Shaping a Nation

Shaping a Nation

A GEOLOGY OF AUSTRALIA

Geoscience Australia

Richard Blewett, Chief Editor

Australian Government

Geoscience Australia

Australian National University

E PRESS

Department of Resources, Energy and Tourism
Minister for Resources and Energy: The Hon. Martin Ferguson, AM MP
Secretary: Mr Drew Clarke

Geoscience Australia
Chief Executive Officer: Dr Chris Pigram

This book is published with the permission of the CEO, Geoscience Australia.

ISBN (print version) 978-1-922103-43-7
ISBN (online version) 978-1-921862-82-3

GeoCat # 73489

Graphic design and production
Marie Lake, Maria Bentley, Alissa Harding, Katharine Hagan, Adrian Yee

Cartography
Silvio Mezzomo, David Arnold, Chris Evenden, Veronika Galinec, Daniel McIlroy

Digital appendix
Julie Silec, James Navin

Editors
Sue Turner; Biotext Pty Ltd

Indexing
Design*emergency* – Tracy Harwood

Project coordination
Suzy Domitrovic, Bobby Cerini

Additional photography
Chris Fitzgerald, Adrian Yee

Printing
Book printing by Paragon Printers, Canberra
DVD printing by Replicat, Melbourne

Published by Commonwealth of Australia (Geoscience Australia) and ANU E Press, Canberra, Australia.

This title is also available online at http://epress.anu.edu.au

Address for correspondence (print edition):
Geoscience Australia
Cnr Jerrabomberra Avenue & Hindmarsh Drive
Symonston ACT 2609
Australia
Email: sales@ga.gov.au

Address for correspondence (online edition):
ANU E Press
R.G. Menzies Building (#2)
The Australian National University
Canberra ACT 0200
Australia
Email: anuepress@anu.edu.au

Citation
Blewett RS (ed.) 2012. *Shaping a Nation: A Geology of Australia*, Geoscience Australia and ANU E Press, Canberra.

Recommended citation for individual chapters
Brodie RS, Lawrie KC & Commander DP 2012. Groundwater—lifeblood of the continent. In: *Shaping a Nation: A Geology of Australia*, Blewett RS (ed.), Geoscience Australia and ANU E Press, Canberra, 332–379.

Cover image
Kings Canyon, Northern Territory. Image by Jim Mason

Printed on Monza Recycled stock. Monza Recycled is certified carbon neutral by the Carbon Reduction Institute (CRI) in accordance with the global Greenhouse Protocol and ISO 14040 framework. Monza Recycled contains 55% recycled fibre and is FSC Mix Certified, which ensures that all virgin pulp is derived from well-managed forests and controlled sources. Monza Recycled is manufactured by an ISO 14001 certified mill.

Minister's introduction

This book holds a mirror to timeless geological processes that have shaped a magnificent country. It shows how our nation has evolved by utilising the natural endowment of this continent that has formed on the back of nearly four billion years of geological processes.

This absorbing documentation of Australia's geological history and how it has shaped our society highlights the value of geoscience information and how it underpins our economy and well-being. A very strong reflection of this history is how plate tectonics has placed Australia adjacent to the fastest growing region on the planet, thus enabling Australians to rise to the challenges that our global economy now faces: energy security, developing cleaner and alternative energy technologies, the impact of climate change and the safety and resilience of communities to recover from devastating natural hazards.

Australia's history has been influenced by the relationship between people and natural resources. This relationship will continue to shape our way of life in Australia, including our families, culture, safety and well-being, society, prosperity and future.

Geoscientists have built a knowledge base that puts Australia at the forefront of resource exploration and development. Future generations will build on this knowledge. They will invent technologies that are perhaps beyond our imagination at this point in time. Currently, geoscientists are exploring and investigating new minerals and energy sources, knowing that we have a responsibility to contribute to the sustainable management of the Australian continent and the planet.

This book takes a journey that covers every landscape, seascape and climate, from the tropics to the Antarctic, across more than twenty seven million square kilometres of our planet. It also takes a journey through deep time, from the oldest evidence of Earth's crust in Western Australia to modern coral reefs in Queensland.

Most of Australia's natural tourism features are products of geological processes that began more than 20 million years ago—the Warrumbungles in New South Wales, the Flinders Ranges in South Australia and the Great Barrier Reef in Queensland. Another significant cultural and tourism site is the Uluru–Kata Tjuta National Park. Uluru, 'the rock', was laid down in an inland sea about 540 million years ago. Uluru and the National Park have a cultural and natural heritage that dates back tens of thousands of years, and which the Anangu traditional owners have looked after over that time.

This exciting book demonstrates the fundamental importance of the study of Earth sciences to our society. It is upon all of us, in every corner of the world, to understand the geological processes of the land on which we live and to continue working together to solve the challenges for our future well-being.

The Hon Martin Ferguson AM MP
Minister for Resources, Energy and Tourism,
Australian Government

Contents

The Pinnacles, Western Australia.

Image by Jim Mason

Out of Gondwana 173

Old, flat and red—Australia's distinctive landscape 227

Living on the edge—waterfront views 277

Groundwater—lifeblood of the continent 333

Foundations of wealth—Australia's major mineral provinces 381

Sustaining Australia's wealth—economic growth from a stable base 433

Deep heat—Australia's energy future? 483

Advance Australia Fair 527

Index 544

Foreword

This book demonstrates not only how the geological history of the Australian continent has shaped where we live, but how it has influenced the way the Australian people have responded to the endowment bestowed by that geological history to build a prosperous and peaceful nation located between the east Indian and southwest Pacific oceans. Our location as an island continent is the product of many hundreds of millions of years of plate tectonic processes, which have left the country in a mid-plate tectonic setting with a latitude band that results in a highly variable climate. The interaction of these tectonic and climatic forces has shaped the landscape and environment of the continent. Our position within middle to low latitudes has left the continent without recent continent-wide glacial activity to replenish and refresh the soils, unlike much of Eurasia and North America. Furthermore, the continent's position within the Australian Plate has meant that the processes that shape it operate at a time-scale that is orders of magnitude slower than the processes that operate at plate boundaries, thereby making it difficult to measure these processes and hence understand their role in shaping the continent. The geological record is, however, clear. Even at very slow rates, enormous changes can and do occur when they have the opportunity to operate over tens to hundreds of millions of years.

The very long geological history preserved within Australia has delivered a great natural endowment of mineral, energy and groundwater resources that have formed and accumulated over close to four billion years. The resources and landscape of Australia have strongly influenced the first Australians—the Aboriginal and Torres Strait Islander people—and, during the past two centuries, the development of Australia as a nation. This influence occurred from the early Asian and European encounters with the west and north coast—which led to a perception of a difficult, inhospitable coastline backed by a commercially uninteresting continent, with little or no potential—to the early exploration of eastern Australia, when the passive margin mountains of the Great Divide initially prevented inland exploration. Once the interior of the continent was being explored, it was often the discovery of mineral resources that led to the establishment of major settlements. As this book recounts, many of the significant changes in Australia during the past 200 years have been driven by major population spurts that were the outcome of minerals booms in one form or another, from the 19th century gold rushes to the modern iron ore- and energy-driven boom.

One decade into the 21st century, what does this view of Australia, shaped by nearly four billion years (4 Ga) of geological processes, tell us about how we should, as geoscientists, tackle the many challenges we face in maintaining a sustainable, vibrant, wealthy and healthy Australia over the next 100 years? Although we have learnt a lot and know a great deal about the evolution and development of our continent, there is much still to learn.

The challenge for the nation is to better understand the processes that have generated the resource endowment that we are currently exploiting,

Windjana Gorge, Western Australia.

Image by Jim Mason

and to identify the location, at depth and under cover, of the next generation of resources that will underpin the nation's economic future. Although we understand many of the first-order factors that drive the processes that affect our endowment, as well as the geological hazards that occur in Australia, we cannot predict or convincingly explain many of the phenomena that we observe. For example, our understanding of intraplate processes is deficient and represents a significant gap in the research focus of our geoscience community.

Much of what we know comes from investigation of the relatively accessible, near-surface geological record. By contrast, we know much less about what lies below that near surface and, perhaps more importantly, how the solid earth behaves in three dimensions through time in response to the processes that shape our world. We are not yet able to answer important questions about why and where the events of Australia's long geological history have placed the next generation of resources that are currently hidden at depth or beneath cover. Although technology has given us tools to image the subsurface in various ways and to various depths (with highly variable resolutions), these tools are measuring a range of physical properties that are surrogates for the geology. We will need to deploy and integrate the full suite of tools and techniques available to provide the necessary geologically meaningful results that will lead to a comprehensive understanding of earth systems processes as they have operated in Australia.

In some geological environments, the tools are superb and give excellent results. In other environments, our ability to interpret and integrate this information in a way that generates meaningful geological information, while developing, still lags well behind where we would like to be. Our capacity to measure the rate of change of the continent's surface down to millimetre-scale accuracy and to image deep into the continent means that we have the opportunity to tackle some of the great geological questions of our time. We will therefore have a very much enhanced prospect of being far more predictive and confident in answering many of the big questions that determine the behaviour of our continent in an ever-changing world. In this way, geoscientists can more confidently contribute to the sustainable management of the planet.

The book is the largely voluntary effort of many people from across the geological community in Australia. Geoscience Australia, in collaboration with The Australian National University, is delighted to bring this publication to fruition in time for the 34th International Geological Congress, and trusts that it will be an enduring contribution to the knowledge of the Australian nation through the lens of its geological evolution.

Dr Chris Pigram
Chief Executive Officer
Geoscience Australia

Kermit's Pool, Hancock Gorge, Karijini National Park, Western Australia.
Image by Jim Mason

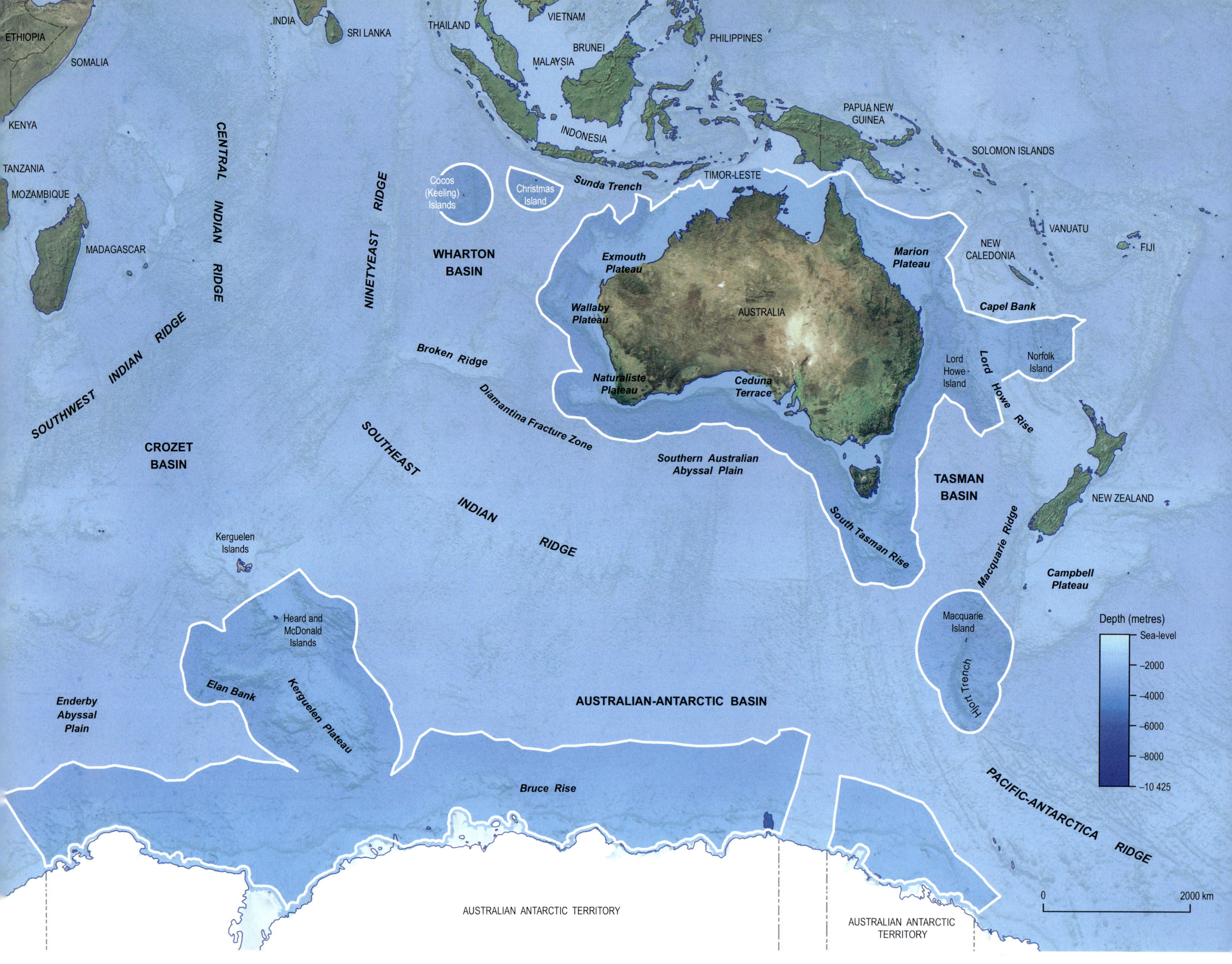

Frontispiece: The limits of Australia. The white line shows the treaty boundaries with adjacent states, combined with Australia's outer limit of extended continental shelf. Major geomorphological features of the surrounding ocean basins are also shown.

Editors' note

This new book on Australia's geology has been prepared for release at the 34th International Geological Congress in Brisbane in August 2012. The book has been co-published by Geoscience Australia and ANU E Press. The book is not intended as a definitive text on all aspects of Australia's diverse geology, nor does it follow the 'traditional' time-based treatment of the topic. Rather, the book tells the story of Australia's geological evolution through the lens of human impacts—illustrating both the challenges and the opportunities presented by the geological heritage of the 'lucky country'.

The book showcases the excellence of Australian geoscience by integrating geoscience disciplines into a systems framework that address many of the 'big questions' relevant to Australians today. It is aimed at geoscientists, but the narrative and messages are relevant to society as a whole. The Editorial Board has tried to bring together a book that is as visually stunning as Australia's geology, and present much new information with minimal scientific jargon.

The book is arranged into 11 chapters, each having a number of breakout boxes and *Did you know?* panels. These are excursions from the main narrative where interesting facts and extra details are presented. Large colour photographs bring the text to life. The opening two chapters set the spatial, temporal and cultural contexts for the book. The following eight chapters are arranged into themes around the various geological influences on Australian society, environment and wealth. These chapters cover the evolution of life in Australia, development of post-Gondwana hydrocarbon systems, evolution of the landscape, the coastal zone, groundwater, minerals and bulk commodities, and, finally, future energy. The concluding chapter considers the major challenges facing the nation and the vital role that geosciences will play in meeting these challenges.

The time-scale we use is the September 2009 chart issued by the International Commission on Stratigraphy. The appendices are a very important component of the book. They are available on a dual-layer DVD bound with the hard-copy printed version. This DVD contains an Australian Common Earth Model and places many of the maps and 3D objects from Appendix 2 into a 3D viewer (NASA's World Wind). Other large-scale maps, reports, animations, tables and documents are arranged by chapter and can be viewed through a browser. The Common Earth Model is, in essence, an atlas of continental-scale maps that were previously published in the 1980s as the BMR Earth Science Atlas. These plates have been updated since, but here they are compiled together in one easy resource.

The book has 54 contributing authors from across Australia and as many external reviewers from across Australia and the world. Many others have also contributed; it has been a community effort (see acknowledgements). The Editorial Board would like to acknowledge all these people for their time, expertise and enthusiasm. We would also like to acknowledge the considerable investment by Geoscience Australia in making this book a reality.

Richard Blewett, Keith Scott, Brian Kennett,
Phil McFadden, Marita Bradshaw, Phil Cummins
(Editorial Board)

Acknowledgements*

J Aitchison, M Alcock, J Alexander, M Allan, H Apps, M Archer, M Barley, A Barnicoat, P Betts, B Birch, K Black, S Blewett, C Boreham, C Brown, P Butler, C Butt, C Carson, L Carson, P Cawood, R Chopping, J Claoue-Long, D Clark, B Collins, K Condie, C Consoli, M Cornelius, R Costello, S Cox, R Cresswell, A Cross, D Curnoe, G Davidson, J Dawson, P De Caritat, P De Dekker, K Derrington, T Eggleton, P English, R Evans, N Exon, S Fishwick, I Fitzsimmons, D Flint, C Foster, G Fraser, J Frazier, E Fredericks, J Gehling, D Giles, A Gill, B Goscombe, J Greenfield, K Grey, L Halas, L Hall, B Handke, A Harris, L Head, P Henson, S Holland, J Hollis, O Holm, D Hopley, M Hutchinson, T Ireland, S Jaireth, L Jaques, N Jarosz, B John, D Kay, M Kendall, J Kennard, B Kohn, R Korsch, I Lambert, R Langford, R Large, D Le Heron, K List, J Long, S Maclaren, J Magee, D Mantle, G Marinelli, J Mason, H Maxwell-Stewart, P McCabe, C McCuaig, A McKay, A McPherson, Y Miezitis, P Milligan, B Minty, T Moore, D Müller, C Murray-Wallace, M Nichol, B Nicoll, M Norvick, Ian Oswald-Jacobs, J Paterson, J Pettigrew, C Pigram, K Piper, P Playford, T Press, T Ransley, A Reading, M Roarty, D Robson, R Rogerson, A Rowett, R Sait, B Salau, M Salmon, M Sandiford, J Scanlon, I Scrimgeour, A Short, K Sircombe, P Southgate, T Stieglitz, A Stewart, G Taylor, T Topper, D Trail, I Tyler, T Tyne, C Vickers, M Walter, M Wenitong, M White, T Whiteway, M Williams, N Williams, E Woehler, J Woodhead, C Woodroffe, T Worthy, G Young and S Zahirovic.

* Authors are listed at the start of each chapter.

1 Australia and the Australian people

The demography, history and culture of the Australian people have been shaped by Australia's size, geographical remoteness, ancient soils and landscape, arid climate, flora and fauna, and bountiful mineral and energy resources. These influences are the result of a rich, and in some ways, unique geological heritage—making Australia a truly 'lucky country'. Australians today number more than 22 million people; we are a diverse nation, having originated from more than 200 other countries. We are custodians of 27.45 million km^2 of Earth, including the continent's mainland and offshore extensions, numerous offshore territories, marine zones and a large part of Antarctica.

Richard S Blewett,[1] Phil A Symonds[1] and Brian LN Kennett[2]

[1]Geoscience Australia; [2]Australian National University

THE LAND THAT SHAPED THE PEOPLE (BOX 1.1)

Heart of Auss

I've seen the Rock at the break of day
And it seems so right for a place to stay
A place in Law to live love and pass away
For the people of my Dreaming

That road to Rock bin bitumen—all-da-way
But worth the wait when you stay
See shards of light that cut bend break and shape
The heart of the land of The People

Long time ago there was nothing else
No plane no car no photocells
And we walk and we talk and believe the land
And bond together in little bands

Then came the pale one and he took stock
Will we paint it will we sell it
It is Our rock to do to have to colour blue
And the people felt old life was through

Now we all live together in this big land
and most of our country is covered with sand
But there in the centre there stands
our heart a sacred place to make a stand

A solid multi-hued rock of our new start
And our heart in the land of the people.

John Michael Wenitong (2010)

My Country

The love of field and coppice,
Of green and shaded lanes.
Of ordered woods and gardens
Is running in your veins,
Strong love of grey-blue distance
Brown streams and soft dim skies
I know but cannot share it,
My love is otherwise.

I love a sunburnt country,
A land of sweeping plains,
Of ragged mountain ranges,
Of droughts and flooding rains.
I love her far horizons,
I love her jewel-sea,
Her beauty and her terror -
The wide brown land for me!

An opal-hearted country,
A wilful, lavish land -
All you who have not loved her,
You will not understand -
Though earth holds many splendours,
Wherever I may die,
I know to what brown country
My homing thoughts will fly.

First two and last stanzas, by Dorothea Mackellar (1904)

© Getty Images [C Wehrmeier]

Australians enjoying a summer day at Bondi Beach, Sydney.

A sense of place

The land of Australia features strongly in the visual art, poetry and songs of the people. This is especially true for the first Australians—the Aboriginal and Torres Strait Islander people—who have a deep connection with the country; the poem of one Indigenous Australian (JM Wenitong) on the most iconic geological and cultural place, Uluru, epitomises this special connection (Box 1.1). To the first Australians, the word *'country'* does not mean just landscape in the usual sense of the rivers, creeks, rocks, hills and waterholes. Rather, it includes all living things; it embraces the seasons, the stories and the creation spirits from the Dreamtime.

More recent arrivals also feel a deep connection to the land. Dorothea Mackellar's homesick elegy, written in London in the early 1900s, conjures vivid images of her adopted land, so different from her English roots (Box 1.1). When Tourism Australia asked what makes this country a special place to live in or to visit, Australians identified the wildlife, beaches, coral reefs, outback, vibrant cities, multicultural people and laid-back lifestyle. By listing these special qualities, and probably without realising it, Australians and their guests identified with the continent's remarkable geology.

This book tells the story of Australia's geology through the lens of its shaping influence on the Australian people and their nation. The story charts our landscape, soils and water, unique flora and fauna, and mineral and energy wealth through a series of interconnected 'pillars' that are deeply rooted in time and space to Australia's geological 'foundations' (Figure 1.1). Without the geological entity—the rocks that form the

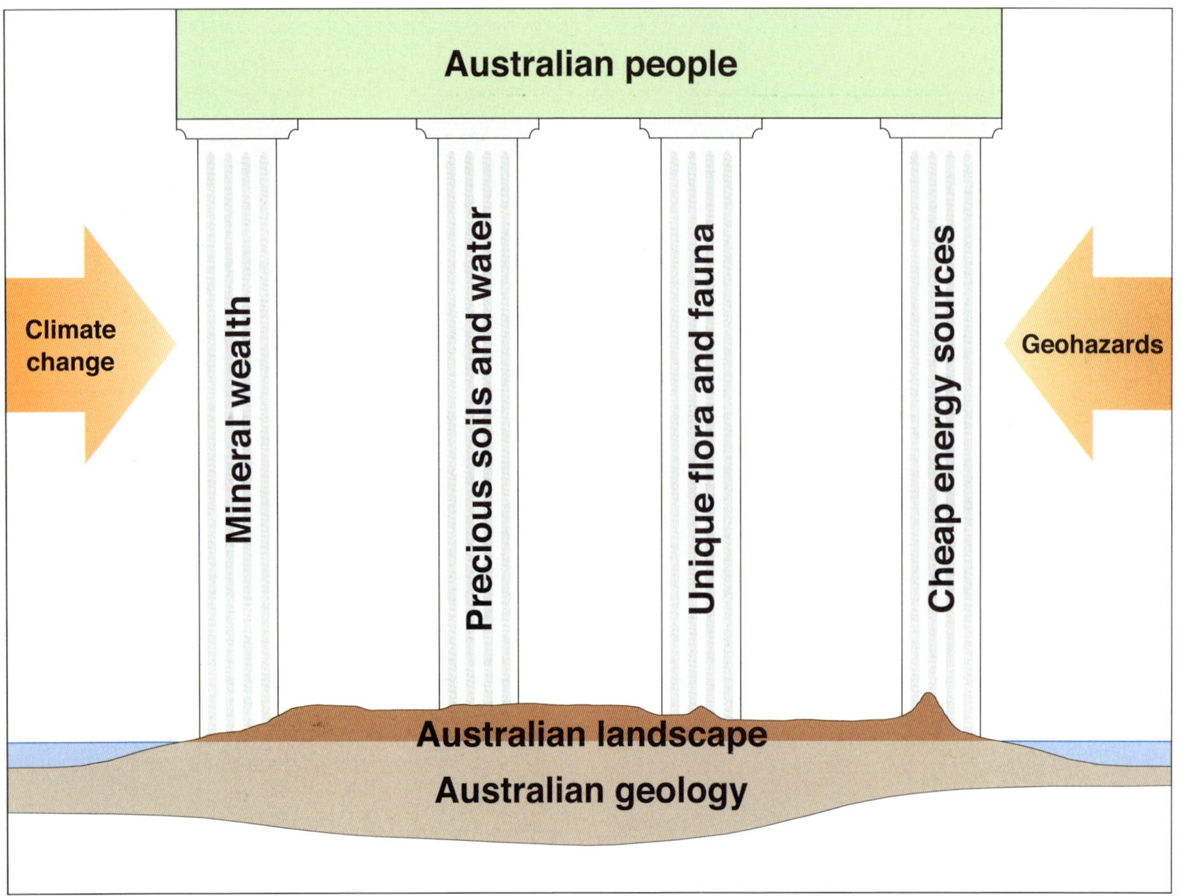

Figure 1.1: Conceptual portrayal of how geology shapes the Australian people. We exist only by 'geological consent' (Durant, 1926). Geology forms the foundation of the landscape within which the soils and water, unique flora and fauna, mineral wealth and energy sources have evolved. These are the rocky 'pillars' that support society, and that climate change and geohazards sometimes undermine.

Australian continent—we would not have such an exciting story to tell. First of all, we give you what makes Australia today. The geological process of continental drift has placed the island continent of Australia into its present climate system south of the equator. This southern climatic regime has acted on ancient geological roots to create a distinctive Australian landscape. The landscape has also been shaped in time by water and soils, which are now scarce and fragile. Australia has evolved unique flora and fauna, and into this environmental system people arrived, possibly as long ago as 60 000 years. Australia's geology also endows the continent with an abundance of mineral and energy resources, first sustaining the people and then making the nation wealthy. Civilisation and the choices made, however, only exist with the 'consent of geology', with geohazards and climate change providing challenges and constraints for the Australian people. Fortunately, this 'lucky country' has provided its citizens with numerous choices for the future.

Let us look closer at the Australian people and their nation.

The Australian people

The Aboriginal and Torres Strait Islander people, the traditional owners, are the first Australians. They arrived from the north up to an estimated 50 000 to 60 000 years ago (ka), via sea and land bridges moulded by climate change, into a country that was very different from today. In terms of the migrations of *Homo sapiens*, the ancestors of today's Indigenous people travelled the furthest and fastest of all the migratory people out of Africa. At sea-levels 60 m below those of today, Australia, New Guinea and Tasmania (Tas.) formed a single continent called Sahul (Figure 1.2). Even at the glacial maximum when sea-level was about 125 m lower than today, this landmass was never connected to the Asian mainland, so island hopping from Sunda across Wallacea to Sahul required several open-ocean crossings, some more than 80 km (Figure 1.2).

Today, Australians originate from a rich variety of cultural, ethnic, linguistic and religious backgrounds. Other than the first Australians, the more than 22.7 million (M) people who call themselves Australians are immigrants, whose ancestors and descendants arrived during the past two centuries from more than 200 countries. Most have come from lands radically unlike Australia—lands with regular and predictable seasons, with flowing rivers and lush fertile plains, not a land with deserts, 'droughts and flooding rains' (Box 1.1).

The Bugis people from Macassar in Sulawesi (Figure 1.2) knew of Australia long before European or North Asian people. They collected bêche-de-mer (trepang) on the northern coast, but did not settle, returning to their apparently more favourable homelands. The first European contact with Australia occurred mostly along the harsh western and northern coast by sailors—Portuguese, Dutch, British, French—intent on reaching the Spice Islands of Indonesia. Ironically, they sailed past or above the mineral and energy riches of the northwest (Chapter 9). The land was regarded as best avoided, with little to offer. Past climate was not always as harsh as it is today (Chapter 5). The interior deserts and the tropical north, largely shunned by Europeans, have been productive regions for Aboriginal people. Linguistic maps illustrate that most of Australia was occupied and, in contrast to European perceptions of the land, these areas gave a good living to the modest numbers of Aboriginal people.

Only much later was the more temperate and fertile east coast of Australia mapped by Captain James Cook. The 18th century British claim of sovereignty was unique in history, as it claimed a continent for a nation and a nation for a continent. Until the advent of long-distance aircraft, all

Figure 1.2: Approximate render of dry land with sea-level 125 m lower than today, showing the outline of Sunda, Sahul and intervening Wallacea. Human arrival into Australia may have taken a northern or a southern path, both necessitating ocean-going vessels. Once humans entered Australia, they were able to travel as far south as Tasmania. The oldest archaeological sites in Sahul are shown with radiocarbon dates as uncalibrated (raw) ages. (Source: after O'Connor, 2009)

KEY MOMENTS IN AUSTRALIA'S POPULATION GROWTH (BOX 1.2)

Image courtesy of John Oxley Library, State Library of Queensland, neg: 62474

Australia is a 'modern' nation in that most of the native-born population is descended from immigrants who arrived in the last 200 years, and it is their cultures that form the basis of a nation that has existed for little over a century. In this regard, Australia is very different from most long-established European and Asian societies. It has a culture that, while already distinctive, continues to develop.

Population growth in Australia was modest for the first 63 years of European settlement, which began with the arrival of the First Fleet at Botany Bay in 1788. It was the discovery of gold in 1851, however, that saw the first mass arrivals of immigrants, with a dramatic jump in the population in the following years (Figure B1.2a). The lure of gold also brought a different mix of people, especially from Asia. The discovery of gold in Victoria, soon after the state (colony at the time) was founded, saw a dramatic change in its fortunes compared with New South Wales (Figure B1.2b). For example, the gold exported from Melbourne in 1860 accounted for 5% of the revenue for the entire British Empire. It was not until 1884 that the population in New South Wales was to surpass that of Victoria (Figure B1.2b). Gold also changed the lives of individuals, with many 'ordinary' people making and losing fortunes.

Australia's population graph shows a dip in the late 1910s, with the tragedy of World War I and subsequent flu epidemic causing the population to decline. The next big jump in population growth followed the end of World War II. Nation building saw mass migration from around the world, at a time when local birth rates were high.

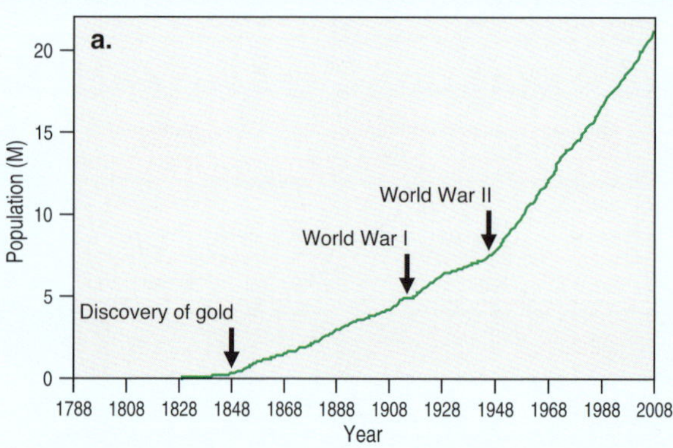

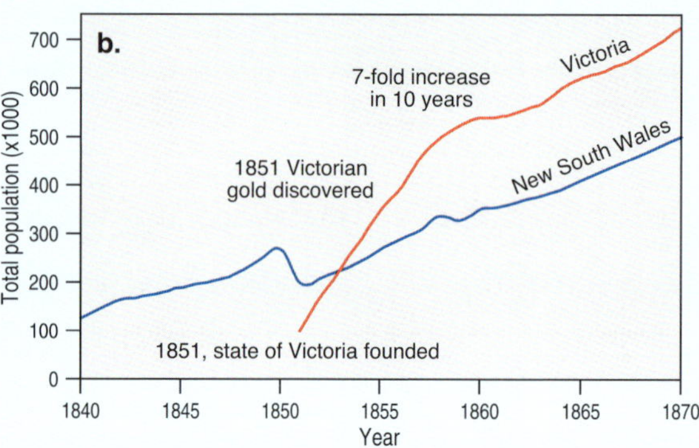

Figure B1.2: (a) Australian population between 1788 and 2008. Note the acceleration in the population after the discovery of gold in 1851 and the post-World War II boom. (Source: Australian Bureau of Statistics). (b) Population of Victoria and New South Wales compared for the period 1840–70. The discovery of gold in the rich fields of Victoria drew people from around the world. The influx of immigrants resulted in Victoria having a greater population than New South Wales for the period up to 1884. (Source: Australian Bureau of Statistics, 2008)

newcomers arrived by sea. Coastal geography and the availability of reliable potable water were key factors in determining early settlement patterns. The capital cities of each state were founded as ports, with maritime trade critical for their individual development and the nation as a whole (Chapter 6). Despite this, there is not a strong maritime tradition in Australia. For example, the modern-day so-called grey nomads circumnavigate Australia not by yacht, but with a four-wheel drive and caravan. People as diverse as artists and adventurers looked inwards, despite the isolating influence of the surrounding seas that so shaped the history of Australia.

Historically, Aboriginal people are from mainland Australia and Tasmania. Torres Strait Islander people come from the islands between the tip of Queensland (Qld) and New Guinea and share many cultural similarities with the people of New Guinea and other south Pacific islands. Australian Indigenous culture is diverse and is the longest continuous living culture on Earth. Since the late 20th century, this culture has been increasingly recognised as an integral and sustaining core to Australia's national identity.

It is estimated that there were between 300 000 and 750 000 Aboriginal and Torres Strait Islander people in Australia at the start of European settlement in 1788. This population declined dramatically during the 19th to mid-20th centuries due to factors including conflict on the frontier of European settlement, the impact of new diseases and other social problems.

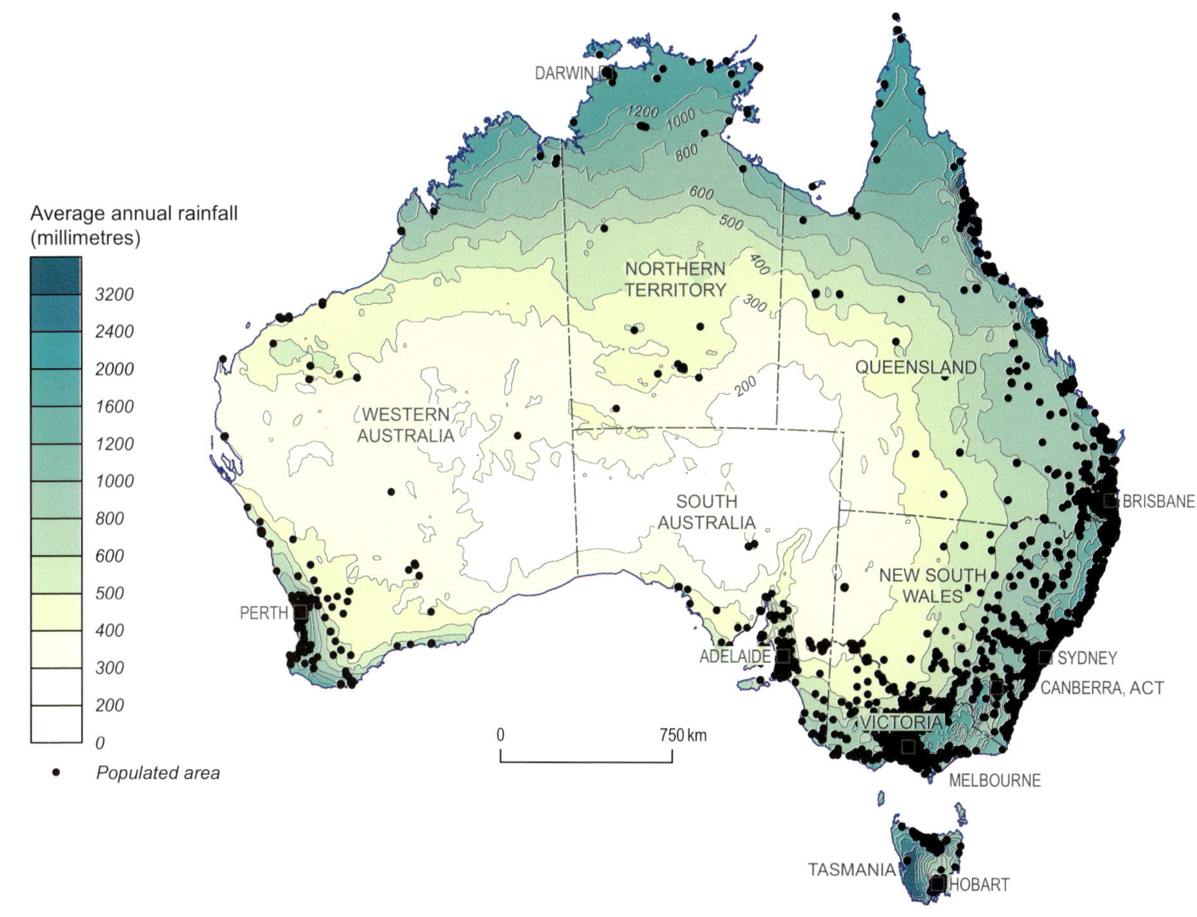

Figure 1.3: Population distribution of Australia, with each dot representing 1000 people. Note the population concentration in the southeast, east and southwest coastal regions of the mainland and around Tasmania, and the scattered and sparse distribution for much of the interior and the north. The boundaries of the states and territories are also delineated. (Source: Australian Bureau of Statistics, 2006. Data are usual residence, mapped to the 2006 Australian standard geographical classification)

Modern immigration began with the British First Fleet in 1788, and continued at a steady pace until the discovery of gold in 1851 (Box 1.2). The population growth rate then jumped to 50 000 immigrants per year during the subsequent gold rushes of the 1850s, with new arrivals from many parts of the world. Asian immigration, in particular by Chinese people, commenced at that time (Chapter 8). By the 1940s, the population reached 7 M, with the bulk of settlers from a British background. Since World War II, more than 6.6 M migrants from more diverse backgrounds have arrived in Australia.

Figure 1.4: Prominent geographical features of mainland Australia, showing major rivers and deserts, population centres and major transport links. Note the remoteness of the centre and its lack of roads or settlement. The boundaries of the states and territories are also delineated. The distance from the coast of the 200 m bathymetric contour varies widely.

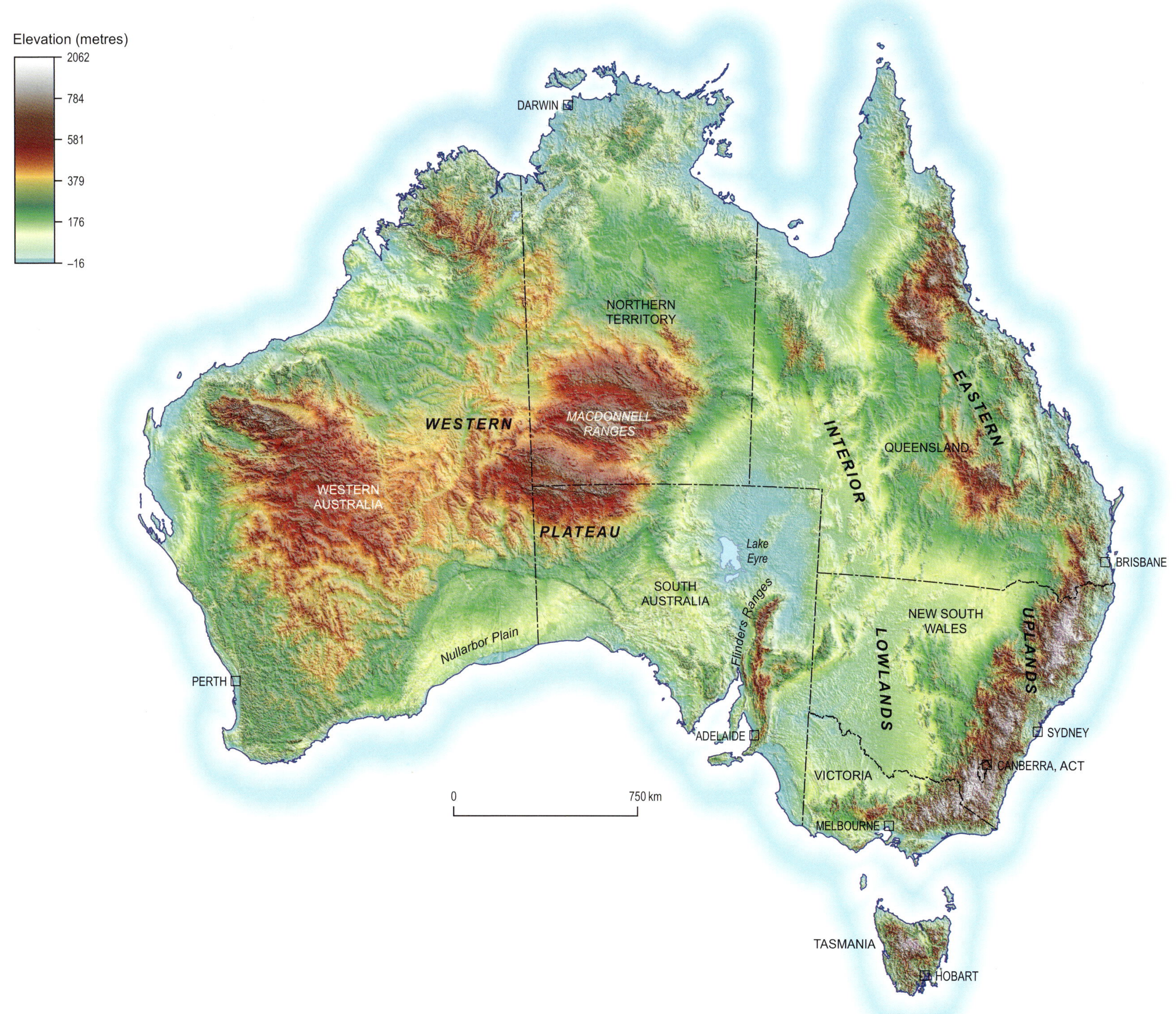

Figure 1.5: Digital elevation model of Australia, showing the main physiographic features of the continent.

Image courtesy of the Royal Flying Doctor Service

The Royal Flying Doctor Service flying over Rottnest Island in Western Australia. The service's motto is 'the furthest corner—the finest care'.

Modern Australia

Ask an Australian what their national dish is and they'll struggle to answer. Ask them about contemporary Australian cooking and they may talk about barbecues, baked dinners or Aussie fusion cuisine—marrying the abundant fresh produce with multicultural influences. Today, almost one in four Australian residents was born overseas. The past 40 years have seen a significant shift in the immigrant country of origin. In the 1960s, 45% of all new settlers were born in the United Kingdom and Ireland. By 2006–07, this had fallen to 17% with increased arrivals from the Asia–Pacific region, Africa and the Middle East. More than 10% of permanent migrants in 2006–07 came from China and, since 1995, more than 200 000 people have come from Africa and the Middle East. The Refugee Council of Australia estimates that more than 750 000 immigrants have come to Australia through refugee or humanitarian programs since 1901.

Australia has a population of more than 22.7 M, with a density of 2.9 people per square kilometre, which is 233rd out of the 239 countries in terms of density ranking. We are unevenly distributed, with around 95% of Australia's population living within 100 km of the coast, which contrasts with the world average of 39% living within 100 km of the coast (Figure 1.3). The central areas of Australia are sparsely populated; the Northern Territory (NT) has a population of around 230 000 in a land area twice that of France, which has a population of more than 62 M! Australia's population growth rate was 1.4% for the year ended 31 March 2011, which was down from a high of 2.2% for the year ended 31 December 2008.

Australia's fundamental settlement pattern has not changed significantly for more than 150 years. Australian society is one of the most urbanised in the world, with the majority living in five large port cities whose populations range from one million to more than four million people. Between them, the metropolitan areas of the state capitals Sydney, Melbourne, Brisbane, Perth and Adelaide house 61% of the population (Figure 1.3). A further 13 smaller cities, each with estimated populations

AUSTRALIA: THE LARGEST ISLAND AND SMALLEST CONTINENT (BOX 1.3)

Australia has a land area of 7.69 M km^2, making it both the largest island and the smallest continent. The coastal outline is distinctive and recognisable and stretches more than 3860 km from north to south and almost 4000 km from east to west. This east–west distance is comparable to that between New York and San Francisco in the United States, Lisbon and Moscow in Europe, and Beijing and New Delhi in Asia.

In area, mainland Australia is the sixth largest nation after Russia, Canada, China, the United States and Brazil (Figure B1.3). Australia has about twice the area of the European Union or the Association of Southeast Asian Nations (ASEAN). Australia's marine jurisdiction—in the top three in the world, along with the United States and France—lies in three oceans and covers nearly 12 M km^2. In fact, Australia is custodian of about 3.8% of the oceans, and about 9.1% of the land, with a full land and marine jurisdiction of 27.45 M km^2 (Table 1.1). Within its exclusive economic zone, Australia has certain sovereign rights to the natural resources (living and non-living) of the water column, seabed and subsoil in areas adjacent to Australia's territorial seas and jurisdiction over the water column, seabed and subsoil.

Australia has a number of large islands in the Pacific, Indian and Southern oceans and the Coral and Timor seas as part of its external territories, as well as several closer to the mainland that are larger than 1000 km^2. Tasmania, with its distinctive heart shape, is Australia's largest island, with an area of 64 519 km^2 and an additional 3882 km^2 of fringing islands. Islands are important for determining the size of a nation's exclusive economic zone. The small Heard and McDonald islands, for example, add some 410 722 km^2 to Australia's exclusive economic zone.

For an island nation, coastlines play an important role in defining national, state and territory boundaries. For example, the 260 or so inhabitants of the low-lying Boigu Island in Torres Strait live only 4 km from Papua New Guinea, our nearest neighbour.

Calculations for Australia's coastline lengths are taken from Geoscience Australia's GEODATA Coast 100K 2004 database. The length of the Australian coastline is 59 736 km, comprising a mainland of 35 877 km and fringing islands of 23 859 km.

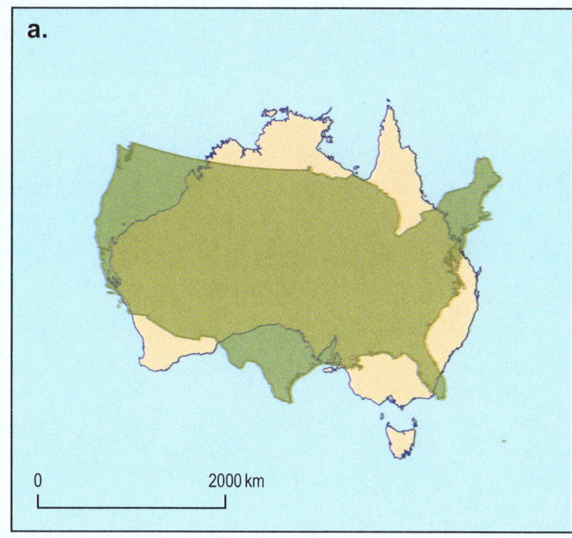

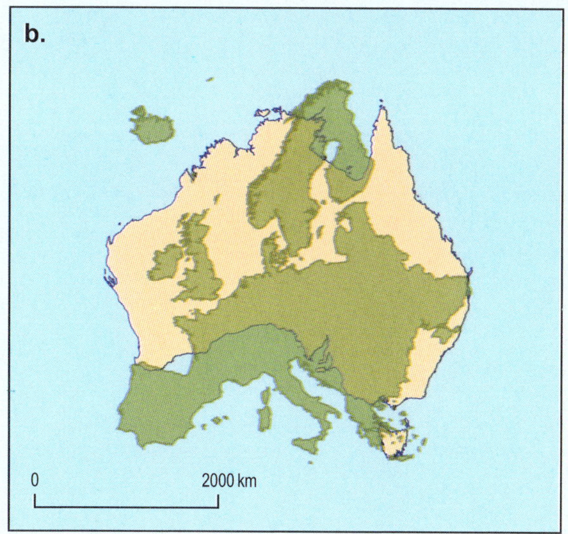

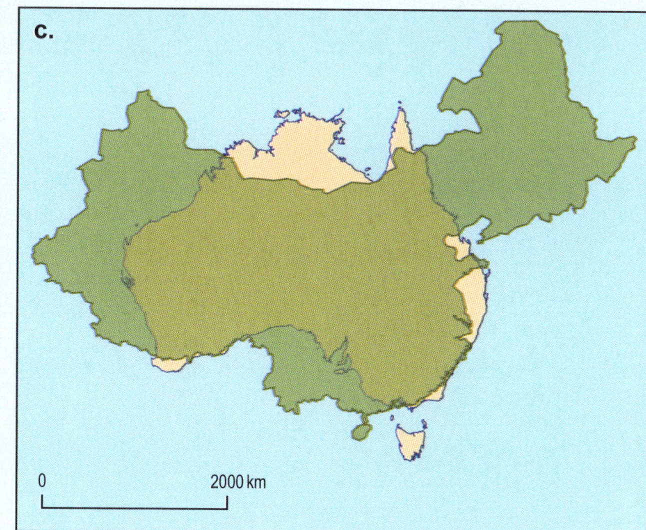

Figure B1.3: Comparative size of continental Australia (population 22.7 M) with: (a) the conterminous United States of America (population >300 M), (b) Western Europe (population >500 M) and (c) China (population >1.3 B).

of more than 80 000 people, accommodate another 14% or so. Even the exploitation of mineral and energy resources in Australia's outback has not changed this pattern; instead, many workers fly-in/fly-out (FIFO) to the remote locations from their urban home base (Chapter 9).

Australia's low population density, the vast distances between settlements (Box 1.3) and the availability of land have all shaped the nation in many ways. The 'quarter-acre block' typified the post-World War II population and construction boom. This expansion was facilitated by the availability of abundant land and cheap energy, which allowed towns and cities to sprawl into the countryside, some of which is Australia's most productive for agriculture (Chapter 10). In the more remote areas, landowners can control enormous parcels of land. Anna Creek Station in South Australia (SA), for example, runs cattle and is the world's largest individual property holding at 34 000 km^2. Graziers on these large properties typically use helicopters for mustering (rounding up) their cattle. Gone are the stock routes and cattle drovers of the late 19th to mid-20th centuries (Chapter 7). Instead, there are now 'road trains', some up to 200 tonnes (t) and consisting of three or four linked trailers.

Road trains ply the long and dusty outback roads of Australia, carrying goods and livestock to and from remote communities and their markets.

The distances in Australia are large; they are measured in days of driving. Football fans have been known to drive the 900 km from Mt Isa to Townsville to watch a rugby league match. Shopping can be 200 km away from home. The low population density means that there is a small tax base to pay for quality highways in such a vast continent. It was not until 1988 that Australia completed the bitumen of the road network encircling the country. The regolith (surface material) cover over much of Australia's inland is of poor quality for use in road building (Chapter 5). The surface readily breaks down, the roads become corrugated, and in places deep, fine-grained 'bull dust' holes develop. The great distances and scattered remote populations in the outback fostered the Royal Flying Doctor Service (established in 1928), whose motto, 'the furthest corner—the finest care', ensures that all Australians are afforded the benefits of 'medivac' access to modern medicine.

Only in the last few years, as global communication has become almost instantaneous, has the sense of remoteness been challenged, although even now the time-zone differences to Europe and North America can prove difficult. The coming of the Internet has revolutionised outback living, with everything now more accessible, from Schools of the Air to checking weather reports of floods and cyclones to sourcing replacement parts, facilitating direct communication for business and citizens. Today, Australia's major trading partners lie to the north, with China, Japan and India being only a time difference of a couple of hours away.

From its foundation as a European-oriented enclave many weeks by sea from the homelands, Australia was forced to rely on its own resources and initiative as far as possible, although high-cost imports were also needed, and paid for by wool and mineral wealth. Local ingenuity was applied to many specific problems, and clever solutions were found without external input. A classic example is the 'stump-jump' plough, invented by Richard Smith in 1876 to counteract the roots of trees left in the ground after clearing of the native vegetation. Even today, items such as the rotary Hills hoist for drying washed clothes in quarter-acre suburban backyards (Chapter 10) are part of the Australian vernacular and seem strange to newcomers.

Less visible is the significant innovative contribution made by the mining technology and services industry. The perfection of the flotation method at Broken Hill in the early 1900s, for the recovery of non-ferrous metals, was a major technological achievement. Prior to this, zinc (Zn) recovery at Broken Hill was only 12%, and recovery of silver (Ag) and lead (Pb) was 50% and 70%, respectively. The evolution of the technique was led by Guillaume Delprat, the General Manager of BHP from 1901, and benefited from the skills of Auguste de Bavay, a Belgian brewer for Fosters beer. This somewhat counterintuitive process selectively floats different sulfide minerals. This technology is seen as the genesis of modern mining, leading to the development of Australia's manufacturing industry. During post-World War II construction of the tunnels for the Snowy Mountains hydroelectric and irrigation scheme (Chapter 5), Australian engineers invented the expanding rock bolt, which allowed that project to proceed quickly and safely. This invention revolutionised underground mining and tunnelling across the world. More recently, Australian innovation in computer software for mining and exploration accounts for around 60% of the global market. This industry generates more export earnings for the nation than the more widely known wine industry.

Australians are proud of their harmonious multicultural society. This developed in part through hardship and necessity, governed by a small population in a large country with many opportunities, with people distant from homelands. The new land, both physically and socially, contrasts greatly with where many settlers originated. The Australian culture and democracy are exemplified by a sense of a 'fair go' (justice for all)

1.1: Australian Coat of Arms

Australian native species feature in national symbols. The supporters on either side of the Australian Coat of Arms, for example, are the red kangaroo (*Macropus rufus*) and the emu (*Dromaius novaehollandiae*). The kangaroo and emu were chosen to symbolise a nation moving forward, reflecting a common belief that neither animal can easily move backwards.

The arms is depicted on a background of sprays of golden wattle (*Acacia pycnantha*), which was proclaimed Australia's national flower in the bicentennial year 1988.

Image courtesy of National Archives of Australia, NAA: A6135

and egalitarian spirit. Few people in a big country meant that labour shortages occurred, driving up wages and enhancing workers' conditions, such as an eight-hour day and weekends free from work. These factors also facilitated the early development of trade unions and universal suffrage. Increased leisure time and money, together with Australia's favourable climate and available space, led to a strong sporting culture and a fierce rivalry with the 'mother country' (e.g. Ashes cricket test matches against England). Sport is an Australian passion, and many of the national sporting teams have faunal names, or their derivatives, such as the Wallabies, Kangaroos, Kookaburras and 'Socceroos'.

A consequence of Australia's small population density is the low number of people available to explore and understand the continent. The fundamental geographical attributes of Australia are the product of the geology—in particular, its continental position, which has influenced the past and present climatic belts, the water balance and flow, the landscape, soil fertility, and the unique flora and fauna. All of this has combined to create the unique continent called Australia, thus shaping the people who live here—the Australians.

Having met the Australian people, let us now consider the role that geology has played in shaping their past and the likely influences on their future.

Australian geology: the foundation that shapes the people

This book is not another 'geology of Australia' with the rocks, minerals and fossils catalogued through time. We wanted instead to concentrate on the geological processes that made this great continent and in turn show how the underlying geology has shaped many aspects of our nation and society (Figure 1.1).

Australia's defining moment occurred around 34 Ma when the continent finally rifted from Gondwana (Antarctica). This geological event set in motion the primary shaping influences on the modern Australian continent. It led to the present-day geographical position ('down under') and tectonic setting, both of which have led to the formation of Australia's unique and ancient landscape (Chapter 5).

Australia's separation from Gondwana released the continent from cold high-latitude southern regions, allowing it to drift rapidly northwards. From the Late Miocene (*ca* 10 Ma) onwards, most of the Australian continent has been positioned where it occurs today, in the horse latitudes (subtropical high) of the Southern Hemisphere, particularly during the warmer months. This simple geographical

Aerial view of longitudinal sand dunes in the Simpson Desert of central Australia.

AUSTRALIA: A LAND OF GEOLOGICAL CONTRASTS (BOX 1.4)

Australia is home to some of the oldest rocks in the world, some dating back to more than 3.7 Ga, and the oldest mineral, a detrital zircon grain from Western Australia with an age of more than 4.4 Ga, only about 150 million years younger than the age of Earth. Surprisingly, some of the oldest parts of Australia are incredibly well preserved, with some 3.5 Ga rocks looking as though they could have formed yesterday. These old rocks have been an important natural laboratory for understanding how Earth worked during its early history: sedimentary rocks provide evidence of earliest life (Chapter 3), volcanic rocks provide evidence of early magmatism and tectonics (Chapter 2), and mineral deposits provide for the operation of mineral systems (Chapter 3).

The oldest known life forms, stromatolites, formed in very old (*ca* 3.5 Ga) sedimentary rocks of the Pilbara of Western Australia and are a feature of many Australian Archean and Proterozoic successions. Stromatolites continue to form today at Shark Bay (Chapter 3). The largest biogenic structures in the world, coral reefs, form a major part of the Canning Basin, and growth of a modern coral reef along the northeast Australian margin has produced the 2900 reefs, 600 continental islands and 300 cays of the largest coral reef system on Earth—the Great Barrier Reef (Fig. 1.12), one of the natural wonders of the world (Chapter 6).

Volcanism has been active in Australia since before 3.5 Ga and has continued intermittently on the mainland up to only a few thousand years ago. Volcanoes are still active today in the remote sub-Antarctic Heard and McDonald islands (Box 1.6). Volcanism and associated magmatism provided energy to drive formation of the oldest ore deposit on Earth, a 3.49 Ga barite deposit in the Pilbara of Western Australia. Similar

The walls of Windjana Gorge, Western Australia, rise sharply from the plains of Lennard River. The gorge is 3.5m km and cuts through Devonian limestone reefs.

Image by Jim Mason

processes are forming ore deposits, so-called 'black smokers', on the seafloor around Papua New Guinea, along the northernmost extension of the Australian Plate (Chapters 2 and 8). Magmatic processes have also enriched certain granites in potassium (K), uranium (U) and thorium (Th) to produce high heat-producing granites that form a potentially vast geothermal resource (Chapter 10).

One of the reasons for the diversity of Australian geology has been complex relationships between stable cratonic elements and intervening orogenic belts. Much of the western two-thirds of Australia consists of cratonic blocks that have been stable since the middle Proterozoic or even the Archean, whereas the eastern third is dominated by a complex orogenic belt associated with the formation of Gondwana and Pangaea (Chapter 2). Most of Australia's mineral (Chapters 8 and 9) and hydrocarbon (Chapter 4) wealth formed as a consequence of the interplay between cratonic and orogenic Australia.

A surprising portion of Australia is covered by Mesozoic basins. In contrast to the older basement terranes, which tend to be hilly or mountainous, these basins can be incredibly flat but contain one of Australia's most important resources, groundwater (Chapter 7). The Great Artesian Basin, which extends from South Australia to northern Queensland, provides the groundwater essential for agriculture, the backbone of the economy of this vast area (Chapter 7).

Some Australian landscapes are very old, perhaps even early Paleozoic in age. Around 80% of the continent is covered by a veneer of regolith—weathered rock, sediments and soils. Australia is a land of geological contrasts that have added variety to our natural history and supported a very diverse economy.

Rock formation before your eyes—the Great Barrier Reef, Queensland.

Spring snow on the Main Range, Snowy Mountains, New South Wales.

Did you know?

1.2: The *SydHarb* is a uniquely Australian unit of measurement

When describing volumes of water, Australian hydrologists tend to talk in megalitres or gigalitres, or even numbers of Olympic-sized swimming pools. They also use the uniquely Australian term, a *SydHarb* (or Sydharb, Sydarb):

- One kilolitre (kL) is 1000 litres or the volume of a cubic metre.
- One megalitre (ML) is one million litres or the volume of 10 cubic metres.
- One gigalitre (GL) is one billion (B) litres or the volume of 100 cubic metres.
- One cubic kilometre (km^3) contains 1000 GL.
- One *SydHarb* is the amount of water in Sydney Harbour, nominally 562 GL.
- One Olympic-sized swimming pool is 50 m long, 25 m wide and 2 m deep, so holds 2500 m^3 or 2.5 ML.

By way of example, the massive rainfall event on 11 January 2011 resulted in about two *SydHarbs* flowing into Brisbane's Wivenhoe Dam in 24 hours.

Image courtesy of Tourism Australia

fact means that much of Australia is arid. Due to the low-latitude position, most of the Australian landscape did not receive the rejuvenating benefits from the Quaternary glaciations to the soils that other continents did (Chapter 2).

Australia's rifting out of Gondwana formed passive margins on three sides of the continent (Chapter 4), positioning Australia distant from destructive plate margins. This is a major reason why Australia is the flattest and lowest of all continents. The lack of mountain building events has also limited the rejuvenation of the landscape and is another reason why Australia's landscape is so old and the soils so leached, poor and salt prone (Chapter 5). The flatness has also prevented the formation of major transcontinental river systems, which, together with the lack of rainfall, means that sediment supply into marine basins is very limited (Chapter 6).

The aridity, together with the infertile ancient soils, has shaped the flora and fauna and the carrying capacity of the land for agriculture and people (Chapters 3, 5 and 7). These factors have thus influenced Australia's demography, with most of the population concentrated near the more fertile and temperate coast in the southeast and far southwest of the continent (Figure 1.3).

Australia's distance from destructive plate boundaries means that the threat from geohazards, especially earthquakes, volcanoes and tsunamis, is different from that of Australia's neighbours. Australia's geographical position, however, means that tropical cyclones, droughts and floods, bushfires and heatwaves are common (Chapter 6).

Rocks and minerals in Australia span most of Earth's geological history (Box 1.4), with more than two-thirds of the continent underlain by Precambrian rocks (Chapters 2 and 10). This great span of geological time means that Australia has benefited from numerous geological events that give us both an iconic geoheritage and vast mineral and energy resources. These resources not only shaped how the nation developed (Chapter 8), but today are the backbone of the nation's economy (Chapters 4, 9 and 10).

We shall now look in more detail at how the geological foundation, and the 'pillars' it supports, have shaped our nation (Figure 1.1).

Australian geography: the land down under

Geology has positioned Australia between the southern trade winds and the northern monsoon belts. This geographical factor has had a profound influence on the continent's climate (Figure 1.6). If Australia were located in the Northern Hemisphere, it would occupy much of the Atlantic Ocean, in a position adjacent to the African Sahara Desert.

With the end of the last ice age, global temperatures increased and sea-levels rose rapidly by 125 m (Chapter 6). This rise flooded land bridges between Tasmania and the Australian mainland (11 ka), and between Australia and New Guinea (9 ka). The sea-level rise also inundated about one-seventh of the larger ice-age continent (Sahul), isolating Tasmania, the Australian mainland and New Guinea (Figure 1.2). Many smaller islands appeared around the evolving continental landmass as rising sea-level isolated high-standing land (e.g. King and Flinders islands between Victoria and Tasmania). New islands were fashioned as reef-building organisms constructed atolls and cays (e.g. Ashmore Reef on the North West Shelf). The rise in sea-level flooded deep river valleys (rias), creating the likes of Sydney Harbour, on the coast of New South Wales. This rise in sea-level ultimately created the Australia we know today (Figure 1.4).

Australia has a distinctive coastal outline (Chapter 6). It has a convex-shaped western and eastern coast, a large bight in the south, and three northward-pointing promontories (Figure 1.4): the Kimberley (WA), Arnhem Land (NT) and Cape York Peninsula (Qld). The submerged extensions of the continent form a differently shaped Australia. Some of these submerged regions (<125 m below sea-level [bsl]) were land bridges and freshwater lakes during the last ice age (Figure 1.2). In total, the continental extensions of Australia are 1.8 times larger than the exposed landmass (Table 1.1).

Because of its geographical position in the Southern Hemisphere, distant from the European and North American powers, Australia has colloquially been known as the land *down under*. Many books and songs have used this term. Geoffrey Blainey wrote about the *Tyranny of Distance*, referring to Australia's geographical position being far from the major powers and economies. In the Oceania region, however, Australia now has a dominant influence; it is the largest and most populous country, and has the largest economy and the most powerful military in the region.

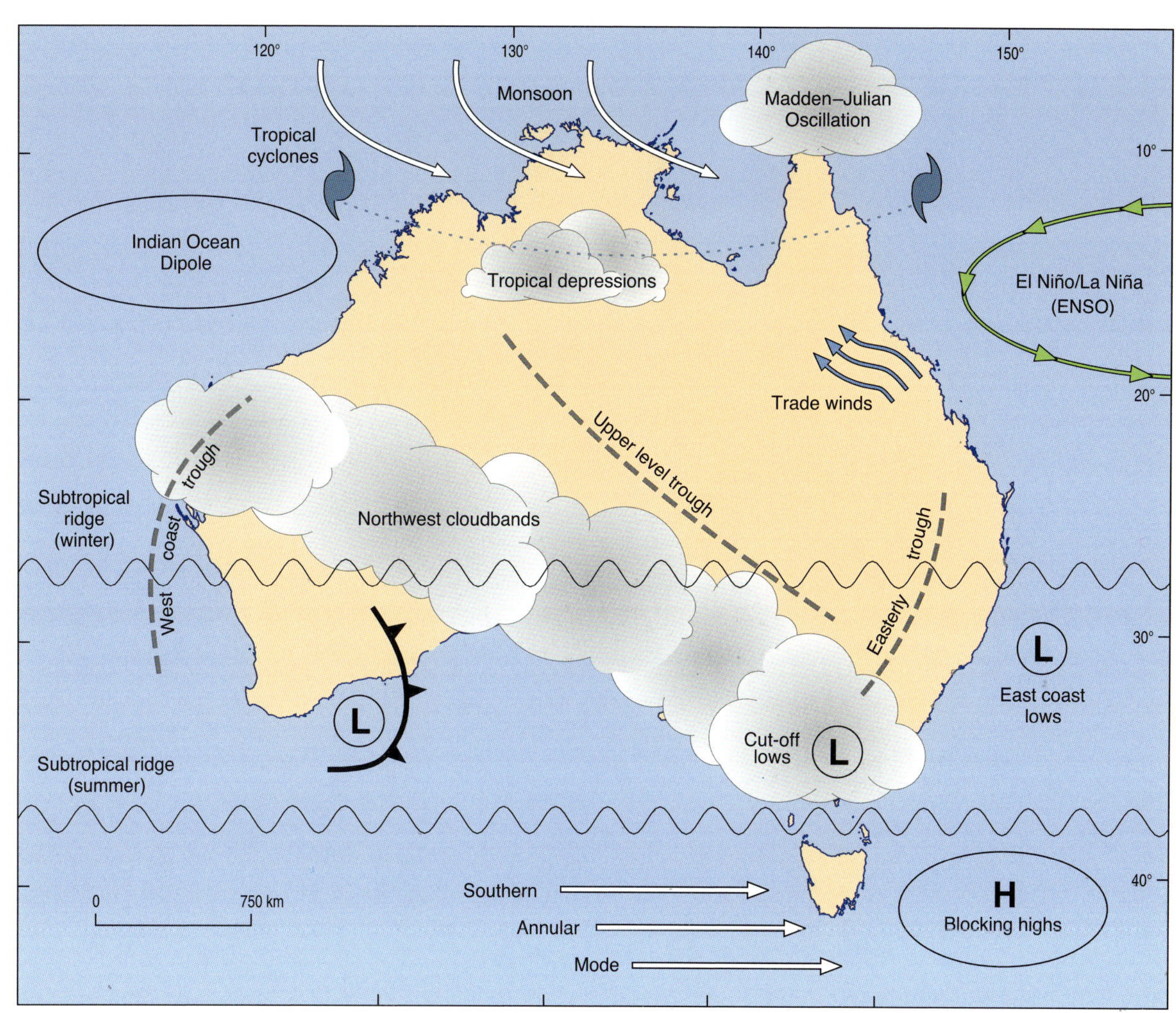

Figure 1.6: Australia's climate can vary greatly from one year to the next. The map shows the present main influences on Australia's climate. These influences have varying levels of impact in different regions at different times of year. (Source: Bureau of Meteorology)

Australian landscape: the flattest on Earth

The topography of the continent of Australia is subdued. With an average elevation of only 325 m, Australia is the lowest of all continents. Without major mountains, Australia's river systems have low and erratic flow rates compared with those on other continents (Chapters 5 and 7). Limited river systems have had a major influence on the availability of surface water, as well as navigable passageways into the continent.

Image courtesy of Tourism Australia

The Twelve Apostles sea stacks are a popular tourist destination along the Great Ocean Road, Victoria.

The major physiographic features of the Australian continent are a large Western Plateau with localised ranges; the Eastern Uplands, with the highest land concentrated along the Great Divide; and an intervening zone of Interior Lowlands (Figure 1.5). The highest point on the continent, Mt Kosciuszko, is only 2228 m asl, and the lowest point is Lake Eyre at around 15 m bsl (Box 1.5).

The lack of elevation across Australia is more than compensated for in landscape variety. In the centre and the west, there are vast stony and sandy deserts; in the east, sweeping plateaus and plains flank narrow coastal slopes. Australia's coastal features vary considerably, with broad, sandy beaches, lush vegetation, stark deserts, steep cliffs, swamps and mangroves, and extensive tidal flats (Chapter 6). These various coastal environments are backed by a great variety of landforms, ranging from the steep cliffs of the Blue Mountains west of Sydney, and the eroded volcanic edifices of the Glasshouse Mountains north of Brisbane, to the monotonous flatness of the Nullarbor Plain on the southern coast. The giant monolith Uluru in the Northern Territory (Box 1.1) and the striking beehive mountains of Purnululu (the Bungle Bungles) in Western Australia attract visitors from around the world, as do the country's beaches and rainforests. Many of these geological features are World Heritage listed.

Australian climate: beautiful one day—perfect the next

Continental drift has positioned Australia in its current latitude, ensuring that it is mostly hot, dry and sunny (Figures 1.6 and 1.7). The Australian Energy Resource Assessment (AERA) estimates that around 58 M petajoules (PJ) of solar radiation fall on the land surface, which is 10 000 times the current energy consumption. Australia has the highest solar radiation per square metre of any continent, and it is especially high in the northwest and centre of the country (Figure 10.1).

The abundant solar resource comes with a sting. The ultraviolet light levels are often extreme in Australia, especially between 10:00 am and 2:00 pm. For a predominantly fair-skinned population, this means that Australians have the highest skin cancer rates in the world. Various government programs have emphasised the slip (shirt)–slop (sunscreen)–slap (hat) strategy to encourage Australians to manage their risk. These programs have been extended to include seek (shade) and slide (sunglasses).

Heat accompanies such a sunny climate. Other than epidemics and war casualties, heatwaves, such as those during the summers of 1895–96 and 1938–39, have been responsible for more deaths (more than 400 each) than any other climate factor or disaster. As recently as February 2009, even with modern air-conditioning, some 374 people perished in the extreme heat. That heatwave was also one of the catalysts for the disastrous Black Saturday fires in Victoria, which killed 173 people and burnt 4500 km^2 of country that had endured a decade of drought.

All regions in Australia enjoy warm summers and relatively mild winters (Figure 1.7); it rarely snows in most state capitals. The highest temperature ever recorded in Australia was 53°C in 1889 at Cloncurry, Queensland. The coldest temperature was –23°C, recorded in 1994 at Charlotte Pass in the Snowy Mountains, New South Wales. The Snowy Mountains are Australia's main winter sport playground, with skiing from June until October. Although Australia's snowfields are not world class in vertical drop or snow quality, they occupy an area of around 1200 km^2 (for two months), which is similar to the area of permanent snow and ice in Switzerland.

As an island continent, Australia, more than any other continent, is subject to the influence of interseason climate variations, especially with respect to rainfall (Figure 1.8). These influences, illustrated in Figure 1.6, include the:

- El Niño/La Niña–Southern Oscillation (ENSO), which is a climate pattern across the tropical Pacific Ocean
- Indian Ocean Dipole (IOD), which is a measure of sea-surface temperatures between the western and eastern Indian Ocean
- Southern Annular Mode (SAM), also known as the Antarctic Oscillation (AAO)
- Madden–Julian Oscillation (MJO), which is a global-scale feature of the tropical atmosphere and influences the monsoon in northern Australia (Figure 1.6).

The ENSO index is calculated from monthly or seasonal fluctuations in the air-pressure difference between Tahiti and Darwin, which is a major

1.3: There are six 'seasons' in the Top End

The traditional owners of the Northern Territory's Arnhem Land describe six distinct seasons. These different times guide the peoples' cultural and spiritual practices:

- *Gudjewg*, from December to March, can be described as the 'true' wet season.
- *Banggerreng*, in April, is the season of clear skies.
- *Yegge*, from May to mid-June, is relatively cool, but humidity remains.
- *Wurrgeng*, from mid-June to mid-August, is the 'cold weather time'; humidity is low.
- *Gurrung*, from mid-August to mid-October, is hot and dry.
- *Gunumeleng*, from mid-October to late December, varies from a few weeks to months, and is the pre-monsoon season of hot weather with progressively more and more humidity.

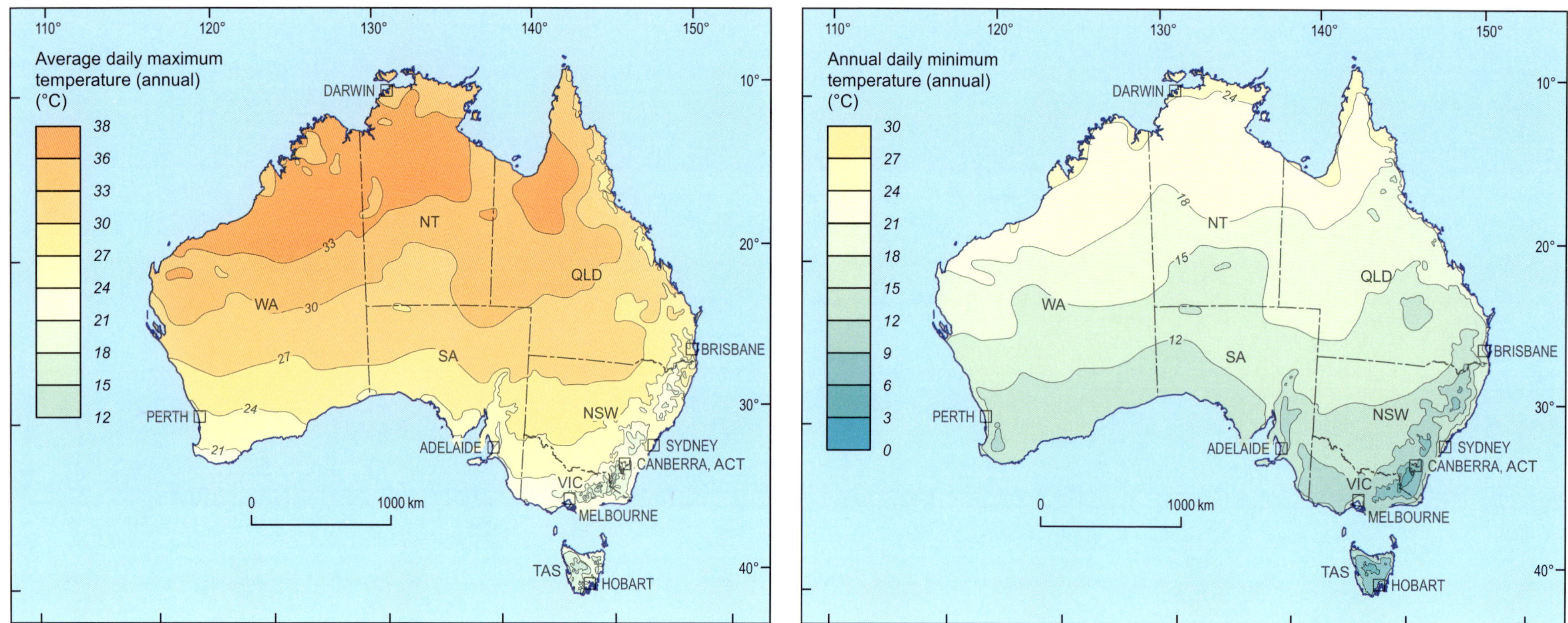

Figure 1.7: Average daily maximum and minimum temperatures. (Source: Bureau of Meteorology)

factor in controlling Australia's rainfall. Sustained negative values indicate El Niño episodes, which are usually accompanied by sustained warming of the central and eastern tropical Pacific Ocean, a decrease in the strength of the Pacific trade winds, and a reduction in rainfall over eastern and northern Australia. Positive values are often associated with La Niña episodes, which lead to stronger Pacific trade winds and warmer sea temperatures to the north of Australia. Waters in the central and eastern tropical Pacific Ocean become cooler. Together, these factors increase the probability of rain in eastern and northern Australia.

The national average annual rainfall of 470 mm varies greatly each year and is distributed unevenly around the continent (Figure 1.8). In fact, the one thing that is predictable about Australia's rainfall is that it is unpredictable. In general, southern Australia has four seasons in terms of temperature, but not necessarily rain. The tropical north of the continent has two dominant seasons, with a relatively warm and dry winter, and a summer monsoon (December–March) known as the 'wet' season (see *Did you know?* 1.3).

The most arid area in Australia is the centre of the continent, particularly the Lake Eyre Basin, which averages less than 125 mm annually (Figure 1.8). Arid areas such as these have very considerable variability in rainfall, with several years' worth of rain falling from a single major weather system, such as in 2009–10. Birdsville in outback New South Wales has an annual average rainfall of 133 mm. In 1917, some 659 mm of rain was recorded, while seven years later, in 1924, only 14 mm was recorded.

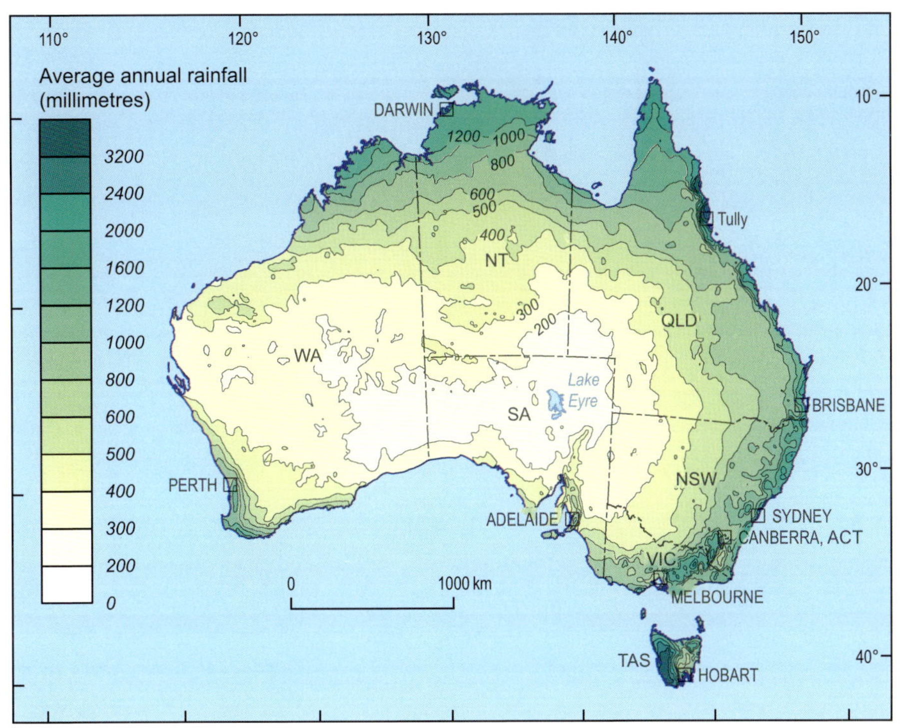

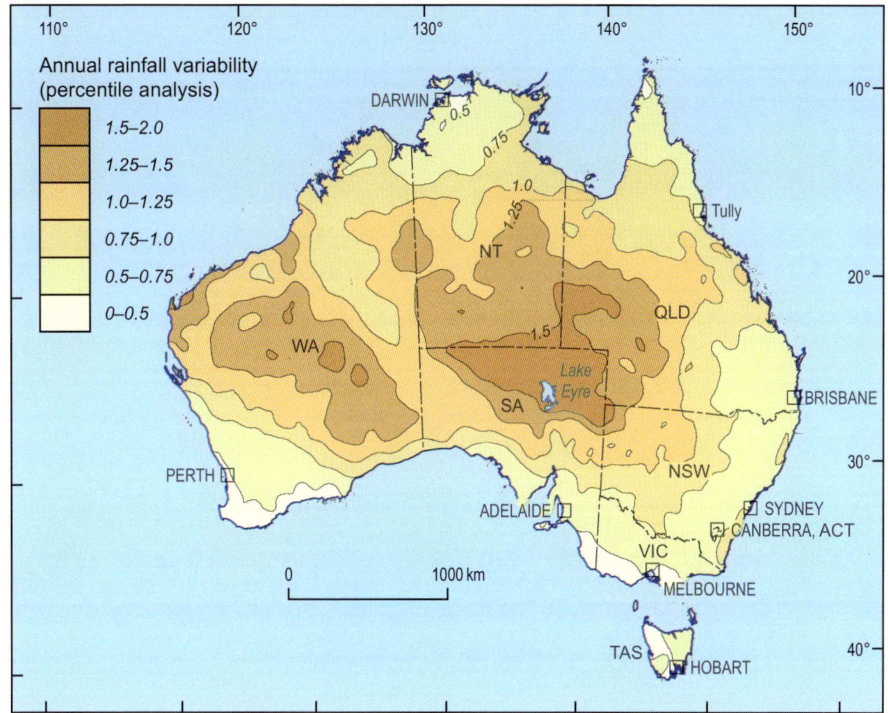

Figure 1.8: Average annual rainfall (1961–90) and rainfall variability. (Source: Bureau of Meteorology)

The desert areas of Australia are much wetter than deserts in Africa and South America, which average less than 50 mm per year. The relatively warm encircling oceans (Chapter 6) and the lack of blocking mountain ranges (Chapter 5) ensure that some precipitation usually falls on inland Australia. Both surface water and groundwater are scarce (Chapter 7). Water is therefore a point of dispute between those wanting to use it, such as for agriculture or mining, and those wanting to preserve it, for environmental flows.

The wettest regions of Australia are located in the tropical northeast and in the far southwest of Tasmania. The mountains in the hinterland of Tully in northeast Queensland have recorded (in 2000) more than 12 m in annual rainfall. The town is Australia's wettest, with average annual rainfall of more than 4 m. Rain on the western and southern coasts of the mainland falls primarily in the winter, whereas the east coast receives rain in all seasons, with more during the summer. The summer of 2010–11 was particularly wet, with catastrophic flooding throughout Queensland and northeast New South Wales, resulting in many fatalities. The summer of 2011–12 repeated this pattern, thankfully without the fatalities. The driver of these events is thought to be a particularly strong La Niña event. These floods followed a decade of drought in eastern Australia. As captured by Mackellar in 1904—'of droughts and flooding rains'—southern Western Australia remained very dry. There were persistent temperatures above 40°C in the summer of 2011, which resulted in bushfires that burned houses in the suburbs around Perth.

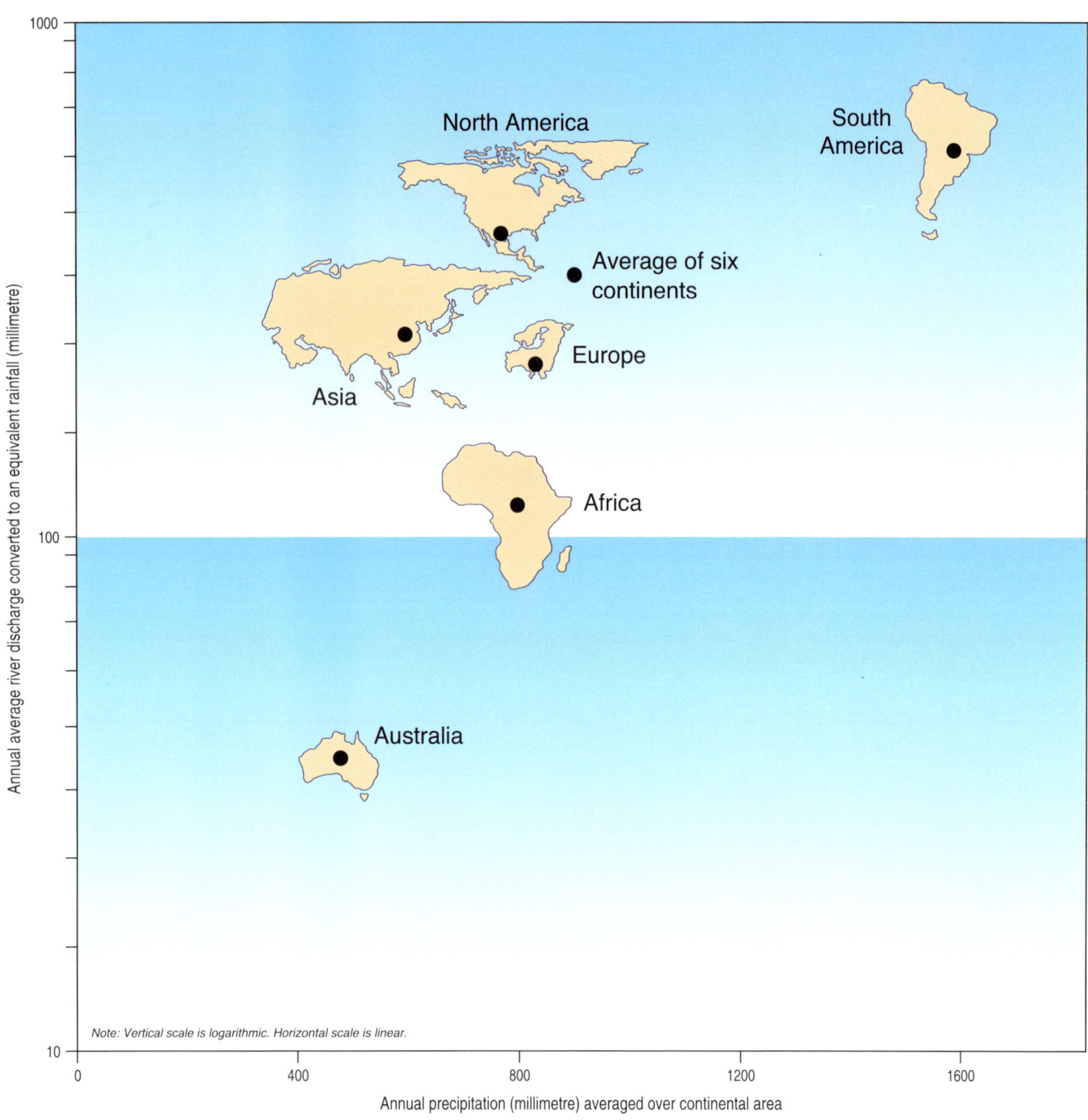

Figure 1.9: Australia's precipitation budget compared with the other inhabited continents. Australia is a standout, with low rainfall and low river discharge. (Sources: Lamont-Doherty Earth Observatory; Bureau of Meteorology, 2009)

Australian water and soils: short but not sweet

In the driest inhabited continent (Figure 1.9), Australia's environment and people face the persistent challenge of securing water. Even with the addition of water, Australia's soils are not very productive (Chapter 5). They are old and fragile and prone to damage and erosion, especially when subjected to European industrial-scale agricultural practices.

About one-third of the Australian mainland lies north of the Tropic of Capricorn, with the remainder stretching to latitude 39°S (Figure 1.4). About 70% of the country is arid or semi-arid, and a large part of the centre is unsuitable for large-scale settlement. More than one-third of the continent is classified as desert. There are 11 principal deserts, the largest of which is the Great Victoria Desert, spanning 348 750 km^2 (4.5% of the continent) through Western Australia and South Australia (Figure 1.4). As a consequence, Australia's rivers have the world's lowest amount of discharge into the sea of any continent (Figure 1.9). This lack of discharge has had a major influence on the physical shaping of the coast's geomorphology, such as ports and harbours, or lack thereof (Chapter 6).

Australia's Murray and Darling rivers make up one of the great river systems of the world in terms of areal drainage. Together, their catchments form the 1 M km^2 Murray–Darling Basin, which covers around 14% of the mainland area. Throughout human history, rivers have provided geographical barriers, as well as corridors for exploration, settlement and development. However, the generally low flow

LAKE EYRE, CONTINENTAL AUSTRALIA'S LOWEST POINT (BOX 1.5)

Early explorers searched in vain for a vast 'inland sea'; the time was not right for them to see one of Australia's most fascinating places. Instead, they found mostly arid country with few people and signs of life. But there are 'inland seas' both past and present. Lake Eyre in South Australia is Australia's largest salt lake and is situated within a major internal river drainage system of the interior lowlands. It actually comprises two modern lakes, North Lake Eyre and South Lake Eyre, which are now connected by a narrow channel. Almost three-quarters of the runoff from Lake Eyre's 1.3 M km^2 catchment finds its way via an intricate network of channels known as the Channel Country through the deserts to its north and east towards Lake Eyre.

Although most of the water is lost through evaporation or absorption, on the rare occasions when the lake fills, it temporarily becomes Australia's largest lake as it spreads out over a vast area and reaches almost 6 m at its deepest point. Then it becomes a wildlife haven and is renowned for the swift return of primitive fairy shrimps, fish, and vast numbers of breeding birds, including pelicans.

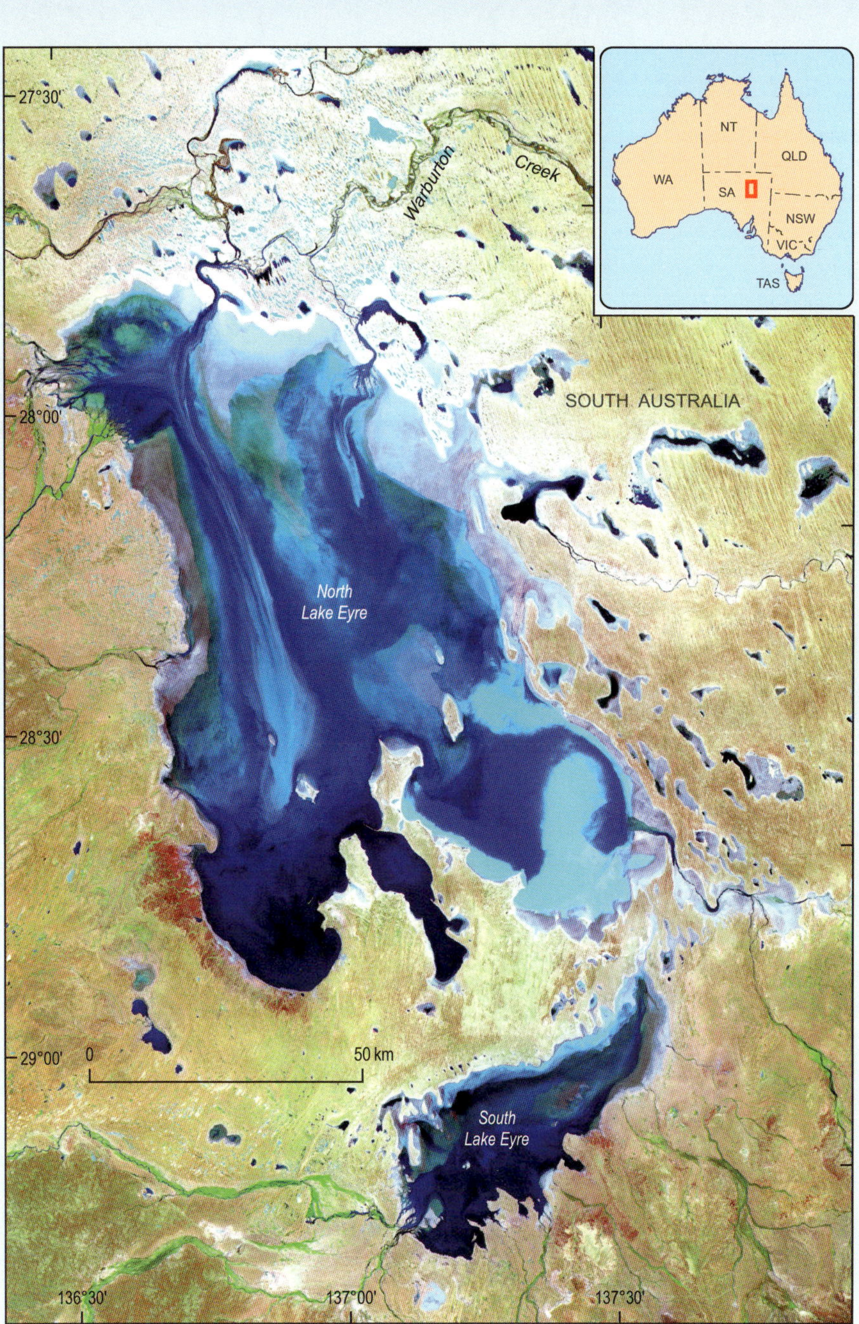

© Getty Images [C Roberts]

Lake Argyle, north Kimberley, Western Australia, is Australia's largest artificial lake. Its stored water is used for irrigation and recreation.

rate in the rivers and high salinity mean that the Murray–Darling system is no Yangtze or Mississippi in terms of highly fertile floodplain soils or even a reliable navigation corridor facilitating development of Australia's arid interior (Chapter 7).

Numerous rivers drain the tropical north, with strong flows in the summer 'wet' season. Capturing some of this flow occurs in dams with more than 19 *SydHarbs* in Lake Argyle on the Ord River in the Kimberley region of northwest Australia, which is second only to the 22 *SydHarbs* held in the Gordon–Lake Pedder hydroelectric dam in Tasmania. Many of Australia's lakes and rivers are either dry or salty, despite being portrayed in blue on maps (e.g. Figure 1.4). The largest lake is Lake Eyre in the central south of the country. This vast salt lake, which is more than 9500 km^2 in area, is dry for lengthy periods (Box 1.5). In fact, Englishman Sir Donald Campbell broke the land-speed record there in July 1964 with an average speed of 648.8 kph. From time to time, particularly following heavy rains as in 2010, the inland draining rivers of the north feed water—via the 'Channel Country' river system—into Lake Eyre. Like all desert regions, life abounds in these infrequent times of plenty.

Not all water is at the surface, and some of it is very old (Chapter 7). The Great Artesian Basin covers much of inland eastern Australia (Chapters 2, 4 and 7), and it was this resource that opened up the arid interior to pastoral development in the late 1800s. The groundwater in the southwest of the basin originated as rain that fell on the uplands of the east up to 2 Ma. Groundwater provides nearly a third of the water that is consumed, but makes up less than one-fifth of the total available freshwater resource in Australia. Geology, landscape and climate dictate the nature of the groundwater resources, with the geology providing the underlying architecture of aquifers and aquitards that define the reservoirs and seals, respectively. Groundwater also provides almost a third of the high-security water that is vital for agriculture and towns during droughts, and is the only source of water for many regional and remote communities and many mining operations. It could be said that groundwater is the lifeblood that sustains Australian communities, industries and habitats, particularly during the tough times (Chapter 7).

Fertile soil for food production is another vital factor that shapes people. The great civilisations of the world were built on floodplains of major river systems whose sources were located in young and active mountain ranges. In much of the Northern Hemisphere, the soils are also young and fertile because they formed after the retreat of major glaciers of the last ice ages. During the Pleistocene, however, Australia occupied a low-latitude position and so ice-age impacts were limited in Australia and did not rejuvenate the old landscape. Weathering of recent volcanic rocks, such as in New Zealand or Indonesia or the western district of Victoria, also creates fertile soils. With recent volcanism restricted to eastern Australia and a position distant from subduction zones, most of Australia's soils lack the nourishment provided by volcanoes. Thus, Australia's landscape is old and covered by soils that, for the most part, are also old. They are also salt prone and nutritionally and organically impoverished (Chapter 5). The poor soils mean that, despite its large land surface area, much of Australia was and is unsuitable for growing food crops to support a large population. Consequently, each wave of incoming humans has struggled to establish an agrarian economy, with Aboriginal people mainly limited to managing the land for hunting and gathering (Chapter 7).

Australian flora and fauna: there's nothing like it

Australia has a remarkably long fossil record that charts the evolution of life on the planet. Examples include the stromatolites and biomarkers from rocks as old as 3500 Ma in the Pilbara (WA), the Neoproterozoic Ediacaran of the Flinders Ranges (SA) and fine specimens of Devonian fish and Plio-Pleistocene megafauna across the continent (Chapter 3). Fossil life not only informs us of our origins, it is a source of great wealth through helping to generate the great banded iron-formations, as well as the coal and hydrocarbon resources that we rely upon (Chapters 4, 8 and 9).

Life in Australia today has a Gondwanan legacy, but the partial isolation of the continent from others over most of the Cenozoic has ensured an evolution that is largely unique. The northwards continental drift, coupled with the end-Pleistocene processes, led to increasing aridity. Adaptation to this aridity is a distinctive feature that has shaped the evolution of Australia's flora and fauna (Chapters 3 and 5). So typical of this country are its *Eucalyptus* trees, with leaves full of aromatic gums, and acacias, the floral emblem of Australia but also known in Africa. One eucalypt, the mountain ash, *Eucalyptus regnans*, is arguably the world's tallest flowering plant.

Spinifex (*Triodia*) grows across much of arid Australia. Compared with non-native grasses (e.g. wheat), gathering seed to make food from these plants is labour intensive.

Wheat is the major winter crop grown in the vast plains of Australia. Australia exported almost $5.5 B worth of wheat in 2011.

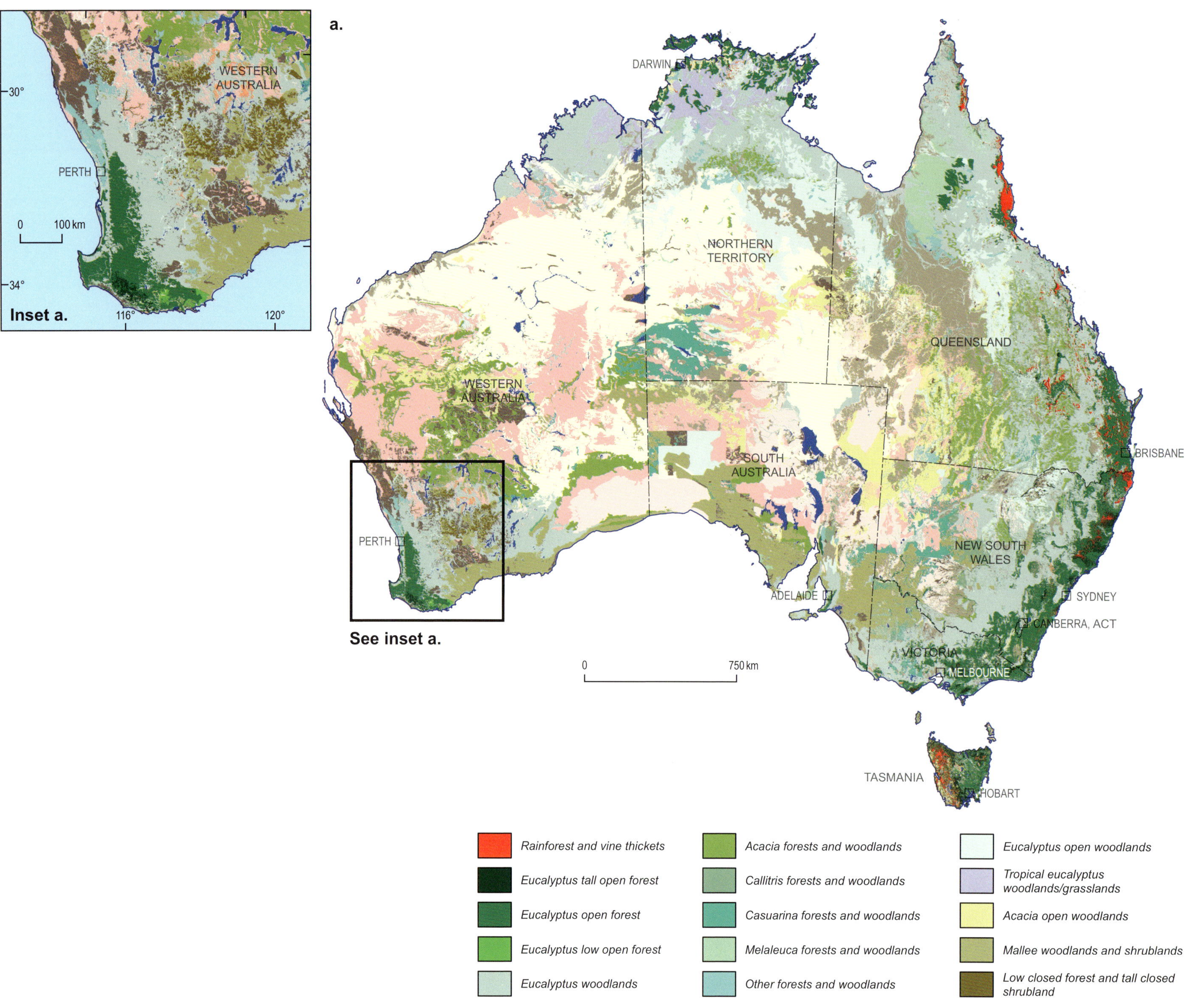

Inset a.
WESTERN AUSTRALIA
30°
PERTH
0
100 km
34°
116°
120°
a.
DARWIN
NORTHERN TERRITORY
QUEENSLAND
WESTERN AUSTRALIA
SOUTH AUSTRALIA
BRISBANE
NEW SOUTH WALES
PERTH
ADELAIDE
SYDNEY
CANBERRA, ACT
See inset a.
0
750 km
VICTORIA
MELBOURNE
TASMANIA
HOBART
Rainforest and vine thickets
Eucalyptus tall open forest
Eucalyptus open forest
Eucalyptus low open forest
Eucalyptus woodlands
Acacia forests and woodlands
Callitris forests and woodlands
Casuarina forests and woodlands
Melaleuca forests and woodlands
Other forests and woodlands
Eucalyptus open woodlands
Tropical eucalyptus woodlands/grasslands
Acacia open woodlands
Mallee woodlands and shrublands
Low closed forest and tall closed shrubland

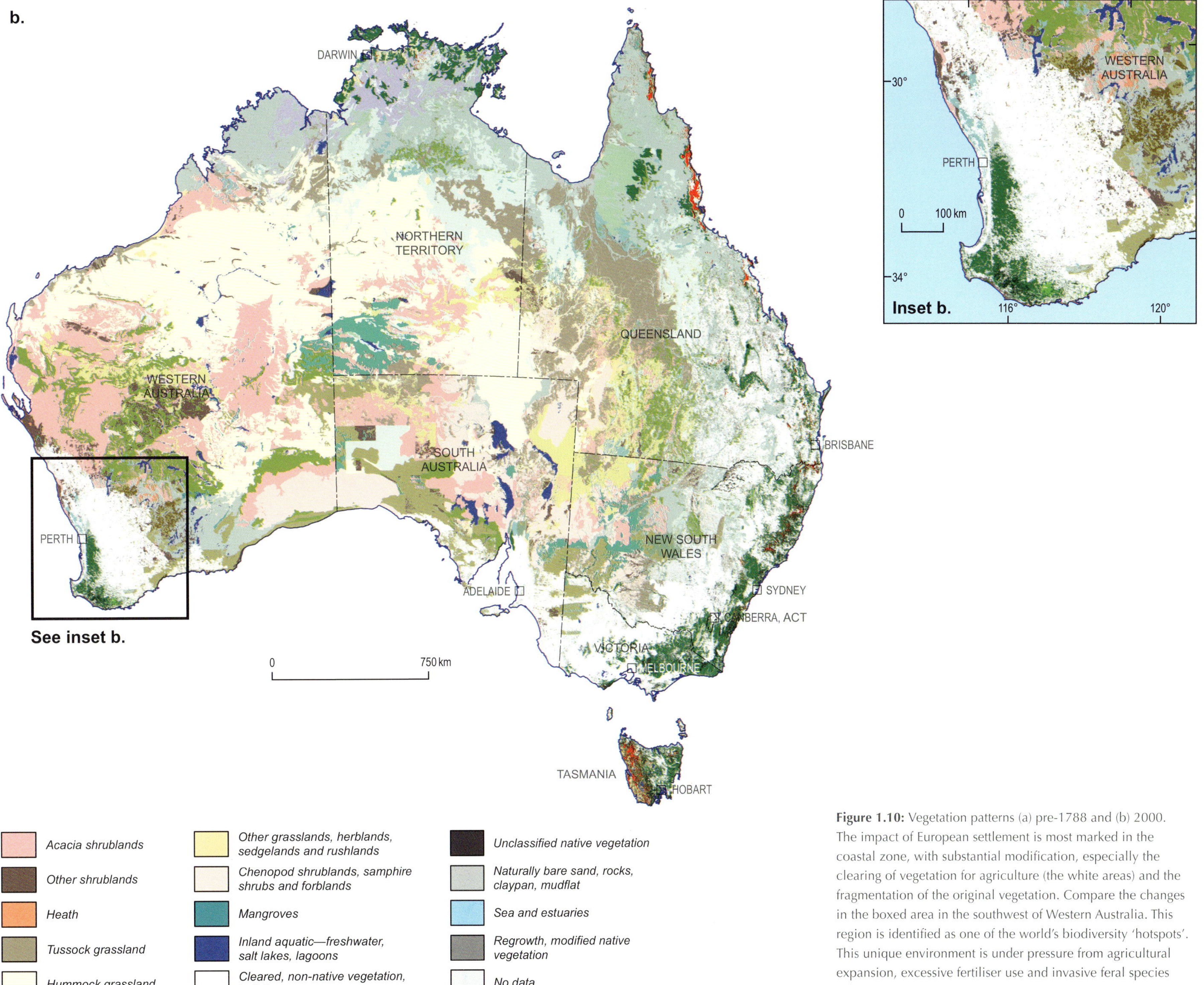

Figure 1.10: Vegetation patterns (a) pre-1788 and (b) 2000. The impact of European settlement is most marked in the coastal zone, with substantial modification, especially the clearing of vegetation for agriculture (the white areas) and the fragmentation of the original vegetation. Compare the changes in the boxed area in the southwest of Western Australia. This region is identified as one of the world's biodiversity 'hotspots'. This unique environment is under pressure from agricultural expansion, excessive fertiliser use and invasive feral species (Chapter 3).

Did you know?

1.4: Unique Australia

- Australia has around 10% of the world's biodiversity.
- Southwest Australia, one of the world's five Mediterranean-type ecosystems, is listed as a unique biodiversity 'hotspot'.
- Of the estimated 20 000 species of vascular plants found in Australia 16 000 are found nowhere else in the world.
- Of the 378 species of mammals in Australia, more than 80% are endemic.
- Of the 869 Australian reptiles, 773 are found nowhere else.
- Around 80% of the marine biota of the Southern coast are endemic.

© Getty Images [W Bibikow]

The capacity of a land's biota for domestication is a prerequisite for a reliable food supply, an essential input for the development of large human populations and complex societies. Contrast the food value of rice, wheat and corn, endemic to other continents, with Australia's spinifex and other native grass seeds. Contrast the utility of mammals such as horses, cattle, sheep and goats, endemic to most other continents, for food and work with Australia's marsupials—kangaroos, koalas and wombats.

Australia is a land like no other, with about a million different native species (Chapter 3), and yet only the macadamia nut tree has been successfully domesticated for human food supply on any large commercial scale. Aboriginal and Torres Strait Islander people occupied a land that not only became increasingly dry, but also had little in the way of natural resources amenable to domestication. Australia's large agricultural industry has been built on imported species of flora and fauna, which have led to many environmental problems.

The partial isolation of the Australian continent has also meant that more than 80% of the continent's flowering plants, mammals, reptiles and frogs are unique to Australia, along with most of its freshwater and even marine fishes and almost half of its birds (Chapter 3). Australia's marine environment is home to 4000 fish, 1700 coral, 50 marine mammal and 110 seabird species, about 75% being endemic. About 80% of marine species found in southern Australian waters occur nowhere else in the world, while only 10% of the known marine species in northern Australia are endemic.

Australia has more than 140 extant species of marsupials, including koalas, wombats and the Tasmanian devil (now found only in Tasmania). Australia (with New Guinea) hosts another unique animal group, the monotremes—egg-laying mammals, often referred to as 'living fossils'. The most distinctive is the platypus, a river-dwelling animal with a duck-like bill, a fur-covered body and webbed feet. This diversity has its drawbacks. Australia is infamous for its highly venomous and human-eating species, including spiders, snakes, sharks, crocodiles, jellyfish and stinging bushes, let alone biting flies and mosquitoes (Chapter 3), many of which curtail people's leisure pursuits, especially in the northern coastal areas.

The vegetation patterns in Australia today are strongly linked to the prevailing rainfall (Figures 1.8 and 1.10). Australian vegetation has evolved and adapted over time to the arid conditions, and so by global standards Australia's deserts are relatively densely vegetated. In the north, the extensive tropical forests grade south into more open scrub, and then into savannah zones where rainfall is less reliable (Figure 1.10). Vegetation patterns have changed during the last million years along with changing climate cycles (Chapters 3 and 5). The use of fire by Aboriginal people to promote open woodland and grasslands is also thought to have contributed to these changes (Chapter 3).

The development of extensive European-style agriculture based on imported species had a substantial impact on the pre-1788 vegetation patterns (Figure 1.10), with wholesale modification of habitats and fragmentation of vulnerable plant communities such as the tropical and temperate rainforests. Although the inland appears to have felt less impact, much is used for raising cattle. Even though cattle densities may sometimes be measured in square kilometres per animal, when the rains fail these hard-hoofed animals, unlike the native species, exact a high toll on fragile soils.

Australia has the world's third largest exclusive fishing zone, covering 11 M km^2 and extending up to 200 nautical miles out to sea (Appendix 1.2.4). The east and west coasts of Australia have intermittent south-flowing warm currents, so the waters are less productive than for other continents, such as the west coasts of southern Africa and South America. Australia therefore only ranks 52nd in world volume of fish landed, with nearly 70% of the catch being exported (Chapter 6).

A relatively small number of Australian farmers feed, clothe and house many millions of people around the world. For example, Australia in 2009–10 produced around 22 Bt of wheat, exporting more than half, with Indonesia, Malaysia and Japan being the largest importers. Other major agricultural exports include cotton, rice, timber products, beef, sheep, wool and fish. These energy-and water-intensive industries partly account for Australia's high per-capita generation of greenhouse gases (Chapter 10). Australia indirectly exports its scarce water through its resources. For example, about 960 L of water are used to make one litre of wine.

Australian mineral and energy resources: wealth and choice

Geology has ensured that Australia is an energy- and mineral-rich country by creating an abundance of coal, iron (Fe), aluminium (Al), lead–zinc (Pb–Zn), gold (Au), copper (Cu), gas, rare earth elements and uranium (U), to list a few (Appendix 8.1.1). There is a positive correlation between the size of a country, the age range of its rocks, and its mineral and energy wealth. This comes about partly because of the greater probability to form and then preserve a resource in a large area and the long period of geological time in which to do this. Australia wins on both counts (Boxes 1.3 and 1.4).

Aboriginal people have mined and traded in ochre and stone tools for millennia. The discovery of Australia's metalliferous mineral resources by new immigrants, however, is what helped to save the early colonies from economic ruin in the 1850s—first with Cu and later Au. The allure of gold especially brought migrants from around the world and changed the social fabric of Australia forever (Box 1.2; Chapter 8).

Today, growth in Australia's economy is one of the highest in the Organisation for Economic Cooperation and Development (OECD), largely because of these mineral and energy resources. Australia's balance of trade is at a 146-year high, with much of this fortune driven by strong demand from the rapidly industrialising Asian economies (Chapter 9). Economic growth and energy use are strongly correlated, and Australians are lucky in having an assortment of energy choices. Coal is readily available and is the primary energy source; its use in electricity generation is one of the major reasons for Australia's high per-capita CO_2 emissions. Uranium, gas in its various forms, and renewable energy, such as geothermal and solar power, provide alternatives (Chapter 10).

Image courtesy of Rio Tinto

Loading iron ore at Parker Point, Dampier, Western Australia.

The favourable location of abundant high-quality coal in particular is a clear example of how geology has provided choices that have shaped the Australian people.

Australian geohazards: living on the edge

Australian people live distant from the active boundaries of the Australian Plate and the geohazards associated with large earthquakes and volcanoes. Australia, of course, is not immune to the destructive power of earthquakes—the fatal event in Newcastle in 1989 being the most significant example (Chapters 2 and 11). Most Australians, and much of Australia's economic infrastructure, is located close to the sea, making distantly generated tsunamis a potential major geohazard (Chapter 6). We do all live in one earth system, and the Chilean volcanic eruptions in mid-2011 grounded airlines in southern Australia, with the ash cloud circumnavigating the globe several times.

The geographical position and tectonic setting of the continent mean that cyclones, floods, droughts, bushfires and heatwaves are the primary geohazards for the Australian people. Climate change modelling indicates that such hazard impacts are likely to be greater in the future, with rising sea-levels also of great concern. As we improve understanding of how our plate responds to stress and unravel the changes in climate from a geological perspective, the Australian people will be better placed to make informed choices regarding their future protection from these geohazards. Of course, this can only happen with the 'consent of geology' (Figure 1.1).

The Geoscience hub of the Joint Australian Tsunami Warning Centre located at Geoscience Australia, Canberra.

Image by Adrian Yee

Knowing Australia's limits

Here we think it is worth making a digression to understand Australia in terms of a legal definition. Most Australians live on mainland Australia or in Tasmania, only small populations live on the many smaller islands scattered about the nation's wider territory (Figure 1.3). Most people think of Australia as a large landmass of deserts and sweeping plains. Australia, however, is really a maritime nation with a confirmed marine jurisdiction of more than 14 M km^2 (Frontispiece), which is almost twice the area of Australia's landmass (Table 1.1 and Appendix 1.2.4).

Today, the country or sovereign nation of Australia consists of:

- mainland Australia and Tasmania, as well as many thousands of small fringing islands and numerous larger ones that rise from the shelf
- a number of remote mid-ocean islands that form parts of Australian states or Australia's external territories (Box 1.6).

Australia also claims the Australian Antarctic Territory, which is an area of nearly 6 M km², or 42% of the total landmass of Antarctica (Table 1.1). The Australian Antarctic Territory has a combined coastline length of about 5500 km. The highest point on the icecap occurs at 4100 m, with the highest recognised mountains being Mt McClintock in the eastern sector at 3490 m and Mt Menzies in the western sector at 3355 m. If the Antarctic Territory is included, Australia is the second largest country in the world.

Australia's seas

The basis for the granting of Australia's confirmed marine jurisdiction is lodged very much in geology (Frontispiece; Figure 1.11; Table 1.1). The geomorphological limits of Australia are defined by the foot of the continental slope or, where present, the continental rise. The continental slope covers around 6.8 M km² (Table 1.1) and contains the continent's greatest variety of geomorphological features, the largest of which are related to rifting on all but the northern margin (Chapter 4). The slope around Australia contains an unusual number of large terraces and trough-bounded marginal plateaus. The geological limit of the continent is defined by the continent–ocean boundary, which marks the change from extended, thinned and modified continental crust to unequivocal oceanic crust.

The margin adjacent to mainland Australia and Tasmania has a shelf of variable width (Frontispiece). In the east, the shelf break often lies within 15 km of the coast, and much of the margin comprises continental slope. The continental shelf is somewhat wider in the Great Australian Bight and in Queensland, where it acts as the base for the Great Barrier Reef. The extensive area of shelf to the north of Australia represents the extension of the Australian continent to the Java Trench and eastwards into Papua New Guinea, parts of the former Sahul landmass.

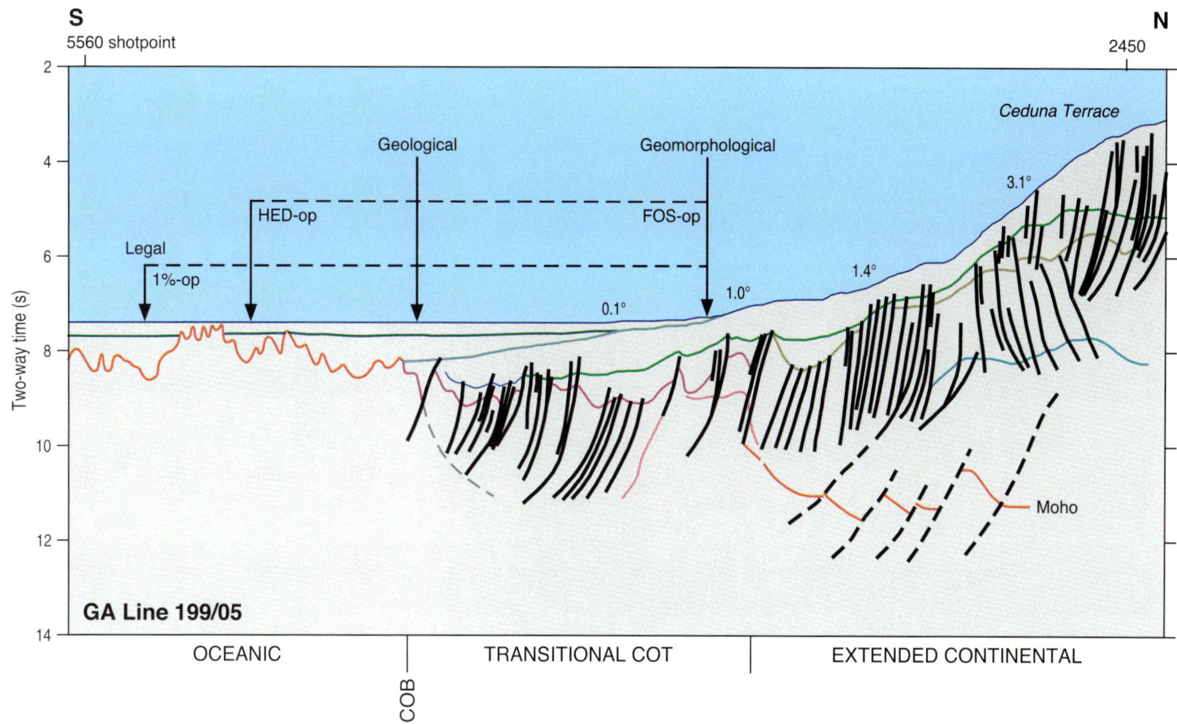

Figure 1.11: Continental margins and their geomorphological, geological and legal limits, explained by this interpreted seismic line across the Great Australian Bight margin of southern Australia. The section is from the outer Ceduna Terrace and underlying Ceduna Sub-basin (Chapter 4), across the continent–ocean transition (COT) zone, continent–ocean boundary (COB) and onto the slow-spreading oceanic crust of the Southern Ocean. The location of the geomorphological, geological and legal limits to the continental margin is illustrated.

Table 1.1: Areas of Australia's land and maritime jurisdiction (see Frontispiece)

Land	Area (million km^2)		% surface area of Earth
Australia and island external territories		7.69	1.51
Australian Antarctic Territory		5.90	1.16
Total Australia/island territory/Australian Antarctic Territory land area		**13.59**	**2.66**
Internal waters			
Approximate total marine area landward of territorial sea baseline (TSB)		**0.26**	**0.05**
Coastal waters (CW)			
Internal waters	0.24		
Area 3 nautical miles seaward of TSB	0.15		
Total area of CW		**0.39**	**0.08**
Territorial sea (TS)			
Australia and island external territories		0.68	0.13
Australian Antarctic Territory		0.17	0.03
State (coastal waters) 3 nautical mile portion of TS	0.15		
Commonwealth portion of TS	0.70		
Total area of TS (full 12 nautical mile zone)		**0.85**	**0.17**
Exclusive economic zone (EEZ)			
Australia and island external territories		8.15	1.60
Australian Antarctic Territory		2.04	0.40
Total area of EEZ		**10.19**	**2.00**
Continental shelf [a]Commission on the Limits of the Continental Shelf			
Continental shelf to 200 nautical miles (200CS)		10.19	2.00
Extended continental shelf (ECS)	**Submitted**	**Confirmed**	
Australia and island external territories (see Frontispiece for locations)	2.69	2.56	0.50
Australian Antarctic Territory—Australia requested the Commission[a] not consider this region for the time being	0.68		
Total area of submitted ECS	**3.37**		
Total area considered by the Commission[a] and yet to be resolved	**0.08**		
Total area of confirmed continental shelf (200CS+ECS)		**12.75**	**2.50**
Australia's confirmed marine jurisdiction (TS+EEZ+ECS)			
Australia and island external territories		11.39	2.23
Australian Antarctic Territory		2.21	0.43
Total area of Australia's confirmed marine jurisdiction		**13.60**	**2.67**
Australia's full confirmed marine jurisdiction (including marine areas landward of TSB to coast *ca* 0.26 M km^2)		13.86	2.72
Australia's maximum possible marine jurisdiction		**14.62**	**2.87**
Australia's complete jurisdiction (land + full confirmed marine)			
Australia and island external territories		19.34	3.79
Australian Antarctic Territory		8.11	1.59
Total area of Australia's complete, confirmed jurisdiction		**27.45**	**5.38**

HEARD ISLAND AND MCDONALD ISLANDS (BOX 1.6)

Australia's islands on the Kerguelen Plateau—Heard Island and the McDonald Islands—are located 4100 km southwest of Perth (WA) and about 1500 km north of Antarctica. Heard Island was discovered by Captain Heard while sailing a great circle route from Boston to Melbourne in 1853. The McDonald Islands were discovered in 1854 by Captain McDonald.

The Kerguelen Plateau, which formed 118–110 Ma, is one of the largest volcanic plateaus in the world and is the largest in the Southern Ocean. It extends for more than 2200 km mostly in water depths of 1000 m to 4000 m, but was emergent or under shallow water for up to 40 Ma, as indicated by wood fragments and coal found in Late Cretaceous sediments.

Heard Island and the McDonald Islands are home to Australia's only active volcanoes, with the most recent episodes of volcanic activity occurring on Heard Island since the mid-1980s, and on the McDonald Islands during the 1990s. The activity on the McDonald Islands group resulted in the island doubling in size and increasing in elevation by about 100 m.

Heard Island has a land area of 368 km^2 and is 43 km long. The island is dominated by the active volcano Big Ben, whose summit, Mawson's Peak, named for Australian geologist and Antarctic explorer Sir Douglas Mawson (Box 2.2), rises to 2745 m asl and is the highest Australian mountain outside of the Australian Antarctic Territory. Heard Island and the McDonald Islands were designated as World Heritage in 1997 because of their pristine sub-Antarctic ecosystems and geological activity.

Figure B1.6: The modern-day Heard Island is dominated by Big Ben, a roughly circular active volcanic cone that rises 2745 m above sea-level. A number of glaciers descend the slopes of the mountain, including the Gotley and Lied glaciers, which are studied for their sensitivity to climate change.

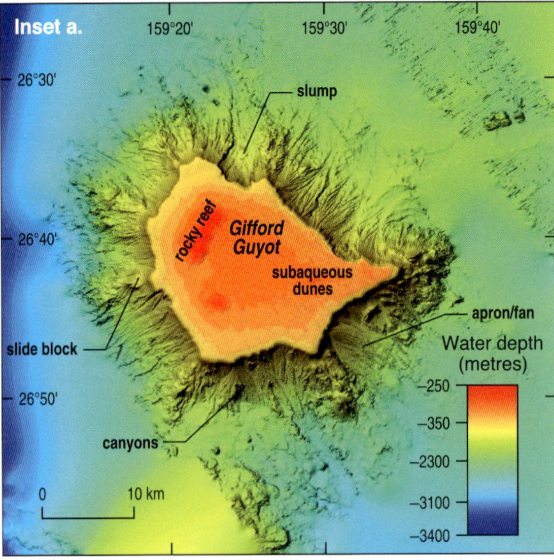

Figure 1.12: Multibeam bathymetry image of the 15.6 Ma Gifford Guyo. The guyot forms part of the Lord Howe Seamount Chain to the east of the Tasmantid Seamount Chain (Figure 2.3). The seamounts are elevated areas of volcanoes generated from mantle hotspots over the northward-moving Australian Plate. The Gifford Guyot has an extensive flat summit *ca* 200 m bsl. Guyots have highly variable seabed habitats, making them an important foundation for diverse marine life (Chapter 6).

The relatively small islands of Lord Howe in the Tasman Sea (Figure 1.12), and Heard/McDonald (Box 1.6) in the Southern Ocean, surmount very large mid-ocean plateaus (Frontispiece). These plateaus, which sit above the level of the deep ocean floor, generate large associated continental margins despite the small area of the exposed landmasses.

The presence of numerous marginal plateaus and terraces within the margins of mainland Australia and Tasmania is a distinctive feature of the Australian margin compared with other continents. Many of the plateaus off eastern and southern Australia are underpinned by rifted/extended continental crust. In contrast, the plateaus off Western Australia are commonly heavily modified by magmatism generated by the breakup of Gondwana (Chapter 4).

Deep ocean-floor provinces occur in water depths of more than 4000 m adjacent to all margins but are absent in the north and around Norfolk Island, off the eastern mainland. Adjacent to mainland Australia, they are most extensive in the east and south of the jurisdiction, particularly where the slope is steep and narrow. Around the Christmas and Cocos (Keeling) islands, most of the surrounding seafloor is deep ocean floor. Similarly deep ocean floor lies close to Macquarie Island. The 6700 m deep Hjort Trench lies adjacent to the southwest portion of the 1600 km long Macquarie Ridge Complex and is the deepest seafloor adjacent to Australian territory (Frontispiece).

Volcanic seamounts and guyots of various sizes are a common aspect of the deep ocean floor. The north-trending Tasmantid Seamount Chain extends along the eastern edge of the north Tasman Basin and Cato Trough off eastern Australia (Figures 1.12 and 2.3). A number of individual seamounts/guyots within this chain rise spectacularly from the relatively flat-lying deep ocean floor of the Tasman Basin at 4500 m depth to within less than 200 m of sea-level (Figure 1.12).

Opportunities and responsibilities within Australia's marine jurisdiction

Australia has sovereign rights for resources across a vast marine jurisdiction. However, many parts of the jurisdiction are remote and poorly known, and lie in water depths of 1000 m to 4500 m. Previous and current petroleum exploration has mostly occurred in the relatively shallow waters. There are very few areas of Australia's marine jurisdiction that can be regarded as mature for petroleum exploration, and most parts are still underexplored by international standards (Chapter 4).

All of Australia's continental margin sedimentary basins with petroleum potential lie within its newly confirmed marine jurisdiction. The deep-water parts of the continental margin include significant frontier petroleum exploration areas. Many of these areas are poorly surveyed and are generally not included in conventional estimates of undiscovered resources. There are many other parts of Australia's jurisdiction, such as the Macquarie Ridge, that do not have any petroleum but do have considerable significance because of their ecological, environmental and scientific values.

The vast marine jurisdiction also provides opportunities for a range of mineral resources (Appendix 1.2.3). The seabed around Australia is known to host a vast array of economic minerals, including manganese (Mn), Au, heavy mineral sands, tin (Sn), tungsten (W), Fe and even phosphate. Industrial commodities such as shell sand, sand and aggregate are currently commercially extracted. The seabed around Tasmania has proved rich in a variety of mineral resources (Figure 1.13).

In addition to the rights to explore and exploit the resources of its marine jurisdiction, Australia has the obligation to protect and preserve the marine environment. This is a significant obligation given that parts of the three large oceans (the Pacific, Indian and Southern) and the Tasman, Coral, Arafura and Timor seas cover all five of the world's temperature zones—tropical, subtropical, temperate, subpolar and polar (Figure 1.4). Consequently, Australia's marine environment is home to a wide biodiversity due to the broad range of habitats afforded by latitude, the characteristics throughout the water column (e.g. nutrients, temperature and light), and the geomorphological and geological variability of the seafloor.

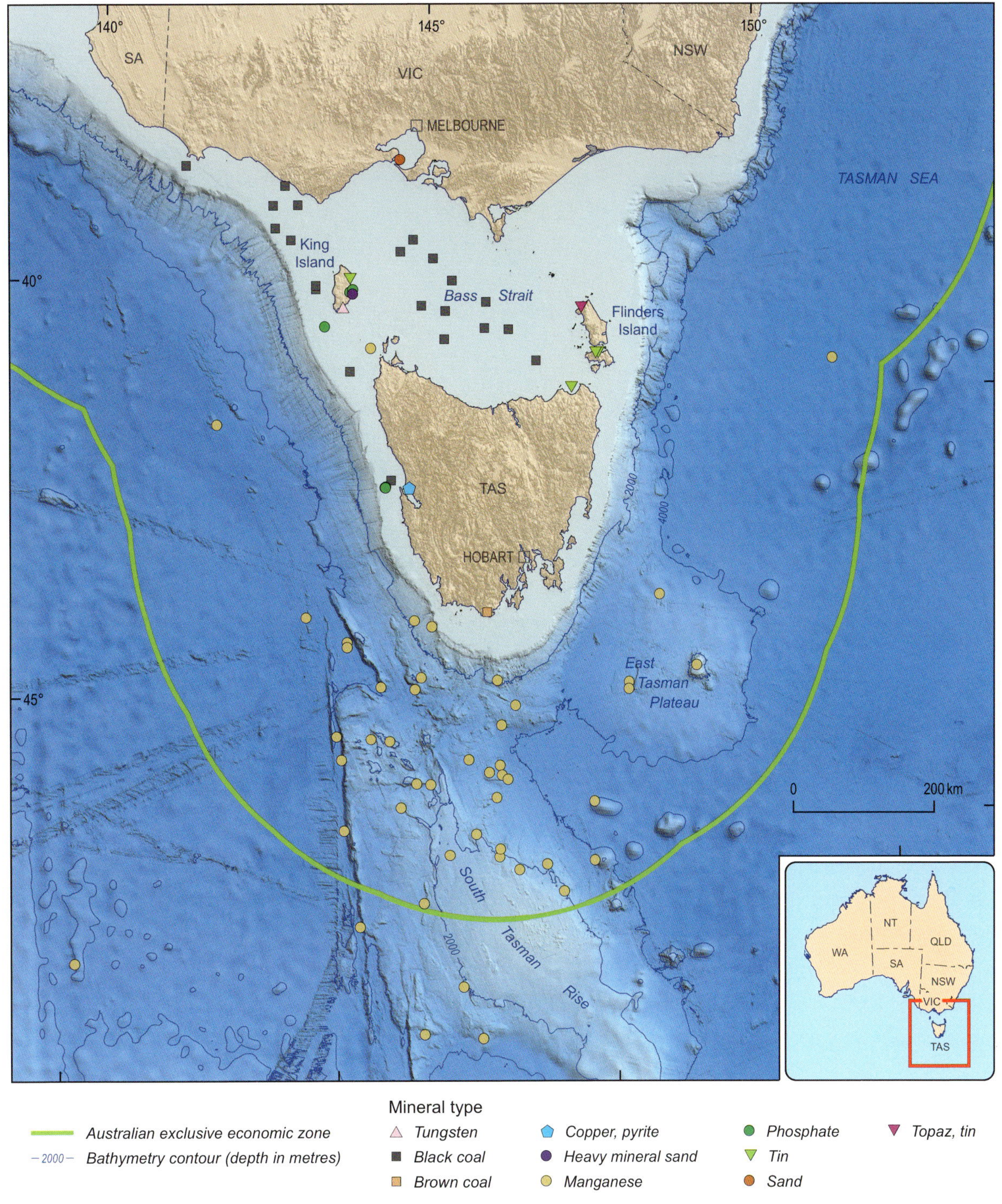

Figure 1.13: Extract around the island of Tasmania from the Australian Offshore Minerals Locations Map (Appendix 1.2.3), showing the array of different mineral commodities and their distribution.

© Getty Images [J & C Sohns]

The red kangaroo (*Macropus rufus*) is superbly adapted to life in Australia's harsh outback.

Bibliography and further reading

The Australian people

Aboriginal Australia Art and Culture Centre, Alice Springs. http://aboriginalart.com.au/culture/dreamtime2.html

Australian Bureau of Statistics 2006. *2006 census of population and housing.* www.abs.gov.au

Australian Bureau of Statistics 2008. *Australian historical population statistics, 2008.* www.abs.gov.au

Blainey G 1975. *The triumph of the nomads*, Macmillan, Sydney.

Blainey G 2001. *The tyranny of distance: how distance shaped Australia's history*, revised edn, Macmillan, Sydney.

Blainey G 2009. *The shorter history of Australia*, Vintage Books, Australia.

Clark M 1987. *A short history of Australia*, 3rd edn, Mentor, New York.

Clark G, Leach F & O'Connor A (eds) 2008. *Islands of inquiry: colonisation, seafaring and the archaeology of maritime landscapes*, Terra Australis 29, ANU E Press, Canberra.

Diamond JM 1997. *Guns, germs and steel*, WW Norton, New York.

Durant W 1926. *The story of philosophy: the lives and opinions of the great philosophers*, Simon and Shorter, New York.

The Economist 2011. The next golden state: a 16-page special report on Australia. *The Economist* 28 May – 3 June. www. economist.com/specialreports

Gammage B 2011. *The biggest estate on earth: how Aborigines made Australia*, Allen & Unwin, Melbourne.

Horne D 1964. *The lucky country: Australia in the sixties*, Penguin Books, Melbourne.

Horton D 1994. *The encyclopaedia of Aboriginal Australia: Aboriginal and Torres Strait Islander history, society and culture*, Aboriginal Studies Press for the Australian Institute of Aboriginal and Torres Strait Islander Studies, Canberra.

Hughes R 1987. *The fatal shore: a history of the transportation of convicts to Australia, 1787–1868*, Collins Harvill, London.

Jupp J (ed.) 2002. *The Australian people: an encyclopaedia of the nation, its people and their origins*, Angus & Robertson, Sydney.

Keneally T 2009. *Australians: origins to Eureka*, Allen & Unwin, Sydney.

Libby R 2007. *How a continent created a nation*, University of New South Wales Press Ltd, Sydney.

Lynch A, Harbort G & Nelson M 2010. *History of flotation*, AusIMM Spectrum Series 18, Australian Institute of Mining and Metallurgy, Carlton.

May SK, Taçon PSC, Wesley D & Travers M 2010. Painting history: indigenous observations and depictions of the 'Other' in northwestern Arnhem Land, Australia. *Australian Archaeology* 71, 57–65.

Mulvaney DJ 1969. *The prehistory of Australia*, Thames & Hudson, London.

O'Connor S 2009. *Earliest seafaring*, AccessScience, McGraw-Hill Companies. www.accessscience.com

O'Connor S & Chappell J 2002. Colonisation and coastal subsistence in Australia and Papua New Guinea: different timing, different modes. In: *Pacific archaeology: assessments and prospects*, Sand C (ed.), Les Cahiers de l'Archéologie en Nouvelle-Calédonie, Nouméa, 15–32.

Pope KO & Terrell JE 2008. Environmental setting of human migrations in the circum-Pacific region. *Journal of Biogeography* 35, 1–21.

Refugee Council of Australia 2011. *Australia's refugee and humanitarian program 2010–11: community views on current challenges and future directions*, Refuge Council of Australia submission to the Australian Government.

Rehman I 2011. *From down under to top center: Australia, the United States, and this century's special relationship*, Transatlantic Academy Paper Series, Transatlantic Academy, Washington.

Rowley CD 1972. *The destruction of Aboriginal society*, Penguin Books Ltd, Australia.

Australian geology: the shaping influence

Australian Government Department of Sustainability, Environment, Water, Population and Communities 2011. *Six seasons of Kakadu*. www.environment.gov.au/parks/kakadu/nature-science/seasons.html

Birrell R et al. (eds) 1982. *Quarry Australia? Social and environmental perspectives on managing the nation's resources*, Oxford University Press, Melbourne.

Bureau of Meteorology 2009. *Climate of Australia: the authoritative climate reference*, Bureau of Meteorology, Melbourne.

Bureau of Meteorology 2011. *Australian climate influences*. www.bom.gov.au/watl/about-weather-and-climate/australian-climate-influences.shtml

Garnaut R 2011. *The Garnaut Review 2011: Australia in the global response to climate change*, Cambridge University Press, Port Melbourne.

Geoscience Australia 2011. *Landforms*. www.ga.gov.au/education/geoscience-basics/landforms.html

Mercer D & Marden P 2006. Ecologically sustainable development in a 'quarry' economy: one step forward, two steps back. *Geographical Research* 44, 183–203.

National Library of Australia 2007. *Australia in maps: great maps in Australia's history from the National Library's collection*, National Library of Australia, Canberra.

Nielsen JN 2008. *Life and landscape*, blog http://geopolicraticus.wordpress.com/2008/11/08/life-and-landscape

Pike D 1970. *Australia: the quiet continent*, 2nd edn, Cambridge University Press, Cambridge.

Price AG 1972. *Island continent: aspects of the historical geography of Australia and its territories*, Angus & Robertson, Sydney.

Seddon G 1997. *Landprints: reflections on place and landscape*, Cambridge University Press, Cambridge.

Knowing Australia's limits

Brown BJ, Müller RD, Gaina C, Struckmeyer HIM, Stagg HMJ & Symonds PA 2003. Formation and evolution of Australian passive margins: implications for locating the boundary between continental and oceanic crust. In: *Evolution and dynamics of the Australian Plate*, Hillis RR & Müller RD (eds), Geological Society of Australia Special Publication 22 and Geological Society of America Special Paper 372, Geological Society of America, Boulder, & Geological Society of Australia, Sydney, 217–237.

Heap AD & Harris PT 2008. Geomorphology of the Australian margin and adjacent seafloor. *Australian Journal of Earth Sciences* 55, 555–585.

Heap AD, Hughes M, Anderson T, Nichol S, Hashimoto T, Daniell J, Przeslawski R, Payne D, Radke L & Shipboard Party 2009. *Seabed environments and subsurface geology of the Capel and Faust basins and Gifford Guyot, eastern Australia*, Geoscience Australia Record 2009/22, Geoscience Australia, Canberra.

Kerr A 2009. *A federation in these seas: an account of the acquisition by Australia of its external territories*, Attorney-General's Department, Canberra.

Symonds PA 2001. Australia's Southern Ocean jurisdiction and opportunities under the Law of the Sea. In: *Looking south—managing technology, opportunities and the global environment*, proceedings of the 2001 Invitation Symposium of the Australian Academy of Technological Sciences and Engineering, Hobart, 20–21 November 2001, 162–180.

Symonds PA & Willcox JB 1989. *Definition of the continental margin using UN Convention on the Law of the Sea (Article 76), and its application to Australia*, Bureau of Mineral Resources record 1988/38, Bureau of Mineral Resources, Canberra.

Symonds PA & Willcox JB 1989. Australia's petroleum potential in areas beyond an exclusive economic zone. *BMR Journal of Australian Geology & Geophysics* 11(1), 11–36.

Australia in time and space

The geology of Australia has exerted a fundamental influence on the welfare and lives of the Australian people and their economic and environmental sustainability. How did Australia's remarkable geology develop, from its deep-time roots in the Archean to the present, and how has this development affected the distribution and abundance of Australia's economic and environmental resources? This chapter provides a summary of Australia's geology in terms of time and space—how Australia was assembled and how it interacted with other continents to produce a rich geology, flora, fauna and landscape. Australia, as a nation continent, has been systematically mapped, and numerous geological and geophysical maps and datasets are available to advance our understanding of the continent.

Richard S Blewett,[1] Brian LN Kennett[2] and David L Huston[1]

[1]Geoscience Australia, [2]Australian National University

Image by Jim Mason

Australia's current tectonic setting

Plate boundaries

Globally, there are 14 large and about 40 small tectonic plates, ranging in size from the Pacific Plate, which comprises 20.5% of Earth's surface, to the Manus Microplate in the Bismarck Sea, which comprises only 0.016% of the surface area. These tectonic plates are thought to be nearly rigid blocks of lithosphere, defined by their boundaries and their trajectory across Earth's surface. The plate boundaries can be convergent, or divergent, or transform. Increasingly evident are regions of diffuse seismicity, with wide, slowly deforming boundaries (as opposed to sharp, discrete ones) between some plates.

Australia was once thought of as simply part of the rapidly moving Indo-Australian Plate, but is now considered as a plate in its own right (Figure 2.1). The Australian Plate appears to be rigid from Cocos Island in the west to Noumea in the east and incorporates the entire Australian continent. The smaller Capricorn Plate is located between the Indian Plate to the northwest and the Australian Plate to the east and southeast. The small Macquarie Microplate is thought to have existed for *ca* 6 Ma. It is a further subdivision of the former Indo-Australian Plate, and is located to the east of the Tasman Fracture Zone.

The boundary forces acting upon the Australian Plate vary, from extension in the south and southwest to compression in the east and north. To the south, an active spreading centre, the Southeast Indian Ridge, separates the Antarctic Plate from the Australian Plate (Figure 2.1). This southern plate boundary, a mid-ocean ridge, was around 100 Myr in the making and developed with the breakup of Gondwana (Chapter 4). Australia's motion was initially to the northwest, but a major plate reorganisation occurred in the Pacific Ocean between 53 Ma and 50 Ma, possibly caused by subduction of the Pacific-Izanagi

The Bungle Bungles in Purnululu National Park, Kimberley, Western Australia.

Image by Jim Mason

The Australian Plate boundary bisects New Zealand, here as the Alpine Fault.

Image by Lloyd Homer, GNS Science

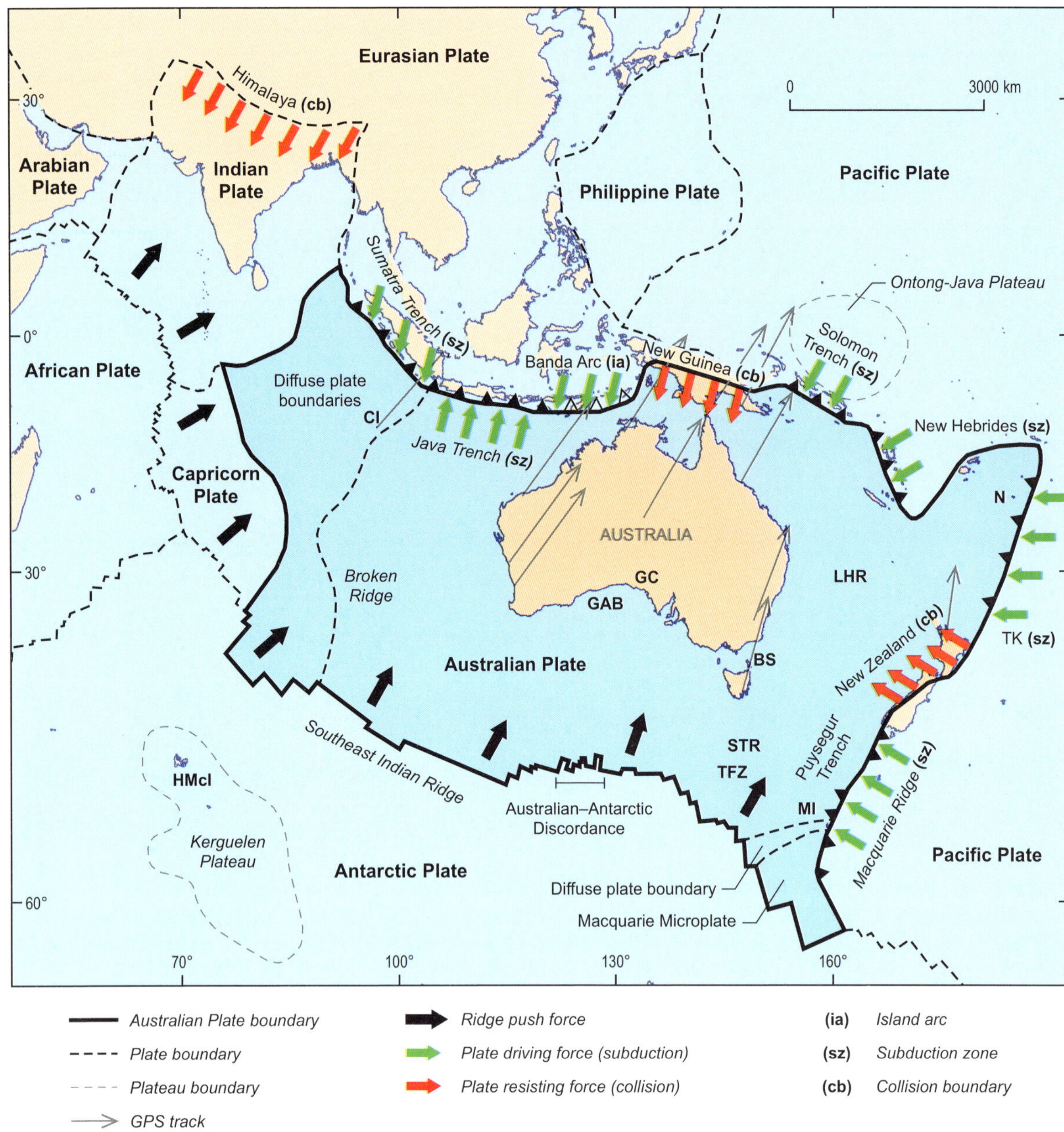

Figure 2.1: Map of the crustal plates and their key boundaries surrounding the Australian Plate. The Capricorn Plate is a zone of diffuse seismicity between the Indian and Australian plates. Red arrows are the plate motion vectors and velocity. GPS measurements show that Australia is moving to the north-northeast at a rate of around 7 cm per year. The Australian–Antarctic Discordance is a region where the mid-ocean ridge between Antarctica and Australia is anomalously deep. (Sources: modified from Royer & Gordon, 1997; Tregoning, 2003; DeMetts et al., 2010)
BS = Bass Strait; CI = Cook Islands; GAB = Great Australian Bight; GC = Gawler Craton; LHR = Lord Howe Rise; HMcI = Heard and McDonald Islands; MI = Macquarie Island; STR = South Tasman Rise; TFZ = Tasman Fault Zone

spreading ridge and subsequent Marianas/Tonga-Kermadec subduction initiation. Whatever the cause, Australia's drift direction changed to its present north-northeast trajectory. Full separation between Australia and Antarctica was not achieved until 34 Ma. Australia has migrated more than 3000 km along this north-northeast path at a rate of 6–7 cm per year, making it the fastest moving continent. The final separation from Gondwana saw the emergence of the island continent that we recognise as Australia (Chapter 4), and marks a significant point in the evolutionary path of our geology and landscape (Chapter 5) and of our distinctive flora and fauna (Chapter 3).

The eastern boundary of the Australian Plate with the Pacific Plate is a collisional zone through New Zealand and a clear subduction boundary along the Tonga–Kermadec Trench north of New Zealand. The Australian Plate is being subducted beneath the Pacific Plate along its northeastern margin at the New Hebrides and Solomon trenches, and again beneath the South Island of New Zealand at the Puysegur Trench. The northern boundary is tectonically complex. Australia is colliding with the Pacific Plate through the island of New Guinea. Further west, the interaction is with the Eurasian Plate, with collision occurring in the Banda Arc region and subduction beneath Indonesia at the Java and Sumatra trenches. The western plate boundary is formed by a diffuse zone of seismicity with the Capricorn Plate (Figure 2.1).

Stress state

The forces generated at the plate boundaries are transmitted across the plates. These forces originate from at least three types of settings: the subduction of oceanic lithosphere, collision zones between continental lithosphere, and seafloor spreading (ridge push). Analysis of earthquakes generated at these boundaries and within the plate can be used to determine the type, magnitude and direction of stress. These earthquakes are also sources of energy that can be used to map the structure of the lithosphere and asthenosphere (see later). Other sources of stress are generated from the gravitational potential inherent in the mass distribution within the plate itself and within the mantle, which is reflected in the shape of the gravitational equipotential surface known as the geoid. Beneath continental Australia, the geoid slopes downwards to the southwest, with a total fall of around 125 m.

Continental Australia is unique among the continents in that the measured stress field is not parallel to the present day north-northeast directed plate motion (Figure 2.2). Much of continental Australia is under a horizontal compressional stress (*Did you know?* 2.1), and most of that stress state is controlled by compression originating from the three main collision boundaries located in New Zealand, Indonesia and New Guinea, and the Himalaya (transmitted through the Indian and Capricorn plates). South of latitude –30°, the stress trajectories are oriented east–west to northwest–southeast. North of this latitude, the stress trajectories are closer to the present-day plate motion, being oriented east-northeast–west-southwest to northeast–southwest. Notably, the main stress trajectories diverge most markedly from one another in north-central New South Wales (Figure 2.2).

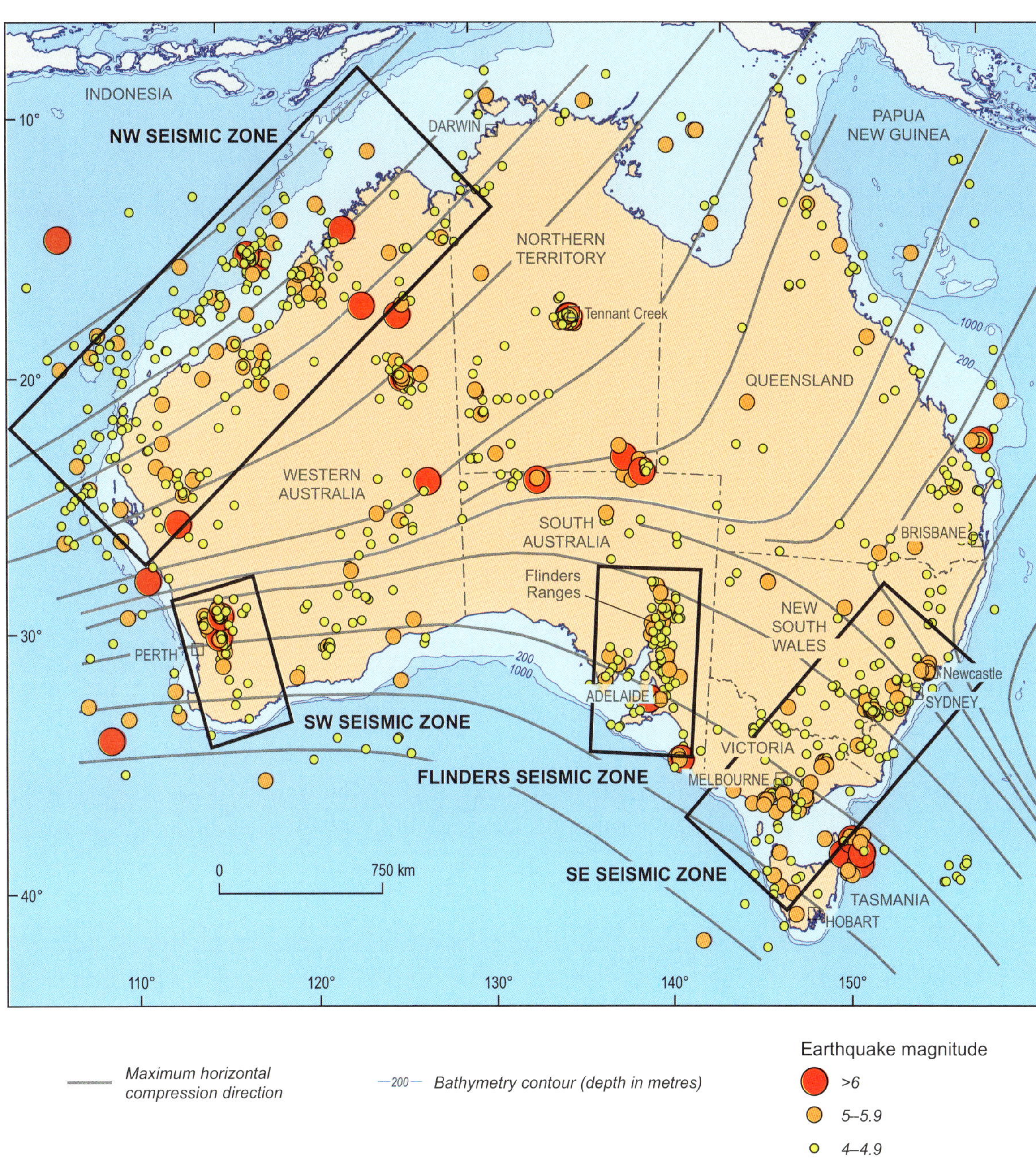

Figure 2.2: Distribution of magnitude M >4 historical earthquake epicentres, mapped neotectonic (young) features (modified from Quigley et al., 2010), and directions of maximum horizontal compression (SHmax), shown as grey lines (after Hillis & Reynolds, 2003). The directions of maximum horizontal compression sweep east–west across the southern half of the continent and diverge to northeast and north-northeast over the northern half of the continent. Australia is divided into four main seismic zones: NW Seismic Zone, SW Seismic Zone, Flinders Seismic Zone and SE Seismic Zone. One of the most active seismic provinces is located in the Flinders Ranges of South Australia. This region coincides with recent uplift of the landscape (Chapter 5) and a zone of enhanced heat flow (Chapter 10).

Image by Dan Clark

Trench exposure, showing Proterozoic Burra Group bedrock (grey) thrust over Pleistocene alluvial sediments (red) across the Williamstown–Meadows (Kitchener) Fault, northern Mt Lofty Ranges, South Australia. The most recent large earthquake event on this fault, some 30–40 ka, resulted in the formation of a scarp >25 km long and 1.5 m high.

For what is generally considered a 'stable' intraplate continental region, Australia experiences a relatively high level of seismicity. The distribution of events with magnitude M ≥5.5 is not uniform across the continent. Rather, the largest earthquakes are clustered into four main regions, which have contrasting basement ages (Archean, Proterozoic and Paleozoic). These seismic zones are characterised by normally low seismic activity punctuated by a period of enhanced seismic activity associated with one or more large earthquakes over decadal time-scales. There is some spatial overlap between such clusters of seismicity and mapped neotectonic reactivation of ancient fault zones and/or regions of elevated crustal heat flow. These spatial relationships suggest that active intraplate deformation in Australia is guided most likely by prior tectonic structures and local to regional thermal weakening of the lithosphere.

The Flinders Seismic Zone in South Australia is one of the most seismically active regions in Australia, with earthquakes of magnitude up to M 6.5 recorded. Interestingly, this region lies within the South Australian Heatflow Anomaly, comprising some of the world's highest concentrations of heat-producing elements (Chapter 10). Major fault displacements, with the uplift of the Mt Lofty Ranges to more than 700 m asl, are a testament to the tectonic activity in this region. In a number of localities, such as the Wilkatana Fault in the central Flinders Ranges, east- and west-directed thrust faults place Precambrian basement rocks over Pleistocene sediments as young as 30 ka. Earthquake focal-mechanism solutions for this zone additionally indicate strike-slip and reverse mechanisms, with a broadly east–west oriented maximum horizontal stress orientation. Quaternary tectonism in southeastern South Australia also resulted in the damming of the Murray River and formation of a large inland lake that persisted for 500 kyr (Chapter 5). In central Australia, the earthquake mechanisms are consistent with north–south compression associated with Australia's ongoing collision with Indonesia/New Guinea.

The recurrence intervals for active faults in continental Australia are not well constrained, as is the case for intraplate regions worldwide, with estimates for large scarp-forming earthquakes (M ≥6.0) between 10 ka and 100 ka. Assessment of seismic hazard is difficult in a region with infrequent large events. Australia's most significant earthquake, in terms of human impacts, occurred in Newcastle (NSW) in 1989, with 13 fatalities and 160 people injured. This notwithstanding, the area had not been identified as particularly significant in the existing hazard maps of the day (*ca* 1979). This example illustrates the difficulty of hazard appraisal in regions where quiescence is the more prevailing status quo. Neotectonic fault movements attest to large earthquakes (M <7.0) in the recent past in proximity to some of Australia's most populated centres in the SE Seismic Zone. Determining the likely recurrence rate for these earthquakes is therefore important for understanding Australia's seismic hazard (Chapter 11).

The eastern and northern margins of the Australian Plate are associated with active volcanism of the 'Pacific rim of fire'. Here, the subduction zones and their overlying environments form one of the most seismically active regions in the world, where around one-third of all earthquakes worldwide occur. The impact on Australia is felt from tsunamis generated by these earthquakes, which can reach Australia's coastline within two to four hours of the seismic event. The Joint Australian Tsunami Warning Centre (JATWC) is operated by Geoscience Australia and the Australian Bureau of Meteorology (BoM). The centre monitors, detects, verifies and warns the community of potential tsunami impacts on Australia's coastline and external territories. This system locates earthquakes using real-time data from 60 seismic stations in Australia and more than 130 international seismic stations, which, together with sonobuoys and tide gauges, calculates the tsunami risk to near-shore regions. The system is integrated with national emergency agencies to issue appropriate warnings within minutes of the initial earthquake. Modelling of hypothetical tsunami events, as well as measured ones, identifies Australia's northwest coast, particularly the Pilbara and Kimberley regions, and the more populated east coast, as having the highest tsunami risk (Chapter 6).

Volcanic activity

There are two active and emergent volcanoes in Australian territory, both lying within the Antarctic Plate on the oceanic Kerguelen Plateau (Figure 2.1). The approximately 2.2 M km^2 Kerguelen Plateau is made of basalts with geochemical characteristics distinct from those of mid-ocean ridges. Mafic volcanism started around 110 Ma and was associated with a hotspot. One active volcano is located on the Big Ben massif of Heard Island. Its most recent lava flow is 2 km long by 50–90 m wide (Box 1.4). The second volcano is located on the McDonald Islands, 44 km to the west of Heard Island—this was dormant for 75 kyr, then erupted in 1992, with activity several times since.

Elsewhere on the Australian Plate, the main belt of volcanoes in the North Island of New Zealand formed as a result of the westward subduction of the Pacific Plate. Here, there are a number of active centres with widespread geothermal activity, with sometimes dangerous consequences for people. Strato-volcanoes include Ruapehu and Tarawera, both of which have claimed lives in the past

Did you know?

2.1: Regional stress is important for hot-rock energy

Australia's unusual horizontal compressional stress state favours the exploitation of the nation's hot-rock geothermal resources (Chapter 10). Hot-rock geothermal systems use a fluid pumped through a closed loop to extract heat energy from surrounding rocks. The loop consists of a down-going well and an up-going well, connected via a fracture network, which is created by pumping fluid to very high pressures, forcing the intervening rock mass to fracture in ways governed partly by the local- to regional-stress state. Horizontal maximum stress encourages horizontal fracture systems, whereas vertical maximum stress encourages vertical networks. Horizontal fracture networks minimise fluid loss in the system, because the fluid, once it leaves the down-going well, can only travel laterally, promoting its intersection with the adjacent up-going well, and thereby recovering the energy. Containing fluid flow in vertical fracture systems is more difficult, as seen in the Rhine Graben in Germany, which is under an extensional (vertical maximum stress) to strike-slip stress state.

Image courtesy of Petrathem

200 years through phreatic eruptions and lahars. Lake Taupo occupies the caldera of a super-volcano, which erupted at about 10 ka, leaving a shroud of tephra across the whole North Island.

The mid-ocean ridge between Australia and Antarctica is also an active volcanic region. This ridge has a significant physiographic influence on marine life, but also influences Earth's climate. The growth of oceanic phytoplankton sequesters around 20–25% of atmospheric CO_2. Carbon storage in the Southern Ocean, however, is below full capacity because these waters lack the iron necessary for vigorous plant growth. Around 5–15%, and in some areas up to 30%, of the iron content of this ocean is derived from hydrothermal processes. The mid-ocean ridge volcanoes provide a relatively constant source of iron over millennial time-scales. These volcanoes also buffer short-term iron fluctuations derived from climate-driven sources such as wind-blown continental dust (Chapter 5) and the resuspended coastal sediments (Chapter 6).

In addition, there are many dormant volcanoes along the eastern margin of Australia (Figure 2.3). These form the Newer Volcanic Group, which is a chain of mostly mafic volcanic rocks that erupted along the Great Divide. These rocks stretch from Cape York Peninsula to southern Victoria

Timor Rock volcanic plug in the Warrumbungle National Park, New South Wales.

and South Australia, and occur as extensive lava fields (e.g. McBride in north Queensland) and as shield volcanoes (e.g. 12 Ma Mt Canobolas in New South Wales). The soils derived from these volcanic rocks are some of the most fertile in Australia, especially where they occur in the higher rainfall regions (Figure 1.5). Diatremes associated with the Newer Volcanic Group also carried gemstones, such as diamonds, sapphires, rubies, garnets and zircons. To the east of the continent lies a series of subparallel seamount chains, with seamounts in the Tasman Sea, in the Lord Howe Rise, and Norfolk Island. They represent eroded volcanoes, such as the 15.6 Ma Gifford seamount (Figures 1.15 and 2.3), which mark the northward track of the Australian Plate over one or more mantle plumes.

The East Australian Plume System originated around 65 Ma, with the rifting of the Coral Sea. As the continent drifted north over the plume system, younger eruptions occurred increasingly to the south. The youngest dormant volcanoes occur in the intraplate region of southeast South Australia to western Victoria. They consist of nepheline hawaiites (a variety of basalt) dated at 4.6 ka, which were derived from partial melts of lower crust to upper mantle mafic rocks. Drill holes into the Gambier Basin produce commercial quantities of CO_2 gas, once thought to be volcanic in origin, but now considered to

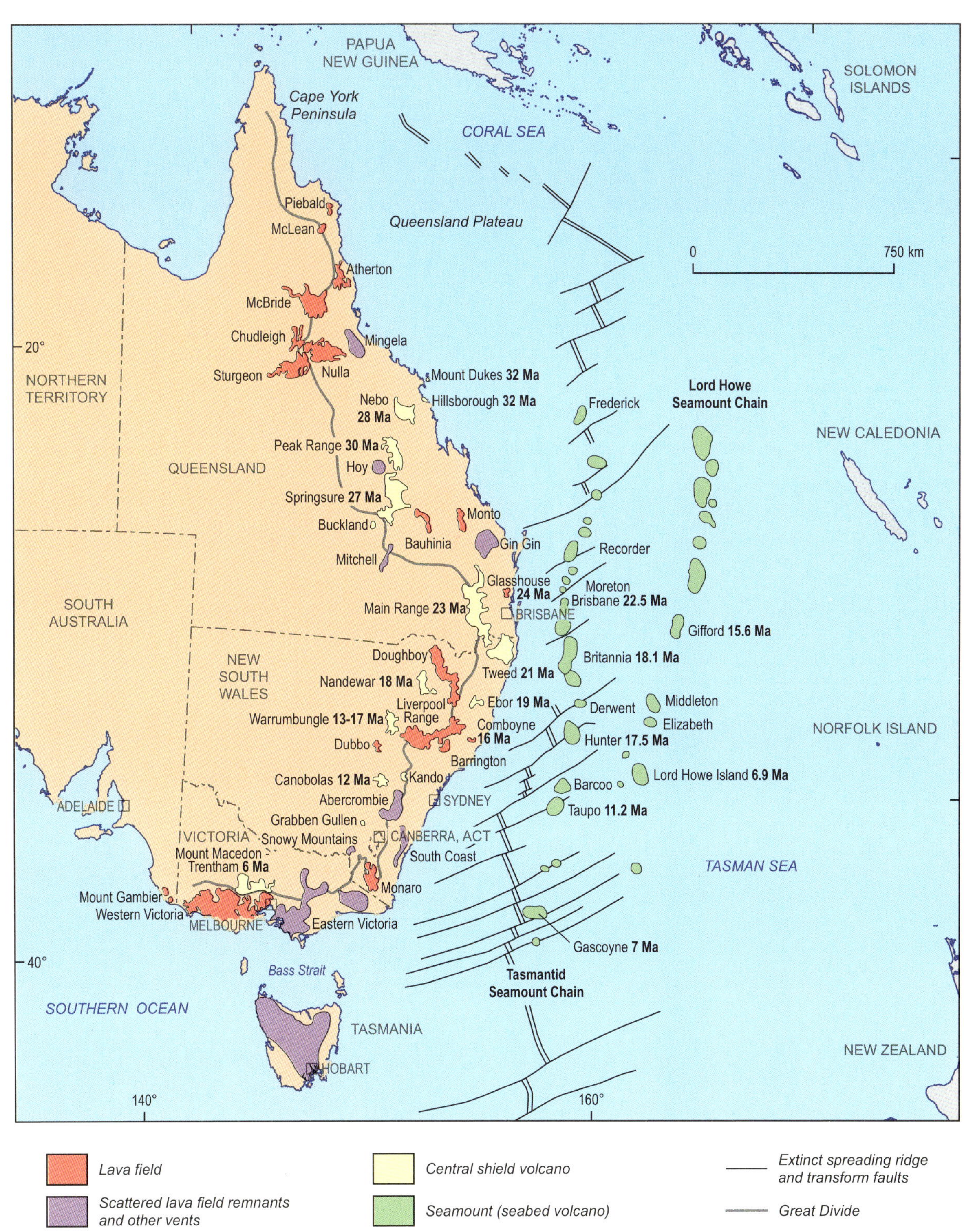

Figure 2.3: Distribution of Cenozoic volcanoes and lava fields in eastern Australia. Note the southward decrease in age of shield volcanoes and seamounts in three parallel tracks. These volcanoes were formed as the Australian Plate moved north-northeast over a series of mantle plumes. (Source: after Johnson, 2009)

be mantle sourced, since the gas includes some of the most primitive neon and xenon gas signatures on Earth. A mantle connection is also implied by the common presence of mantle xenoliths in these Newer Volcanic Group rocks. The age progression of the volcanism suggests that the next volcano on the Australian continent is likely to erupt in or around Bass Strait (Figure 2.3).

Mapping Australia

Aboriginal and Torres Strait Islander people have been mapping their country for millennia. Their religious art expresses social and spiritual relationships rather than precise geography. Topographic information is commonly portrayed as plan-view maps in rock or woodcarvings or as sketches on the ground. These maps often included the location of water and food resources (Figure 7.3).

Terra Australis appeared on European maps as early as 1499. The oldest known map to use the title 'Australia' is the woodcut map in Johann Honter's *Astronomia* from 1545. Dutch sea captains are credited as the first to begin to understand the shape and size of the island continent in the early 1600s, working for Dutch East India Company interests but undoubtedly with earlier Portuguese charts to hand. In 1605, Janszoon was commissioned

The highlands of New Guinea formed by the collision of the Australian and Pacific plates. The Bismarck Range in the central western highlands, Papua New Guinea, reaches over 4500 m. Mount Wilhelm is the highest peak; it is often snow capped.

to discover and chart 'unknown southlands'. By 1623, Carstensz had named parts of Cape York Peninsula. From the 1600s to the 1800s, various European explorers, including the Dutch, French and British, had charted Australia's coastline and added to the map. Louis de Freycinet and Matthew Flinders published the first complete maps of Australia in 1811 and 1814, but James Cook had noted geological features such as the Glasshouse Mountains on the southeast Queensland coast in 1770. European settlement, migration and inland exploration provided further geographical details, and geological exploration began.

Unlike all the other continents on Earth, Australia is governed as a single sovereign nation—a political factor that has facilitated the systematic mapping of the continent. With each technological advance, Australia has been mapped in ever-increasing sophistication and detail. Australia is well served by high-quality geological and geophysical maps. The data form a vital input to Australia's pre-competitive data inventory, with the Australian Government, together with state and territory geological surveys, promoting the nation as an attractive destination for mineral and energy explorers. Many others have since copied this lead, but no other country has such a comprehensive coverage, which is freely available for download (www.ga.gov.au/gadds). The maps and results illustrated in the following section include representations of continental-scale topography and bathymetry, remote sensing, soils, gamma-ray spectrometry, potential fields (gravity and magnetics) and lithospheric mapping. Further presentations of these results are in Appendix 2. Definitions of the spatial and temporal terms are given in Table 2.1.

Table 2.1: Definition of the spatial and temporal terms used

Term	Definition
Supercontinent	A large continent formed by the amalgamation of most or all of Earth's continental landmasses.
Supercraton	A large, ancestral (largely Archean) landmass consisting of two or more cratons.
Continent	One of Earth's major landmasses (or former major landmasses).
Element	Part of a continent that has some shared broad-scale geological history; often an interpreted proto-continent or collection of such continental fragments (including cratons) that now forms part of an extant continent.
Craton	A part of Earth's continental crust that has attained stability and has been little deformed for a prolonged period.
Province	A large geological region showing similarities in its geological history to adjacent provinces, but with a different geological history.
Superterrane	A collection of two or more terranes.
Terrane	A region with essentially similar geology and geological history.
Domain	A usually fault-bounded region of similar geology.
Superbasin	A group of temporally and genetically related basins.
Basin	A low area at Earth's surface of tectonic origin in which sediments have accumulated.
Inlier	An exposure of basement rocks completely surrounded by younger basinal rocks.
Seismic province	A discrete volume of middle to lower crust, which cannot be traced to the surface, whose seismic reflectivity is different from that of adjoining provinces, either laterally or vertically.
Orogen	An often linear or arcuate region that has been subjected to one or more common episodes of deformation and metamorphism (orogenies).
Orogeny	Geological event or genetically and temporally closely related events involving rock deformation.
Movement	Geological event or genetically and temporally closely related events involving rock deformation. Used where previously defined orogenies (e.g. Alice Springs) are found to include unrelated orogenic events.
Event	A temporally and, commonly, spatially restricted occurrence of a geological process or related geological processes. This can include magmatism (igneous event), deformation (deformational event, orogeny or movement) or mineralisation (mineralisation event).

Surface relief

Flying from Europe or Asia over the Australian continent to the east coast, it is easy to see why the expression 'old, flat and red' to describe Australia is so apt (Chapter 5). The large cratonic blocks of Precambrian rocks in the western two-thirds of Australia form the core of the continent. The ochre-red colours of the 'outback' are a testament to the aridity, driven by Australia's plate position with respect to the monsoonal belts in the north, and the westerly winds in the Southern Ocean. Australia is the lowest and flattest continent, and has been the most slowly eroded and deeply weathered on Earth (Figure 1.5).

The average elevation of the continent is about 330 m, with a maximum local topographic relief typically less than 1500 m. Other than the upland areas, the bedrock erosion rates are typically less

© Getty Images [J Edwards]

Meandering river system, lined by trees, dissects the arid and flat Lake Eyre Basin between Oodnadatta and William Creek, South Australia.

than 1–10 m per Myr. The interior of Australia has been largely devoid of major mountain building for the past 200 Myr, with most uplift restricted to the eastern margin. There are, however, local areas of elevation that are relatively young in terms of continental Australia's ancient past. Some, such as the Flinders Ranges, have been uplifting intermittently at rates of 10–50 m per Myr. The old landscape is dotted with impact craters, some of which now rise above the low plains (Box 2.1).

Most of Australia's jurisdiction lies beneath the ocean (Chapter 1), and it is here that most of the surface relief is found. The east–west spine of submarine ranges forming the active spreading centre of the Southeast Indian Ridge in the Southern Ocean is a legacy of Australia's release from Gondwana (Figure 1.5). The ridge is dissected by swarms of north–south-trending transform faults, with a marked zone of anomalous depth at the Australian–Antarctic Discordance. Continental Australia's older geology influenced the Gondwanan breakup and resultant marine bathymetry. The pronounced gap in transform faults along strike from the Archean Gawler Craton marks the influence of the old, strong lithosphere. The conjugate southern margins of Australia along the Great Australian Bight and the Antarctic coast, created by the opening of the Southern Ocean, show slightly wider continental shelves. Huge deltas drained the interior of Australia into the extending basins along this evolving southern margin (Chapter 4). The shallow sea of Bass Strait, once a series of freshwater lakes during lower sea-level periods, now crosses the end of the long pathway of some of the earliest and farthest travelled humans—from Africa to Tasmania.

To the east, rifted fragments of continental Australia provide frontier opportunities for energy resources in the basins of the Lord Howe Rise, which at its southern end becomes part of New Zealand (Figure 2.1). Between here and the coast lie the seamounts, which are a legacy of hotspot activity of the Australian Plate's northwards move (Figures 1.8 and 2.3). The broad continental shelf off Queensland, left behind after the Coral Sea opened, forms a foundation for Earth's largest single living entity—the Great Barrier Reef, which extends 2300 km and is made up of more than 2900 individual reefs. This iconic symbol of Australia is estimated by Tourism Australia to be worth around $7 B per year for the Australian economy (Chapters 3 and 6).

IMPACT CRATERS (BOX 2.1)

The collision of extraterrestrial bodies with Australia is inevitable, given the great age of the geology and landscape. Australia has a total of 35 confirmed impact sites from the Late Precambrian onwards, 22 unconfirmed impacts and 10 sites with identified impact ejecta. The size of the impacts ranges from a few hundred metres to more than 100 km in diameter.

Australites are the glassy tektites (ejecta) found across southern Australia. Examples from Victoria are around 800 ka. They were reputedly used as cutting tools or sacred objects by Aboriginal people. Other impacts are concealed beneath the ocean or in sedimentary basins—these are inferred from their geophysical signatures. For example, a large feature beneath the Cooper Basin in South Australia is interpreted as a major impact that might have been partly responsible for generating the heat anomaly. The area coincides with an elliptical-shaped magnetic high, which is around 100 km in diameter. Shock-textured quartz from Carboniferous granites recovered from deep drillholes is consistent with an impact.

Other impacts, such as the impressive 300 ka Wolf Creek crater, are exposed as eroded circular-shaped remnants in the landscape. The smaller Veevers crater in Western Australia is thought to be <20 ka, and this impact may well have been felt by nearby Aboriginal people. In the Northern Territory, rocks fractured into a regular 'shatter cone' pattern by intense, sudden pressure have been found at the 5 km diameter and 150 m high Gosses Bluff crater west of Alice Springs in central Australia. This impact occurred *ca* 142.5 Ma. The site is sacred to the local Western Arrernte Aboriginal people. They name it *Tnorala* and relate its formation to a cosmic impact during the Dreamtime:

> *A group of celestial women were dancing as stars in the Milky Way. One of the women became tired and placed her baby in a wooden basket. The women continued to dance and the basket with baby fell to the Earth, the impact forced the ground upward, forming the circular mountain range. The baby's parents, the evening and morning star (Venus), continue to search for their baby.* (Thornton, 2007)

Aerial view of Gosses Bluff meteorite crater, Northern Territory.

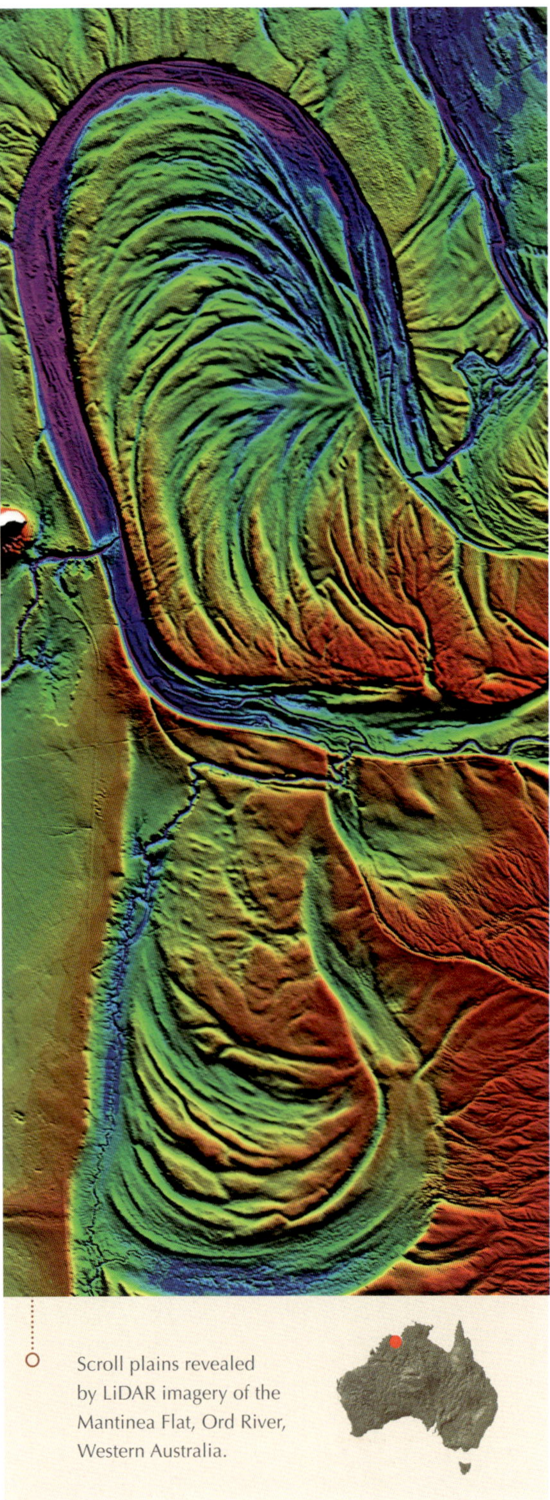

Scroll plains revealed by LiDAR imagery of the Mantinea Flat, Ord River, Western Australia.

The wide continental shelf between the coast of northern Australia and Indonesia, Timor and New Guinea is a relatively fertile fishing ground and also host to great hydrocarbon wealth. The extended continental North West Shelf contains large natural gas resources, and is marked by complex embayments and salients left over from the breakup of Gondwana (Chapter 4). The continental slope on all margins is deeply incised, with steep-sided canyons up to 2 km deep. Australian margins, like those in Antarctica, are unusual because deep river-associated canyons do not dissect their continental shelves. More than 80% of Australia's canyons are blind, probably because of Australia's low rainfall, low relief and lack of continentally derived sediment transport to the oceans (Chapter 6). Some of the larger canyons influence ocean currents and attendant fertility—the head of the Perth Canyon off southwest Western Australia, for example, is a haven for whales.

Topography and bathymetry

The representation of surface topography is now carried out with a digital elevation model (DEM) based on the use of point elevation data; commonly, a regular grid of elevation points is employed. Such grids can be directly observed, but generally they are computed from other elevation information such as contours or irregularly spaced spot heights. High-resolution digital terrain maps can be generated from the National Aeronautics and Space Agency's (NASA, in the United States) Shuttle Radar Topography Mission (SRTM). NASA has released the SRTM dataset for Australia, New Zealand and many South Pacific islands. Research agencies in Australia have combined to increase the resolution to a one-second grid, or around a 30 m pixel size. The new image has many benefits; for example, water catchment boundaries can be better mapped, and terrain data can be used for modelling surface/groundwater interactions, as well as landscape evolution. Underlying geological features are also revealed in the elevation data (Figure 1.5).

For more local high-resolution elevation information, the LiDAR (light detection and ranging) technique can be employed. LiDAR maps are generally acquired over smaller areas than the SRTM technique. These high-resolution digital elevation maps have led to significant advances in geomorphology, where very subtle topographic features are readily resolvable. For example, near the mouth of the Murray River in eastern South Australia, ancient river terraces with scrolls and channel banks are defined in great detail (Chapter 5). Combining aircraft-based LiDAR and satellite geodesy provides another tool for detecting faults and measuring uplift across a region.

Most of the Australian Plate is under the sea, and bathymetric images reveal a highly complex surface, reflecting the formative geological processes. Geoscience Australia has been collecting bathymetry data routinely from seismic and sampling surveys around the Australian margin since 1963. More recently, Geoscience Australia has taken on the role of national bathymetry custodian for all holdings within the Australian marine jurisdiction. These data consist of ship-track and swath bathymetries, digitised soundings from hydrographic charts and laser airborne depth-sounder data from more than 1400 surveys. The data were acquired by

Geoscience Australia, as well as by other scientific institutions, oil exploration companies and academic organisations. In addition, some data have been sourced from the National Geophysical Data Centre (United States), to which various institutions have contributed (Figures 1.5 and 4.9).

Satellite remote sensing

Satellite remote sensing has proved very valuable in characterising features of the Australian continent. There are two basic types of sensor systems mounted on orbiting satellites:

- active systems, such as Synthetic Aperture Radar (SAR) that generate their own electromagnetic radiation, measuring a return signal response
- passive systems, such as Landsat, that use an array of detectors to record electromagnetic radiation, which is reflected and/or emitted from Earth's surface.

The data are transmitted to ground stations, such as at Alice Springs (NT). These data, when processed, can be used to create images of Earth's surface. Satellite images differ from high-altitude aerial photographs in two main ways:

- firstly, the acquisition of a broader bandwidth of the electromagnetic spectrum, such as infra-red, enables improved identification and assessment of surface features
- secondly, the regular updates from each satellite pass enable time-based analysis of the surface. Being a digital product, satellite remote sensing imagery can be easily integrated with other spatial imagery.

The Landsat 7 Picture Mosaic of Australia was produced by the Australian Greenhouse Office as part of its National Carbon Accounting System (Figure 2.4). The mosaic consists of 369 individual Landsat satellite scenes, acquired between July 1999 and September 2000, and utilises the 2, 4, 7 spectral bands (as blue, green, red colours). The mosaic clearly shows the arid interior, in brown and yellow colours, contrasting with the more watered coastal zone of eastern Australia and southwest Western Australia, in shades of green. The major ephemeral lake systems in central Australia mark the lowest point in the landscape, and these capture occasional surface-water runoff from the Channel Country to the northeast (Figure 1.4). The productive 'wheat belt' in southwest Western Australia is clearly visible from space, demonstrating the impact that people can have on the landscape (Figure 2.4). To the east and north, the desert and semi-desert of the interior has saline lakes, remnants of past freshwater river and lake systems that drained into the Southern Ocean. The classic Archean granite-cored domes are well exposed near the Pilbara coast. These domal map patterns may have formed under an early-Earth tectonic system, one that was different from modern plate tectonics (see later).

Land cover is defined as the observed biophysical cover on Earth's surface, including trees, shrubs, grasses, soils, exposed rocks and waterbodies, as well as anthropogenic elements such as plantations, crops and built environments. Changes occur to land cover from factors such as seasonal weather, severe weather events, fires and human activities, including mining, agriculture and urbanisation.

Did you know?

2.2 Ocean drilling

The Ocean Drilling Programme (ODP) was an international partnership involving 22 countries and hundreds of researchers to understand Earth's evolution. Projects were focused on all aspects of the geosciences. Australia participated from 1988 until ODP finished in 2003. Up to 2001, 17 survey legs were completed, many in Australia's jurisdiction. Leg summaries and their locations in Australian waters are presented in Appendix 2.

The Integrated ODP is the current successor organisation (www.oceandrilling.org). Australia, in partnership with New Zealand, joined in 2005, with funding guaranteed until 2012. Expedition 325, completed in mid-2010, drilled three key localities on the outer edge of the Great Barrier Reef in water depths of 42 m to 167 m. The area was chosen to study sea-level rise and ocean chemistry for the past 20 kyr, because it is located in a tectonically stable region distant from major ice sheets.

Image courtesy of Neville Exon

Figure 2.4: True colour mosaic (bands 2, 4 and 7) of Landsat satellite imagery over Australia, highlighting the differences in vegetation as a reflection of rainfall (Chapter 1). The boxed area in the Pilbara is enlarged in Figure 2.13a. (Source: Australian Greenhouse Office, 2005)

Satellites are able to map land cover and how it changes through time. Australia's land cover map was made from time-series satellite data using the Moderate Resolution Imaging Spectroradiometer (MODIS) system from NASA. The time series includes 186 snapshots of vegetation greenness for each 250 m by 250 m area across the continent over the period April 2000 to April 2008. The land cover map provides a baseline for reporting on change and trends in vegetation cover and extent. Information about land cover dynamics is essential to understanding and addressing challenges such as drought, salinity, water availability and ecosystem health.

Soils

Australian soils, like the landscape, are the products of the long-term evolution of the climate, life, topography and geological parent materials (Chapter 5). Australian soils tend to be old, salty and clayey, except in the west of the continent, where they tend to be sandy, acidic, and nutritionally and organically impoverished, and pose structural challenges to building construction. Compared with those in the Northern Hemisphere, Australian soils generally have less organic matter and poor structure, and tend to be clay-rich near the surface. These clayey characteristics tend to restrict water drainage, impede root growth and, due to their 'shrink and swell' nature, impact engineering and farming. The clays adsorb heavy metals and even pesticides, concentrating them in the environment until chemical changes release them.

The agricultural landscapes of Australia support a great range of soils of varying thickness. Most are ancient, strongly weathered and infertile. Others are younger and more fertile. This variety, along with the natural limitations of many soils and their interactions with climate, has made it difficult to develop sustainable agricultural systems. Limitations to productivity have also been induced through human impacts on soils. While some forms of degradation such as nutrient deficiencies can be corrected, others, such as soil erosion, are difficult to remedy. Remediation is impeded because the rate of soil formation is slow in most areas.

Large areas of Australian soils are affected by salt and have various nutrient and physical limitations for plant growth and therefore agriculture. Their iron-rich nature means that such soils are susceptible to phosphorus adsorption, impeding the uptake of added fertiliser to crops. Soil fertility and usability, together with net water budgets, have had a profound influence on determining the distribution of the Australian population (Figure 1.3). These factors highlight the contrast between Australia's agricultural development and that of Europe and North America.

Gamma-ray spectrometry

All rocks and soil contain some of the major radioactive elements uranium (U), thorium (Th) and potassium (K). The decay of these elements gives rise to a natural gamma-ray flux, which can be measured with a suitably equipped aircraft. The energy distribution of the gamma rays is specific to the decay chain for the particular elements, and hence the relative contributions can be measured and extrapolated back to the concentrations of radioactive elements (radioelements) at Earth's surface. This process requires corrections for background radiation, the height of the aircraft above the ground, and the response and sensitivity of the detector.

The normal mode of display of the results is via a three-colour image, with the K concentration on the red channel, Th on the green channel and U on the blue channel. The ternary image is essentially a chemical map of the near-surface distribution of these three elements and is strongly correlated with surface geology and the rate of erosion or deposition in the landscape. Areas that are low in all three radioelements appear as dark hues (ultramafics, quartzites and sandstones), and areas that are high in all three elements appear as white

Dead trees on edge of salt lake—salinity damage due to rising water-table between Beacon and Bencubbin, Western Australia.

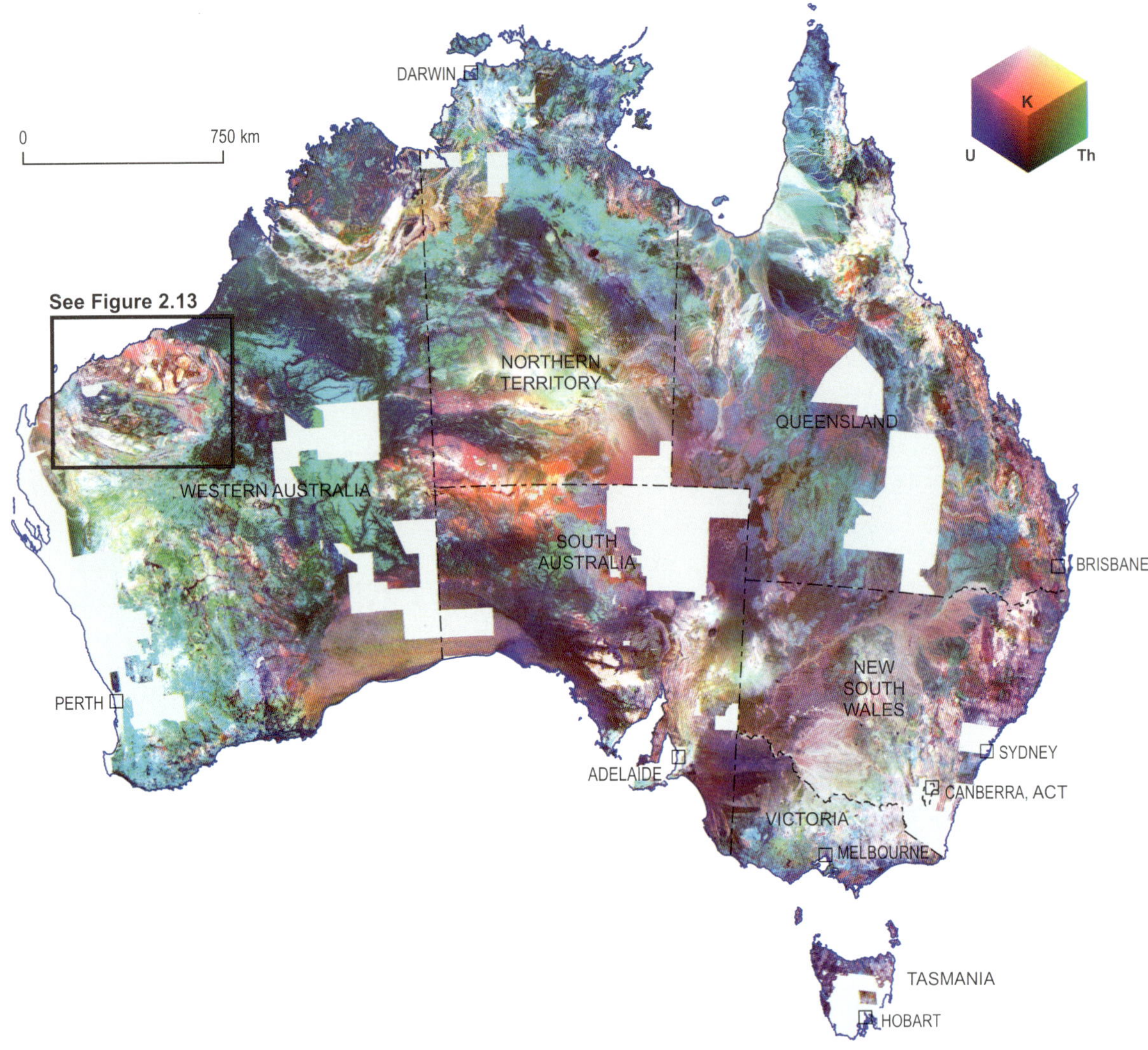

Figure 2.5: Gamma-ray or spectrometric map of Australia synthesised by combining results from many individual airborne surveys (after Geoscience Australia). The gamma-ray map shows the distribution of potassium (red), thorium (green) and uranium (blue) in the top 30 cm of the surface. Dark coloured and black regions are low in total counts of the radioelements—these are commonly quartz-rich sand-covered areas. The white regions are high in total radioelements and are commonly associated with areas of active erosion, especially of felsic (granitic) rocks. The pale-grey areas are data acquisition gaps. The boxed area in the Pilbara is enlarged in Figure 2.13b.

hues (felsic volcanics and granites). Weathering, erosional and depositional processes play a large part in forming the radiometric response. Many of the green and green-blue areas, representing high Th but low K, are highly weathered surfaces rich in iron.

Most gamma-ray surveys are carried out over a limited area, and careful processing is required to link together the results from separate surveys. A major effort has been made in recent years to link the results at the continental scale, and the spectacular outcome is shown in Figure 2.5. This map is the largest single gamma-ray dataset collected at this resolution anywhere on Earth.

Localised high concentrations of the radioelements are clearly visible in the granites of the Pilbara and the eastern Yilgarn cratons (WA), in central Australia and in the New England region of northern New South Wales. Other concentrations occur in the Flinders Ranges (SA), which is also noted as a zone with enhanced geothermal heat flux, most likely associated with a high concentration of these heat-producing elements in the crust (Chapter 10).

Many geological features and boundaries are well delineated in the ternary image. For example, the Mesozoic shoreline on the northern margin of the Nullarbor Plain (WA) is clearly distinguished from the Th-rich Yilgarn Craton. The actively eroding fold-belts surrounding the Kimberley region are prominent against the more sombre tones of the Kimberley itself. There is also a very clear image of the remote Canning Basin in northwestern Western Australia, much of it forming the Great

Sandy Desert between the Kimberley and the Pilbara regions. The mineral provinces around Mt Isa and Broken Hill have a distinctive signature. The granites along the eastern margin are bright red and white colours and shades, consistent with their high concentrations of radioelements.

In Australia, the acquisition of airborne gamma-ray data has become routine when magnetic data are being acquired. The data are used specifically for geological and landscape mapping and in assessment of alteration associated with mineralisation—especially U. Linking the measurements from rock-sample geochemistry and the gamma-ray data has shown that Australia is 'missing' a huge amount of U (Chapter 10). The data have become one of the key means of prediction to map specific soil properties (e.g. clay content). The results have been used for land-use planning, such as mapping soils, vegetation (even viticulture) and animal ecology.

Surface geochemistry

Australia has a near complete coverage of multi-element surface geochemistry. Because of the varied landscape and climate of Australia (Chapter 5), a specific sampling methodology was developed. The coverage consists of 1315 samples collected near the outlet of 1186 catchments, covering 81% (6.2 M km^2) of Australia. All samples were analysed for up to 68 elements by as many as three complementary methods. The average sampling density was around 1 site per 5200 km^2. A geochemical atlas and dataset are freely available from Geoscience Australia's website. The geochemical maps show continental-scale patterns (Figure 2.6). They also provide the geochemical

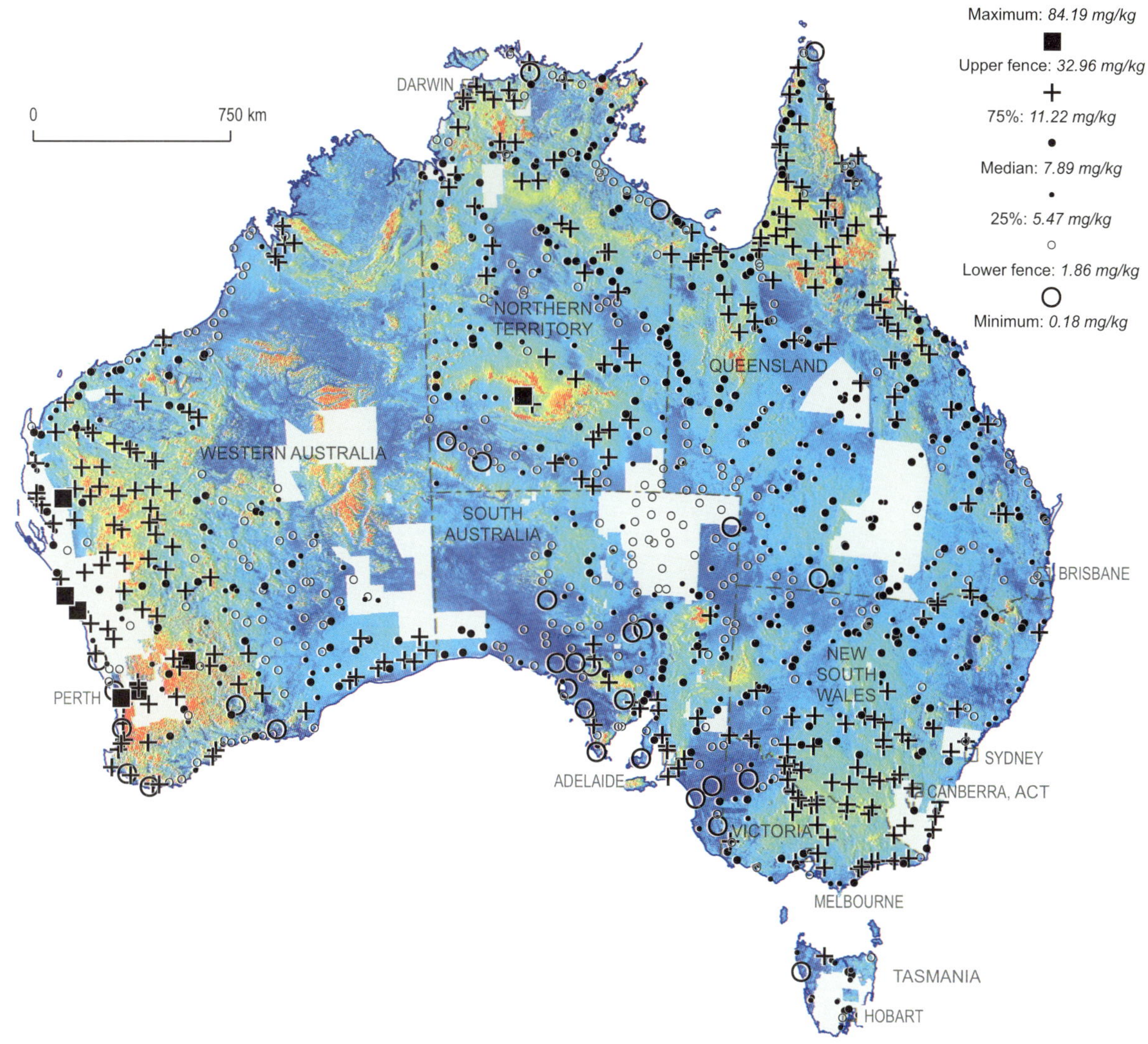

Figure 2.6: Map of soil Th concentration from the top 10 cm of the surface from 1190 sites distributed across Australia, overlain on an airborne radiometric Th channel image. Although the data density is low, the map provides the first consistent picture across most of Australia of the distribution of the elements analysed. (Source: de Caritat & Cooper, 2011)

Airborne radiometric Th channel data (background colour ramp, from blue for low Th to red for high Th)

PIONEERS OF GEOLOGICAL MAPPING IN AUSTRALIA (BOX 2.2)

Mapping after the 1830s, by explorers such as J Lhotsky, PE Strzelecki, T Mitchell, JD Dana and L Leichhardt, noted areas likely to contain coal, limestone or metalliferous deposits, as well as fossils to date rocks, potential dam sites and the variability of soils. These tantalising glimpses of potential led to the colonial governments establishing geological surveys. In terms of national geological maps, the Strzelecki 1845 map of New South Wales and southeast Australia and the *First sketch of a geological map of Australia including Tasmania*, published in 1875 by R Brough Smyth, were the first geological maps of the continent. The Smyth map was issued at a scale of 110 miles to an inch in two sheets, which are in the National Library of Australia.

In 1887, the map *Continental Australia: from the most recent information and materials supplied by the survey departments of the several colonies geologically colored [sic] by Arthur Everett* was published. The map was sourced from state geological maps, with the base map supplied by the Victorian Lands Department. It was one of the largest geological maps published in Australia at that time, and, priced at £3.10.0 (more than $500 in 2012 terms), was the most expensive map issued by the department.

The 1932 *Geological map of the Commonwealth of Australia* and a volume of *Explanatory notes,* the culmination of several decades of work by TW Edgeworth David, were one of the most significant early contributions to Australian geology. In the late 1940's, the Bureau of Mineral Resources (now Geoscience Australia) and state geological surveys began 1:250 000 scale mapping of the map sheets of Australia; there were more than 600 of these maps. This marked

Extract of Edgeworth David's geological map, showing southeast Australia.

Image courtesy of National Library of Australia, MAP RM 4019 the beginnings of a comprehensive understanding of Australia's geology. This mapping was built on the observations and recording of all those earlier generations of geoscientists who covered vast distances under difficult physical conditions and limited basic information.

The unknown frontier today is depth. The challenge for the various mapping agencies is to map the 3D geology of Australia. Advances in geophysics are integral to this aim (Chapter 11).

Some of the early pioneers in geological mapping are noted below.

Image courtesy of State Library of Victoria. Image no. H83.86/1

Alfred Richard Cecil Selwyn (1824–1902)

Appointed Geological Surveyor of Victoria in 1851, Selwyn mapped large tracts of the colony, helping to establish the stratigraphy. Under his direction (1853–69), the Geological Survey of Victoria issued more than 60 geological maps and numerous reports. At the time, they were compared favourably with the best in the world. In 1865, he published a geological map of Victoria in eight sheets.

Image courtesy of LINC Tasmania

Charles Gould (1834–93)

Gould was Tasmania's first Geological Surveyor (1859–69). His primary task was to discover gold. Gould's surveys covered much of the colony and added greatly to the geographical knowledge of western Tasmania. He established the Ordovician to Lower Devonian stratigraphy, correctly deduced the succession of Permian and Triassic coals and Jurassic dolerite, and suggested mining for coal under dolerite.

Image courtesy of State Library of South Australia. Image number B7102

Henry Yorke Lyell Brown (1843–1928)

Brown arrived in Australia in 1865 and worked in Victoria, Western Australia and New South Wales. In 1882, he was appointed Government Geologist of South Australia. His major achievement was the publication of the geological map of the colony in 1899 (including the Northern Territory). On his death, it was said that 'he knew every mineral belt from Darwin to Mt Gambier'.

Image courtesy of John Oxley Library, State Library of Queensland, Neg: 19233

Robert Logan Jack (1845–1921)

As Geological Surveyor for northern Queensland (1876), Jack contributed greatly to the geological knowledge of Queensland. His work covered gold, tin, gemstones and coal. Jack's work was of outstanding quality and quantity, and remarkable for its accurate and detailed observation. Perhaps his greatest contribution was his work on artesian water and the construction of the first government bore in the Great Artesian Basin (1887).

Image courtesy of John Oxley Library, State Library of Queensland, Neg: 19233

Sir Tannatt William Edgeworth David (1858–1934)

David is arguably Australia's most widely accomplished geologist. His mapping assisted the tin and coal industries in New England and Hunter regions. He was Professor of Geology at the University of Sydney (1891–1924). He led the first expedition to reach the South Magnetic Pole (1909) and distinguished himself in World War I. His *Geological map of the Commonwealth of Australia* and *Explanatory notes* were published in 1932.

Henry William Beamish Talbot (1874–1957)

Talbot joined the gold rush at Kalgoorlie (1893). He worked for the Geological Survey of Western Australia (1899–1920), typically spending 7–8 months a year exploring remote areas of the state. He explored for oil in the Canning Basin (1921–30), and worked for Western Mining Corporation (1933–47). Government Geologist AG Maitland said 'Few men have contributed more to our knowledge of the inaccessible and arid regions of the state'.

Image courtesy of National Library of Australia, image number: nla.pic-an10932811-47

Sir Douglas Mawson (1882–1958)

Mawson's major contributions to Australian geology were in the area of Precambrian stratigraphy and glaciation, particularly the rocks of the Flinders Ranges. Under his leadership, the University of Adelaide became renowned in geology. He also led the scientifically successful Australasian Antarctic Expedition (1911) and the British, Australian and New Zealand Antarctic Research Expedition (1920).

Image courtesy of National Archives of Australia, NAA: A1200,L46934

Sir Harold George Raggatt (1900–68)

Raggatt saw Australia as a country with great natural resources. He also recognised the importance of systematic geological mapping and mineral resources development. As Geological Adviser to the Australian Government (1940), he compiled a comprehensive inventory of Australia's mineral resources. He was the first Director of the Bureau of Mineral Resources, Geology and Geophysics (BMR) in 1946 (now Geoscience Australia).

Did you know?

2.3: Geological surveys in Australia

Since the mid-19th century, most states have had a Geological Survey office that produced maps and publications. Before World War II, there was a Commonwealth Geologist and Palaeontologist advising government, but the national geological survey, the Bureau of Mineral Resources of Geology and Geophysics (BMR), was only created by Sir Harold Raggatt in 1946 after a wartime push to find petroleum and useful minerals.

The BMR took responsibility for distant territories such as Papua New Guinea and the Australian Antarctic Territory and Commonwealth waters. The BMR changed its name to the Australian Geological Survey Organisation (AGSO) in 1992. Geoscience Australia was formed in mid-2001, following a merger between AGSO and the Australian Surveying and Land Information Group (AUSLIG). Geoscience Australia has a very broad role in meeting the geoscience needs of the nation to build Australia's future.

context for higher resolution regional geochemical studies and present a baseline for comparison for future geochemical surveys. Different processes emerge at the continental scale compared with the regional or local scale. In Australia, prolonged weathering and relative landscape stability have resulted in higher median silicon dioxide (SiO_2) and zirconium (Zr) concentrations than in Europe, for instance.

Surface geology

The knowledge contained in the Surface Geology of Australia map (Figure 2.7) was compiled by numerous geologists and geophysicists over the past 150 years (Box 2.2). The first impression when examining this map is of the predominant yellow or light-green colours. These Mesozoic and younger sedimentary rocks make up more than 80% of the geology at the surface and illustrate the general tectonic stability of much of the continent from this time onwards. It is on this platform that Australia's regolith and landscape evolved (Chapter 5). Although Figure 2.7 shows the extent of Cenozoic cover to the shoreline, seismic and other data indicate that this cover extends offshore, forming basins that contain sediments in excess of 15 km in thickness, which accumulated during the breakup of Gondwana (Chapter 4).

The eastern seaboard, including the island of Tasmania, is a multicoloured patchwork of Paleozoic metamorphic, sedimentary and igneous rocks. They are revealed as highlands, due to the uplift generated by the formation of the Tasman and Coral seas. Eastern Australia is where the I-type (igneous sourced) and S-type (sediment sourced) granite classification concept was developed. These granites and associated rocks built the eastern margin of the old continent through a process of subduction rollback throughout the Phanerozoic. This rollback process, together with intermittent collision and orogeny, created the meridional grain of eastern Australia's geology. The Flinders Ranges, a Y-shaped region of uplifted Neoproterozoic to Paleozoic sedimentary rocks north of Adelaide (SA), attest to the influence of regional horizontal compression across the Australian Plate from the far eastern margin in New Zealand.

Across northern Australia, large areas of mostly Proterozoic metasedimentary rocks occur in the Kimberley, Pine Creek, Macarthur and Mt Isa areas. These basins were filled with vast sandsheets during a time when Earth's land surface was devoid of the stabilising influence of life (Chapter 3). The basins became the containers for major base metal and U mineral systems (Chapter 8). A series of east–west mountain spines in central Australia is the remnant of the Devonian–Carboniferous Alice Springs Orogeny. Much later, this landscape shaped the first Australians and their beliefs (Box 1.1), with the development of ranges such as the spectacular Uluru (Ayers Rock) monolith—an iconic postcard example of outback Australia (Chapter 5).

Geological regions

Stripping away the young Cenozoic cover from the surface geology map reveals the main geological regions of Australia (Figure 2.8). Geological regions are represented by the dominant outcropping to subcropping geology—for example, the Mt Isa

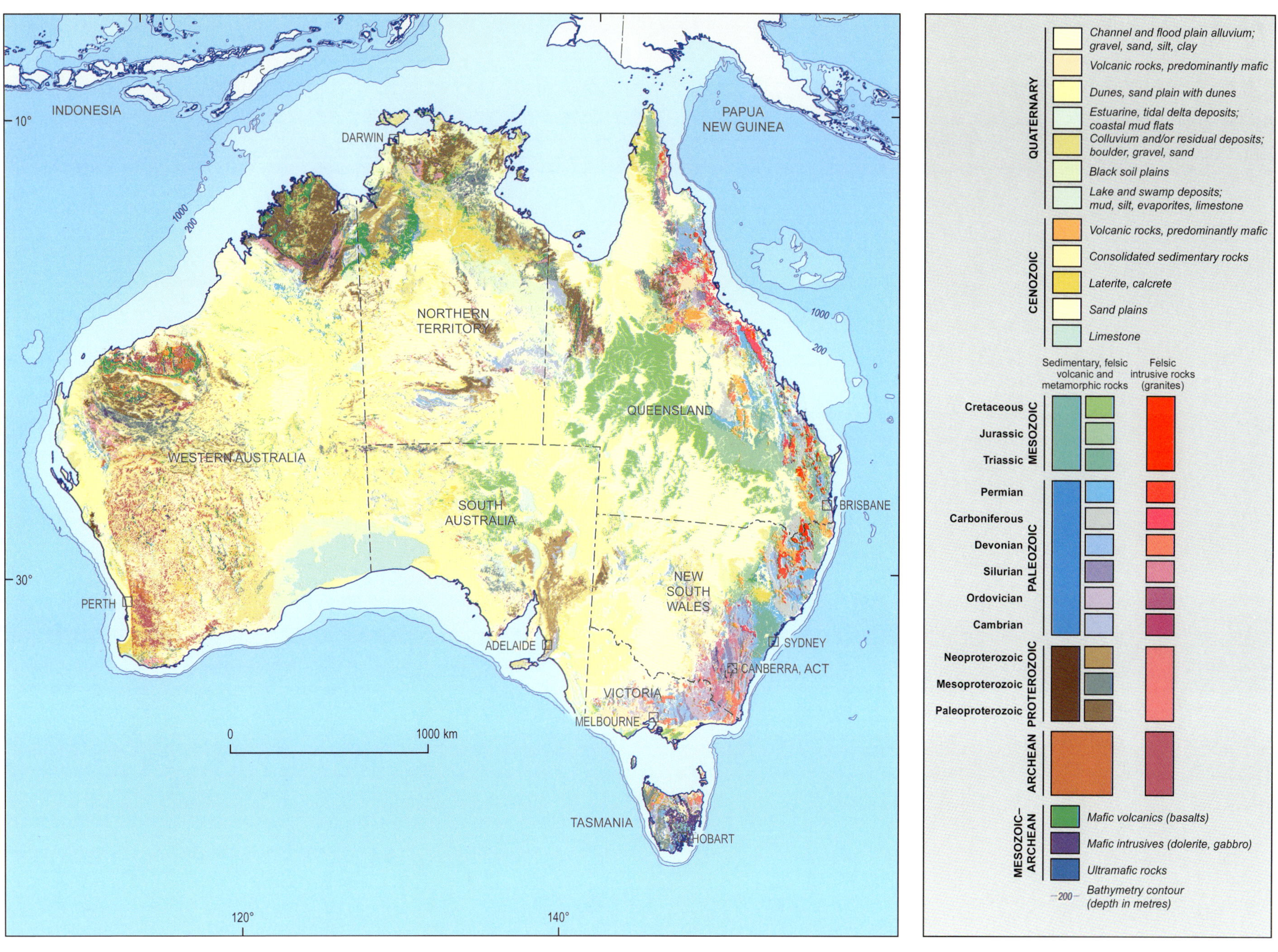

Figure 2.7: Surface Geological Map of Australia. Note the vast areas of pale-yellow rocks of Cenozoic cover and regolith (Chapter 5).

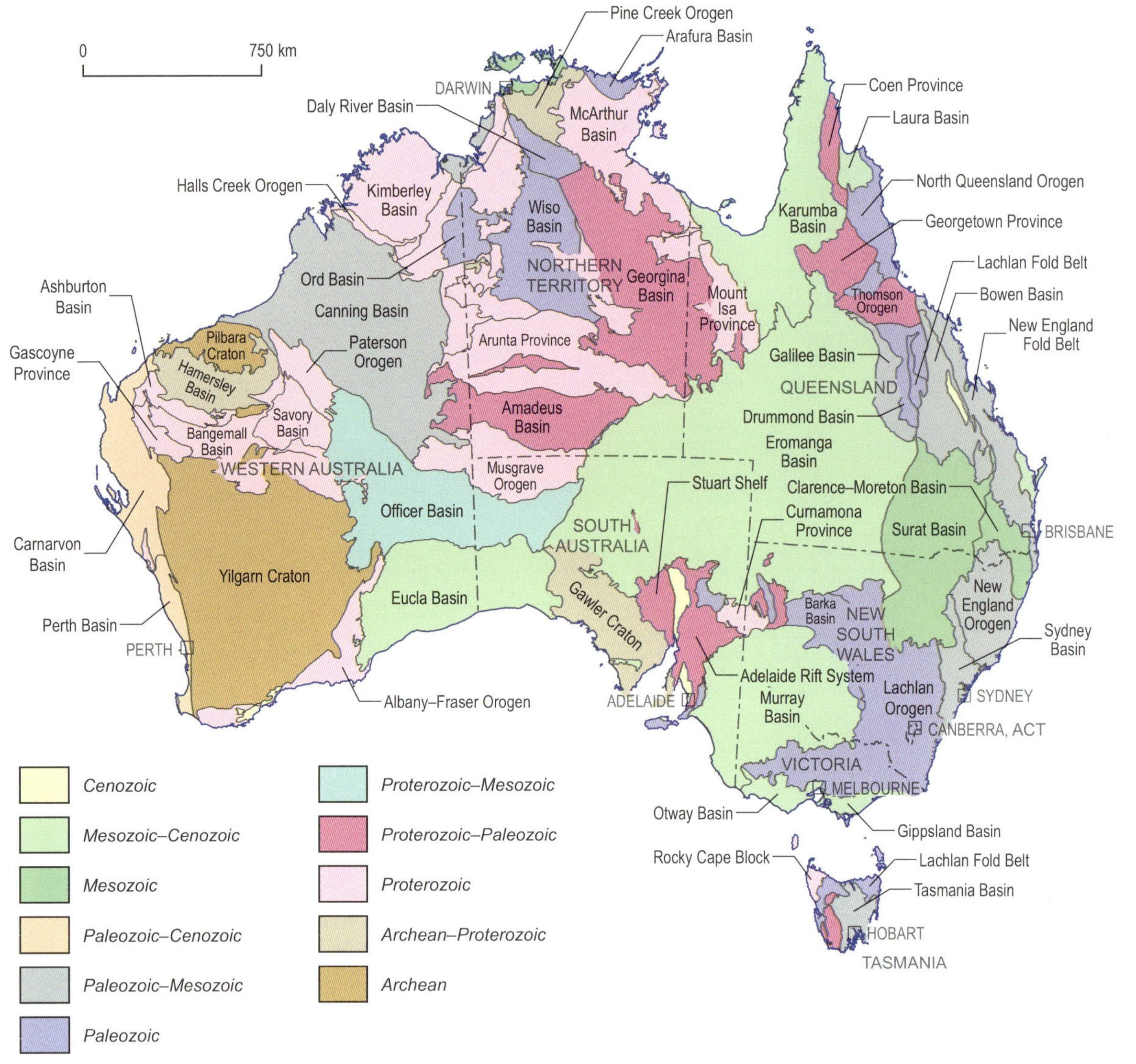

Figure 2.8: The basic building blocks of the continent are revealed in this map of major onshore geological regions of Australia, with the major cratons, inliers, orogenic belts and Phanerozoic sedimentary basins shown.

Inlier, Yilgarn Craton and Eromanga Basin. The geological regions are convenient geographic subdivisions of Australia. In a broad sense, a region may be thought of as the intersection of a three-dimensional geological province with Earth's surface.

The Archean rocks occur mostly in the west, with additional small exposures in the south and north. Proterozoic rocks extend as far east as the northeast coast, western Tasmania and Broken Hill. The Centralian Superbasin is a collection of late Proterozoic to Paleozoic basins that cover much of the centre and central west of the continent. The Paleozoic orogens (Lachlan to New England) in eastern Australia are partially covered by the vast Mesozoic Eromanga Basin, which covers around a fifth of the country. Unlike much of the regolith veneer visible in Figure 2.7, the Cenozoic Murray, Karumba and Eucla are discrete basins of significant thickness (up to several kilometres) and are thus retained on this map. The map illustrates the general trend of younger rocks in the east of the Australian continent.

Basins

Since the late Proterozoic breakup of the supercontinent of Rodinia, Australia has been the locus of numerous events that have resulted in the development of extensive and, in places, deep sedimentary basins (Figure 2.9). The process of basin formation is varied. Some basins formed in foreland settings as the various elements of the continent were shortened and mountain belts formed (e.g. Amadeus Basin in the centre, and the Sydney and Bowen basins in the east). Others are associated with rifting and continental separation of Australia and Antarctica (e.g. Otway and Gippsland basins). Some basins originated because of the lithospheric response to asthenospheric flow in the mantle. For example, the sinking of dense subduction slabs, or counter-flow from upwelling mantle plumes, can draw down the lithosphere. The deepening of certain eastern Australian basins, such as the Early Cretaceous Eromanga Basin and the Cenozoic Murray Basin, are viewed this way.

Many of the onshore basins that are located on the outer edge of the landmass have significant offshore extensions. Notable among these are the Bight, Northern Carnarvon and Arafura basins, all of which have much larger offshore dimensions than onshore (Figure 2.9). The basins of the North West Shelf region (WA) and the basins between Victoria and Tasmania host most of Australia's hydrocarbon production (Chapter 4). Some of the basins are under-explored, especially the vast Bight Basin, which stretches across the southwestern half of the country, and is the largest south-facing shelf on Earth. This coastal zone is exposed to the full force of the Southern Ocean, and is also a region of unique marine biodiversity (Chapter 6).

The central part of eastern Australia is a region with a vast number of overlapping basins that are concealed or partly concealed beneath the Mesozoic Eromanga Basin (Figure 2.9). The region has clearly been one of ongoing subsidence and basin formation since the mid-Paleozoic. These basins are important for Australia's coal and onshore hydrocarbon resources (Chapters 4 and 9), and are also becoming a focus for geothermal energy exploration (Chapter 10).

Figure 2.9: Major pre-Cenozoic basins of onshore and offshore Australia. In this map, the older basins beneath the Eromanga Basin are shown across eastern Australia. The onshore to offshore connections of the basins are also shown. Many of these named basins are hosts to Australia's producing, as well as potential, hydrocarbon and coal resources (Chapters 4 and 9).

Figure 2.10: The 2008 definition of major crustal elements in Australia (after Hoatson et al., 2008 revision of Shaw et al., 1995). The oldest is the West Australian Element and the youngest is the Tasman Element. The zone between the West Australian, South Australian and North Australian elements is made up of Proterozoic- to Paleozoic-aged orogens, which are the 'geological glue' created during the assembly of Australia (see later). The bulk of the continent was built from west to east. Mesozoic or younger ocean floor bounds the outer edges of the crustal elements, except in the north. Here the map extends beyond Australia to the active plate edge in New Guinea, while to the south it stretches to the South Tasman Rise (Figure 2.1).

Crustal elements

Beneath the late Proterozoic and younger basins (Figure 2.9) lie the fundamental building blocks, or major crustal elements, of continental Australia. The interpretation of these elements and their boundaries has evolved over the past 100 years, and will continue to evolve as new data are acquired.

Edgeworth David produced the first tectonic map for Australia in 1911, which he updated and revised in 1931. In 1960, a solid geology map of Australia showed the continent with Cenozoic rocks removed (similar to Figure 2.8). Around 1970, Russian geologists published a tectonic map of Australia in their *Great soviet encyclopedia*. The Geological Society of Australia produced a 1:5 M scale tectonic map of Australia and Papua New Guinea in 1971. As part of an earth science atlas of Australia, the Bureau of Mineral Resources published the *Major structural elements of Australia*. This 1979 map divided the continent into five main groups of cratons and so-called covers: the Archean Craton, the North Australia Craton, Northeast orogens, Central Australian Mobile Belts and the Tasman Fold Belt.

In 1995, the *Australian crustal elements map* was produced, utilising the improvements in national-scale geophysics, geochemistry and geochronology. The map identified eight coherent 'mega-elements'. These mega-elements represent crust with similar geological and geophysical characteristics, within a common set of boundaries, which were interpreted to mark crustal-scale changes in composition and/or structural patterns.

AGSO's (Geoscience Australia's) Aero Commander geophysical survey aircraft VH–BGE, which flew airborne magnetic and gamma-ray spectrometric surveys across the continent between 1990 and 1999.

The Australian continent is grouped into six major elements (Figure 2.10). This interpretation is similar to the 1979 version, with the addition of the Pinjarra Element in the west. The Tasman Line was first defined in 1951 as the most easterly outcrop of Precambrian rocks. This line has been redrawn many times, mostly on the basis of interpretation of geophysical datasets. Proterozoic rocks occur east of most proposed Tasman Lines, such as in northwest Tasmania and north-central Queensland. They are likely to be the basement to much of the Tasman Orogen, or as rifted remnants of the margins of Rodinia, a Neoproterozoic supercontinent (see below and Box 2.5) to which Australia belonged. Studies of the seismic structure in the mantle beneath Australia show a broad correspondence with these crustal elements, but the boundary of the Tasman Element in the mantle appears to lie significantly further east (see Figure 2.17).

The Central Australian Element is a zone of reworking of the older elements north and south, together with juvenile input during a series of rifting events. This element includes the Albany–Fraser Orogen and the Musgrave–Paterson Orogen. The North Australian Element, from the Kimberley region in the west to Cape York Peninsula in the east, consists of mostly Proterozoic rocks that were cratonised by 1550 Ma. The West Australian Element consists of the Archean Yilgarn and Pilbara cratons, which are separated by Proterozoic

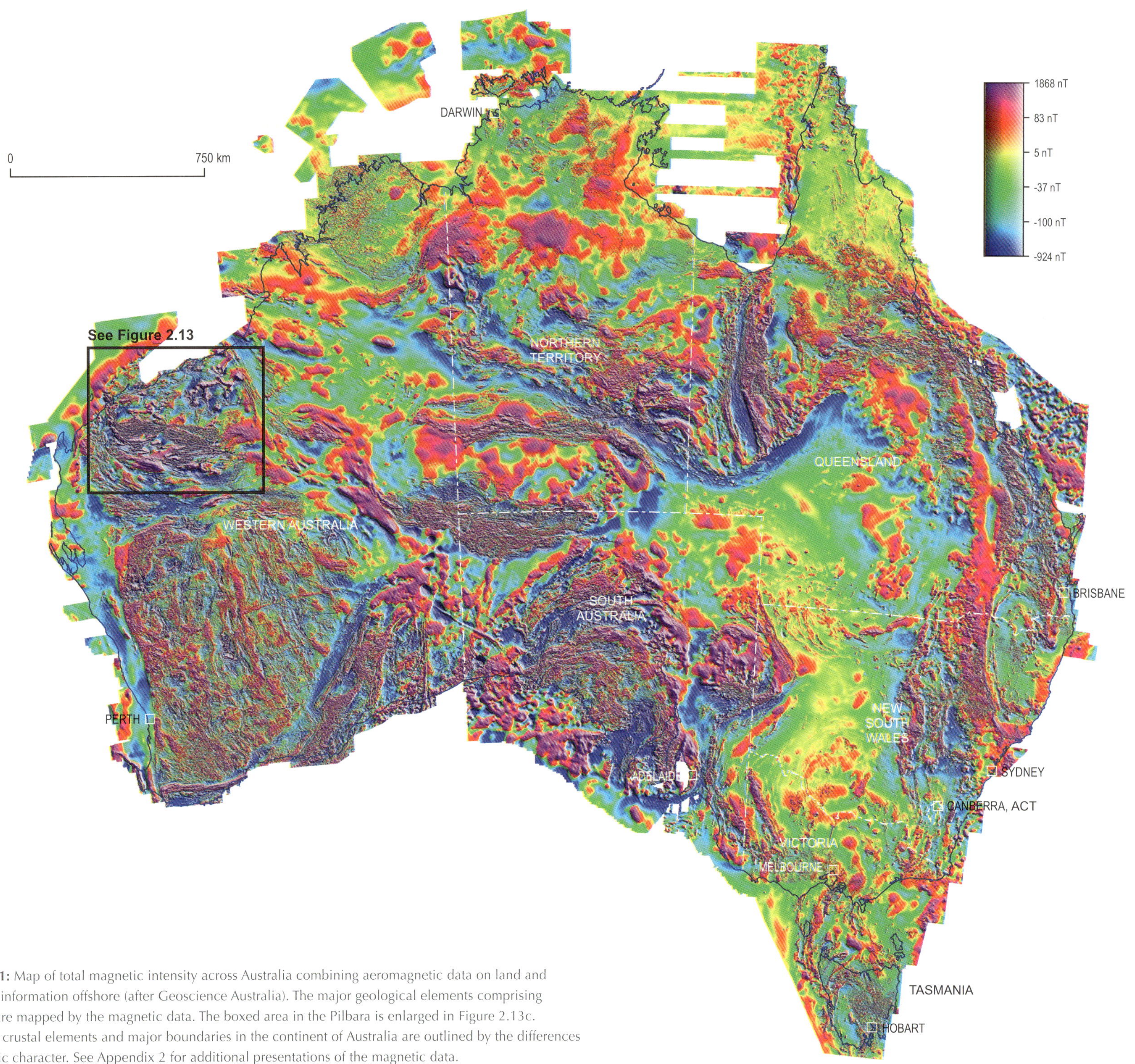

Figure 2.11: Map of total magnetic intensity across Australia combining aeromagnetic data on land and ship-track information offshore (after Geoscience Australia). The major geological elements comprising Australia are mapped by the magnetic data. The boxed area in the Pilbara is enlarged in Figure 2.13c. The major crustal elements and major boundaries in the continent of Australia are outlined by the differences in magnetic character. See Appendix 2 for additional presentations of the magnetic data.

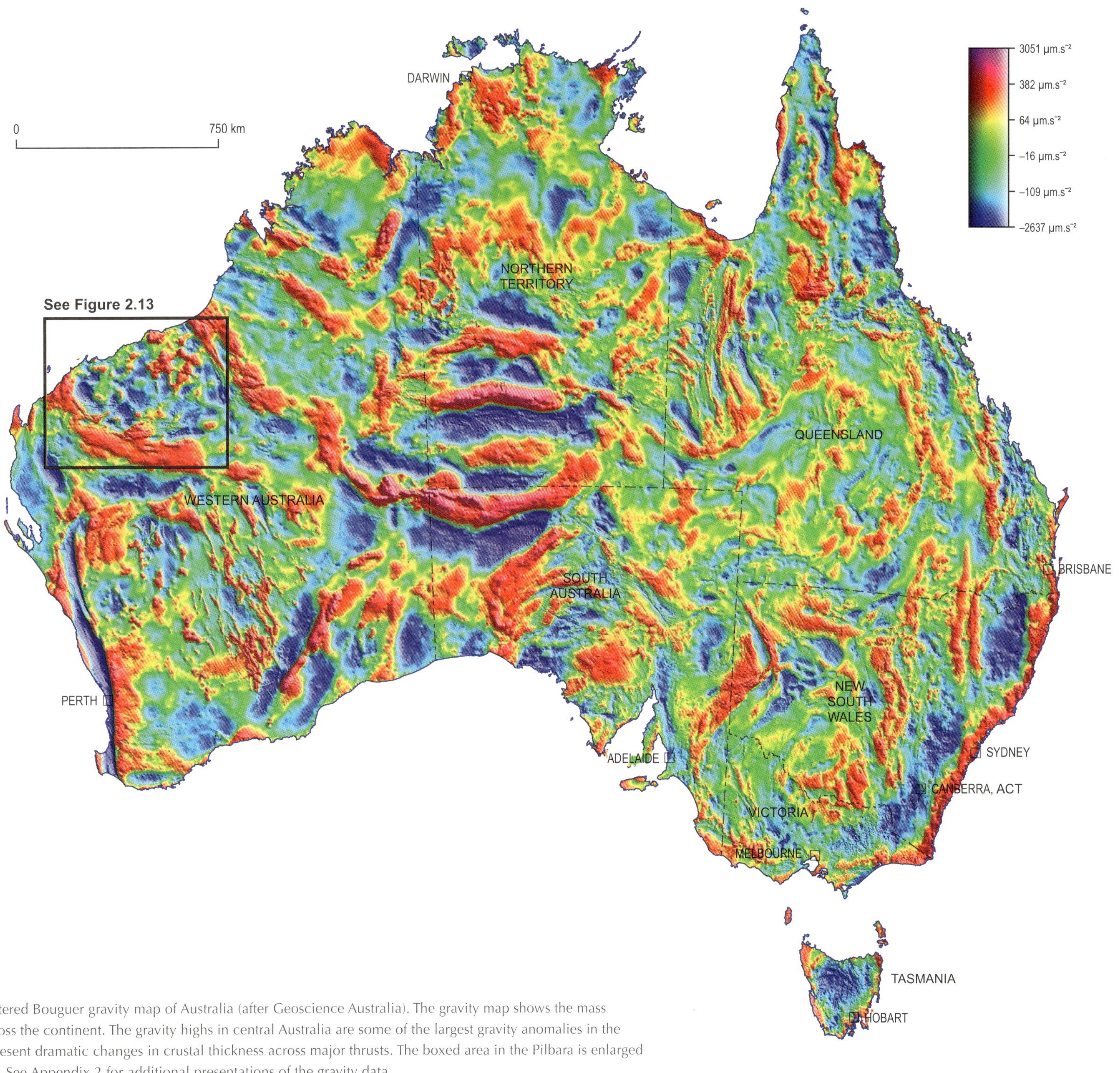

Figure 2.12: Filtered Bouguer gravity map of Australia (after Geoscience Australia). The gravity map shows the mass distribution across the continent. The gravity highs in central Australia are some of the largest gravity anomalies in the world, and represent dramatic changes in crustal thickness across major thrusts. The boxed area in the Pilbara is enlarged in Figure 2.13d. See Appendix 2 for additional presentations of the gravity data.

mobile belts. The Pinjarra Element comprises Proterozoic and younger reworking of the western margin of the continent. The South Australian Element consists of the Archean Gawler Craton and Proterozoic Curnamona Province, which have some geological affinities with the element to the north. The Tasman Element consists of a series of eastward younging orogens accreted onto the older western two-thirds of the continent, which include the Delamerian, Lachlan, Thomson, North Queensland and New England orogens that are collectively known as the Tasmanides.

Measuring absolute gravity at Mawson Station, Australian Antarctic Territory.

Image courtesy of Ray Tracey, Geoscience Australia

Australia's potential fields

Around 80% of Australia is covered by surficial material (Figure 2.7) called regolith (Chapter 5). Its presence makes potential field geophysics an essential tool set for mapping the buried geology beneath. This extensive regolith cover has driven the collective efforts of the state and territory geological surveys and Geoscience Australia to acquire high-quality continent-wide airborne magnetic data (Figure 2.11). The gravity information is not as densely sampled, but many parts of the continent are sampled on a 1–2 km grid (Figure 2.12). The magnetic and gravity data are complementary, mapping different fundamental bulk rock properties (susceptibility for magnetism and density for gravity).

The potential fields reflect both the spatial distribution and the strength of variations in physical properties, so that longer spatial wavelengths are sensitive to anomalies at a greater depth. The gravity signal diminishes with the square of the depth, while the magnetic signal diminishes with the cube of the depth, which means that gravity is a 'deeper looking' technique. No single presentation satisfies all needs, and filtering the data is required to extract the different depth and textural information (see Appendix 2 for different presentations).

Magnetics

Magnetic data measure the variation in Earth's magnetic field, which is caused by differences in the magnetic susceptibility of the underlying rocks. Such data typically provide information on the structure and composition of the magnetic basement, which commonly has a distinctive magnetic signature characterised by the magnitude, heterogeneity and fabric of the magnetic response. When calibrated with known geology, terranes can be mapped under a cover of sedimentary rock and/or water, a technique that has proved valuable in Australia, given its extensive and deep cover (Figures 2.8, 2.9 and 2.10).

The total magnetic intensity map (Figure 2.11) shows a number of features in common with the filtered gravity map (Figure 2.12), but also a more distinct representation of texture linked to the major geological features. Major basement structures can be interpreted from consistent discontinuities and/or pattern breaks in the magnetic fabric. We see, for example, the distinct oroclinal swirl of structure around the southern end of the New England Orogen on the east coast. There are still a few areas with thick sedimentary cover where flight-line spacing is sparse, notably around the Simpson Desert in northeast South Australia, so that loss of detail here may not just be associated with cover. The sharp, jagged magnetic boundaries along the western borders of New South Wales and Queensland are often interpreted as the preserved salients and embayments created by the breakup of Rodinia. To the west, the old craton and province of the Gawler and Curnamona are circular in shape, and are fringed by younger mobile belts. The Yilgarn Craton makes up much of southern Western Australia, and it too has sharp magnetic boundaries from younger events, such as the Pinjarra and Albany–Fraser orogens, on its margins.

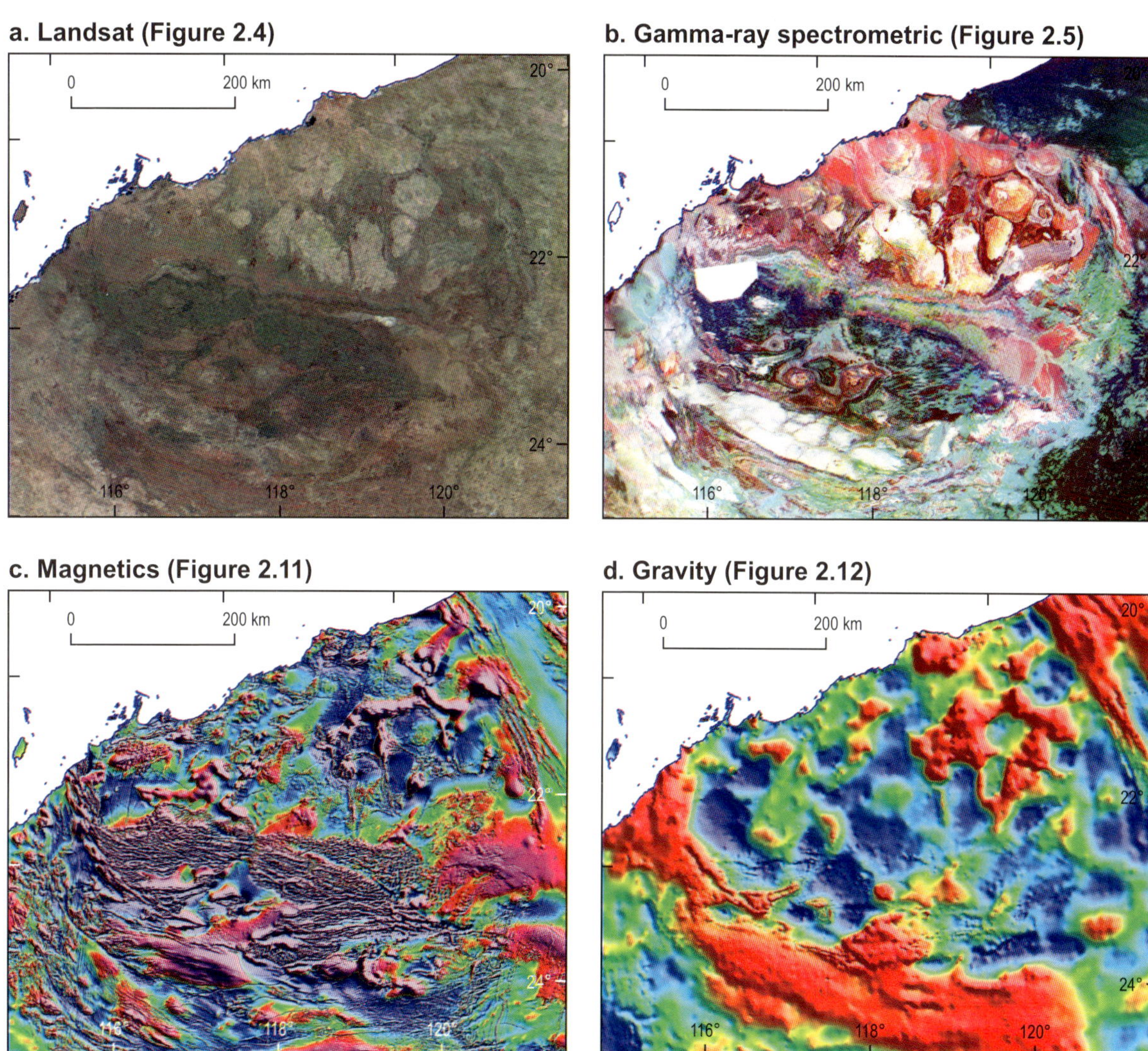

Figure 2.13: The North Pilbara Craton of Western Australia is a classic dome-and-basin map pattern for Mesoarchean geology. The exposed granite-cored domes are up to 50 km in diameter; they are enveloped by predominantly basaltic (greenstone) volcanic rocks. These rock types have distinctly contrasting properties, visible in the images. In contrast to the greenstones, the granites have high albedo (reflectance of light); high radioelement concentrations, particularly K (coloured red to white); low magnetisation; and low density. Covering the southern and eastern region are the Neoarchean volcanic and sedimentary rocks of the Fortescue and Hamersley basins (Figure 2.9).

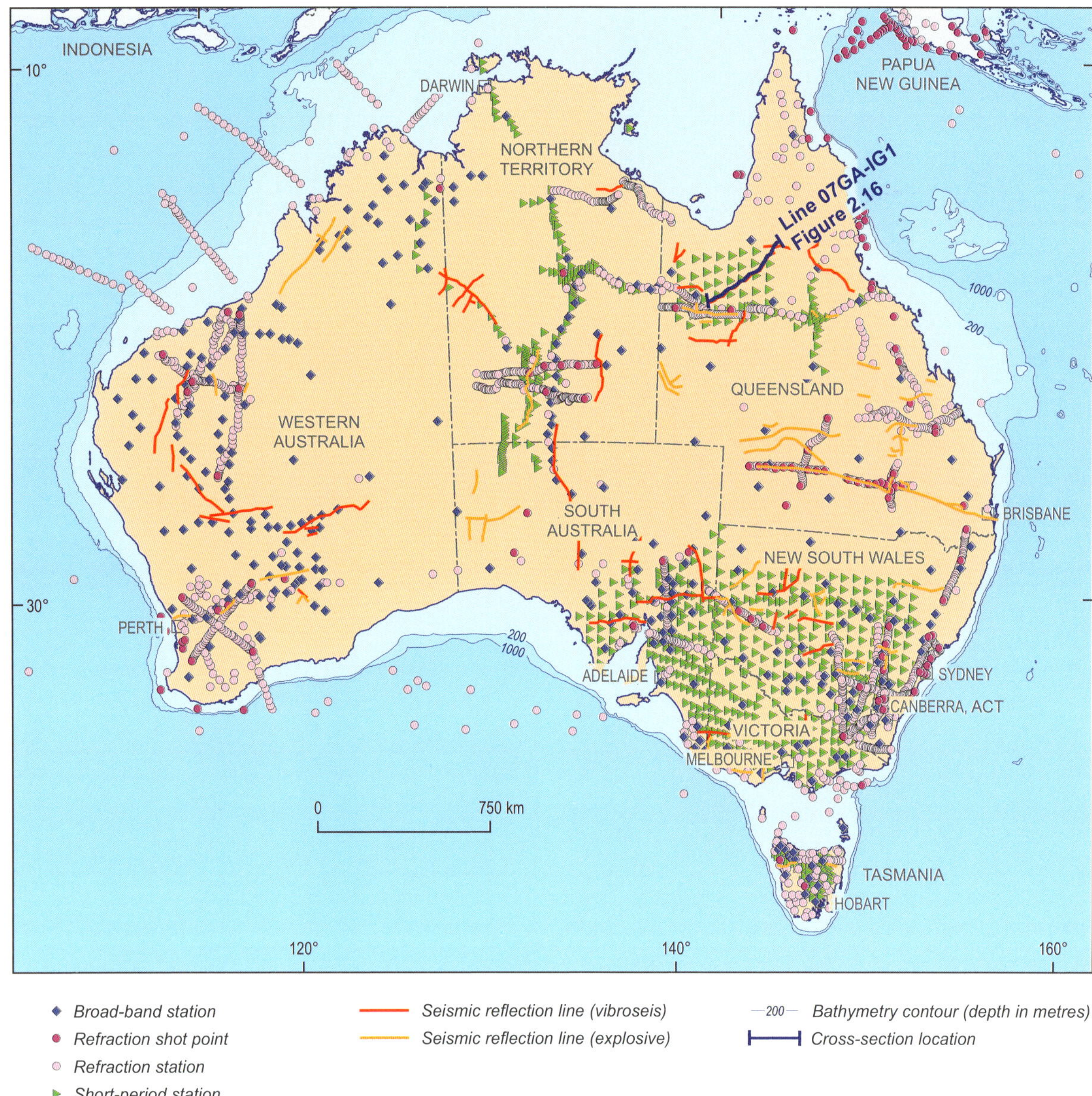

Figure 2.14: Both controlled-source and passive seismic techniques are systematically and regularly deployed across Australia. The major seismic reflection lines cross most of the major elements and their boundaries. A selection of the reflection profiles is presented in the Australian Common Earth Model (on DVD). Many of these onshore surveys have been acquired in the past decade, which has seen improvements in the processing and acquisition of data. The location of the seismic line in Figure 2.16, which crosses part of north Queensland, is also shown.

Gravity

Gravity data map subtle changes in Earth's gravitational field caused by variations in the density of the underlying rocks. The spatial resolution of gravity data tends to be lower than that of aeromagnetic data, but its ability to image anomalies at depth is better than magnetic data (Figure 2.12). Continental Australia has a gravity station spacing coverage of 11 km. South Australia, Tasmania and part of New South Wales are covered at a spacing of 7 km, and Victoria at approximately 1.5 km. Over the past 15 years, the Australian, state and Northern Territory governments have funded exploration initiatives for the systematic infill of the continent at a grid spacing of 2 km, 2.5 km or 4 km to provide improved coverage for areas of scientific or economic interest.

The current gravity grid was produced from 1.4 million onshore gravity stations. Data for the offshore region were extracted from a satellite altimetry dataset provided by the Scripps Institution of Oceanography, California. The gravity image links the Bouguer field on the continent to free-air gravity anomalies offshore. The full gravity field clearly displays the contrasts associated with the various cratonic elements, and considerable substructure is revealed when the longer wavelength gravity anomalies are removed by filtering (Figure 2.12). The group of alternating bands of east–west-trending low and high gravity anomalies in central Australia dominate the gravity image, and are associated with substantial localised changes in crustal thickness in the zone affected by the Alice Springs Orogeny. Similar bands of alternating north–south-trending gravity anomalies in eastern Australia are interpreted to be associated with a series of island arcs and a succession of Paleozoic accretionary events onto the evolving eastern margin of the continent. The major mineral province of Mt Isa (western Qld) has strong positive gravity anomalies that trend north–south, a tectonic grain imparted during the Mesoproterozoic. These trends are truncated in the south, seen by a sharp gravity gradient that most likely formed during the latest Proterozoic breakup of the supercontinent Rodinia, probably when Laurentia (North America) rifted from Australia to form the palaeo-Pacific Ocean. The contrasting Archean map patterns within the Pilbara Craton (ovoid) and the Yilgarn Craton (elongate) are evident in the gravity response to the distribution of dense mafic greenstone rocks.

Australian lithosphere

The nearly rigid outer skin of Earth is termed the lithosphere, which in Australia ranges in thickness from more than 200 km, beneath the Archean cratons of Western Australia, to very thin at the active mid-ocean ridges of the Southern Ocean. With crustal and upper mantle components, the lithosphere sits upon the more ductile asthenosphere. The Australian lithosphere is moving more quickly than the asthenosphere beneath, driven by ridge push from the mid-ocean ridge in the Southern Ocean and slab-pull along the subduction zones to the north and east.

Crust to mantle

The uppermost layer of the lithosphere is the crust. A sustained program of active seismology using human-made sources, as well as passive

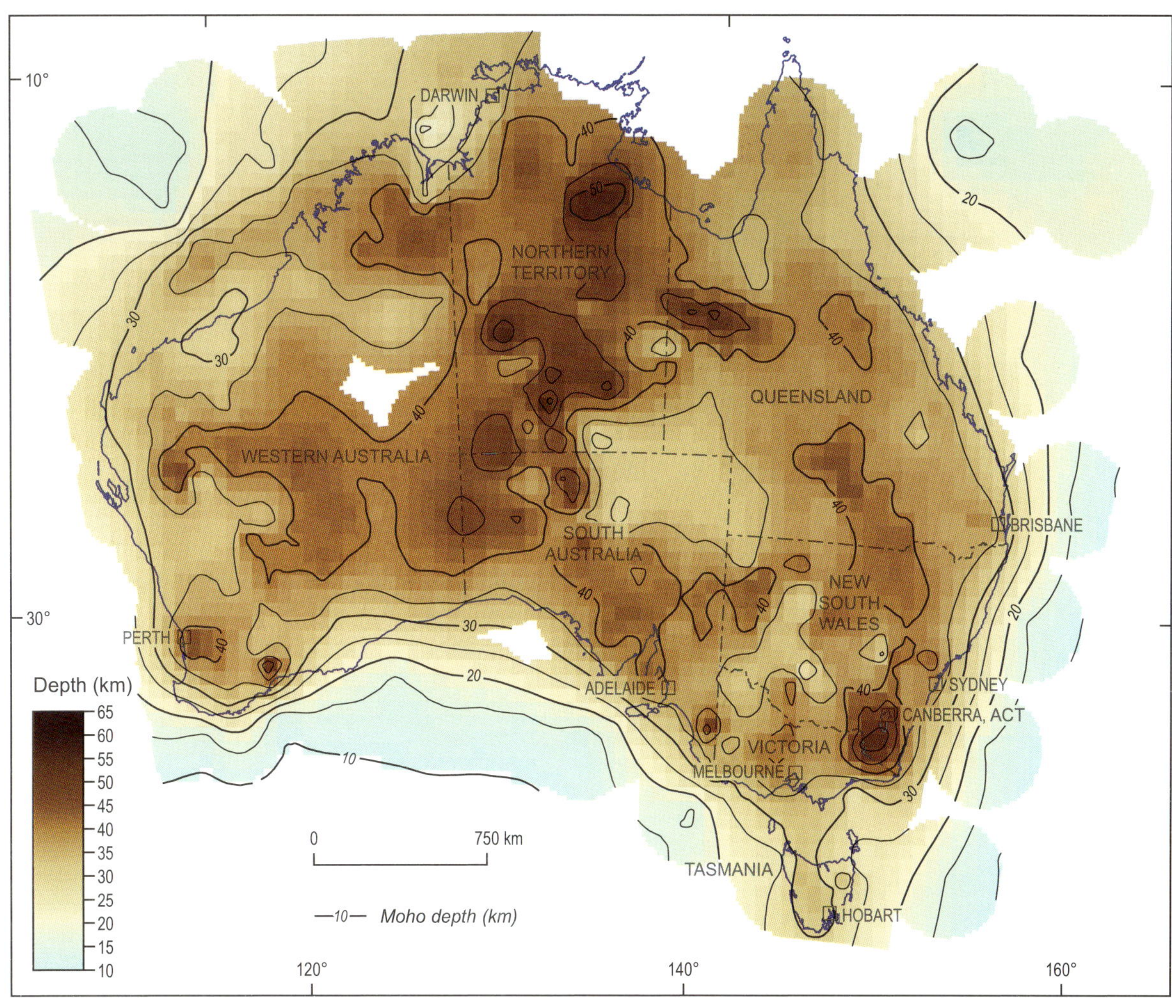

Figure 2.15: Contour map of crustal thickness (depth to Moho) in kilometres. The data are derived from seismic reflection, refraction and receiver function methods (Figure 2.14). The continent has an average crustal thickness of 38 km, with a range in thickness from 24 km to 59 km. There are a number of large gradients in crustal thickness, particularly in central Australia around longitude 132°E. The Moho changes from nearly 60 km in the west to around 30 km in the east over a distance of only about 200 km. The crust around Mt Isa and Tennant Creek in the North Australian Element is unusually thick (50 km), as is the crust beneath the 'high country' of the Snowy Mountains in southeast New South Wales.

LAND-BASED DEEP CRUSTAL REFLECTION PROFILING (BOX 2.3)

Australia was a pioneer in the use of land-based seismic reflection methods to probe the deep crust, with short reflection profiles conducted as early as the late 1950s. An ongoing acquisition program has been maintained over the decades by Geoscience Australia and its predecessors, with increasing sophistication in styles of experiment and recording.

The AuScope infrastructure initiative, launched in 2007 by the Australian Government, has enabled extra seismic reflection coverage to be acquired in South Australia, Victoria and Western Australia. More than 17 600 km of deep seismic reflection profiles have been acquired across the continent (not including marine surveys).

Explosions were used as the energy source until 1998, with major surveys across southern Queensland and in central Australia. The establishment of the ANSIR Major National Research Facility for Earth Sounding in 1997 provided heavy-duty vibrator trucks and digital recording equipment, which were employed in a number of projects across the country to explore lesser known geological provinces.

The survey along the remote and regolith-covered region of north Queensland shows a remarkable range of structures through the crust, including major changes in seismic character, unknown sedimentary basins, possible sutures and a clearly defined sharp Moho (Figure 2.16).

The result of this work is an extensive national coverage across many of the continent's major structures. Such information helps in understanding the crustal architecture in the context of resource exploration, and provides depth control on the evolution of the Australian continent.

Processing deep-crustal seismic data is different from processing conventional sedimentary basin seismic data. The hard rock environment has relatively high P-wave velocities at shallow depths, and commonly has steeply dipping geology. Reflections from the deep crust also tend to be less coherent. The improvements in imaging have been made by use of better imaging algorithms, as well as more data channels per source location (data fold).

Vibroseis trucks acquiring deep seismic reflection data in the Flinders Ranges, South Australia.

sources from earthquakes generated at subduction zones surrounding Australia and further afield, has provided a comprehensive view of the thickness and internal structure of the crust across Australia (Box 2.3). The extensive coverage of the continent is indicated in Figure 2.14, which shows the locations of seismic stations for reflection, refraction and passive seismic studies across Australia.

Various styles of seismic data analysis were employed to generate a new map of the crustal thickness across the continent. For consistency with earlier studies, the seismic Moho is taken at the base of the zone where seismic wavespeeds reach mantle values (e.g. above 7.8 km/s for P-waves). This definition may not correspond to the petrological Moho in terms of mantle rocks, and the transition from crust to mantle is gradational in a number of places across the continent. The combination of results from numerous deep seismic reflection, refraction and receiver function profiles has yielded the new map (Figure 2.15). The Australian continental crust, excluding the thinned margins, has an average thickness of 38 km, with a range from 24 km to 59 km. The clarity of the crust–mantle boundary in the seismic data varies. In some terranes, like the Proterozoic region between Cloncurry and Croydon, north Queensland, the boundary is 'knife' sharp (Figure 2.16). A similar clear Moho is evident beneath the Archean Yilgarn Craton. In other areas, the boundary is difficult to define, such as in the Proterozoic Curnamona Province and the Archean–Proterozoic Gawler Craton (Figure 2.8). The crust–mantle boundary can have significant localised topography, with distinct offsets of more than 10 km due to large displacements on presumed faults (e.g. in north Queensland (Figure 2.16) and central Australia). There are also steep horizontal gradients in crustal thickness, particularly in central Australia.

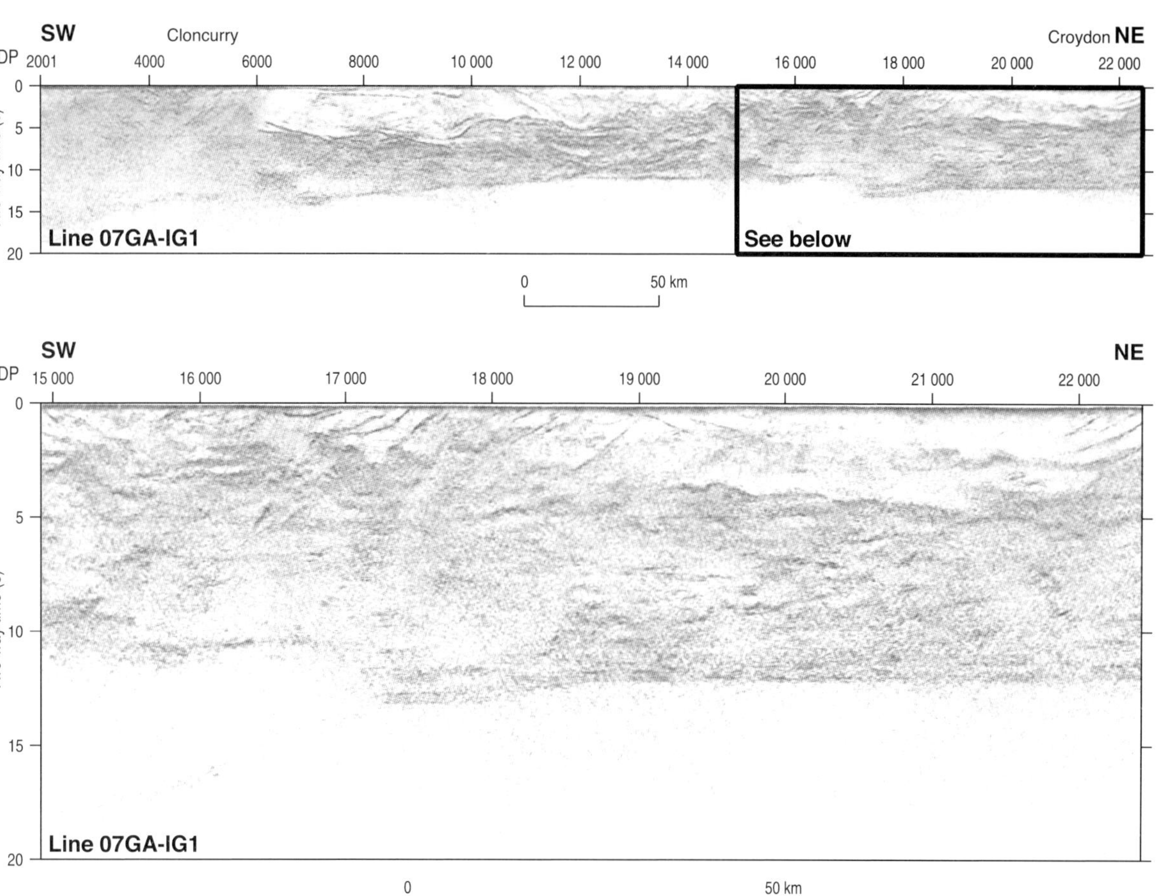

Figure 2.16: Full crustal reflection profile between Cloncurry (left) and Croydon (right) in north Queensland carried out with vibrator sources (seismic line 07GA–IG1). Lower image shows a detailed view of the region in the box. The frame displays 20 seconds of record (i.e. approximately 60 km in depth). The migrated section shows contrasting seismic character of different crustal blocks. The Mt Isa Province has a seismic character that is diffuse, and the depth to Moho appears gradational or is poorly defined; the crust is at least 50 km thick (Figure 2.15). To the northeast, a major crustal boundary is crossed and the Numil–Abingdon seismic province is imaged with a sharply defined Moho and a well-developed series of seismic features. Note the Moho step towards the northeast end (detailed lower image); this is where a reflector extends into the mantle and the Moho is offset by *ca* 10 km. The features are interpreted as a fossil subduction zone. For more images and interpretations, see the Australian Common Earth Model (on DVD).

The general pattern for post-Archean crust is towards thinner crust with decreasing age. This relationship has been interpreted to be associated with magmatic underplating, with the crust being accreted successively from below. This age–thickness relationship does not hold for Archean crust, much of which is currently relatively thin, being around 30 km beneath the Pilbara Craton

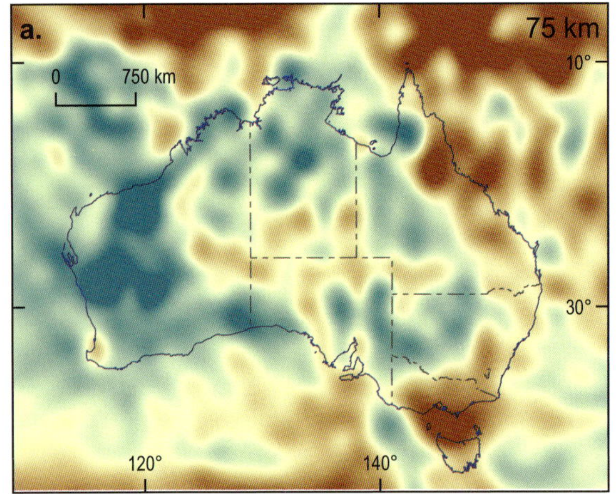

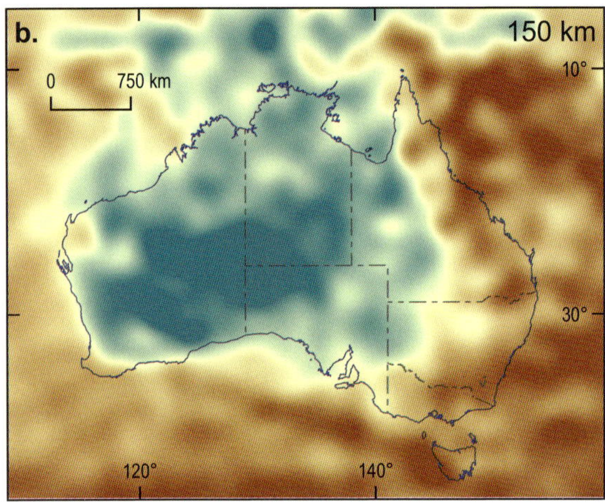

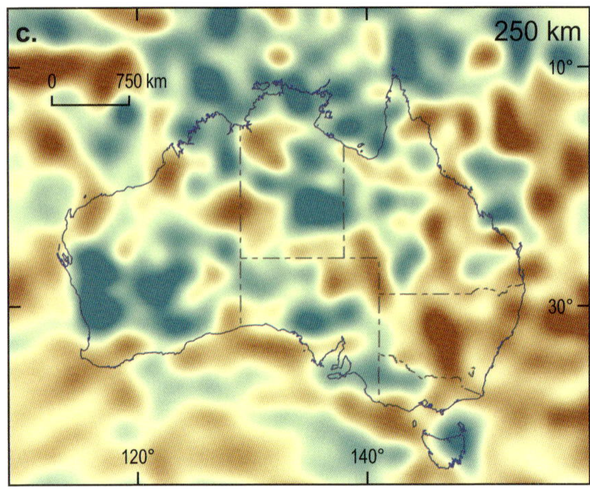

and around 38 km beneath the Yilgarn Craton. The thickest crust is under the North Australian Element. Interestingly, the Mt Isa region has anomalously thick crust of around 55 km, and it also has a large volume of high-density material (Figure 2.12). The thinnest crust occurs beneath Tasmania (Figure 2.15).

Over the past 30 years, a series of long reflection seismic transects has been acquired (Figure 2.14). One series of surveys transects the crust from the Lord Howe Rise in the east some 2000 km to central-west Queensland. Many of the land surveys were high-quality vibroseis sourced with up to 22 seconds two-way-time recording that provided information on the crust and uppermost mantle. The total line length for onshore deep-seismic reflection surveys in Australia is 17 600 km.

More recently, magnetotelluric (MT) data have been added along many of the major seismic transects. This passive technique measures natural electrical and magnetic fields of the crust and mantle and is used to map major boundaries and different crustal domains. A deep conductive zone extends north–south from east of Mt Isa in the north to the eastern Flinders Ranges in the south. This zone approximates the boundary between the mostly older western and central Australian crust and the generally younger eastern Australia crust. Anomalous zones of enhanced crustal conductivity have been mapped beneath some of the major mineral systems, such as Olympic Dam and Kalgoorlie (Chapter 8).

Figure 2.17: A representation of the 3D variation in seismic wavespeed maps of the upper mantle through perspective view of depth slices at 75 km, 150 km and 250 km. The three maps are displayed with a common seismic velocity scale, with the neutral tone representative of continental upper mantle structure. The red–brown zones have slower wavespeeds, probably due to the effects of higher temperatures or the presence of volatiles. The blue areas have faster wavespeeds and reveal the extent of the deep lithospheric keels of the continent.

Each new deep seismic and magnetotelluric survey discovers something. Major crustal boundaries are imaged through the entire crust, and some penetrate the Moho, leaving signatures interpreted as fossil subduction zones (e.g. Figure 2.16). Some of these deep structures are also spatially associated with Australia's major mineral deposits, such as beneath Mt Isa, Olympic Dam and Kalgoorlie (Chapter 8). The quest for mineral wealth is one of the principal drivers behind these deep surveys.

The regular release of offshore acreage is a key part of the Australian Government's strategy to encourage investment in petroleum exploration. The acreage release involves the delivery of seismic data and their interpretations in selective regions of the Australian margin. The 2011 release areas were located in Commonwealth waters offshore from the Northern Territory, Western Australia and Victoria (Arafura, Bonaparte, Browse, Roebuck, Carnarvon, Perth, Otway and Gippsland basins).

The offshore north and northwest region is a world-class gas province with significant oil accumulations. The region includes the offshore and marginal basins of the Arafura Sea (Carpentaria, Arafura and Money Shoal basins) and the North West Shelf (Bonaparte, Browse, Offshore Canning, Roebuck and Carnarvon basins and Wallaby Plateau) (Figure 2.9). Cumulatively, the North West Shelf basins comprise the Westralian

Superbasin, a thick Late Paleozoic, Mesozoic to Cenozoic sedimentary succession related principally to the fragmentation of Gondwana (Chapter 4).

Under the UN Convention on the Law of the Sea, regional marine seismic data, together with potential field data, have been acquired across Australia's entire marine claim. These data were instrumental in defining Australia's limits (Chapter 1). They also reveal how the continent and its margins developed and evolved (Chapter 4).

Upper mantle and lithosphere–asthenosphere boundary

The Australian Plate is bounded by belts of earthquakes along the Indonesian arc, through New Guinea and then down through the Tonga–Kermadec arc and New Zealand to join the mid-ocean ridge system south of Macquarie Island (Figure 2.1). These earthquakes have been exploited in a variety of studies of the structure beneath Australia, using both refracted body waves and the large late surface waves recorded at the few permanent and many temporary stations across the continent. The frequency of earthquakes is greatest to the north and east, but the southern events provide very useful additional information.

The earthquake signals are exploited in seismic tomography to extract images of the 3D shear wavespeed distribution beneath the Australian region. The reliability of the results depends on the density of paths between sources and receivers, and especially on a sufficient number of crossing paths. The result is reliable for most of the Australian continent and the Tasman Sea to the east, but the absence of western sources means that coverage to the west of Australia is limited. Thanks to systematic coverage of the continent with portable seismic instruments undertaken since 1993 by The Australian National University, Canberra, the resolution of mantle structure is of the order of 200 km horizontally and 30 km vertically.

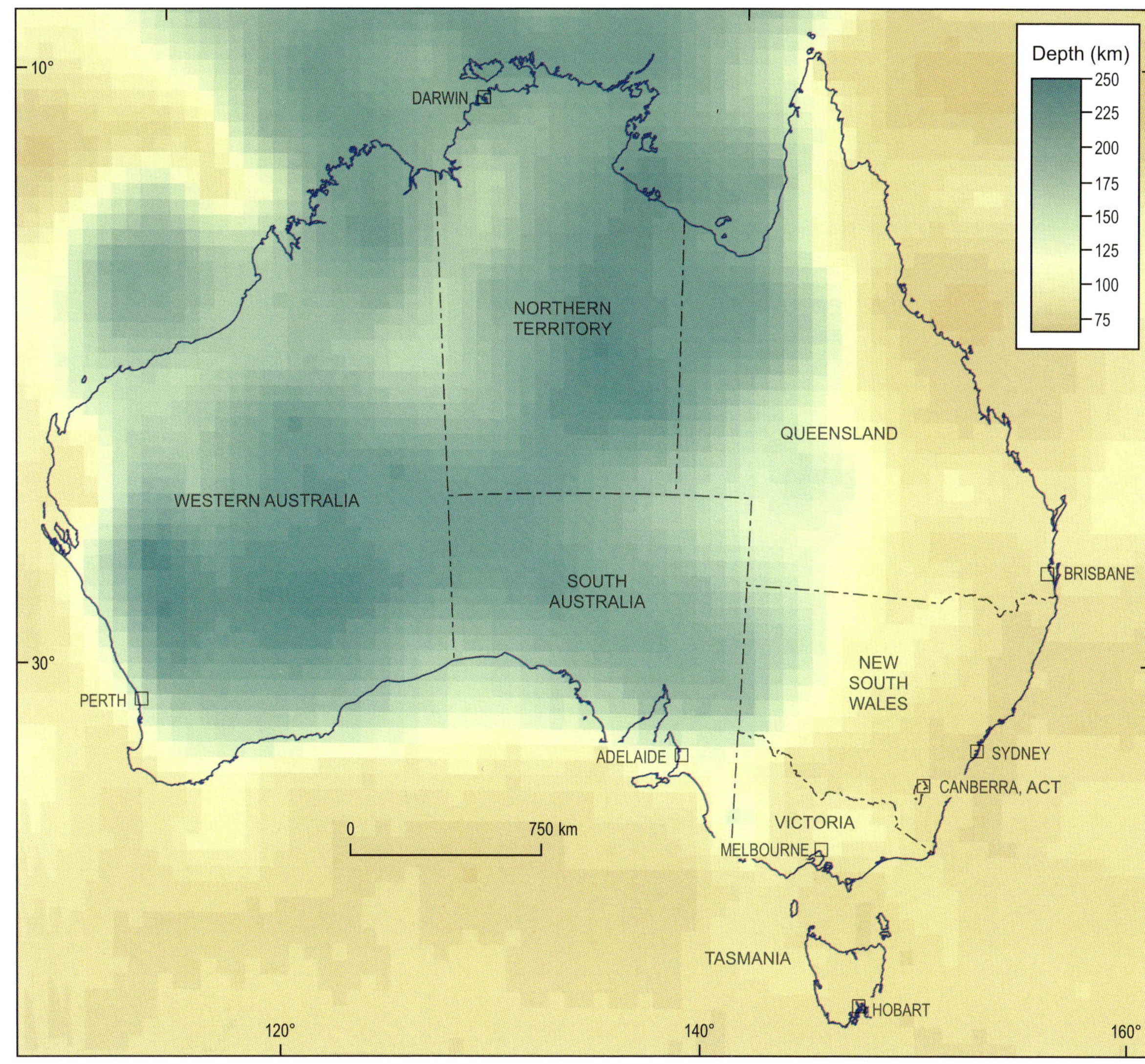

Figure 2.18: Estimate of the depth to the base of the lithosphere derived from various seismological data. The map shows the mostly *ca* 200 km-thick lithosphere in the western two-thirds of the continent. The thickness of lithosphere beneath the Tasman Element in the east is *ca* 100 km or less, which is similar to the global average for continents.

Did you know?

2.4: Deep carbon

Diamond is the high-pressure allotrope (crystal form) of carbon. Most diamonds come from areas located above the edges of thick lithospheric keels of Archean cratons, such as the Kaapvaal Craton in South Africa and the Slave Province in Canada. If lithospheric depth is one main control on diamond formation, then does the thick lithosphere under Proterozoic-aged central Australia suggest that these regions and along their edges may be the place to look for them?

Image courtesy of Argyle Diamonds

Argyle, in the Kimberley of Western Australia, produces more than 90% of the world's pink diamonds. These rare and precious gems have an intensity and vibrancy of colour not seen in any other diamonds.

The shear wavespeed (or velocity) is strongly dependent on temperature, and the base of the lithosphere is generally assumed to occur at mantle temperatures around 1300°C. A striking feature in the 3D images of shear wavespeed is the strong contrast between fast wavespeeds (blue-green in Figure 2.17) in central and western Australia and slower wavespeeds (brown in Figure 2.17) to the east. This divide is most clear between 125 km and 200 km depth. The fast zones lie beneath the ancient cratons, whereas slower wavespeeds are seen beneath the eastern Phanerozoic belts and particularly beneath the Tasman Sea. At shallower depths, the fast wavespeed domains in the uppermost mantle are broken up in the areas of more recent deformation (Paleozoic), such as central Australia. The slow wavespeeds across the centre at 75 km depth are difficult to explain, and the suitable combination of thermal or compositional structure remains a question of considerable scientific interest. This zone also corresponds with the former extent of the Centralian Superbasin, which formed following rifting during the Neoproterozoic breakup of the supercontinent Rodinia at around 850 Ma (see below).

Australia is underlain by mostly thick lithosphere, which, at around 200 km, is about double the global continental average. This is important because the thickness controls a continent's tectonics and seismicity, basin evolution, resource accumulation, topography, and landscape. The lithosphere appears to be thick beneath the Yilgarn Craton (Figure 2.18). The thickest lithosphere, however, occurs beneath the central Australian Proterozoic mobile belts, which were tectonically active in the Paleozoic. The observation of thick lithosphere and relatively young orogenesis raises interesting questions regarding the age of the lithosphere (Paleozoic or younger) and the possible process or processes of its formation.

Thick subcontinental lithosphere is strong and difficult to melt, although it can be mechanically weakened by stretching, leading to basin formation. The way in which sedimentary basins evolve through time, from rifting to thermal subsidence, is strongly influenced by the thickness and structure of the lithosphere. Thick lithosphere conducts heat more slowly than thin lithosphere, so that the thermal sag sedimentation phase following rifting of thick lithosphere is delayed and prolonged. Australia's intracontinental basins are broad features that were filled over a prolonged interval from 850 Ma to the Pliocene, with many basins overlying older basins (Figure 2.9).

The overall tectonic stability of much of Australia is also governed by the generally thick lithosphere. A consequence of this tectonic stability has been the preservation of ancient landscapes and the creation of the generally flat relief over most of the continent (Chapter 5). This lack of relief, however, has had a major impact on the development of rivers, which are generally sluggish and have low rates of discharge into the sea (Chapter 6).

Measuring Australia's record of 'deep' time

Since the very beginning of geology as a science, obtaining a fix on geological, or 'deep', time has always been a goal for geoscientists. This knowledge allows for comparison and correlation, as well as estimates of the rates of geological processes. The earliest attempts were by palaeontologists, who recognised that fossil assemblages varied systematically through stratigraphic columns. Moreover, they were able to establish that these patterns varied systematically, in many cases, across the world. These fossil assemblages, through a discipline called 'biostratigraphy', could be used to establish relative ages of rocks (Chapter 3). The next challenge was to establish absolute ages.

The absolute age of Earth, and of rocks, is a quest that has interested humans—notably theologians and scientists—throughout time. One of the first estimates of the age of Earth was by Bishop Ussher, who in 1650 put it at 4004 BC (about 6000 years BP) based upon biblical records. Early naturalists and scientists also made estimates. In the latter part of the 18th century, Georges Buffon made one of the earliest estimates, at *ca* 75 ka, based on the inferred cooling rates of Earth. In the middle of the 19th century, the age of Earth was thought to be of the order of a few hundred million years; Charles Darwin suggested 300 Myr based on denudation rates, and William Thomson (later Lord Kelvin) gave a range of 10–400 Myr based on thermodynamic considerations. It was not until the discovery of radioactivity in 1896 by Henri Becquerel, and the subsequent recognition that radioactive elements decayed at a constant, measurable rate, that geochronology as we understand it today began.

The discovery in 1913 of isotopes, followed by the persistence of Arthur Holmes, established radiometric dating, particularly using the U–Pb system, as the best tool for determining the age of rocks and of Earth. Since then, a large number of radiogenic isotope systems—U–Pb, C, Rb–Sr, K–Ar (including Ar–Ar), Sm–Nd, Hf–Lu, Re–Os—have been used as geological (and archaeological) chronometers. More importantly, the limitations of these systems have been established, and correlations of ages determined by independent systems have been established. Absolute ages determined radiometrically have also been used to calibrate relative age scales. For instance, radiometric 'gold spikes' have been used to anchor relative biostratigraphic ages and, occasionally, biostratigraphic assemblages can give more accurate and precise dates than radiogenic ones. Moreover, the radiometric ages have also been used to calibrate other geochronological techniques. Temporal variations in palaeomagnetic data, anchored by radiometric ages, have been used to provide accurate and precise ages of mineralising events, and variations in the stable isotope ratios of oxygen, hydrogen, sulfur and carbon can be used in certain cases to estimate the age of rocks and minerals that contain them.

Radiogenic (and stable) isotope ratios can also be used as tracers of geological processes. An example is variations in neodymium (Nd) and lead (Pb) isotope ratios that can indicate the juvenile or evolved character of crust and identify major crustal boundaries. Such boundaries are

Image by Geoff Fraser

The Cooyerdoo Granite, dated at 3157 ± 2 Ma, is South Australia's oldest known rock.

important controls on the distribution of ore deposits in Australia and elsewhere (Chapter 8). The Ar–Ar technique is used to date metamorphic and mineralisation events (Box 8.2). The Re–Os isotopic system has also been used to date mineral deposits. Other techniques are deployed for younger rocks or investigating younger geological events, such as landscape evolution (Chapter 5) or coral growth (Chapter 6). They include fission-track, multiple isotopic systems, electron-spin resonance and various luminescence dating techniques.

Earth is now thought to be around 4540 Ma old (Figure 2.19). Starting as a molten body that began to differentiate into layers according to elemental density and chemistry, the crust only began to develop around 150 Myr later. Evidence for the world's oldest continental crust is preserved in the Jack Hills region (WA), which has metasedimentary rocks containing *ca* 4404 Ma detrital zircons. The oldest known rocks on the Australian continent are the 3731 ± 4 Ma Meeribie Gneiss from the Murchison district (WA). The oldest rock within Australia's jurisdiction, however, is the Mt Sones Orthogneiss from the Napier Complex in Antarctica, which is dated at 3927 ± 10 Ma. These old rocks were all dated using the U–Pb isotopic system by the Sensitive High-Resolution Ion MicroProbe (SHRIMP), developed at The Australian National University (Box 2.4). The applications of SHRIMP have revolutionised the study of the old rocks of Australia and elsewhere, and SHRIMP has been used to date the oldest rocks in each state and territory (Table 2.2). One of the amazing features of SHRIMP, however, is its ability to accurately and precisely date the full spread of geological time (Figure 2.20).

Table 2.2: Age of oldest known rocks from each state and territory of Australia

State/territory	Rock unit and location	Rock type	Age (millions of years)	Geological time (era)
Australian Antarctic Territory	Mt Sones Orthogneiss, Enderby Land	Granodioritic gneiss	3927 ± 10	Eoarchean
Western Australia	Meeberrie Gneiss, Murchison District	Gneiss	3731 ± 4	Eoarchean
South Australia	Cooyerdoo Granite, Eyre Peninsula	Granite	3157 ± 2	Mesoarchean
Northern Territory	Woolner Granite, Pine Creek Orogen (under Mesozoic cover near Darwin)	Granite	2674 ± 3	Neoarchean
Queensland	Yaringa Metamorphics, Mt Isa	Orthogneiss	1874 ± 4	Paleoproterzoic
New South Wales	Redan Gneiss, Broken Hill	Gneiss	1710 ± 4	Paleoproterozic
Tasmania	Surprise Bay Formation, King Island	Metasedimentary	*ca* 1300	Mesoproterozoic
Victoria	Glenelg River Metamorphic Complex Ceres Gabbro	Ultramafics-Serpentinite Gabbro	> *ca* 520 *ca* 600?	Early Cambrian Neoproterozoic?
Australian Capital Territory	Adaminaby Group, Brindabella Ranges	Metasedimentary	*ca* 470	Ordovician
Heard and McDonald Islands	Limestone on Lauren's Peninsula	Limestone Basalt?	*ca* 50 *ca* 120??	Eocene Cretaceous?
Norfolk Island	Ball Bay Basalt	Basalt	*ca* 3.2	Pliocene
Ashmore-Cartier Islands	Cartier Formation	Limestone	*ca* 30	Oligocene
Cocos (Keeling) Islands	Undated limestone	Limestone	*ca* 50?	Eocene by correlation with Christmas Island?
Christmas Island	Basalt, age greater than Eocene sediments	Basalt	*ca* 60	Paleogene?

Eon	Era	Period	Epoch	Stage	Age (Ma)
Phanerozoic	Cenozoic	Quaternary	Holocene		0.0117
			Pleistocene	Tarrantian	0.126
				Ionian	0.781
				Calabrian	1.806
				Gelasian	2.588
		Neogene	Pliocene	Piacenzian	36.00
				Zanclean	5.332
			Miocene	Messinian	7.246
				Tortonian	11.608
				Serravallian	13.82
				Langhian	15.97
				Burdigalian	20.43
				Aquitanian	23.03
		Paleogene	Oligocene	Chattian	28.4 ± 0.1
				Rupelian	33.9 ± 0.1
			Eocene	Priabonian	37.2 ± 0.1
				Bartonian	40.4 ± 0.2
				Lutetian	48.6 ± 0.2
				Ypresian	55.8 ± 0.2
			Paleocene	Thanetian	58.7 ± 0.2
				Selandian	*ca* 61.1
				Danian	65.5 ± 0.3
	Mesozoic	Cretaceous	Upper	Maastrichtian	70.6 ± 0.6
				Campanian	83.5 ± 0.7
				Santonian	85.8 ± 0.7
				Coniacian	*ca* 88.6
				Turonian	93.6 ± 0.8
				Cenomanian	99.6 ± 0.9
			Lower	Albian	112.0 ± 1.0
				Aptian	125.0 ± 1.0
				Barremian	130.0 ± 1.5
				Hauterivian	*ca* 133.9
				Valanginian	140.2 ± 3.0
				Berriasian	145.5 ± 4.0

Eon	Era	Period	Epoch	Stage	Age (Ma)
					145.5 ± 4.0
Phanerozoic	Mesozoic	Jurassic	Late	Tithonian	150.8 ± 4.0
				Kimmeridgian	*ca* 155.6
				Oxfordian	161.2 ± 4.0
			Middle	Callovian	164.7 ± 4.0
				Bathonian	167.7 ± 3.5
				Bajocian	171.6 ± 3.0
				Aalenian	175.6 ± 2.0
			Early	Toarcian	183.0 ± 1.5
				Pliensbachian	189.6 ± 1.5
				Sinemurian	196.5 ± 1.0
				Hettangian	199.6 ± 0.6
		Triassic	Late	Rhaetian	203.6 ± 1.5
				Norian	216.5 ± 2.0
				Carnian	*ca* 228.7
			Middle	Ladinian	237.0 ± 2.0
				Anisian	*ca* 245.9
			Early	Olenekian	*ca* 249.5
				Induan	251.0 ± 0.4
	Paleozoic	Permian	Lopingian	Changhsingian	253.8 ± 0.7
				Wuchiapingian	260.4 ± 0.7
			Guadalupian	Capitanian	265.8 ± 0.7
				Wordian	268.0 ± 0.7
				Roadian	270.6 ± 0.7
			Cisuralian	Kungurian	275.6 ± 0.7
				Artinskian	284.4 ± 0.7
				Sakmarian	294.6 ± 0.8
				Asselian	299.0 ± 0.8
		Carboniferous	Pennsylvanian	Gzhelian	303.4 ± 0.9
				Kasimovian	307.2 ± 1.0
				Moscovian	311.7 ± 1.1
				Bashkirian	318.1 ± 1.3
			Mississippian	Serpukhovian	328.3 ± 1.6
				Visean	345.3 ± 2.1
				Tournaisian	359.2 ± 2.5

Eon	Era	Period	Epoch	Stage	Age (Ma)
					359.2 ± 2.5
Phanerozoic	Paleozoic	Devonian	Late	Famennian	374.5 ± 2.6
				Frasnian	385.3 ± 2.6
			Middle	Givetian	391.8 ± 2.7
				Eifelian	397.5 ± 2.7
			Early	Emsian	407.0 ± 2.8
				Pragian	411.2 ± 2.8
				Lochkovian	416.0 ± 2.8
		Silurian	Pridoli		418.7 ± 2.7
			Ludlow	Ludfordian	421.3 ± 2.6
				Gorstian	422.9 ± 2.5
			Wenlock	Homerian	426.2 ± 2.4
				Sheinwoodian	428.2 ± 2.3
			Llandovery	Telychian	436.0 ± 1.9
				Aeronian	439.0 ± 1.8
				Rhuddanian	443.7 ± 1.5
		Ordovician	Late	Hirnantian	445.6 ± 1.5
				Katian	455.8 ± 1.6
				Sandbian	460.9 ± 1.6
			Middle	Darriwilian	468.1 ± 1.6
				Dapingian	471.8 ± 1.6
			Early	Floian	478.6 ± 1.7
				Tremadocian	488.3 ± 1.7
		Cambrian	Furongian	Stage 10	*ca* 492
				Stage 9	*ca* 496
				Paibian	*ca* 499
			Series 3	Guzhangian	*ca* 503
				Drumian	*ca* 506.5
				Stage 5	*ca* 510
			Series 2	Stage 4	*ca* 515
				Stage 3	*ca* 521
			Terreneuvian	Stage 2	*ca* 528
				Fortunian	542.0 ± 1.0

	Eon	Era	Period	Age (Ma)
				542.0 ± 1.0
Precambrian	Proterozoic	Neo-proterozoic	Ediacaran	*ca* 635
			Cryogenian	850
			Tonian	1000
		Meso-proterozoic	Stenian	1200
			Ectasian	1400
			Calymmian	1600
		Paleo-proterozoic	Statherian	1800
			Orosirian	2050
			Rhyacian	2300
			Siderian	2500
	Archean	Neoarchean		2800
		Mesoarchean		3200
		Paleoarchean		3600
		Eoarchean		4000
	Hadean (informal)			*ca* 4600

Figure 2.19: Geological time-scale modified from 2009 International Stratigraphic Chart (IUGS International Commission on Stratigraphy).

SHRIMP—AUSSIE INGENUITY (BOX 2.4)

The Sensitive High-Resolution Ion MicroProbe, or SHRIMP, designed and built by The Australian National University, measures the age of rocks by analysing the isotopic ratios of daughter products from the natural decay of radioactive minerals. The most commonly used mineral for ancient rocks is zircon (zirconium silicate, $ZrSiO_4$), which is a relatively stable mineral that is common in felsic igneous rocks and provides a natural radioactive clock due to its relatively high U content.

Zircons can preserve a long geological history. For example, a zircon extracted from a particular granite may have:

- an old core, which is a relic of the source rock that melted to form the granite, that is
- encased in younger magmatic zircon that crystallised with the granite melt, which is
- rimmed by even younger zircon, a consequence of post-magmatic metamorphism.

Zircon is also resistant to weathering, so it is a common detrital mineral in sediments.

The zircon crystals are extracted from rock samples with careful crushing and mineral separation. Target zircons are then placed on a mount alongside a standard zircon, such as the Temora standard. A 30 µm beam of ionised O_2 is focused onto a spot within a single zircon grain to blast off atoms, which are collected and measured. This process allows individual sites of the crystal to be dated and the full history of the rock to be determined. The U–Pb isotopic system is well understood. Uranium decays to Pb, and it is the ratio of the isotopes of U and Pb that is measured by the mass spectrometer and used to calculate the age.

Other minerals can be dated with SHRIMP using the U–Pb isotopic system. These include monazite, apatite, baddelyite, titanite and allanite, as well as carbonates. SHRIMP can also analyse stable isotope ratios, including those of S, C, O and B, which provide information about scores and processes affecting these elements and can indirectly provide age data.

Today, SHRIMP machines are built in Canberra and operate in nine countries across five continents.

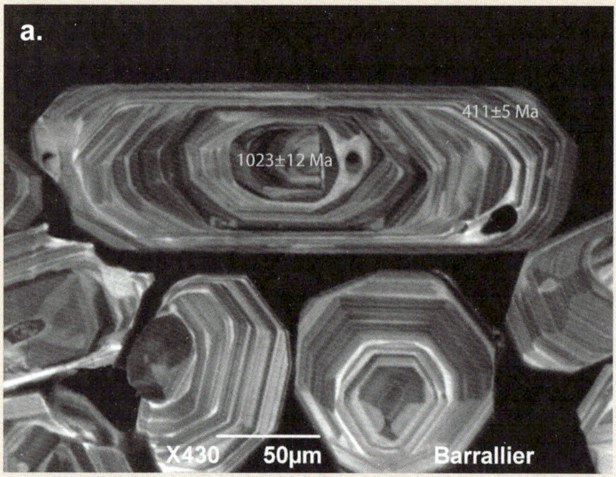

Images by Keith Sircombe

(a) Different regions of a zircon can be dated, telling us about the evolution of a particular rock. (b) Geoscience Australia's SHRIMP II dates around 150 rocks per year from regions of Australia.

Australia through time: a summary of the tectonic evolution

Australia became the island continent *ca* 34 Ma when it finally broke from Antarctica. This was one of the last events in the breakup of Pangaea, the youngest of the world's supercontinents. Supercontinents form when most or all of Earth's continents amalgamate into one large landmass (Box 2.5). The make-up and even the existence of supercontinents through geological time has been the subject of much debate, with only the configuration of Pangaea generally agreed.

Although the existence and configuration of ancient supercontinents might seem to be a topic of debate only among academics, in truth, much of the geology and important mineral wealth, as well as the evolution of life in Australia, are the consequence of the amalgamation and breakup of supercontinents and smaller amalgamations (supercratons) through geological time—Vaalbara, Kenorland, Nuna (Columbia), Rodinia and Pangaea (including Gondwana) (Figure 2.20). Supercontinent history has not only controlled the distribution of many resources, but has surprising influences on geological processes that, in many ways, make Australia remarkable among the continents. With the important exception of Rodinia, peaks in the emplacement ages of igneous rocks closely correspond with periods of supercontinent assembly and breakup, in both Australia (Figure 2.20) and the rest of the world. This suggests that supercontinent processes involve extensive reworking of old continental crust and the formation of new continental crust.

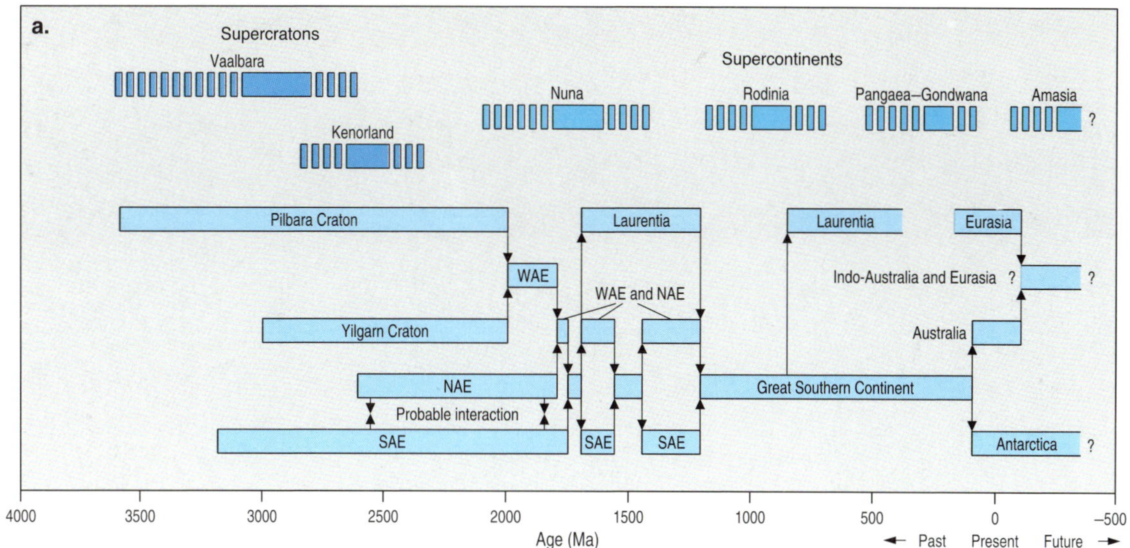

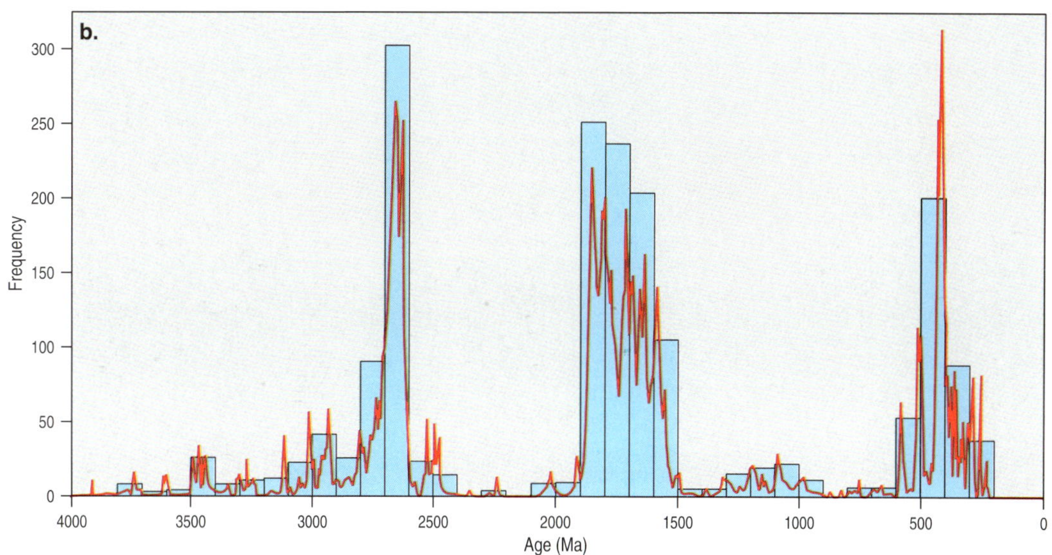

In simple terms, continental Australia grew from west to east. Archean rocks mostly occur in the west, Proterozoic rocks in the centre and Phanerozoic rocks in the east (Figure 2.8). The western two-thirds of the continent consist of three Precambrian elements: the West Australian, North Australian and South Australian elements,

Figure 2.20: (a) Assembly and breakup of the cratonic elements that constitute Australia from 4000 Ma and their relationship to supercontinent cycles. (b) Histogram showing the distribution of the age of igneous rocks from Australia as determined using SHRIMP and other U–Pb analytical techniques. In (b), the blue bars are igneous rock ages binned in 100 Myr intervals, and the red line indicates the combined probability density distribution for the same igneous rocks. (Sources: Dr Keith Sircombe, Geoscience Australia and OZCHRON database)
NAE = North Australian Element; SAE = South Australian Element, WAE = West Australian Element

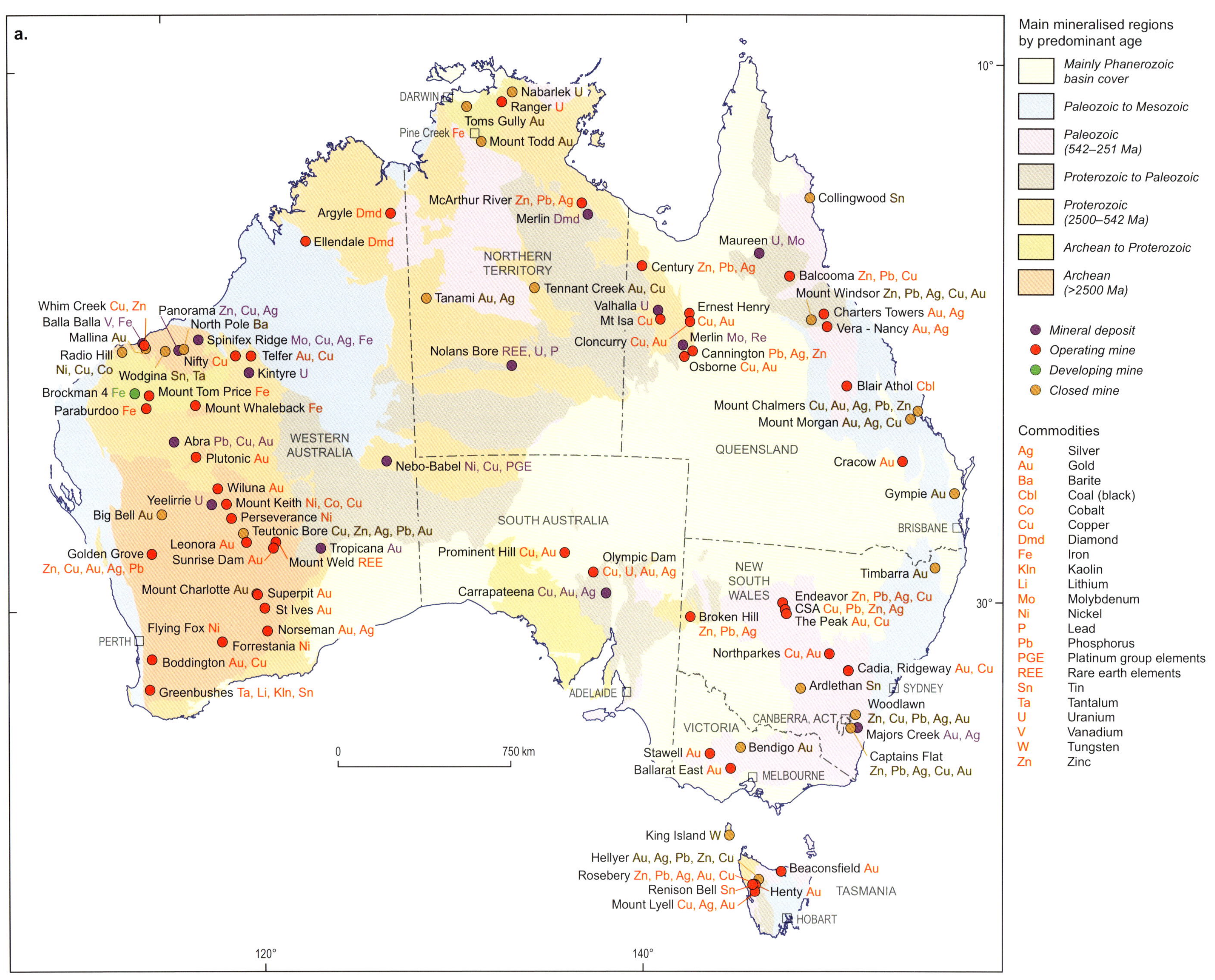

Figure 2.21: (a) Important mineral resources in 'hard rock' Australia. (b, opposite) Important mineral and energy resources in basins of Australia.

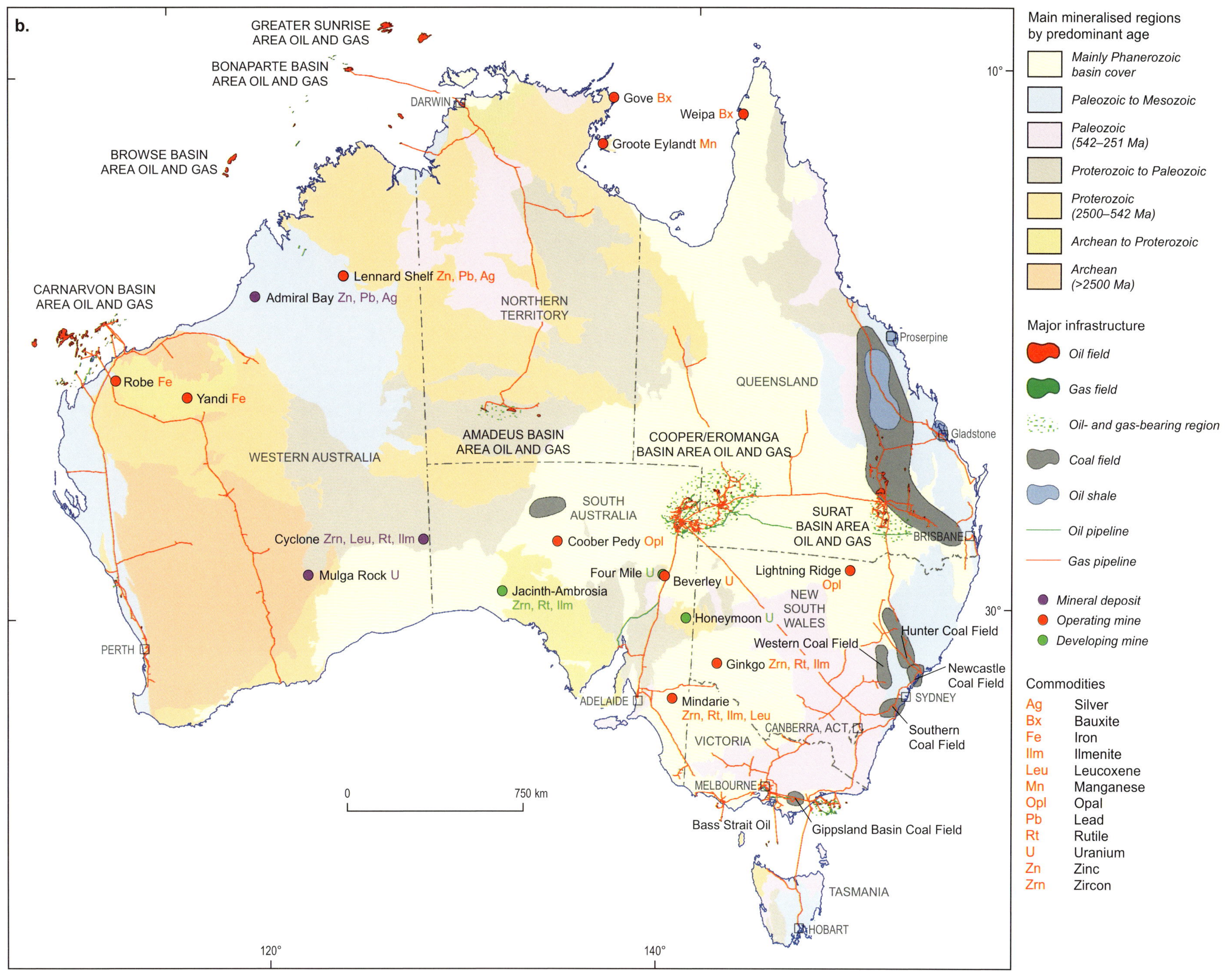

b.
GREATER SUNRISE AREA OIL AND GAS
BONAPARTE BASIN AREA OIL AND GAS
BROWSE BASIN AREA OIL AND GAS
CARNARVON BASIN AREA OIL AND GAS
DARWIN
Gove Bx
Weipa Bx
Groote Eylandt Mn
Lennard Shelf Zn, Pb, Ag
Admiral Bay Zn, Pb, Ag
NORTHERN TERRITORY
Robe Fe
Yandi Fe
WESTERN AUSTRALIA
AMADEUS BASIN AREA OIL AND GAS
COOPER/EROMANGA BASIN AREA OIL AND GAS
QUEENSLAND
Proserpine
Gladstone
SOUTH AUSTRALIA
SURAT BASIN AREA OIL AND GAS
BRISBANE
Cyclone Zrn, Leu, Rt, Ilm
Coober Pedy Opl
Mulga Rock U
Four Mile U
Beverley U
Lightning Ridge Opl
Jacinth-Ambrosia Zrn, Rt, Ilm
NEW SOUTH WALES
Honeymoon U
Hunter Coal Field
Western Coal Field
PERTH
Ginkgo Zrn, Rt, Ilm
Newcastle Coal Field
SYDNEY
ADELAIDE
Mindarie Zrn, Rt, Ilm, Leu
CANBERRA, ACT
Southern Coal Field
VICTORIA
0
750 km
MELBOURNE
Bass Strait Oil
Gippsland Basin Coal Field
TASMANIA
HOBART
10°
30°
120°
140°
Main mineralised regions by predominant age
Mainly Phanerozoic basin cover
Paleozoic to Mesozoic
Paleozoic (542–251 Ma)
Proterozoic to Paleozoic
Proterozoic (2500–542 Ma)
Archean to Proterozoic
Archean (>2500 Ma)
Major infrastructure
Oil field
Gas field
Oil- and gas-bearing region
Coal field
Oil shale
Oil pipeline
Gas pipeline
Mineral deposit
Operating mine
Developing mine
Commodities
Ag Silver
Bx Bauxite
Fe Iron
Ilm Ilmenite
Leu Leucoxene
Mn Manganese
Opl Opal
Pb Lead
Rt Rutile
U Uranium
Zn Zinc
Zrn Zircon

WHAT ARE SUPERCONTINENTS? (BOX 2.5)

Supercontinents are amalgamations containing nearly all Earth's continental blocks. The concept for a geological entity originated with the coining of the term Gondwana by geologists in India in the late 1800s. First recognised from the late Paleozoic floral *Glossopteris* assemblage, found only in the southern landmasses of South America, Africa, Madagascar, India and Australia (and later Antarctica), this widespread distribution could only be explained at first if the countries were linked by now-absent 'land bridges' (Chapter 3). Since Alfred Wegener's 1912 continental drift hypothesis and work in the Southern Hemisphere, it has become apparent that continental blocks have amassed together and subsequently dispersed at several times in the Proterozoic and Paleozoic since their earliest formation. Supercontinent amalgamation commonly overlaps with dispersal, and thus the period of maximum packing of continental blocks defines the age span.

The interplay of continental blocks and mantle dynamics is fundamental to supercontinent formation and breakup, since it is likely that continents move away from locations of mantle upwelling and move towards downwelling mantle cells. Dispersion can occur through a number of mechanisms. There can be uplift and lateral stresses due to warming of the mantle beneath the supercontinent, since subduction is not able to cool the upper mantle within the interior of the large landmass. Alternatively, dispersion can be driven by thermal plumes linked to reorganisation of mantle flow. These two mechanisms may be linked.

Supercontinent formation and breakup have a significant influence on global climate because the grouping and location of continents affects wind patterns and ocean circulation. Eustatic sea-level is also affected; rapid seafloor spreading can decrease the volume of the ocean basins, as in the Eocene. This is because young oceanic crust is relatively buoyant and elevated compared with old oceanic crust (think of the mid-ocean ridges). There are also feedbacks between global tectonics, climate and sea-level. Times of continental amalgamation have been linked to major glacial accumulations ('icehouse' episodes, as in the end-Precambrian, end-Ordovician, end-Carboniferous) and hence lowered global sea-level. Dispersal of a supercontinent correlates with greenhouse regimes and flooded continents, as in the Silurian and mid-Cretaceous (Box 4.3).

The meridional distribution of continents in Pangaea in the Triassic induced an intense monsoonal system, and the seasonal aridity is recorded as red beds in many continents. The opening of circumpolar circulation can drive polar glaciation, such as in the Cenozoic rifting of Australia from Antarctica, and the opening of the Drake Passage, leading to the Antarctic ice sheets. Glaciation in the last 1–2 Myr in the Northern Hemisphere is also probably linked to the closure of equatorial circulation between the Pacific Ocean and the Caribbean Sea.

Image by Adrian Yee

The tongue-shaped leaves of *Glossopteris* are found across Gondwana.

whereas the eastern third is made up mostly by the Paleozoic Tasman Element (Figure 2.10). The link between the spatial and temporal patterns is consistent with supercraton and supercontinent growth, particularly Kenorland, Nuna and Pangaea. The distribution of Australia's energy and mineral resources (Figure 2.21) is also governed by this broad pattern, with each of the four major cratonic elements characterised by distinctive deposit assemblages, both in time and in composition.

The Australian continent evolved in five broad but distinct time periods: 3800–2100 Ma, 2100–1300 Ma, 1300–700 Ma, 700–160 Ma and 160 Ma to the present.

The first period saw the growth of nuclei about which cratonic elements grew, whereas the latter four periods involved the amalgamation and dispersal of Nuna, Rodinia and Pangaea–Gondwana, respectively. Below we present a history of the growth of the present-day Australian continent using this framework, noting that there is often significant uncertainty and disagreement about specific details. This 'geohistory' provides context for events that have shaped and changed Australia and Earth, including the evolution of life and changes in the composition of the atmosphere and hydrosphere.

As mineral deposits have been increasingly linked to geodynamic processes, the evolution of Australia's mineral and petroleum systems provides a parallel framework for the evolution of the continent as a whole. Appendix 8.1.1 provides a summary of the major mineral provinces through time.

Image courtesy of Anthony Harris, University of Tasmania

Figure 2.22: Photographs from the Pilbara Craton, Western Australia, showing (a) association between stromatolites and hydrothermal barite in the Dresser Formation, and (b) the Strelley Pool Formation angular unconformity. In (b), the unconformity is located at the base of the prominent outcrop along the ridgetop. The trend of the dark outcrop in the centre–right indicates the trend of bedding in the units below the unconformity.

3800–2200 Ma—growth of cratonic nuclei

The Narryer Terrane of the Yilgarn Craton contains both the oldest known rock in continental Australia and the oldest mineral known on Earth—3731 ± 4 Ma and *ca* 4404 Ma, respectively. The Jack Hills zircons are only about 150 Myr younger than the age of Earth, and their study has important implications for the characteristics of the earliest period of Earth's history. The existence of zircons indicates that continental crust was formed very early. Oxygen isotope characteristics of these old zircons indicate that they crystallised from a magma that melted from a source that had previously interacted with seawater, telling us that both continents and oceans were present on Earth during the Hadean.

The Yilgarn and Pilbara cratons form the nuclei of the West Australian Element and are the most extensive exposures of old rocks in Australia. Over the past five years, old rocks have been increasingly

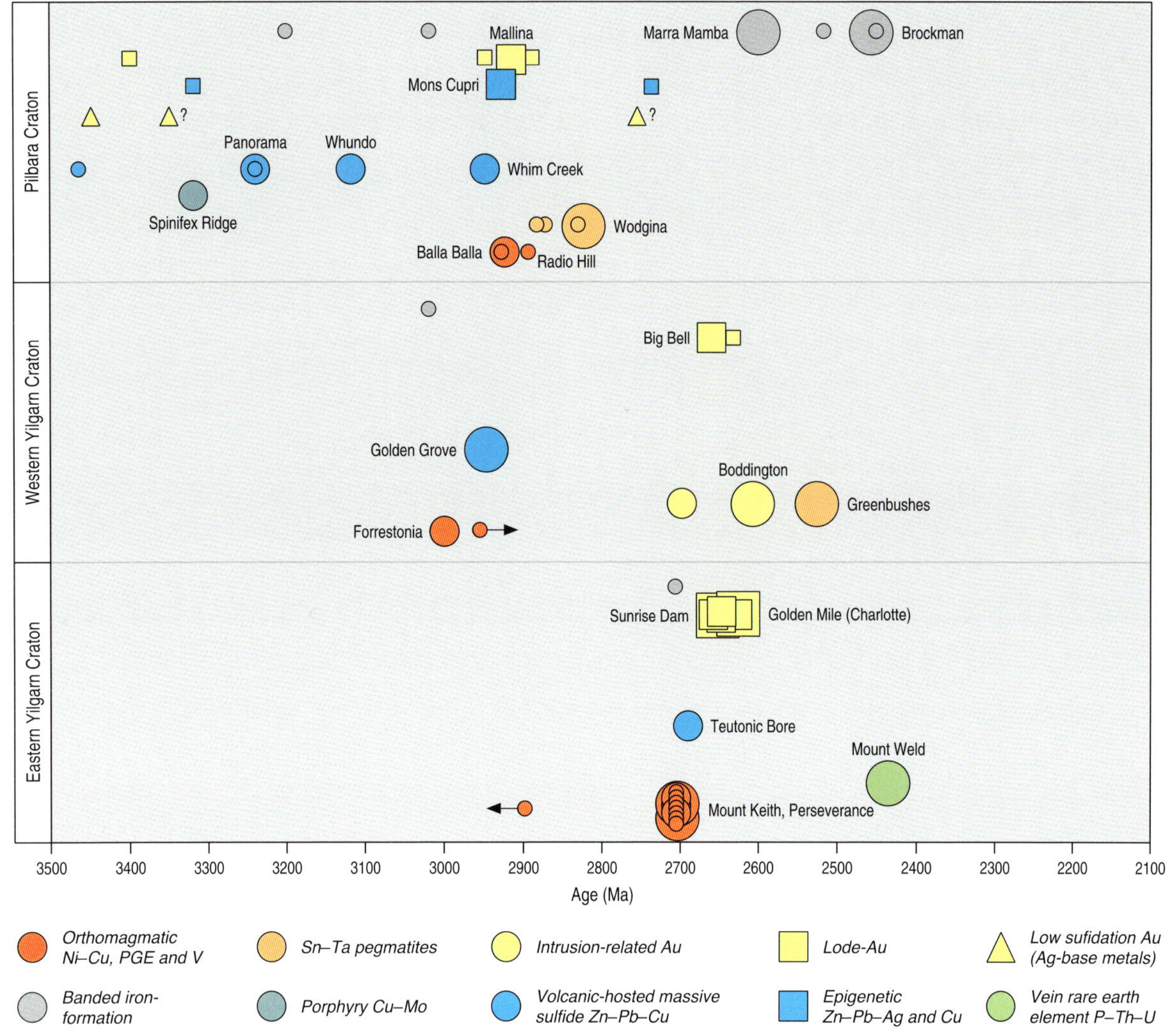

Figure 2.23: Major mineral deposits of Australia that formed between 3500 Ma and 2100 Ma; see Figure 2.21a for localities. Note that all deposits of this age occur in the West Australian Element. Symbol size broadly indicates deposit size relative to deposits with similar commodities.
PGE = platinum group elements

identified within the North and South Australian elements, forming the nuclei upon which these elements were built.

The 3655–2840 Ma Pilbara Craton is a classic Archean granite–greenstone terrane. Due to its remarkable preservation and exposure, it contains the earliest examples of many geological processes and so has been the focus of numerous studies by Australian and international researchers.

The tectonic regime that formed the older units of the stratigraphy (>3200 Ma) has been a particular focus of scientific debate. Some people suggest that these old rocks developed within a thick oceanic plateau as the consequence of mantle plumes in a hotter early Earth, and that vertical tectonic processes were dominant (as opposed to plate tectonic processes). The rock types, their associations and the map patterns are certainly unlike anything we observe today. In younger units of the stratigraphy, such as the *ca* 3120 Ma Whundo greenstone belt in the western part of the Pilbara Craton, some of the oldest examples of modern plate tectonics are preserved. The volcanic and intrusive rocks of this region have distinctive rock associations and geochemical characteristics that are very similar to those found in modern arc environments. The Pilbara Craton also preserves the world's oldest unconformity (Figure 2.22b). This is dated at *ca* 3430 Ma and is marked by the Strelley Pool Formation, which is a regional chert overlying steeply dipping Warrawoona Group rocks.

The Pilbara Craton contains the oldest examples of many types of mineral deposits (Figure 2.23), including volcanic-hosted massive sulfide, lode gold, and epithermal and porphyry copper deposits. The Dresser Formation hosts the world's oldest ore deposit, *sensu stricto*, the North Pole barite deposit, which produced 129 000 tonnes for use in drilling mud. These ancient mineral deposits share many features with geologically young and modern examples, indicating that many mineralising processes have persisted throughout most geological time. The oldest known hydrocarbons, within ore-related fluid inclusions, are associated with *ca* 3240 Ma volcanic-hosted massive sulfide deposits in the Panorama district (Figure 2.23).

The Pilbara Craton also contains the oldest unequivocal evidence of life on Earth, preserved as stromatolites in the *ca* 3480 Ma Dresser Formation (Chapter 3). Stromatolites grow today in the shallow, warm waters of Shark Bay, about 700 km to the southwest of their ancient 'ancestors' (Figure 3.2). Stromatolites are commonly closely associated with hydrothermal vents, as seen by their association with barite in the North Pole area (Figure 2.22a). These associations support the idea that life on Earth might initially have evolved in a hydrothermal environment.

The Pilbara Craton was probably a constituent of the oldest supercraton, Vaalbara (Figure 2.20a). This supercraton, which was made up of the Kaapvaal (southern Africa) and Pilbara cratons, began to form by 3600 Ma, and broke up at about 2800 Ma. Supercratons, which probably were smaller than most modern continents, were more long lived (Table 2.1, Box 2.5), possibly reflecting the stability of smaller cratonic blocks in an evolving Earth.

With the *ca* 2800 Ma breakup of Vaalbara, the Fortescue and Hamersley basins developed on the old crust of the Pilbara Craton, as some of Earth's earliest preserved examples of a passive-margin sequence. Thick mafic and felsic volcanic rocks, and sedimentary rocks dominate the basin fill, the most important being the banded iron-formations. These 2590–2450 Ma deposits (Figure 2.23) formed when reduced Fe^{2+}-rich bottom waters were oxidised as they welled up onto the wide passive margin, depositing the iron. The vast majority of Australia's and the world's banded iron-formations were deposited between *ca* 2600 Ma and *ca* 1800 Ma, during a period when Earth's hydrosphere and atmosphere were oxygen poor (Chapter 3), an environment very unlike today.

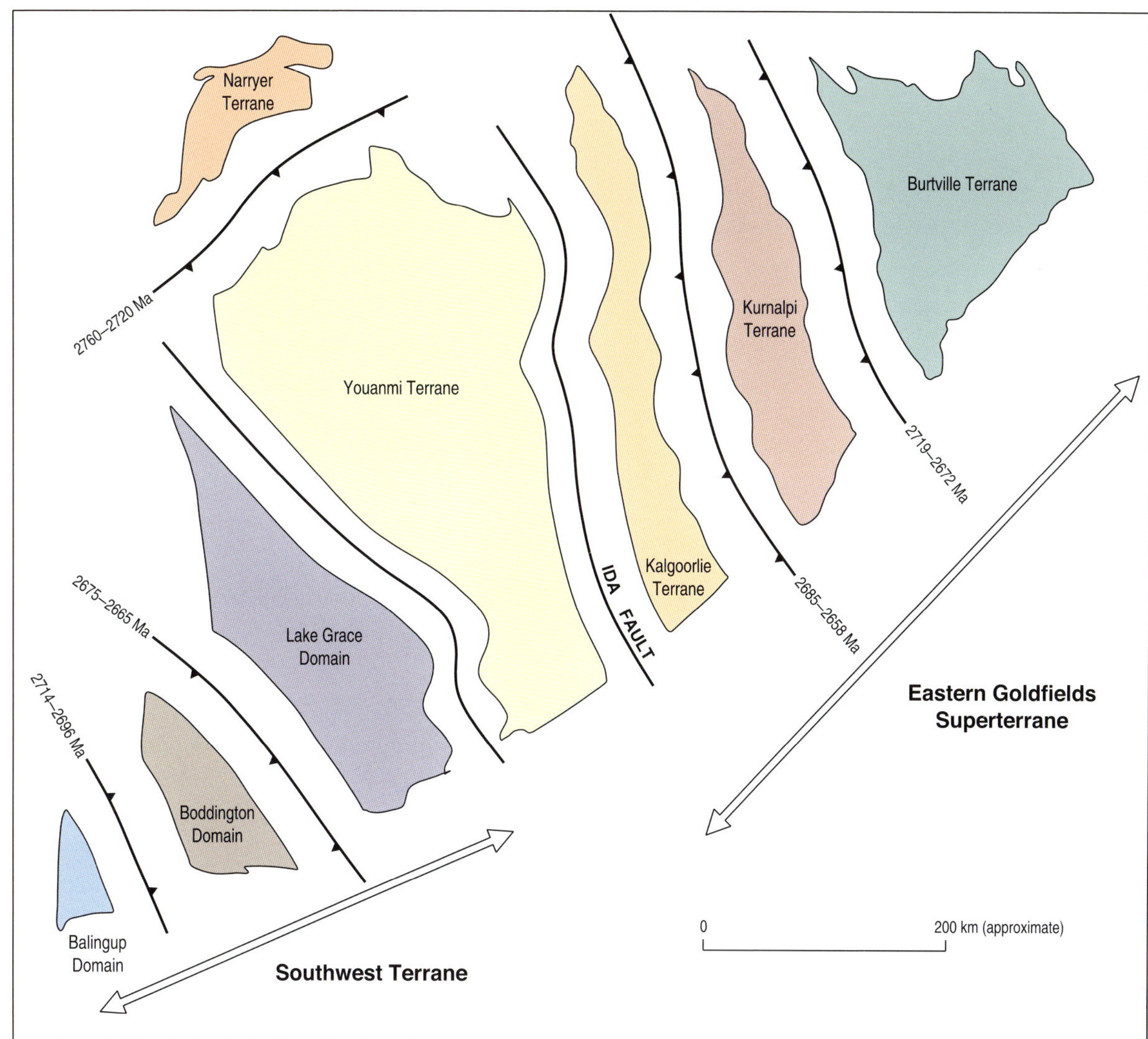

Figure 2.24: Neoarchean assembly of the terranes comprising the Yilgarn Craton of Western Australia.

The Yilgarn Craton, the larger of the two West Australian Element nuclei, differs in a number of important ways from the Pilbara Craton. Although both cratons have extended geological

Figure 2.25: Photographs comparing (a) spinifex grass (*Triodia*) growing near Laverton with (b) spinifex-textured komatiite from a *ca* 2700 Ma ultramafic lava near Leonora, both in the Eastern Goldfields of Western Australia. The magnesium-rich lavas were erupted at high temperature, and the spinifex textures result from the cooling of olivine and pyroxene (crystals to 30 cm in this example).

histories, the Yilgarn is characterised by the presence of short-duration, even catastrophic, crust-forming events. This is shown particularly well in Figure 2.20b, in which a short, sharp peak in the ages of igneous rocks corresponds to the final assembly of the Yilgarn Craton (and Kenorland) between 2720 Ma and 2655 Ma. In contrast, the smaller Pilbara Craton is marked by several indistinct peaks over a 750 Myr period (3500 Ma to 2850 Ma), which likely reflects a slower rate of overall crustal growth.

The Yilgarn Craton is older in the west than the east. The Narryer, Youanmi and Southwest terranes formed between 3730 Ma and 2900 Ma in the west; and the Kalgoorlie, Kurnalpi, Burtville and Yamarna terranes, which constitute the Eastern Goldfields Superterrane, formed between 2940 Ma and 2660 Ma in the east. Assembly of these terranes probably occurred through a series of east-dipping subduction zones that closed intervening basins between 2780 Ma and 2655 Ma. The processes were probably broadly similar to modern subduction, with the formation of backarc basins and major orogenic events as fragments collided (Figure 2.24). These processes produced laterally continuous crust-penetrating shear zones that accessed the mantle, which was previously fertilised by subduction. These shear zones were important conduits for gold (Au) mineralisation (Chapter 8), and may be one of the keys to the gold riches of the Eastern Goldfields Superterrane.

The Eastern Goldfields Superterrane is one of the two largest global Archean lode-Au provinces, with total resources of more than 8500 tonnes (Chapter 8). The discovery of Au changed the fortunes of Western Australia and encouraged mass migration of people in the 1890s (Chapter 1). The region is also a major nickel (Ni) province. Individual deposits are hosted by komatiites, which are mantle-derived ultramafic volcanic rocks, a product of a hotter Archean Earth. Komatiites are also known for their 'spinifex' texture (Figure 2.25), which is caused by the quenched crystallisation of olivine and pyroxene minerals. The texture resembles needle-like 'spinifex' (*Triodia*), which is a characteristic grass that grows across more than 20% of the Australian continent (Chapter 3).

Like the Pilbara Craton, the Yilgarn Craton appears to have been a constituent of a supercraton—in this case, Kenorland—which likely included the Abitibi Subprovince in Canada (Figure 2.20a). These provinces formed over the same time period, share some isotopic similarities and are the two most richly mineralised Archean provinces known. Kenorland, which had amalgamated by *ca* 2660 Ma and began to break up at 2480 Ma, is associated with different continental fragments from Vaalbara, suggesting the presence of two distinct, relatively long-lived supercratons in the Archean.

The recent identification of *ca* 3150 Ma granites in the South Australian Element extends its geological history by about 600 Myr. Recent dating has greatly increased the extent of known Archean rocks in the North Australian Element (Table 2.2). Taken together, these new data indicate a much more significant and prolonged Archean history in both the North and South Australian elements—a history that will become clearer as more data are collected.

2200–1300 Ma—amalgamation and breakup of Nuna

The Paleo- to Mesoproterozoic evolution of the Australian continent is controversial. There are a number of tectonic models that fall into two broad groups: 'fixist' models that suggest there was relatively little lateral movement between crustal blocks, and 'mobilist' models that suggest relatively large lateral movements between crustal blocks. Below we present a mobilist model that implies that the three major Precambrian elements of Australia were assembled in the Paleo- to Mesoproterozoic as the supercontinent Nuna was constructed. There is, however, still significant difference of opinion regarding the details of this assembly; references to alternative ideas are presented in the bibliography.

The Pilbara and Yilgarn cratons were amalgamated by a series of orogenies (Capricorn Orogen) from *ca* 2215 Ma to 1950 Ma to form the West Australian Element, which is one of the earliest building blocks of Nuna (Figures 2.20 and 2.26). Most of the North Australian Element formed before 1840 Ma as a consequence of the amalgamation of the combined Tanami–Tennant–Mt Isa Province with the combined Kimberley–Pine Creek Province. The Aileron Province accreted from the south before 1840 Ma, and the Numil–Kowanyama–Abingdon seismic province accreted from the east before 1850 Ma (Figures 2.20 and 2.26). Proto-Australia was probably linked with Laurentia—basically present-day North America—to the east.

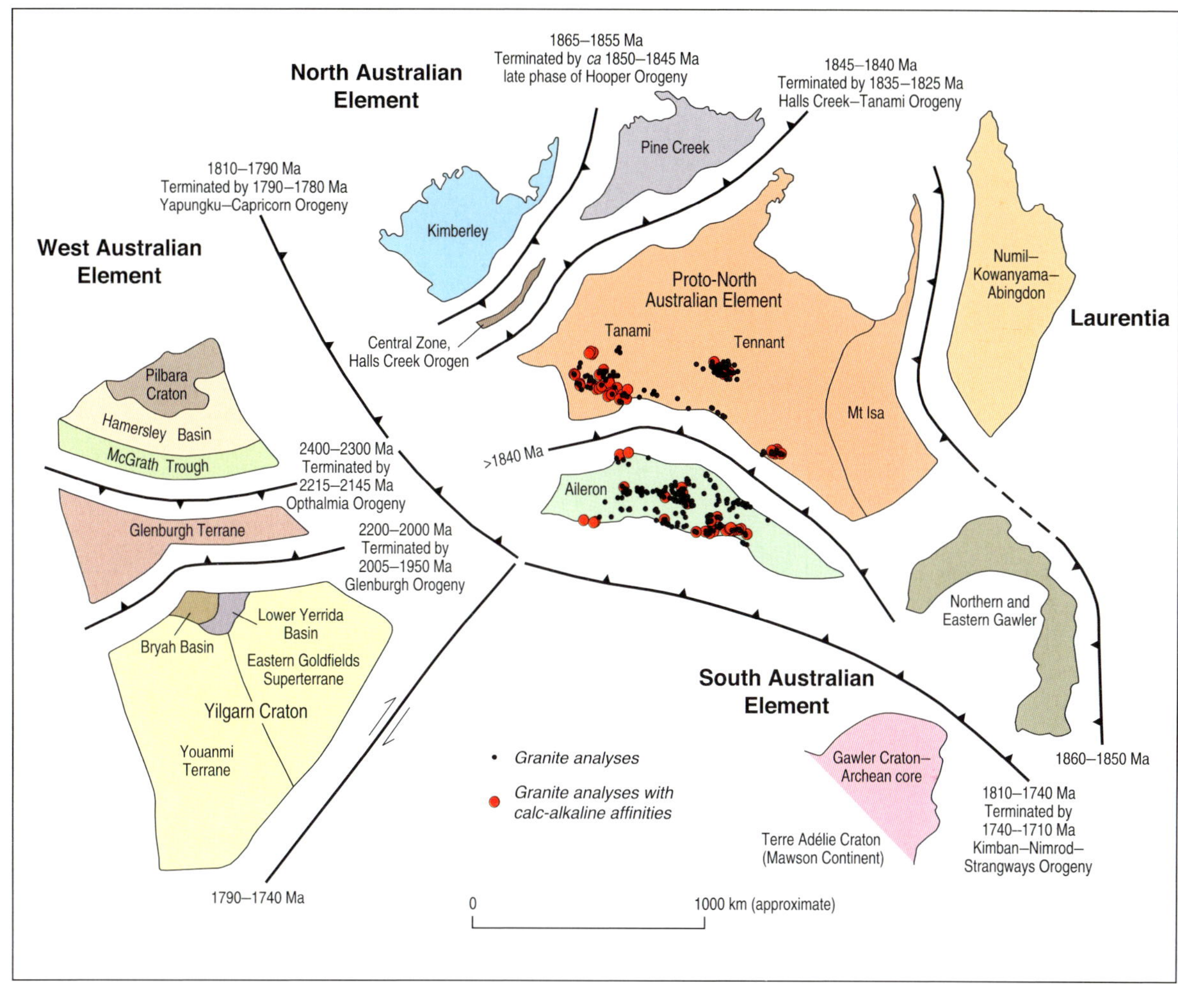

Figure 2.26: Assembly of proto-Australia. Proto-Australia consists of the three elements—the West, North and South Australian elements—which were assembled from smaller provinces in the mid-Paleoproterozoic. During the late Paleoproterozoic (*ca* 1810 Ma), these elements began to amalgamate, with final closure by *ca* 1740 Ma during the Kimban–Nimrod–Strangways Orogeny. It is likely that Laurentia was joined to proto-Australia on the east, and that the Terre Adélie Craton in Antarctica joined the Gawler Craton to the south. Parts of the North and South Australian elements may have been joined earlier (*ca* 2500 Ma and *ca* 2020 Ma; Figure 2.20) as they show common histories at this time. However, the presence of abundant 1810–1760 Ma granites, many with calc-alkaline affinities (large red circles), along the southern margin of the North Australian Element, combined with the virtual lack of similar-aged magmatism in the South Australian Element, suggests that these elements were separated but likely converging at this time. Alternative interpretations can be found in the bibliography.

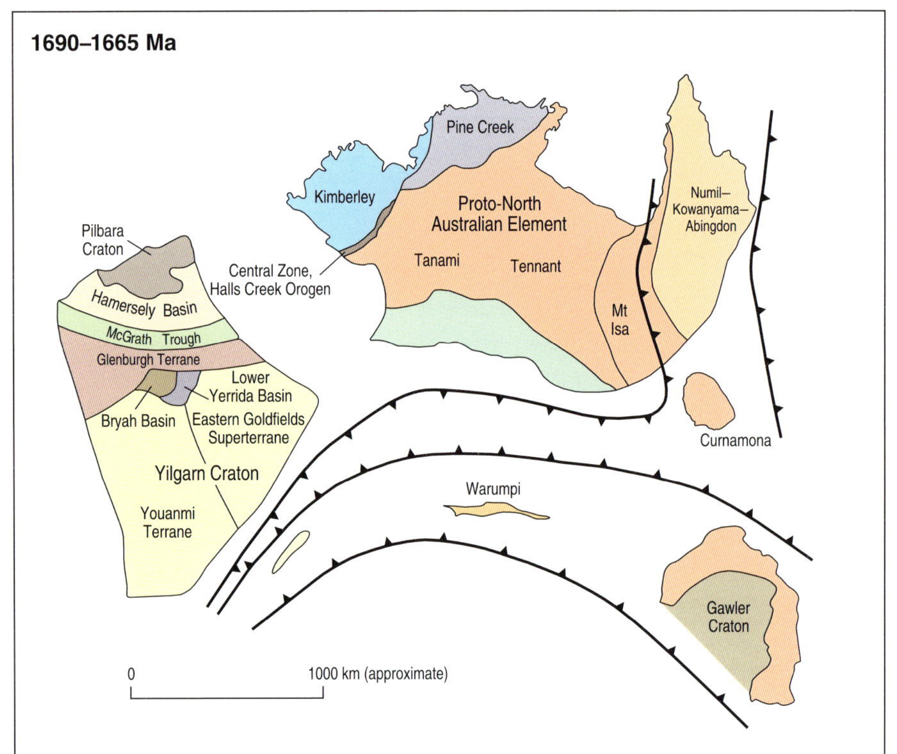

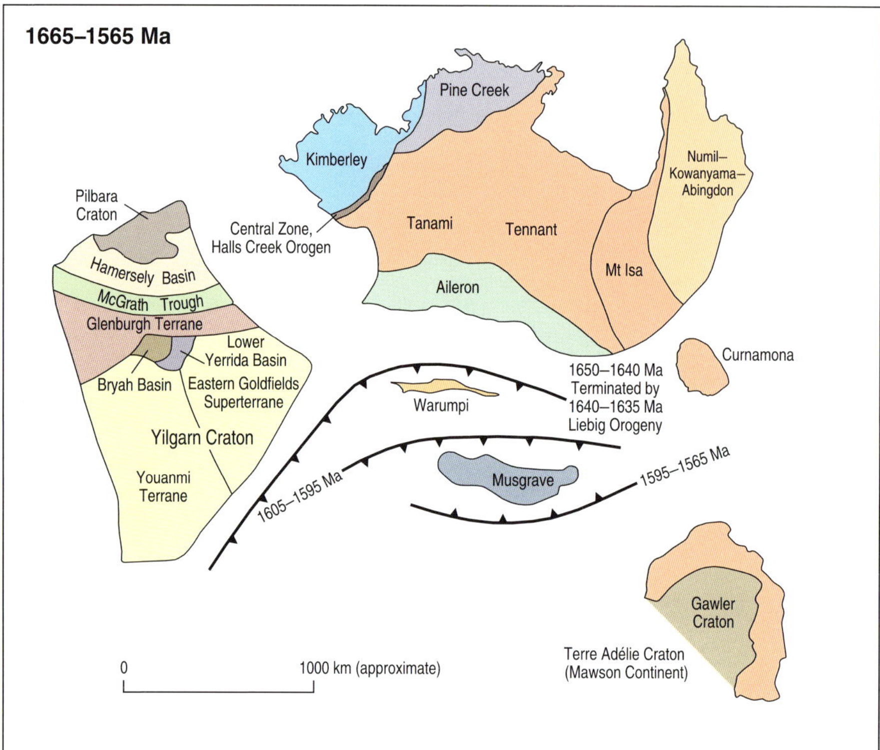

Figure 2.27: Diagram illustrating an interpretation of the evolution of proto-Australia from 1690 Ma to 1565 Ma. This involved extension along the eastern margin of proto-Australia and separation from Laurentia (1690–1665 Ma) and convergence along the southern margin of the combined North and West Australian elements (1665–1565 Ma).

The northern and eastern parts of Gawler Craton, which forms the core of the South Australian Element, grew, at least in part, as a consequence of subduction from the east, and may have originally been part of the North Australian Element (Figures 2.20 and 2.26). This interpretation is based on the provenance of sediments from the northern and eastern Gawler Craton, which matches the Mt Isa Province but not the older Archean core of the Gawler.

By about 1840 Ma, most of the West and North Australian elements, and large parts of the South Australian Element, had been assembled. Although the North and South Australian elements may have been together intermittently since *ca* 2500 Ma (Figure 2.20), the period between 1810 Ma and 1710 Ma saw the assembly of the three Paleoproterozoic to Archean elements into the proto-Australian continent (Figure 2.26), which was one of the building blocks of Nuna. Initially, the North and West Australian elements amalgamated at 1790–1780 Ma during the Yapungku–Capricorn Orogeny, a union that has remained largely intact until today.

Beginning at about 1810 Ma, north- to northeast-directed subduction along the southern margin of the North Australian Element resulted in convergence between the North Australian and West Australian elements and the Archean core of the Gawler Craton. The West Australian Element was first to dock with the North Australian Element at *ca* 1790 Ma during the Yapungku–

Capricorn Orogeny. After this collision, north-directed subduction continued underneath the eastern part of the North Australian Element, with a convergent margin along the southeast edge of the West Australian Element (Figure 2.26). This period of convergence concluded when the Archean core of the Gawler Craton (Figure 2.26) was accreted onto the combined North and West Australian elements during the Kimban–Nimrod–Strangways Orogeny at 1740–1710 Ma. During this time, the Gawler Craton probably joined to the Terre Adélie Craton to the south (now located in Antarctica).

Shortly thereafter, Nuna began to break up. This process, however, was complicated and involved not only divergence of Laurentia to the east, but development of a backarc basin system along the southern margin of Proterozoic Australia as subduction stepped to the south (Figure 2.27). The development of basins filled with turbidites and emplacement of tholeiitic mafic rocks along the eastern margins of the South Australian Element (forming the Curnamona Province) and the North Australian Element suggests divergence along the eastern margin of Proterozoic Australia beginning at about 1690 Ma (Figure 2.27). During this period, or shortly thereafter, felsic magmatic rocks and minor sedimentary rocks (now ortho-and paragneisses) were emplaced and deposited to form the oldest known rocks in the Warumpi Province. In addition, convergence along the southeast margin of the West Australian Element resulted in the development of backarc basins between 1710 Ma and 1665 Ma.

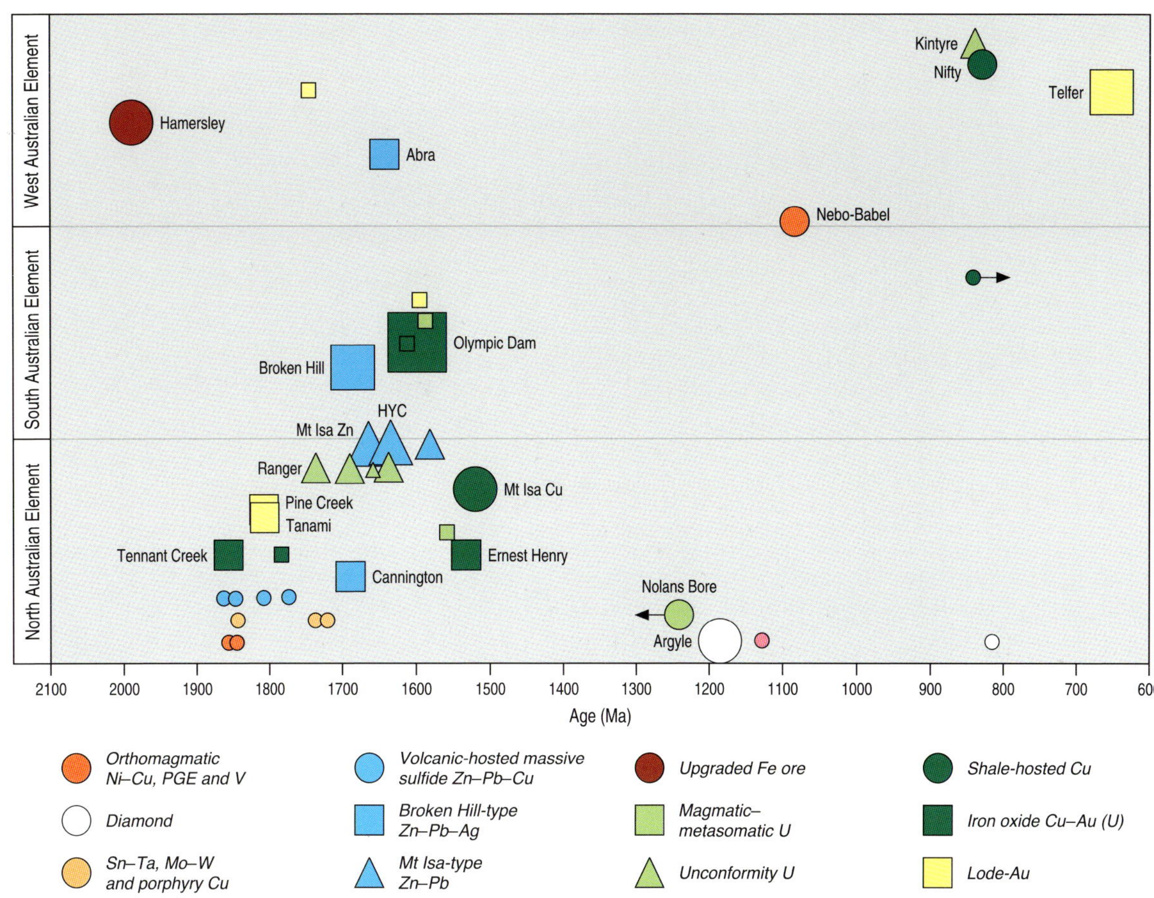

Figure 2.28: Major mineral deposits of Australia that formed between 2200 and 700 Ma. Symbol size broadly indicates deposit size relative to deposits with similar commodities. See Figure 2.21 for localities.
PGE = platinum group elements

At about 1660 Ma, this backarc basin system began to close through a series of both south-and north-directed subduction systems. These initially accreted the Warumpi Province at *ca* 1640 Ma during the Liebig Orogeny, followed by the Musgrave Province at *ca* 1590 Ma, and the Gawler Craton at *ca* 1560 Ma, to the North Australian Element. The Riversleigh inversion and the early phase of the Isan Orogeny in the Mt Isa Province may relate to accretion of the Warumpi and Musgrave provinces, respectively. At some time after the *ca* 1560 Ma event, the Mawson Continent (now in Antarctica), together with

Figure 2.29: Cauliflower chert from *ca* 1650 Ma platform carbonates in the Mt Isa Province, Queensland. Cauliflower chert forms by the replacement of sulfate minerals during evaporation of seawater. These brines are important as they form mineralising fluids with a high capacity to carry metals (Chapter 8).

the Gawler Craton, rifted from the combined West–North Australian Element, as these elements appear to have converged later during the assembly of Rodinia (see below). The extensive magmatic provinces comprising the 1595–1565 Ma Hiltaba Suite in the Gawler Craton and the 1545–1500 Ma Williams Suite in the Mt Isa Province may be related to hotspot activity.

Much of Australia's mineral wealth, particularly iron ore and base metals, formed during the amalgamation and breakup of the Nuna Supercontinent (Figure 2.28). Iron ore deposits in the Hamersley Basin were upgraded (Chapter 9), and lode-Au deposits of the Pine Creek–Tanami formed, during amalgamation (Figure 2.26). In contrast, zinc (Zn)–lead (Pb)–silver (Ag) deposits of the Australian Proterozoic Zn belt formed during breakup, and deposits of the Olympic Dam and Cloncurry mineral provinces may be related to hotspot activity (Chapter 8). The Mesoproterozoic was also a time of abundant high heat-producing granite and volcanic rock production. These rocks contribute to Australia's crustal heat production today, providing widespread potential for geothermal power or, in a few places, U and possibly Th resources for fission reactors (Chapter 10). Another curiosity of this time is the development of the world's oldest oil play, although in uneconomic quantities, in the Upper Roper Group of the Mesoproterozoic Macarthur Basin, northwest of Mt Isa.

A major chemical change occurred in the oceans during the Proterozoic. The process of oxidation of the atmosphere, which began *ca* 2460 Ma, culminated in the oxidation of the oceans at 1900–1800 Ma. This had a number of important consequences. First, as the oxidised form of iron, Fe^{3+}, is highly insoluble, the ocean waters were scrubbed of their iron, which resulted in a major period of banded iron-formation deposition during this time. Since this time, virtually no more banded iron-formations have been deposited, except during the Neoproterozoic icehouse period. Another consequence was an increase in oceanic sulfate concentrations, particularly in the upper part of the ocean, resulting in an increasing abundance of sulfate minerals, both evaporative (Figure 2.29) and hydrothermal, in the geological record.

The oxidation of the atmosphere had a dramatic effect on ore deposition in Australia and the rest of the world, a prime example being iron ore deposits. In the Hamersley Basin, giant iron ore deposits formed through the superposition of two geological systems (Figure 2.23). A 2590–2450 Ma sedimentary system deposited the original banded iron-formation during global anoxia, and then a *ca* 2000 Ma hydrothermal system, related to convergence between the Pilbara Craton and the Glenberg Terrane (Figure 2.26), upgraded the banded iron-formation into iron ore. This upgrading event was one of the earliest consequences of the amalgamation of Nuna, as described above. Hamersley Basin iron ore is Australia's largest export earner today (Chapter 9).

1300–700 Ma—amalgamation and breakup of Rodinia

Compared with other periods of time, Australia was relatively quiescent between 1300 Ma and 700 Ma, possibly because the landmass was largely in an intraplate setting during this time. Recent

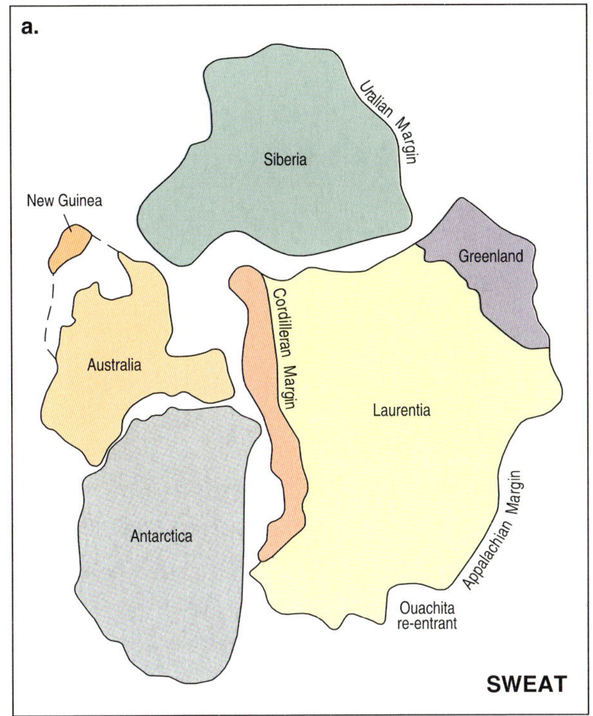

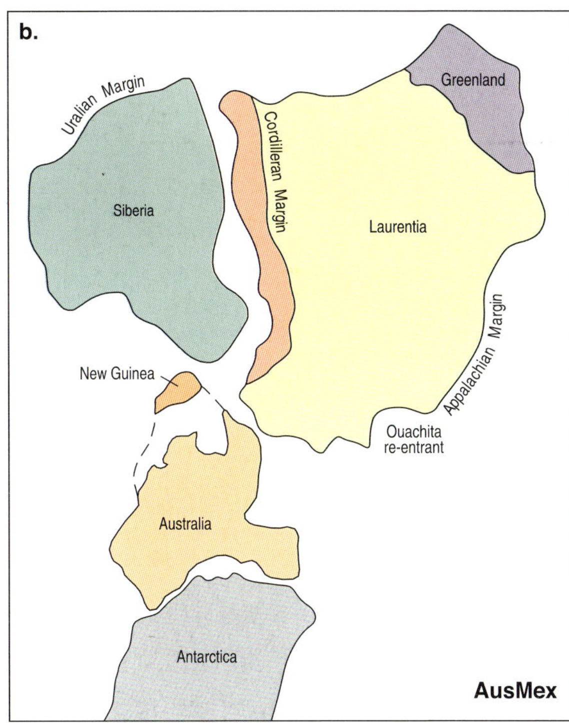

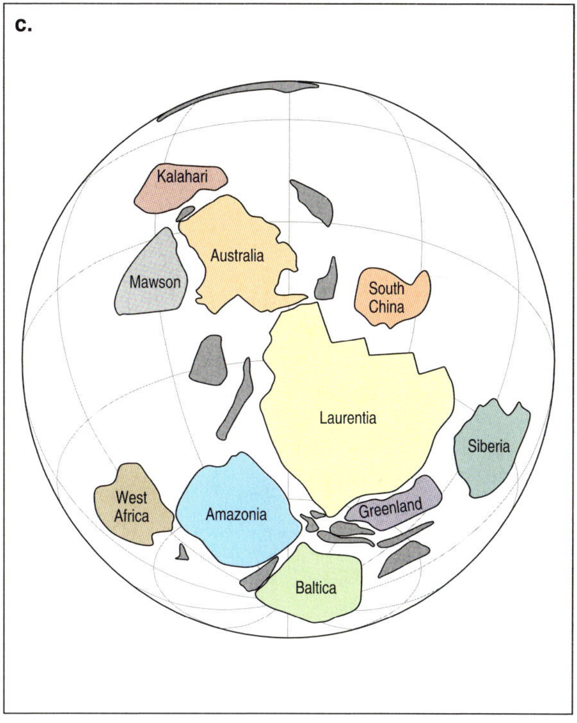

Figure 2.30: Alternative reconstructions of Rodinia. (a) The SWEAT (Southwest (US) – East Antarctic; Hoffman, 1991) reconstruction. (b) AusMex (Australia–Mexico; Wingate et al., 2002) reconstruction. (c) Modified AusMex (Pisarevski et al., 2003) model. These are some of the many variants of the Rodinian reconstructions.

mineral discoveries and new geochronological data, however, highlight the tectonic activity of this period, which appears to have been associated with a different assemblage of mineral systems from those before. The supercontinent of Rodinia was assembled from 1300 Ma to 900 Ma. Of the various reconstructions of Rodinia, most place eastern Australia (without the Tasman Element, which formed later) adjacent to Laurentia (Figure 2.30), with some placing North China in between.

During the time of Rodinian assembly, the West Australian Element was reworked by Meso- to Neoproterozoic orogenies along its northeastern (Paterson Orogen), southern (Albany–Fraser Orogen) and western (Pinjarra Orogen) margins. The Albany–Fraser Orogen and correlated deformation in the Musgrave Province mark the collision at 1345 Ma to 1140 Ma between the combined West–North Australian Element and the Mawson Craton. The Tropicana Au deposit lies along this margin (Figure 2.21a).

The earliest mineral deposits of this interval are a diverse suite, including diamond, rare earth elements and orthomagmatic Ni–copper (Cu)–platinum (Pt) group elements (PGE) (Figure 2.28). All of these deposits are associated with alkalic rocks, including the >1240 Ma Nolans Bore rare earth element deposit and the *ca* 1180 Ma Argyle diamond pipe, which might have resulted from an intracratonic plume event. The abundance of base metal and Au deposits that characterise the assembly of earlier (Kenorland and Nuna) and later (Pangaea–Gondwana) supercontinents is missing, both in Australia and overseas. In central

Figure 2.31: Photograph of the MacDonnell Ranges, Northern Territory, showing the effect of geology on topography. Resistant rocks, such as the Heavitree Quartzite, form ridges, whereas softer rocks, such as the Bitter Springs Formation, form valleys.

and western Australia, the Warrakuna Large Igneous Province crops out over 1.5 M km^2, and was emplaced around 1075 Ma. It is associated with mafic–ultramafic intrusions such as the Giles Igneous Complex and associated Ni–Cu–PGE deposits such as Nebo–Babel (Figure 2.28).

The onset of rifting leading to breakup of Rodinia commenced *ca* 850 Ma, leaving the jagged Tasman Line as the mark of the successful separation, probably, from Laurentia (Figures 2.10 and 2.30). One of the earliest manifestations of this breakup in Australia was the northwest-trending Gairdner Large Igneous Province (830 Ma), which was intruded into the South and North Australian elements and into the Paterson Orogen in the northwest. Extension associated with Rodinia breakup was likely to have been the immediate driver for the formation of the Centralian Superbasin. This basin system, which initiated at *ca* 850 Ma, and continued to the Devonian, affected most of central Australia and includes the Officer, Amadeus, Georgina and Yeneena basins (Figure 2.9). These basins include the earliest known major salt deposits, in the Bitter Springs Formation of the Amadeus Basin, several periods of glaciation, the first flowering of multicellular life and later the emergence of vertebrates (Chapter 3). The deposition of salt (and subsequent formation of salt deposits) has lowered the salinity of seawater by a factor of two or more since the Mesoproterozoic. Towards the bottom of these basins lie the Heavitree Quartzite and its correlatives. This quartzite makes up the spectacular red cliffs of the MacDonnell Ranges and Kings Canyon (NT), which are special places to the Aboriginal owners, and are famous as a tourist destination (Figure 2.31). Sandstone sheets like the Heavitree Quartzite are tens to hundreds of metres thick and extend for hundreds of kilometres laterally. They were only able to achieve this vast lateral extent because the Proterozoic land surface was not yet covered with vegetation (Chapter 3), so erosional rates were likely to have been enormous.

The global ice ages of the Neoproterozoic have left clear traces in a band across central Australia, from the Kimberley region (WA) to the Flinders Ranges (SA). Clear evidence of glacial dropstones can be found in the arid deserts of central Australia. Indeed, South Australian localities have given their names to these Sturtian and Marinoan glaciations (Figure 2.32). Over this long time span, the complexity of life increased and, towards the end of this period, the first clear fossils of multicellular life appeared. The type locality is in the Ediacaran Range slightly to the west of the Flinders Ranges (Chapter 3).

Uranium and Cu deposits in the Yeneena and the Adelaide basins formed between 840 Ma and 790 Ma (Figure 2.28); these deposits were related to basin formation and/or inversion. The revised AusMex reconstruction of Rodinia (Figure 2.30c) places Australia adjacent to the Kalahari Craton, which contains the highly productive Zambian Copper Belt. This belt is hosted by a basin of similar age and with similar fill to the Centralian Superbasin, and the ages of mineralisation overlap.

700–250 Ma—amalgamation of Gondwana and Pangaea

The Phanerozoic components of Australia were mostly assembled in a sequence of tectonic cycles, beginning with extension and ending in orogeny. To the west, the Pinjarra Element divides East Gondwana into Australo–Antarctic and Indo–Antarctic domains, which are distinct continental fragments with different Proterozoic histories that were juxtaposed by oblique collision at 550–500 Ma during the assembly of Gondwana.

Between 650 Ma and 600 Ma, deformation was accompanied by granite emplacement and related Cu–Au and tungsten (W)–Cu–Zn mineralisation in the Paterson Province (Figure 2.33). The *ca* 550 Ma Petermann Orogeny was a major intraplate event in central Australia. This event caused crustal thickening with around 20 km of vertical motion (eclogites are exposed today) and the formation of a large mountain belt with adjacent foreland basins, such as the Amadeus Basin. This mountain belt is part of a proposed 'Supermountain range' that may have stretched more than 8000 km through Gondwana, perhaps reaching Himalayan proportions. The weathering and erosion of this unvegetated mountain belt released a huge volume of sediment and nutrients (such as iron, phosphorus, strontium and calcium) into the marginal oceans and seas, possibly triggering the evolution of animal life (Chapter 3). In central Australia, the arksoic sandstone of the Uluru monolith (Ayers Rock) and the conglomerate of Kata Tjuta (the Olgas) formed in a foreland basin adjacent to this vast mountain range.

To the east, the Tasman Element was assembled successively onto the older Precambrian continental core to the west (Figure 2.34). These tectonic cycles produced Australia's most significant volcanic-hosted massive sulfide (VHMS), epithermal and porphyry Cu–Au

Image by Helen Dulfer

Figure 2.32: Dropstone in the Merinjina Tillite, Adelaide Rift Basin, South Australia. Dropstones form in shallow seas when icebergs melt, dropping contained stones onto muds on the seafloor. This example is from the Sturtian glaciation, a record of so-called icehouse Earth, when much of Earth's land surface was affected by glaciers.

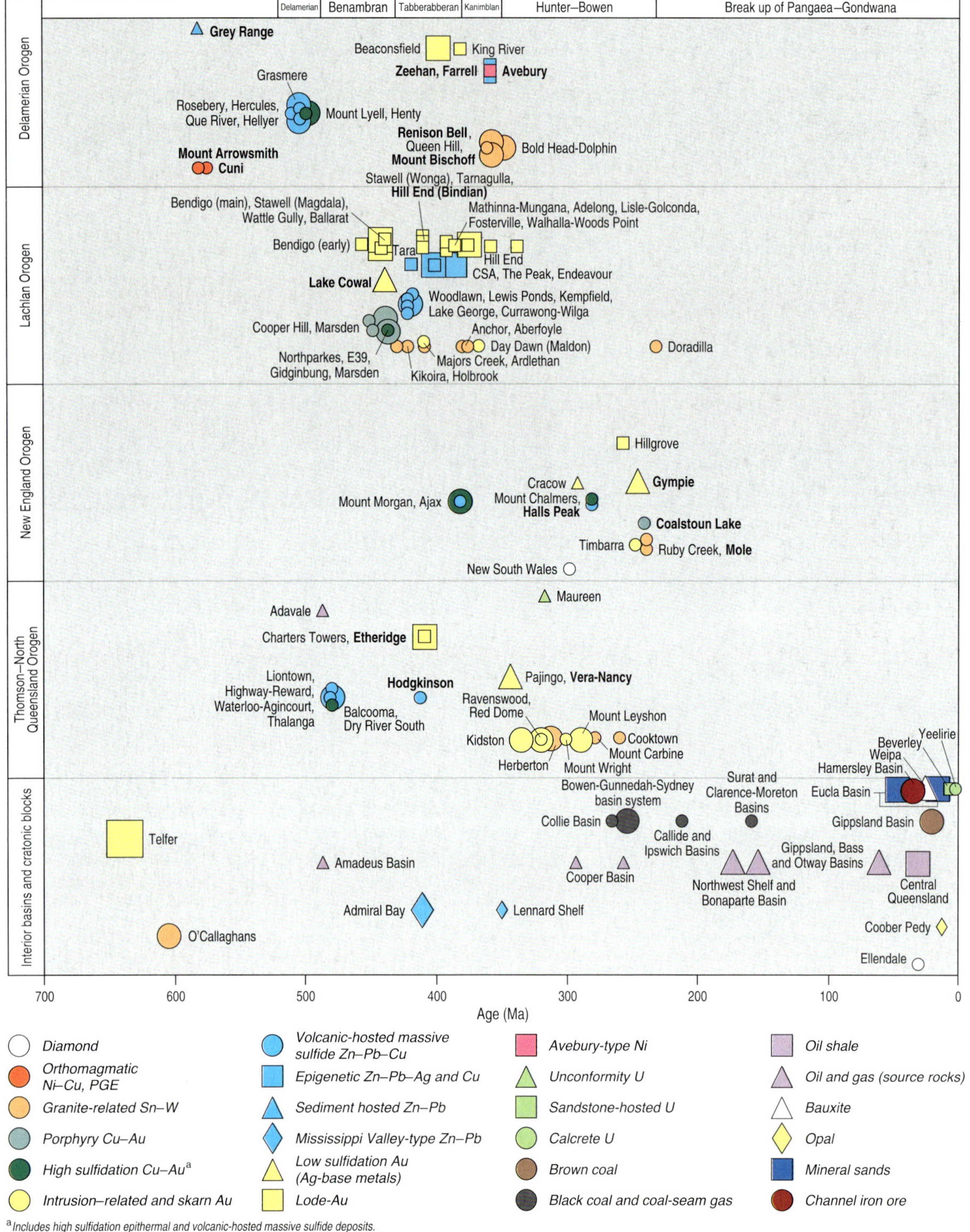

deposits, and major lode-Au and magmatic tin provinces (Figure 2.33 and Appendix 8.1.1). Central Australia underwent a series of intraplate extensional and contractional events, many being time equivalent to those in the east.

Following the breakup of Rodinia, a passive margin with local rifts formed on the eastern edge of Australia, which now faced the proto-Pacific Ocean (Figure 2.34). By 515 Ma, or even earlier (560 Ma), subduction started in southeastern Australia, which saw the crust/lithosphere of eastern Australia grow via a series of episodic tectonic cycles. These cycles lasted between 30 Myr and 130 Myr, beginning with a period of extension associated with the formation of backarc basins, and terminating with contractional deformation events, when continental slivers and island arcs were accreted back onto the Australian continent (Figure 2.34). There were five such cycles: the Delamerian (515–490 Ma), Benambran (490–440 Ma), Tabberabberan (440–380 Ma; includes the Bindian), Kanimblan (380–350 Ma) and Hunter–Bowen (350–220 Ma). The younger cycles overprint the older cycles, but, in general, the cycles become younger to the east (Figure 2.34).

There is a general progression in the types of mineral deposits formed within individual cycles (Figure 2.33). Mineral deposits formed in arcs

Figure 2.33: Major mineral deposits of Australia formed during the period 650–0 Ma, in particular during the assembly and breakup of Pangaea–Gondwana. Symbol size broadly indicates deposit size relative to deposits with similar commodities. See Figure 2.21 for localities.
PGE = platinum group elements

(e.g. porphyry and epithermal deposits) and backarcs (e.g. volcanic-hosted massive sulfide deposits) form early in cycles, whereas mineral deposits formed mostly during contractional deformation (e.g. lode-Au and structurally hosted base metal deposits) came late in the cycle.

While accretion was occurring in the east, parts of Gondwana were being rifted from the northwest. The first of these rifting events may have been related to the emplacement of the Kalkarindji Large Igneous Province in Western Australia and the Northern Territory, which is known from basalts dated at 520 Ma. A progression of younger rifting events led to excision of a succession of slices off northwest Australia, with components now found in north China (380 Ma), Sibumasu in South-East Asia (*ca* 290 Ma), Lhasa in Tibet (210 Ma) and Argo Land in Burma (162–154 Ma). From the Carboniferous, thick sedimentary successions accumulated in rift basins along the evolving northwest margin and in the Perth Basin (Chapter 4).

Activity related to convergence on the eastern seaboard also influenced the continent's interior. Interior events that link temporally to events in eastern Australia include development of the Larapintine Seaway and the Alice Springs Orogeny (which actually includes three separate orogenic events). The Larapintine Seaway is a concept that goes back to the 1930s, linking the

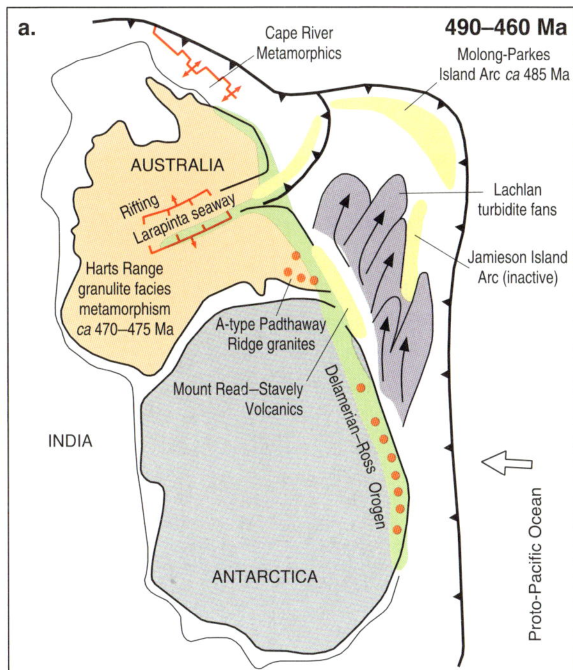

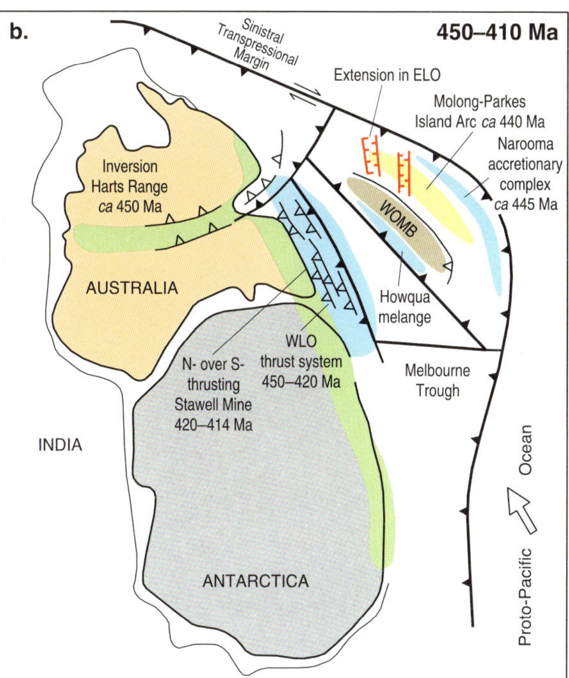

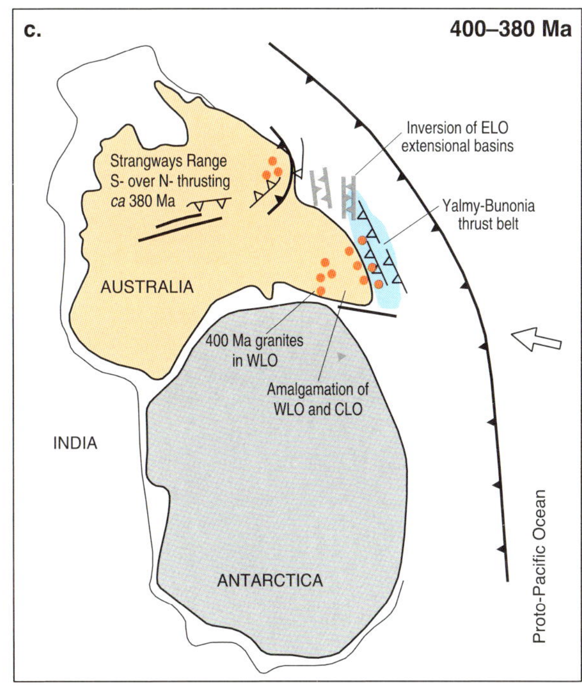

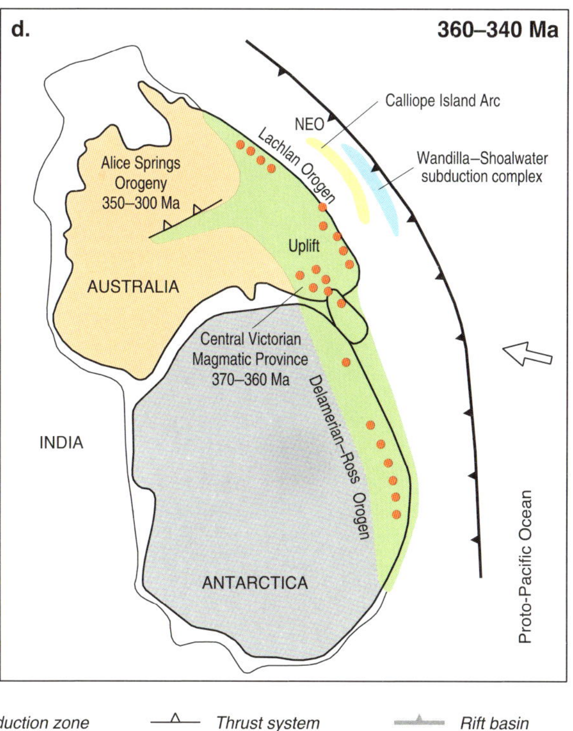

Figure 2.34: Schematic diagrams showing the tectonic evolution of Australia between 490 Ma and 340 Ma.
CLO = Central Lachlan Orogen; ELO = East Lachlan Orogen; NEO = New England Orogen; WLO = West Lachlan Orogen; WOMB = Wagga-Omeo Metamorphic Belt

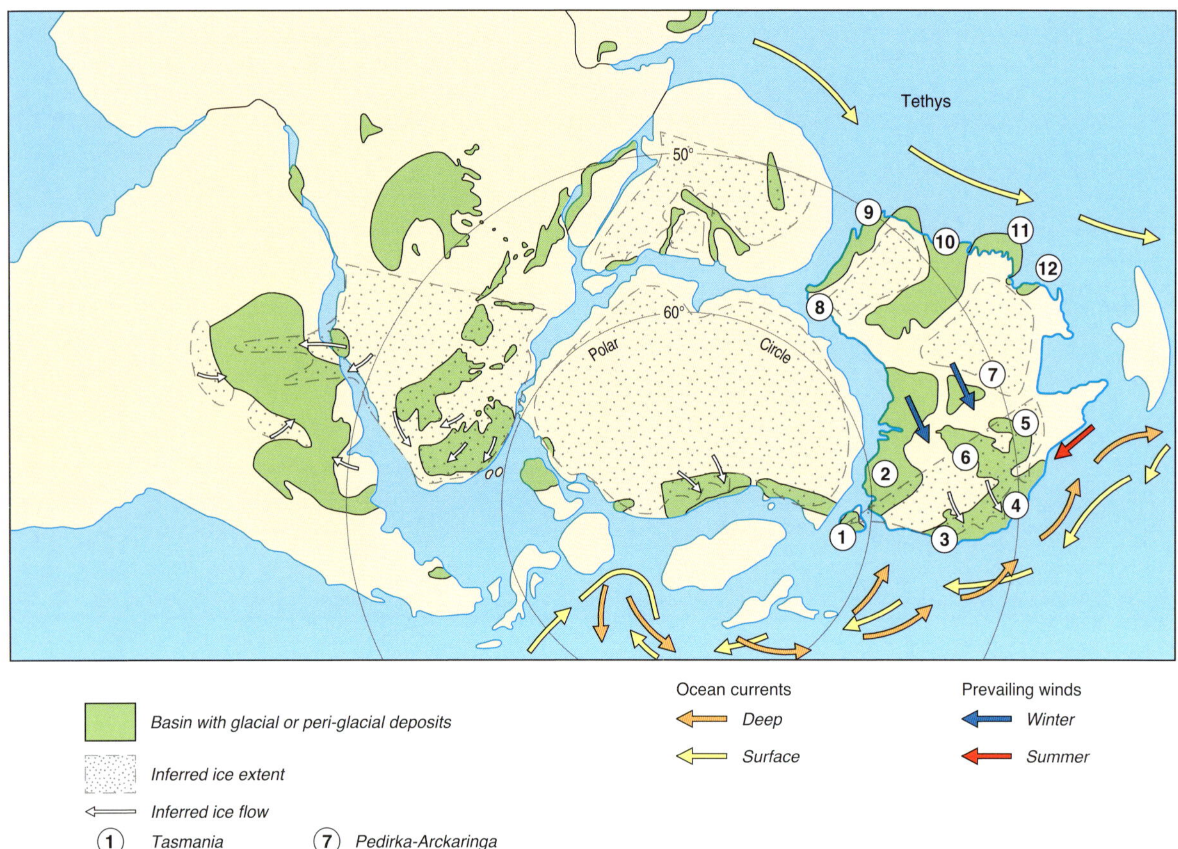

Figure 2.35: Gondwana reconstruction from the late Carboniferous to early Permian. Sedimentary basins formed across much of Australia at this time, some being the sites of major coal and coal-seam gas resources (Chapter 9). Australia and Antarctica were tightly coupled in East Gondwana. Australia was completely separated from the Antarctic vestige of Gondwana at 34 Ma, becoming an island continent in the geologically recent past (Figure 2.20). (Source: after López-Gamundí, 2010)

Ordovician (and later) marine sediments of the Canning, Wiso, Georgina and Amadeus basins through to the deep-water margin in eastern Australia (Figure 2.9). The evocative name of this seaway derives from the Aboriginal name for the Finke River in central Australia, the sinuous course of which is the track of the Dreamtime rainbow serpent (Chapter 4). Ordovician sediments, including limestone, sandstone and organic-rich marine shale, were deposited across tropical Gondwana to Tarim, which is now located in western China. Deep-water Ordovician sediments marking the edge of Gondwana are found in West Papua and eastern Australia, where they host the Victorian goldfields (Chapter 8).

Larapintine has also been used to describe the particular petroleum systems related to oil source rocks deposited in warm, shallow seas during the early Paleozoic in onshore Australia (Chapter 4). The largest commercial Larapintine petroleum accumulations are the Palm Valley gas field and Mereenie oil field in the Amadeus Basin in central Australia (Figure 2.21b). Ordovician organic-rich marine sediments provide the source rock and seal for the gas and oil in underlying sandstone reservoirs. Giant anticlines, formed during the Alice Springs Orogeny, provide the structural architecture for the hydrocarbon traps. However, in comparison with younger systems (see below and Chapter 4), the Larapintine system was very minor (<10% of Australia's total hydrocarbon resources).

In central Australia, the intraplate Alice Springs Orogeny was a succession of contractional events that inverted large parts of the Larapintine Seaway. Prior to this, the basins of the Centralian Superbasin (Amadeus, Georgina, Wiso and Ngalia) were contiguous. The orogeny was intense: deep crustal rocks were exhumed some 30 km and juxtaposed with undeformed rocks of similar age. As part of this orogeny, the south-directed Redbank Shear Zone offset the Moho, creating one of the largest gravity gradients on Earth (Figure 2.12). The spectacular MacDonnell Ranges at, and near, Alice Springs comprise Amadeus Basin sediments tilted as a consequence of exhumation associated with the Redbank Shear Zone (Figure 2.31). The Alice Springs Orogeny also produced topographic relief. This topography resulted in the development of foreland basins and possibly the topographic head that drove hydrothermal fluid flow in the Mississippi Valley-type Pb–Zn deposits of the Canning Basin (Figure 2.33).

The assembly of the continents into Pangaea led to extensive glaciations in the late Carboniferous and Permian (Figure 2.35). Streams of ice entered Australia from the uplands of Antarctica to the south, bringing glacial debris into the sedimentary record, with extensive deposition in central Australia and in the rift sequences along the western margin, in the Perth, Carnarvon and Bonaparte basins. The Permian glaciation also removed or modified much of the arid zone's pre-existing landscape, although large tracts were ice free. An example is the deep oxidation, visible to more than 60 m depth, in the Au deposits of the Tanami Desert region in the Northern Territory (Figure 5.3b). Deep Permian-aged oxidation was likely enhanced by the fact that the atmosphere may have contained 50% more oxygen than today. A further example of Permian weathering is the karst system developed at 345 Ma in the Jenolan Caves to the west of Sydney, making these the oldest currently open cave system in the world. It is possible that some of these old landscapes have been formed, buried and more recently exhumed, rather than being continuously exposed for hundreds of millions of years (Chapter 5).

Gondwana breakup seen by Jurassic (*ca* 175 Ma) dolerite tors on Mt Wellington, Tasmania.

The vast Permian coal deposits of the Sydney–Bowen–Gunnedah basin system in eastern Australia formed in a foreland basin behind the continental volcanic arc system of the New England Orogen. Counter-flow in the mantle wedge above the subduction zone is thought to have created the slowly subsiding accommodation space for the thick accumulation of the coal measures. The cool high-latitude plate position of Australia at this time, an environment similar to the modern Russian taiga, permitted the accumulation of vast amounts of peat from *Glossopteris* and other species. These peat seams, accumulated over hundreds of thousands of years, were compacted into black coal, with the largest seam at Blair Athol being 31 m thick (Chapter 9).

250–160 Ma—breakup of Pangaea

After final amalgamation around 250 Ma, Pangaea lasted less than 100 Myr before it started to break apart. This breakup initiated at around 180 Ma as the Atlantic Ocean and Tethys Sea began to form, separating Gondwana (which included Australia, Antarctica, South America, Africa and India) from Laurasia (which included North America and Eurasia). During the Triassic, the western margin underwent a series of transpressional events, known collectively as the Fitzroy Movement. Ongoing subduction occurred along the eastern margin of the continent, with widespread volcanoclastic detritus being deposited in the Eromanga and Surat basins. The emplacement of voluminous *ca* 175 Ma tholeiitic dolerite sills, which form part of the Karoo–Ferrar Large Igneous Province in Tasmania, presaged the breakup.

One of Australia's iconic rocks, the Hawkesbury Sandstone of the Sydney Basin, was deposited as Pangaea broke up. This quartz-rich sandstone was both a boon and a drawback to the growing colony of New South Wales. Ideally suited for building because it is both durable and easily worked, it was used to construct many public buildings in Sydney, including Government House and the older buildings at the University of Sydney. However, soil that developed on the Hawkesbury Sandstone was poor, and escarpments formed from it were a major impediment to exploration and settlement across the Blue Mountains west of Sydney. On the western margin of the continent, vast delta systems poured sediment onto the North West Shelf, creating the Triassic-aged reservoir rocks for Australia's vast hydrocarbon resources (Chapter 4).

160–65 Ma—breakup of Gondwana and creation of Australia

Gondwana began to fragment in the Jurassic, as East and West Gondwana separated with seafloor spreading between Africa and Madagascar, and the Atlantic Ocean started to open between Africa and South America. A triple point of extension, centred just southwest of Perth (WA), cut the jigsaw between India, Antarctica and Australia. The northwest margin of Australia had formed by 154 Ma, as the last of a series of continental slivers, Argo Land in Burma, rifted away. By 120 Ma, the breakup had extended far into Gondwana, separating Australia from Greater India, and the entire west coast of Australia faced the Indian Ocean. A series of panels illustrating the breakup is given in Chapter 4.

Australia unzipped from Antarctica via an easterly-propagating rift system. Rift basins had begun to form as far east as the Polda Basin in South Australia at 155 Ma. By 145 Ma, rifting in the Gippsland Basin, Victoria, had occurred. By 120 Ma, volcanic detritus was flooding into eastern Australia (Eromanga Basin), and rifting stopped in the Bass Basin. The Large Igneous Province of Kerguelen started to form around 118–110 Ma and the Broken Ridge around 95 Ma (Figure 2.1), while volcanism continued on the Lord Howe Rise (now submerged off eastern Australia).

During the Cretaceous, a large inland sea formed across much of eastern Australia (Box 4.3). This Eromanga Basin is up to 1200 m thick and covers several older basins. Some workers have suggested that remnants of the west-directed subduction zone slab, which dominated the tectonics of eastern Australia, continued to exert an influence by descending in the mantle and drawing down the lithosphere, creating the accommodation space for the Eromanga Basin sediments. This concept helps explain why much of eastern Australia was inundated by sea in the Early Cretaceous but largely high and dry in the Late Cretaceous when global sea-levels were at their highest (Chapter 4). Along the southern margin, the giant Ceduna delta partly filled the seaway between Antarctica and Australia with continental detritus.

By 95 Ma, the longstanding west-directed Palaeopacific Ocean subduction zone beneath eastern Australia retreated to the east. This event coincided with uplift and formation of the eastern highlands of Australia (95–85 Ma), including uplift and erosion of the Eromanga Basin. By this time, a significant seaway had developed between Australia and Antarctica, though there was still connection through Tasmania. Breakup could have occurred to the north of Tasmania through the Bass Strait region where the rift basins of the Otway, Bass and Gippsland lie. The dolerite massif of Mt Wellington, which stands above present-day Hobart, is a legacy of earlier attempts at rifting. The thinning of the crust was accommodated by basin formation in the Gippsland and Bass Strait areas, creating one of Australia's main hydrocarbon provinces (Chapter 4).

Rifting along the east coast began before *ca* 85 Ma in the south, and propagated northwards, resulting in the opening of the Tasman Sea, taking the Lord Howe Rise and New Zealand to the east. The former links to the other parts of Gondwana remain evident through associations of flora and fauna (Chapter 3). The Southern beech, *Nothofagus*, is today found in Tasmania, New Zealand and southernmost South America. Australia even has a native rhododendron whose nearest relatives are in the Himalaya. The distinctive marsupial populations of Australia are a thriving remnant of a distribution that once spread across the southern continents (Chapter 3).

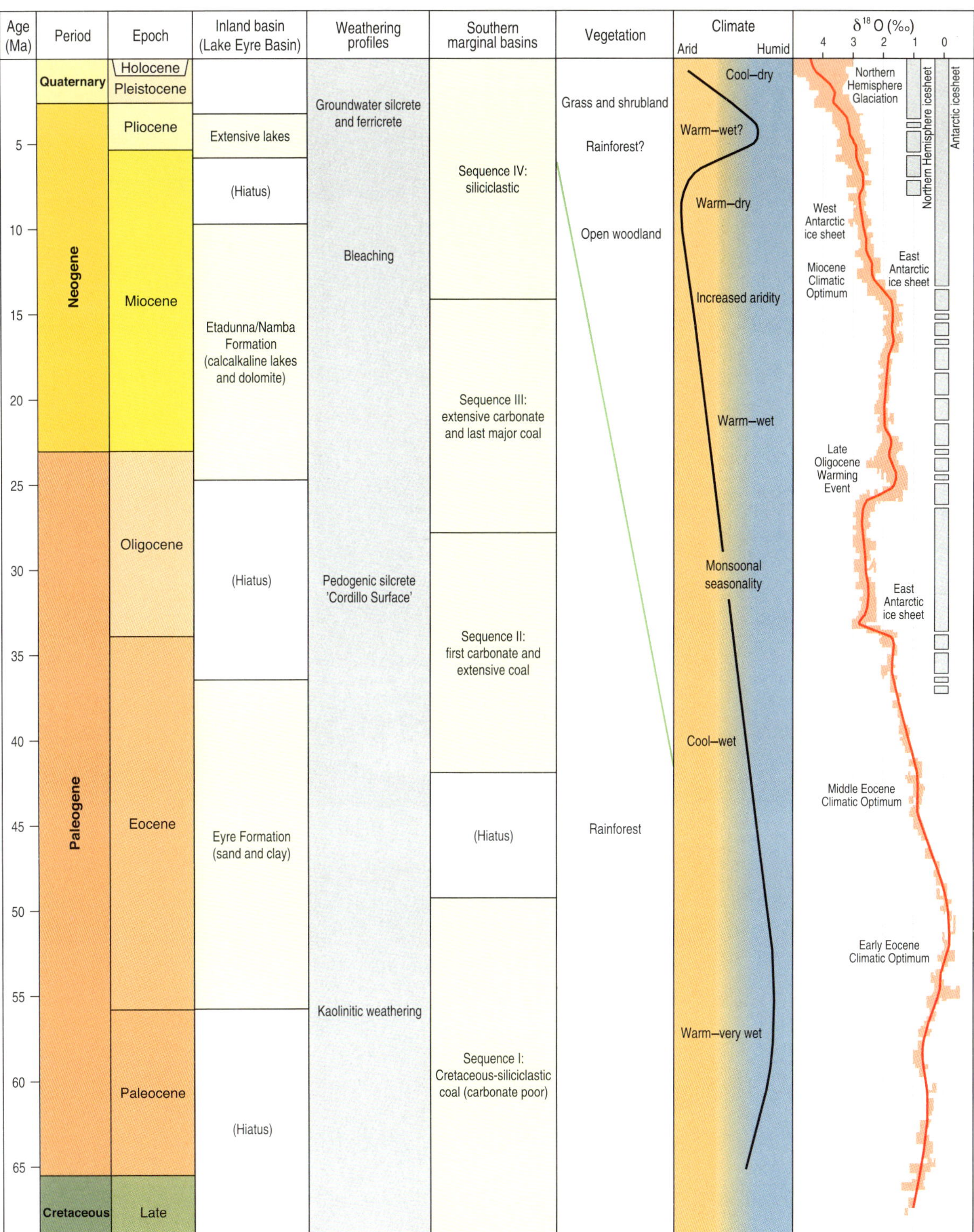

Figure 2.36: Summary of the climate of Cenozoic Australia, inferred from sedimentological and palaeontological evidence in inland and southern marginal basins and from weathering regimes. Oxygen isotopes are proxies for ocean temperature, data sourced from deep-sea drilling cores. The low $\delta^{18}O$ values correspond to warmer periods, and the onset of cooling and glaciation is marked by a sharp rise in the $\delta^{18}O$ values. (Sources: after Fujioka & Chappell, 2010, and references therein)

Image courtesy of Undara Volcanic Park

Lava caves in the Undara Volcanic National Park in north Queensland. The lava caves are the remnants of conduits through which lava flowed. When volcanism ceased, the conduits remained, free of lava, to form tube-like caves. The Undara Volcanic National Park is part of the 190 ka McBride Volcanic Province (Figure 2.3).

Mesozoic weathering and landscapes continued to develop, especially on granitic rocks in eastern Australia. By the end of the Cretaceous, the fundamental physiography and the palaeovalley architecture of today's arid Australia was established. These palaeovalley systems formed as a network of wide and shallow valleys in the landscape, but without sediment infill. Many of these palaeovalleys are important groundwater aquifers (Chapter 7). Some of these landscapes are as old as the Cambrian, but they are unlikely to have remained continuously exposed at the surface for this duration. Fission-track dating suggests that these very ancient landscapes were exposed, then buried and finally more recently exhumed (Chapter 5).

65–2.6 Ma—Australia girt by sea

The past 65 Ma have left an extensive imprint on the Australian continent, with around 80% of the surface geology of Australia comprising Cenozoic rocks of some kind (Figure 2.7). Most of this cover consists of thin aeolian sand, lakes and fluvial deposits. Locally, discrete basin entities developed, such as the Eucla Basin in Western Australia, the Murray Basin in southeastern Australia and the Karumba Basin in northern Australia, as well as their offshore extensions (Figure 2.9). Knowledge of Cenozoic Australia comes from the combined study of landscape, stratigraphy, palaeontology, isotope geochemistry and geochronology (Figure 2.36).

By 65 Ma, oceanic crust floored the Southern Ocean as far east as western Victoria. Further to the east and the south, the crust was very thin at this time, but a connection with Antarctica remained as sediments sourced from the south were deposited in the sag phase of the Sorell Basin off the west coast of Tasmania. The Ceduna delta system was abandoned at this time. Seafloor spreading started with the opening of the Coral Sea and the Queensland Plateau at around 60 Ma. Following breakup, Australia's initial drift was to the northwest. But a ridge subduction in the northwest Pacific at around 52 Ma resulted in termination of Tasman Sea spreading and a change in the Australian Plate vector to its present-day northerly direction. Seafloor spreading in the Southern Ocean accelerated at around 45 Ma, and it was not until 34 Ma that full separation of Australia and Antarctica occurred. Australia, as a separate continent, was released from the remnants of Gondwana and commenced the northward march to meet Asia.

The formation of the passive margins not only created most of Australia's hydrocarbon plays, but also preserved the huge energy reserves in coal and coal-seam gas along the eastern margin. If the Pacific–Australian Plate subduction margin, now in New Zealand and Fiji, had remained just offshore of New South Wales and Queensland, then these easily won energy resources may have been destroyed or rendered uneconomic (Chapter 9). Had this rollback of the subduction zone not occurred, maybe the mountains, and thus ski fields, of southeast Australia might have been world class.

From about 45 Ma to 10 Ma, Cenozoic sediments were deposited in the Lake Eyre (360 m), Eucla (750 m onshore, to 6 km offshore), Karumba (up to 1700 m offshore) and Murray (600 m) basins. The climate was warmer and wetter than today, and rainforests were extensive (Figure 2.36). During this time, the sea made several incursions onto the land. The Eocene (42.5–34 Ma) and Miocene–Pliocene (15–2.6 Ma) shorelines preserved along the Eucla Basin margin are good examples of these repeated incursions. These marine incursions have left a legacy of salt, which today is being mobilised into the present-day rivers and groundwater. These incursions have also deposited sulfur in some coal and lignite deposits, rendering them less suitable to use. Most of the salt burden contained in Australian soils is, however, a function of aeolian salt carried directly from the sea or as dissolved salt carried in rainwater. Salt remains in the Australian landscape because the continent's river systems either do not drain into the sea or, if they do, are sluggish and have low discharge rates (Chapter 5).

Major palaeovalley systems provided detritus to the Cenozoic basins and their offshore extents (Figure 2.37). Some palaeovalleys were shaped by older landscape features. The general pattern of these palaeovalleys across Australia is an early Cenozoic fluvial succession of sand, overlain by a mid-Cenozoic succession of fine-grained lacustrine sediments. These palaeovalleys are known for a variety of resources, including iron ore, U and groundwater (Chapter 7). For example, a significant proportion of the Hamersley iron ore production in the Pilbara (WA) comes from pisolitic iron that formed in the Robe and Yandi palaeochannels, which were active at around 30 Ma (Figure 2.33). In this region, late-stage goethite mineralisation has been dated at 18 Ma near the surface, and 5 Ma at depth, reflecting the progressive lowering of the water-table as northern Australia became more arid (Chapter 5).

It can snow any time of year in the high country of southeast Australia. Small glaciers carved the landscape during the last ice age. The meltwater from these snows is now piped across the Great Divide into the west-flowing Murray–Darling system (Chapter 5).

High-stand palaeoshorelines of the Eucla Basin are elevated, now sitting at 250 m asl on the Nullarbor Plain. Around Mt Gambier, South Australia, shorelines occur 180 m asl, equating to uplift rates of 65 m per Myr. This is also the region of the youngest volcanoes on the Australian continent. Interestingly, there are no comparable-aged marine shorelines exposed on the north coast. Possibly, there was a northward tilting of the plate, caused by the subduction pull of the northern margin (Chapter 6). Some of these Cenozoic palaeoshorelines contain economic concentrations of rutile, zircon, ilmenite, and minor leucoxene and monazite (Figures 2.33 and 2.37).

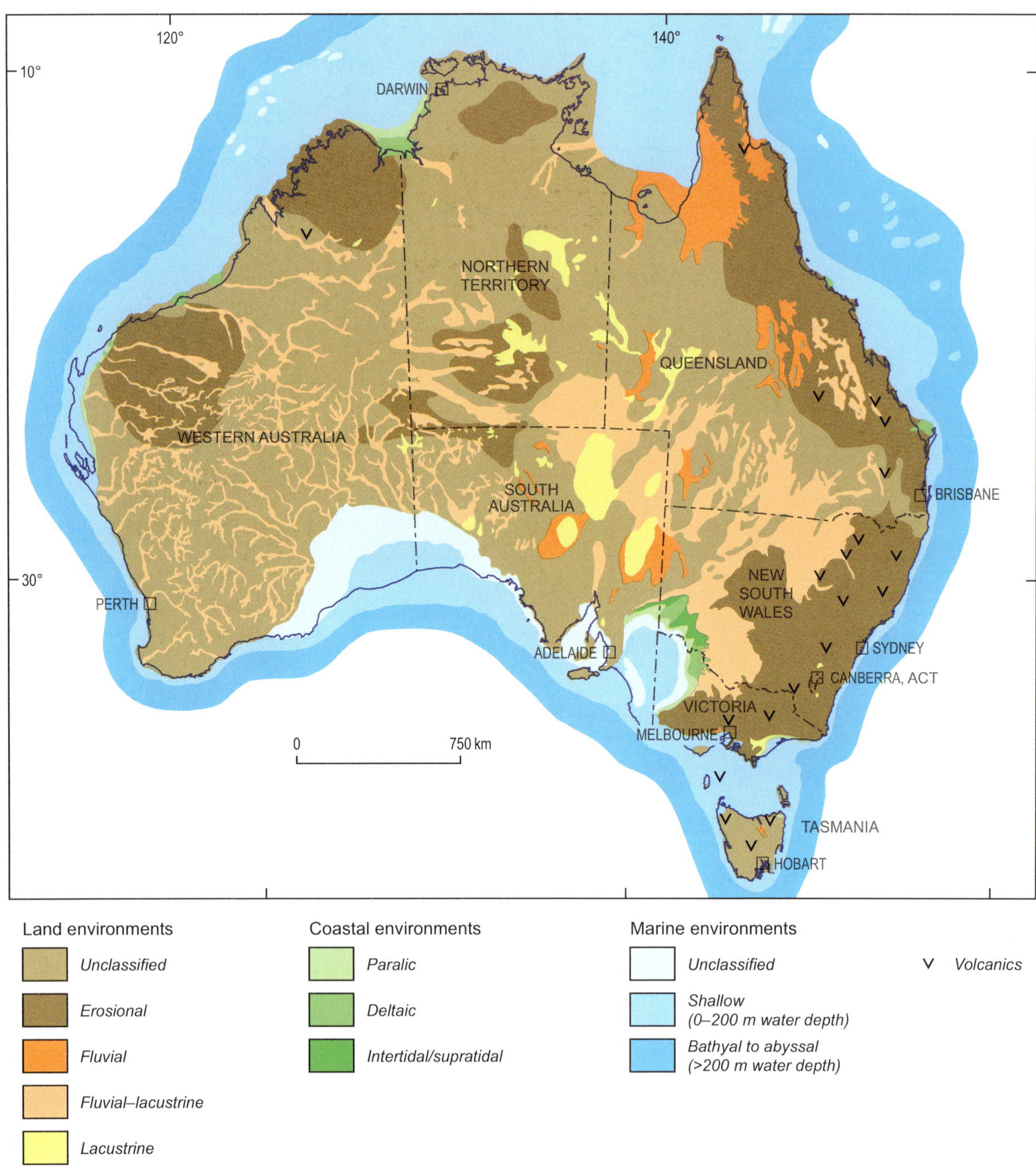

Figure 2.37: Map of the environments during Cenozoic Australia, between 30 Ma and 10 Ma.

The brown coals of the La Trobe Valley in southeast Victoria, which are burnt in power stations today, are up to 130 m thick. They were formed between 30 Ma and 20 Ma, when the climate was humid. Similarly, the oil-shale resources near Gladstone in Queensland (Figure 2.21b) and diatomite deposits from the Warrumbungles (NSW) into southeast Queensland were also formed in lacustrine basins at this time.

The Ellendale lamproite field (WA) was emplaced around 25 Ma, probably as a result of a short-lived mantle hotspot. More than 80 separate pipes cross the northern margin of the Canning Basin, which marks a persistent zone of continental weakness adjacent to the very resistant margin of the Kimberley Craton. Along the eastern margin of the continent, gemstones were carried in eruptions associated with the Newer Volcanic Group as the Australian Plate moved northwards over mantle hotspots. At 32 Ma, the Mount Dukes Intrusive Complex in tropical northeast Queensland contains the youngest granite known on the Australian mainland (Figure 2.3). This granite was comagmatic with the shield volcanoes common to the east coast.

Australia and Antarctica—and South America and Antarctica—had separated sufficiently at around 25 Ma for full circulation between the Southern and Pacific oceans to occur. Although polar glaciation had started in West Antarctica around 34 Ma, the great ice sheet on Antarctica began to develop around 15–10 Ma in the east and 10–6 Ma in the west. The formation of the Antarctic ice sheet fundamentally altered Earth's marine and terrestrial climate and sea-levels. The impacts on Australia were large, as the meridional temperature gradients steepened and the subtropical monsoon contracted northwards, which placed much of the continent in the drier mid-latitudes (Figure 2.36).

The late early Miocene (*ca* 16 Ma) was the climax of a Neogene warm period, followed by cooling and increasing aridity. Opal began to be deposited, such as in Coober Pedy (SA), at this time and during punctuated intervals thereafter. Deep oxidation and weathering occurred in southeast Australia in the Miocene. Between 16 Ma and 12.9 Ma, there were major short-term climate fluctuations. This was the period of the stabilisation of the East Antarctic ice sheet. The boundaries between climatic zones strengthened, which led to the increased aridification of mid-latitude continental regions in Australia, Africa, and North and South America. Grasslands developed, stimulating the evolution of grazing mammals (Chapter 3), and in Australia the rainforests retreated (Figure 2.36).

By 10 Ma (Late Miocene), Australia had moved northwards into a latitudinal position similar to that of today. Sea-levels were lower, so that much of the continent was exposed to weathering and erosion (Chapter 5). Pollen evidence shows that the period between *ca* 5 Ma and *ca* 2 Ma was wet in Australia, with rainforests again supported (Figure 2.36). The extensive cave systems of the Nullarbor Plain formed in the uplifted marine sediments of the Eucla Basin. Recent dating has shown a range in ages of these speleothems, from 7.5 Ma to *ca* 1 Ma, with a distinct peak in cave formation between 4.1 Ma and 3.5 Ma.

The Cenozoic tectonic history of Australia was not entirely associated with breakup. The active convergence that is still occurring today along Australia's north commenced at about 40 Ma. Timor is now accreted onto the Australian Plate and is moving together with Australia. Collision with proto-Indonesia and related fragments caused the cessation of carbonate shelf deposition and initiation of widespread siliciclastic sedimentation on top of the Australian continental basement at about 12 Ma. The collision emplaced the Irian (West Papua) ophiolite and created the present mountainous topography that forms the spine of the island of New Guinea. A well-developed fold-thrust belt has grown over the Australian Plate margin in the southern part of New Guinea; this active seismic region has folds developing today. The foreland basins created by Australia's collision also host the giant accumulations of hydrocarbons (such as the Tangguh and Hides gas fields). The collision and underlying structure of the Australian Plate also control the location of the region's giant Cu–Au deposits (such as Grasberg and Ok Tedi).

Convergence and major dextral strike-slip faulting commenced around 5 Ma across the Australian Plate boundary in New Zealand. This resulted in far-field shortening across Australia, and renewed uplift along the eastern highlands and the Flinders Ranges (SA). Present-day convergence along the eastern plate margin contributes to the continent's stress regime and earthquake hazard (Figure 2.2).

Towards the end of the Pliocene, the climatic zones now governing Australia had become established, although the climate was often wetter than today. The pace of climate change increased, as the Quaternary glacial cycles commenced, including the expansion of ice caps in the Northern Hemisphere around 3 Ma. In the African savanna, along the widening eastern rift valleys, early hominins were evolving.

2.6–0 Ma—evolution of modern Australia

One of the difficulties in understanding the Australian landscape is that it is a complex mosaic of many superimposed landscapes that developed during vastly differing climatic conditions. For example, many of the features of Quaternary Australia were inherited from landscapes as old as the Cambrian (Chapter 5). The preservation of these ancient landscapes and its regolith is a function of the tectonic stability afforded by the old cratonic roots of the continent and the plate position during the Pleistocene to Holocene. Most of Australia avoided the physical rigours of the last ice ages, with glaciers limited to the southeastern highlands.

The Pleistocene and Holocene ice age impacts in Australia were driven by cycles of aridity, which were linked to the glacial–interglacial cycles. In Tasmania, the maximum ice extent occurred at *ca* 1 Ma, with later advances in the past 100 kyr being less extensive. During these glacial episodes, atmospheric circulation intensified, with a reduction in atmospheric moisture and an increase in 'continentality' because of lower sea-levels (Box 2.6). The net water budget reduced, due to decreased precipitation and increased wind-enhanced evaporation. Aeolian processes

OCEANS TO ICE—QUATERNARY CLIMATE AND SEA-LEVEL CYCLES (BOX 2.6)

Global changes in climate over several million years are recorded in oxygen (O) isotopes preserved in the tests of fossil Foraminifera that have been buried in deep-sea sediments. Figure B2.6a shows a time series of these data for the last 4 Myr, derived from several sediment cores. The lower values in this graph reflect periods of warm global temperature when there was a greater relative abundance of the lighter O isotope, ^{16}O, in seawater and the calcium carbonate ($CaCO_3$) tests of the foraminiferans.

Conversely, higher values reflect periods of cool global temperature, when more of this O isotope was locked up in ice caps. Likewise, sea-level was higher in the warm periods and lower in the cool periods. The trend line displays the decline in global temperatures and sea-level since the Late Miocene and the onset of lower, more variable global temperature and sea-level in the Quaternary (last *ca* 2.6 Myr), especially during the last 800 kyr.

Large, regular fluctuations in sea-level have occurred over the Quaternary as the ice sheets grew and then melted, approximately every 100 kyr (Figure B2.6b). The record of sea-level change for the last glacial cycle (130 ka to present, Figure B2.6c) reveals well the second-order fluctuations in sea-level that occur within a glacial cycle. These data show that sea-level is rarely stable over geological time-scales and has seldom sat at its present high position. Note that the arrival of humans between 55 ka and 21 ka occurred when sea-level was 60 m to 125 m lower than today.

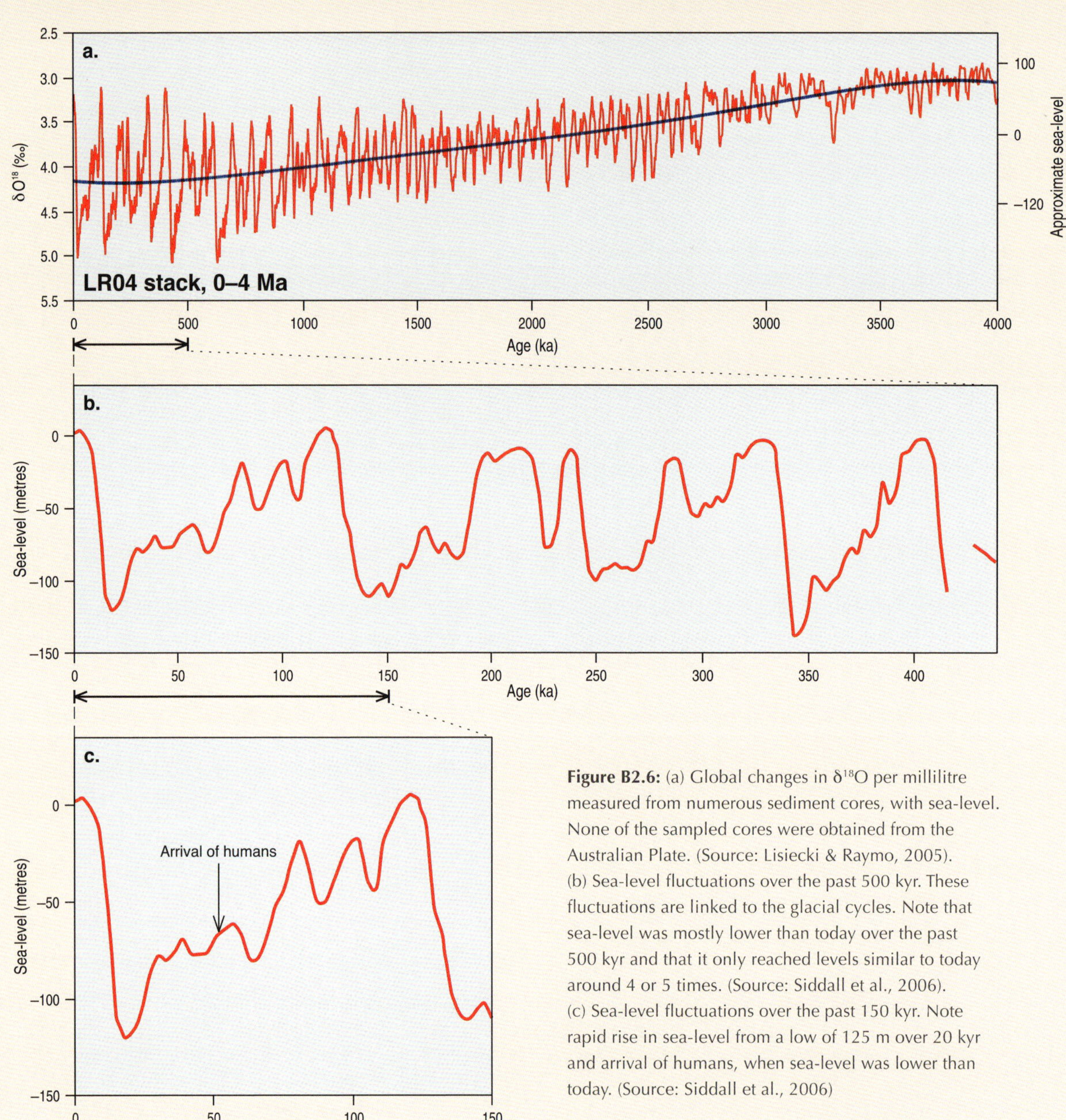

Figure B2.6: (a) Global changes in $\delta^{18}O$ per millilitre measured from numerous sediment cores, with sea-level. None of the sampled cores were obtained from the Australian Plate. (Source: Lisiecki & Raymo, 2005). (b) Sea-level fluctuations over the past 500 kyr. These fluctuations are linked to the glacial cycles. Note that sea-level was mostly lower than today over the past 500 kyr and that it only reached levels similar to today around 4 or 5 times. (Source: Siddall et al., 2006). (c) Sea-level fluctuations over the past 150 kyr. Note rapid rise in sea-level from a low of 125 m over 20 kyr and arrival of humans, when sea-level was lower than today. (Source: Siddall et al., 2006)

became increasingly important. They led to areas of deflation and the formation of vast areas of sand dunes, and the transport of aeolian silt and dust. Today, aeolian sand dunes and sand plains mantle around 40% of the Australian arid zone (Chapter 5).

The Quaternary was also the period when many valuable industrial materials formed, including the sand, gravel and gypsum from which Australian cities and buildings have been constructed. Australia's regolith is the only material available for road construction; much of it is not optimal for the purpose (Chapter 5). Salt is a curse for Australia's agricultural lands, but salt from brines in the arid-zone salt lakes is a commodity being harvested and exported or used in the domestic market. Quaternary groundwaters accumulated in vast quantities in the Great Artesian Basin (as old as 2 Ma) and other aquifers all over the continent; they continue to be exploited today (Chapter 7).

The record of drying in central Australia is found in the change from perennial to ephemeral playa lake sediments. The deposits have been dated using palaeomagnetic reversal techniques. They reveal that aridity started in Australia around *ca* 900 ka, and, by *ca* 500 ka, had extended to the edges of the continent. Dust was deposited in the arid far regions of western New South Wales before 500 ka. Around this time, the nature of the weathering changed in South Australia from oxide to calcareous (calcrete). By 350 ka, dust from the interior began to be deposited in the Tasman Sea. This dust deposition continues today on the east and southeast coasts, in New Zealand, and even further east, and provides a significant source of iron for plankton growth. Aeolian dunes started to form in the lowlands and arid-zone basin areas around 300 ka, and the most recent period of major activity was at the height of the last glacial maximum, at 25 ka to 15 ka. Longitudinal dunes are oriented across the continent in an anticlockwise whorl, largely consistent with the subtropical high-pressure atmospheric circulation. The dunes were avoided (and largely ignored) by the early explorers and settlers, as water was more easily found in gorges and rocky waterholes of the upland areas (Chapter 7).

Lake Eyre reached its maximum water levels 130–110 ka and was progressively lower after this, particularly at 95–80 ka and at 65–62 ka. The period between 50 and 44 ka was wet, with high lake levels and an expanded range of tropical

Unusual floodwater between dunes, seen from 'Big Red', Simpson Desert's biggest dune, Simpson Desert, South Australia.

forests. This was the continent that the first Australians settled. The forests declined rapidly from about 45–42 ka, accompanied by an increase in native grasses (Poaceae) and biomass burning, all of which may be linked to human arrival and disturbance (Chapter 3).

The last glacial maximum saw Lake Eyre completely dry and exposed to the wind, which scoured its base and deepened the lake floor. Transport of dust was particularly active at 22 ka and was likely active for a few thousand years before and after this event. Aeolian dust also mantled many of the hills and slopes of arid Australia, covering bare rocks and slopes and changing surface-water runoff. Towards 22 ka, continental Australia was 10°C cooler than today. It was also much drier and windier, and sea-level was around 125 m lower (Box 2.6).

The Sahul and Bass Strait land bridges connected the mainland with New Guinea and Tasmania. Perennial lakes existed in the Gulf of Carpentaria and Bass Strait. The river mouths were located at the edge of the continental shelf, which meant that the rivers were longer than today. Many of the rivers carved deep canyons into the outer continental slope, such as the Murray River (SA) and Perth (WA) (Chapter 6). The Murray Canyons are still active conduits for shelf-derived sediment transport to the abyssal plains, and the Perth Canyon is an important breeding ground for the endangered blue whale.

A spectacular 90 m high waterfall on the Yarra River crowned the entrance to what is now Port Phillip Bay near Melbourne. Aboriginal stories say it showered Point Nepean in a permanent mist. The local Boonwurrung Aboriginal people were known as the 'people-of-the-mist', and the traditional name for the Yarra was *Birrarung*, meaning 'river of mists'.

The climate warmed, sea-levels rose, and the land bridges were swamped around 6 ka. Palynology records the return of warmer and wetter climatic conditions, which saw the tropical rainforests expand between 8–6 ka and again at 3 ka (Chapter 3). Lake levels in western Victoria reached a maximum at 7.2 ka. After 5.5 ka, these lakes started to become more saline. Continental Australia still had volcanoes, the last being active around 4.6 ka in Mt Gambier.

During the glacial maximum, drought and cold, together with the lower CO_2 levels, would have had a negative impact on Australian trees and forests (Chapter 3). Grasslands and savanna became more dominant at this time. The widespread loss of trees and their associated deep roots led to rising groundwater tables, which brought salt closer to the land surface (Chapter 7).

Aboriginal and Torres Strait Islander people have occupied Australia at least since 55–45 ka. They have survived the many major climatic upheavals that have beset the country over these millennia. There are many examples of episodic occupation by Aboriginal people at specific sites. These patterns were driven by availability of water and food and sea-level. For example, at sites around Shark Bay on the mid-western Australian coast, there are records of two periods of occupation: firstly, during the late Pleistocene between 30 ka and 18 ka, and secondly, during the Holocene between about 7 ka and 6 ka. Rock shelters at the nearby Eagle Bluff and the Zuytdorp Cliffs were occupied around 4.6 ka. Other sites at Monkey Mia and Useless Loop show that the current period of occupation began about 2.3 ka. When people returned to Shark Bay in more recent times, they came primarily to hunt turtles, dugongs and land mammals.

To sum up

The geological evolution of Australia is a fascinating microcosm of the evolution of planet Earth. The fundamental features of Australia's geology, including the climate, environment, wealth and landscape, have shaped the Australian people in many ways. The first Australians—the Aboriginal and Torres Strait Islander people—whose beliefs, customs and extraordinary culture extend back millennia, have also been shaped by Australia's remarkable geology. In turn, they shaped the landscape to their needs and adapted to the remarkable changes in climate over the past 60–50 kyr. The Anthropocene is an epoch of dramatic change. How Australians living today will deal with change will, in many ways, be shaped by the geological heritage of this continent (Chapter 11).

In the next chapter, we go back to the beginning and look further at the astonishing changes in the life forms in Australia through geological time.

Image by Jim Mason

The Otway Ranges, a fine example of Cool Temperate Rainforest in southwest Victoria. Over 80% of this bioregion still has a cover of native vegetation, and 22% is in formal reserves.

Bibliography and further reading

Australia's current tectonic setting

Braun J, Dooley J, Goleby B, van der Hilst R & Klootwijk C (eds) 1998. *Structure and evolution of the Australian continent*, Geodynamics Series 26, American Geophysical Union, Washington.

DeMets C, Gordon RG & Argus DF 2010. Geologically current plate motions. *Geophysical Journal International* 181, 1–80.

Ferrett R 2005. *Australia's volcanoes*, Reed New Holland, Chatswood.

Graham I (ed) 2008. *A continent on the move: New Zealand geoscience into the 21st century*, Geological Society of New Zealand, Wellington.

Hillis RR & Reynolds SD 2003. In situ stress field of Australia. In: *Evolution and dynamics of the Australian Plate*. Hillis RR & Müller RD (eds), GSA Special Publication 22 and GSA Special paper 372, 49–60.

Quigley MC, Clark D & Sandiford M 2010. Tectonic geomorphology of Australia. In: *Australian landscapes*, Bishop P & Pillans B (eds), Geological Society, London, Special Publications 346, 243–265.

Royer J-Y & Gordon RG 1997. The motion and boundary between the Capricorn and Australian Plates. *Science* 277, 1268–1274.

Tagliabue A, Bopp L, Dutay J-C, Bowie AR, Chever F, Jean-Baptiste P, Bucciarelli E, Lannuzel D, Remenyi T, Sarthou G, Aumont O, Gehlen M & Jeandel C 2010. Hydrothermal contribution to the oceanic dissolved iron inventory. *Nature Geoscience* 3, 252–256.

Thornton W 2007. *Tnorala–baby falling*. CAAMA Collection, Ronin Films, Canberra.

Tregoning P 2002. Plate kinematics in the western Pacific derived from geodetic observations. *Journal of Geophysical Research* 107, 1–8.

Tregoning P 2003. Is the Australian Plate deforming? A space geodetic perspective. In: *Evolution and dynamics of the Australian Plate*, Hillis RR & Müller RD (eds), Geological Society of Australia Special Publication 22, 41–48.

Mapping Australia

Aitken ARA 2010. Moho geometry gravity inversion experiment (MoGGIE): a refined model of the Australian Moho, and its tectonic and isostatic implications. *Earth and Planetary Science Letters* 297, 71–83.

Australian dictionary of biography, 2006–11, National Centre of Biography, Australian National University, Canberra. http://adb.anu.edu.au

Australian Greenhouse Office 2005. *National Carbon Accounting System*, Canberra. www.greenhouse.gov.au/ncas/

Birch WD 2003. *Geology of Victoria*, Geological Society of Australia Special Publication 23.

Branagan D 1986. Strzelecki's geological map of southeastern Australia; an eclectic synthesis. *Historical Records of Australasian Science* 64, 375–393.

Branagan D 2009. The Geological Society on the other side of the world. In: *The making of the Geological Society of London*, Lewis CLE & Knell SJ (eds), Geological Society, London, Special Publication 317, 341–371.

Bureau of Mineral Resources, Geology and Geophysics 1979. *Earth science atlas of Australia*, Bureau of Mineral Resources, Geology and Geophysics, Canberra.

Burrett CF & Martin EL 1989. *Geology and mineral resources of Tasmania*, Geological Society of Australia Special Publication 15.

Day RW, Whitaker WG, Murray CG, Wilson IH & Grimes KG 1983. *Queensland geology: a companion volume to the 1:2,500,000 scale geological map (1975)*, Geological Survey of Queensland Publication 383.

de Caritat P & Cooper M 2011. *National geochemical survey of Australia: the geochemical atlas of Australia*, Geoscience Australia Record 2011/20, Geoscience Australia, Canberra.

Doutch HF 1979. Major structural elements of Australia. In: *Earth science atlas of Australia*, Bureau of Mineral Resources, Geology and Geophysics, Canberra.

Drexel JF, Preiss WV & Parker AJ 1993. *The geology of South Australia*, Geological Survey of South Australia, Bulletin 54.

Earth Sciences History Group 2010. *Thematic issue: snapshots of the geological program covering the whole of Australia*, Geological Society of Australia ESHG Newsletter 41.

Edgeworth David TE 1911. *Notes on some of the Chief Tectonic Lines of Australia* [part of a Presidential Address]. Journal and Proceedings of the Royal Society of New South Wales: 45, 4–60.

Finlayson DM 2008. *A geological guide to Canberra region and Namadgi National Park*, Geological Society of Australia (ACT Division), Canberra.

Geological Society of Australia 1971. *Tectonic map of Australia and New Guinea, 1: 5M*, Geological Society of Australia, Sydney.

Geological Survey of Western Australia 1990. *Geology and mineral resources of Western Australia*, Memoir 3, Geological Survey of Western Australia, Perth.

Geoscience Australia 2001. *Palaeogeographic atlas of Australia*. www.ga.gov.au/meta/ANZCW0703003727.html

Gunn P (ed) 1997. *Thematic issue: airborne magnetic and radiometric surveys*, AGSO Journal of Australian Geology and Geophysics 17.

Harris PT & Whiteway T 2011. Global distribution of large submarine canyons: geomorphic differences between active and passive continental margins. *Marine Geology* 285, 69–86.

Hoatson DM, Claoué-Long JC & Jaireth S 2008. Guide to using the 1:5 000 000 scale map of Australian Proterozoic mafic–ultramafic magmatic events, Geoscience Australia Record 2008/15, Geoscience Australia, Canberra.

Johnson DP 2009. *The geology of Australia*, 2nd edn, Cambridge University Press, Port Melbourne.

Lymburner L, Tan P, Mueller N, Thackway R, Lewis A, Thankappan M, Randall L, Islam A & Senarath U 2011. *250 metre dynamic land cover dataset of Australia*, version 1, Geoscience Australia, Canberra.

McKenzie N, Jacquier D, Isbell R & Brown K (ed) 2004. *Australian soils and landscapes: an illustrated compendium*, CSIRO Publishing, Collingwood.

National Library of Australia 2007. *Australia in maps: great maps in Australia's history from the National Library's collection*, National Library of Australia, Canberra.

Plumb KA 1979. The tectonic evolution of Australia. *Earth Science Reviews* 49, 205–249.

Ross S 2010. A chronicle of change: AusGeo News celebrates 20 years. *AusGeo News*, December 2010, 100. www.ga.gov.au/ausgeonews/ausgeonews201012/chronicle.jsp

Scheibner E 1996, 1998. *Geology of New South Wales—synthesis*, Geological Survey of New South Wales Memoir 13, Geological Survey of New South Wales, Maitland.

Shaw RD, Wellman P, Gunn P, Whitaker AJ, Tarlowski C & Morse M 1995. *Australian crustal elements (1:5,000,000 scale map) based on the distribution of geophysical domains (v 1.0)*, Australian Geological Survey Organisation, Canberra.

Australian lithosphere

Clitheroe G, Gudmundsson O & Kennett BLN 2000. The crustal thickness of Australia. *Journal of Geophysical Research* 105, 13697–13713.

Kennett BLN, Salmon M, Saygin E & AusMoho Working Group 2011. AusMoho: the variation of Moho depth in Australia. *Geophysical Journal International* 187, 946–958.

Hillis RR & Müller RD (eds) 2003. *Evolution and dynamics of the Australian Plate*, Geological Society of Australia Special Publication 22.

Australia through time

Australian Journal of Earth Sciences 2008. *Thematic issue: geochronology of Australia*, Australian Journal of Earth Sciences 55(6/7).

Betts PG & Giles D 2006. The 1800–1100 Ma tectonic evolution of Australia. *Precambrian Research* 144, 92–125.

Betts PG, Giles D, Schaefer BF & Mark G 2007. 1600–1500 Ma hotspot track in eastern Australia: implications for Mesoproterozoic continental reconstructions. *Terra Nova* 19, 496–501.

Bishop P & Pillans B (eds) 2010. *Australian landscapes*, Geological Society of London Special Publication 346.

Cawood PA 2005. Terra Australis Orogen: Rodinia breakup and development of the Pacific and Iapetus margins of Gondwana during the Neoproterozoic and Paleozoic. *Earth-Science Reviews* 69, 249–279.

Cawood PA & Korsch RJ 2008. Assembling Australia: Proterozoic building of a continent. *Precambrian Research* 166, 1–396.

Champion DC, Kositcin N, Huston DL, Mathews E & Brown C 2009. *Geodynamic synthesis of the Phanerozoic of Eastern Australia and implications for metallogeny*, Geoscience Australia Record 2009/18, Geoscience Australia, Canberra.

Fujioka T & Chappell J 2010. History of Australian aridity: chronology in the evolution of arid landscapes. In: *Australian landscapes*, Bishop P & Pillans B (eds), Geological Society, London, Special Publication 346, 121–139.

Giles D, Betts PG & Lister GS 2004. 1.8–1.5-Ga links between the North and South Australian Cratons and the Early–Middle Proterozoic configuration of Australia. *Tectonophysics* 380, 27–41.

Gray DR & Foster DA 2004. Tectonic review of the Lachlan Orogen: historical review, data synthesis and modern perspectives. *Australian Journal of Earth Sciences* 51, 773–817.

Great Barrier Reef Marine Park Authority 2007. *Great Barrier Reef Climate Change Action Plan 2007–2011*. www.gbrmpa.gov.au/corp_site/key_issues/climate_change/management_responses

Hawkesworth CJ, Dhuime B, Pietranik AB, Cawood PA, Kemp AIS & Storey CD 2010. The generation and evolution of the continental crust. *Journal of the Geological Society* 167, 229–248.

Hoffman P 1991. Did the breakout of Laurentia turn Gondwanaland inside-out? *Science* 252, 1409–1412.

Huston DL, Blewett RS & Champion DC 2012. Australia through time: a summary of its tectonic and metallogenic evolution. *Episodes*: in press.

International Commission on Stratigraphy 2009. *International stratigraphic chart*. www.stratigraphy.org/upload/ISChart2009.pdf

Korsch RJ, Kositcin N & Champion DC 2011. Australian island arcs through time: geodynamic implications for the Archean and Proterozoic. *Gondwana Research* 19, 716–734.

Laurie J (ed) 2009. *Geological timewalk*, Geoscience Australia, Canberra.

Le Heron DP 2011. Sea ice-free conditions during the Sturtian glaciation (early Cryogenian), South Australia. *Geology* 39, 31–34.

Lisiecki LE & Raymo ME 2005. A Pliocene–Pleistocene stack of 57 globally distributed benthic $\delta^{18}O$ records. *Paleoceanography* 20, PA1003.

López-Gamundí OR 2010. *Late Paleozoic glacial events and postglacial transgressions in Gondwana*, Geological Society of America Special Papers 468.

Moss PT & Kershaw AP 2000. The last glacial cycle from the humid tropics of northeastern Australia: comparison of a terrestrial and a marine record. *Palaeogeography, Palaeoclimatology, Palaeoecology* 155, 155–176.

Myers JS, Shaw RD & Tyler IM 1996. Tectonic evolution of Proterozoic Australia. *Tectonics* 15, 1431–1446.

Neumann NL & Fraser GL (eds) 2007. *Geochronological synthesis and time-space plots for Proterozoic Australia*, Geoscience Australia Record 2007/06, Geoscience Australia, Canberra.

Pisarevsky SA, Wingate MTD, Powell CM, Johnson S & Evans DAD 2003. Models of Rodinia assembly and fragmentation. In: *Proterozoic East Gondwana: supercontinent assembly and breakup*, Yoshida M, Windley BE & Dasgupta S (eds), Geological Society, London, Special Publication 206, 35–55.

Siddall M, Chappell J & Potter E-K 2006. Eustatic sea level during past interglacials. In: *The climate of past interglacials*, Sirocko F, Litt T, Claussen M & Sanchez-Goni M-F (eds), Elsevier, Amsterdam, 74–92.

Squire RJ, Campbell IH, Allen CM & Wilson CJL 2006. Did the Transgondwanan supermountain trigger the explosive radiation of animals on Earth? *Earth and Planetary Science Letters* 250, 116–133.

van Ufford AQ & Cloos M 2005. Cenozoic tectonics of New Guinea. *AAPG Bulletin 89*, 119–140.

Veevers JJ (ed) 2000. *Billion-year earth history of Australia and neighbours in Gondwanaland*, GEMOC Press, Sydney.

Veevers JJ 2006. Updated Gondwana (Permian–Cretaceous) earth history of Australia. *Gondwana Research* 9, 231–260.

Wingate MTD, Pisarevsky SA & Evans DAD 2002. A revised Rodinia supercontinent: no SWEAT, no AUSWUS. *Terra Nova* 14, 121–128.

3

Living Australia

Although the major features of the evolution of life have shaped the flora and fauna on whichever continent one explores, each has its own peculiarities. These are largely because of differences in each continent's tectonic and climatic history, coupled with the timing of evolutionary events. So, in its own way, each continent is unique. But of all the continents, Australia is special, at least in part because of its recent relative isolation from other large continental masses—an isolation that is only now coming to an end. The changing biosphere has also had a profound effect on the atmosphere, hydrosphere and lithosphere.

John Laurie,[1] Brian Choo,[2] Stephen McLoughlin,[3] Suzanne Hand,[4] Peter Kershaw,[5] Glenn Brock,[6] Elizabeth Truswell,[7] Walter Boles[8] and John Long[9]

[1]Geoscience Australia, Canberra; [2]Institute of Vertebrate Paleontology and Paleoanthropology, Beijing; [3]Swedish Museum of Natural History, Stockholm; [4]University of New South Wales, Sydney; [5]Monash University, Melbourne; [6]Macquarie University, Sydney; [7]Australian National University, Canberra; [8]Australian Museum, Sydney; [9]Natural History Museum of Los Angeles County, Los Angeles

Image by Adrian Yee

Figure 3.1: Map of Australia, showing the localities mentioned in the text.

Life in Australia

Life has had a profound effect on our planet and on Australia. It has changed the composition of the atmosphere, of the oceans, and of the sediments laid down on the deep ocean floor, on continental shelves, in lakes and on floodplains. In so doing, it has dramatically altered the historical trajectory of this continent by giving rise to three of our most important and valuable mineral resources (coal, gas and iron ore), which have contributed enormously to the economic wellbeing of this nation over the past century or so.

In turn, the geological history of the continent has had an effect on the life that inhabits it. We have a Gondwanan legacy, but the partial isolation of the continent from others over most of the Cenozoic has allowed the evolution of a flora and fauna that are largely unique. The idea of a long-sustained isolation of an Australian landmass is coming under increasing challenge from molecular phylogenetic studies. These, combined with an increased understanding of the geology of the oceans surrounding Australia and the potential of small landmasses to offer stepping stones for biotic interchange, suggest that the isolation of Australia was less complete than once believed. However, even if limited, this isolation also delayed the arrival of humans until much later than for the islands to our northwest. The drift of the continent northwards towards the tropics, coupled with the development of the Pleistocene ice age, led to the continent drying out, so that it is now by far the driest inhabited continent (Chapter 5). Adaptation to this aridity is a distinctive feature of the flora and fauna, and may have in part mandated a nomadic existence for the first human inhabitants of the continent; however, in the temperate southeast, there is some evidence for seasonal settlements.

The beginning

The origin of life on the planet is shrouded in mystery and lacking in fossils. However, there is considerable evidence from meteorites that organic molecules may be quite common throughout the solar system. In September 1969, a rare form of meteorite, a carbonaceous chondrite, landed near the town of Murchison in Victoria. This meteorite was found to contain many organic molecules, including amino acids, which are the building blocks of protein, of which most organisms are largely made. Chondrites are considered to represent material from the asteroid belt, and those that have been dated radiometrically are about 4550 Ma.

Although fossils are rare in very old rocks, Australia has large areas of relatively undeformed crust. It is on one of these (the Pilbara Craton) that the oldest known possible fossils are found. These are stromatolites from the Strelley Pool Formation, near Marble Bar, Western Australia (Figures 3.1 and 3.2) and are approximately 3500 Ma in age. However, it is possible, but unlikely, that they are simply chemical precipitates. Younger (3430 Ma) examples from the same area are more convincingly biogenic (Figure 3.2). The oldest stromatolites of confirmed microbial origin are from the Tumbiana Formation, also in the Pilbara, which are 2700 Ma in age.

Stromatolites are layered structures formed in shallow water by the trapping and cementation of sedimentary grains by films of microorganisms, especially mats of photosynthetic cyanobacteria. Stromatolites may be conical, stratiform, branching, domal or columnar in shape. They are a major constituent of the fossil record for the Late Archean and most of the Proterozoic eons. They peaked in abundance about 1200 Ma, and then declined in importance and diversity. The most probable explanation for their decline is decimation by grazing organisms, implying that reasonably complex multicellular animals existed by about 1000 Ma. The best modern examples of stromatolites can be found in intertidal environments of Hamelin Pool, a hypersaline arm of Shark Bay, Western Australia (Figure 3.2).

These Pilbara stromatolite fossils set a minimum age limit for the origin of life on Earth. Some chronological analyses of divergence in DNA and protein sequences (yielding a 'molecular clock') suggest that life arose about 4200 Ma. However, such clocks need to be calibrated against the fossil record and, where the fossil record is sparse, as in the Archean, calibration is difficult at best. It is most likely that life had appeared on Earth at least by the Paleoarchean (3600 Ma).

Some hydrocarbon molecules are restricted in their modern occurrence to particular types of organisms, and these molecules ('biomarkers') can be found in trace amounts in some rocks. Application of this method indicates that all three domains of life (Bacteria, Archaea and Eukarya) were in existence by the Neoarchean (2800 Ma) or Paleoproterozoic (2500 Ma).

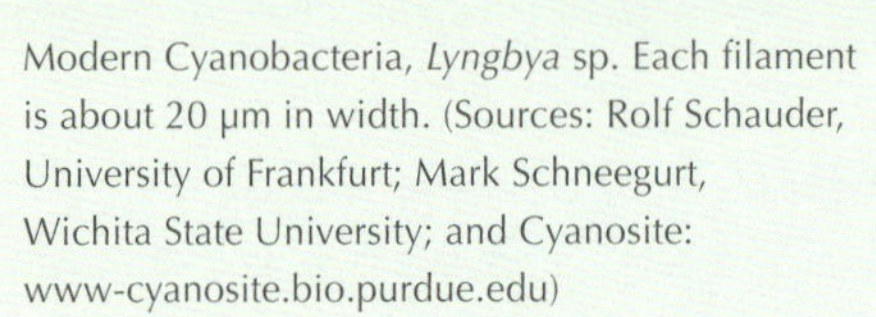

Modern Cyanobacteria, *Lyngbya* sp. Each filament is about 20 µm in width. (Sources: Rolf Schauder, University of Frankfurt; Mark Schneegurt, Wichita State University; and Cyanosite: www-cyanosite.bio.purdue.edu)

(Top image) © Getty Images [P Siver]

Figure 3.2: Ancient and modern stromatolites. (a) Intertidal stromatolites in Hamelin Pool, Shark Bay, Western Australia. (b) Underwater view of stromatolites with jellyfish, Shark Bay, Western Australia, a scene probably similar to that of 600 Ma. (Image courtesy of Martin van Kranendonk, Geological Survey of Western Australia). (c) *Baicalia burra*, probably from the Kanpa Formation, near Lake Throssell, Western Australia. (Source: Kath Grey, Geological Survey of Western Australia) (d) Small, conical, probable stromatolites (*ca* 3500 Ma), Strelley Pool Formation, near Marble Bar, Western Australia. (Image courtesy of Martin van Kranendonk, Geological Survey of Western Australia).

Oxygenic photosynthesis commenced with the evolution of the precursors of modern cyanobacteria, perhaps during the Mesoarchean or Neoarchean (*ca* 2800 Ma). This process converts carbon dioxide into organic compounds (mostly sugars) using sunlight as an energy source, and releases oxygen as a waste product. Photosynthesis changed the atmosphere forever.

Before oxygenic photosynthesis evolved, there was very little or no free oxygen in the atmosphere or the ocean. However, the oxygen initially produced in the ocean by these photosynthesisers would have been removed almost immediately by the oxidising of reduced compounds, especially those of iron and sulfur dissolved in the ocean. This oxidation (the 'mass rusting') led to the deposition of huge thicknesses of iron-rich rocks (banded iron-formations) in sedimentary basins throughout the world. The wide continental shelves that developed on the Pilbara Craton (Figure 3.1) in the Neoarchean (2800–2500 Ma) provided sites for deposition of the enormous volumes of iron ore in the overlying Hamersley Basin (Chapter 9).

By about the end of the Archean, free oxygen began to accumulate in the atmosphere and ocean. This event, called the Great Oxygenation Event (Figure 3.3), would have restricted anaerobic microbes to isolated environments away from the deadly oxygen. The Great Oxygenation Event changed the prospects for life profoundly, because aerobic metabolism is considerably more efficient than anaerobic pathways. The presence of oxygen created new possibilities for life to explore. An atmospheric oxygen concentration of about 2% is required for the formation of compounds such as collagens, which are employed by all multicellular animals to provide structure to their cellular makeup and connective tissue to complex organ systems. Without free oxygen in the atmosphere, complex animal life would have been impossible.

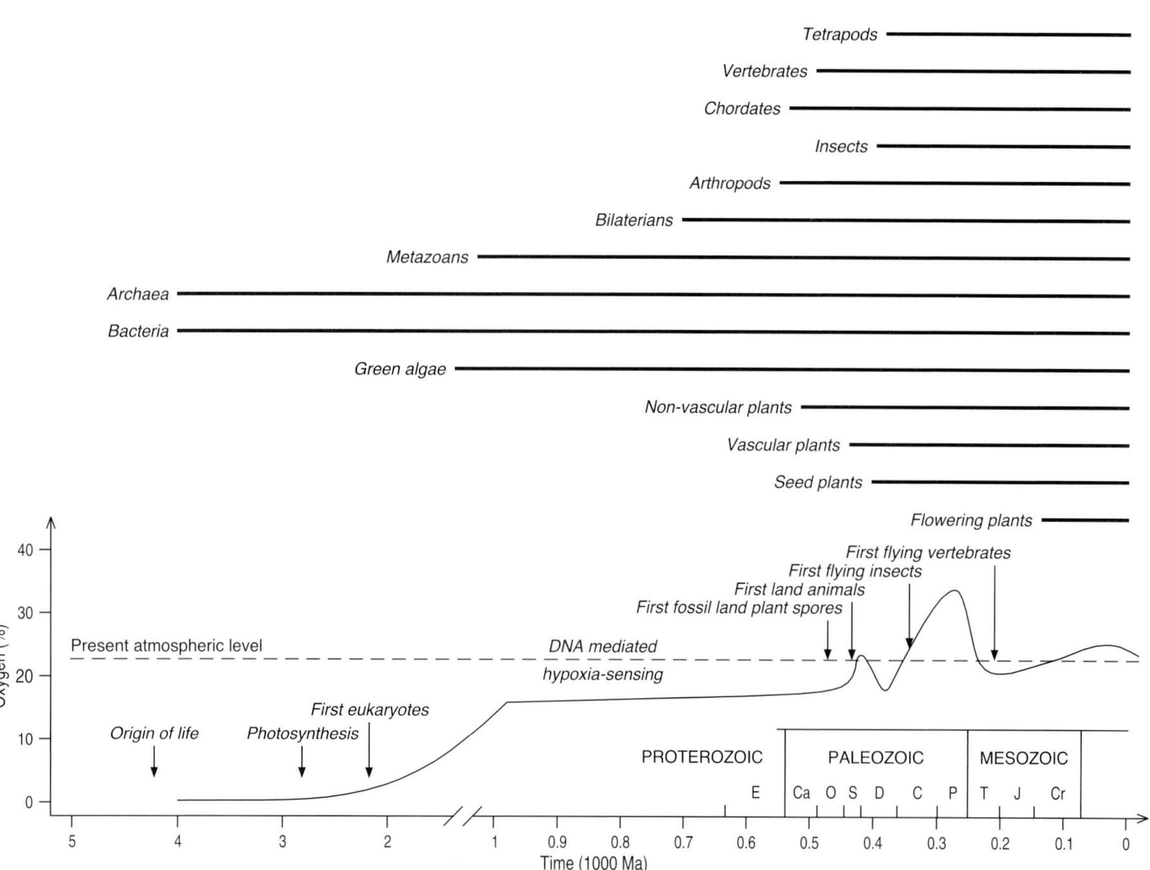

Figure 3.3: Relationship between geological time, atmospheric oxygen concentration and appearance of some life forms. (Source: modified from Flück et al., 2007)

Complexity and cooperation

Eukaryotes (ancestors of 'protists'—plants, animals and fungi) differ from Bacteria and Archaea in that their cells have a membrane-bound nucleus, among other distinctive characteristics. Most also contain other organelles such as mitochondria and chloroplasts. The rise of the eukaryotes allowed the development of predation, adding a level of complexity to ecosystems that was not previously possible. It also allowed the development of sexual reproduction, which increased the variation upon which natural selection could operate. Sexual reproduction is also more efficient at removing deleterious mutations from the population and provides a mechanism for repair of damaged DNA.

It is unclear when the first eukaryotes appeared, but the oldest fossils suspected to be eukaryotes are the probable multicellular algae (*Grypania*) which come from rocks about 2100 Ma. There is some debate about these fossils because they show very little structure, leading some researchers to consider that they are simply bacterial colonies. However, their size and regular spiral shape suggest

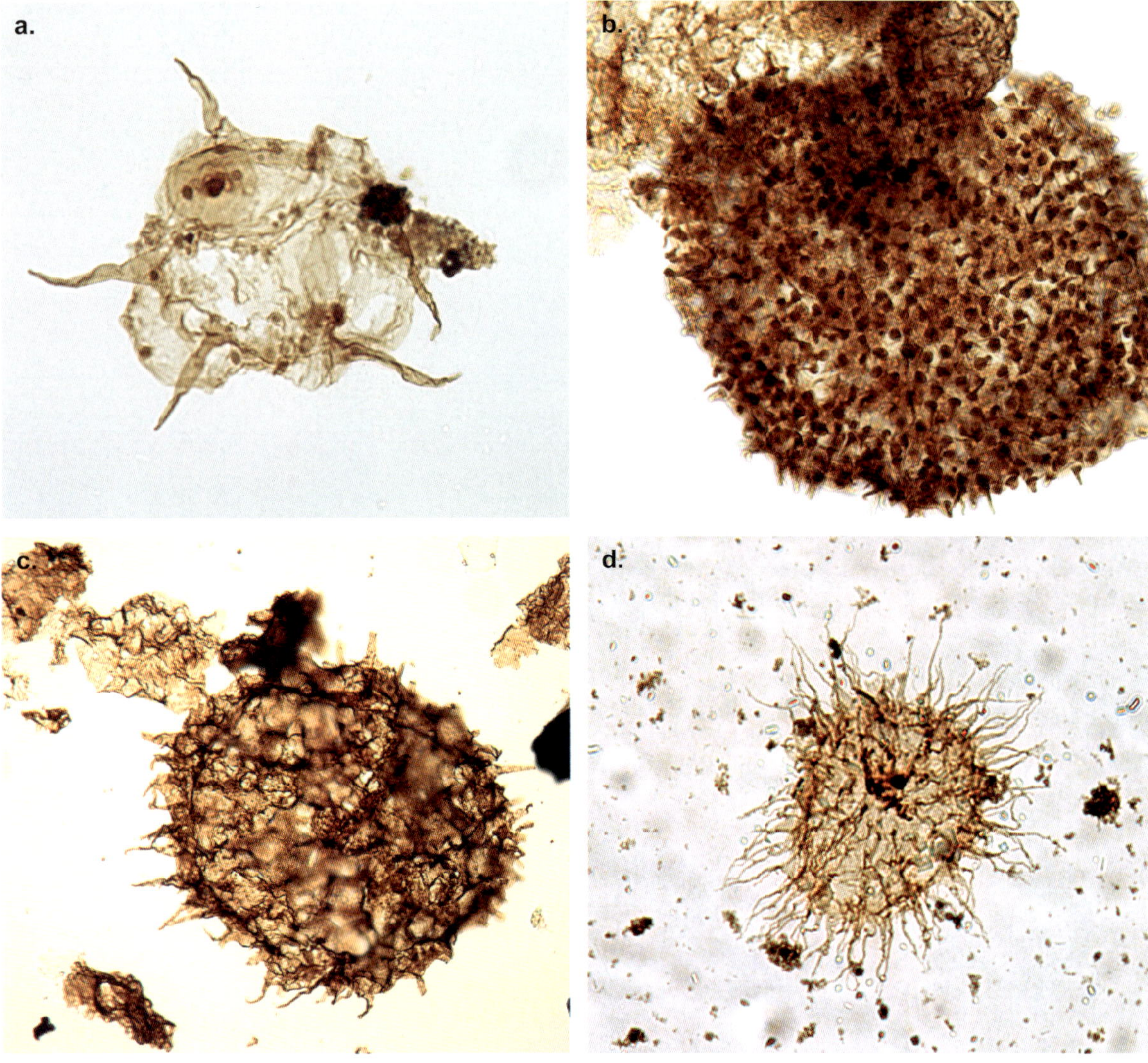

Figure 3.4: Ediacaran acritarchs. (a) *Tanarium paucispinosum*, Wilari Dolomite, Observatory Hill 1 well, Officer Basin, South Australia. (b) *Taedigerasphaera lappacea*, lower Ungoolya Group, Observatory Hill 1 well, Officer Basin, South Australia. (c) *Gyalosphaeridium multispinulosum*, Munta 1 well, Officer Basin, South Australia. (d) *Tanarium pycnacanthum*, Tanana Formation, Munta 1 well, Officer Basin, South Australia. (Source: Kath Grey, Geological Survey of Western Australia)

that they are more likely to be eukaryotes. Recently discovered organic-walled microfossils in 3200 Ma rocks in South Africa have been interpreted as either Eukaryota or colonial Cyanobacteria. Large and complex microfossils from 3000 Ma and possibly even 3400 Ma rocks from the Pilbara Craton in Western Australia might also be eukaryotes. If they are eukaryotes, then surprisingly, they predate the Great Oxygenation Event by hundreds of millions of years and are much older than the oldest certain eukaryotes. The latter are acritarchs (Figure 3.4) from rocks of about 1800 Ma in China. Acritarchs are small (10–250 μm), organic-walled, acid-insoluble microfossils of uncertain origin and are particularly well represented in the Ediacaran of Australia (Figure 3.4). However, most of them probably represent the resting cysts of some types of marine algae. They are known in the fossil record from the Mesoarchean and are most abundant in the Proterozoic and Paleozoic. They suffered a significant extinction during the Neoproterozoic glaciations, but quickly diversified thereafter.

Multicellularity arose independently in at least 17 groups of eukaryotes, and in some cases numerous times within the same group. It also arose in several groups of Bacteria, including Cyanobacteria, as well as in some Archaea. Even though most of the organisms on Earth are still unicellular, if all of the multicellular eukaryotes suddenly disappeared, to the naked eye the planet would look almost as barren as Mars.

It is most likely that multicellularity began simply as a symbiotic association of several cells of the same species; such associations have been seen in many modern organisms. Multicellularity allows organisms to increase in size beyond the limits imposed by a single cell; this gives them a greater ability to survive predation, as almost all single-celled predators consume by engulfing, and it is difficult to engulf something much larger than oneself. In addition, multicellularity gives organisms a greater ability to survive environmental change by producing a regulated internal environment; it also allows them to have increased mobility and permits the groups of cells to communicate.

Multicellularity allows individual cells in the organism to specialise, thereby becoming more efficient at the tasks required for life. It is likely that cell differentiation appeared fairly early in the evolution of multicellularity, because some of the simplest modern multicellular organisms show such differentiation, usually into somatic and germ (reproductive) cells. Multicellularity also facilitates sexual reproduction, and it allowed the evolution of development from the fertilised egg to the mature individual. In this way, multicellularity permitted a huge increase in the possible complexity of individual organisms.

Working out when the first multicellular organisms evolved is very difficult, because it is likely that most would have left no fossil record. As noted above, multicellularity arose in numerous lineages at many different times until quite recently. An example of the most recent such event is the evolution of multicellularity in some freshwater algae, which probably happened only in the last 50 Myr.

The rise of animals

Not all animals are multicellular, but all multicellular animals are referred to the Metazoa, and it seems that true multicellularity evolved only once in the animals. The oldest convincing metazoan fossils are found in rocks from the Ediacaran, although there are some possible examples from the Cryogenian (*ca* 750 Ma). These organisms subsequently diversified, and their fossils are found in many places around the world in rocks from the latter half of the Ediacaran (Figure 3.5).

The actual timing of the evolution of metazoans is problematic, with body fossils possibly indicating a middle Cryogenian age (*ca* 750 Ma), and some trace fossils in rocks from the Stirling Ranges, Western Australia (Figure 3.1), dated at more than 1200 Ma—the latter being consistent with molecular sequence divergence dates of Ectasian to Stenian (*ca* 1200 Ma). It is possible that the

Figure 3.5: Ediacaran fossils from the Flinders Ranges, South Australia. (a) *Spriggina floundersi*. (b) *Dickinsonia costata*. (c) *Parvancorina minchami*. (Source: J Gehling, South Australian Museum)

first metazoans were sponge-like organisms and that they evolved from the free-living or colonial single-celled choanoflagellates, as these organisms bear a striking similarity to the choanocyte cells of sponges. In sponges, choanocytes are the cells that cooperatively move their flagella to create the feeding and respiratory currents that flow through the sponge's pores. However, recent molecular data indicate that the Placozoa, a small metazoan group having a very simple body plan, may have preceded the sponges.

Late in the Proterozoic (Cryogenian), two major glaciations overtook Earth. These are named the Sturtian and Marinoan glaciations, after rock successions in the Flinders Ranges in South Australia (Figure 3.1) where the glacial sediments are well preserved. Their cause is unknown, but the excessive burial of organic matter because of the particular tectonic conditions of the time is one possibility. Some researchers have suggested that these glacial events may have been so extensive that ice covered the entire planet (the Snowball Earth hypothesis); this would have had a devastating effect on the biota. However, although there was a significant decrease in diversity of acritarchs, there is little evidence that the biota suffered to any great degree, so the more extreme interpretations of the extent of these glaciations must be considered unlikely. They were, however, quite severe. This is indicated by the fact that the extensive glacial deposits in the Flinders Ranges were, at the time of their deposition, less than 10 degrees from the equator.

While their relationships are still debated, the first widespread fossils of multicellular animals are found late in the Ediacaran (*ca* 600 Ma). These fossils, first found in large numbers in the Flinders Ranges (Figure 3.1), are an enigmatic group of organisms of discoidal, tubular, frond-like or bag-like forms (Figure 3.5). Up to 100 genera have been found in rocks of this age on most continents, and they range in size from a few millimetres to a couple of metres in length. It has been suggested that some of these organisms represent jellyfish, sponges, worms, early arthropods, sea pens and even chordates. However, although it is clear that there are sponges, jellyfish and possibly annelid worms in the fauna, the relationship of some of the other species to modern phyla has been shown to be unlikely. At best, their relationships to modern animals are obscure, and it has even been proposed that they represent a 'failed experiment' in multicellular life, very distantly related to modern phyla and now completely extinct.

The Cambrian explosion

The Cambrian explosion is perhaps the single most dramatic event documented in the fossil record. Milestone evolutionary innovations associated with this bioevent include the evolution of most new (mainly bilaterian) animal body plans (recognised today as distinct phyla), the development of true eyes, the advent of biomineralised hard parts, increased ecological complexity and the development of the first animal reef builders. Significantly, these major evolutionary changes are broadly coincident with perhaps the most extreme shifts in ocean chemistry recorded during geological time.

The Cambrian explosion thus heralds more than just the rise of skeletons—it heralds the emergence and rapid diversification of bilaterian animals, which today make up more than 99% of all living animals. Although the relationships of living animal clades can be reconciled using molecular phylogenetics, and the timing of important lineage splits can be estimated using increasingly accurate molecular clocks, the fact remains that the fossil record provides the most direct means of accessing important macroevolutionary data from the archive of the Cambrian explosion. This includes detailed investigation of both exceptionally preserved soft-bodied assemblages, as well as the more commonly preserved shelly faunas.

The lower Cambrian (542–510 Ma) succession of sedimentary rocks exposed in the Flinders Ranges of South Australia represents perhaps the most complete time capsule of Earth's biosphere during the momentous events associated with the Cambrian explosion. These rocks have yielded abundant, diverse and superbly preserved lower Cambrian marine fossils, indicating that tropically situated South Australia was a global hotspot of biotic diversity during the Cambrian explosion (Box 3.1).

By the middle Cambrian (510–501 Ma), the ocean teemed with animal life, and many modern phyla were already in existence. The trigger for the Cambrian explosion is still obscure, and probably multifaceted; increased atmospheric oxygen, development of new genetic and developmental pathways and increased ecological complexity all probably played a role. The acquisition of numerous, diverse shells across a wide spectrum of largely unrelated animal groups was seemingly protective in nature, and probably in response to escalating macroscopic predation.

THE CAMBRIAN EXPLOSION (BOX 3.1)

By the middle 1800s, geologists in England realised that there appeared to be an abrupt appearance of fossils in rocks of early Cambrian age. At the time, there were few records of possible fossils in older rocks, and the appearance of the Cambrian fauna with no apparent antecedents led Charles Darwin to realise that this could be used as an objection to his theory of evolution by natural selection. Since that time, many macroscopic fossils (mostly sponge and cnidarian grade) have been found in pre-Cambrian rocks. The early Cambrian saw the emergence and rapid diversification of more complex bilaterian grade animals that ultimately evolved into the diagnostic anatomical designs recognised today as major phyla. This Cambrian explosion of life also coincides with the development of the first animals with eyes, rapid movement, efficient burrowing and predatory life habits—in essence the development of complex ecological systems. Many easily recognised forms are represented, including brachiopods, trilobites and other arthropods, molluscs, priapulids, nematomorphs, lobopodians, echinoderms and chordates. However, the earliest skeletal remains in the early Cambrian are often represented by tiny mineralised plates, tubes and spines that are simply termed 'small shelly fossils'. Some of these represent the shells of complete miniature animals such as molluscs, but many others are elements from composite exoskeletons that were prone to disarticulation after death. The superbly preserved fossils from the lower Cambrian rocks of South Australia are allowing detailed study and clarification of the relationships of some of these odd little fossils, and some have proved to be early members of modern phyla.

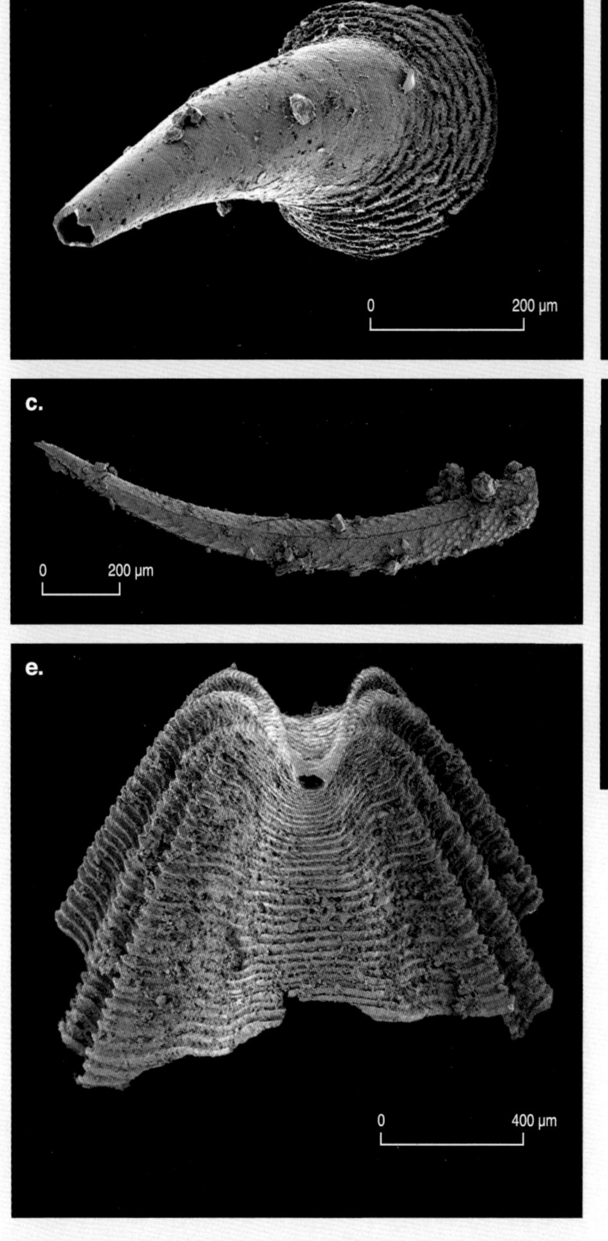

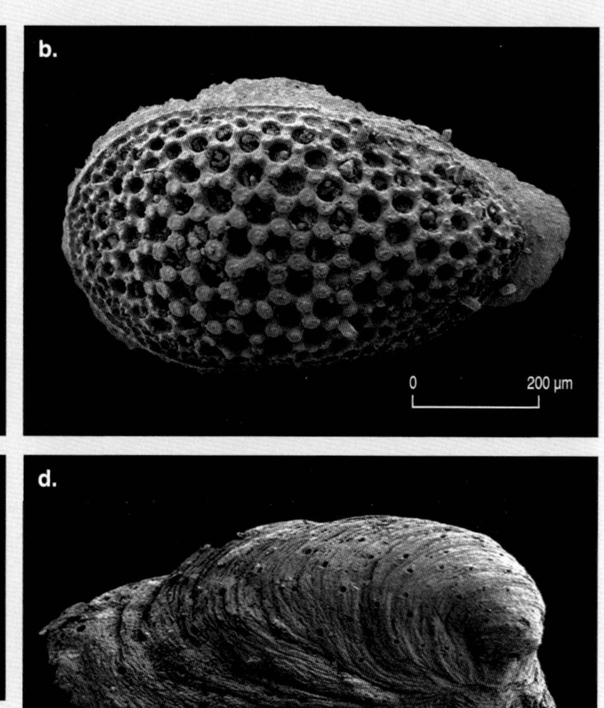

Figure B3.1: Scanning electron micrographs of some small shelly fossils from the lower Cambrian of South Australia. (a) *Stoibostrombus crenulatus*. (b) Lobopodian plate referred to *Microdictyon* sp. (c) Spine of the bradoriid arthropod *Mongolitubulus* sp. (d) The tommotiid *Micrina etheridgei*. (e) The camenellan tommotiid *Dailyatia ajax*.

Figure 3.6: Fossils from the Emu Bay Shale, Kangaroo Island, South Australia. (a) Bivalved arthropod *Tuzoia australis*. (b) Grasping appendage of arthropod predator *Anomalocaris briggsi*. (c) Possible nematomorph *Palaeoscolex* sp. overlying two specimens of the small trilobite *Estaingia bilobata*. (d) Large trilobite *Redlichia takooensis*. (e) Arthropod *Emucaris fava*. (Source: John Paterson, University of New England)

One of the peculiarities of the Cambrian fossil record is the unusually large number of sites preserving soft tissues of organisms. The most famous is the legendary Burgess Shale in Canada, but there are similar localities in various parts of the world, including China (Chengjiang), Greenland (Sirius Passet) and Kangaroo Island in South Australia (Figure 3.1). This last occurrence has been known for many years, but it has only recently been the subject of intensive excavation and study. The fauna includes several species of trilobite, numerous soft-shelled arthropods (some with eyes and antennae preserved), sponges, assorted 'worms', rare brachiopods and molluscs, and other strange, problematic animals (Figure 3.6).

The extraordinary biological events associated with the Cambrian explosion were largely played out in the oceans—there is relatively little evidence of complex life occupying the exposed Cambrian continents. However, the events that eventually changed the face of the Australian continent started some time before the Cambrian, with the first venture onto land by cyanobacteria, algae, fungi and lichen.

Invading the land

There is some evidence for cyanobacteria-dominated mats developing on soil surfaces as long ago as the Neoarchean (*ca* 2650 Ma), and these would almost certainly have been limited to areas adjacent to bodies of water. It is possible that cyanobacterial mats occurred in drier conditions, as desert crusts do today all over central Australia, but there is little evidence to support this because traces of these environments are rarely preserved in the geological record. The cyanobacteria in these mats were probably the only terrestrial organisms during the later Archean and most of the Proterozoic. However, carbon-isotopic studies indicate that, by the Cryogenian (*ca* 850 Ma), there was an extensive photosynthetic land flora. This may have consisted of lichens (symbiotic associations of algae and fungi) and free algae. It is also possible that free-living fungi had begun to inhabit the land at this time.

In the early Middle Ordovician (*ca* 470 Ma), the first land plant spores appear, indicating that the land was at least partly covered by liverwort-grade non-vascular plants. In Australia, the earliest known occurrence of land plant spores comes from the Caribuddy Formation in the Canning Basin of Western Australia. There, trilete spores held in tetrads are typical of those from a range of sites in the Northern Hemisphere and indicate a Late Ordovician to earliest Silurian age. Recently, fossil liverworts have been reported from the Lower Ordovician in Argentina. These early spore-producing plants were probably rapid colonisers, with some tolerance to drying out, but were confined to habitats that were periodically damp. This covering of the land by non-vascular plants persisted until the Late Ordovician (*ca* 455 Ma), when spores of the type characteristic of early vascular plants first appeared. However, it is not until much later, in the middle Silurian (*ca* 425 Ma), that the first macrofossils of vascular plants are found. By the end of the Silurian, vascular plants were quite diverse and were many centimetres in height. The clubmoss *Baragwanathia longifolia* from the late Silurian of Yea, Victoria, is a prime example (Figure 3.7). Individual stems of *Baragwanathia* may reach 5 cm in diameter and 2 m in length. The stems are clothed in loose spirals of simple spine-like leaves. They probably inhabited delta wetlands where they were prone to being washed out to sea during flood events and preserved alongside marine organisms.

It is likely that some arthropods and molluscs from the Early Ordovician (*ca* 480 Ma) occasionally ventured onto land, but they spent most of their time in the water. There are, however, fossil burrows of Late Ordovician age, which have been interpreted as those of millipedes.

The presence of coprolite-like masses of cryptospores in Ordovician terrestrial sediments from Scandinavia (then also in equatorial latitudes) also suggests that at least some animals were exploiting the new liverwort-grade vegetation as a food source. However, it was only in the middle Silurian that the first arthropods with the ability to breathe air appear in the fossil record. These include herbivorous millipedes, and predatory centipedes and trigonotarbids (extinct spider-like animals) that indicate that a multitiered food web was already in existence.

Late in the Ordovician, Gondwana drifted over the South Pole. It might have been this that triggered the Andean–Saharan ice age, which lasted into the early Silurian. This decrease in global temperature has been suggested as a cause of the end-Ordovician mass extinction, which is one of the 'big five' extinction events (Figure 3.8). This extinction saw about half of all genera become extinct.

The rise of vertebrates

The oldest fossil fish-like animals are small, finned creatures from the early Cambrian of China. In Australia, shark-like scales indicate that fishes having true bony structures had appeared by the late Cambrian, these being the oldest record of such fossils. By the Middle Ordovician, the first jawless armoured fishes had evolved; by the Silurian, fishes had diversified, and the first jawed vertebrates (gnathostomes), represented by sharks and armoured placoderms, had evolved. These are largely represented in Australian Silurian deposits by isolated scales and teeth. However, by the start of the Devonian, there was a diverse fauna of early jawed fishes, including large (4 m) placoderms, small bony fishes ancestral to the predominant modern ray-fins, and a wide range of primitive marine lungfish. Australia has the oldest fossil coelacanths and onychodontid fishes, both primitive kinds of lobe-finned fishes (sarcopterygians). Many superbly preserved fish fossils have been recovered from Lower Devonian (*ca* 405 Ma) limestones near Canberra. By the Middle to Late Devonian (*ca* 385 Ma), fishes had greatly diversified to inhabit both tropical marine reefs and freshwater environments. The fish

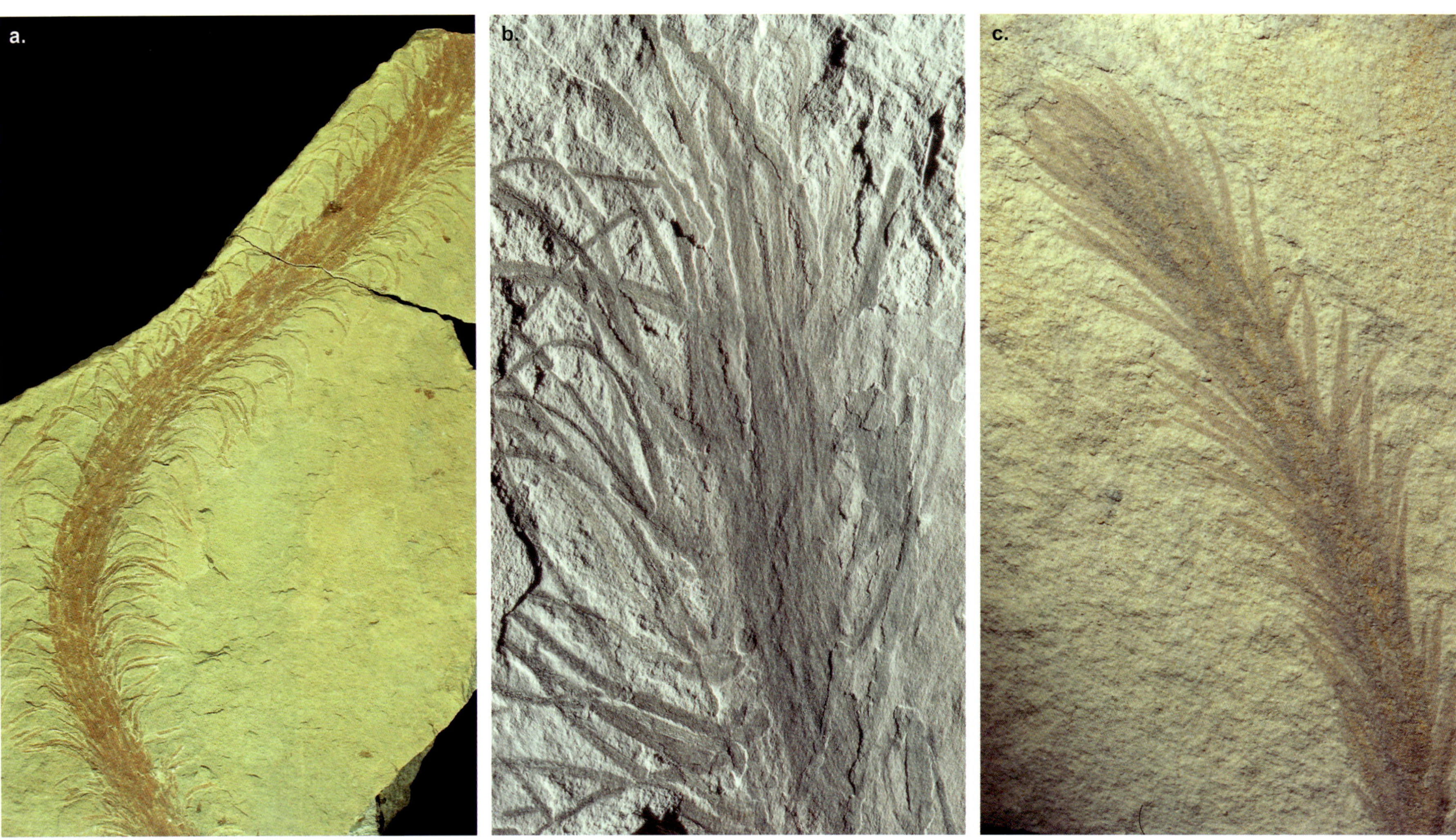

Figure 3.7: *Baragwanathia longifolia* from the Yea area, Victoria. (a) Part of a mature stem with long leaves. (b) Tip of a mature stem. (Source: White, 1986; photographs by Jim Frazier). (c) Tip of a young stem. (Source: White, 1986; photographs by Jim Frazier)

fauna from Mt Howitt, central Victoria, has an abundance of lungfishes, the first to show clear adaptations for air breathing. This seemingly occurred in response to the depression of oxygen levels in the ocean in the late Middle Devonian. Perhaps the most significant collection of fossil fish from the Late Devonian is that from the Gogo Formation in the Canning Basin of northern Western Australia (Box 3.2). After the extinction near the end of the Devonian (Figure 3.8), the world's fish faunas would never be the same. Placoderms and other archaic groups died out; their place was taken by sharks and ray-finned fishes, which today constitute some 99% of all living fishes.

The first forests

Land plants continued to diversify rapidly during the Early Devonian (*ca* 405 Ma) and, by the end of the period, the landscape had been transformed.

FISH FOSSILS FROM THE GOGO FORMATION, CANNING BASIN WESTERN AUSTRALIA (BOX 3.2)

The late Devonian fishes of the Gogo Formation, Canning Basin, represent the world's most significant collection of fishes of this age, all perfectly preserved in three dimensions and many with preserved, mineralised soft tissues such as muscle bundles, nerve cells and internal organs. These fishes, along with a great diversity of invertebrates such as crustaceans, early ammonoids, gastropods, brachiopods, corals, trilobites and echinoderms, all lived in and around a great reef system built mostly of algae and sponge-like stromatoporoids, which is so well exposed in Windjana Gorge, Western Australia.

More than 50 fish species have been recorded, about half of these being predatory arthrodires or joint-necked placoderms, with the others comprising boney fishes (Osteichthyes) such as lungfishes (Dipnoi), early ray-finned fishes (Actinopterygii), an onychodontid (*Onychodus*), a tetrapodomorph (*Gogonasus*), and rare sharks (Chondrichthyes) and acanthodians (an extinct group of jawed spiny fishes). Gogo fishes have in recent years been instrumental in resolving major problems in early vertebrate evolution such as elucidating the anatomical transformation of fishes to land animals (tetrapods). Studies of Gogo placoderms have shown that they possess true teeth with pulp cavities, that they exhibit sexual dimorphism in the pelvic fins, and that some bore live young supported by maternal umbilical structures. The origins of complex sexual reproduction, with males internally fertilising females by copulation, was first confirmed in our ancestral line from these fossils.

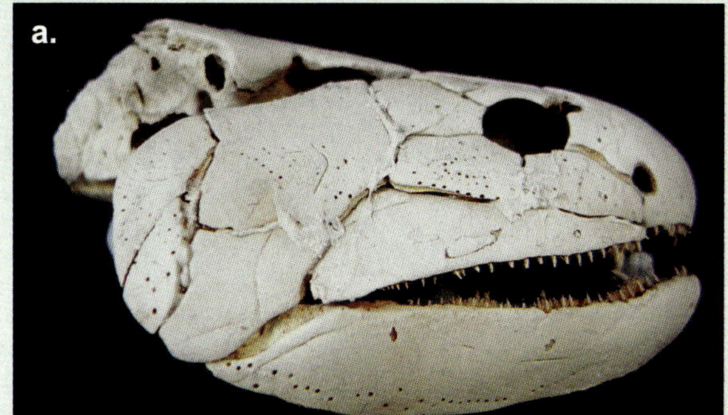

Figure B3.2: (a) The skull of *Gogonasus andrewsae*, a lobe-finned fish from Gogo that bridges the gap between fishes and land animals. (b) The bony armour of *Mcnamaraspis kaprios*, a placoderm fish from Gogo, which is today recognised as the official state fossil emblem of Western Australia. (c) The skull of *Onychodus jandemarrai*, a predatory 'dagger-toothed fish' that hunted among the reef crannies like a modern moray eel.

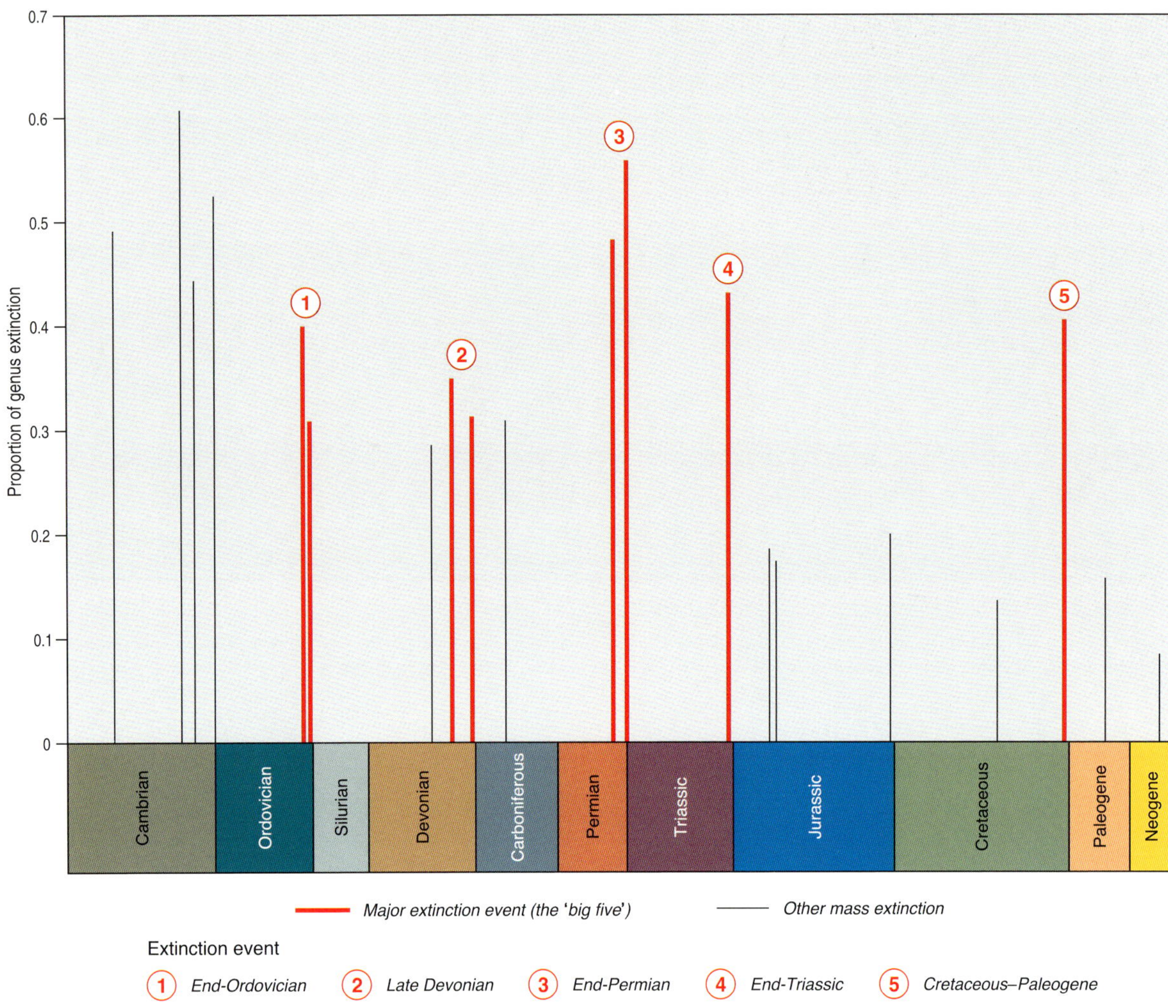

Figure 3.8: Extinction events in the Phanerozoic. This is a graph of proportion of generic extinction at a substage level against time. The 'big five' extinction events are labelled. These occurred at the end of the Ordovician, in the Late Devonian, at the end of the Permian, at the end of the Triassic and at the end of the Cretaceous. (Source: modified from Bambach, 2006)

At the beginning of the Devonian, there were scattered shrub-sized lycophytes and smaller plants; by the end of the period, there were forests containing tree-sized lycophytes, ferns, horsetails and early seed plants. Plants are characterised by an alternation of generations. That is, they have a gametophyte stage of their life cycle, during which the male and female gametes are produced, and a sporophyte stage that grows from the fertilised egg cell and, in turn, produces spores that germinate to grow into the succeeding gametophyte generation. The earliest land plants were essentially restricted to moist habitats, since they required water for the transfer of the male gametes to the egg cell. Simple free-sporing plants such as mosses, lycophytes, horsetails and ferns are still mostly restricted to such wet habitats today.

During the Devonian, one plant lineage developed two significant innovations—the dispersal of the male gametophyte (in a desiccation-resistant waxy coat that we call pollen), and the retention and protection of the female gametophyte stage within the tissues of the parent sporophyte to produce the first seeds. These innovations influenced the future development of life on the land, since they allowed plants to liberate themselves from their dependence on free water for reproduction and to colonise drier inland and upland habitats that were mostly barren previously. Animals followed the advance of the vegetation, creating complex ecosystems that now covered the entire land surface. This dramatic change in the landscape is termed the Devonian Transformation. Not only did this change the appearance of the landscape, it also altered the style of weathering, allowing chemical weathering to become relatively more important and deeper soils to develop. This partly changed the composition of sediments transported by rivers and deposited in the ocean, in lakes and on floodplains. Thicker soils and vegetation-stabilised soil surfaces even changed the nature of river systems around the globe, from a dominance of sandy, braided rivers to one of mud-rich, meandering rivers. The growth of so much plant material also led to its accumulation and the generation of the first coals and non-marine petroleum source rocks (Chapter 9).

As the vegetation changed during the Devonian, so did the fauna. By late in the Devonian (*ca* 375 Ma), there were numerous groups of arthropods living on land. These included harvestmen, mites, centipedes, millipedes, springtails, scorpions, pseudoscorpions, spiders, isopods, insects and trigonotarbids. The Middle Devonian also saw the evolution of the first tetrapods from lobe-finned fishes. These early tetrapods were probably fully aquatic animals, with the limbs and digits more akin to paddles than legs, and most likely unable to bear the weight of the animal out of water. These limbs were probably an adaptation to ambush predation in shallow water, whereas the air breathing may have been a response to oxygen-poor water (see below).

In the Late Devonian (*ca* 375 Ma), there was a complex extinction, in which several closely spaced events had a devastating effect on the marine fauna. This is considered to have been caused partly by the dramatic increase in land plants and the rapid increase in nutrient-rich runoff, which may have led to oceanic anoxia. Most marine organisms suffered during this series of events, particularly reef-building organisms, which were almost completely wiped out. The large, magnificent Devonian reef complex exposed in Windjana Gorge in Western Australia (Figure 3.1) is a superb example of the structures that suffered during this extinction event. Major reef-building did not re-emerge for more than 100 Myr with the evolution of modern types of corals. Land plants and animals, including freshwater animals, were only slightly affected by this extinction, another indication that it was an oceanic event.

Image courtesy of Phil Playford

Devonian reef in cross-section, Windjana Gorge, Western Australia.

During the early part of the Carboniferous (*ca* 340 Ma), the land flora was similar to that of the Late Devonian, with existing groups of plants continuing to diversify and the largest tending to get larger. However, during the period, the first cycads and conifers appeared. The huge increase in the diversity, extent and density of plants in forests during the Late Devonian and early Carboniferous led to an increase in atmospheric oxygen and an enormous sequestration of carbon in the form of coal deposits in the palaeotropics (North America, Europe and parts of Asia), such that by late in the Carboniferous free oxygen formed up to 35% of the atmosphere, about two-thirds more than current levels. This led to a degree of gigantism among arthropods and amphibians, probably because of the effective increase in efficiency of their respiration. The Carboniferous also saw the first flying insects, the first animals to develop this ability. On the negative side, such high oxygen levels, coupled with the presence of dense forests, contributed to a highly combustible landscape, and the presence of large quantities of fossil charcoal in Carboniferous sediments indicates that forest fires had become a regular occurrence.

Amphibians diversified and increased in size dramatically during the Carboniferous, but perhaps the greatest vertebrate innovation during the period was the evolution of the amniotic egg—an egg with a waterproof coating. This allowed tetrapods to lay their eggs on dry land, dispensing with the aquatic larval stage characteristic of amphibians and, therefore, allowing them to become fully terrestrial. These first fully terrestrial, amniotic vertebrates were the first reptiles. They very soon separated into two lineages, the synapsids and the diapsids, in the late Carboniferous (*ca* 310 Ma). The synapsids (mammal-like reptiles) were to give rise to the mammals, whereas the diapsids were to give rise to the dinosaurs, modern reptiles and birds.

Early in the Carboniferous, the supercontinent of Euramerica collided with Gondwana, followed soon after by Siberia colliding with Euramerica to form the supercontinent of Pangaea (Chapter 2). Through the Carboniferous, this supercontinent drifted further south to cover a large proportion of the high latitudes of the Southern Hemisphere, including the South Pole. This may have been the trigger that led to a decrease in global temperatures, eventually plunging the planet into the Karoo Ice Age during the late Carboniferous to early Permian. The decreasing temperature of the late Carboniferous had a dramatic effect on the Australian flora; the flora of large lycophytes disappeared, to be replaced by the impoverished *Nothorhacopteris* and *Botrychiopsis* seed-fern floras. During the Karoo Ice Age, ice sheets covered large parts of southern Gondwana, including Australia (Chapter 4).

Seed ferns dominate

As the Karoo ice sheets receded in iterative pulses, the continent was partly flooded, and marine sediments were deposited in several large basins around its periphery. The warming climate triggered an explosion of plant life, and the *Glossopteris* (Figure 3.9) flora spread across Gondwana (*Did you know?* 3.1). *Glossopteris* was a genus of tall seed plants with distinctive air chambers in their roots and a deciduous habit that enabled them to thrive in the vast high-latitude (temperate) lowland swamps of the Gondwanan Permian. The vast peatlands dominated by the *Glossopteris* flora provided the raw material for the huge coal deposits found in the Permian (299–251 Ma) of the Gondwanan continents (Chapter 4).

Figure 3.9: Representatives of the *Glossopteris* flora. (a) (opposite) Leaf whorls from the horsetail *Phyllotheca australis*, from Newcastle, New South Wales. (b) A cross-section of a silicified tree-fern trunk (*Palaeosmunda* sp.) with two main stems, from Blackwater, Queensland. (c) Leaves of *Glossopteris duocaudata* from Cooyal, New South Wales. (d) A small branch of a conifer (*Walkomiella australis*) terminating in a cone, from Ulan, New South Wales. (e) A cross-section of a silicified tree trunk 6 cm in diameter, with closely spaced annual rings indicating a cold climate, from the Werrie Basin, New South Wales. (f) A fern, *Neomariopteris* sp., from Newcastle, New South Wales. (Source: White, 1986; photographs by Jim Frazier)

Did you know?

3.1: The distribution of *Glossopteris*—evidence for continental reconstruction

The distribution of fossil *Glossopteris* leaves across the Southern Hemisphere continents (Antarctica, Australia, southern Africa, Madagascar and South America) and India was one of the first firm pieces of evidence used, in 1861, by the Viennese geologist Eduard Suess to deduce that all of these lands were once united into a single supercontinent. He called this supercontinent Gondwanaland, which is now more correctly shortened to Gondwana. Suess did not, at the time, understand that the continents moved across the surface of the earth; he thought that sea-level fell to expose land bridges between the present-day continents.

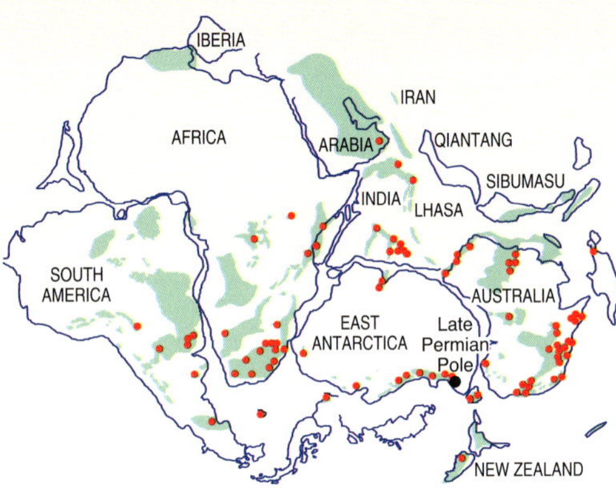

Figure DYK3.1: Distribution of Glossopteris flora on a reconstruction of Gondwana, showing where it is confidently recorded (red dots) and where Permian sedimentary basins are located (green regions).

In 2011, Australia exported more than $58 B worth of metallurgical and thermal coal, mostly of Permian age, making coal one of the country's most significant export commodities (Chapter 9). *Glossopteris* plants survived in high-latitude habitats with several months of darkness per year by being deciduous (shedding their leaves) and retreating into a relatively dormant state during the winters. Very few fossils of terrestrial vertebrates are found associated with the *Glossopteris* flora, but a broad range of fish, insects and other invertebrates are known from lake deposits containing *Glossopteris* and associated plants. In addition to the glossopterid gymnosperms, the Permian peatlands generating these coal deposits were also inhabited by horsetails, ferns, lycopods and mosses; the drier uplands may have hosted conifers, horsetails, cycads and ferns, but these are less commonly represented in the major fossil deposits that typically accumulated in lowland areas.

The fossil record of Permian terrestrial vertebrates in Australia is sparse and consists almost entirely of temnospondyl amphibians found in association with the Sydney Basin coal deposits. Similarly, the Australian fossil record of Permian terrestrial invertebrates is very limited, with only one major locality, in the Sydney Basin coal measures, containing a diverse fauna, including scorpionflies, beetles, lacewings, caddisflies and crickets.

The *Glossopteris* flora and its attendant fauna persisted until the end of the Permian, at which time Earth's most severe extinction event occurred (Figure 3.8). This event, the end-Permian extinction, wiped out up to 90% of marine species, 70% of all terrestrial animal species and half of all land plant species. It is believed to have been caused by the eruption of the Siberian Traps, one of the largest volcanic events to have occurred in the Phanerozoic. The *Glossopteris* flora disappeared completely during this event, as did a large proportion of the insect and vertebrate fauna.

This extinction event was so severe, with so much biodiversity lost, that it took many millions of years for ecosystems to recover, much longer than for any other extinction event. It was the closest land plants and animals came to complete extermination since they had begun to invade the land several hundred million years before. This event was the extinction that fundamentally changed Earth's ecosystems, bringing to an end the Paleozoic world and allowing the remaining organisms to begin to construct the world with which we are familiar. The plants and animals currently occupying our world still reflect those that survived the catastrophe at the end of the Permian (Figure 3.8).

On land, this extinction event resulted in a severely impoverished flora and fauna. The diverse glossopterids and their associates, which formed the huge Gondwanan peatlands in the Permian, had all disappeared and were replaced by a sparse flora of lycophytes, seed ferns, horsetails and conifers. Compared with the lush, diverse flora of the late Permian, the Early Triassic (251–245 Ma) flora was one of smaller opportunistic plants commonly having small leaves with thick cuticles and sunken gas-exchange pores (stomata)—adaptations to warmer and drier conditions. The ability of the flora to generate the volume of vegetation to form peatlands (and therefore coals) had gone, partly

Figure 3.10: Representatives of the *Dicroidium* flora. (a) *Dicroidium zuberi*, from Brookvale, Sydney, New South Wales. (b) *Dicroidium odontopteroides* and *Dicroidium elongatum* (long, narrow leaves), from Ipswich, Queensland. (c) Delicate twigs of the conifer *Voltziopsis wolganensis* with lax terminal cones and needle-like leaves. (d) The ginkgophyte *Ginkgoites semirotunda*, from Benolong, New South Wales. (e) The horsetail *Neocalamites* sp., from Brookvale, Sydney, New South Wales. (Source: White, 1986; photographs by Jim Frazier)

because of its impoverishment and partly because the climate had warmed. This Early Triassic interval is called the 'Coal Gap' and represents the only period during the last 350 Myr when no coal was accumulating anywhere on Earth.

It would be several million years before even the thinnest coals would reappear. Despite coals being significant enough for mining from the Late Triassic (*ca* 210 Ma) of southeast Queensland, South Australia and Tasmania, the Jurassic of southeast Queensland and the Paleogene–Neogene of Victoria (Chapter 9), they were never again as extensive in Australia as they were in the Permian.

By the Middle Triassic, the seed fern *Dicroidium* (Figure 3.10) had taken over most lowland habitats and had diversified extensively, in much the same way as the *Glossopteris* flora had done in the Permian. *Dicroidium* was a woody seed-bearing plant with foliage that was superficially fern-like. The microscopic gas-exchange pores (stomata) on its leaves were in many cases strongly sunken or protected by overarching cuticle lobes, signalling adaptations to warmer climates and periodic drought. *Dicroidium* exploited a similar geographic range to the preceding *Glossopteris* flora and was also predominantly a plant of wet lowland habitats. Later in the Triassic, in Australia, its fossil remains contributed to the economic coal deposits at Ipswich, Tarong and Callide in Queensland and at Leigh Creek in South Australia (Figure 3.1).

The Triassic was also a period of extensive diversification of the conifers, such that all the major extant groups of conifers appeared in this period. The Australian land vertebrate fossil record is limited but was dominated by temnospondyl amphibians, some primitive archosaurs and rare mammal-like reptiles.

Conifers dominate

At the end of the Triassic period, another mass extinction event occurred. This was probably caused by the eruption of the basalts of the Central Atlantic Magmatic Province in western Pangaea, and it wiped out most of the crurotarsans, a group of crocodilian-like reptiles that were direct competitors with the early dinosaurs. This left the way open for the latter to become the dominant land vertebrates for the remainder of the Mesozoic. The extinction event also wiped out most of the temnospondyl amphibians, with only two families surviving in China and Gondwana.

The *Dicroidium* flora disappeared and was replaced by a flora dominated by conifers, cycad-like bennettitaleans, ginkgoes, ferns and herbaceous lycophytes. By the Jurassic (*ca* 200 Ma), all extant conifer families were already in existence and, in Australia, conifers of the Araucariaceae, Podocarpaceae and extinct Cheirolepidiaceae dominated the flora. Gingkoes, several orders of seed ferns (Caytoniales, Pentoxylales, Corystospermales) and cycads were the other dominant midstorey plants. Apart from a couple of aquatic groups, all major extant fern families had evolved by the Jurassic and, together with small but abundant and diverse lycophytes and horsetails, they dominated the ground stratum of the vegetation.

Cheirolepidiacean conifers were particularly prominent in the Early Jurassic (200–176 Ma) of Australia, as determined by the great abundance of their distinctive pollen. Bennettitaleans and Caytoniales were also prominent seed plants of the Early Jurassic, while osmundacean, dipteridacean, dicksoniacean, matoniacean and gleichenacean ferns were well represented herbaceous groups. Some floristic differences are apparent between western and eastern Australia in the Early Jurassic, and this is interpreted to result from generally drier and warmer conditions that prevailed at middle latitudes (Western Australia) at that time, compared with the cool, moist climate experienced by eastern Australia at a latitude of 40–75°S.

The marine transgression near the end of the Pliensbachian (*ca* 183 Ma) and the early Toarcian (*ca* 180 Ma) oceanic anoxic event coincided with a change in the Australian flora. The araucarian conifers became more common at the expense of their relatives, the cheirolepidiacean conifers. Podocarp conifers formed a lesser part of the flora, and bennettitaleans, lycophytes and a range of ferns were also common. The climate was seasonal, and the flora in the west was much the same as that in the east, as the latitudinal differentiation was no longer clearly evident.

Later in the Middle Jurassic, a slight cooling trend again altered the flora. The podocarp conifers became more common at the expense of the araucarian conifers, while some fern families, which are now restricted to the wet tropics (e.g. Dipteridaceae), became quite rare. The abundance and diversity of other ferns and lycophytes indicate that the climate was seasonal and humid, with sufficient dry spells to inhibit peat formation, thus preventing the formation of extensive coal deposits.

Reconstruction of *Rhoetosaurus brownei*.

The record of the Jurassic vertebrate fauna is limited to a few isolated specimens, but comprises sauropod and theropod dinosaurs and a few temnospondyl amphibians. The temnospondyl amphibians became extinct elsewhere in the Late Jurassic, but survived into the Early Cretaceous in Australia. No mammal fossils are known from the Australian Jurassic. The insect record from the Australian Jurassic is slender, with only a few beetles, cicadas and leafhoppers, among others, being recorded from the Talbragar Fossil Fish Bed (Figure 3.1), which, together with fish, plants and molluscs, provides a snapshot of a Late Jurassic (*ca* 153 Ma) ecosystem. The deposit contains several horizons preserving mass-mortality events, where hundreds of fish, insects and plants were rapidly buried and are exquisitely preserved (Figure 3.11). It contains one of the finest mid-Mesozoic continental fossil assemblages in the world. Particularly well represented are the fish *Cavenderichthys,* the robust araucariacean conifer *Podozamites jurassica* (distantly related to the modern bunya, hoop, kauri and Wollemi 'pines'), and a range of now-extinct seed ferns.

The flowering

Molecular phylogenetic studies suggest that the angiosperms (flowering plants) had evolved by the Middle Jurassic (*ca* 175 Ma) and underwent a gradual initial diversification through to the beginning of the Late Jurassic (*ca* 161 Ma). After this, a rapid diversification occurred, so that by the end of the Early Cretaceous five major lineages had been established, most of which probably had a herbaceous or shrubby habit and preferred wet environments.

It is unclear from where the angiosperms evolved, but it has been suggested that their close relatives are one or other of the groups of seed ferns, and recent studies favour the extinct Caytoniales, Bennettitales or Pentoxylales as their closest relatives. To date, the Jurassic component of the lineage leading to angiosperms lacks a clear fossil record. The oldest fossil pollen, leaves and flowers of angiosperms come from the late Valanginian to early Hauterivian (*ca* 135 Ma), but the diversity of these early records suggests that the group as a whole had a much earlier origin. The oldest flower fossil (*Archaefructus*), from Liaoning in northern China, is an aquatic plant whose position within the early angiosperms remains a subject of debate.

The dominantly coniferous forests of the Jurassic continued into the earliest Cretaceous. The earliest Cretaceous flora comprised forests in which araucarian and podocarp conifers formed the canopy with a few members of the now extinct Cheirolepidaceae. The middle and lower strata of the forests consisted of bennettitaleans, ferns, seed ferns, lycophytes and bryophytes. There was some regional differentiation in the flora, with that from Western Australia being more like India in having more diverse araucarian conifers and ferns, while that of eastern Australia was more like western Antarctica and New Zealand.

In the Valanginian to early Barremian (*ca* 138–128 Ma), podocarps became more common in the canopy at the expense of the araucarians, while seed ferns, bennettitaleans, ferns and lycophytes continued to constitute the midstorey and understorey. At this time, ginkgoaleans had little representation in the flora. Whereas ginkgoaleans were essentially absent from the Australian Jurassic, they returned in the Barremian–Aptian (*ca* 125–120 Ma). Their presence signifies an interval of significant cooling, which is corroborated by the presence of cold-water sedimentological indices, such as glendonites and ice-rafted dropstones in southern Australian marine strata of this age.

Together with pentoxylaleans, ginkgoaleans formed a significant deciduous component of the Early Cretaceous forests, which were dominated for the most part by the same conifer families that had dominated the Jurassic. Growth-ring analysis also indicates that the climate was cool, with marked seasonality. Riparian vegetation included horsetails,

liverworts, ferns, lycophytes and early angiosperms. Regional variation was still a feature of the vegetation, with high frequencies of *Sphagnum*-like spores in Queensland suggesting extensive development of moss bogs, whereas in Victoria lycophytes were more prominent. Similarly, liverworts, hornworts and riparian ferns were more common in Queensland than in the southeast.

There was a major vegetational change near the Aptian–Albian boundary (*ca* 111 Ma). In Victoria, a new suite of araucarian and podocarp conifers appeared, with some Cupressaceae and

Cheirolepidaceae. Associated with the conifers in the canopy was a small-leafed *Ginkgo*, which had replaced *Ginkgoites*. Also very common in the understorey was the osmundaceous fern *Phyllopteroides*.

In Queensland, horsetails, cycads and *Ginkgo* were present; the pentoxylalean *Taeniopteris* was common in Queensland, but was absent in the southeast. In both regions, angiosperms were more widely distributed and increasingly diverse, with assemblages from the north being more diverse than those from the southeast.

During the Cenomanian (*ca* 98 Ma), uplift commenced along the eastern margin of the continent, and subsidence began along the western margin. At about the same time, Australia started to separate from Antarctica, and marine environments extended into Victoria from the west (Chapter 4). This was accompanied by profound changes in the vegetation of the southeast. The canopy largely comprised araucarian conifers, with a small-leafed *Ginkgo* and a minor component of angiosperms, but the latter had become much more diverse. In the lower storey, there was an increased diversity of clubmosses and leptosporangiate ferns.

Figure 3.11: Representative fossils from Talbragar, New South Wales. (a) (opposite) Araucariacean conifer, *Podozamites jurassica*. (b) Seed fern, *Komlopteris* sp. (Source: S McLoughlin). (c) Twigs of a small podocarp conifer with stems covered in scale-like leaves and three small cones at the tip of each stem. (d) A conifer spur shoot showing rhombic leaf detachment scars. (e) Small freshwater fish, *Cavenderichthys talbragarensis*. (Source: except where noted, White, 1986; photographs by Jim Frazier)

© Getty Images [R Blakers]

Ancient myrtle beech trees (*Nothofagus cunninghamii*) grow in the Vale of Belvoir, northwestern Tasmania.

In Queensland, the flora was more diverse, with the canopy containing several araucarians, a podocarp and a cupressacean, along with a *Ginkgo*. Also present were the pentoxylalean seed fern *Taeniopteris*, the bennettitalean *Ptilophyllum* and at least eight angiosperm species (based on macrofossils, although the pollen flora suggests the presence of many more). The latter group became co-dominant elements of the flora for the first time—a change that fundamentally altered the structure of the world's vegetation to the present day.

At about the Cenomanian–Turonian boundary (*ca* 94 Ma), there were profound vegetational changes, with forms readily recognisable in the modern flora being introduced. Lineages that include the modern Huon pine (*Lagarostrobos*) and *Dacrydium* (another podocarp) evolved in southern high latitudes, together with the distinctive austral angiosperms Proteaceae and *Ilex* (holly).

By the Santonian (*ca* 86 Ma), the podocarp/araucarian forests had proteacean angiosperm associates in the canopy. The Proteaceae includes the modern *Banksia*, *Grevillea*, *Hakea*, *Telopea* (waratah) and *Macadamia*. Early in the Campanian (*ca* 80 Ma), *Nothofagus* (the southern beech) evolved in southern high latitudes, and its early diversification probably occurred in the southern South America–Antarctic Peninsula region. Appearing just before the ultimate breakup phase of eastern Gondwana, this genus has remained a key component of the forests of Australia, New Zealand, New Caledonia, New Guinea and South America through the Cenozoic. Its fossils are also known from Antarctica, indicating that this continent hosted extensive forests in the Cretaceous and Paleogene, which probably played an important role in biotic interchange before the breakup of the southern continents. The Campanian also saw the diversification of the Proteaceae, which, in Victoria, includes such forms as *Adenanthos* (woolly bush) and *Persoonia* (geebung). The record there suggests a vegetation with both rainforest and sclerophyll taxa, with podocarps, araucarians and a range of flowering plants. The diversification of the Proteaceae is particularly evident in Maastrichtian (*ca* 68 Ma) pollen floras from central Australia.

By the Maastrichtian, the Winteraceae (e.g. modern mountain pepper), Ericaceae (e.g. modern common heath), Trimeniaceae (e.g. modern bitter vine) and Ranunculaceae (e.g. modern small-leafed clematis) had also appeared in the Australian flora, and the true grasses (Poaceae) had evolved in Gondwana (probably in South America and Africa), although the latter remained a very minor component of the vegetation until well into the Cenozoic.

The record of Cretaceous terrestrial vertebrates is extensive but fragmentary in Australia. The continent had a diverse fauna of sauropod, theropod and ornithopod dinosaurs, which at lower palaeolatitudes grew to enormous size, similar to those on other continents. In higher palaeolatitudes, the fauna was apparently diminutive and dominated by herbivores adapted to high-latitude seasonality. In compensation for the relatively poor skeletal fossil record, there are some important dinosaurs trackways recorded from Australia. The Lark Quarry trackway (near Winton in Queensland; Figure 3.1) is perhaps the best known. It comprises thousands of footprints made by a herd of Early Cretaceous herbivores that have been stampeded, slipping and sliding in the mud, by the appearance of a large predatory theropod or two. Recently, a dinosaur trackway of similar age has been reported from the Otway Basin in Victoria. In addition to dinosaurs, birds, monotremes, other archaic mammals, pterosaurs, crocodilians and the last of the temnospondyl amphibians were also present on the continent. All temnospondyls on other continents had become extinct by the Late Jurassic, but they persisted in southeastern Australia.

With the demise of the last of the temnospondyl amphibians in the Cretaceous, only the Lissamphibia (modern amphibians) remained, comprising the Anura (frogs and toads), Caudata (salamanders and newts) and Apoda (caecilians). Of these, only the frogs and toads currently inhabit Australia. The Eocene fossil locality at Murgon, Queensland, reveals that salamanders may have once lived in Australia, along with leiopelmatids, an archaic frog family that still survives in New Zealand. The absence of the worm-like caecilians in Australia is striking, given their wide distribution (fossil and living) in other Gondwanan fragments at lower latitudes.

Living in the shadow of their archosaur relatives were some of the precursors of modern reptile groups. The diversity of Australian Cretaceous chelonians included early sea turtles, sinemydids (freshwater turtles also found in Cretaceous of central Asia) and meiolaniids (a Gondwanan family of land turtles). Curiously, the side-necked turtles (Pleurodira) that fill Australia's modern freshwater habitats were absent. Crocodyliforms were also present, including *Isisfordia* from the mid-Cretaceous of Queensland. This is the earliest and most primitive known eusuchian, the group that includes modern crocodiles and alligators and hints at a Gondwanan origin for these predators. A diagnostic fossil record for Australia's terrestrial squamatans (lizards and snakes; remains of the extinct marine mosasaurs are known) does not extend into the Cretaceous, but several groups would certainly have been present. Lizards, including the geckos, agamids (commonly called dragons), skinks and varanids (goannas and monitors) are all Late Mesozoic lineages, first appearing in the Jurassic or Cretaceous of Asia, but not appearing in the Australian fossil record until much later.

Several species of archaic mammal are known from the Lower Cretaceous of Victoria. Some are clearly monotremes; another may be allied to the multituberculates, a primitive mammal group; whereas others are of uncertain affiliation, but with some similarities to archaic placentals.

A few scattered bird fossils are known from the Cretaceous of Australia, and these indicate the presence of enantiornithid birds on the continent. This group of birds was very diverse in the Cretaceous, but did not survive the extinction event at the end of that period (Figure 3.8).

One of the most important fossil localities from the Cretaceous (Aptian: *ca* 120 Ma) in Australia is the Koonwarra locality (Figures 3.1 and 3.12). It is a lake deposit formed in what was then a rift valley along Australia's southern margin associated with continental breakup from Antarctica (Chapter 4). The deposit contains several species of fish and plants, including liverworts, lycophytes, horsetails, ferns, seed ferns, ginkgoes, araucarian and podocarp

Figure 3.12: Representative fossils from Koonwarra, Victoria. (a) Twigs of a podocarp conifer, *Bellarinea richardsii*. (b) A frond of a seed fern, *Rintoulia variabilis*. (c) A delicate fern, *Aculea bifida*. (d) A deeply dissected leaf of the ginkgophyte *Ginkgoites australis*. (e) A veliid water strider insect, *Duncanovelia extensa*. (Sources: White, 1986; White, 1990; photographs by Jim Frazier)

conifers, and Australia's oldest flower. The relationships of this flower are obscure, but it may belong to an early magnoliid lineage. Numerous insects are also preserved, including mayflies, dragonflies, cockroaches, beetles, fleas, flies and wasps, together with other arthropods, including a spider. In addition, a few isolated feathers have been found; these are the oldest evidence for birds (or feathered dinosaurs) on Gondwana.

The Koonwarra deposit represents an important window into the evolution of Australia's terrestrial biota, since this interval marks the first appearance in the continent's fossil record of flowering plants—a group that now dominates the world's vegetation. It also provides insights into the structure of Cretaceous high-latitude communities, given that the Gippsland region was located at about 80°S at this time.

Beginning to build the modern world

Among the vertebrates, the Cretaceous–Paleogene extinction event exterminated the non-avian dinosaurs, enantiornithid and some other archaic birds, pterosaurs and marine reptiles, but groups like crocodilians, turtles, amphibians and mammals suffered relatively little. Ironically, it was this extinction event that saw the marsupials disappear from North America (the extant opossums were Pliocene immigrants from South America). This extinction event also saw the demise of the cheirolepidacean conifers, which had formed a major part of the forest canopy for many millions of years.

The end of the Cretaceous (65.5 Ma) marks a significant and convenient boundary in the history of life in Australia. All the major groups of organisms that dominate our modern biota (lizards, snakes, birds, mammals and flowering plants) had evolved, and the Cretaceous–Paleogene extinction event had removed many of their main competitors or, in the case of seed ferns, led to their serious decline. This boundary, therefore, marks the beginning of the rise of the modern biota, with mammals and birds subsequently diversifying into many of the niches previously occupied by their competitors, while the flowering plants came to dominate the flora, with many of the previously dominant forms (e.g. conifers) becoming more restricted in their distribution.

Freshwater fishes of two genera persisted from the Mesozoic: the Queensland lungfish (*Neoceratodus forsteri*) and the two saratogas (genus *Scleropages*) from southeastern Queensland and the far north. Saratogas are osteoglossids (bony-tongues), archaic predatory teleosts that are found throughout the tropical freshwaters of the world. Although the Australian fossil record does not extend into the Mesozoic, genetic comparison with Asian congeners indicates that the genus must have been present in the Early Cretaceous when the Asian and Australian lines diverged, likely due to the separation of India from the rest of Gondwana (which transported the genus to Asia).

The remainder of Australian freshwater fishes evolved more recently from marine lineages during the Cenozoic; indeed, the majority of local freshwater families also contain marine or

Image courtesy of Jean Joss, Macquarie University

Queensland lungfish, *Neoceratodus forsteri*.

estuarine species. On other continents (excluding Antarctica), freshwater faunas are dominated by ostariophysians, the teleost group that includes the cyprinoids (carp-like fishes), characoids and catfishes. In Australia, the only native freshwater ostariophysians are representatives of two mostly marine catfish families (the eel-tailed plotosids and the fork-tailed ariids).

It is likely that marsupials evolved in Asia during the Early Cretaceous, dispersing to North America not long after. They entered South America during the Late Cretaceous to early Paleocene, and Australia by the late Paleocene. The marsupial-dominated fauna of Australia is perhaps the most distinctive mammalian fauna on any continent. Apart from Antarctica, all other continents (South and North America, Eurasia and Africa) have exchanged terrestrial placental mammals, including camels, lions, anteaters, sloths, armadillos and elephants. This mixing has not involved Australia, largely because of Australia's relative isolation throughout much of the radiation of the placentals. Australia entered this period of isolation as it completely separated from Antarctica by the Oligocene (34 Ma). As a consequence, land animal immigration was completely curtailed, and the marsupial fauna was allowed to diversify with little interference from other mammal lineages. It was only as the northward drift of the continent uplifted New Guinea late in the middle Neogene that extensive contact with the outside world resumed. Unfortunately, the early marsupial fossil record in Australia is patchy, so it is difficult to be certain how the early diversification of the marsupial fauna occurred.

Paleogene climatic optimum

The wet forests persisted from the Late Cretaceous in the southeast of the continent during the early Paleogene, and were dominated by araucarian and podocarp conifers, and angiosperms. Several of the latter, especially members of the Proteaceae, became extinct at this time. However, the angiosperms recovered quickly from the changes at the Cretaceous–Paleogene transition, and ancestral *Nothofagus* forests became dominant over large areas of the continent by late in the Paleocene. In central Australia, although dating remains poor, it has been suggested that a more variable temperate rainforest, dominated by Cupressaceae, Cunoniaceae (e.g. coachwood) or Proteaceae, was present. In northwestern Australia, the forests were dominated by she-oaks (Casuarinaceae) and Proteaceae, whereas, in northeastern Australia, marshland-adapted burr-reeds (Sparganiaceae) and the mangrove palm (*Nypa*) are present.

As the global temperature rose during the early Eocene (*ca* 53 Ma), Casuarinaceae and Proteaceae came to dominate Australian coastal rainforests; away from the coast, the rainforests were more variable, and subtropical rainforest taxa covered much of the continent. These forests also probably included the earliest representatives of *Eucalyptus* (Myrtaceae) and *Banksia* (Proteaceae)—perhaps growing on poorer soils in rainforest fringes, where their leathery (sclerophyllous) leaves made them pre-adapted to the more arid conditions that would arise later in the Cenozoic. The *Nothofagus* (*Brassospora*) forests, which had been more widespread in the southeast, seem to have become confined to the Tasmanian highlands. At the same time, the mangrove palm ranged as far south as western Tasmania, while some pantropical palms reached central Australia. Their descendants can still be found in such places as Palm Valley, to the west of Alice Springs (Figure 3.1).

The middle Eocene (*ca* 45 Ma) saw the maximum extent of rainforest across the continent. *Nothofagus* (*Brassospora*) was widespread in these forests, even in central Australia (Box 3.3). This rapid expansion of *Nothofagus*-dominated forests may reflect global cooling. A decrease in global temperatures is apparent from the mid- to late Eocene, and this was reflected in the Australian vegetation. Causes of the temperature decline remain uncertain. A conventional view has been that the development of the Antarctic Circumpolar Current close to the Eocene–Oligocene boundary (*ca* 34 Ma) was related to the separation of Antarctica from Australia and South America, isolating Antarctica from warmer waters to the north. Lately, however, this view of the dramatic cooling event has been challenged; the separation of southern South America may have been a later event, and the onset of the Antarctic Circumpolar Current may not have caused the onset of Antarctic glaciation. Other factors, such as decreased levels of CO_2, may be implicated.

True crocodiles first evolved in the Early Cretaceous, possibly in Gondwana. For most of the Cenozoic, a diverse Australian crocodile fauna existed throughout much of the continent, with most fossil assemblages containing three or more species. Most belonged to a regionally endemic subfamily of crocodylids, the Mekosuchinae, which first appear in the Eocene. These reptiles

STUART CREEK SILCRETE FLORA (BOX 3.3)

Nothing illustrates the aridification of Australia better than comparing the Eocene fossils of the Stuart Creek silcrete flora from the Lake Eyre Basin with the modern flora of that region, which is dominated by *Acacia*, chenopod shrubs, *Casuarina*, *Xanthorrhoea*, spinifex and samphire. The Stuart Creek silcrete flora is probably Middle Eocene in age and contains tree ferns, podocarp conifers, araucarian conifers, cypress conifers, putative *Eucalyptus*, paperbark (*Melaleuca*), flame trees (*Brachychiton*) or kapok trees (*Cochlospermum*), she-oaks (*Gymnostoma*), palms and Proteaceae.

Today, *Brachychiton* and *Cochlospermum* occur mainly in the tropics or along the humid east coast; *Melaleuca* is usually associated with watercourses and swamps; tree ferns are confined to tropical, subtropical or temperate rainforest; and podocarp and araucarian conifers are mostly restricted to tropical, subtropical or temperate rainforests or alpine areas. *Gymnostoma* is now restricted to tropical rainforests. These indicate that, at the time these plants were growing, the climate was warm and humid, probably with permanent rivers, billabongs and swamps.

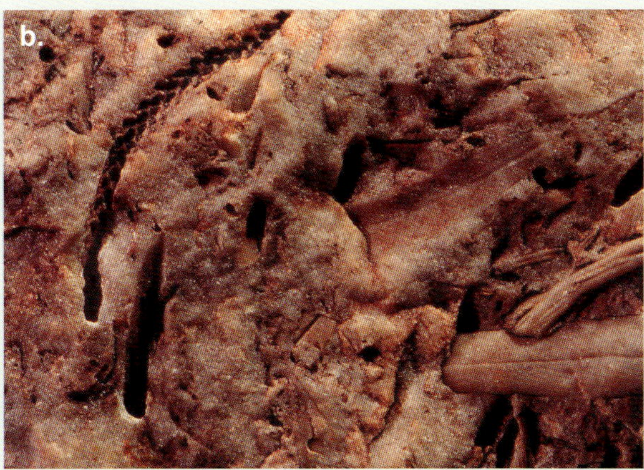

Figure B3.3: Representatives of the Stuart Creek flora, South Australia. (a) Three different angiosperm leaves. (b) Impression of a conifer twig. (c) Part of a palm frond. (d) A leaf with narrow lamina belonging to the Proteaceae. (e) Large leaf of a flame tree or kapok tree (*Brachychiton* or *Cochlospermum*). (Source: M White, 1994; photographs by Jim Frazier)

occupied a much greater range of niches than their modern counterparts; they included dwarf, perhaps even arboreal, forms (*Trilophosuchus*), terrestrial carnivores (*Quinkana*) and 'conventional' large semiaquatic predators (*Pallimnarchus*).

The earliest significant Cenozoic bird fossils come from the Eocene and include a giant penguin; other large, flightless birds (probably dromornithids); a stone curlew–like bird (but probably not closely related to modern stone curlews); and a songbird (passerine). The latter is the oldest passerine fossil known anywhere in the world and, coupled with molecular evidence, it seems most likely that this most successful group of birds evolved in Gondwana during the Paleocene (*ca* 65.5–56 Ma) and only reached the northern continents during the Oligocene (*ca* 28 Ma).

The passerines now account for more than half the world's bird species. In Australia, they include the lyrebirds, fairy wrens, pardalotes, currawongs, bowerbirds, willie wagtails, Australian magpies, honeyeaters, Australian finches and numerous others. Some later dromornithids were up to 3 m tall and may have weighed well over 300 kg. Originally thought to be related to the ratites, they are now known to be most closely related to waterfowl (ducks and geese).

The oldest Australian marsupial fossils are from the early Eocene (*ca* 54 Ma) of Murgon in southeastern Queensland (Figure 3.1), and these include three animals much like early marsupials from South America, together with one that is distantly related to the group that includes modern bandicoots and bilbies (the Peramelemorphia). This is roughly concurrent with the proposed origin of the group based on molecular evidence. There is only one extant species of bilby, the greater bilby, which is a nocturnal omnivore of the western desert. The lesser bilby, from the central Australian desert, has probably been extinct since the 1950s. Bandicoots are much more diverse and widespread, with about 20 species inhabiting desert to rainforest habitats. They are found throughout mainland Australia, Tasmania and New Guinea, having entered the latter recently as the close approach of Australia and lower sea-levels permitted.

There is increasing evidence that non-volant (non-flying) placentals had arrived in Australia by the Eocene, with fragmentary dental, cranial and post-cranial skeletal remains now known from the early Eocene of Murgon (Figure 3.1). If such animals did exist on this continent in the Eocene, they left no descendants. The oldest well-represented fossil placentals from Australia are early Eocene bats from the same locality. Nearly 25% of all mammal species are bats, and they are essentially worldwide in distribution. Although the oldest fossil bats are from the early Eocene, molecular evidence suggests that they may have evolved as early as the Late Cretaceous. The earliest fossil bats from Australia represent an archaic group, although it is clear that they could use echolocation. It seems likely that these early Australian bats entered from South America via Antarctica. The place of origin of bats is unknown, mostly because of the poor fossil record, but it seems likely that the 80 species of extant bats in Australia comprise a mixture of endemic groups (some 'evening bats' and 'leaf-nosed bats'; Figure 3.18) and more recent arrivals (flying foxes).

The cooling

Cooler temperatures in the southeast during the late Eocene (*ca* 37 Ma) caused mangrove palms to disappear from Tasmania, while the presence of *Ginkgo* and deciduous species of *Nothofagus* suggest cooler winter temperatures. Several more sclerophyllous and xerophytic genera, such as the *Banksia serrata* and *Dodonaea triquetra* (modern hop bushes) lineages, appear at this time. Mangrove palms still survived in the warm shelf waters of South Australia, while a more megathermal mangrove flora persisted in southern Queensland. This global cooling created intensifying climatic gradients, in both rainfall and temperature, resulting in increasingly provincial and heterogeneous vegetation into the Oligocene. The appearance of what is considered the characteristic Australian genus, *Acacia* (the wattle), may have been related to the expansion of drier conditions in some regions. Although the family Mimosaceae is widespread globally, pollen records suggest its first appearance in the early Oligocene (*ca* 30 Ma) in Australia, but indicate that the genus did not become common until the Miocene.

In the southeast, the thick accumulations of brown coal in the Gippsland Basin reflect the heterogeneous nature of Oligocene and Miocene vegetation communities. Within the coal seams of the Latrobe Valley, three main peat-swamp vegetation types have been identified: a *Lagarostrobus* (Huon pine)-dominated swamp forest, a Myrtaceae–Elaeocarpus (e.g. blueberry ash) swamp forest, and an open-canopied swamp forest distinguished by conifer and sclerophyll taxa. *Nothofagus* forest grew on the higher ground. Today, these coals provide most of Victoria's electricity (Chapter 9).

Some modern fish genera had evolved by the Oligocene (34 Ma), and several species have remained virtually unchanged since the Miocene. An example of this is the modern Murray cod (*Maccullochella peelii*), which is also represented by superbly preserved fossil specimens in middle to late Miocene (*ca* 12 Ma) diatomite near Bugaldie (Figure 3.1) in northern New South Wales. Today, Australia has a relatively low diversity of freshwater fishes, with about 280 species, but the fauna is highly distinctive.

Both widespread modern Australian frog families—the Hylidae (tree frogs, of the endemic Australo-Papuan subfamily Pelodryadinae) and the Myobatrachidae (Australasian ground frogs)—were well established in the Oligocene–Miocene rainforests of Riversleigh (with more than 20 species, arguably the most diverse fossil frog assemblage in the world). Many modern genera were already present, including the tree frog *Litoria* and several myobatrachids.

Due to their fragile osteology, the fossil record of snakes in Australia (and the rest of the world) is sparse. The archaic madtsoids were almost certainly present here in the Cretaceous (owing to their fossil record in other Gondwanan sites). Whereas this group had disappeared from other continents by the mid-Cenozoic, they continued to flourish in Australia, producing anaconda-sized constrictors (*Wonambi*), before their extinction in the late Pleistocene. By the Oligocene–Miocene (*ca* 23 Ma), blindsnakes, pythons and elapids had made their local appearances alongside the madtsoids. Filesnakes and colubrids are comparatively recent arrivals from Asia, with correspondingly low endemicity.

Murray cod (*Maccullochela peelii*) is the largest Australian freshwater fish, reaching 113.6 kg and 1800 mm in length. The species is important in Aboriginal mythology: a huge Murray cod is believed responsible for creating the Murray River and all its fishes.

Australasian ratites (emus and cassowaries) stem from an old Gondwanan lineage, which probably evolved during the Cretaceous, as they now occupy the southern continents and adjacent islands (e.g. rhea in South America, ostrich in southern Africa, kiwi in New Zealand and, until recently, elephant bird in Madagascar). The oldest fossil ratites in Australia are from the Oligocene (*ca* 28 Ma). Today, the emu occupies most of the mainland, whereas the cassowary is restricted to northeast Queensland.

The emu (*Dromaius novaehollandiae*) is Australia's largest bird, and is the world's second largest after the ostrich.

It was also during the Oligocene and Miocene that the first representatives of the waterfowl, grebes, cormorants, pelicans, flamingos, cranes, rails, stone curlews, birds of prey and pigeons also appear in the Australian fossil record.

The oldest diverse snapshot of the Australian marsupial fauna comes from Oligocene to Miocene limestone deposits of Riversleigh, in northwestern Queensland (Figure 3.1), where numerous early representatives of extant and extinct lineages have been discovered. The oldest thylacinid fossils are of late Oligocene age (*ca* 25 Ma). By the early Miocene (*ca* 20 Ma), at least four genera inhabited the rainforests of northern and eastern Australia; by 8 Ma there were three species; and by 5 Ma there was one, the modern Tasmanian tiger, although it was distributed throughout Australia and New Guinea.

The oldest fossil vombatiforms (of which wombats and koalas are the only living members) come from the late Oligocene (*ca* 25 Ma), but molecular evidence indicates an origin early in the Eocene. The koalas were a very diverse group in the Miocene, when the climate was wetter, and rainforest and suitable woodland covered more of the continent. In the Miocene, vombatimorphians (which include only wombats), like the koalas, were much more diverse, with as many as 11 genera being recorded. Some of the Miocene to Pleistocene forms were as much as three times the dimensions of the largest modern species (i.e. 2 m in length).

The diprotodontoids were the giants of the marsupial lineage, with the largest being just under 3 tonnes in weight. This group of herbivores was very diverse, with 20 or more genera having been recorded. Like many other groups of marsupials, the oldest fossils of this group come from the late Oligocene, but it is likely that their origins are older.

The largest of the thylacoleonids (marsupial lions) were the top predators of their day and reached the size of a modern lioness. They probably used their elongated canine-like lower incisors to tear out the throat of their prey, in much the same way as the sabretooth cats of the Northern Hemisphere used their oversized upper canines. Again, the

FIRE AND THE AUSTRALIAN BIOTA (BOX 3.4)

Fire is a key element in the functioning of Australian ecosystems, affecting the distribution of this continent's diverse flora and fauna. Adaptations shown by much of the modern flora suggest a long history of exposure to fire regimes of varying frequency and intensity. Adaptations to fires may be behavioural or morphological. Behavioural responses include the spectacular flowering of some species immediately after fire—grass trees (*Xanthorrhoea*) are a well-known example. Another behavioural trait is the triggering of seed release by fire—species of *Banksia* and *Eucalyptus* show this. Morphological adaptations offering protection to fire include the presence of subterranean buds (lignotubers), or above-ground buds concealed beneath thick bark. Although adaptations to fire are unlikely to be preserved as fossils, there are direct indications of fire in the Australian sedimentary record. In the Murray Basin, for instance, charcoal conglomerates are known from Lower Cretaceous sediments. Widespread rainforest vegetation in the Paleogene in Australia is probably linked with a lower fire frequency, but drier conditions and more open vegetation types characterise the Neogene and correlate with a general increase in burning. In the Gippsland Basin, the Miocene Yallourn Coal Seam contains a high proportion of fossil charcoal (fusinite).

But it is in the Quaternary that the record of fire and its impact on the biota is best documented. A general increase in charcoal within records from the Late Quaternary is considered to reflect drier and more variable conditions. Evidence for the acceleration of this trend in some records within the last glacial–interglacial cycle has been interpreted to reflect an increase in ignition resulting from the arrival and impact of Aboriginal people on the continent. By contrast, analysis of a large compilation of charcoal records covering the last 70 ka has failed to detect a notable human influence, except during the period of European occupation. It is proposed that fire activity on both orbital and millennial scales has been higher during wetter periods, probably as a result of higher levels of plant biomass. However, debate over the relative importance of people and climate on fire activity is active in the literature, globally as well as in Australia.

Figure B3.4: (a) Australian bushfire. (b) Aftermath of fire. (c) Bush regenerating after fire.

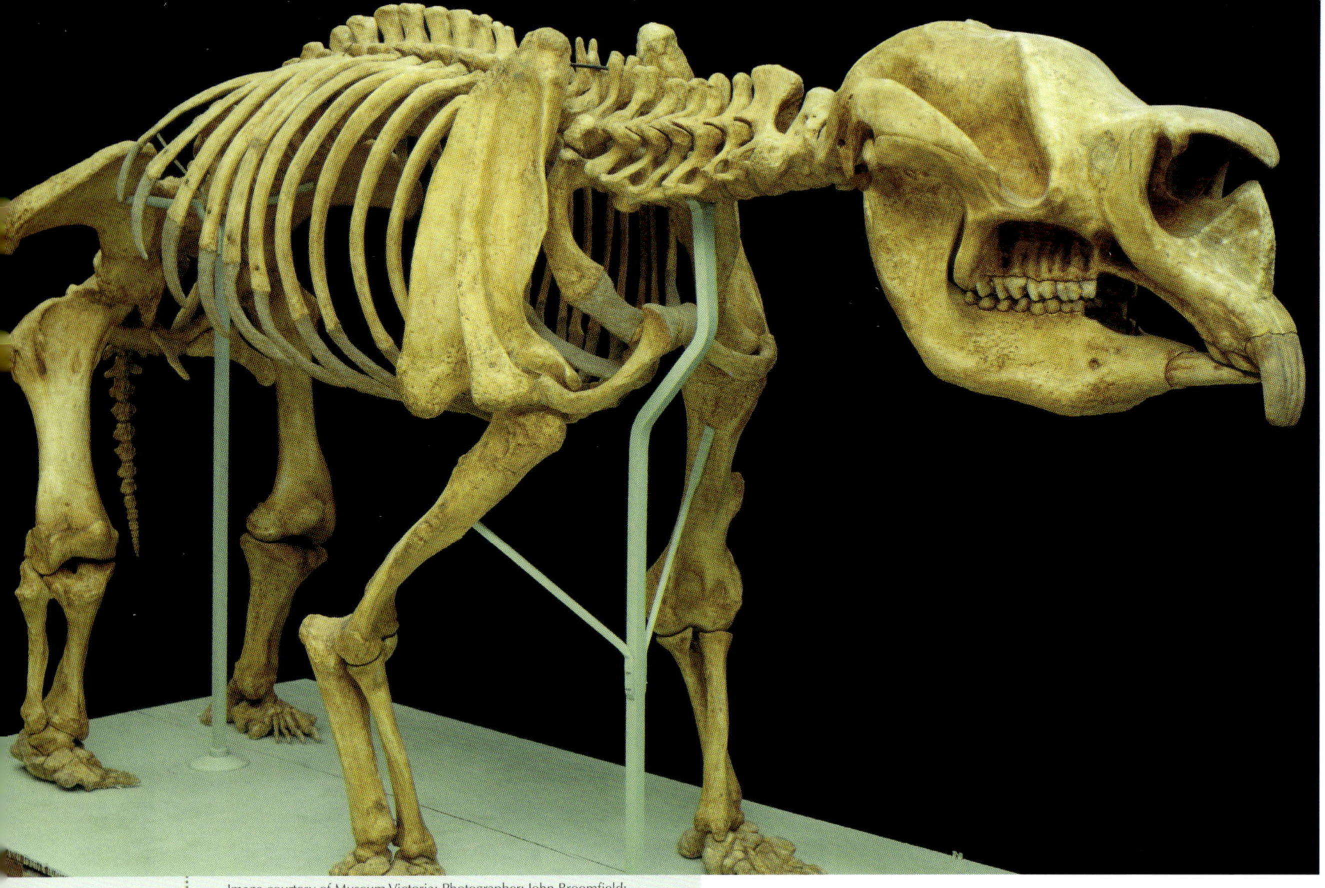

Image courtesy of Museum Victoria; Photographer: John Broomfield; Copyright Museum Victoria 1997

Skeleton of the hippopotamus-sized *Diprotodon optatum;* the largest marsupial that ever lived.

oldest thylacoleonid fossils are known from the late Oligocene, but their origins are probably much older.

The phalangeriforms include the plethora of possums, pygmy possums, cuscuses and gliders. The oldest members of this group in the fossil record are from the late Oligocene (*ca* 25 Ma). The distribution of some lineages altered dramatically with the aridification of the continent in the Neogene, with the ranges of many that were adapted to cooler, wetter environments decreasing as the continent dried out. An example of this is the mountain pygmy possum lineage; the fossil record indicates that this was once widespread in inland Australia, but it is now found only in the Mt Hotham and Mt Kosciuszko areas (Figure 3.1). It is also clear that the group spread into New Guinea and Sulawesi as the close approach of Australia and lowered sea-level allowed (Chapter 1).

The extant kangaroos are most likely descended from a phalangeriform, and molecular evidence indicates that this may have occurred in the Eocene. The macropods are divided into two main groups: the kangaroos and wallabies (Macropodidae), and the rat kangaroos (Potoroidae and Hypsiprymnodontidae). Early in the evolution of the group, in the late Oligocene to early Miocene (*ca* 25–17 Ma), the extinct kangaroo family Balbaridae was more diverse than any of the modern lineages, but, as the continent dried out and forest was replaced by grasslands and woodlands, balbarids declined in diversity and abundance, and the kangaroos and rat kangaroos diversified.

The warming

By the end of the Miocene (*ca* 5.5 Ma), strong contrasts between the coastal and inland vegetation were well established along similar, but less intense, gradients to those of today (Figure 3.13). Rainforest and some sclerophyll taxa that now live in different communities coexisted in a mosaic landscape, while araucarian forest was still widespread, together with conspicuous Podocarpaceae, along the eastern seaboard and southern Western Australia. By this time, the range of *Nothofagus* (*Brassospora*) had shrunk so that it probably only survived in highland areas along the eastern seaboard.

Together with drying of the landscape (Figure 3.13) and expansion of sclerophyllous vegetation, fire became an increasingly important component of the environment through the Neogene (Box 3.4). Many Australian plant groups show particularly strong adaptations to surviving wildfire (e.g. regrowth from epicormic buds and lignotubers, and production of fire-resistant seeds and capsules). Indeed, some groups have become strongly dependent on fire to release their seeds or to provide an open landscape for establishment of seedlings. The wildfires that regularly devastate large tracts of the Australian bush and human infrastructure are essential for the survival of many native species (Box 3.4).

Cockatoos and swiftlets are first found in the early Miocene (*ca* 17 Ma) of northern Australia, as are the first lyrebirds, logrunners, orioles, storks, kingfishers and owlet-nightjars. Owlet-nightjars are an exclusively Australasian group, but related forms (apodiforms) have been recorded from the Paleogene of Eurasia. Kingfishers (e.g. modern kookaburra) are thought to have originated in South-East Asia and dispersed to Australia during the Miocene.

The Oligocene–Miocene rainforests that grew at Riversleigh, Queensland, harboured many forms of lizard, including geckos, legless lizards (which, phylogenetically, are also geckos), skinks (including the modern blue-tongued lizard genus, *Tiliqua*) and agamids (including the modern water dragon genus, *Physignathus*). The goannas and monitors (Varanidae) have truly flourished in Australia (excluding Tasmania) and include both imposing apex predators and unique pygmy forms. They appear in the Australian fossil record in the Miocene, and today Australia plays host to half of the world's varanid diversity.

The fossil record of monotremes is very sparse, so it is difficult to interpret the phylogeny and biogeography of the group. It is most likely that they are an early offshoot of the Mammalia, separating from the lineage leading to modern mammals between the Early Jurassic and Early Cretaceous. The oldest monotreme fossils are from the Early Cretaceous of Australia and represent a variety of groups, including platypus-like, but not echidna-like, forms. The oldest echidna fossils are Miocene in age. A platypus-like fossil from the Paleocene of Argentina indicates that the monotremes were a Gondwanan lineage, but that they became extinct on the other Gondwanan continents, only surviving in Australia, and spreading to New Guinea as proximity and lowered sea-level allowed.

Dasyurids are mostly nocturnal carnivores and insectivores. They include the modern quolls, dibblers, false antechinuses, marsupial mice,

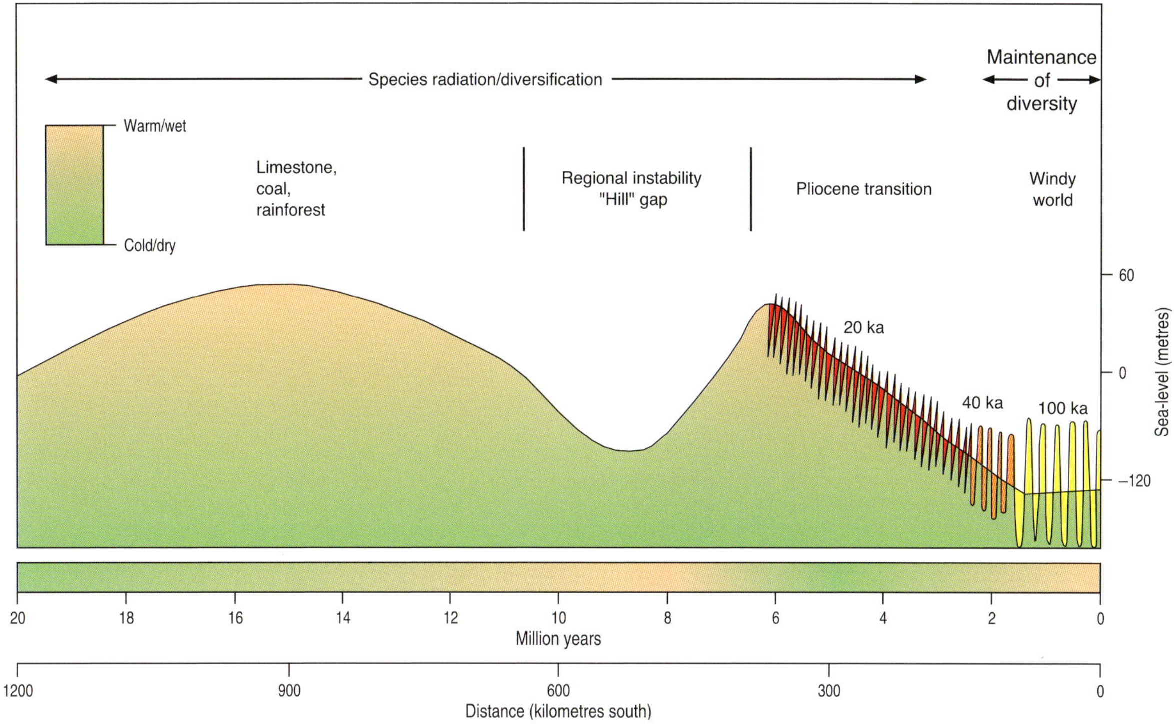

Figure 3.13: Summary of the palaeoclimatic conditions in Australia during the evolution of the arid-zone biota from 20 Ma to present. The vertical axis represents sea-level, but is not to scale. The horizontal axis represents, at the top, time in millions of years before present and, at the bottom, distance that the continent of Australia was further south than at present. Development of cycling of climatic conditions and sea-level changes is evident through the Pliocene and Pleistocene. Shading indicates warm/wet vs cold/dry climatic conditions. (Source: modified from Byrne et al., 2008)

Fossil skull of a Miocene platypus, *Obdurodon dicksoni*, from Riversleigh, western Queensland.

kowaris, mulgaras and the Tasmanian devil. The oldest fossil dasyurids are from the early Miocene (*ca* 17 Ma), with molecular data indicating a middle to late Oligocene origin, with dasyuromorphs as a whole (dasyurids, numbats and thylacinids) diverging from other marsupials in the Paleocene to Eocene. Miocene fossils indicate that the marsupial moles evolved in rainforest habitats and were pre-adapted for digging in sandy soil as the continent dried out in the late Cenozoic. The two remaining marsupial mole species are today restricted to the continent's northern and central sandy deserts.

The drying

In pockets of southwestern and southeastern Australia, there was a transient resurgence in *Nothofagus* (*Fuscospora*, *Lophozonia*), accompanied by peaks in rainforest conifers, during the early Pliocene (*ca* 4.5 Ma), but, otherwise, the trend towards *Acacia*, *Casuarina*, *Eucalyptus* and low, open communities dominated by Asteraceae (e.g. daisies), Chenopodiaceae (e.g. samphire), Gyrostemonaceae (e.g. corkybark) and Poaceae (true grasses) continued (Figure 3.13). An alternation of cool rainforest communities with low, open communities, suggesting climatic swings, is evident in probable Pliocene sediments underlying Lake George, near Canberra. Two xerophytes (*Eucalyptus spathulata* and *Stirlingia*), which first appear across southern Australia in the late Miocene, had become confined to the southwestern Western Australia by the end of the Pliocene.

At about the end of the Pliocene (*ca* 2.6 Ma), across southeastern Australia there were marked declines in cool temperate rainforest and wet heath taxa (Figure 3.13). Many of the taxa that disappeared from the southeastern Australian pollen record survived in the northeast, where the changes to the flora were not as pronounced. However, *Nothofagus* (*Brassospora*) seems to have disappeared from the northeast at this time. Almost all of the taxa that declined in southeastern Australia have relatives whose predicted temperature ranges should have permitted them to survive at these high latitudes. It is, therefore, unlikely that a temperature decrease alone led to these extinctions, and most likely that increased climatic variability and reduced precipitation were involved. It has been suggested that there was a switch from predominantly summer to winter rainfall at this time in southeastern Australia, due to intensification and northward movement of the mid-latitude high-pressure belt with the continued development of the Antarctic icecap, but this event is now considered to have occurred later, during the mid-Pleistocene transition (Figure 5.5). It also appears that the decline in rainforest taxa in southeastern Australia at the end of the Pliocene has been exaggerated, since one recently analysed early Pleistocene pollen site has revealed the local survival of rainforest as diverse as any in the Australian Cenozoic, although it was clearly surrounded by sclerophyll forests.

The discovery, in 1994, of a small stand of the araucarian *Wollemia nobilis* (Figure 3.14)—a taxon unknown from the spore-pollen record since the Pleistocene—in the Blue Mountains of New South Wales demonstrates the ability of the rainforest to survive in small patches protected from fire through considerable periods of time.

Pollen indistinguishable from that of the modern Wollemi pine, and referred to the fossil genus *Dilwynites*, has long been known from the fossil record, and occurs from the Late Cretaceous through most of the Cenozoic of Australia, Antarctica and New Zealand. There are close links too with macrofossils, such as the lanceolate linear leaves known from the mid-Cretaceous Winton Formation of Queensland (Figure 3.14). In northeastern Australia and probably along most of the eastern seaboard, the vegetation in the early Pleistocene continued to be dominated by drier araucarian forest; wetter rainforest was restricted to coastal and montane areas. Little is known about the vegetation in the more arid centre of the continent, but records from the similarly arid northwest indicate that it was probably hummocky grassland with a sparse, open *Acacia* canopy, or savanna grassland with a *Casuarina* canopy.

Toward the end of the Early Pleistocene (*ca* 1 Ma), there were marked changes in the vegetation of the northeast. There was an increase in importance of mangroves, which seems to have been facilitated by strong progradation at the end of glacial cycles, a suggestion supported by the close relationship between peaks in mangrove pollen and the glacial–interglacial transitions. Mangrove growth may also have been facilitated by increased sea-surface temperatures in the Coral Sea and associated development of the Great Barrier Reef (Chapter 6), which would have provided protection for mangrove expansion. By contrast, higher temperatures, resulting from an increasing latitudinal thermal gradient in combination with increasing amplitude of climate cycling between warm, wet interglacial and cool, dry glacial intervals, may have placed sufficient stress on the cool, temperate conifers to bring about their gradual elimination.

The first evidence of Quaternary vegetation in monsoonal northwestern Australia indicates the dominance of eucalypt savanna through at least the last 500 ka. Changes in tree density resulted from monsoon intensity related to a combination of Northern Hemisphere–driven glacial–interglacial cyclicity and regional Southern Hemisphere insolation. Sustained changes in vegetation under progressively drier and/or more variable climatic conditions and altered fire regimes are recorded broadly over the continent within the last 200 ka or so. The major response was in the northeast, where araucarian forest began a dramatic decline relative to eucalypt woodland from around 130 ka, while wetter forest elements, such as palms and ferns, were substantially replaced by grasses along the coastal plains about 175 ka. This influence extended to northwestern Australia, where there is evidence for a shift to a more open canopy in savanna woodland and a loss of drier rainforest, while, down the east coast of the continent, the most marked change was the demise of Casuarinaceae and minor rainforest elements relative to *Eucalyptus* from about 130 ka. In addition, Cunoniaceae (e.g. modern Christmas bush) and *Quintinia* (possumwood) disappeared from Tasmania, while *Phyllocladus* (celery-top pine) and *Gunnera* (Tasmanian mudleaf) disappeared from the mainland and are now restricted to Tasmania. The geographical pattern of decline tends to reflect the initiation or intensification of the El Niño–Southern Oscillation (ENSO) climate phenomenon around this time (Figure 1.5).

Figure 3.14: Foliage of *Wollemia nobilis* and a close relative from the Cretaceous. (a) Foliage of *Wollemia nobilis*, showing transition from juvenile foliage (in two ranks) to adult foliage (in four ranks). (b) Cretaceous fossil juvenile twig of an araucarian conifer from the Winton Formation, western Queensland.

Did you know?

3.2: Ten of the most venomous snakes in the world are native to Australia

The LD50 test is a scale that measures the potency of a snake's venom. It refers to the amount of venom that, when injected into mice, kills 50 per cent of those injected. So the smaller the amount of venom to reach LD50 (median lethal dose), the more potent is the venom. Using this as the criterion, the 10 most venomous snakes in the world are native to Australia, according to the Australian Venom Research Unit at the University of Melbourne. Of these, the five most venomous are:

1. inland taipan (fierce snake)—*Oxyuranus lepidotus*
2. eastern brown snake—*Pseudonaja textilis*
3. coastal taipan—*Oxyuranus scutellatus*
4. tiger snake—*Notechis scutatus*
5. black tiger snake—*Notechis ater*.

The most toxic venom of snakes not native to Australia is that of the Asian cobra. However, the median lethal dose of cobra venom is 22.5 times that of the inland taipan. Despite the extreme toxicity to mice of the venom of Australian snakes, far fewer people (2–3 per year) are killed by Australian snakes than by snakes on other continents, probably because of the widespread availability of anti-venom, our sparse human population, or the more urbanised nature of Australia. It has been estimated that worldwide there may be as many as 125 000 deaths from snakebite per year. The reason why Australian elapid snakes are so venomous is unknown.

Although recent analyses of charcoal records in Australia have tended to indicate a strong relationship between fire and climate, many changes—the most conspicuous being the rapid demise of *Araucaria* in the northeast and the disappearance of several conifers on the southeastern mainland around 40 ka—suggest that burning by Aboriginal people had some impact, especially given the weight of archaeological evidence for continental colonisation at this time and the lack of correlation with a major global climate event. As a result of the combination of events during the late Quaternary period, it is likely that environmental conditions were most extreme during the height of the last glaciation, around 25–20 ka, and that vegetation recovery during the present interglacial (Holocene) has been restricted by the loss of some components of the plant landscape and an increase in fire-tolerant and fire-promoting communities (Box 3.4).

Among the fishes, and besides the Cretaceous relicts and catfishes, the majority of the larger species in Australian freshwaters are various forms of percoid (perch-like teleosts). South of the tropics, including in the Murray–Darling Basin, these are predominantly percichthyids—notable forms include the Murray cod, golden perch, Australian bass and river blackfishes, as well as the smaller pygmy perches. In warmer areas, the coastal drainages abound with diverse terapontids (grunters, spangled perch) alongside archerfishes, jungle perch and the iconic barramundi (*Lates calcarifer*, actually a widespread catadromous giant perch that is distributed throughout tropical Indo–West Pacific). Major groups of smaller fishes include colourful goby-like forms (true gobies and eleotrid gudgeons) and endemic radiations of non-percoids like the atheriniforms (hardyheads, Australasian rainbowfishes and blue-eyes) and, in cooler areas, the osmeriforms (Australian smelt, galaxiids and the endemic salamanderfish).

Away from the major drainages, the predominantly arid Australian landscape presents special challenges for freshwater fishes, with local species exhibiting striking adaptations for life under extreme conditions. In Witjira National Park, on the western fringe of the Simpson Desert in South Australia, gobies, catfish and hardyheads live in spring-fed billabongs with temperatures rising to more than 40°C. The salamanderfish (*Lepidogalaxias*) of southwestern Australia dwells in acidic ephemeral pools where the pH may be as low as 3.0, aestivating in the sand when the water evaporates over summer. Beneath the Cape Range Peninsula of Western Australia, blind milyeringid cave-gudgeons navigate through sunless subterranean lakes that experience extreme shifts in salinity, low oxygen levels and layers of noxious H_2S.

The largest known Australian frog was the cane toad–sized *Etnabatrachus maximus* from the Pleistocene of Mt Etna, Queensland. It is provisionally assigned to the Hylidae, of which the largest living Australian frog (white-lipped tree frog, *Litoria infrafrenata*) is also a member. More recent arrivals from Asia are the microhylids and one species of ranid that have colonised the wetter parts of the tropical north. The bufonids (true toads) were unknown in Australia until the introduction of the cane toad (*Bufo marinus*) in the 1930s, with disastrous consequences for the native fauna.

The environmental extremes of Australia have forced many frogs to adopt innovative survival and reproductive strategies. The water-holding frog (*Litoria* [*Cyclorana*] *platycephala*—actually an aberrant 'tree frog') survives drought by cocooning itself underground, with large amounts of water stored in its bladder, and aestivating for several years, if necessary. Some myobatrachids dispense with free-swimming tadpoles, raising their young in hip 'pockets' (*Assa*) or even inside the stomach (*Rheobatrachus*). Much of Australia's unique frog fauna is in decline, with several species critically endangered or feared extinct. Threats include habitat destruction, feral animals, climate change and infection from chytrid fungus.

For much of the Cenozoic, soft-shell turtles (family Trionychidae) were abundant in the region, but they vanished from Australia in the Pliocene (remaining abundant elsewhere in the tropics). The meiolaniids grew into horned behemoths (*Meiolania*, *Ninjemys*), but perished with most of the megafauna in the Late Pleistocene. The final representatives of this ancient family persisted on remote Pacific islands until well into the Holocene. Side-necked turtles (family Chelidae) appeared in the Australian record during the Miocene, presumably having arrived from South America (where they appeared in the Cretaceous) via Antarctica, and became the dominant freshwater turtle group.

Living Australian cryptodires (those that retract their necks into the shell in a vertical plane) comprise six species of sea turtle (including the regionally endemic flatback turtle) and the freshwater pig-nosed turtle (*Carettochelys insculpta*) of the Northern Territory and southern New Guinea. The latter is the sole survivor of a family that was diverse and widespread in the Eocene. Although much of the arid interior is devoid of turtles, the side-necked turtles are abundant in most freshwater habitats in the remainder of the continent, with major centres of diversity in the 'Top End' of the Northern Territory and the Fitzroy–Dawson drainage system of eastern Queensland. Some have evolved highly specialised habits, like the critically endangered western swamp tortoise (*Pseudemydura umbrina*) that lives in ephemeral pools and aestivates underground through the baking summer, and the Fitzroy turtle (*Rheodytes leukops*) that rectally acquires oxygen from the surrounding water via a highly vascularised cloaca, hardly ever surfacing for air.

The last Australian mekosuchine crocodiles vanished in the late Pleistocene, but the group persisted on remote Pacific islands (New Caledonia, Vanuatu) until the arrival of humans in those lands only a few thousand years ago. The two living crocodile species of the Australian north belong to the pantropical genus *Crocodylus* and are comparatively recent additions to the local fauna. The saltwater crocodile (*C. porosus*), the largest living reptile, is widespread throughout the tropical Indo-Pacific and arrived in Australia in the Pliocene or latest Miocene. The endemic freshwater crocodile (*C. johnstoni*) appeared in the Pleistocene of Queensland, probably evolving from ancestral *C. porosus* stock.

Prudent Australian bushwalkers are all too aware of the presence of the venomous Elapidae in Australia (*Did you know?* 3.2), the family that includes most of Australia's iconic snakes (taipan, brown snakes,

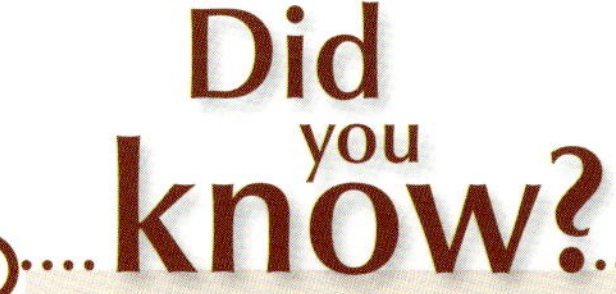

3.3: Goannas and dragons are venomous too

It has recently been discovered by researchers at the University of Melbourne that agamid (including the bearded dragon) and varanid lizards (including Mitchell's water monitor and lace monitor) are venomous. Bites from some species can cause swelling, pain, disruption of blood clotting and a decrease in blood pressure and may paralyse small prey.

© Getty Images [P Nevin]

Figure DYK3.3: The bearded dragon (*Pogona* sp.) is known as ngarrang by the Dharawal people from the Sydney region.

Two intertwined coastal taipans (*Oxyuranus scutellatus*).

tiger snakes, death adders, etc.), along with the sea snakes (as well as cobras, kraits, mambas and coral snakes on other continents). Australians can take pride in knowing that theirs is the only continental herpetofauna where the venomous outnumber the non-venomous snakes, in both abundance and diversity. Molecular evidence indicates that Australian elapids and sea snakes form a group exclusive of the rest of the family, with the sea snakes evolving from terrestrial Australian ancestors.

About half the world's python species (Pythonidae) are found in Australia. The largest Australian snake (the scrub python, which exceeds 5 m in length) is of this family. Most of the locals are much smaller, however, including the world's smallest python, the 50 cm long *Antaresia perthensis*.

Of the remaining families, the blindsnakes (Typhlopidae) are a widespread, primitive family of burrowers that resemble oversized earthworms, with more than 30 Australian species (*Ramphotyphlops* spp.). Colubrids, locally including a collection of tree snakes and water snakes, are the most diverse family in the world, but are represented in Australia by only a handful of species, most of which are shared with New Guinea and Indonesia. They include both non-venomous and rear-fanged (as opposed to the front-fanged elapids) venomous species. Filesnakes (Acrochordidae) are flabby, aquatic species; the distributions of two widespread South-East Asian species extend to northern Australia.

Most Australian gecko species belong to a regionally endemic radiation, comprising the diplodactyline geckos (knob-tailed, leaf-tailed and velvet geckos) and the Australasian legless lizards (pygopods), which have clearly been here for a very long time. The legless lizards are the most heavily modified of this group (traditionally separated into their own family), having lost their functional limbs, with some evolving a highly kinetic skull to assist in swallowing large prey. Also in Australia are the more globally widespread gekkoines that arrived in the region more recently.

With great endemic diversity, some of the most distinctive Australian lizards are agamids (Figure 3.15), including the frill-necked lizard (*Chlamydosaurus*), the thorny devil (*Moloch*) and bearded dragons (*Pogona*; *Did you know?* 3.3). Many of the local forms are arid adapted, while surprisingly sparse in rainforest (given their abundance in tropical Asian jungles), being represented here by only two species of angle-heads (*Hypsilurus*; Figure 3.15).

The skinks (Scincidae) are the most diverse saurian group on the continent (and globally), with well over 300 Australian species described. These adaptable reptiles are found everywhere, from baking deserts to the summit of Mt Kozsciusko. The familiar blue-tongues and shinglebacks (*Tiliqua* species; Figure 3.15) are among the largest skinks in the world. The Australian assemblage is highly endemic; few genera extend beyond southern New Guinea, and much of the diversity is probably locally derived rather than the result of repeated invasions.

At well over 2 m in length, the perentie (*Varanus giganteus*; Figure 3.15) is the biggest lizard in the modern outback. Its larger cousin, the Komodo dragon (*Varanus komodoensis*) had its origins in Australia before migrating to Indonesia, as it is present in Queensland Pliocene deposits. Both were dwarfed by the Megalania (*Varanus priscus*), a 5.5 m–long colossus that was the largest terrestrial predator of Pleistocene Australia and may have survived long enough to terrorise the first human arrivals.

Most of the bird lineages that appear in the fossil record in the Oligocene to Miocene still occur in Australia. The story of the flamingos is unusual, in that they were quite diverse in Australia during the Miocene to Pleistocene, but became locally extinct as increasing aridity in central Australia led to the large waterbodies, which they frequented, becoming ephemeral rather than permanent.

By the Pliocene (*ca* 5.5 Ma), the number of birds invading from the north had increased, with the black-necked stork (sometimes called jabiru; Figure 3.16), ibis, swamphen and bustard, among others, appearing in the fossil record. At this time, the avifauna had essentially developed its modern character (Figure 3.16). The radiations of the inherited Gondwanan taxa comprised about 80% of the fauna, and the remainder comprised post-Oligocene immigrants. The Gondwanan forms had diversified greatly, but the immigrant species did not diversify much (except for the grassfinches), probably because of their relatively recent arrival coupled with the prior occupation of most of the available niches.

The monotremes are now known from only three genera: the platypus (*Ornithorhynchus*) and two genera of echidna (*Tachyglossus* and *Zaglossus*). The platypus is restricted in its range to the eastern mainland and Tasmania, whereas the short-beaked

Figure 3.15: Representative Australian lizards. (a) Perentie (*Varanus giganteus*), Australia's largest varanid. (b) Spiny-tailed gecko (*Strophurus spinigerus*). (c) Gippsland water dragon (*Physignathus lesueurii howittii*), an agamid. (d) Bobtail skink (*Tiliqua rugosa rugosa*). (Source: © Getty Images [J Cancalosi]). (e) Boyd's forest dragon (*Hypsilurus boydii*), an agamid. (Source: © Getty Images [M Harvey]). (f) Burton's legless lizard (*Lialis burtonis*), a pygopodid.

echidna (*Tachyglossus*; Figure 3.17) is known from Tasmania, mainland Australia and New Guinea. The several species of long-beaked echidna (*Zaglossus*) are known only from New Guinea and from fossils on mainland Australia.

The last remaining thylacine species, the Tasmanian tiger, still survived on the mainland until about 4 ka and lingered on in Tasmania until the 1930s.

Of the previously very diverse vombatiforms, only one species of koala survives in the woodlands of South Australia, Victoria, New South Wales and Queensland. The wombats are currently represented by three species in two genera; they are mostly restricted to the eastern and southeastern mainland and Tasmania.

The diverse and large diprotodontids completely disappeared in the late Pleistocene, possibly only a few thousand years after the arrival of humans on the continent. Similarly, their main predator, the marsupial lion *Thylacoleo carnifex*, also became extinct at about the same time.

The phalangeriforms are widespread in Australia, New Guinea and Sulawesi, having spread to the latter two as Australia approached from the south. Some of this group have proven to be very adaptable, with the common brushtail possum being found over most of the continent, from arid woodland in central Australia to cool temperate forests and even in tropical rainforests along the east coast. It is also commonly found in suburbia and is the marsupial most commonly seen by inhabitants of Australian cities. It was introduced to New Zealand in the late 1800s and, without any natural predators, its population increased dramatically, such that it is considered a pest there.

There are nearly 80 species of Macropodidae (wallabies, kangaroos, wallaroos, pademelons, *Dorcopsis*, tree kangaroos, potoroos and the quokka; Figure 3.17), and they are widespread throughout the continent, having diversified at the expense of their extinct relatives the Balbaridae.

The numbat (Figure 3.17) is the only living species of myrmecobiid, and there are no pre-Pleistocene fossils of the group; those from the Pleistocene belong to the modern species. Molecular data indicate a much older origin, of about late Oligocene, for this group.

The rodents (Figure 3.18) are the planet's most spectacularly successful group of mammals, with the number of extant species roughly equalling that of all other groups of mammals combined. The oldest rodent fossils are from the late Paleocene (*ca* 58 Ma) of North America, and it is most likely that they evolved in Laurasia (modern North America, Europe and Asia). They had spread to Africa and South America by about the end of the Paleogene (*ca* 23 Ma), but, because of Australia's isolation, they did not reach this continent until the late Miocene or early Pliocene (*ca* 6 Ma), presumably invading from the north as Australia approached South-East Asia. Australian rodents now represent about 25% of all mammal species on the continent, and all come from the family Muridae (true rats and mice); they include hopping mice (Figure 3.18), tree rats, rock rats and water rats.

a.

e.

Figure 3.16: Representative Australian birds.
(a) (opposite) Rainbow lorikeets (*Trichoglossus haematodus*). (Source: Shane Cridland and the Australian Government Department of Sustainability, Environment, Water, Population and Communities). (b) Black-necked stork (sometimes called jabiru in Australia; *Ephippiorhynchus asiaticus*). (Source: © Getty Images [J&C Sohns]). (c) Splendid fairy wren (*Malurus splendens*). (d) Kookaburra (*Dacelo novaeguineae*). (Source: © Getty Images [S Turner]). (e) (opposite) Sulphur-crested cockatoo (*Cacatua galerita*). (f) Southern cassowary (*Casuarius casuarius*). (g) Gouldian finch (*Erythura gouldiae*). (Source: except where noted, Brian Furby collection)

Figure 3.17: Representative Australian monotremes and marsupials. (a) Short-beaked echidna (*Tachyglossus aculeatus*). (Source: © M Cohen, Lonely Planet Images). (b) Fat-tailed dunnart (*Sminthopsis crassicaudata*). (Source: © Getty Images [T Mead]). (c) Numbat (*Myrmecobius fasciatus*). (Source: A White, University of New South Wales). (d) Bilby (*Macrotis lagotis*). (Source: Arid Recovery). (e) Tiger quoll (*Dasyurus maculates*). (Source: A White, University of New South Wales). (f) Eastern grey kangaroo with joey in pouch (*Macropus giganteus*). (Source: © Getty Images [G Lewis])

Arrival of humans

The first humans to leave Africa were members of *Homo erectus*, reaching China by perhaps as early as 2 Ma and Java by as early as 1.6 Ma. Although *H. erectus* reached Java and Flores, there is no evidence for their presence on Borneo, Sulawesi, New Guinea or Australia (Figure 1.2). It has been suggested that this is because *H. erectus* was only capable of crossing small ocean barriers, perhaps where the destination island could be seen from the departure point. This was perhaps why travelling from the Asian mainland down through the Sunda Arc was possible, but travelling to the larger islands to the north (Borneo, Sulawesi) or the east (New Guinea, Australia) was not.

Homo sapiens (modern humans) appeared in Africa about 250 ka and initiated a wave of migration about 100 ka. They reached India by 70 ka, Europe by 50 ka (where they encountered *Homo neanderthalensis*), Australia and the large islands of South-East Asia by 60–40 ka (having encountered *Homo floresiensis* along the way), North America by 30 ka and South America by 15 ka. From this rapid dispersal, it is clear that *Homo sapiens*, unlike *H. erectus*, was capable of undertaking significant ocean journeys.

Aboriginal people reached Australia during the last glacial phase, which, at its subsequent climax around 20–18 ka, saw sea-levels fall to around 125 m below present levels, making ocean barriers less intimidating and exposing the seafloor between New Guinea and Australia, such that travel between them could be on foot (Figure 1.2).

Aboriginal people mostly remained nomadic hunter gatherers until well after Europeans arrived in Australia, although it is clear that, in the southeast, along the Murray and Darling River systems (Victoria–New South Wales) and along some of the coastal rivers in Victoria, there were seasonal settlements based around fish and eel trapping.

Although it is clear that the megafauna of Australia was in serious decline through much of the Pleistocene, largely because of the drying out of the continent, it is possible that the *coup de grace* was delivered to many species belonging to the megafauna by the arrival of the first humans on the continent, as a result of either hunting or concomitant changes to the fire regime and hence the vegetation. Firestick farming, as it has been called, involved regular use of fire in certain seasons to facilitate hunting and foraging, and to change the composition of plant and animal species in the area being burnt. It has been considered that this changed the vegetation of the continent, with an increase in more fire-tolerant species, the spread of fire-dependent species and the transformation of scrublands into grasslands, leading to a decline in browsers and an increase in grazers. However, as previously mentioned, the evidence for these fire impacts from analysis of charcoal is very limited temporally, with fire activity being closely related to climate change. Indeed, evidence of a major and sustained decline in megafaunal dung fungi just before the increase in charcoal, and the beginning of the replacement of araucarian rainforest by sclerophyll vegetation in a recent high-resolution record from Lynch's Crater in north Queensland from around 43 ka, suggests strongly that the increase and impact of burning were a result of megafaunal demise, presumably through overkill.

The arrival of humans on the continent led to the concomitant introduction of many and varied species of plants and animals. Perhaps the first of these was the dingo (Figure 3.18). The dingo is related to the wild dog of South-East Asia and is genetically very similar to most domestic dogs. Molecular studies indicate that a small founding group of dingoes arrived in Australia about 5 ka. They are now the largest terrestrial predator on the continent and are widespread on the mainland, but are not found in Tasmania. It is thought that

Figure 3.18: Representative Australian placentals. (a) Dingo (*Canis lupus dingo*). (Source: Arid Recovery). (b) Large-footed myotis (*Myotis macropus*). (Source: Arthur White, University of New South Wales). (c) Spinifex hopping mouse (*Notomys alexis*). (Source: Arid Recovery)

dingoes may have been one of the factors leading to the extinction of the preceding largest terrestrial predator, the thylacine, on the mainland about 4 ka; in Tasmania, where there are no dingoes, the eventual extinction of the thylacine in the 1930s was caused predominantly by bounty hunting.

More humans

The first European landfall on the continent of Australia was made by Willem Janszoon on the west coast of Cape York Peninsula, Queensland, in 1606, during an exploratory voyage in his ship *Duyfken*. However, it was not until 1788 that Europeans began to arrive in numbers, when a fleet of nine transports accompanied by two Royal Naval escorts arrived in Botany Bay.

Apart from the dingo, most of the remaining invasive species have been introduced deliberately since European settlement. Invasive species are non-indigenous organisms that adversely affect the ecosystems they invade. Invasive animals can impact native species by parasitism, predation, competing for food and nest sites, damaging habitat and spreading disease. Invasive plants can impact native species by parasitism, hybridisation, displacement, altering the fire regime, altering soil properties and decreasing water availability (Figure 3.19).

In 1859, about two dozen rabbits imported from England were released on a farm near Winchelsea in Victoria for the purposes of hunting. Within 10 years, as many as 2 million were shot or trapped annually and, by 1950, the population was estimated at about 600 million. Rabbits eat plant species that are also part of the diet of such native animals as bilbies, and will sometimes take over their burrows. Biological control measures have proven to be the most effective, with the myxoma virus in the 1950s and the calicivirus in the 1990s dramatically decreasing the rabbit population. However, the population is again on the increase.

Perhaps a more well-meaning introduction was that of the cane toad. A native of Central and South America, it was introduced from Hawaii in 1935 to combat the cane beetle, which was attacking sugar cane crops on the eastern seaboard. Unfortunately, it did not have a significant effect on the cane beetle, but its spread across much of tropical Queensland and the Northern Territory, coupled with the toxicity of its skin secretions, has led to significant decreases in the populations of some varanid lizards (e.g. Mertens' water monitor), the Johnston River crocodile, and marsupial predators such as the northern quoll. Other land animals that have had a significant impact on Australia—either by predation on native species, or damage to native vegetation, crops or pasture—include foxes, mice, cats, dromedary camels, goats, donkeys, pigs and water buffalo (Figure 3.19).

Since European arrival, several imported species of freshwater fish have become established, either deliberately introduced for sport (trout, carp, redfin perch) or pest control (*Gambusia*), or else liberated from aquaria as escapees or discards (cichlids). Nearly all have had a negative impact on the local biota, particularly the European carp, whose benthic feeding habits have degraded many freshwater habitats in the southeast. Native Australian species have generally fared poorly, owing to habitat destruction, disease and competition from introduced fishes. Several endemic species, including the trout cod, barred galaxias and red-finned blue-eye, are in danger of extinction without strong conservation measures.

Similarly, invasive bird species such as the Indian myna (Figure 3.19), starling, sparrow and common pigeon have had a damaging effect on native bird species.

Invasive plants are numerous in Australia; some of the worst are alligator weed, blackberry, lantana, mimosa, Paterson's curse, prickly pear, giant salvinia, serrated tussock, water hyacinth and willow. Many of these were deliberately imported as ornamental plants. Although some have been largely eradicated by biological control, weed control still costs the Australian economy about $4 B per year.

Invasive arthropods are also present in Australia. They have either been introduced accidentally (e.g. red fire ant, Argentine ant, Portuguese millipede, European wasp—Figure 3.19) or have been introduced on purpose and have developed feral populations (e.g. the honey bee). Most of these compete with, or are predators on, native arthropods, but do not have a direct economic impact. However, there are occasional economic costs. The Portuguese millipede was accidentally introduced into South Australia in the 1950s and has spread across the southern half of the continent. It has no natural predators and can reach plague proportions, to the extent that it has occasionally disrupted train services, the large number of squashed carcasses preventing drive wheels from gaining sufficient traction on railway lines.

Figure 3.19: Some invasive species. (a) Indian myna (*Acridotheres tristis*). (Source: © Getty Images [C van Lennep]). (b) European red fox (*Vulpes vulpes crucigera*). (Source: Arthur White, University of New South Wales). (c) Giant salvinia (*Salvinia molesta*), with cane toad. (Source: Buck Salau). (d) Cane toad (*Bufo marinus*). (Source: D McRae). (e) European wasp (*Vespula germanica*). (Source: Local Government Association of South Australia). (f) Feral pig (*Sus scrofa*). (Source: Buck Salau)

Extinctions

Extinctions have always occurred; indeed, the overwhelming majority of the species that have ever existed are now extinct (Figure 3.8). In modern Australia, the rate of extinctions seems to be increasing. Since European settlement, more than 50 vertebrate species have become extinct. These include both species of gastric-brooding frog, King Island emu, paradise parrot, desert rat kangaroo, Toolache wallaby, lesser bilby, thylacine and numerous others. The precise reasons for each of these extinctions are uncertain, but it is most likely that clearing for agriculture, hunting, predation by foxes and cats, and changes to fire regimes are the prime causes for the extinction of the birds and mammals, while the introduced chytrid fungus is suspected in the frog extinctions.

Since European settlement, nearly 50 plants have become extinct. These include *Acacia kingiana*, Diel's wattle, Daintree's river banana, short spider-orchid, robust greenhood and many others. The reasons for each of these extinctions are uncertain, but it is suspected that land clearing for agriculture, competition with introduced species, grazing and changes in the fire regime are the most likely causes.

Several invertebrates are known to have become extinct, but, as less than half the invertebrate species have been described, let alone studied in detail, it is likely that many more than we know have disappeared.

In addition to those plants and animals that are already extinct, another 130 vertebrate species and more than 600 plant species are threatened with extinction. It is almost impossible to estimate how many of the remaining 225 000 Australian invertebrates are endangered.

The living economy

Because there are no large-seeded grasses native to Australia and no legumes that have been suggested as more than a supplement to pasture in marginal environments, and because of the aridity of most of the continent, it is not surprising that agriculture did not originate on the Australian continent, as it did in some other parts of the world early in the Holocene. Indeed, broadacre and other forms of agriculture, despite having been in Australia for more than 200 years, still use almost entirely imported species for large-scale farming. The only native species now used as a large-scale commercial food crop is the macadamia nut.

The top 10 export commodities, which together provide 65% of the $230 B in export earnings for Australia, include six variously related to living systems. The most closely related of these are beef and wheat, which are grown from introduced species by means of broadacre agriculture. Three of the commodities are fossil fuels (coal, natural gas, crude petroleum) derived from the anaerobic decay of mostly Ordovician to Eocene plant and algal

The extinct thylacine (*Thylacinus cynocephalus*), also known as the Tasmanian Tiger, was a carnivorous marsupial that weighed up to 30 kg. It was the only member of the family Thylacinidae to survive into modern times. The last known thylacine died in Hobart Zoo in 1936.

remains. The last of the six, iron ore, was laid down in huge deposits in Western Australia as a direct result of the oxygenation of the atmosphere brought about by the evolution of photosynthetic bacteria.

Of the remaining commodities exported, about one-third are derived from living material. They include alcoholic beverages, meats other than beef, wool, dairy products, live animals, woodchips and paper, barley, fish, crustaceans, fruit, nuts, vegetables, leather and hides, pearls, cotton and numerous others.

Final comments

Life has left an indelible mark on the planet and on Australia. Photosynthetic bacteria oxygenated the atmosphere, giving rise to the huge iron ore deposits in Western Australia and on other continents. The oxygen in the atmosphere allowed complex, multicellular life to evolve. Several hundred million years later, some of these multicellular organisms formed huge peatlands that gave us most of our enormous energy resource of coal, oil and gas. Many millions of years later, one of these complex species studies the geological history of the continent and writes books about how that history has shaped us. Although Australia has not been part of Gondwana for many millions of years, life on our continent has a Gondwanan heritage. The tectonic history behind that heritage is outlined in the next chapter.

Koala (*Phascolarctos cinereus*) was once widespread across eastern Australia. Hunting and land clearing practices in the 19th and 20th centuries saw numbers decline dramatically. In eastern Australia, they are now classified as vulnerable and have been added to the threatened species list.

Image by Jim Mason

Bibliography and further reading

Anderson M, Kool L & Hann C 1999. *Dinosaur dreaming*, Monash Science Centre, Melbourne.

Bambach RK 2006. Phanerozoic biodiversity mass extinctions. *Annual Reviews of Earth and Planetary Sciences* 34, 127–155.

Benton MJ 2005. *Vertebrate palaeontology*, 3rd edn, Blackwell Science Ltd, Malden.

Benton MJ & Harper DAT 2009. *Introduction to paleobiology and the fossil record*, Wiley-Blackwell, Chichester.

Byrne M, Yeates DK, Joseph L, Kearney M, Bowler J, Williams MAJ, Cooper S, Donnellan SC, Keogh JS, Leys R, Melville J, Murphy DJ, Porch N & Wyrwoll KH 2008. Birth of a biome: insights into the assembly and maintenance of the Australian arid zone biota. *Molecular Ecology* 17, 4398–4417.

Clarkson ENK 1998. *Invertebrate palaeontology and evolution*, 4th edn, Wiley-Blackwell, Malden.

Cowen R 2000. *History of life*, 3rd edn, Blackwell Science Inc., Malden.

Fedonkin MA, Gehling JG, Grey K, Narbonne GM & Vickers-Rich P 2007. *The rise of animals: evolution and diversification of the kingdom Animalia*, Johns Hopkins University Press, Baltimore.

Flück M, Webster KA, Graham J, Giomi F, Gerlach F & Schmitz A 2007. Coping with cyclic oxygen availability: evolutionary aspects. *Integrative and Comparative Biology* 47, 524–531.

Hill RS (ed.) 1994. *History of the Australian vegetation: Cretaceous to recent*, Cambridge University Press, Cambridge.

Laurie JR (ed.) 2009. *Geological TimeWalk*, Geoscience Australia, Canberra.

Long JA 2011. *The rise of fishes: 500 million years of evolution*, 2nd edn, University of New South Wales Press, Sydney.

Merrick JR, Archer M, Hickey, GM & Lee MSY 2006. *Evolution and biogeography of Australasian vertebrates*, Austscipub, Oatlands.

Ogg JG et al. 2008. *The concise geologic time scale*, Cambridge University Press, Cambridge.

Playford PE, Hocking RM & Cockbain AE 2009. Devonian reef complexes of the Canning Basin, Western Australia. Geological Survey of Western Australia, Bulletin 145, 444 P

Prothero D 2004. *Bringing fossils to life: an introduction to paleobiology*, 2nd edn, McGraw Hill, New York.

Rich PV et al. 1985. *Kadimakara: extinct vertebrates of Australia*, Pioneer Design Studio, Melbourne.

Stewart WN & Rothwell GA 2010. *Paleobotany and the evolution of plants*, Cambridge University Press, New York.

Of the flora and fauna species that have been recorded in the Otway Ranges, Victoria, 92 species are considered rare or threatened, including 46 plants, 9 mammals, 32 birds, 2 fish and 3 invertebrates.

Taylor TN, Taylor EL & Krings M 2008. *Palaeobotany: the biology and evolution of fossil plants*, Academic Press, Amsterdam.

Valentine JW 2004. *On the origin of phyla*, University of Chicago Press, Chicago.

White ME 1984. *Australia's prehistoric plants and their environment*, Methuen Australia, North Ryde.

White ME 1986. *The greening of Gondwana*, Reed Books, Frenchs Forest.

White ME 1988. *Australia's fossil plants*, Reed Books, Frenchs Forest.

White ME 1990. *The nature of hidden worlds*, Reed Books, Balgowlah.

White ME 1991. *Time in our hands*, Reed Books, Balgowlah.

White ME 1994. *After the greening: the browning of Australia*, Kangaroo Press, Kenthurst.

Woodford J 2005. *The Wollemi pine*, Text Publishing, Melbourne.

4

Out of Gondwana

This is the story of how the island continent of Australia emerged from the supercontinent of Gondwana, and the far-reaching impacts of this history on Australian resources, economy and society. We owe our energy resources of coal, oil and gas to Gondwana and its breakup, and hence an Australian economy dependent on fossil fuels. The isolation imposed by the breakup has been fundamental to how life developed in Australia and has shaped Australian society in response to the tyranny (or opportunity) of distance. This chapter focuses on how Gondwana broke up, why the fossil fuels are where they are, why their discovery and exploitation have followed the pattern that they have, and why all that is important to Australians.

Marita T Bradshaw, Irina Borissova, Dianne S Edwards, George M Gibson, Takehiko Hashimoto, Gabriel J Nelson, Nadege Rollet, Jennifer M Totterdell

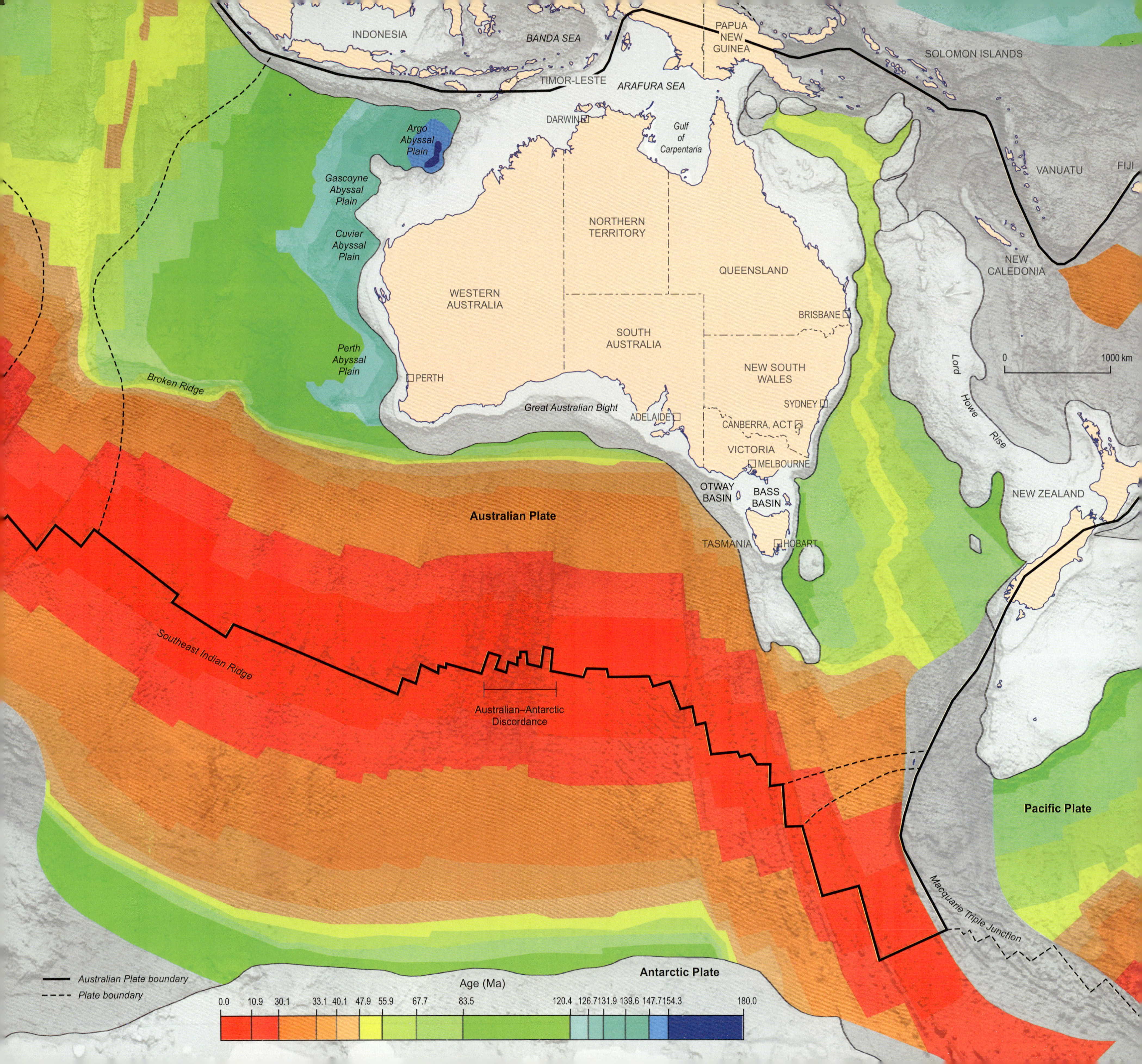
INDONESIA
BANDA SEA
PAPUA NEW GUINEA
SOLOMON ISLANDS
TIMOR-LESTE
ARAFURA SEA
DARWIN
Gulf of Carpentaria
Argo Abyssal Plain
Gascoyne Abyssal Plain
Cuvier Abyssal Plain
Perth Abyssal Plain
VANUATU
FIJI
NORTHERN TERRITORY
QUEENSLAND
NEW CALEDONIA
WESTERN AUSTRALIA
SOUTH AUSTRALIA
BRISBANE
NEW SOUTH WALES
0
1000 km
Lord Howe Rise
PERTH
Broken Ridge
Great Australian Bight
ADELAIDE
SYDNEY
CANBERRA, ACT
VICTORIA
MELBOURNE
OTWAY BASIN
BASS BASIN
NEW ZEALAND
Australian Plate
TASMANIA
HOBART
Southeast Indian Ridge
Australian–Antarctic Discordance
Pacific Plate
Macquarie Triple Junction
Antarctic Plate
Australian Plate boundary
Plate boundary
Age (Ma)
0.0
10.9
30.1
33.1
40.1
47.9
55.9
67.7
83.5
120.4
126.7
131.9
139.6
147.7
154.3
180.0

Gifts of Gondwana

The coal, oil and gas resources of Australia are the 'gifts of Gondwana'. The quality, quantity and distribution of these fossil fuels were determined by the geological evolution of Australia as it emerged from the supercontinent. First we will chart this dramatic history from the Early Paleozoic tropical seaways to the Late Paleozoic glaciation, through Permian coal-forming forests to the breakup of Gondwana and the creation of the island continent with its hydrocarbon-rich marginal basins, now on its journey north. In the second section of the chapter, we will examine the history of Australia's search for oil using three case studies. Combined, these form a story about why the fossil fuels are where they are, why their discovery and exploitation have followed a particular pattern, and why they are important to Australians today and in the future.

Australia's journey out of Gondwana is recorded in the patterns of seafloor spreading in the surrounding oceans (Figure 4.1) and the sediments that were deposited in the onshore and offshore basins of the continent (Figure 4.2). The Mesozoic marginal basins created as Gondwana broke apart contain around 90% or more of Australia's conventional oil and gas. The majority of the known gas accumulations are on the North West Shelf, and most of the oil was originally in the Gippsland Basin (Figure 4.2), but it has now been largely produced. Australia has sufficient gas resources to support growing domestic and export demands for liquefied natural gas (LNG) over many decades, especially when unconventional resources such as coal-seam gas (CSG) are also considered (Chapter 9); but finding a future oil supply is more of a challenge.

Early energy self-sufficiency with wood and coal in the age of steam was replaced by dependency on imported oil once the internal combustion engine became crucial to keeping Australians moving

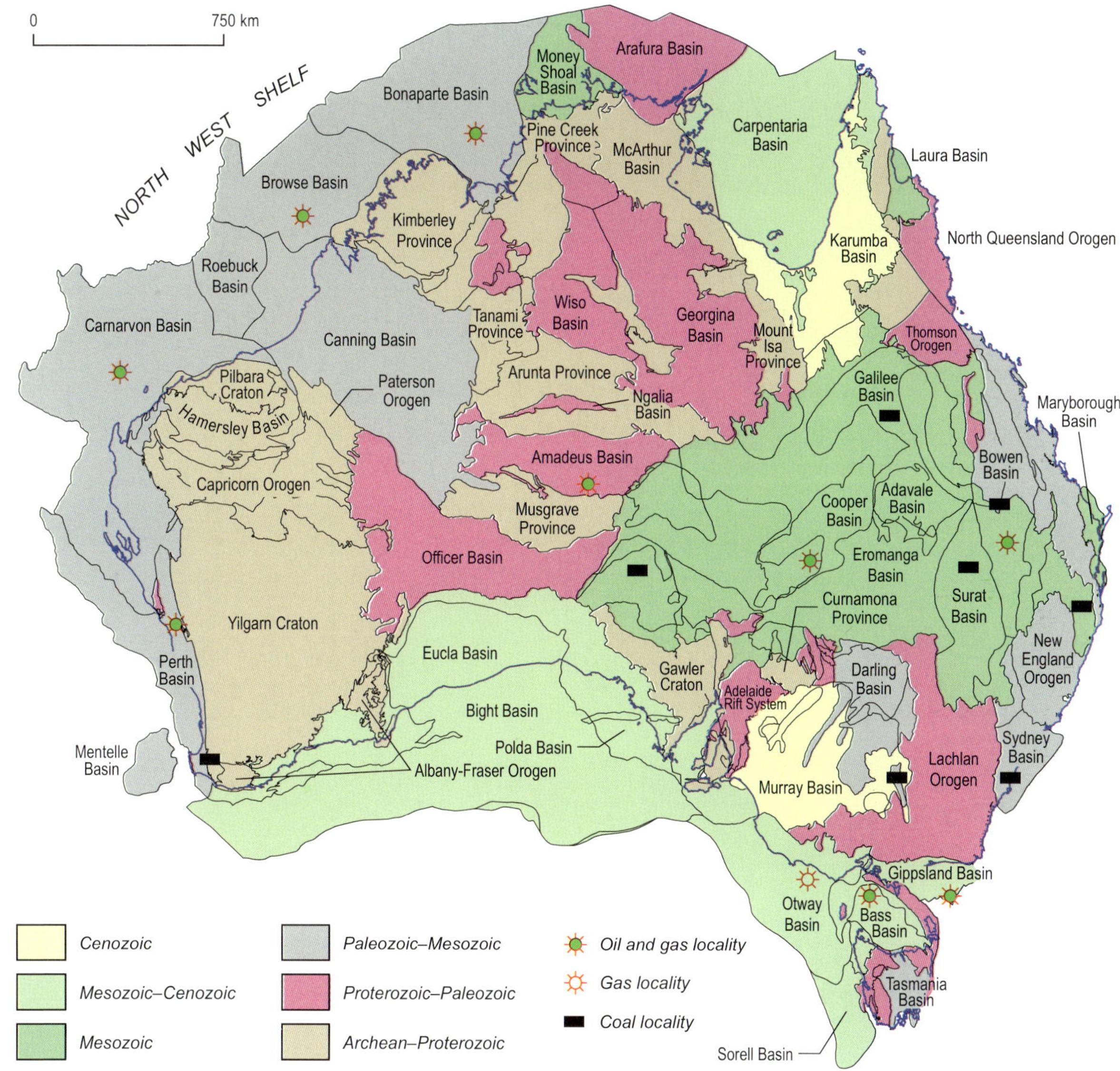

Figure 4.2: Australian basins map, showing basin age and major oil, gas and coal occurrences.

Figure 4.1 (opposite): Age of ocean floor around Australia. Note the rapid spreading in the Southern Ocean after about 40 Ma. Light blue regions are thinned continental crust below sea-level. Grey regions are oceanic crust; age not asssigned. (Source: modified from Muller et al., 2011)

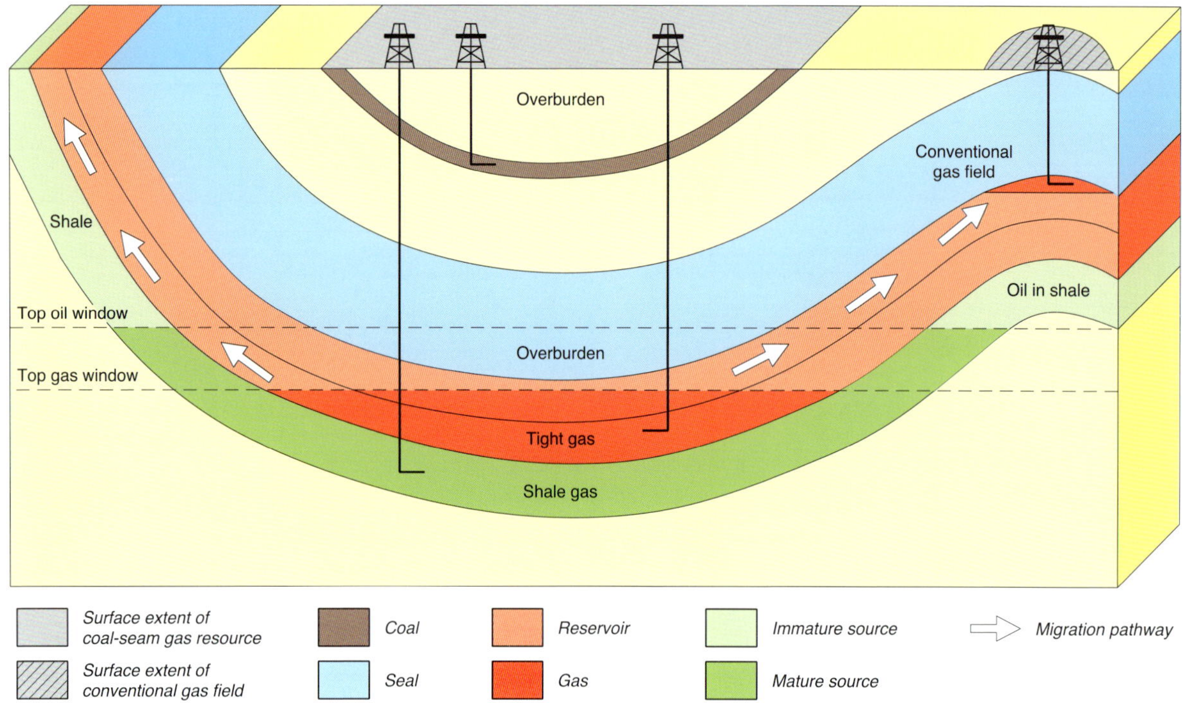

Figure 4.3: Schematic block diagram of a petroleum system, showing both conventional and unconventional hydrocarbon resources. (Source: modified from Schenk & Pollastro, 2001)

across the vast distances by the early 1920s. Energy self-sufficiency returned in the 1960s with the discovery of giant oil fields in the offshore Gippsland Basin and more oil finds on the North West Shelf. These fields supplied most of Australia's crude oil needs for 30 years, with production peaking in 2000. Since then, crude oil production has declined and, by 2030, imports are expected to supply three-quarters of our oil demand. There still remains the chance that a major new oil province awaits discovery in the deepwater frontier basins around Australia's margin, formed as Gondwana broke apart.

Hydrocarbon exploration is guided by a concept known as the petroleum system, which brings together all the essential elements for oil and gas accumulations to occur. Let us look at what this system entails.

Petroleum systems and hydrocarbon resources

Petroleum is formed by the thermal and chemical alteration of organic matter buried in sedimentary basins. Time and temperature act on organic matter, generating buoyant gas and oil. As they are lighter than water, they move upwards, following permeable migration pathways through the sedimentary sequence until trapped by a permeability barrier. For petroleum accumulations to form and be preserved, a complex series of processes occurring over millions of years is needed. The crucial elements of a petroleum system are:

- the depocentre—a basin container influenced by the underlying basement architecture
- the source—an organic-rich rock, such as a coal or oil shale
- the reservoir—porous and permeable rock, such as sandstone
- the seal—an impermeable rock such as a shale
- the trap—a subsurface structure that contains the accumulation, such as a fault block or anticline
- the overburden—overlying sediments that are required to bury the source rock to depths hot enough for thermal maturation to occur
- the migration pathways—to link the matured source to the trap (Figure 4.3).

In addition to these static elements, the actual processes involved—trap formation, hydrocarbon generation, expulsion, migration, accumulation and preservation—must occur, and in the correct order, for the petroleum system to operate successfully and for oil and gas accumulations to be formed and preserved. A mineral system is a similar conceptual process (Chapter 8), but with a critical difference. In a mineral system, the fluids are not trapped but are fluxed through a 'throttle' or smaller rock mass where metals in the fluid are deposited and the residue fluid is expelled.

Unconventional oil and gas resources are those that are more difficult and costly to extract than conventional accumulations. In a conventional field, the hydrocarbons generated from organic-rich source rocks have migrated into a reservoir from which they can be fairly readily produced. Unconventional oil and gas are often hosted within the source rock (shale oil, shale gas and coal-seam gas), in a poor-quality reservoir (tight gas), or is of poor quality (tar sands). Oil shale is an example, where a thermally immature source rock has not generated and expelled hydrocarbons (e.g. Early Cretaceous Toolebuc Formation in the Eromanga Basin; *Did you know?* 4.1). Oil or tar sands occur where conventional crude oil has failed to be trapped at depth and has migrated near to the surface and become degraded by evaporation, biodegradation and water washing to produce a viscous, heavy, oil residue. In the case of coal-seam gas, the gas molecules are still within the coal source rock, absorbed onto the surface of the coal. Coal-seam gas can only flow to the surface once the confining pressure has been decreased by dewatering the coal seam (Chapter 9).

Another key difference is that conventional oil and gas resources occur in discrete accumulations within structural closures, whereas many unconventional hydrocarbon resources are described as 'continuous'. Examples include basin-centred gas, which occupies large areas in a basin axis, contained in tight gas sands, shale and/or coaly source rocks; and coal-seam gas that is throughout a coal formation, often covering hundreds of square kilometres (Figure 4.3).

The petroleum resource pyramid (Figure 4.4) describes how a smaller volume of easily extracted conventional gas and oil is underpinned by larger volumes of unconventional gas and oil, which are more difficult and more costly to extract. For the unconventional hydrocarbon resources, additional technology, energy and capital have to be applied to extract the oil or gas, replacing the natural action of the geological processes of the petroleum system. Technological developments and commodity price rises can make the lower parts of the resource pyramid accessible and economic to produce. The recent development of oil sands in Canada, and of shale gas in the United States, are examples where rising energy prices and technological development have facilitated the exploitation of large unconventional hydrocarbon resources lower in the pyramid.

With an understanding of a petroleum system and the nature of these hydrocarbon resources, let us now look briefly at how Australia has evolved in the Paleozoic and, in particular, how the continent came *out of Gondwana*.

Australian petroleum supersystems

The dramatic shifts in climate and tectonic regime experienced as Australia evolved within, and then separated from, Gondwana have produced

4.1: Australia's largest potential oil resource is the Toolebuc oil shale

There are an estimated 1.5 trillion barrels of shale oil in the Toolebuc Formation in western Queensland. The Toolebuc is a widespread but very low-grade oil shale. This unconventional hydrocarbon resource requires mining, crushing and heating to unlock the oil and hence has high cost and carbon emissions. The geological processes of the petroleum system have failed to generate, expel, migrate and accumulate the oil in a sealed reservoir. Instead, the immature source rock of the Toolebuc Formation remains exposed at the surface or mostly buried beneath a few hundred metres of overburden, insufficient to push it down into the oil-generating zone (Figure 4.3). These reducing successions in an otherwise oxidised sedimentary package are favourable sites for metal deposition; the formation is anomalous in uranium and vanadium, and also copper, lead, cobalt, nickel and manganese.

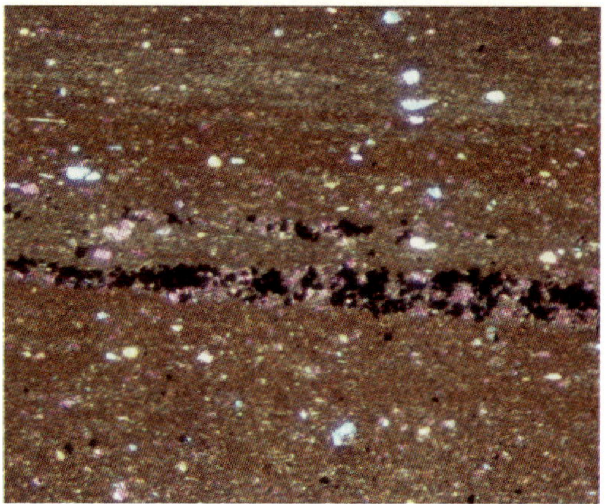

Polarised light micrograph of oil shale. Magnification 14× at 35 mm.

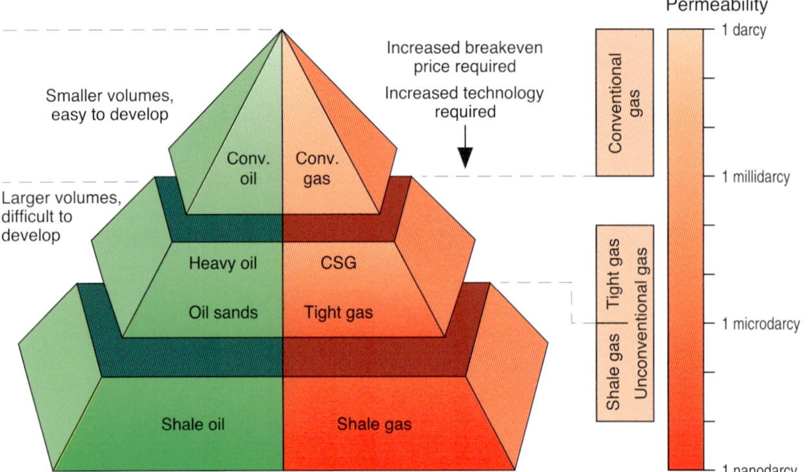

Figure 4.4: Petroleum resource pyramid, showing how resource quality varies with permeability. The pores in a conventional sandstone reservoir (a) are several orders of magnitude larger than those in a shale gas reservoir (b). (1.3 mm horizontal field of view in both thin section photomicrographs) (Sources: Geoscience Australia; photomicrographs SE Phillips, Phillips-Gerrard Petrology Consultants)

a great variety of individual petroleum systems with widely differing characteristics. Source rocks range in age from the Mesoproterozoic to the Cenozoic, and, along with the reservoir and seal facies, almost every sort of depositional environment is represented, from desert salt lakes and tropical reefs to glacial outwash plains and deep-sea fans. However, a pattern emerges among this extreme diversity when continent-wide studies of palaeogeography, hydrocarbon habitat and oil and gas geochemistry are considered.

The individual systems have been classified into a number of large groupings or petroleum supersystems with shared geological characteristics (Appendix 4.1.1). These are the McArthur, Urapungan, Centralian, Larapintine, Gondwanan, Westralian, Austral, Capricorn and Murta supersystems (Figures 4.5 and 4.6). The spatial pattern of these supersystems is reflected and validated in the geochemistry of Australian oil and gas, which again fall into a number of distinct oil families. Oil families match petroleum supersystems because oil and gas composition is controlled by the depositional environment of the source-rock facies and the biological evolution of the algae, bacteria, plants and animals, all of which contribute the initial organic matter. The oil families in Australia can be geographically represented in a dendrogram based on a cluster analysis, showing the relationship between them. For example, oils from the Papuan, Bonaparte (Vulcan and Sahul), Browse and Carnarvon (Dampier, Barrow and Exmouth) basins cluster together as they are derived from Late Jurassic marine shales. This Westralian oil family shares more chemical affinity with Perth Basin oils sourced from Triassic marine rocks than with oils derived from land plants found in the Gippsland, Otway and Eromanga basins (Figure 4.6).

Most of Australia's conventional oil and gas resources are within the Larapintine, Gondwanan, Westralian and Austral supersystems (Figure 4.5), although there is the potential for significant

Figure 4.5: Generalised time-space plot of Australian petroleum supersystems, illustrating the main Phanerozoic sedimentary units across the continent and the stratigraphic position of the gas and oil resources. Early Paleozoic sediments include carbonates and evaporites; oil and gas occurs in Ordovician and/or Devonian rocks in the Amadeus and Canning basins (Larapintine Supersystem). Carboniferous deposition records a change into clastic-dominated sequences. Glacial sediments were deposited in many basins in the Late Carboniferous and Early Permian. Coal, oil and gas are found in Permian of the Cooper, Perth, Canning, Bonaparte and Bowen basins (Gondwanan Supersystem). Oil and gas occurs in the Mesozoic clastic sediments of the marginal basins (Westralian and Austral supersystems). Early Cretaceous marine shales in offshore and onshore basins mark the inundation of the continent and form the seal for many petroleum systems. There is a progressive change from clastic to carbonate sediments from the northwest to the southeast after the Late Cretaceous. See Appendix 4 for detailed stratigraphic columns.

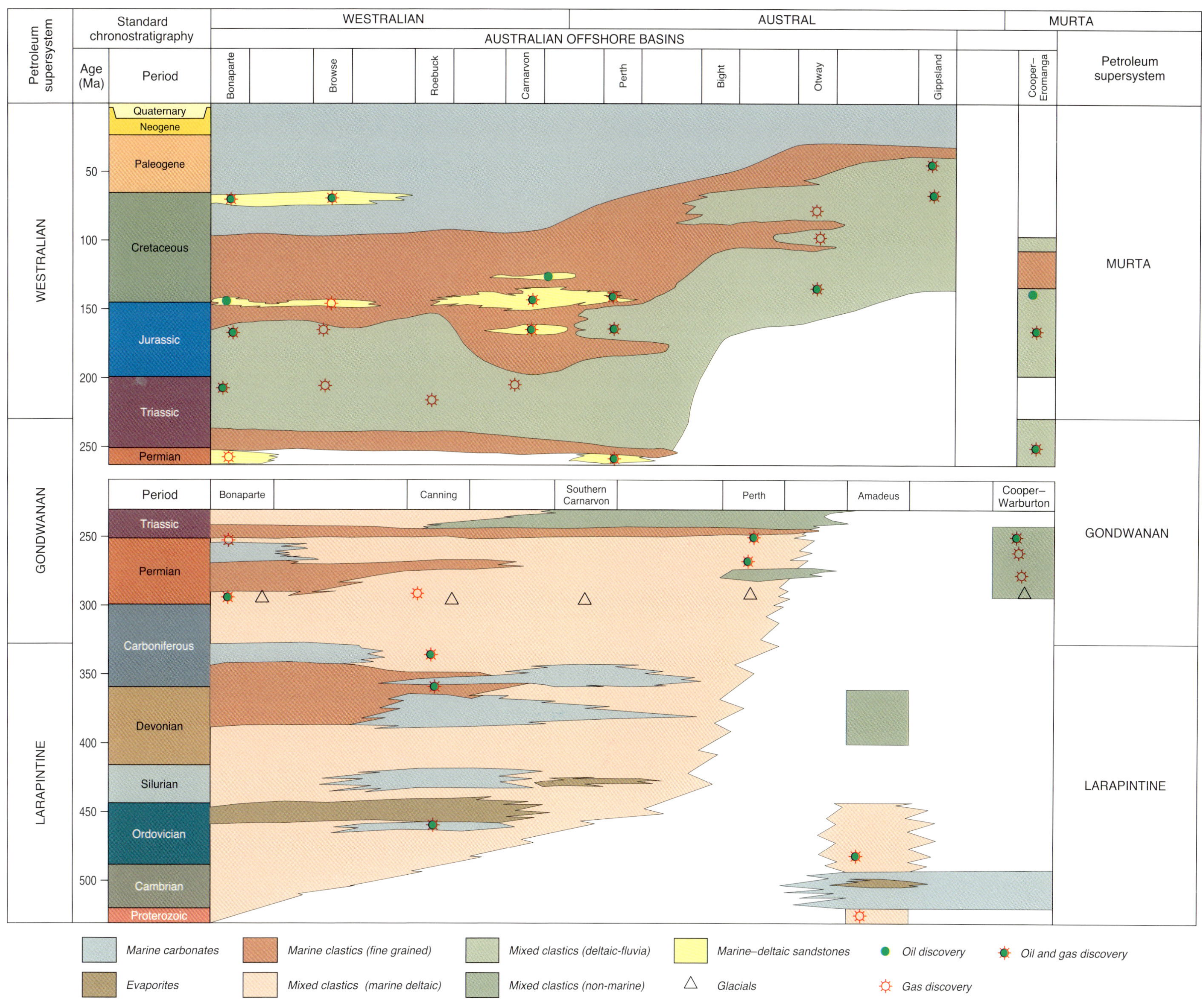
Petroleum supersystem
Standard chronostratigraphy
Age (Ma)
Period
WESTRALIAN
AUSTRAL
MURTA
AUSTRALIAN OFFSHORE BASINS
Bonaparte
Browse
Roebuck
Carnarvon
Perth
Bight
Otway
Gippsland
Cooper–Eromanga
Petroleum supersystem
Quaternary
Neogene
Paleogene
Cretaceous
Jurassic
Triassic
Permian
50
100
150
200
250
WESTRALIAN
MURTA
Period
Bonaparte
Canning
Southern Carnarvon
Perth
Amadeus
Cooper–Warburton
Triassic
Permian
Carboniferous
Devonian
Silurian
Ordovician
Cambrian
Proterozoic
250
300
350
400
450
500
GONDWANAN
LARAPINTINE
Marine carbonates
Marine clastics (fine grained)
Mixed clastics (deltaic-fluvia)
Marine–deltaic sandstones
Oil discovery
Oil and gas discovery
Evaporites
Mixed clastics (marine deltaic)
Mixed clastics (non-marine)
Glacials
Gas discovery

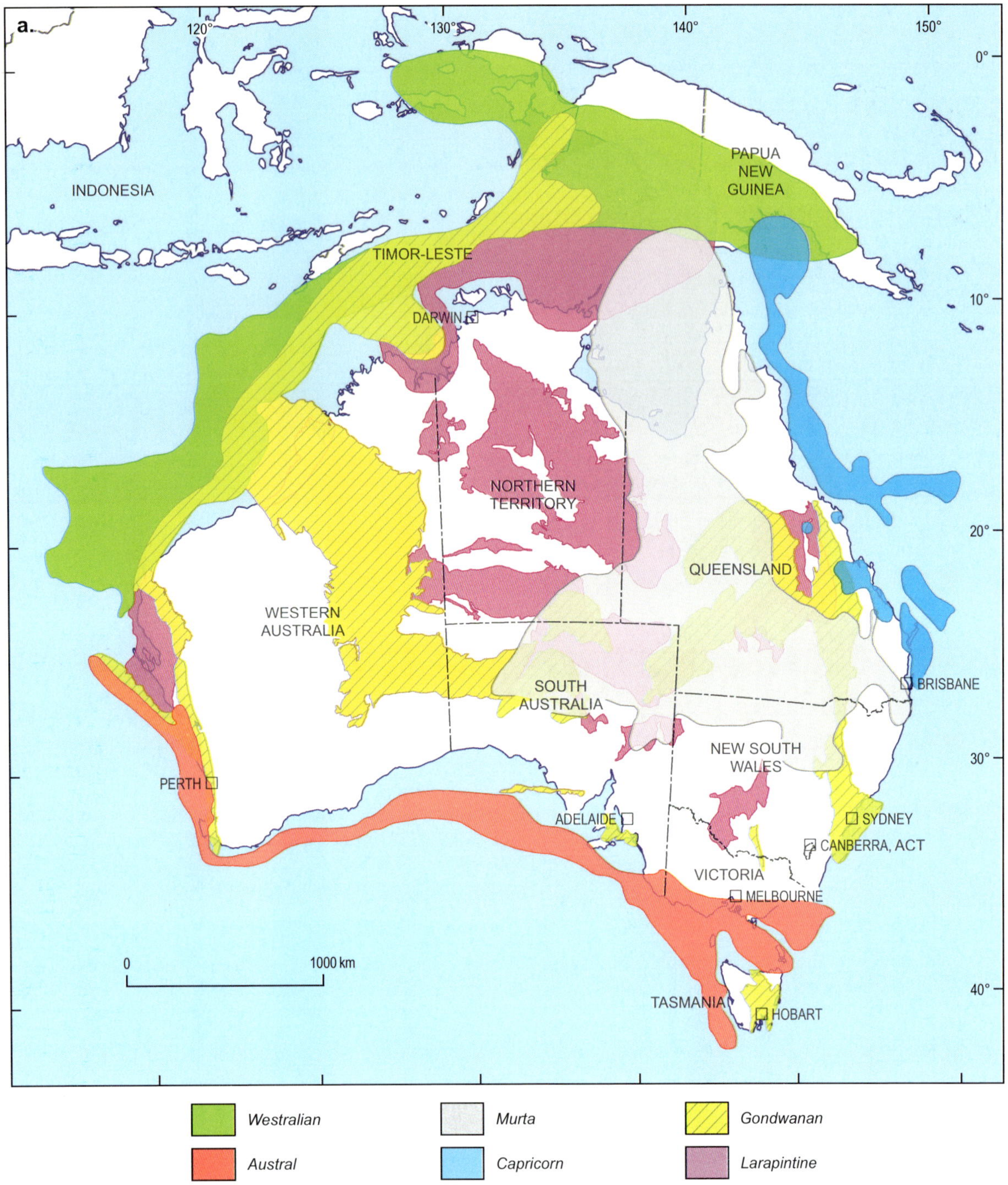

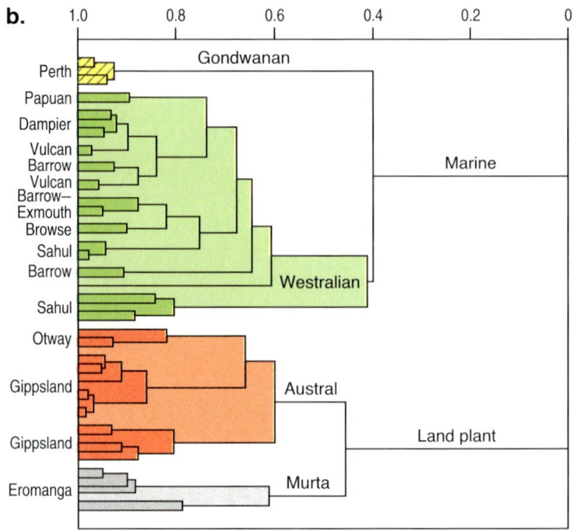

Figure 4.6: (a) Map of Australian Phanerozoic petroleum supersytems. (b) Dendrogram showing the chemical relationships between selected oils from the Gondwanan, Westralian, Austral and Murta petroleum supersystems. (Sources: modified from Bradshaw, 1993; Bradshaw et al., 1994)

unconventional shale gas resources in the Proterozoic McArthur Supersystem, as well as oil shale deposits in the younger Murta and Capricorn supersystems.

Gondwana to island continent

The fragments of continental crust that are now Australia have undergone several amalgamation and fragmentation supercontinent cycles (Chapter 2). However, the heritage of Gondwana has remained indelible in the landscape, in the flora and fauna, and in the distribution of fossil fuel resources, and so has shaped the Australian economy and society. The formation of the continent's margins gave the island nation of Australia its shape, *girt by sea*, and the boundary conditions within which the coastal realm developed. This is where most Australians live (Chapter 1).

Tectonics and climate are the big drivers that play out in the supercontinent cycles. Driven by the processes of plate tectonics, the continental blocks wander over Earth's surface, changing latitude; landmasses group together and break apart, altering climate and sea-level as the world shifts from greenhouse to icehouse and back again (Box 2.5). During the Permian, the climate was paramount, Earth being under the 'bulldozer' of ice, followed by the greening thaw. In the Jurassic and Cretaceous, tectonic processes dominated during the breakup. Mostly there is an intricate interaction between tectonics and climate. These forces are also identified as the big drivers in landscape evolution (Chapter 5).

The prelude to this tale begins in the Ordovician, when Australia was located in tropical latitudes in the Northern Hemisphere and locked into the recently formed jigsaw of Gondwana. Antarctica lay to the south, Greater India to the west and various Asian continental blocks to the north and west (Figure B4.1a). Sea-level was high, Australia was criss-crossed by shallow seas and a series of volcanic arcs formed the eastern seaboard of the continent. Oil source rocks were deposited in the seaways, and some of the gold endowment of eastern Australia was created by this tectonic activity (Chapter 8).

By the Late Devonian, Australia, within Gondwana, had shifted southwards and lay in the Southern Hemisphere. Tropical conditions still held sway across the north, with reef-fringed marine embayments on the western margin (Carnarvon, Canning and Bonaparte basins), and on the eastern margin (Broken River, Adavale basins). The Ordovician Larapintine Seaway (Chapter 2) was severed, and central Australia, as today, was under desert sands. North China and other Asian

Devonian outcrop, Canning Basin, Western Australia.

FROM THE TROPICS TO THE POLE AND BACK AGAIN (BOX 4.1)

Over the past half a billion years, dramatic changes have shaped the Australian continent as it travelled from the tropics to the pole and back again. Australia commenced this journey within Gondwana, then emerged as the island continent and is now in collision on its northern margin with Asia. This history can be unravelled from the record preserved in the sedimentary basins, the evidence cut and etched into the landscape, and the patterns of seafloor spreading in the surrounding oceans (Figure 4.1).

The detailed stratigraphy of basins such as the Canning (Appendix 4.3.7) captures this epic story—the Ordovician Larapinta seaway retreats, as limestones are replaced with salt and desert sands, which in turn are overlain by Permian glacial sediments. Then these are capped by the mud and sand of the Early Cretaceous inland sea.

In the Kalbarri cliffs (left), Ordovician sandstones deposited in fluvial to coastal environments when Australia was in the tropics are overlain by the Cretaceous sediments of the rising seas, deposited when the continent was in polar latitudes.

Using such information from outcrop and well data from across the whole continent, a series of palaeogeographic maps from the Cambrian to the Quaternary were produced for Australia (BMR Palaeogeographic Group, 1992). The 70 time-slice maps show how the pattern of environments shifted across the continent through the Phanerozoic (Appendix 2.3.5). The more recent products of the Earthbyte Group at the University of Sydney are needed to see Australia in its plate tectonic context (Appendix 4.4.1). (Sources: BMR Palaeogeographic Group, 1992; www.earthbyte.org)

Sea cliffs at Kalbarri, Western Australia, showing the Ordovician Tumblagooda Sandstone overlain by Cretaceous sediments.

Figure B4.1a: Middle Ordovician (*ca* 470 Ma) global plate reconstruction, showing Australia (highlighted in white) within Gondwana with Antarctica, India and other Asian fragments. The palaeo-equator (marked in red) bisected Australia. Shallow tropical marine environments extended from the North China Block to eastern Australia, with the Larapintine Seaway linking the Canning and Amadeus basins to the deepwater convergent margin. (Source: modified from PALEOMAP Project, 2008)

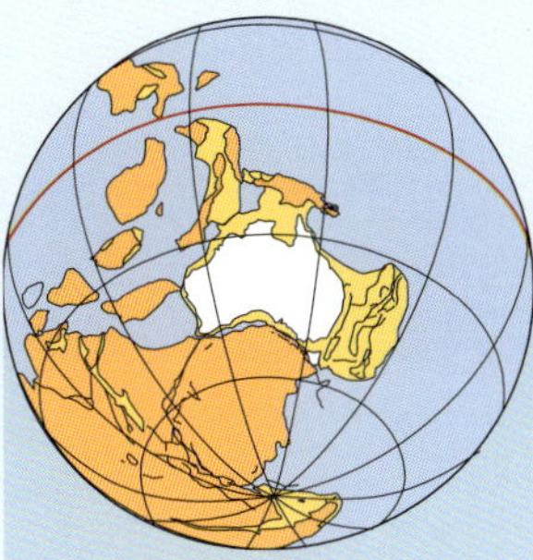

Figure B4.1b: Late Devonian (*ca* 360 Ma) global plate reconstruction, showing Australia within Gondwana. Note that Australia has shifted south but still edges into the tropics. The seas have retreated from the interior as Gondwana begins to collide with Laurentia. A large marine embayment occupies the Canning Basin, with a fringing coral reef developed along its northern margin. A deep marine seaway has opened as North China and Tarim separate from Gondwana. (Source: modified from PALEOMAP Project, 2008)

Figure B4.1c: Early Permian (*ca* 280 Ma) global plate reconstruction. Australia has continued its southward journey within Gondwana and now lies partially within the polar circle. Most of the continent is covered by an icecap, like modern-day Antarctica. The Asian continental fragments have parted company with Gondwana and are heading north, while Antarctica, India, southern Africa and South America remain in the cold south with Australia. (Source: modified from PALEOMAP Project, 2008)

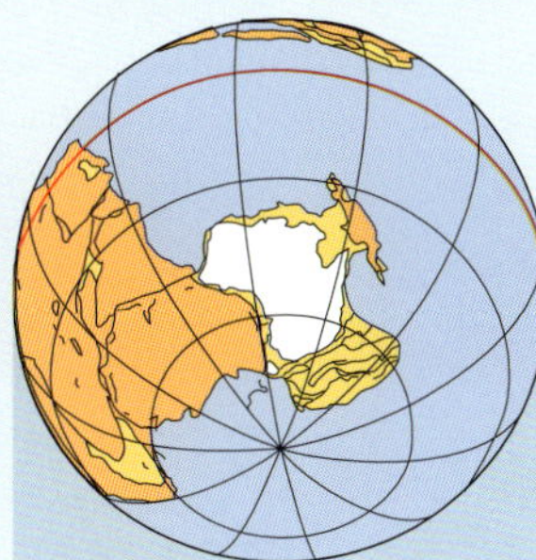

Figure B4.1d: Late Jurassic (*ca* 160 Ma) plate reconstruction for the Australian region, showing the continent still within southerly latitudes. The northwest margin is facing onto the Tethys Ocean, and organic-rich sediments accumulate in restricted marine rifts. The Australian–Antarctic rift valley remains non-marine at this time. Lush forests in eastern Australia were the site of coal deposition, some of these coals are now the source of coal-seam gas. (Source: modified from PALEOMAP Project, 2008)

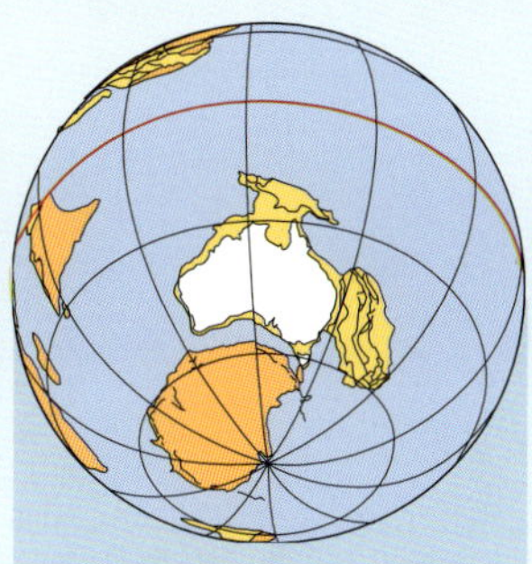

Figure B4.1e: Late Cretaceous (*ca* 70 Ma) plate reconstruction for the Australian region, showing a developing marine gulf between Australia and Antarctica. The Lord Howe Rise is also separating from the eastern margin as the Tasman Sea is formed. The southwest margin is fully marine, and India has left Gondwana and is moving north towards its rendezvous with Asia. The Tasman Sea has begun to open. (Source: modified from PALEOMAP Project, 2008)

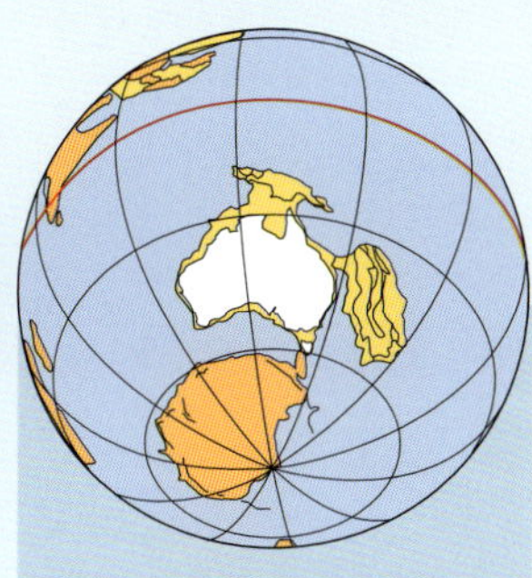

Figure B4.1f: Late Eocene (*ca* 34 Ma) plate reconstruction for the Australian region, showing Australia starting its journey back to the north. Australia is now the island continent, having recently separated from Antarctica along the Tasman Fracture Zone. The opening of this seaway, together with that between South America and Antarctica, allowed the circumpolar current to flow. (Source: modified from PALEOMAP Project, 2008)

© J Wark, Lonely Planet Images

Aerial view of a modern-day tropical sea, the Bahamas, Caribbean Sea. Similar shallow-water carbonate environments are interpreted to have occurred in the Ordovician Larapinta Seaway.

© Getty Images [P Chesley]

Aerial view of the modern-day Great Barrier Reef, Australia. In the Late Devonian, similar reef environments occurred in the Canning and Bonaparte basins.

© Getty Images [B Webber]

Modern-day icefield. Large areas of Gondwana were similarly mantled with ice in the Early Permian.

© B Winebrenner, Lonely Planet Images

Modern-day fluvial environment in high latitudes. Similar environments are interpreted to have occurred along the Australian–Antarctic rift valley in the Late Jurassic.

© W Walton, Lonely Planet Images

Coastline at western end of the Great Australian Bight, near Esperance, Western Australia, where a broad seaway was established between Australia and Antarctica by the Cretaceous.

© D Mayfield, Lonely Planet Images

Aerial view of the Nullarbor Plain and cliffs, Great Australian Bight. Neogene limestones are exposed along this southern edge of the island continent.

Outcrop of Permo–Triassic Beacon Supergroup, Beaver Lake, Antarctica.

blocks rifted from the northwest margin of the supercontinent at this time, and their geological history began to diverge from that of Australia, which continued a southward journey with its partners in Gondwana (Figure B4.1b).

Gondwana's heritage

The distinctive Permo-Carboniferous Gondwana facies was deposited as basal glacial sediments overlain by coal measures across Australia, Antarctica, India, Arabia, Madagascar, Africa and South America. This shared stratigraphy was an important part of the early evidence pointing to continental drift and hence plate tectonics (Box 4.2). In Australia, sediments record the history of the Late Paleozoic ice age, followed by coal deposition in foreland basins along the convergent eastern Gondwanan margin, creating a legacy of major Permian coal resources that underpin Southern Hemisphere fossil energy supplies (Chapter 9). In contrast, the main episode of Northern Hemisphere coal deposition was in the appropriately named, and preceding, Carboniferous period.

Gondwana glaciation

Australia, within Gondwana, continued on a journey south during the Carboniferous, shifting rapidly through about 40° of latitude from the tropics to the polar regions. As Pangaea came together in the mid- to Late Carboniferous, there were continuing collisions between parts of Gondwana and Euramerica, causing the Variscan or Hercynian Orogeny in Europe, and the Orogeny Alleghenian in North America. At the same time in central and eastern Australia, the Alice Springs and Kanimblan orogenies (Chapter 2) formed vast uplands, where seasonal snow cover had a better chance of becoming permanent ice than at lower altitudes. Once established, the ice sheets reflected more heat back into space. Compounding the cooling effect was the enhanced removal of CO_2 from the atmosphere due to accelerated erosion and weathering induced by this episode of mountain building. In this way, the increase in elevation and the meridional arrangement of landmasses helped move Earth into an icehouse regime, and glaciation took hold across the Gondwanan continents arrayed around the South Pole (Figure B4.1c). The ice sheets appear to have been centred in what are now the Gambertsev and Transantarctic mountains in Antarctica, and glaciers radiated out into southern Australia and the other conjugate Gondwana continents (Figure 2.35).

There were three main episodes of glaciation from the Carboniferous through to the Permian, which ranged across Gondwana from South America to Australia. The third episode, in the Late Carboniferous to Early Permian, was the most extensive and the one that deposited glacial sediments in many Australian basins (Figure 2.35). Hundreds of metres to several kilometres of diamictite and other glacially influenced sediments were dumped into the Carnarvon, Canning, Officer and Cooper basins around the edges of three ice sheets that may have joined together to cover half of Australia at the glacial maximum. The clean and wind-winnowed sandstones deposited at the edge of the ice are reservoirs for oil, gas and groundwater in the Cooper and Canning basins.

GONDWANA—WHAT'S IN A NAME? (BOX 4.2)

The name 'Gondwana' is derived from the Sanskrit gondavana, meaning the forest of the Gonds, an ancient indigenous people of India. The term Gondwana series was used by the Geological Survey of India from the 1870s to denote plant-bearing sedimentary sequences with characteristic *Glossopteris* flora. By the 1900s, Gondwanaland had entered the global literature as the term for the southern supercontinent, fragments of which, although now dispersed by continental drift, still preserve glacial sediments deposited when the fragments were united around the South Pole. Southern Hemisphere geologists (Alex Du Toit and Lester King in South Africa and Sam Carey in Australia), surrounded by this evidence, championed the idea, although it was not until the plate tectonics revolution of the late 1960s that the mechanism of continent dispersal by seafloor spreading was understood (Chapter 2).

Today Gondwana has passed into the general consciousness, especially in Australia where it is found in literature, in song and in company names for clothing lines and tourist ventures. The name conjures up remnants of something ancient and original. Even the origin of the name, from one of the earliest Indo-European languages, seems apt as an echo of a lost world.

© O Strewe, Lonely Planet Images

© R Barnett, Lonely Planet Images

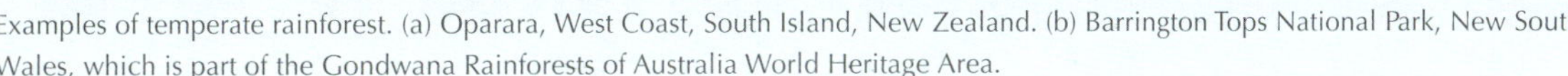

Examples of temperate rainforest. (a) Oparara, West Coast, South Island, New Zealand. (b) Barrington Tops National Park, New South Wales, which is part of the Gondwana Rainforests of Australia World Heritage Area.

Image courtesy of Mary White

Glossopteris, the iconic fossil of Gondwana.

Image by Mining Photo

Gondwanan Permian coal measures, Meandu mine, Bowen Basin, Queensland.

In eastern Australia, mountain glaciers scoured the highlands along the convergent margin of Gondwana where the ice was established earlier and lingered longer.

The *Glossopteris* forests of Gondwana: the origins of Australia's energy base

When the ice sheets melted, sea-level rose, and marine to brackish water embayments extended into Gondwana, which was now re-mantled by thick vegetation, including *Glossopteris* forests of (Chapter 3). These forests may have resembled the high-latitude peat-swamp forests of today's taiga in Siberia. Thick coal measures accumulated on all the Gondwanan continents in areas of basin subsidence. In eastern Australia, the Sydney–Gunnedah–Bowen basin system formed as a foreland depocentre behind the convergent margin (Chapter 2).

Black coal, conventional gas and coal-seam gas are the major resources that were formed while Australia was part of Gondwana during the Permian. The organic matter that accumulated in Gondwanan swamps now powers Australia. Coal has been mined for more than 200 years and forms the cornerstone of Australia's domestic economy; it is our largest energy resource, has a low cost of extraction and is located close to population centres. High-quality Australian black coal is also a major contributor to the trade balance, with Australia being the world's largest coal exporter (Chapter 9).

In addition to the traditional mineable coal, the Permian basins also host major coal-seam gas and conventional oil and gas resources. Conventional gas from the Cooper and Bowen basins has supplied the domestic gas needs of much of eastern Australia for more than 40 years. These resources are the products of the widespread Gondwanan petroleum supersystem that is characterised by clastic terrestrial facies and coaly source rocks (Figure 4.5).

Production from conventional fields is now being supplemented by the new coal-seam gas industry. This resource is now considered large enough to rival the giant conventional gas fields on the North West Shelf. In Queensland, several coal-seam gas to LNG projects are progressing to development, although not all stakeholders are happy with the potential intrusion into agricultural land. The development of unconventional hydrocarbon resources does bring environmental and other challenges, including understanding the impact on groundwater and competing land uses (Chapters 7 and 11). The economic benefits, however, may outweigh environmental concerns.

Challenges also lie ahead in a carbon-constrained world for an Australia dependent on coal for power generation and export income; late 2011 saw the Australian Parliament pass a new law to tax carbon emissions. Gas-fired electricity is less carbon intensive, and carbon capture and storage offers a low-emission way to continue to reap the benefits from Gondwanan coal and gas. The eastern Australian coal basins are being actively studied for possible CO_2 storage sites (Chapter 11). Coal-to-liquids and gas-to-liquids are other technologies that may form part of the future energy mix, again using the gifts of Gondwana to power Australia.

Pangaea at its zenith

With the coalescence of most landmasses into the one huge supercontinent of Pangaea, the remaining building blocks of Gondwana were placed mostly south of the equator. Australia, attached to Antarctica as East Gondwana, was moving southwards in the farthest southern latitudes. The new configuration allowed land flora and fauna to achieve rapid and vast distributions. Throughout the Late Permian, the forests were laying down coal widely across Australia, from the Laura Basin in Cape York to Tasmania in the south and across to the Perth Basin (Figure 4.2). Following the end-Permian global extinction event (Chapter 3), the sedimentary record preserves a very different world, often referred to as the 'coal gap'. The supercontinent Pangaea reached its maximum size in the Triassic, altering the global climate to one dominated by semi-arid conditions and a Pangaean monsoonal circulation pattern. Triassic red beds overlie Permian coal-bearing sediments in many locations across Gondwana (southern Africa, Antarctica, India and Australia), as well as in north China.

In eastern Australia, Permian coal measures are overlain by Early Triassic red beds and fluvial sandstones; marine deposits are rare. In contrast, along the western margin, thick marine shales mark a major Early Triassic inundation (Figure 4.5) The Blina Shale crops in the Canning Basin, and the Locker Shale is a distinctive bland seismic package in the offshore Carnarvon Basin. The Mount Goodwin Shale in the Bonaparte Basin is the seal facies for gas accumulations contained in Late Permian sandstones (Petrel gas field); in the Perth Basin, the Kockatea Shale is both a seal and an oil-rich source rock for oil and gas accumulations (Figure 4.7).

From the Permian and into the Triassic, eastern Australia was part of the active Gondwana convergent plate margin. Triassic igneous activity occurred along this margin from South America through Antarctica, New Zealand and eastern Australia (New England Orogen, Chapter 2) to Papua New Guinea. A fragmentary sedimentary record for onshore Australia in the Triassic reflects deformation and erosion related to major tectonic events, including the Hunter–Bowen Orogeny in the east and the Fitzroy Movement in the west (Chapter 2). Widely scattered coal deposition had returned by the Late Triassic—in narrow rift basins in Queensland and northern New South Wales, and in small fluvio-lacustrine basins in South Australia and Tasmania.

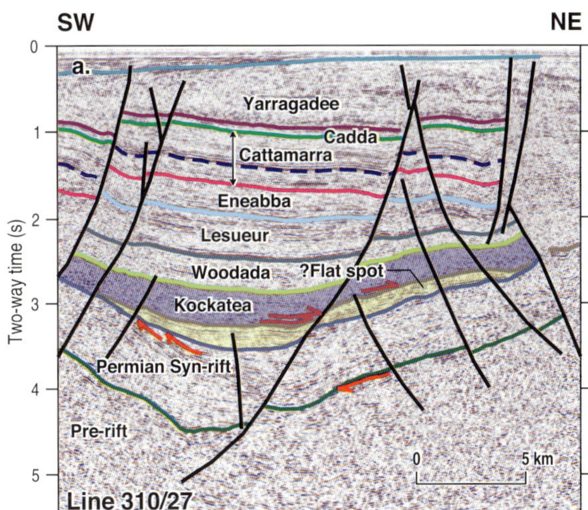

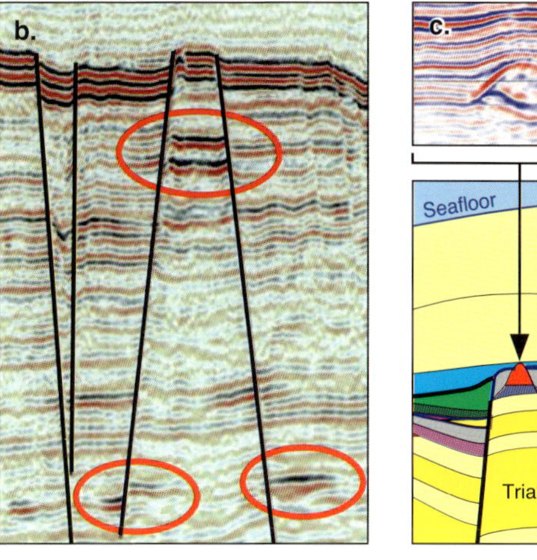

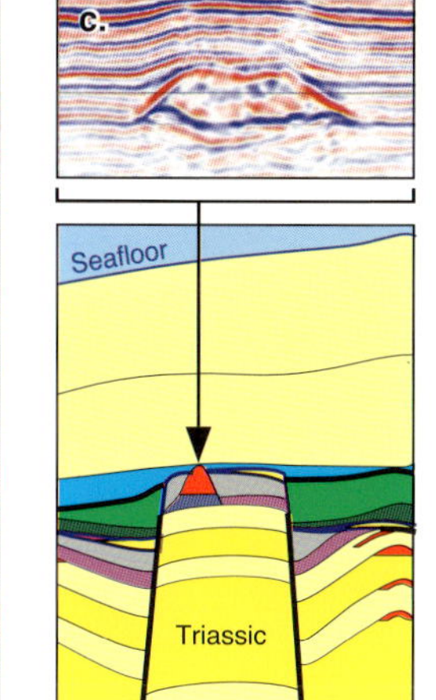

Figure 4.7: Seismic sections of Triassic features. (a) Perth Basin, showing the distinctive Kockatea Shale with a bland texture in the seismic data on-lapping older units. (b) Carnarvon Basin—Triassic fault blocks with high amplitudes circled, indicating possible gas sands. (c) Exmouth Plateau—seismic image and a schematic diagram of a Late Triassic carbonate pinnacle reef. (Sources: Geoscience Australia; Woodside, 2010)

In the Middle Triassic through to the Early Jurassic, major deltas up to 5 km thick were built along a stretch of more than 3000 km on the northwest margin, in the Carnarvon, Roebuck, Browse and Bonaparte basins. Contributions to this vast sediment pile were delivered by rivers that rose in the central Australian highlands, and from the south, from the Pinjarra Orogen and perhaps from elsewhere in Gondwana (Antarctica and Greater India). The coaly sediments and fluvial sandstones that accumulated on these delta plains form the source rocks and reservoirs, respectively, for many of the giant gas fields on the North West Shelf (Figures 4.2 and 4.5). They provide the lion's share of Australia's conventional gas resources and support a major export LNG industry (Chapter 9). The petroleum systems of the Carnarvon Basin will be one of the case studies discussed later in the chapter, where we will look in detail at the Triassic delta of the Mungaroo Formation.

Although the sandstone gas reservoirs on the North West Shelf are a key part of Australia's energy-security future, other Triassic sandstones also occupy a place in our consciousness. The earliest convict settlements were hewed from the Hawkesbury Sandstone in Sydney and the Ross Sandstone in Hobart. Both building stones are Early to Middle Triassic quartzite deposited from a major river system that flowed from Antarctica. The rugged sandstone escarpment of the Blue Mountains was a barrier to the early explorers (Chapter 5) and is now a scenic escape from Sydney. Before the convicts laboured in 'The Rocks' or expressways were sliced through, Aboriginal people carved their stories into the sandstone. The clean, quartz sands were well winnowed in their journey down transcontinental rivers, to be reworked and deposited in the tidal delta environments of the Hawkesbury Sandstone. Later diagenesis has produced a well-cemented quartzite that provides strong foundations for Sydney's skyscrapers and preserves Aboriginal rock art.

What was beyond the shore?

As we have seen, Australia in the Triassic was largely high and dry; except for the Early Triassic transgression along the western margin, non-marine to deltaic sedimentation dominated. But beyond the delta fronts, at the continent edges, the fine-grained materials carried by the giant rivers of Gondwana were deposited as marine silts and muds. Further out on the northwest margin, there were carbonate facies, including pinnacle reefs lapped by the tropical waters of the Tethys Ocean (Figure 4.7c). On nearby Timor, thick Triassic marine shales and limestones, deposited as the distal edge of the North West Shelf sequence, have been caught up in the Cenozoic collision between Asia and Australia.

We can only speculate on the fate of the finer grained sediments that were shed by Gondwana's rivers and deltas. On the eastern margin, did the Hawkesbury Sandstone delta deposit fine-grained sandstones on the Lord Howe Rise? Are the muds of the Mungaroo delta now somewhere in South-East Asia?

Now, let's move into the Jurassic and see how these depositional systems were dismembered as Gondwana broke apart and the continental fragments were dispersed.

Top: Sandstone buildings, Salamanca Place, Hobart,Tasmania (© Getty Images [W Fogden]). Bottom: Aboriginal rock carvings on cliffs at North Bondi, New South Wales, depict marine scenes (© CA Wiley, Lonely Planet Images). Right: The entrance to Sydney Harbour, New South Wales, is protected by the Triassic sandstone cliffs of the Hawkesbury Sandstone (© Getty Images [D Messent]).

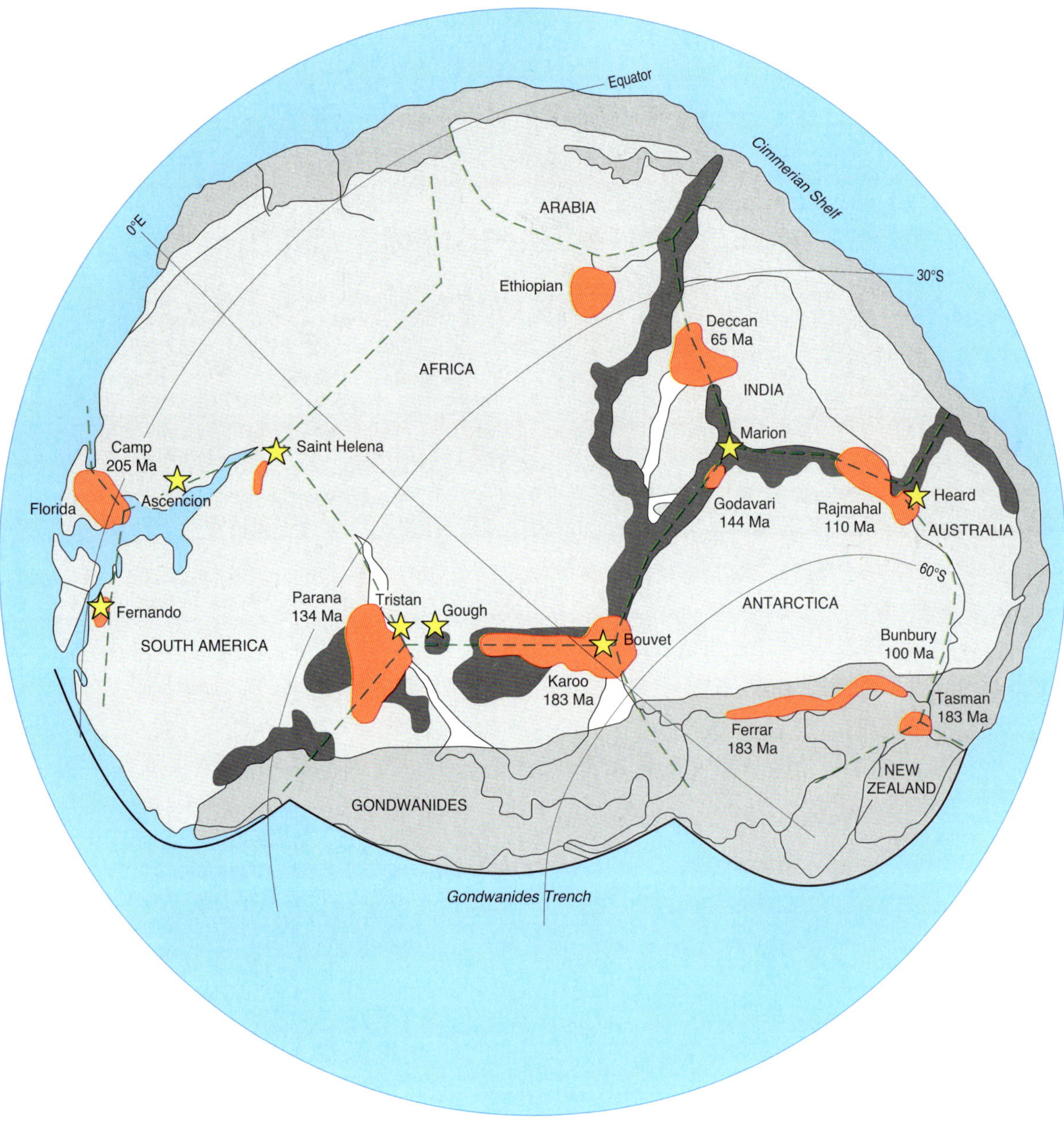

Figure 4.8: Map of Gondwana, showing the location of Jurassic–Cretaceous large igneous provinces (LIPs) (orange) with intrusion and eruption dates shown. Yellow stars are major hotspot volcanoes; coal-bearing Permian rifts are shown in black. Note that the Tasman, Ferrar and Karoo dolerites are all of Early Jurassic age. (Source: modified from Sears, 2007)

Gondwana breakup

By the Middle Jurassic, Gondwana had started to break up. The Gondwana series in India, Argentina, South Africa and Tasmania were capped with dolerites in response to mantle upwelling associated with the breakup (Figure 4.8). The spectacular landscapes of Tasmania record the rending of Gondwana as doleritic magma was intruded into the Permo–Triassic sediments (*Did you know?* 4.2). Although Tasmania did not break away, elsewhere around Australia the process went to completion with the formation of new seafloor. Deep-blue sea (4 km and deeper) encircles Australia on three sides—to the west, south and east—floored by oceanic crust (Figure 4.1). These are the divergent passive margins of the continent. To the north, the margin is currently actively convergent (although it too was divergent and passive in the Mesozoic).

The pattern of seafloor ages (Figure 4.1) indicates that new oceanic crust was formed progressively off the northwest, followed by the southwest, south, southeast and northeast margins, as Australia unzipped from the 'jacket' of Gondwana. The last clasp opened as Tasmania, which remained with mainland Australia, slid past Antarctica in the latest Eocene. Rapid seafloor spreading since the Middle Eocene then shifted Australia back towards the tropics and into collision with Asia and the Pacific Plate (Chapter 2). Ongoing seafloor spreading from the South East Indian Ridge continues to move Australia northward.

Each of the Australian margins differs in character, reflecting varying histories and controlling factors. The nature of the margin (rifted or sheared), the amount of volcanism, the influence of ancient basement structures on the pattern of breakup, and post-breakup modifications (later convergence, growth of deltas or carbonate build-ups) are all-important criteria differentiating the various margins of Australia:

- The ragged northwest margin of submerged plateaus and failed rifts is the product of the successive continental slivers launched towards Tethys and Asia over hundreds of millions of years from the Early Paleozoic to the Cretaceous (Figures 4.9a and 4.9b).
- The southwest margin is dominated by the Wallaby–Zenith Fracture Zone and records the impact of the Kerguelen Large Igneous Province it shared with India. This margin had a dextral transtensional sense of opening.
- The southern margin marks the break from Antarctica along Proterozoic and Paleozoic lines of weakness, with its shape modified by progradation of the Late Cretaceous Ceduna delta (Figures 4.9c and 4.9d).
- The eastern margin is the complex Tasman borderland, which has evolved out of the Paleozoic convergent margin of Gondwana.
- The northern margin is now in collision, forming the highlands of New Guinea and the swirl of the islands through the Banda Arc in eastern Indonesia and Timor.

The continent's breakup history from Gondwana has had a fundamental control on the distribution of Australia's major hydrocarbon resources. The identified petroleum resources are dominated by widely distributed natural gas, but the major known accumulations of crude oil are restricted to the Gippsland Basin and five 'oily' sub-basins along the northwest margin (Figure 4.10). This distribution is controlled by the occurrence of deep, narrow troughs containing mature oil source rocks. In the case studies at the end of this chapter, the operation of the proven petroleum systems of the Gippsland and Carnarvon basins are discussed, along with the potential in the Great Australian Bight. But before we delve into the oil kitchens and how they have influenced the Australian economy, let's paint a bit more of the broader context by looking at each margin in turn, since they delimit the edges of Australia and reveal the inheritance of Gondwanan breakup.

Australia's elusive conjugate margins

The neat fit of Africa to South America demonstrates that they are conjugate passive margins now separated by seafloor spreading from the Mid Atlantic Ridge. In contrast, Australia's conjugate margins are far less obvious, as many are 'hidden' continental fragments. The strongest evidence for Argo Land—which separated from Australia in the Late Jurassic when the Argo Abyssal Plain formed (Figure 4.1)—is the requirement for another side of the rifted northwest and southwest margins of Australia. West Burma, Sumatra and other destinations in South-East Asia have all been proposed for Argo Land. Greater India, which once lay to the west of the Perth Basin, is now considered to be hidden beneath the Himalaya. The Wallaby–Zenith Fracture Zone marks the northern limit of Greater India (Figure 4.11). The Gascoyne Terrane is postulated to be a separate

Did you know?

4.2: Dolerite in Tasmania

Tasmania has the largest exposure of dolerite in the world, with 30 000 km^2 and a volume of 15 000 km^3 of the rock. The dolerite has been described both as a curse, where it has destroyed the utility of the Permian coal measures, and a blessing, where it has produced rich and fertile soils. In 1642, Abel Tasman saw the huge fluted cliffs of columnar jointed dolerite as he approached the island that was to bear his name.

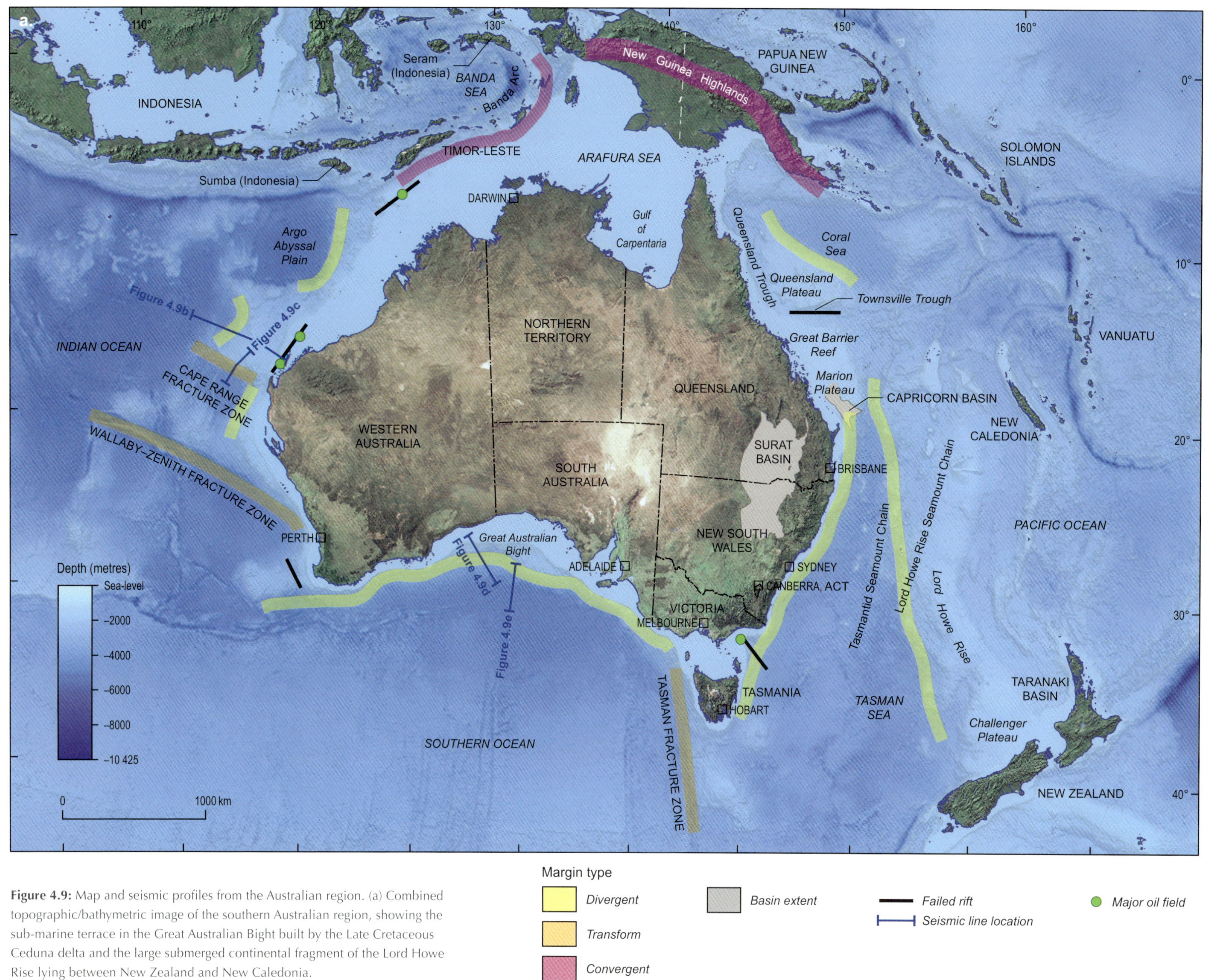

Figure 4.9: Map and seismic profiles from the Australian region. (a) Combined topographic/bathymetric image of the southern Australian region, showing the sub-marine terrace in the Great Australian Bight built by the Late Cretaceous Ceduna delta and the large submerged continental fragment of the Lord Howe Rise lying between New Zealand and New Caledonia.

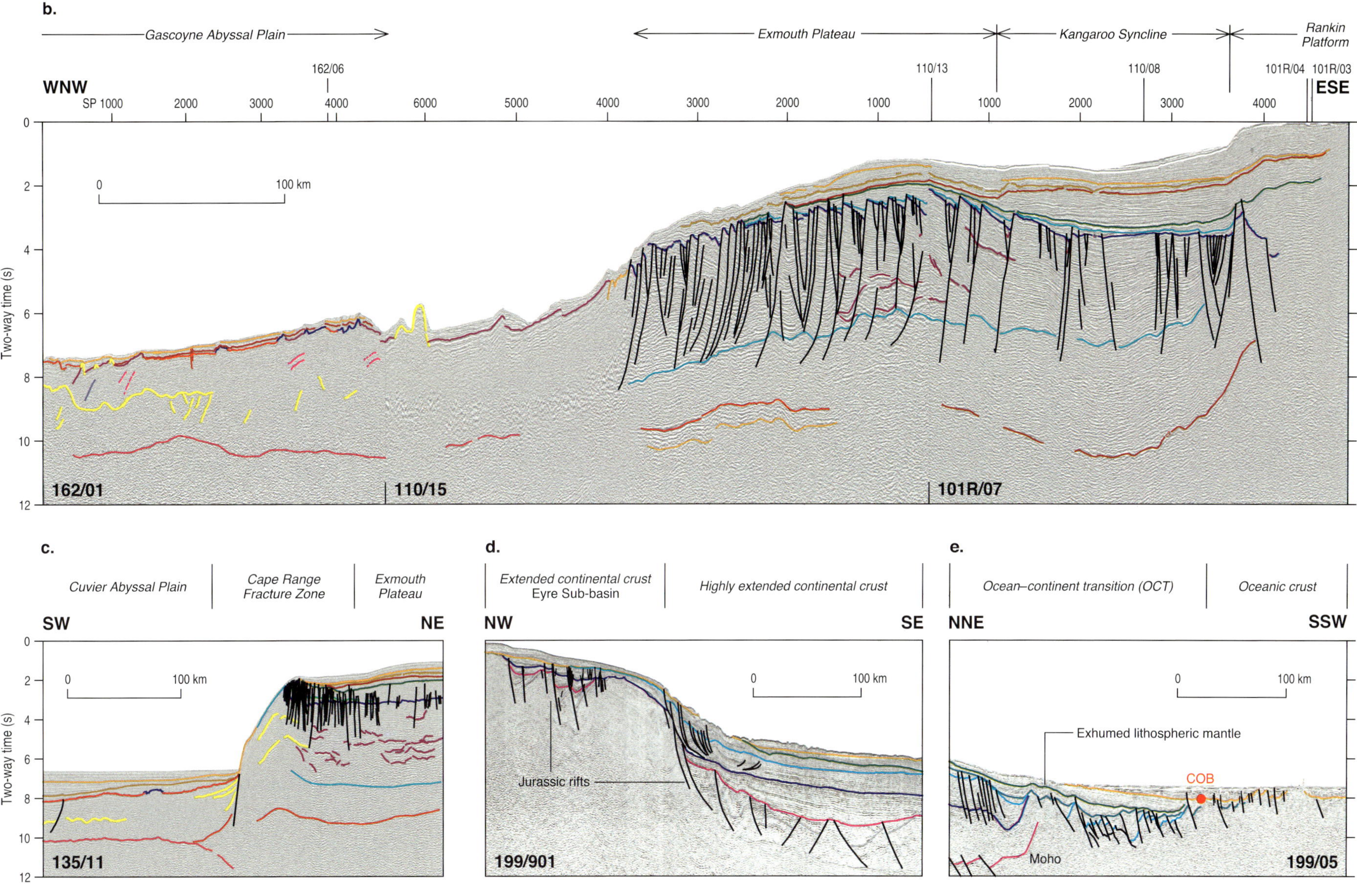

Figure 4.9 continued: (b) Seismic section across the wide and complex volcanic western margin of the Exmouth Plateau, showing seamounts on the Gascoyne Abyssal Plain. (c) Seismic section across the narrow sheared southern margin of the plateau along the Cape Range Fracture Zone stepping down to the ocean floor of the Cuvier Abyssal Plain. The block-faulted Triassic sediments of the Exmouth Plateau host giant gas fields and are overlain by a drape of younger sediments. (d) and (e) Seismic section across the Bight Basin rifted margin, showing the perched Jurassic rift of the Eyre Sub-basin and the highly extended continental crust beneath the Bight Basin sediments, extending out to the continent–ocean boundary (COB). (Sources: modified from Alcock et al., 2006; Symonds et al., 2000; Totterdell et al., 2000; Stagg et al., 2004)

block between Greater India and Argo Land. The terrane broke away in the Early Cretaceous, with the formation of the Gascoyne and Cuvier abyssal plains. The submerged Lord Howe Rise and other continental fragments in the Tasman and Coral seas are the poorly known conjugates of the eastern Australian margin. Only on the southern margin do we have an obvious conjugate pair. This southern conjugate is found in Antarctica, although now it is mantled by kilometres of ice and Cenozoic glacial sediments.

Northwest margin

Looking to the northwest of the continent, we see the Argo Abyssal Plain, a deep-blue 'window' on the map in contrast to the complex collision zone to the north and east (Figure 4.9a). This plain contains the oldest seafloor around Australia and some of the oldest preserved seafloor on Earth (Figure 4.1). This is the area where Argo Land, perhaps now located in West Burma, parted company with the Australian continent *ca* 156 Ma (Oxfordian) as eastern Gondwana fragmented. The record of Jurassic seafloor further to the northeast has been lost to subduction and collision in the Cenozoic convergent margin. Argo Land was one of a series of Asian continental blocks and slivers that were successively rifted from the Gondwanan northwest and northern margins and transported across Tethys to be accreted onto Eurasia since the Early Paleozoic (Chapter 2).

The embayed configuration of the northwest margin reflects the interplay of two structural trends: the dominant northeast–southwest Westralian trend established in the Permian and seen in the alignment of the margin and the offshore basin depocentres; and an older northwest–southeast grain, which controls the distribution and orientation of the Early Paleozoic basins and the northern margin of the Exmouth Plateau (Figure 4.10). These older basins may have been exploiting lines of weakness established in the Proterozoic along former collision zones such as the Paterson Orogen (Figure 2.10).

During Late Jurassic rifting, the Westralian trend controlled the pattern of deep-marine depocentres that now host the source of the major oil fields on the North West Shelf. The Exmouth, Barrow and Dampier sub-basins are failed rifts inboard of the Exmouth Plateau, and the Vulcan Sub-basin occupies a similar structural position inboard of the Ashmore Platform to the north. The oil-rich Nancar–Flamingo–Sahul area is where the Westralian trend intersects the older northwest–southeast grain of the Petrel Sub-basin, again creating a deep trough where organic-rich Upper Jurassic oil source rocks accumulated (Figure B4.1d).

Early Cretaceous seafloor spreading west and south of the Exmouth Plateau formed the Gascoyne and Cuvier abyssal plains (Figures 4.1 and 4.9b) and further broke apart Gondwana. The rifted conjugate continental fragment may be the Gascoyne Terrane, now located in South-East Asia, and considered to have docked in Burma some 20 Myr after Argo Land. The Barrow delta was the Gascoyne Terrane's parting gift to Australia. This terrane was the conduit and one of the sources of sediment to the north-flowing fluvial system that built the Barrow Group as a series of delta lobes out across the southern half of the Exmouth Plateau and the inboard rifts in the earliest Cretaceous. Locally, the delta provided enough sediment load to push the underlying Late Jurassic organic-rich marine shales in the Exmouth and Barrow sub-basins into the hydrocarbon-generating window, as well as clean reservoir sands to host the expelled oil and gas.

The Barrow delta was severed from its hinterland when the Gascoyne Terrane broke away, sliding along the transform margin of the southern edge of the Exmouth Plateau as the Cuvier Abyssal Plain was formed. Comparison of the western and southern edges of the Exmouth Plateau provides insight into the processes of continental breakup (Figure 4.9). The volcanic rifted western edge contrasts with the sheared southern margin that follows the old northwest–southeast trend, which is picked up again in the Wallaby–Zenith Fracture Zone further to the south and was crucial in shaping the southwest margin. The seismic profile in Figure 4.9c depicts the steep and linear southern transform margin along the underlying Cape Range Fracture Zone. There is an abrupt change from a thick sedimentary pile over the Exmouth Plateau to ocean-floor basalts. The transition zone is a narrow intruded and uplifted rim that forms a steep slope mantled with basaltic

Figure 4.10: Structural elements map of the North West Shelf region, showing the thick Mesozoic depocentres trending northeast–southwest and the Paleozoic basins extending from onshore along a northwest–southeast trend.
OPGGSA = *Offshore Petroleum and Greenhouse Gas Storage Act 2006* (Cwlth)

Field outlines are provided by Encom GPinfo, a Pitney Bowes Software (PBS) Pty Ltd product. While all care is taken in the compilation of the field outlines by PBS, no warranty is provided regarding the accuracy or completeness of the information, and it is the responsibility of the reader to ensure, by independent means, that those parts of the information used by them are correct before any reliance is placed on them.

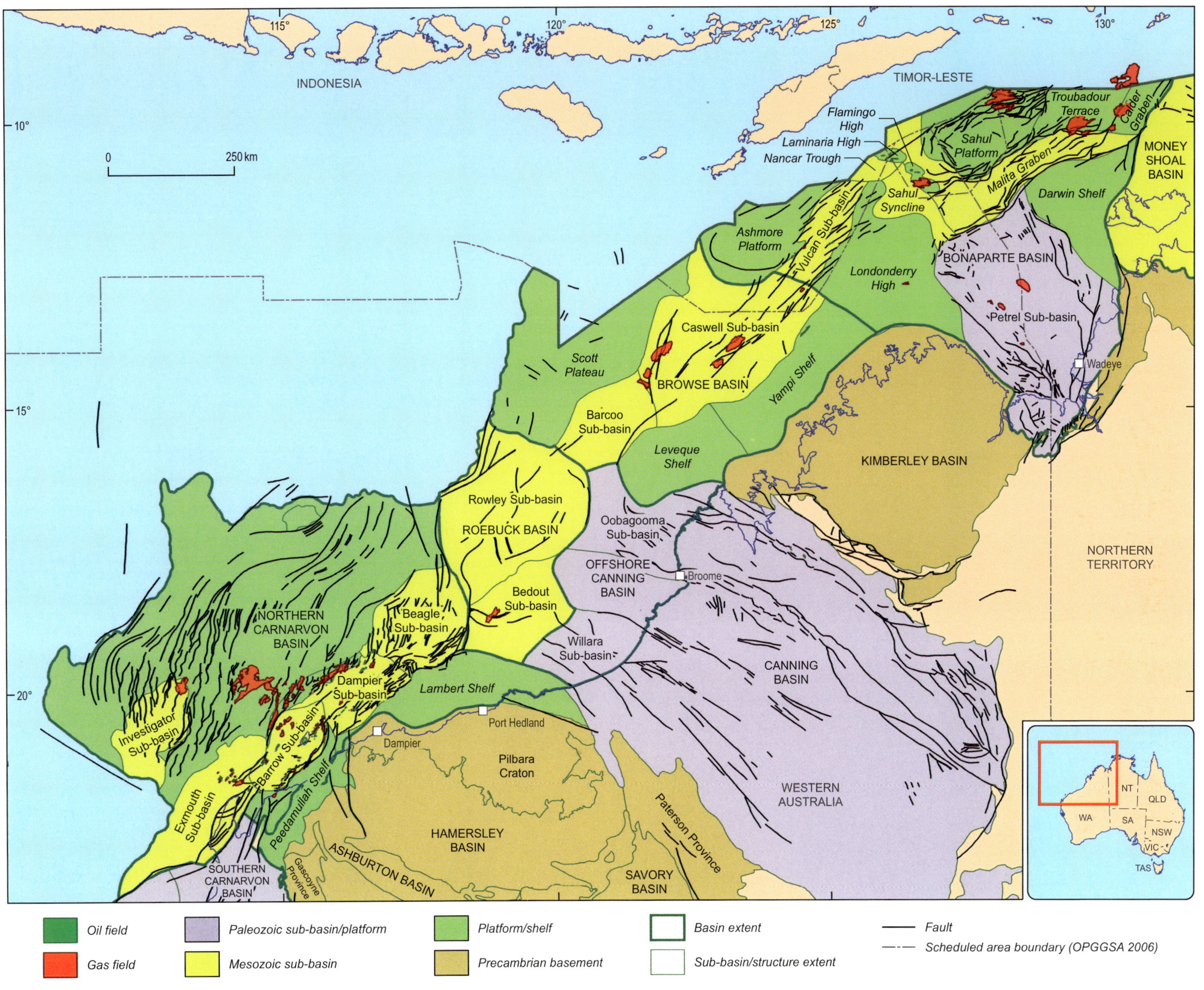
115°
120°
125°
130°
10°
15°
20°
INDONESIA
TIMOR-LESTE
0
250 km
Flamingo High
Laminaria High
Nancar Trough
Troubadour Terrace
Calder Graben
Sahul Platform
Malita Graben
Sahul Syncline
MONEY SHOAL BASIN
Darwin Shelf
Ashmore Platform
Vulcan Sub-basin
Londonderry High
BONAPARTE BASIN
Petrel Sub-basin
Caswell Sub-basin
Scott Plateau
BROWSE BASIN
Yampi Shelf
Wadeye
Barcoo Sub-basin
Leveque Shelf
KIMBERLEY BASIN
Rowley Sub-basin
ROEBUCK BASIN
Oobagooma Sub-basin
OFFSHORE CANNING BASIN
Broome
NORTHERN TERRITORY
Bedout Sub-basin
NORTHERN CARNARVON BASIN
Beagle Sub-basin
Willara Sub-basin
CANNING BASIN
Dampier Sub-basin
Lambert Shelf
Investigator Sub-basin
Port Hedland
Dampier
Barrow Sub-basin
Pilbara Craton
Peedamullah Shelf
Exmouth Sub-basin
Paterson Province
WESTERN AUSTRALIA
HAMERSLEY BASIN
SOUTHERN CARNARVON BASIN
Gascoyne Province
ASHBURTON BASIN
SAVORY BASIN
NT
QLD
WA
SA
NSW
VIC
TAS
Oil field
Gas field
Paleozoic sub-basin/platform
Mesozoic sub-basin
Platform/shelf
Precambrian basement
Basin extent
Sub-basin/structure extent
Fault
Scheduled area boundary (OPGGSA 2006)

CRETACEOUS INLAND SEA (BOX 4.3)

During the breakup, not all the action was on Australia's developing margins; there were also important depositional events under way in the interior, as vast seaways once again invaded deep into the continent. Twice in the Early Cretaceous, marine environments held sway from the Gulf of Carpentaria through the Eromanga Basin to the Great Australian Bight and from the Canning Basin in the west to the Maryborough Basin in the east. The first Cretaceous marine transgression occurred in the Valanginian (from *ca* 135 Ma) at the same time as the breakup between Greater India and Australia's southwestern margin, and was probably caused by widespread cooling following the onset of seafloor spreading.

The second marine flooding event occurred later in the Early Cretaceous, during the Aptian and Albian. The last time the continent was under such extensive marine conditions was about 400 Myr previously in the Ordovician (see Chapter 2). As then, fine-grained sediments, some rich in organic matter, were widely deposited. These Early Cretaceous marine shales are now crucial seal facies in petroleum and groundwater systems. In onshore basins, they have not been buried deeply enough to have generated hydrocarbons (see *Did you know?* 4.1). However, Cretaceous marine shales of the inland sea remain important as the confining beds for the Great Artesian Basin (Chapter 7) and as major elements in the landscape (Chapter 5). Some of Australia's most exciting fossils are giant marine reptiles and sharks from the Cretaceous inland sea (Chapter 3).

The high sea-level phases of the Cretaceous are associated with global oceanic anoxic events, and at least one of these is recognised in Australia—the late Albian (*ca* 100 Ma), coinciding with the widespread deposition of the oil shales of the Toolebuc Formation in the Eromanga Basin (Figure B4.3). The organic-rich shale is finely interlaminated with shell beds containing *Inoceramus sutherlandi*, filter-feeding bivalves that were tolerant of oxygen-poor bottom water conditions in the inland sea.

Elsewhere, there was also major inundation of the continents in the Cretaceous and deposition of oil source rocks—in North and South America, north Africa and the Middle East. But looking more closely at the global palaeography for the Early and Late Cretaceous, we see that Australia is out of step; the Turonian (93 Ma) is the highest sea-level globally but, for Australia, highstand occurs nearly 30 Myr earlier in the Aptian. Why?

The global eustatic signal of high sea-level in the Late Cretaceous reflects the low volumetric capacity of the young ocean basins, formed as Gondwana breaks apart. The flooding of Australia in the Early Cretaceous has a more local tectonic cause. As Australia moved eastward, it passed over the sinking subducted slabs on the convergent Gondwana margin, and this higher density material in the mantle induced widespread subsidence and, consequently, the inland sea. By the Turonian, the interior of the continent had passed over the slab and rebounded to be high and dry—an example of dynamic topography.

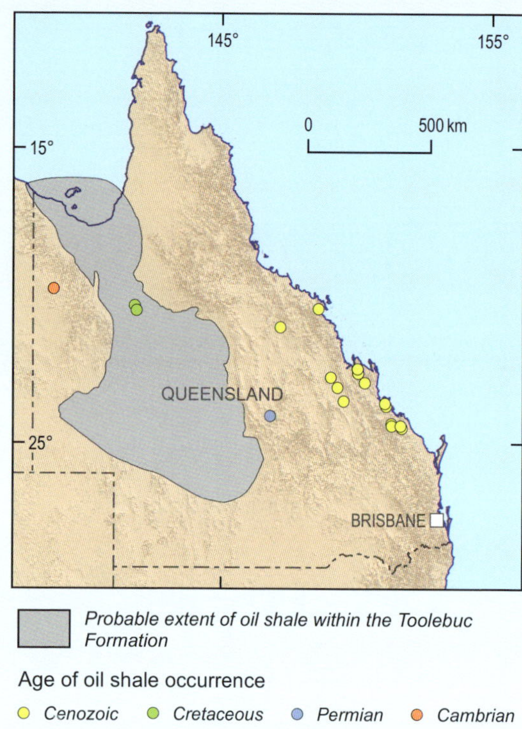

Figure B4.3: Map of oil shale localities in Australia and the probable area of the Toolebuc Formation resource. The abundant Cenozoic oil shale deposits represent the organic-rich sediments of tropical lakes, part of the Capricorn petroleum supersystem (see Figure 4.6). (Sources: Geoscience Australia; Ozimic & Saxby, 1983)

flows (Figure 4.9c). In contrast, the western volcanic rifted margin of the Exmouth Plateau steps down across a rugged and broad slope to the Gascoyne Abyssal Plain (Figure 4.9b). A complex pattern of volcanic ridges, flows and buildups, seaward-dipping reflector sequences and igneous intrusions was created during the breakup as the Cuvier spreading centre cleared the southwestern corner of the plateau and overprinted the earliest formed Gascoyne oceanic crust.

Following continental breakup and seafloor spreading *ca* 135 Ma (Valanginian or Hauterivian), there was a major marine transgression as the northwest margin subsided in response to thermal collapse or cooling of the stretched lithospheric mantle from Perth to New Guinea. This transgression was followed by a period of condensed, deepwater sedimentation, which lasted from the Valanginian to the Barremian. Fine-grained marine sediments accumulated in deepwater environments, forming a regional shale blanket, the Muderong Shale and its equivalents, which is the seal facies to most of the North West Shelf's oil and gas resources (Figure 4.5). By the Late Aptian, deposition of radiolarian-rich claystones records the establishment of open ocean circulation. Contemporaneous flooding occurred on much of onshore Australia, making the Early Cretaceous inland seas (Box 4.3).

On the northwest margin, marine shales and siltstones gave way to carbonate sediments by the Late Cretaceous. The initial deposits were pelagic marls and chalks that graded into coarser calcarenites as the carbonate pile built up into shallower water and prograded seaward, eventually

Figure 4.11: Southwest margin map, showing its sedimentary basins and surrounding ocean floor, hydrocarbon fields and pipelines. The Wallaby and Naturaliste plateaus are largely volcanic features, underpinned by extended continental crust and formed during the final stages of the breakup between Greater India and Australia. The Perth Basin has several small oil and gas fields that are part of the Gondwanan petroleum supersystem. The deepwater frontier Mentelle Basin is still to be tested. (Source: modified from Bradshaw et al., 2003)

Well symbol information is sourced either from 'open file' data from titleholders (where it is publicly available as at 1 December 2011) or from other public sources. Field outlines are provided by Encom GPinfo, a Pitney Bowes Software (PBS) Pty Ltd product. While all care is taken in the compilation of the field outlines by PBS, no warranty is provided regarding the accuracy or completeness of the information, and it is the responsibility of the reader to ensure, by independent means, that those parts of the information used by them are correct before any reliance is placed on them.

creating the modern bathymetric North West Shelf. From the mid-Miocene, coral reefs were common, and they continue to flourish, reflecting Australia's journey north back into tropical waters (Chapter 6). The prograding Cenozoic carbonate wedge can be up to 3 km thick and, in several areas, (e.g. Dampier Sub-basin, Carnarvon Basin, Vulcan Sub-basin, Bonaparte Basin) provides the crucial sedimentary load to push the underlying Jurassic source rocks into the oil window.

In the later Miocene, the architecture of the passive margin was modified by a major regional compressive event as the Australian and Eurasian plates collided (Chapter 2). Pre-existing normal faults were reactivated and inverted, new reverse-fault trends created and anticlines formed; some of these structures now trap hydrocarbons. In the Carnarvon Basin case study, we will see that this is only the last in a long sequence of events required for the oil and gas fields to form on the North West Shelf.

Southwest margin

In contrast to the petroleum-rich northwest, the southwest margin contains only one small oil field offshore and a scattering of modest oil and gas accumulations onshore in the Perth Basin (Figure 4.11). In fact, the major gas supply for the city of Perth is piped more than 1000 km from the giant North West Shelf fields. Why is this contrast so stark? Let's consider the story of this pivotal point within Gondwana to see if we can find an answer.

The southwest margin formed during the breakup as oceanic crust was created in the Perth Abyssal Plain from *ca* 135 Ma (Valanginian) onwards;

the Wallaby Plateau may be a fragment of the Gascoyne Terrane left behind along the transform margin. Greater India slid out along the transform margin of the Wallaby–Zenith Fracture Zone, following the same ancient northwest–southeast trend that controlled the southern and northern edges of the Exmouth Plateau (Figure 4.10); control probably came from the Proterozoic and Archean shear zones that cut the adjacent Yilgarn Craton. However, the structural trend of the Permo-Carboniferous Perth Basin is north–south, reflecting its origin as part of the Gondwanan rift system (Figures 4.8 and 4.11) and the underlying grain of the Neoproterozoic basement (Pinjarra Orogen, Figure 2.10). The dispersal of Gondwana was aided by a mantle plume that was located close to the triple junction between Australia, Antarctica and India, producing large igneous provinces of the Kerguelen Plateau (*ca* 115 Ma), Broken Ridge (*ca* 95 Ma) and Naturaliste Plateau, as well as extensive volcanism in the Mentelle Basin (Figure 4.11).

The Perth Basin contains Permian coal measures, which are mined at Collie (Chapter 9; Figure 4.11) and are a source for some of the gas accumulations (e.g. Whicher Range, in the Bunbury Trough). However, the main source rock for the small oil and gas fields (Cliff Head, Dongara, Mt Horner) is the Early Triassic Kockatea Shale, deposited in restricted marine environments (Figure 4.7) as the sea flooded a landscape created by earlier Permian rifting. The remainder of the Triassic and much of the Jurassic is dominated by sandstones, including red beds with only minor marine and lacustrine shales and coal measures. The coarse sands are good groundwater aquifers (Chapter 7), but are lean in the organic matter needed to source hydrocarbons. There is no sign of the thick pods of Late Jurassic marine shales that are oil sources along the North West Shelf. The time-equivalent strata in the Perth Basin are the coarse fluvial sandstones of the Yarragadee Formation, which were probably derived from Antarctica to the south. There may be a different story in the offshore frontier Houtman Sub-basin and the yet-to-be-drilled Mentelle Basin, which in part is a failed rift between the Australian mainland and the Naturaliste Plateau (Figures 4.11 and 4.12). Very little is known about these two remote deepwater regions.

As with the northwest margin, after the breakup the southwest subsided, and marine shales followed by carbonates were deposited as the passive margin sequence. In their early post-rift history, the Wallaby and Naturaliste plateaus and western Mentelle Basin were affected by extensive volcanism, triggered by the proximity of the Kerguelen Large Igneous Province. Several hundred metres of igneous and volcaniclastic rocks were emplaced and deposited, masking the original geometries, including any pre-existing depocentres. This volcanic episode, which continued for at least 10 Myr after the breakup, adversely impacted their hydrocarbon potential in comparison with the

Semisubmersible offshore drilling rig (in the foreground) and an oil rig construction vessel on the North West Shelf, Western Australia. The North West Shelf is one of Australia's premier hydrocarbon provinces.

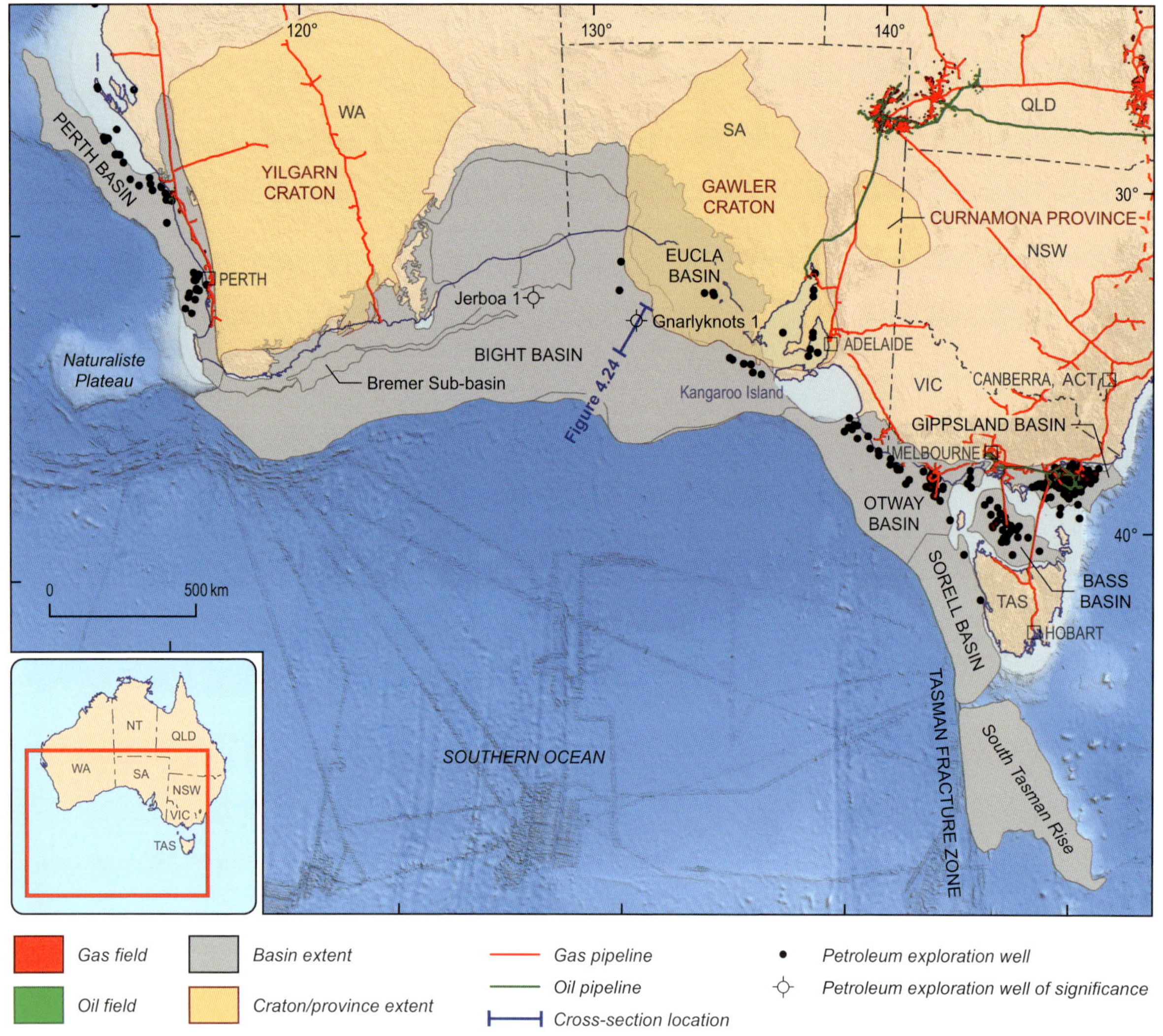

Figure 4.12: Map of the southern Australian margin, showing sedimentary basins, oil and gas fields, exploration wells and pipelines. Note the extensive drilling and infrastructure in the Gippsland, Bass and Otway basins in contrast to the west of Kangaroo Island or on the South Tasman Rise.

Well symbol information is sourced either from 'open file' data from titleholders (where it is publicly available as at 1 December 2011) or from other public sources. Field outlines are provided by Encom GPinfo, a Pitney Bowes Software (PBS) Pty Ltd product. While all care is taken in the compilation of the field outlines by PBS, no warranty is provided re the accuracy or completeness of the information, and it is the responsibility of the reader to ensure, by independent means, that those parts of the information used by them are correct before any reliance is placed on them.

gas-rich Exmouth Plateau, where intense volcanic activity was limited to its western margin. Elevated heat flow and associated hydrothermal fluids can destroy reservoirs, and cook away oil or gas already trapped, and intrusions can physically disrupt a petroleum system. Volcanic rocks can create impermeable flows that cut through a migration pathway, although in rare cases they can act as seals, as in the Kipper field in the Gippsland Basin.

Southern margin

Australia's break with Antarctica created the southern margin, the longest south-facing coastline in the world, stretching more than 3500 km from the Naturaliste Plateau to the South Tasman Rise (Chapter 6). At its eastern end are the hydrocarbon-producing basins of Bass Strait. But from Kangaroo Island to the west is a large region with only around 12 exploration wells, making this area very much a frontier province (Figure 4.12).

Antarctica has been Australia's longest partner in Gondwana, for close to a billion years (Chapter 2), and the process of separation between the two continents took more than 100 Myr to complete (Figure 4.13). The apparent clean break of the current coastline belies the complex story that commenced with a Middle Jurassic to Early Cretaceous rift valley (*Did you know?* 4.3). The crust thinned along the rift, which contained a system of lakes, rivers and high-latitude forests that stretched from the Naturaliste Plateau to the Gippsland Basin. The eastern highlands uplifted in the late Albian (*ca* 100 Ma), with reorganisation of the drainage networks to flow southwest and the beginning of the growth of the Ceduna delta, which was well established by 95 Ma (Cenomanian).

The margin evolved through repeated episodes of extension and thermal subsidence leading up to, and following, the commencement of seafloor spreading between Australia and Antarctica. Breakup progressed from west to east, with a well-established initial age of seafloor spreading of 83 Ma (Santonian) for the central Bight Basin, although perhaps starting as early as *ca* 90 Ma in the Bremer Sub-basin. Propagating to the east, seafloor spreading off the Otway Basin commenced at *ca* 55 Ma, and off the Sorell Basin at *ca* 47 Ma (Figure 4.1). In general, breakup was not accompanied by significant magmatism, and the margin is classified as a magma-poor (or 'non-volcanic') rifted margin in contrast to those on the Exmouth Plateau. As in the northwest, however, there are both rifted and transform components (Figures 4.9c and d).

Initial northwest–southeast ultra-slow to slow seafloor spreading (latest Santonian to mid-Eocene, 83–45 Ma) was followed by north–south fast spreading. This history, and the pre-existing basement architecture, has resulted in segmentation into:

- an oblique to normally rifted margin that extends from the westernmost Bight Basin to the central Otway Basin
- a transform continental margin in the east (western Tasmania – South Tasman Rise) along the Tasman Fracture Zone
- a transitional zone between those end-members (southern Otway–Sorell basins) (Figures 4.9, 4.12 and 4.13).

Seafloor spreading thus was slow in the Late Cretaceous, but by the mid-Eocene (45 Ma), a widening seaway had appeared between Australia and Antarctica as seafloor spreading accelerated. Tasmania stayed with the mainland, and both separated from Antarctica along the Tasman Fracture Zone. The final break came in the Late Eocene at *ca* 34 Ma (Figure 4.13).

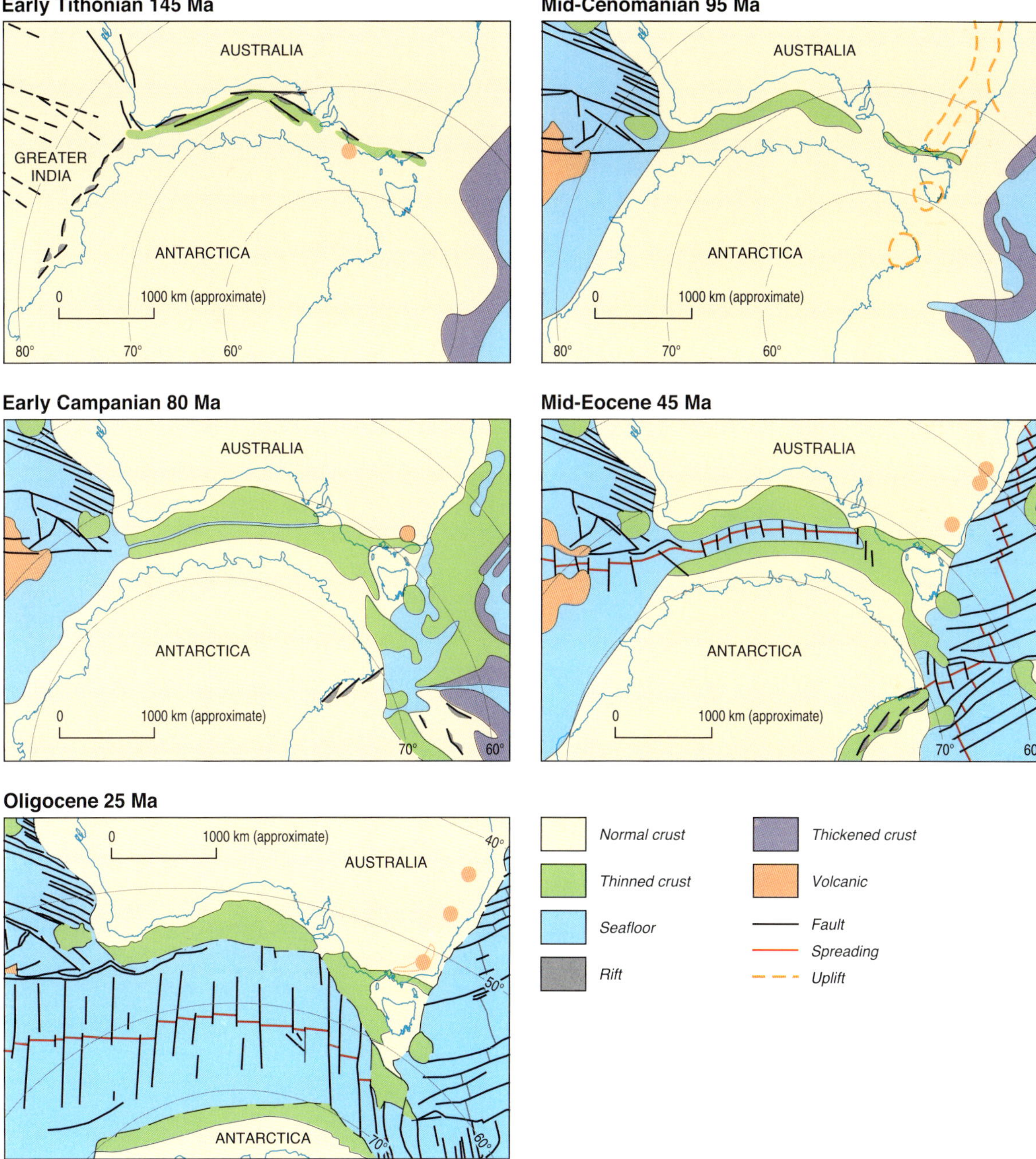

Figure 4.13: Schematic structural evolution of the southern margin with reconstructions from the Late Jurassic (Early Tithonian, 145 Ma) to the latest Oligocene (25 Ma). In the west, the rift valley marks the site of the future break (probably along the southern edge of the Naturaliste Plateau), whereas, in the east, Bass Strait overlies the area where the rifting failed to go to seafloor spreading. The Oligocene reconstruction shows the final configuration with the sharp edge of the South Tasman Rise. (Sources: modified from Norvick & Smith, 2001; Norvick, 2005)

Did you know?

4.3: Why did Australia and Antarctica break apart where they did?

As we have seen, ancient basement structures played a role in determining how and where these two last remaining partners in Gondwana separated. Another idea is that the arc of the southern margin may have followed the shortest or least-work path of breakup.

Or was the late Douglas Adams onto something …

> *Australia is a very confusing place, taking up a large amount of the bottom half of the planet. It is recognisable from orbit because of many unusual features, including what at first looks like an enormous bite taken out of its southern edge; a wall of sheer cliffs which plunge deep into the girting sea. Geologists assure us that this is simply an accident of geomorphology and plate tectonics, but they still call it the 'Great Australian Bight' proving that not only are they covering up a more frightening theory, but they can't spell either …*

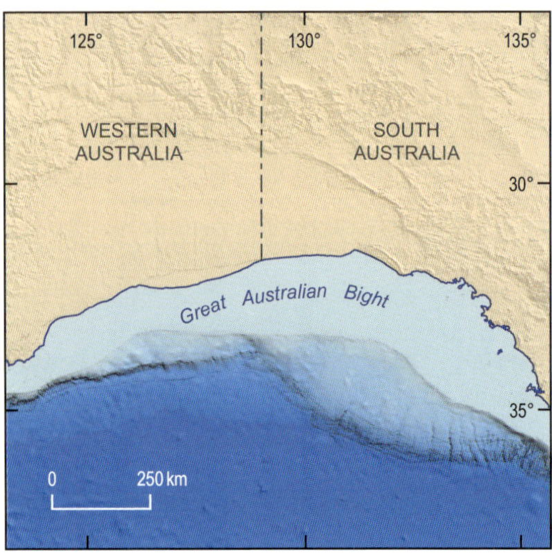

The oblique- to normally-rifted margin is characterised by a broad zone of crustal thinning and thick extensional basin development. In the Bight and Otway basins, a well-developed distal ocean–continent transition zone includes basement highs interpreted as exhumed subcontinental lithospheric mantle (Figure 4.9e). In the eastern part of the margin, where transcurrent stresses controlled deformation, lithospheric thinning is not as marked, and the continent–ocean boundary is probably a combination of rift and transform elements. In the southern Sorell Basin and the South Tasman Rise, the Tasman Fracture Zone forms a transcurrent continent–ocean boundary (Figures 4.1, 4.9a and 4.12).

Basement terranes such as the Mesoproterozoic Albany–Fraser Orogen, Mesoarchean–Proterozoic Gawler Craton and Lower Paleozoic Delamerian Orogen influenced the breakup of the southern margin (Figure 4.14; Chapters 2 and 8). The Shipwreck Trough off western Tasmania marks an abrupt change in Early Cretaceous rift basin geometry from northwest–southeast to dominantly north–south- or northeast-trending structures. This switch is also evident onshore and coincides with the west-dipping, crustal-scale Paleozoic Avoca Fault, a structure that forms the western boundary of a poorly exposed Proterozoic basement terrane in central Victoria, the Selwyn Block (Figure 4.14), and continues southward into the Sorell Fault and the Tasman Fracture Zone. Crustal weaknesses inherited from the underlying basement rocks (Chapters 2 and 8) evidently persisted into Mesozoic times, when they were reactivated under extension during the Australia–Antarctica separation.

The characteristic sedimentary sequence in the southern margin basins is Late Jurassic to Early Cretaceous fluvio-lacustrine facies with a pervasive volcaniclastic component, overlain by Late Cretaceous to Paleogene deltaic and marine sediments with a cap of Cenozoic carbonates (Figure 4.5). This sequence records the progression from the rift valley deep within Gondwana to a passive margin facing the ever-widening sea between Australia and Antarctica. The mid-Cretaceous uplift of the eastern highlands was a prelude to the breakup of that margin. The Ceduna delta was so large that it modified the bathymetry of the margin (Figures 4.14 and 5.14) as it built out an apron of sediment more than 5 km thick over an area of more than 100 000 km^2. Which raises the question of why such an enormous delta was deposited when the river that fed it is no longer seen.

The accommodation space provided by the eustatic high sea-levels of the Late Cretaceous (Box 4.3) and the subsidence of the underlying highly extended continental crust (Figure 4.9d) may have allowed the sediments gleaned from half a continent to prograde out into the narrow seaway between Australia and Antarctica. A dramatic reduction in sediment supply at the end of the Cretaceous, and perhaps switching of the principal river system feeding it, may have caused the abrupt abandonment of deltaic deposition. Regional uplift resulted in some erosion, but, as the delta sank with Early Cenozoic thermal subsidence, it was preserved as a deepwater terrace. We will return to consider the Ceduna delta and its hydrocarbon potential in the frontier petroleum system case study. After Australia and Antarctica severed their last connection along the Tasman Fracture Zone and the circum-Antarctic current was established at the end of the Eocene, cool-water carbonate deposition became dominant.

Australia's southern continental margin is home to both rich hydrocarbon resources and huge potential. Most of Australia's oil has come from one small basin located at the eastern end of the Australia–Antarctic rift system—the Gippsland Basin. Here, all the elements and processes of the petroleum system have converged to produce Australia's only billion-barrel oil fields. Non-marine coaly sediments are source rocks for the oil and gas in the Otway, Bass and Gippsland basins; deltaic to shallow marine sandstones are the reservoirs; and marine mudstones and marls provide sealing facies.

Eastern margin

Eastern Australia is the ancient convergent margin of Gondwana (Chapter 2) and, in the mid-Cretaceous, the plate boundary with the palaeo-Pacific Ocean shifted eastwards. Pre-existing crustal weaknesses along possible former terrane boundaries were exploited as the breakup process opened the Tasman Sea as a backarc basin and created the narrow shelf along the New South Wales coast (Figure 4.14). Here, the Late Cretaceous seafloor of the deep Tasman Sea lies a few tens of kilometres from the shore. On the other side of the spreading ridge, the continental fragment of Lord Howe Rise now lies many hundreds of kilometres to the east and is linked to New Caledonia and New Zealand, also former parts of Gondwana (Box 4.4). Initial seafloor spreading began in the south between New Zealand and Tasmania in the Santonian (*ca* 85 Ma) and propagated irregularly northwards, finishing in the Coral Sea in the Early Eocene (*ca* 52 Ma). A number of sedimentary basins without drillhole evidence are of probable Jurassic to Cretaceous age, and these occur on the Lord Howe Rise; some may be prospective for petroleum, as are the once-neighbouring Gippsland Basin and Taranaki Basin (New Zealand).

North from the apex of the Tasman Sea, the Queensland margin opens out into a broad and complex borderland (Figure 4.9a). Again, as in the northwest, thick sedimentary basins (Capricorn

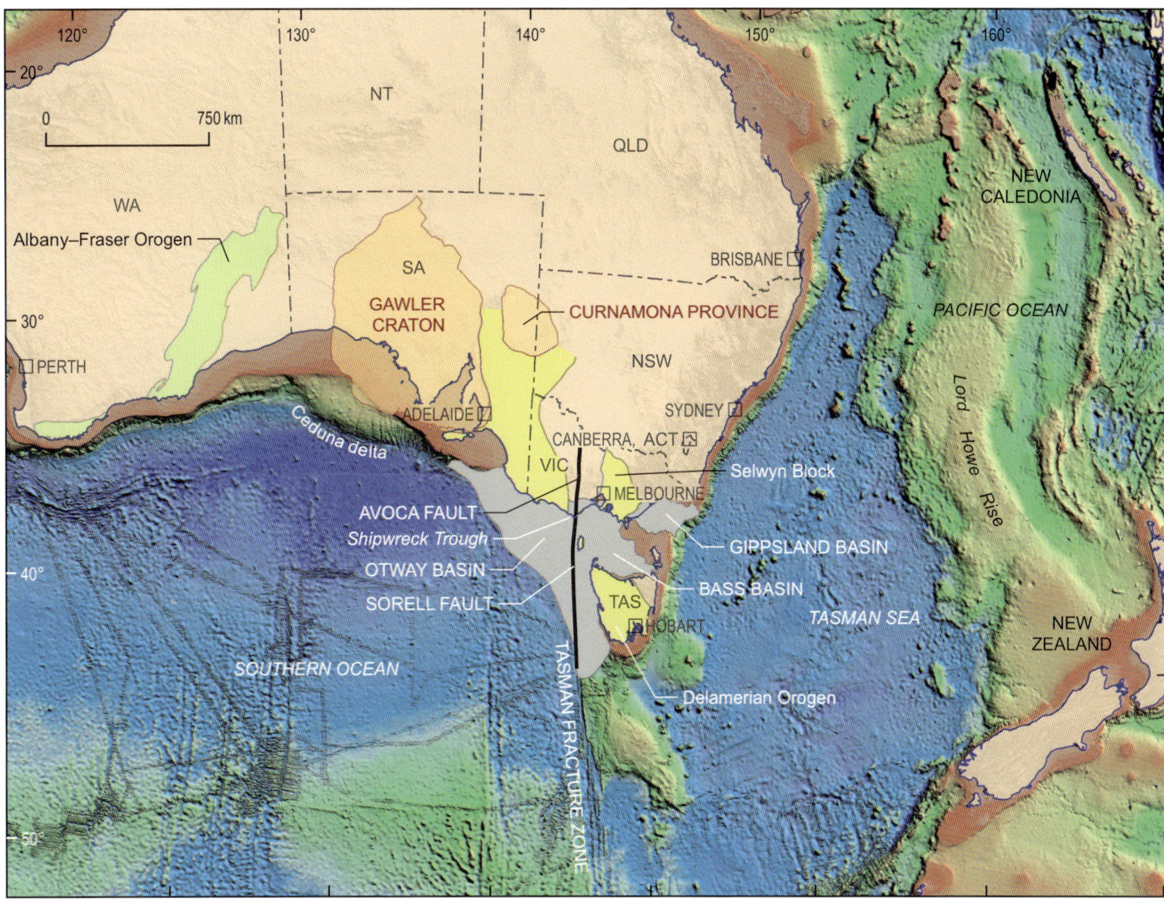

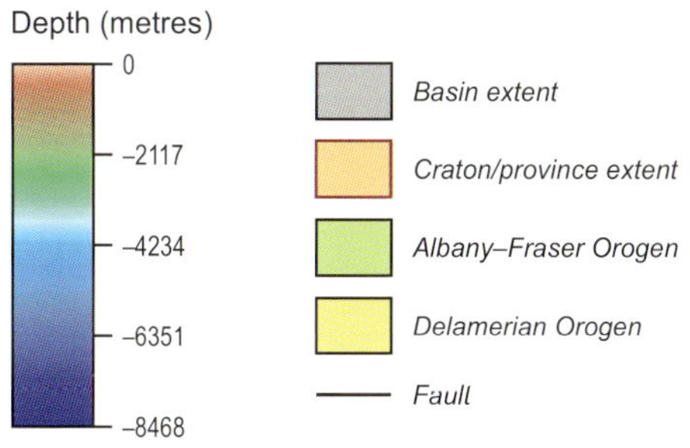

Figure 4.14: Combined topographic and bathymetric image of southern Australia and New Zealand, showing the major features of the Ceduna delta, South Tasman Rise and Lord Howe Rise. Also shown is the relationship of the Avoca and Sorell faults to the basement features. The Tasmanian segment of the Delamerian Orogen extends northward at depth beneath the Bass Basin into central Victoria (Selwyn Block).

LORD HOWE RISE (BOX 4.4)

The Lord Howe Rise is a major submerged plateau (1000–3000 m of water depth) located east of the Australian continent. The Lord Howe Rise extends almost 2000 km from southwest of New Caledonia to the Challenger Plateau, west of New Zealand, and has an area of about 1.5 M km^2. The central portion of the Lord Howe Rise lies within the Australian marine jurisdiction as defined by the United Nations Convention on the Law of the Sea (Chapter 1). Australia now shares political boundaries with France–New Caledonia and New Zealand.

The rise is surrounded by a complex of continental fragments, ocean basins, volcanic arcs and accretionary prisms, as far as the Tonga and Kermadec trenches and New Zealand, where the present-day active plate margin now lies (Chapter 2). The deepwater region off eastern Australia formed during the Cretaceous to Recent, by continental rifting, magmatic arc formation, subduction rollback and backarc spreading. The resulting continental crust is therefore a rheologically weak collage of accreted Paleozoic to Early Cretaceous arc and related material, which broke up into numerous irregular pieces.

The Lord Howe Rise may have originally consisted of at least two separate continental pieces that rifted from eastern Australia during the Cretaceous breakup of the East Gondwana margin and the Tasman Sea opening (85–52 Ma). Rock samples and tectonic reconstructions suggest that the Lord Howe Rise is underlain by offshore continuations of the Paleozoic New England and Lachlan orogens (Chapter 2) and Mesozoic basins of eastern Australia (Clarence–Moreton and Maryborough basins), New Zealand and New Caledonia.

Subparallel geological provinces that extend along its length dominate the basin framework on the rise: the Lord Howe Platform, the Central Rift Province, the Western Rift Province and the Western Ridge System. Depocentres on the Lord Howe Rise have the potential to generate and accumulate petroleum, comparable with other East Gondwanan Cretaceous rift basins, such as the Taranaki, Gippsland and Clarence–Moreton basins. Evidence for Cenozoic post-rift igneous activity is widespread throughout the Lord Howe Rise region. The most prominent are the north–south seamount hotspot chains, the Tasmantid and Lord Howe seamount chains (Figure 4.9a; Chapter 2). They developed to the west of the rise from the Miocene to Recent.

The Lord Howe Rise is a unique, isolated and large continental fragment with volcanic islands and exceptional wildlife. Lord Howe Island was officially inscribed on the UNESCO World Heritage List on 14 December 1982 because of the outstanding universal values of its terrestrial and marine species and ecosystems. The island is covered in beautiful banyan, pandanus and Australian rainforest species and fringed by the world's southernmost coral reef (Chapter 6). Although this tiny World Heritage speck is only 11 km long and 3 km wide, it is home to numerous endangered endemic species. Some species endangered in other places (for example, masked owls in Tasmania) now thrive in densities unheard of elsewhere following their introduction to Lord Howe Island. In the Pleistocene, it was also a refuge for species such as the giant horned turtle, *Meiolania* (Chapter 3).

Basin, Townsville and Queensland troughs) and submerged marginal plateaus (Marion and Queensland) formed as the rifted margin developed, leading to the opening of the Coral Sea in the Paleogene (Figure 4.1). The Marion and Queensland plateaus now support coral reefs, and, to landward, the Great Barrier Reef has grown during the Quaternary (Chapter 6). Despite early exploration, the environmental value of this region precluded the development of the petroleum potential in these basins. Seafloor spreading between about 60 Ma and 52 Ma created the Coral Sea, the last new ocean basin to open around the Australian margin as it emerged from Gondwana.

The next act in the drama is the formation of the convergent northern margin, as the island continent proceeds on its journey north.

Northern margin

Between the Argo Abyssal Plain in the northwest and the Coral Sea in the northeast are the island chains of eastern Indonesia and Timor, the island of New Guinea and the shallow Arafura Sea and Gulf of Carpentaria (Figures 4.1 and 4.9a). This collage of geomorphological and geological elements is home to active volcanoes, equatorial jungle swamps and towering mountains capped by a glacier.

Travelling rapidly north in the Eocene, Australia collided with the Pacific and Eurasian plates, forming the tectonically active northern margin. The structural impact of this collision, which has lasted for about 30 Myr, has been profound on the buckled and sheared northern margin of the plate, but has also had an influence far into the continent and continues to do so today (Chapter 2). There is an enrichment of the flora and fauna as the old Gondwana meets Asia across the Wallace and other biogeographical lines (Chapters 1 and 3). Many of Australia's oil and gas fields are in traps formed or reactivated by the Neogene collision.

The complex northern margin can be considered in two parts: the New Guinea fold belt, running from the Aure Trough to the Bird's Head, and the outer Banda Arc, running from Seram around to Timor and Sumba (Figure 4.9a). The island of New Guinea, extending northwards to the fold belt, is geologically contiguous with Australia. This region reflects a similar east–west contrast between a Late Paleozoic active magmatic margin in the east and a region of extension with Gondwana breakup to the west. Strong Mesoproterozoic crust and lithosphere and thick Neoproterozoic to Paleozoic sedimentary rocks occur in the southwest (West Papua), an extension of the North Australian Element and the overlying Arafura and Bonaparte basins. To the east of the (projected) Tasman Line in New Guinea, the oldest basement is formed by Permian metasediments intruded by Triassic granites. Triassic and Jurassic rifting created a Cretaceous passive margin, with promontories of extended continental crust and embayments of oceanic crust similar to those preserved on the northwest margin. Although similar stratigraphy and shared oil geochemistry show the affinity of the Papuan Basin to the North West Shelf, the original New Guinea passive margin has been severely modified by terrane accretion and strike-slip faulting.

West of New Guinea is the horseshoe-shaped Banda Arc, part of a wide and complex suture zone in eastern Indonesia. There is an outer non-volcanic arc (Timor, Seram) where deformed Mesozoic sediments, which are distal equivalents of the North West Shelf sequences, crop out. There is also a younger, inner volcanic arc, which has been active for the past 10 Myr. The Banda Arc formed where an embayment of oceanic crust within the northward-moving Australian Plate reached the subduction zone around 12 Ma. The old, cold and dense Jurassic ocean lithosphere fell away rapidly, causing the Banda subduction hinge to roll back to the south and east, forming the arc, opening the Banda Sea and, by about 3 Ma, bringing into collision the inner volcanic arc with the Australian continent in the region of Timor.

The island continent

The last episode in the story of Australia's release from Gondwana began with the final clearance of Tasmania from Antarctica in the latest Eocene. Australia was for the first time the island continent on its northward path into Asia. This final separation changed ocean currents and global climate, including the start of glaciation in Antarctica. With changing latitude, cold-climate rainforests were eventually replaced by deserts. However, along the northeastern monsoonal margin of Australia, new Asian rainforest floral elements began to invade this part of the continent. The Gondwanan flora and fauna then evolved through millions of years, mostly in isolation (Chapter 3). Aridity and deserts developed as Australia moved up into the mid-latitudes in the Late Miocene.

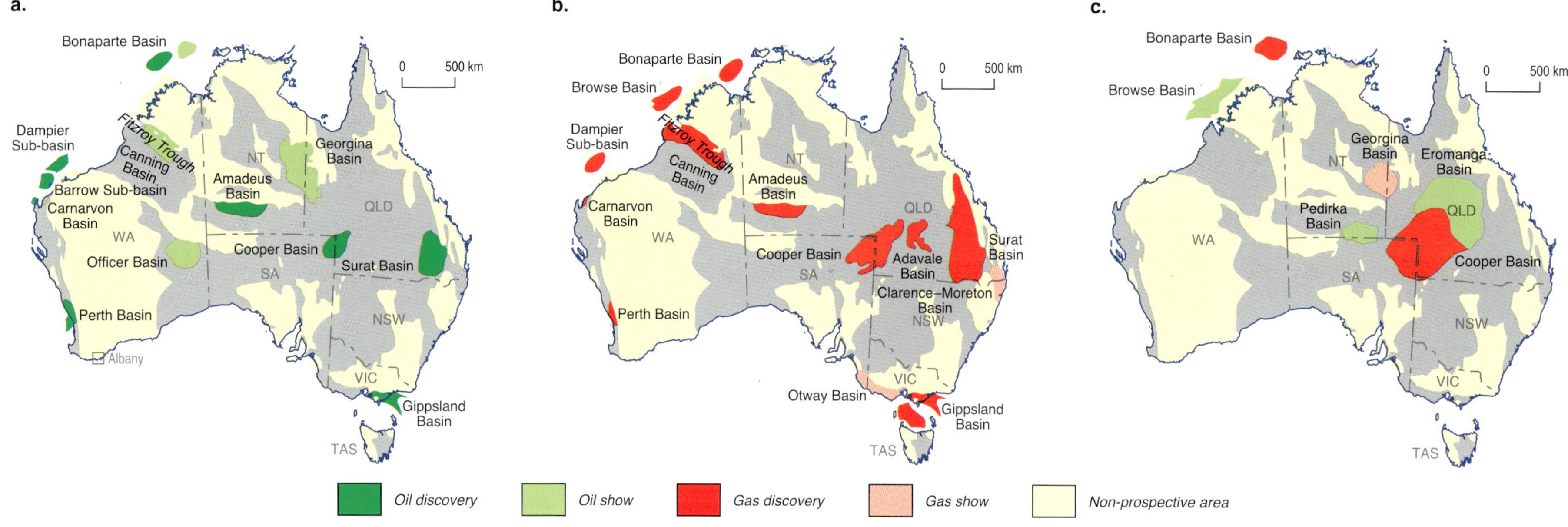

Figure 4.15: Maps of hydrocarbon shows and discoveries in Australian basins over time. (a) Oil shows and discoveries 1960–72. (b) Gas shows and discoveries 1960–72. (c) Oil and gas shows and discoveries 1973–77 (in basins without previous hydrocarbon shows and/or discoveries). (d) Oil and gas shows and discoveries 1978–88 (in basins without previous hydrocarbon shows and/or discoveries). (e) Oil and gas shows and discoveries 1989–98 (in basins without previous hydrocarbon shows and/or discoveries). (f) Oil and gas shows and discoveries 1999–2011 (in basins without previous hydrocarbon shows and/or discoveries). (Source: updated and modified from Bradshaw et al., 1999)

We will see in Chapter 5 how the impact of this change to aridity affected the Australian landscape and people.

The search for Australia's petroleum

The pattern of hydrocarbon discovery in Australia has followed a path from onshore to offshore, and from Bass Strait to the North West Shelf (Figure 4.15; Appendix 4.1.2). Today, with the development of coal-seam gas and other unconventional hydrocarbons, the activity is shifting back onshore. The main drivers of this exploration activity have been market demand, perceived prospectivity, technological change, government behaviour and geological endowment, along with the strong positive market feedback of significant discoveries encouraging further exploration. In recent years, there has been a fundamental shift in the market for gas in the Asia–Pacific region, with the industrialisation of China and India, and this is pushing along the current boom in gas exploration and development (Chapter 9).

Petroleum exploration began in Australia in the 1860s after there were shows of gas and oil in different places. Activity was initially low, with only a handful of wells drilled from 1900 to 1950, and exploration technology was rudimentary by today's standards. Initially, the wrong clues were followed—the obvious surface oil seeps that directed early exploration efforts elsewhere were rare in Australia. False guides, such as beach stranding of bitumen and algal mats in coastal lagoons, inspired the first wells. The great depth of 9 m was reached at Salt Creek in the Coorong in 1866, targeting the source of 'Coorongite', now known to be the remains of modern algal blooms and dead fish. The first 'offshore' well was drilled

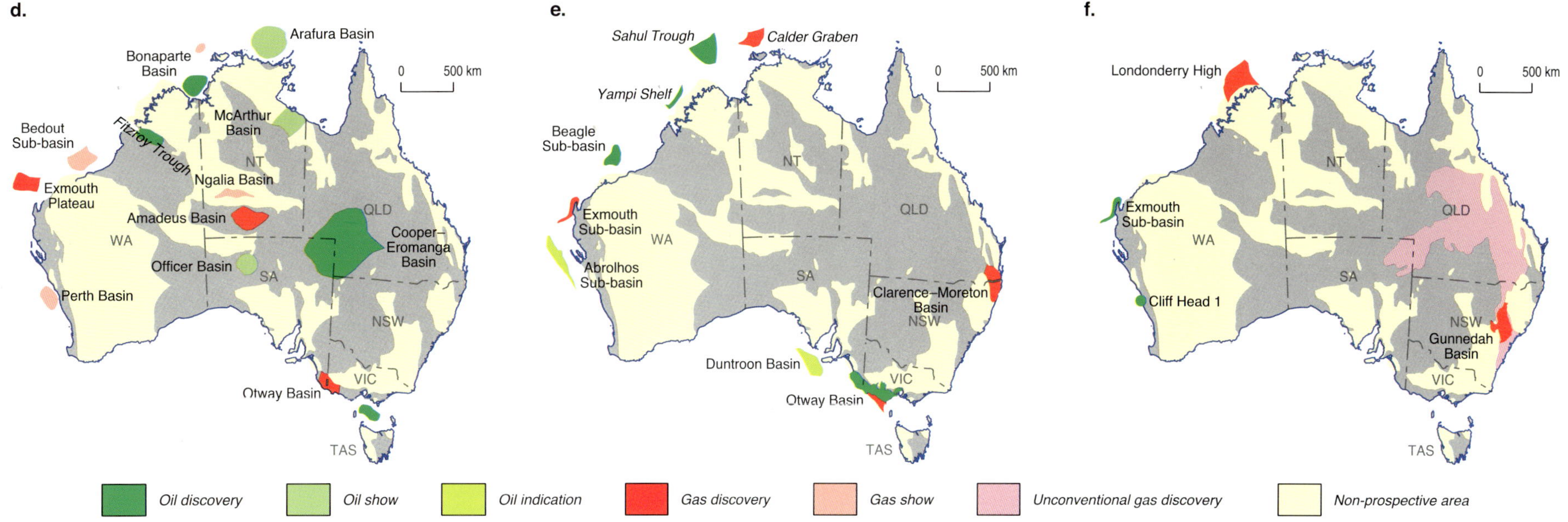

in Albany Harbour in 1907 (Figure 4.15a), looking in the Mesoproterozoic Albany–Fraser Orogen basement terrane for the source of the bitumen strandings common along the southern margin. But, as we shall see, these strandings may in the end lead us to a major new oil province.

Gas and oil shows in water bores (Appendix 4.1.2) prompted a low and sporadic level of drilling through the first half of the 20th century. Most activity was concentrated around Roma in the Surat Basin, where gas was eventually found (Figure 4.15b). Drilling was restricted to onshore areas and consequently was largely dependent on viable Paleozoic source rocks for success. The technology was not available to access the more prospective and younger offshore basins. Prospectivity was perceived by overseas experts to be low, as the Australian continent was considered too old to have preserved major oil accumulations. Despite this predominance of negative factors, some minor and sporadic production of hydrocarbons had been achieved from the Gippsland and Surat basins by the 1940s.

After nearly a hundred years of largely unrewarded effort, the first flow of oil to the surface in Australia occurred in 1953, with the discovery of a small oil accumulation at Rough Range 1 in the onshore Carnarvon Basin. In 1957, the Australian Government introduced a 50% subsidy scheme for exploration drilling, and in 1959, the fledging industry body APEA (the Australian Petroleum Exploration Association) was formed to counteract 'waning public, government and even professional interest in the Australian oil search'. By the early 1960s, the momentum of exploration had built and, with increased drilling, discoveries came. First was onshore in the Surat, Cooper, Adavale, Amadeus and Perth basins, and on Barrow Island in the Carnarvon Basin, followed by offshore in the Gippsland Basin (Figure 4.15a; Appendix 4.1.2). The giant Gippsland oil discoveries in Bass Strait encouraged more exploration and further discoveries, with the result that, by 1972, all the major basins and petroleum systems that are today producing hydrocarbons had been found (Figure 4.15a).

Australian oil production from the 1960s to the 1990s was dominated by the three giant fields in the offshore Gippsland Basin and the Barrow Island field that produced at a steady but much lower rate (Figure 4.16). As production from the Gippsland fields peaked in the late 1980s, dozens of smaller fields on the North West Shelf (Carnarvon and Bonaparte basins) progressively came on stream to make up the shortfall and maintain Australian self-sufficiency in oil (Figure 4.16). Today, production is in decline and the oil search continues, now

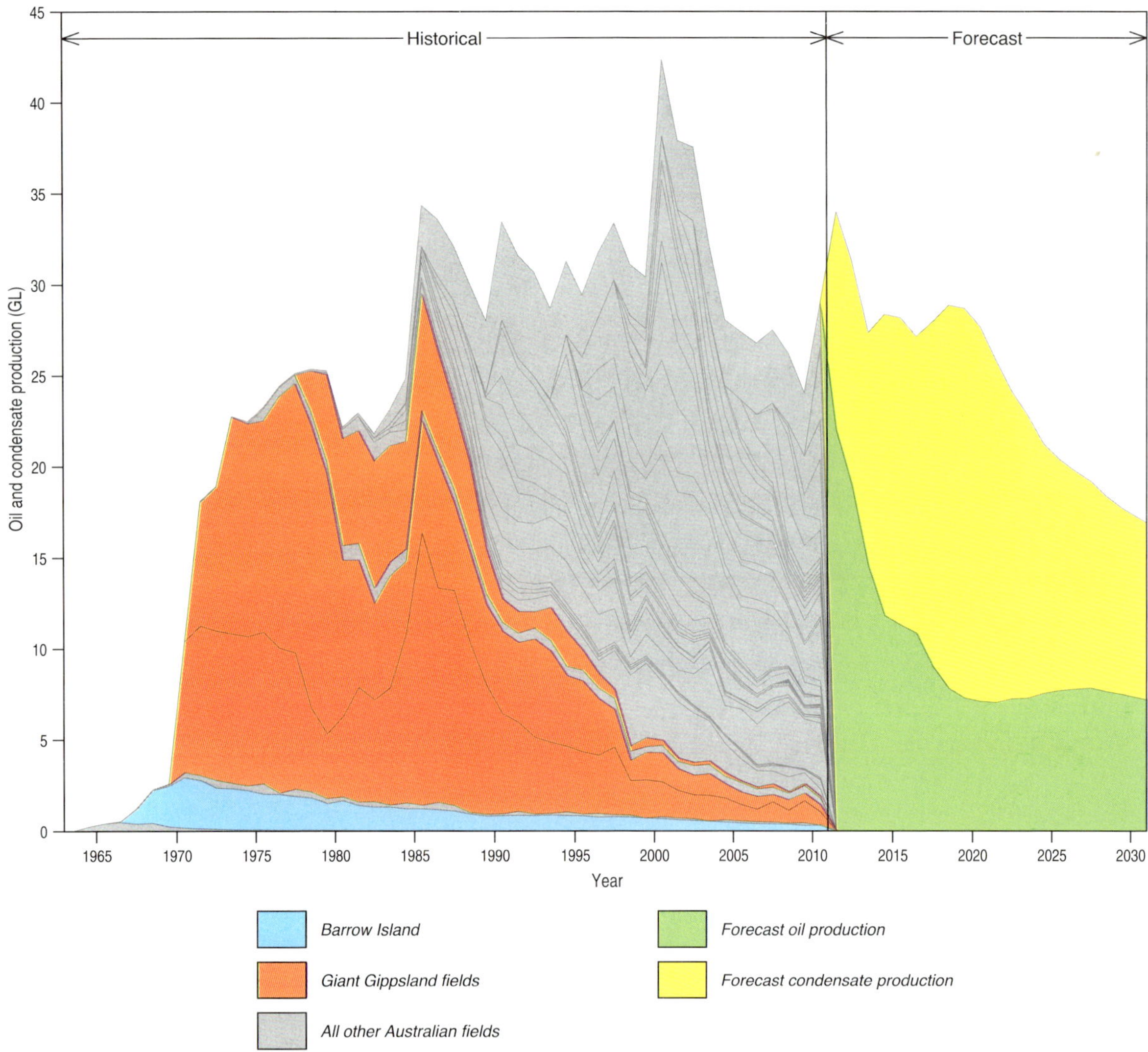

Figure 4.16: Production profiles of individual Australian fields. (Source: updated and modified from Powell, 2001)

moving into deeper offshore Australian waters, further exploring for 'gifts of Gondwana' in the basins formed by its breakup.

Gippsland Basin, Victoria

The Gippsland Basin is Australia's prolific, world-class oil province. It has provided most of Australia's oil and has underpinned the economy (Figure 4.16). It is the most heavily explored basin and one of the best-known petroleum systems in Australia. Lake Bunga No. 1, drilled in 1924 near Lakes Entrance (Figure 4.17), is considered our first oil discovery. The Latrobe Valley brown coal deposits are another major energy resource (Chapter 9) in this relatively small basin of 46 000 km^2, nestled between the Paleozoic basement terranes of the Lachlan Orogen and the Cretaceous seafloor of the Tasman Sea (Figure 4.14).

Discovery and development

As happened elsewhere in onshore Australia, traces of oil and gas were found in a flow of artesian water from the first well sunk onshore near the town of Lakes Entrance. At 370 m, the bore intersected a 13 m interval of oil-saturated Oligocene glauconitic conglomerates. The oil was very viscous and would not flow to the surface; it was 'heavy oil' that had been biodegraded and water-washed. The shallow and cool oil reservoir with its intrusion of freshwater was an ideal environment for bacteria to grow, consuming the shorter chain hydrocarbons and leaving behind a heavier, sticky oil. More than 60 other wells were drilled in following decades, which delineated an extent of around 20 km^2 for the shallow oil sands. Although the oil would not flow to the surface under its own buoyancy, it was shallow enough to be pumped or bailed up out of the reservoir, and around 3000 barrels of oil was produced in this fashion by 1941.

During World War II, amid concerns about security of oil supply via shipments from overseas, the Australian Government initiated the 'Lakes Entrance Oil Shaft Project'. A 3 m-wide shaft was sunk down some 365 m, and then, from an expanded work chamber at the bottom, horizontal wells were drilled into the oil-bearing sandstones. The oil that collected at the bottom of the shaft was then hauled to the surface in buckets. Close to another 5000 barrels of oil were produced, a negligible addition to the fuel supply in a time of petrol rationing. It is ironic that such desperate measures were employed, when the giant oil fields nearby in the offshore were eventually found to have excellent production characteristics, with light oil flowing at a rate of tens of thousands of barrels per day. But their discovery and development had to wait until the 1960s.

The Australian oil search was in full swing in the 1960s, encouraged by finds from Western Australia to Queensland and government subsidies for exploration. On the advice of Lewis Weeks, retired chief geologist of the Standard Oil Company of New Jersey (forerunner of Esso–Exxon), BHP Ltd acquired a number of leases in 1960, including some offshore in Bass Strait, where they focused their search on the thick, young sediments adjacent to onshore oil indications. Initial seismic surveys were conducted in 1962 by BHP Ltd, which then formed a joint venture with Esso Australia Pty Ltd in 1964 over the offshore Gippsland Basin. The seismic surveys revealed the Central Deep and six large anticlinal closures. The first two wells, Barracouta 1 (1964–65) and Marlin 1 (1965–66), were large gas-condensate discoveries, but the third, Kingfish 1 (drilled 1967), struck Australia's largest oil field (1.2 B barrels recoverable), and Australia was at last on the world oil map (Figure 4.17; *Did you know?* 4.4).

By the beginning of the 1970s, the first five of 11 newly discovered fields were on production, with gas and oil pipelines connected to onshore processing facilities (Figure 4.17). The hydrocarbons were reservoired in the coarse-grained clastics at the top of the Latrobe Group (Figure 4.18). Many more, but smaller, discoveries were to follow, and exploration added new play types such as the Kipper gas field, the first significant intra-Latrobe find in 1986 (Figure 4.18). In 1990, the oil province was extended into deep water, with the Blackback discovery in 400 m of water, off the shelf edge (Figure 4.17). Since the late 1990s, high oil and gas prices, combined with the application of 3D seismic technology, have continued to drive exploration efforts in the Gippsland Basin. More than 300 exploration wells have now been drilled, and there is a dense grid of 2D and 3D seismic coverage with an inventory of more than 30 fields.

Bass Strait oil discovery: among the most pivotal events in Australia's history?

The discovery of a giant oil field in one of the first offshore exploration wells drilled in Australian waters revolutionised our energy security by removing the vulnerability clearly demonstrated during World War II. Important economic benefits flowed from the next decades of oil self-sufficiency following the Gippsland Basin discoveries. The development of Australia's oil and gas industry improved the balance of trade and helped underpin the nation's economic prosperity and growth. The discoveries were the crucial kick-start for the $28 B/year ($25.6 B in 2010–11) oil and gas industry, which, in 2008, contributed 58% of Australia's primary energy, 2.5% of Australia's

4.4: Australia's largest oil field was nearly missed when first drilled

The sedimentary sequences in the offshore Gippsland Basin have been cut by large submarine channels and canyons, particularly in the Neogene carbonates (Figure 4.18). Less consolidated channel fill and areas of dolomitic cementation create abrupt lateral changes in density and thus in seismic response, as seismic data record the travel time of soundwaves through the sediments, with a faster and shorter travel time through denser material. The resulting complex velocity variations in the upper section can mask underlying structures. Such a velocity anomaly from channelling in the Seaspray Group occurs over the top of the Kingfish field. The conversion from travel time to depth, based on the 1960s vintage seismic data, was so misleading that the billion-barrel oil field was almost missed by the drill. Today's explorers have the advantage of 3D seismic data and visualisation technology.

Image courtesy of ExxonMobil

Geoscientist reviewing exploration data.

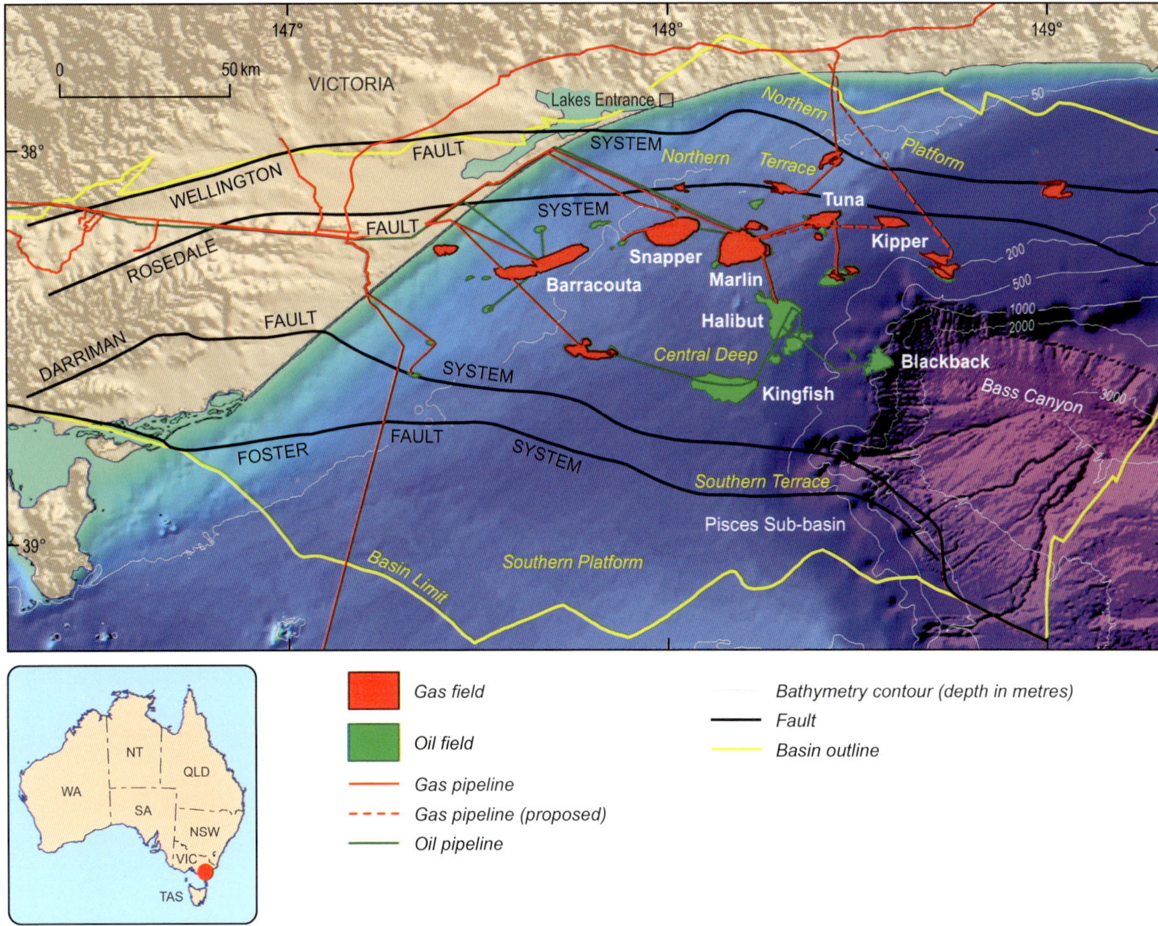

Figure 4.17: Digital terrain image of the Gippsland Basin, showing the location of oil and gas fields, pipelines, bathymetric features and structural elements.

Field outlines are provided by Encom GPinfo, a Pitney Bowes Software (PBS) Pty Ltd product. While all care is taken in the compilation of the field outlines by PBS, no warranty is provided regarding the accuracy or completeness of the information, and it is the responsibility of the reader to ensure, by independent means, that those parts of the information used by them are correct before any reliance is placed on them.

gross domestic product (GDP), and almost $9 B in direct tax payments. Because of this, the Bass Strait finds have been assessed as one of the 10 most important events in recent Australian history.

The training of a generation of Australian petroleum industry professionals was an additional benefit, as infrastructure and industries grew with the development of the rich hydrocarbon resources close to the major population centres in southeastern Australia. The Bass Strait finds showed that there was significant oil in Australia and so sustained the search into the other offshore basins formed during the breakup of Gondwana, such as the Carnarvon Basin on the North West Shelf.

Petroleum systems

Why so much oil in this one small basin?

As we have shown in the story of Gondwanan breakup, the Gippsland Basin is in a special tectonic position. Sitting at the eastern end of the Early Cretaceous rift between Australia and Antarctica, the basin is also on the rifted eastern margin that developed between Australia, the Lord Howe Rise and New Zealand.

As part of the great southern rift zone (Figures 4.13 and 4.14), the Gippsland Basin began as a series of Early Cretaceous northeast–southwest-trending half-grabens. In the Late Cretaceous, as the Tasman Sea opened, continued extension generated a broader rift in what is now the offshore part of the basin, flanked by fault-bounded platforms and terraces to the north and south (Figures 4.17 and 4.18). This Central Deep became the main depocentre, which filled with large volumes of material eroded from the uplifted basin margins. The thick succession included organic-rich rocks of lacustrine, coastal plain and marine facies (Figure 4.19), and thus the Central Deep became the main hydrocarbon kitchen when pushed into the oil window by the overburden load of more than 3 km of Cenozoic sediments.

The relatively thick Cenozoic section is another feature that distinguishes the Gippsland from other Australian basins. The section thickness

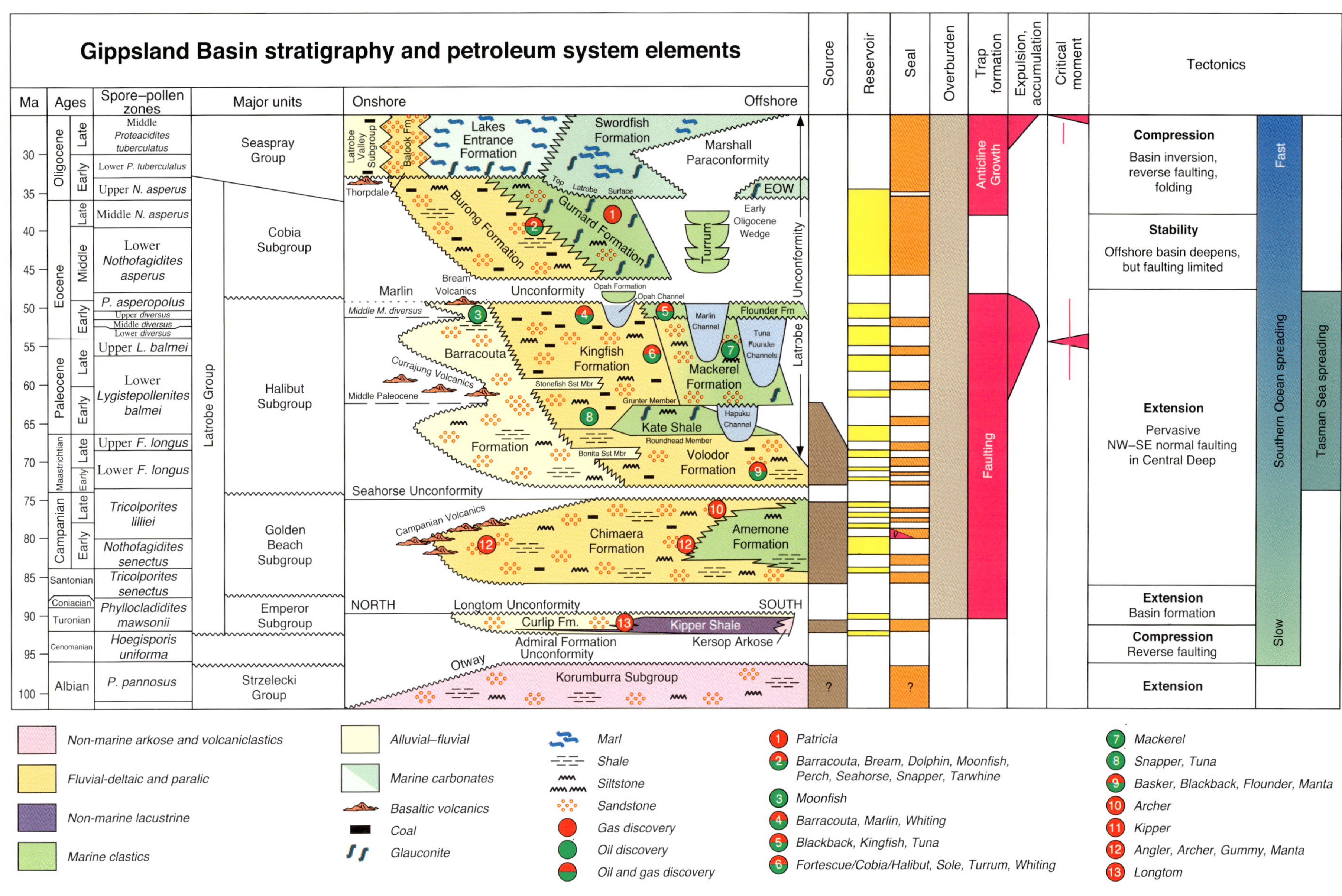

Figure 4.18: Gippsland Basin stratigraphic chart, showing sedimentary units, petroleum system elements and tectonic events. (Source: modified from Bernecker et al., 2006)

a.

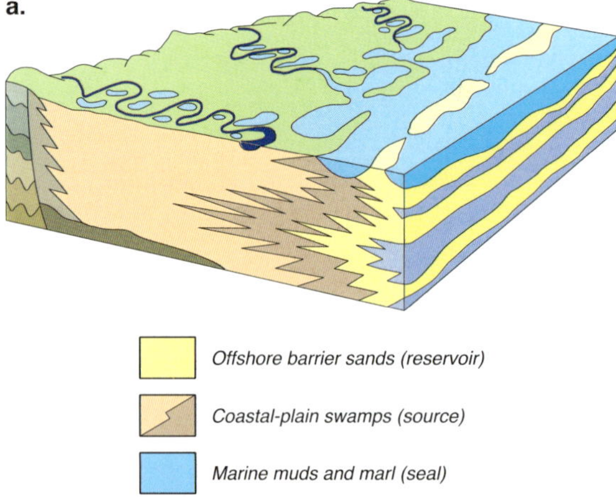

Figure 4.19: (a) Block diagram of the sedimentary packages in the Gippsland Basin. (b) Modern sedimentary environments, Lakes Entrance, Victoria. (Sources: modified from Johnstone et al., 2001; Bernecker et al., 2006)

reflects ongoing clastic supply from the rugged and well-watered hinterland (Chapter 5) and the Neogene accumulation of cool-water carbonate sediments in the offshore. The large anticlinal traps that now contain the giant oil fields are the product of compression and partial basin inversion from Eocene to Quaternary as the Australian Plate moved into a convergent regime. The late loading and generation means that only traps formed in this most recent episode of the 'out of Gondwana' story can be filled with hydrocarbons. There is also less time for trap breach, leakage, gas displacement and biodegradation to affect oil accumulations, although, as we will see in the analysis of the petroleum system, some of these destructive processes are under way.

The thick 10 km sedimentary succession in the Gippsland Basin records the last 100 Myr of the 'out of Gondwana' story in southeastern Australia (Figure 4.18). The Early Cretaceous volcaniclastic and coaly Strzelecki Group was deposited during initial Australia–Antarctica rifting. Renewed crustal extension occurred during the Late Cretaceous in an abandoned rift branch or aulacogen, associated with the opening of the Tasman Sea. The Central Deep was established, infilling with copious material eroded from the uplifted basin margins (Latrobe Group). A series of large, deep lakes developed, which were rapidly filled by the lacustrine Kipper Shale in the Turonian

The environment that formed the coaly source rocks of the giant Gippsland fields may have looked like this modern-day swamp in Victoria.

Image courtesy of ExxonMobil

Offshore oil production platform Gippsland Basin, Bass Strait, Victoria.

(Emperor Subgroup). The first marine incursion into the Gippsland Basin occurred in the Late Santonian (Anemone Formation, Golden Beach Subgroup) in the eastern part of the basin. Rift-related extension and sagging continued until the Early Eocene, with a sequence of alluvial–fluvial, deltaic and marine sediments deposited across the basin (Halibut Subgroup).

By the Middle Eocene, seafloor spreading had ceased in the Tasman Sea, and there was a period of basin sag, during which the lower coastal-plain, coal-rich Burong Formation was deposited, followed by the transgressive shallow to open marine and partly condensed Gurnard Formation (Cobia Subgroup). Marine deposition expanded from the Early Oligocene, with the deposition of the Lakes Entrance Formation (Seaspray Group). These onlapping, marly sediments provide the principal sealing unit across the basin (Figures 4.18 and 4.19). At the same time in the onshore part of the basin, the major brown coal resource (Latrobe Valley Coal Measures) was deposited during the Oligocene and Miocene (Chapter 9).

The Neogene deposition of the thick Gippsland Limestone, particularly during the Late Miocene and Pliocene, provided the critical overburden load for late generation of hydrocarbons from source rocks in the deeper Latrobe and Strzelecki groups (just as Neogene carbonates play a similar role in the operation of the petroleum systems on the North West Shelf). The major anticlines hosting the large oil and gas accumulations (Barracouta, Tuna, Kingfish, Snapper and Halibut) were formed by compression and basin inversion, starting in the Early Eocene and finishing by the end-Eocene. A younger onshore inversion event formed the anticlines that constitute the Strzelecki Ranges, which are Late Miocene to Recent in age.

Most of the hydrocarbons in the Gippsland Basin are located offshore in the Central Deep, with gas in the north and big oil fields in the southeast, although there are some smaller fields on the terraces and onshore (Figure 4.18). Stratigraphically, most of the oil and gas is at the top of the Latrobe Group, trapped under the complex Latrobe Unconformity (Figure 4.18), which ranges in age from latest Cretaceous to Eocene and younger. In addition to giant anticlinal traps in the top Latrobe Group, there are also hydrocarbon accumulations within the group, sealed by local shales and, in some cases, by volcanic rocks. Erosion and fill of ancient submarine channels in the Eocene has created trap geometries; again, the modern environment of the Gippsland Basin reflects the past (see the Bass Canyon in Figure 4.17).

The hydrocarbon fields in the Gippsland Basin are classified as part of the Austral Petroleum Supersystem (Figure 4.6; Appendix 4.1). There are several source units, including a gas source. These are mainly in the onshore in the Early Cretaceous (Austral 2), which provides geochemical evidence of a contribution from marine source rocks in the Late Cretaceous (Anemone Formation). The dominant and most prolific source, however, is the coals and carbonaceous shales within the Latrobe Group (Austral 3). There were several episodes of generation and expulsion, but the first major hydrocarbon charge was probably oil in the Late Miocene, which was partially displaced

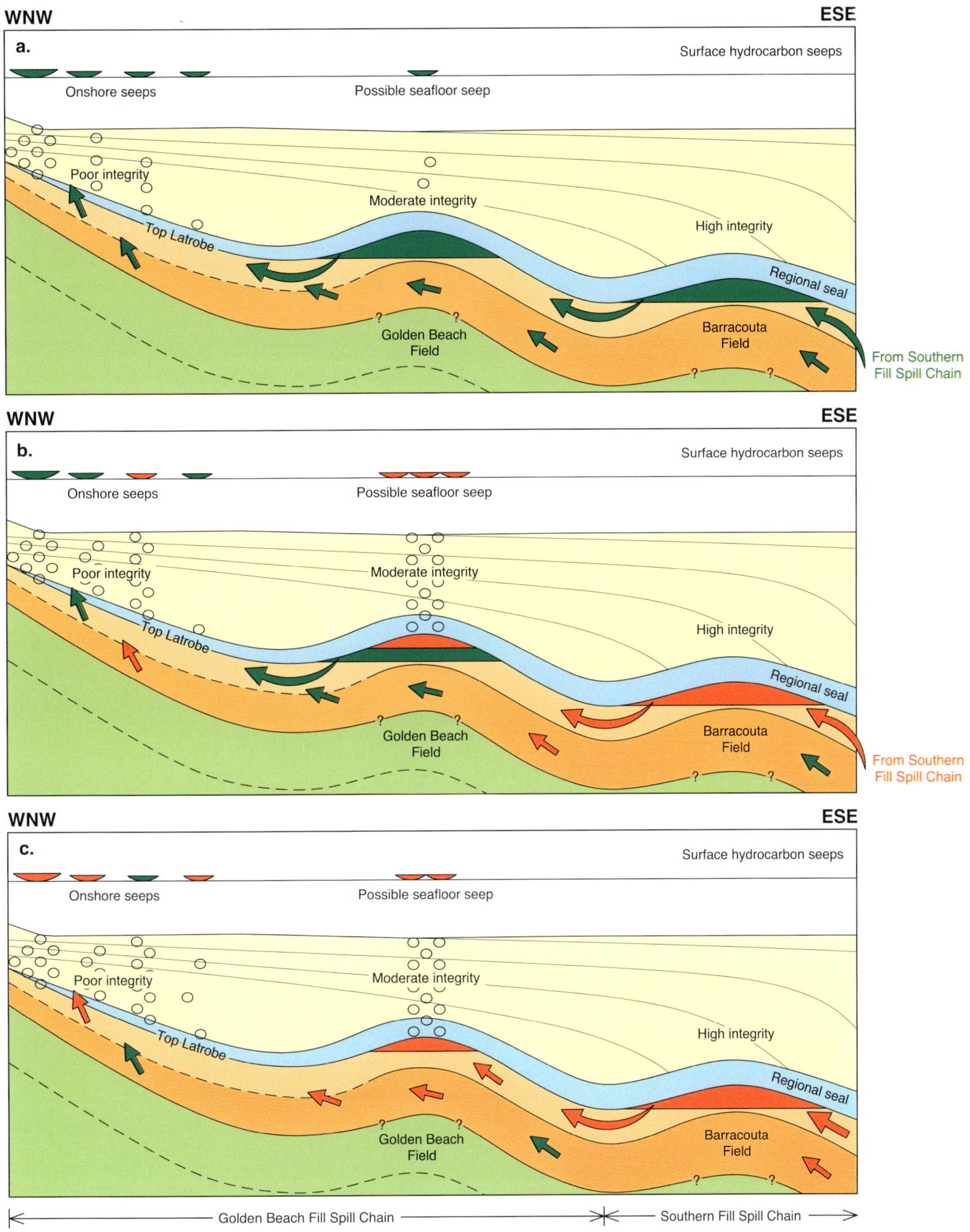

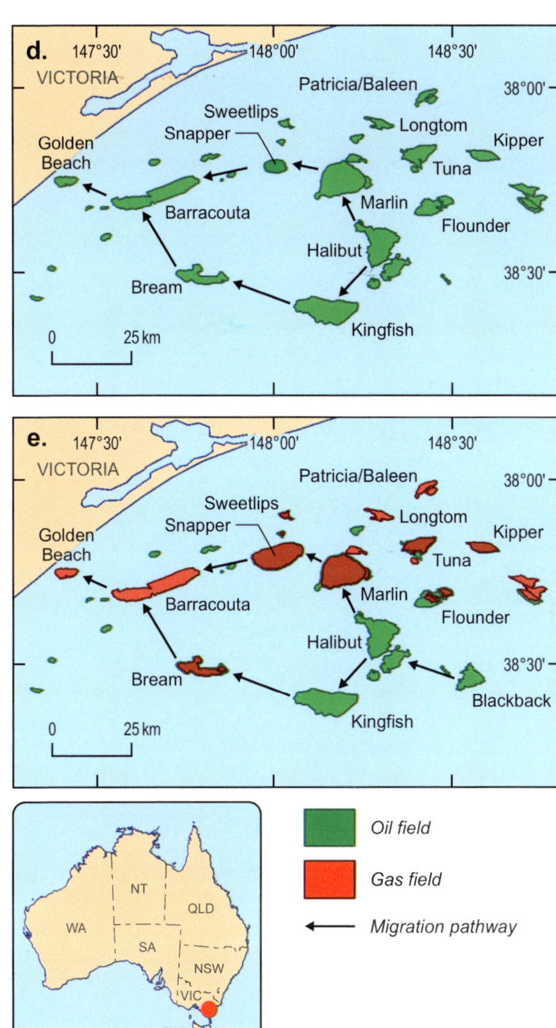

Figure 4.20: Gippsland Basin schematic cross-section, showing hydrocarbon leakage and seepage through the seal during (a) Miocene peak oil charge, (b) Pliocene peak gas charge and (c) present day. (d) Gippsland Basin, showing the postulated distribution of hydrocarbon accumulations and migration pathways in the Miocene. (e) Present-day Gippsland Basin, showing the postulated distribution of hydrocarbon accumulations and migration pathways. (Sources: modified from O'Brien et al., 2008; Tingate et al., 2011)

Field outlines are provided by Encom GPinfo, a Pitney Bowes Software (PBS) Pty Ltd product. While all care is taken in the compilation of the field outlines by PBS, no warranty is provided regarding the accuracy or completeness of the information, and it is the responsibility of the reader to ensure, by independent means, that those parts of the information used by them are correct before any reliance is placed on them.

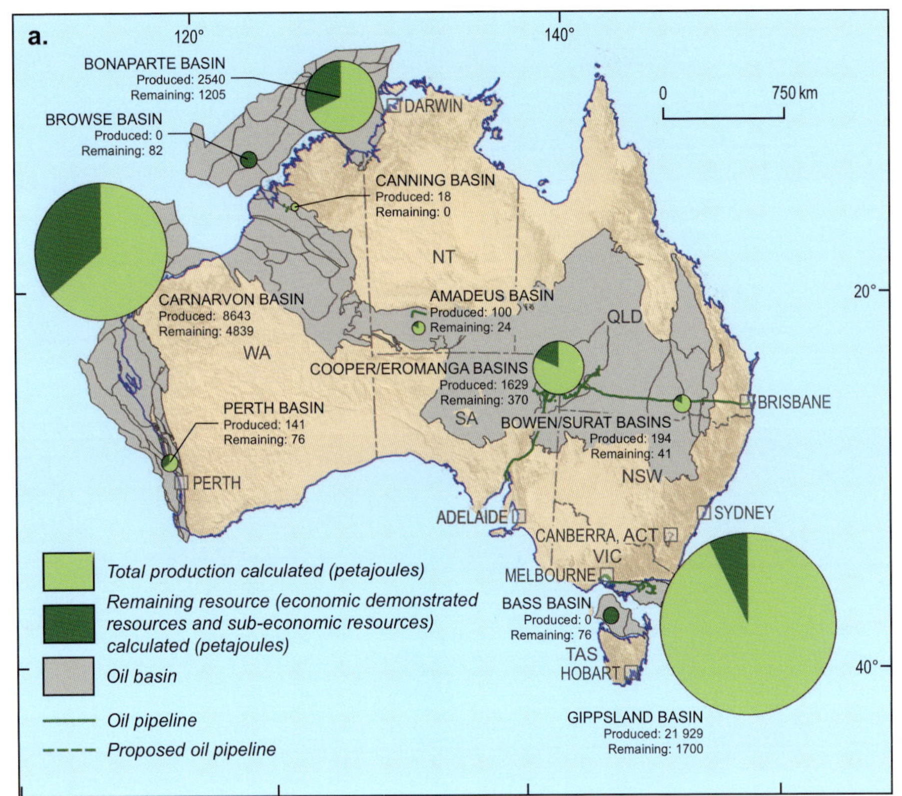

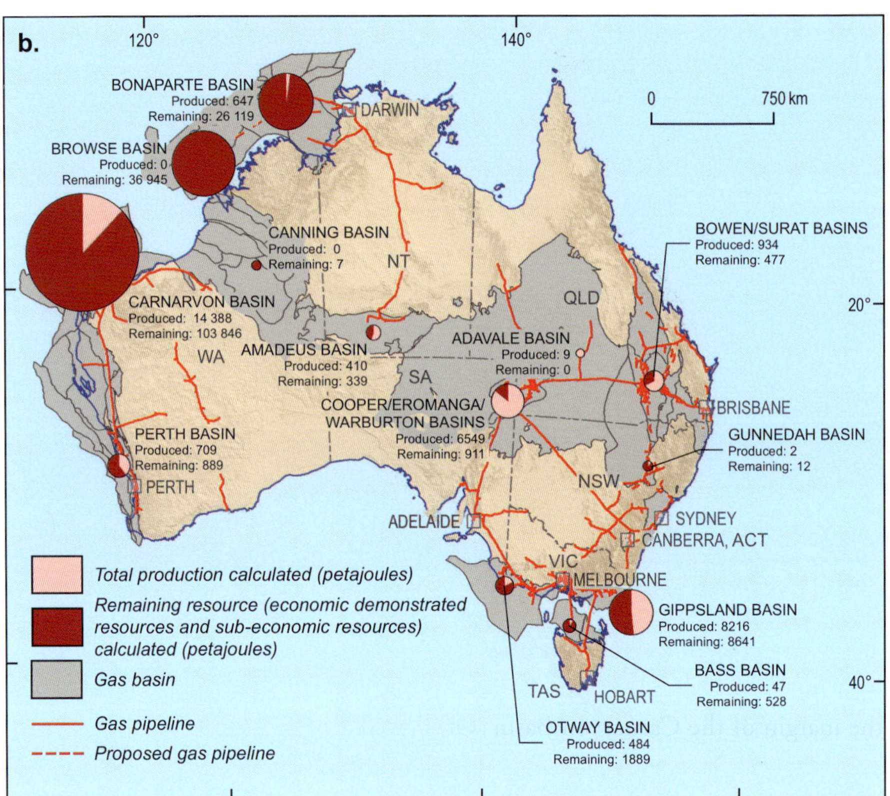

Figure 4.21: Maps showing Australia's conventional (a) oil and (b) gas resources and infrastructure. (Sources: modified from Geoscience Australia & Australian Bureau of Agricultural and Resource Economics, 2010)

by gas in the Pliocene in the northern fields. Analysis and modelling of the Gippsland Basin petroleum system have revealed a complex history of hydrocarbon generation, expulsion, migration and accumulation, followed by gas flushing and leakage (Figure 4.20).

Why is there an anomalous distribution of oil and gas in the Gippsland Basin?

Most petroleum systems operate so that oil fields are located towards the basin edges; gas fields tend to be in the centre of the basin, overlying the highest temperature regions of the hydrocarbon kitchen (Figure 4.3). This is because the early-generated oil migrates to the basin margin, driven along by the later generated gas, which fills the traps in the middle of the basin. The distribution of oil and gas in the Gippsland Basin is anomalous, in that the gas is on the northern basin margin and the oil towards the centre (Figure 4.17). A buoyancy-driven migration model explains this distribution of hydrocarbon fields. Two main migration pathways or fill-spill chains link the hydrocarbon kitchen of the Central Deep to the oil and gas fields (Figure 4.20). In the Miocene, the traps were mostly filled by oil, but, as generation continued, further burial moved the area of the Central Deep between Kingfish and Barracouta into the gas window. Gas moving along the northern fill-spill chain displaced oil, which leaked away at the basin margins where the seal facies is thin and sandier. Some of this oil was trapped onshore as oil sands in Cenozoic sediments (Lakes Entrance Formation—hence the World War II Lakes Entrance Oil Shaft Project). Today, the southern fill-spill chain continues to receive oil generated from the more recently buried southeastern part of the basin.

Carnarvon Basin, Western Australia

Discovery and development

The first flow of oil to the surface in Australia was from Rough Range No. 1, drilled in 1953 onshore near Exmouth Gulf. This, the first petroleum exploration well drilled in Australia after World War II, targeted a surface anticline at the margin of the Carnarvon Basin. Oil flowed at 500 barrels per day from the Lower Cretaceous Birdrong Sandstone, but it proved to be a very small accumulation, and further drilling on the same anticline within a few hundred metres of the first well failed to replicate the initial success. As with the Gippsland Basin, the real hydrocarbon potential was in the offshore part of the basin.

Decades earlier, émigré geologist Curt Teichert had inferred that the limited outcrop of marine Jurassic sediments was the feather-edge of a large Mesozoic basin offshore. This proved to be correct when oil exploration ventured offshore with an island drilling campaign by West Australian Petroleum (WAPET). In 1964, the giant Barrow Island oil field was discovered in a surface anticline. More than a billion barrels of oil are in place in multiple reservoirs, and more than 300 M barrels have been produced, providing a sustained base level of oil production for Australia over many decades (Figure 4.16). Also, both Curt Teichert and later Woodside chief geologist Nicholas Boutakoff suggested that the bathymetry of the North West Shelf might reflect a series of anticlinal ridges propagated from Timor and viewed this as prospective because of Timor oil seeps. In 1963, on Boutakoff's advice, Woodside Petroleum Ltd leased almost all of the rest of Western Australia's offshore areas north of WAPET's existing permits. One of the earliest offshore wells, Legendre 1, drilled in 1968 by Woodside, discovered oil in the Dampier Sub-basin. Several giant gas discoveries were later made along the Rankin Platform in the 1970s and early 1980s—North Rankin, Goodwyn and Gorgon, a gas field with a significant CO_2 content (Figure 4.22). In 1984, the North West Shelf Venture, led by Woodside, commenced domestic gas production from the North Rankin field and, in 1989, the first LNG cargo was shipped to Japan. Since then, the project has contributed $70 B to Australia's GDP (Chapter 9).

In the period 1979–80, Australian exploration took the brave step out into the deepwater Exmouth Plateau to test the large structures imaged on seismic surveys (Figure 4.9b). These initial exploration programs, operated by Esso Standard Oil Co. and Phillips Australian Oil Co., began when no proven technology existed to develop a deepwater oil field. Disappointingly in one way, rather than the hoped-for oil charge, only gas was found in the 11 deepwater wells drilled in 740–1375 m water depths, but this did include the giant Scarborough field. In the 1980s,

Did you know?

4.5: The colourful history of Barrow Island

Whereas man-made destruction prevented West Australian Petroleum (Wapet) from exploring Barrow Island in the 1950s to 1960s, it was nature's destructive forces that delayed the spudding of its first well on the island. Britain used the Monte Bello Islands, located 20 km north of Barrow Island, as an atomic testing site between 1952 and 1956. Not until May 1963 did the Australian Government lift its ban on access to Barrow Island, which Wapet field parties had earlier recognised to be a very broad anticline. In April 1964, while Wapet was preparing to move the rig onto the island for the first well, a late-season cyclone wiped out the newly constructed airstrip and beach landing. Barrow Island-1 was eventually spudded on 7 May.

Barrow Island will soon be home to the Gorgon LNG facility and the site of a carbon-capture and storage project, to demonstrate the technology and to mitigate the problems of the high CO_2 content of the gas (Chapter 11).

Image courtesy of Chevron Australia

The WAPET landing on Barrow Island, Western Australia.

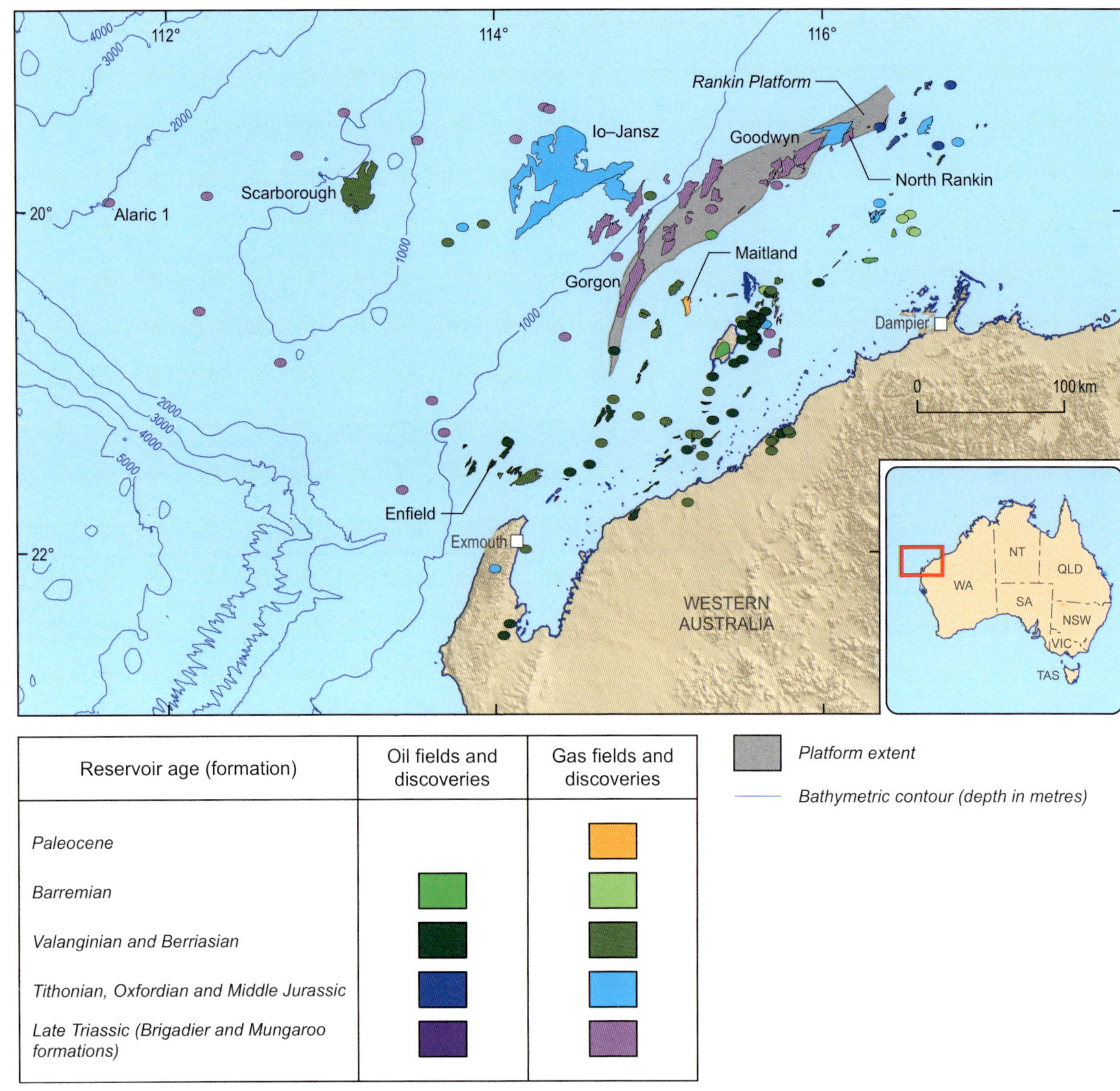

Figure 4.22: Carnarvon Basin map, showing major fields and discoveries, with the age of the reservoir indicated by colour.

Field outlines are provided by Encom GPinfo, a Pitney Bowes Software (PBS) Pty Ltd product. While all care is taken in the compilation of the field outlines by PBS, no warranty is provided regarding the accuracy or completeness of the information, and it is the responsibility of the reader to ensure, by independent means, that those parts of the information used by them are correct before any reliance is placed on them.

such 'stranded' gas fields littered the North West Shelf (Figure 4.15), awaiting a change in the global energy market to make their development commercially viable (Chapter 9).

Exploration in the offshore Carnarvon Basin then shifted inboard, back to the 'oily' sub-basins and away from the home of the gas giants. From the early 1980s to the mid-1990s, a number of significant, mostly medium-sized oil (and gas) discoveries were made in the Barrow and Dampier sub-basins, as a result of the application of dense 2D seismic surveys. More recently 3D seismic and amplitude *vs* offset (AVO) technology has contributed to an improvement in the success rates. In 1999, the Enfield discovery was made in the Exmouth Sub-basin. A string of nearby oil finds followed, which were brought into production from 2006.

Growing demand for LNG has stimulated exploration along the Rankin Platform and a return to the deepwater Exmouth Plateau in recent years, and dozens of major gas discoveries have resulted. Accompanying exploration success has been a series of new, large-scale development projects and associated investment in infrastructure (*Did you know?* 4.5).

Petroleum systems

Major hydrocarbon fields occur all along the North West Shelf, from the Exmouth Sub-basin in the southwest, to the Malita Graben and Sahul Platform in the northeast (Figure 4.10). Most of these fields are part of the Westralian Supersystem (Appendix 4.1.1). The offshore Carnarvon Basin, however, led the way in discovery and development and currently contains the largest resources (Figure 4.21). There are major oil fields in the failed rift marked by the Exmouth, Barrow and Dampier sub-basins, and giant gas fields on the western rim of the rift (the Rankin Platform) and all the way out across the Exmouth Plateau (Figures 4.10 and 4.11).

The offshore Carnarvon Basin, at 535 000 km^2, is more than 10 times the size of the Gippsland Basin. The basin is also more complex and relies on older Mesozoic petroleum systems—a product of Pangaea and Gondwana breakup. The key elements are:

- sources—Triassic and Early Jurassic coaly gas-prone rocks (Westralian 1) and restricted marine Late Jurassic oil kitchens (Westralian) in the failed rift
- reservoirs—Triassic fluvial reservoirs in the Early Cretaceous Barrow delta and also marine sands in the Jurassic, Cretaceous and Paleocene (Figure 4.22)
- regional seal—Early Cretaceous deep-marine shale
- overburden—prograding Cenozoic carbonate wedge to bury the underlying elements, providing thermal maturation (Figure 4.23).

The geological history of the offshore Carnarvon Basin means that oil-prone source rocks are restricted to the Late Jurassic failed rifts, but a very extensive gas-prone source system stretches for more than 500 km across the Exmouth Plateau—even to Alaric 1, a recent large gas find close to its western edge (Figure 4.22). The major source of all this gas is the deeply buried coals and carbonaceous claystones of the Triassic Mungaroo Formation. These are the delta plain deposits of transcontinental rivers that prograded across the plateau when Australia was still locked into the Pangaea–Gondwana supercontinent. The thick, clean sands of the fluvial channels provide the reservoir facies, and the typical hydrocarbon trap is a fault block formed in the Late Triassic – Early Jurassic and sealed by Early Cretaceous or older shales (Figure 4.23). Some of the largest gas fields, however, are stratigraphically trapped in reservoirs of other ages. Two such fields are Io–Jansz in Late Jurassic shallow marine sands, reworked from the uplifted western rim of the failed rift, and Scarborough, an Early Cretaceous basin-floor fan deposited in deep water in front of the Barrow delta (Figures 4.22 and 4.23).

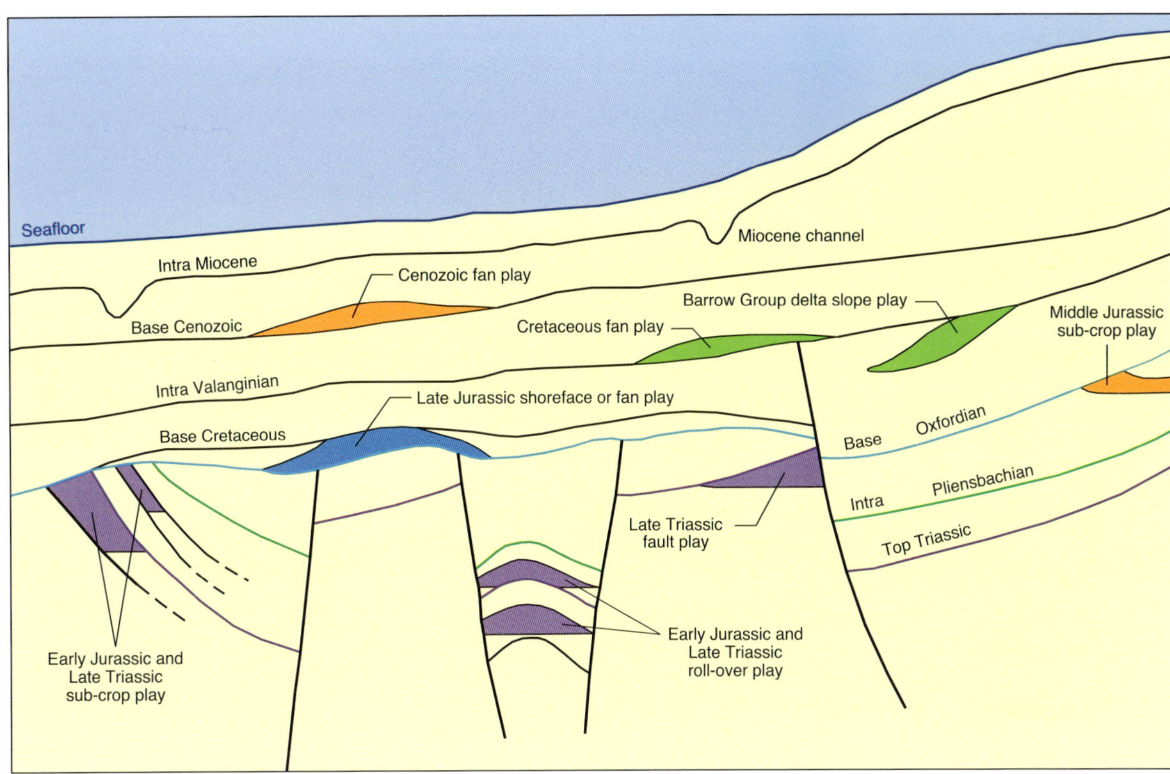

Figure 4.23: Carnarvon Basin petroleum play diagram.

Did you know?

4.6: The tyranny of distance can be an opportunity

Geoffrey Blainey termed the phase the 'Tyranny of Distance' as a unifying theme for Australian history. It refers to the vast distances within Australia and the distance of Australia from the centres of European civilisation. Returning 'home', as Britain was viewed by many in the 19th and early 20th centuries, was a trip requiring months at sea; even after the advent of passenger aircraft, the trip still requires a considerable investment of time and money.

But distance also brings opportunities and benefits. These facts of geography mean that Australians grow up feeling that they live in a big but sparsely populated country (Chapter 1), where there is room to move and explore. And, being a long way from the centres of civilisation, there is also a need to get out and explore the wider world—hence, Australians are great travellers and outward looking. The export-focused Australian economy (see Chapter 9) also reflects this mindset.

The tyranny of distance meant that Australia was the continent last settled by the European empires and hence guided by the ideals of Enlightenment rather than those of earlier thinking. There is also the feeling of being far from the troubles of other lands; immigrants can come and forget the conflicts of the 'old world', as hundreds of thousands did following World War II and continue to do today (Chapter 1).

Jurassic and Cretaceous shallow marine to deltaic sands are the main oil reservoirs located within the inboard rifts overlying or close by the Dingo Claystone oil kitchens. Especially important are the various sands related to progradation and then reworking of the Barrow delta as sea-level rose with breakup and the creation of new seafloor in the Early Cretaceous (Figure 4.23). The first phase of oil generation was in the Early Cretaceous in the Exmouth Sub-basin and southern parts of the Barrow Sub-basin due to the loading of the Barrow delta. Oil accumulations formed at this time required preservation for about 100 Myr. Not all survived intact, and there are examples of biodegradation, trap destruction and displacement of oil by later gas migration. In the Dampier Sub-basin to the northeast, where the stratigraphic equivalent of the Barrow delta is a thin, distal marine shale, the main phase of hydrocarbon generation was in the Cenozoic when the progradation of the carbonate shelf provided sufficient burial to shift the Jurassic source rocks into the oil window (Figure 4.3). The same is true beyond the basin, with geochemically similar oils recognised in the Carnarvon, Bonaparte and Papuan basins, all derived from Late Jurassic marine source rocks. Compare how this successful petroleum system has evolved with that of the less successful Toolebuc Formation (*Did you know?* 4.1).

The majority of the hydrocarbons discovered to date in the offshore Carnarvon Basin are in highly porous sandstone reservoirs beneath the Early Cretaceous Muderong Shale. This thin but effective regional seal makes a major contribution to exploration success in the basin. One notable exception is in the Barrow Island oil field, where the oil-bearing Windalia Sandstone (top Muderong Shale) is top-sealed by the Aptian Windalia Radiolarite, an open-ocean siliceous unit deposited at the height of Australian inundation, when the inland sea stretched across the continent and the Carnarvon Basin was far from shore (Box 4.3). Another exception is the Maitland gas accumulation, in which a Paleocene sandstone is the reservoir (Figures 4.22 and 4.23).

The main trap styles in the basin are anticlines, horsts, fault roll-over structures and stratigraphic pinch-outs beneath the regional seal (Figure 4.23). Extensional faulting in the Triassic and Jurassic rift sequences set up the fundamental architecture of the basin and the geometries of many of the fault block traps. The Cenozoic collision at the northern plate boundary caused tilting, inversion and renewed faulting. This latter event destroyed some accumulations but also produced another suite of structural traps to capture late-generated hydrocarbons.

Why is Australia gas rich and oil poor?

Australia was located in the high southerly latitudes and largely emergent during the key global source-rock depositional episodes of the Permian, Jurassic and Cretaceous (Box 4.1). Many of the richest oil-prone source rocks—in the Middle East, northern South America and west Africa—formed in marine palaeotropical environments. These are regions of high algal productivity, where oil-prone organic matter is linked to the fundamental driver of high solar-energy input near the equator. In the marine environment, the source rocks can also be deposited as basin-wide blankets. In contrast, depending on the vegetation, non-marine coaly source rocks, which dominate Australian petroleum systems, are often inherently gas prone. There are some 'sweet spots' in the delta systems where marine influence and bacterial reworking of terrestrial organic matter produced a more oil-prone source rock. This was the case in the Gippsland Basin.

In addition, the maturation history of many Australian basins has been less than ideal for generating and preserving oil charge. Oil accumulations sourced from Early Paleozoic tropical marine shales of the Larapintine petroleum supersystem (Appendix 4.1.1) had a high likelihood of being destroyed in the mid-Carboniferous Alice Springs Orogeny (Chapter 2). And as we have seen, the Toolebuc Formation oil shale has not been buried sufficiently to generate hydrocarbons (*Did you know?* 4.1). Also, the Toolebuc Formation was deposited far from the plate boundary. The general strength of the Australian continental lithosphere (Chapters 2 and 5) has meant that there are no inland (intraplate) areas that have subsided fast enough to receive a thick enough sedimentary overburden load in the Cenozoic, such as in the oil and gas-rich Gippsland Basin closer to the continental margin. Many Australian petroleum systems rely on preserving oil generated in the Cretaceous or early Cenozoic. Later gas flushing has spilled and displaced many accumulations.

There is a large Late Cretaceous delta in the Bight Basin and evidence of Cretaceous marine oil shales that may have been buried enough to generate hydrocarbons. Let us now look at this frontier basin.

The Bight Basin: Australia's largely untested southern frontier

One of the largest unexplored frontier provinces in Australia occupies the deep water of the Great Australian Bight, including the Late Cretaceous Ceduna delta, which underpins part of the Bight Basin (Figures 4.12 and 4.14). Remarkably few wells have been drilled as yet, and these mostly lie on the shelf, in water depths less than 250 m. One exception is Jerboa 1, drilled in 761 m of water by Esso Ltd in 1980 to test the Eyre Sub-basin. It was drilled with the deepwater rig brought to Australia for the Exmouth Plateau campaign and provided a welcome break in that effort, allowing time to assimilate the initial results of finding gas rather than oil on the Exmouth Plateau.

The only well to investigate the 'real prize' in the Great Australian Bight—the delta you can see on the bathymetry—has been Gnarlyknots 1, drilled by Woodside and partners more than 20 years after Jerboa 1 in 2003 in 1313 m of water (Figures 4.12 and 4.24). This well was plagued by mechanical problems and was eventually abandoned due to bad weather before the target depth was reached and some 1500 m above the planned total depth. It was an expensive exercise with an inconclusive result, and exploration stalled. Evidence of a viable petroleum system would be needed to encourage another try.

How can we reveal the petroleum system in the Bight Basin?

Collecting seismic and other geophysical data, dredging samples from the seafloor and remote sensing are all ways to understand this deepwater

The island continent of Australia that emerged Out of Gondwana was colonised by Europeans in the late eighteenth century in the age of tall ships. The *Bounty* with Opera House at sunset, Sydney, New South Wales.

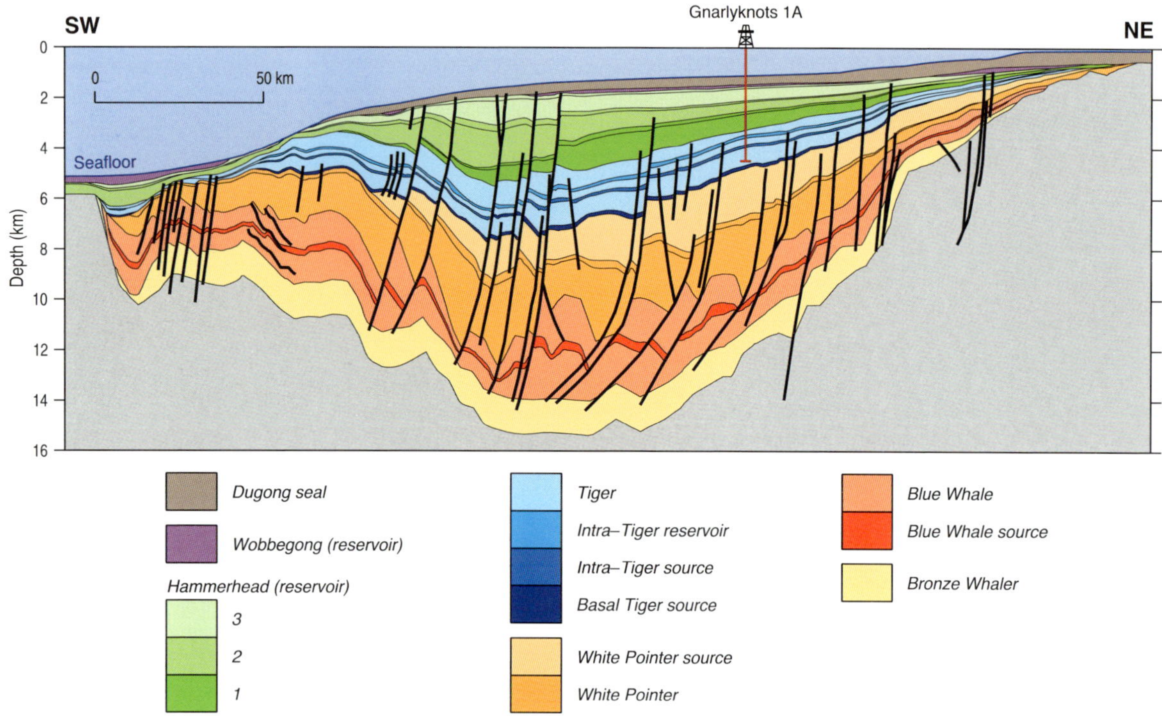

Figure 4.24: Bight Basin schematic cross-section of the Ceduna delta, showing reservoir, seal and source units.

frontier. Geoscience Australia and its predecessors have a long history of research in the Bight Basin, conducting several gravity and magnetic surveys and acquiring more than 28 000 km of regional seismic data. The geochemistry of the bitumen washed ashore along the southern margin beaches was studied, as well as satellite data indicative of natural seepage of oil. Using all these data and the limited well information, a detailed analysis was undertaken. Models of the possible petroleum system and plays were produced. These studies informed a marine survey in 2007, which recovered potential source rocks of Late Cenomanian to Early Turonian age from the northwestern edge of the Ceduna Sub-basin. Exploration is now under way in the deepwater Bight Basin, with the awarding of new petroleum exploration permits in 2011. Drilling is planned in 2013 or 2014.

So, is Australia really oil poor? We can only answer once this, and other frontier basins, have been more fully tested.

Postscript

From the story of 'Out of Gondwana', we see that geology has directed the track of the past and constrains our future choices, a Phanerozoic story with modern consequences. How much of a role will oil and gas play in the future energy mix; how will the development of unconventional resources be balanced against environmental protection; will there be a new oil province in the Great Australian Bight?

In the distant future, Australia will once again join a supercontinent; this time it will be Amasia, linking Australia back with its Asian partners from Gondwana. A collision zone of Himalayan proportions will have destroyed the hydrocarbon fields of the North West Shelf. But new oil and gas accumulations will have formed, perhaps where the Toolebuc Formation has been buried deeply by sediments shed from the mountains. Will oil be of any relevance to whoever remains 50 Myr in the future? Perhaps oil will still be a convenient, energy-dense substance, capturing and concentrating as chemical energy the sunlight that warmed ancient seas.

We will now look at the journey Australia made out of Gondwana, but from the perspective of the old, flat and red landscape …

Image by Jim Mason

Natures Window, Kalbarri, Western Australia.

Bibliography and further reading

Gifts of Gondwana

BMR Palaeogeographic Group 1992. *Australia: evolution of a continent*, Commonwealth of Australia, Canberra.

Bradshaw MT 1993. Australian petroleum systems. *PESA Journal* 21, 43–53.

Bradshaw MT, Bradshaw J, Murray AP, Needham DJ, Spencer L, Summons RE, Wilmot J & Winn S 1994. Petroleum systems in west Australian basins. In: *The sedimentary basins of Western Australian 2*, Purcell PG & Purcell RR (eds), Proceedings of the Petroleum Exploration Society of Australia Symposium, Perth, 1998, 93–118.

Brakel AT & Totterdell JM 1990. *The Permian palaeogeography of Australia*, BMR record 1990/060, Bureau of Mineral Resources, Canberra.

Cook PJ & Totterdell JM 1991. *Palaeogeographic atlas of Australia, volume 2: Ordovician*, Bureau of Mineral Resources, Canberra.

Edwards DS & Zumberge JE 2005. *The oils of Western Australia*, Australian Geological Survey Organisation/ Geomark Research Ltd, Canberra and Houston.

Fielding CR, Frank TD, Isbell JL, Henry LC & Domack EW 2010. Stratigraphic signature of the late Palaeozoic Ice Age in the Parmeener Supergroup of Tasmania, SE Australia, and inter-regional comparisons. *Palaeogeography, Palaeoclimatology, Palaeoecology* 298, 70–90.

Langford RP 1991. *Permian coal and palaeogeography of Gondwana*, BMR record 1991/95, Bureau of Mineral Resources, Canberra.

Magoon LB & Dow WG 1994. The petroleum system. In: *The petroleum system—from source to trap*, Magoon LB & Dow WG (eds), AAPG Memoir 60, American Association of Petroleum Geologists, Tulsa, 3–24.

McCabe PJ 1998. Energy resources—cornucopia or empty barrel? *American Association of Petroleum Geologists Bulletin* 82, 2110–2134.

Metcalfe I 2011. Palaeozoic–Mesozoic history of SE Asia. In: *The SE Asian gateway: history and tectonics of the Australia–Asia collision*, Hall R, Cottam MA & Wilson MEJ (eds), Geological Society, London, Special Publication 355, 7–35.

Müller RD, Gaina C, Tikku A, Mihut D, Cande SC & Stock JM 2000. Mesozoic/Cenozoic tectonic events around Australia. *Geophysical Monograph*, 121, 161–188.

Ozimic S & Saxby JD 1983. *Oil shale methodology: an examination of the Toolebuc Formation and the laterally contiguous time equivalent units, Eromanga and Carpentaria Basins*, Bureau of Mineral Resources and CSIRO research project.

PALEOMAP Project 2008. http://scotese.com/Default.htm

Schenk CJ & Pollastro RM 2001. *Natural gas production in the United States*, National Assessment of Oil and Gas series, USGS Fact Sheet FS-113-01, US Geological Survey, Denver.

Sears JW 2007. Lithospheric control of Gondwana breakup: implications of a trans-Gondwana icosahedral fracture system. Geological Society of America Special Paper 430, 593–601.

Stampli GM & Borel GD 2002. A plate tectonic model for the Paleozoic and Mesozoic constrained by dynamic plate boundaries and restored synthetic ocean isochrons. *Earth and Planetary Science Letters* 196, 17–33.

Stewart JA 1990. *Drifting continents and colliding paradigms: perspectives on the geoscience revolution*, Indiana University Press, Bloomington.

Veevers JJ (ed.) 2000. *Billion-year earth history of Australia and neighbours in Gondwanaland*, GEMOC Press, Sydney.

White ME 1986. *The greening of Gondwana*, Reed Books, Sydney.

Pangaea at the zenith

Brakel AT, Wilford GE, Totterdell JM & Bradshaw M 1995. Palaeogeographic settings of Australian coal measures. In: *Geology of Australian coal basins*, Ward CR, Harrington HJ, Mallett CW & Beeston JW (eds), Geological Society of Australia Coal Geology Group Special Publication 1, 1–15.

Norvick MS 2002. *Palaeogeographic maps of the northern margins of the Australian Plate: final report*, unpublished report for Geoscience Australia.

Sircombe K 1999. Tracing provenance through the isotope ages of littoral and sedimentary detrital zircon, eastern Australia. *Sedimentary Geology* 124, 47–67.

Southgate P, Sircombe K & Lewis C 2011. New insights into reservoir sand provenance in the Exmouth Plateau and Browse Basin. Extended abstract, 2011 Australian Petroleum Production and Exploration Association (APPEA) journal and conference proceedings (DVD).

Veveers JJ 2006. Updated Gondwana (Permian–Cretaceous) earth history of Australia. *Gondwana Research* 9, 231–260.

Woodside 2010. Half yearly results briefing. www.woodside.com.au/investors-media/announcements/documents/18.08.2010%202010%20half%20year%20results%20briefing%20slide%20pack.pdf

Gondwana breakup

Alcock MB, Stagg HMJ, Colwell JB, Borissova I, Symonds PA & Bernardel G 2006. *Seismic transects of Australia's frontier continental margins*, AGSO record 2006/04, Australian Geological Survey Organisation, Canberra.

Leaman D 2002. *The rock which makes Tasmania*, Leaman Geophysics, Hobart.

Seymour DB, Green GR & Calver CR 2007. The geology and mineral deposits of Tasmania: a summary. Tasmanian Geological Survey Bulletin 72, Mineral Resources Tasmania.

Symonds PA, Eldholm O, Mascle J & Moore GF 2000. Characteristics of continental margins. In: *Continental shelf limits: the scientific and legal interface*, Cook PJ & Carleton CM (eds), Oxford University Press, New York, 25–60.

Northwest margin

Gibbons A, Whittaker J & Müller RD 2010. Revised plate tectonic history of the west Australian margin reveals how the Gascoyne Terrane docked at West Burma. ASEG Extended Abstracts, ASEG2010 21st Geophysical Conference, Sydney, 2010.

Heine C & Müller RD 2005. Late Jurassic rifting along the Australian Northwest Shelf: margin geometry and spreading ridge configuration. *Australian Journal of Earth Sciences* 52, 27–40.

Keep M & Moss SJ (eds) 2002. *The sedimentary basins of Western Australia 3*, Proceedings of the Petroleum Exploration Society of Australia Symposium, Perth, 2002.

Norvick MS 2002. *Palaeogeographic maps of the northern margins of the Australian Plate: final report*, unpublished report for Geoscience Australia.

Purcell PG & Purcell RR (eds) 1994. *The sedimentary basins of Western Australia 2,* Proceedings of the Petroleum Exploration Society of Australia Symposium, Perth, 63–76.

Purcell PG & Purcell RR (eds) 1998. *The sedimentary basins of Western Australia 2*, Proceedings of the Petroleum Exploration Society of Australia Symposium, Perth, 1998.

Stagg HMJ, Alcock MB, Bernadel G, Moore AMG, Symonds PA & Exon NF 2004. *Geological framework of the outer Exmouth Plateau and adjacent ocean basins*, AGSO record 2004/13, Australian Geological Survey Organisation, Canberra.

Southwest margin

Bradshaw BE, Rollet R, Totterdell JM & Borissova I 2003. *A revised structural framework for frontier basins on the southern and southwestern Australian continental margin*, Geoscience Australia Record 2003/03, Geoscience Australia, Canberra.

Crostella A & Backhouse J 2000. *Geology and petroleum exploration of the central and southern Perth Basin, Western Australia*, Western Australia Geological Survey report 57, Geological Survey of Western Australia, Perth.

Thomas BM & Barber CJ 2004. A re-evaluation of the hydrocarbon habitat of the Northern Perth Basin. *APPEA Journal* 44, 59–92.

Southern margin

Müller RD, Gaina C & Clarke S 2000. Mesozoic/Cenozoic tectonic events around Australia. In: *The history and dynamics of global plate motions*, American Geophysical Union Monograph 121, 161–188.

Norvick MS 2005. *Plate tectonic reconstructions of Australia's southern margins: final report*, AGSO record 2005/07, Australian Geological Survey Organisation, Canberra [electronic resource].

Norvick MS & Smith MA 2001. Mapping the plate tectonic reconstruction of southern and southeastern Australia and implications for petroleum systems. *APPEA Journal* 41, 15–35.

Stagg HMJ, Cockshell CD, Willcox JB, Hill AJ, Needham DJL, Thomas B, O'Brien GW & Hough LP 1990. *Basins of the Great Australian Bight region, geology and petroleum potential*, Bureau of Mineral Resources, Continental Margins Program Folio 5.

Totterdell JM, Blevin JE, Struckmeyer HIM, Bradshaw BE, Colwell JB & Kennard JM 2000. A new sequence framework for the Great Australian Bight: starting with a clean slate. *APPEA Journal* 40, 95–117.

Eastern margin

Gaina C, Müller DR, Royer J-Y, Stock J, Hardebeck J & Symonds P 1998. The tectonic history of the Tasman Sea: a puzzle with 13 pieces. *Journal of Geophysical Research* 103(B6), 12413–12433.

Norvick MS 2005. *Plate tectonic reconstructions of Australia's southern margins: final report*, AGSO record 2005/07, Australian Geological Survey Organisation, Canberra [electronic resource].

Norvick MS, Langford RP, Rollet N, Hashimoto T, Higgins KL & Morse MP 2008. New insights into the evolution of the Lord Howe Rise (Capel and Faust basins), offshore eastern Australia, from terrane and geophysical data analysis. In: *Eastern Australasian Basins Symposium III: energy security for the 21st century*, Blevin JE, Bradshaw BE & Uruski C (eds), Petroleum Exploration Society of Australia Special Publication, 291–310.

Sdrolias M, Müller RD & Gaina C 2003. Tectonic evolution of the southwest Pacific using constraints from backarc basins. In: *Evolution and dynamics of the Australian Plate*, Hillis RR & Müller RD (eds), Geological Society of Australia Special Publication 22 and Geological Society of America Special Paper 372, 343–359.

Van de Beuque S, Stagg HMJ, Sayers J, Willcox JB & Symonds PA 2003. *Geological framework of the northern Lord Howe Rise and adjacent areas*, Geoscience Australia Record 2003/01, Geoscience Australia, Canberra.

Northern margin

Crowhurst PV, Maas R, Hill KC, Foster DA & Fanning CM 2004. Isotopic constraints on crustal architecture and Permo–Triassic tectonics in New Guinea: possible links with eastern Australia. *Australian Journal of Earth Sciences* 51, 107–122.

Hall R 1997. Cenozoic plate tectonic reconstructions of SE Asia. In: *Petroleum geology of Southeast Asia*, Fraser AJ, Matthews SJ & Murphy RW (eds), Geological Society Special Publication 126, 11–23.

Hall R 2002. Cenozoic geological and plate tectonic evolution of SE Asia and the SW Pacific: computer-based reconstructions, model and animations. *Journal of Asian Earth Sciences* 20, 353–431.

Hall R, Cottam MA & Wilson MEJ (eds) 2011. *The SE Asian gateway: history and tectonics of the Australia–Asia collision*, Geological Society, London, Special Publication 355.

Hall R & Smyth HR 2008. Cenozoic arc processes in Indonesia: identification of the key influences on the stratigraphic record in active volcanic arcs. In: *Formation and applications of the sedimentary record in arc collision zones*, Draut AE, Clift PD & Scholl DW (eds), Geological Society of America Special Paper 436, 27–54.

Hill KC & Hall R 2003. Mesozoic–Cenozoic evolution of Australia's New Guinea margin in a west Pacific context. In: *Evolution and dynamics of the Australian Plate*, Hillis RR & Müller RD (eds), Geological Society of Australia Special Publication 22 and Geological Society of America Special Paper 372, 265–290.

The Australian petroleum search

Bein J & Taylor ML 1981. The Eyre Sub-basin: recent exploration results. *APEA Journal* 21, 91–98.

Bernecker T & Partridge AD 2001. Emperor and Golden Beach subgroups: the onset of Late Cretaceous sedimentation in the Gippsland Basin, SE Australia. In: *Eastern Australasian Basins Symposium: a refocused energy perspective for the future*, Hill KC & Bernecker T (eds), Petroleum Exploration Society of Australia Special Publication, 391–402.

Bernecker T, Thomas JH & O'Brien GW 2006. Hydrocarbon prospectivity of Areas V06-2, V06-3 and V06-4, Southern Offshore Gippsland Basin, Victoria, Australia; 2006 Acreage Release, VIMP Report 88, Department of Primary Industries, Victoria.

Blainey G 2001. *The tyranny of distance: how distance shaped Australia's history*, Macmillan, Sydney.

Bradshaw MT, Foster CB, Fellows ME & Rowland DC 1999. The Australian search for petroleum—patterns of discovery. *APPEA Journal* 39, 1–18.

Brown BR 1986. Offshore Gippsland silver jubilee. In: *Second SE Australia Oil Exploration Symposium*, Glennie RC (ed.), Petroleum Exploration Society of Australia, 29–56.

Eccleston R 2007. The fateful shore. *The Bulletin*, 24 April.

Geoscience Australia & Australian Bureau of Agricultural and Resource Economics 2010. *Australian energy resource assessment*, Geoscience Australia & ABARE, Canberra.

Johnstone EM, Jenkins CC & Moore MA 2001. An integrated structural and palaeogeographic investigation of Eocene erosional events and related hydrocarbon potential in the Gippsland Basin. In: *Eastern Australasian Basins Symposium: a refocused energy perspective for the future*, Hill KC & Bernecker T (eds), Petroleum Exploration Society of Australia Special Publication, 403–412.

Johnstone MH 1979. A case history of Rough Range. *APPEA Journal* 19, 1–6.

Longley IM, Bradshaw MT & Hebberger J 2001. Australian petroleum provinces of the 21st century. In: *Petroleum provinces of the 21st century*, Downey M, Threet J & Morgan W (eds), AAPG Memoir 74.

Norvick MS, Smith MA & Power MR 2001. The plate tectonic evolution of eastern Australasia guided by the stratigraphy of the Gippsland Basin. In: *Eastern Australasian Basins Symposium: a refocused energy perspective for the future*, Hill KC & Bernecker T (eds), Petroleum Exploration Society of Australia Special Publication, 15–23.

O'Brien GW, Tingate PR, Goldie Divko LM, Harrison ML, Boreham CJ, Liu K, Arian N & Skladzien P 2008. First order sealing and hydrocarbon migration processes, Gippsland Basin, Australia: implications for CO_2 geosequestration. In: *Eastern Australasian Basin Symposium III: energy security for the 21st century*, Blevin JE, Bradshaw BE & Uruski C (eds), Petroleum Exploration Society of Australia Special Publication, 1–28.

Powell TG 2001. Understanding Australia's petroleum resources, future production trends and the role of the frontiers. *APPEA Journal* 41, 273–288.

Tingate P, Campi M, O'Brien G, Miranda J, Goldie-Divko L, Liu K, Mills D, Volk H, Hall D & Hamilton J 2011. Understanding the plumbing of the Gippsland Basin: new results on fluid migration and reservoir quality. Extended abstract, 2011 APPEA Conference, Perth.

Totterdell JM, Stuckmeyer HIM, Boreham CJ, Mitchell CH, Monteil E & Bradshaw BE 2008. Mid–Late Cretaceous organic-rich rocks from the eastern Bight Basin: implications for prospectivity. In: *Eastern Australasian Basins Symposium III: energy security for the 21st century*, Blevin JE, Bradshaw BE & Uruski C (eds), Petroleum Exploration Society of Australia Special Publication, 137–158.

Walker TR 2007. Deepwater and frontier exploration in Australia—historical perspectives, present environment and likely future trends. *APPEA Journal* 47, 1–15.

5

Old, flat and red—Australia's distinctive landscape

Most of Australia's landscape is stark, and the climate is harsh. The landscape is also remarkably flat, with an average elevation of only about 325 m and local relief never more than 1500 m. As a result, many of Australia's major rivers are slow flowing, commonly into salt lakes in the arid interior. A thick regolith also blankets much of the continent. Sustained weathering has resulted in the formation of abundant iron oxides in the regolith, giving the Australian landscape its distinctive red colour. The regolith holds great wealth for Australia, but can also present challenges for mineral exploration. Human occupation over the past 50 000 years, and especially the last 200 years, has left a significant mark on the landscape.

Colin F Pain,[1,3] Brad J Pillans,[2] Ian C Roach,[3] Lisa Worrall[4,3] and John R Wilford[3]

[1]University of Seville, [2]Australian National University, [3]Geoscience Australia, [4]Zeus Uranium Ltd

Image by Mark Gray

Figure 5.1: Locality and features map, also showing Australia's deserts, rivers and lakes.

Regolith

The Australian landscape is remarkably flat, and around 80% is covered by a thick, and in places ancient, blanket of red-coloured regolith (Figure 5.1, Box 5.1). The regolith forms as a result of the interaction between rocks, water, air and life (Figure 5.2a), which drive processes operating at or near the surface, such as weathering, erosion, transportation, sedimentation and cementation (Figure 5.2b). Regolith can be thought of as everything between fresh rock and fresh air; it is somewhat synonymous with what has been called the 'critical zone'. Regolith dictates the nature of the Australian landscape, and has had a profound effect on the development of Australia and the way people have lived, and continue to live, on the continent.

Regolith may vary in thickness from a few centimetres to hundreds of metres and can have a complex architecture. Components of the regolith range in age from very young—for example, sediments deposited by a recent flood (Figure 5.3a)—to very old—for example, Permo–Carboniferous oxidation of iron in weathering profiles exposed by mining in the Tanami region of the Northern Territory (Figure 5.3b). The fact that ancient landscapes have been preserved is linked to the overall tectonic stability of the continent. This stability and landscape longevity have resulted in a continent that is deeply weathered and soils that are nutritionally poor and fragile.

Geography and plate position are driven by tectonics. After the breakup of Pangaea–Gondwana, Australia drifted northwards through the Cenozoic into a geographical position that has made it the most arid inhabited continent. The aridity, combined with the flatness of the landscape, has created river systems that are slow flowing. The slow-moving rivers, many draining into the interior, have struggled unsuccessfully to remove the wind-blown salt from the continent. As a result, the nutritionally poor soils are also prone to salinisation.

The soils and landscape differ markedly between Australia and much of the Northern Hemisphere. A key determinant of these differences was the role (or lack) of glaciers. The glaciers of the last ice ages transformed the Northern Hemisphere's landscape and soils, but the ice was restricted to small upland areas in southeastern Australia. This lack of glacial reworking and regeneration of soil and landscape in Australia was due to the low overall elevation and the latitudinal position of the continent at the time (Chapter 2).

Australia's landscape and regolith have been key determining factors in shaping where and how Australian people live. The regolith underpins our economic, social and infrastructure systems; we live on it, we grow our food in it, it is the foundation of many major engineering works, and much of our water supply is stored in it (Chapter 7). Regolith also hosts or hides valuable mineral deposits, and mineral exploration through and within the regolith is one of the greatest challenges facing mineral explorers in Australia in the 21st century (Chapter 11).

The richness and diversity of Australia's regolith makes a fascinating story, so let us first look back to how the continent's landscape and climate were transformed after the continent was finally released 'out of Gondwana' (Chapter 4).

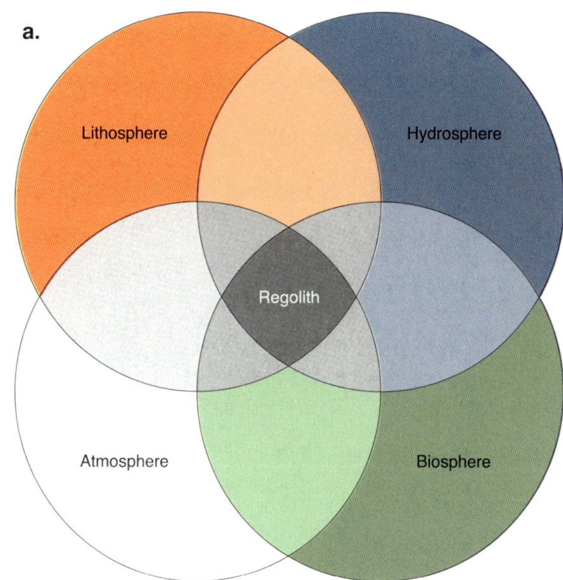

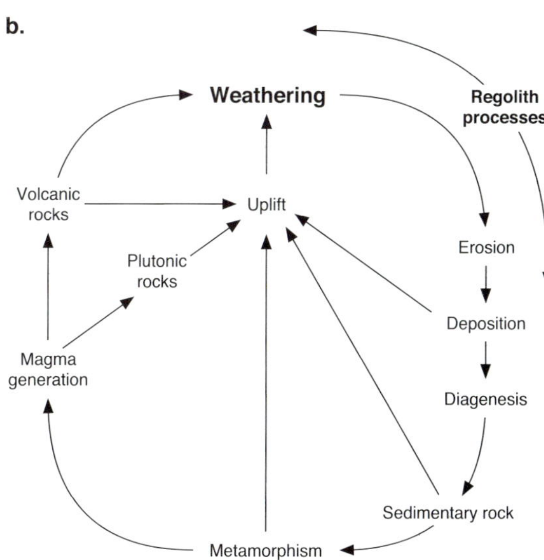

Figure 5.2: (a) The regolith is the critical interface between four of Earth's spheres: the lithosphere, hydrosphere, biosphere and atmosphere. (Source: Taylor & Eggleton, 2001). (b) Role of regolith processes in the geological rock cycle. (Source: Wilson, 2004)

Figure 5.3: (a) Sediments deposited by recent floods on the Darling River Floodplain near Menindee, western New South Wales. (b) Weathering profile (more than 30 m deep) exposed in the wall of an openpit gold mine, Tanami region, Northern Territory.

Cenozoic climate changes of the Australian landscape

Australia is the second driest continent in the world, Antarctica being the driest. There is ample evidence of dramatically different climates in the past compared with the present (Chapter 3), with the strongest imprint on the landscape occurring during the Cenozoic. The period from 65 Ma to about 35 Ma (broadly equivalent to the Paleocene and Eocene epochs) saw a significantly warmer world, with no Antarctic ice sheet, sea-levels perhaps some 60 m higher than present and atmospheric CO_2 concentrations more than three times that of modern levels. During that time, the Australian continent lay well to the south of its current position, and the dominant vegetation was warm to cool temperate rainforest, even in central Australia (Chapter 2).

From about 35 Ma (latest Eocene), the Antarctic ice sheet progressively increased in size, global temperatures decreased, sea-levels fell, and central Australia became progressively more arid as the Australian continent moved northwards to straddle the zone of subtropical high pressure that is characterised by mid-latitude deserts on both sides of the equator. Pollen evidence suggests that seasonal aridity may have been the norm in parts of central Australia as early as 40 Ma (middle Eocene). Nevertheless, around 30 Ma the moisture-loving ancestors of Wollemi pines and southern beech trees were growing in the now semi-arid Pilbara in Western Australia, and around 25 Ma large central Australian freshwater lakes, forerunners to Lake Eyre, had formed, inhabited by crocodiles, dolphins and giant flamingos. Indeed, fully desert conditions were probably not established until around 3 Ma (Late Pliocene). For example, from cosmogenic ^{21}Ne and ^{10}Be dating, it is clear that stony deserts in northern South Australia formed at 2–4 Ma, while major dunes formed somewhat later, from around 1 Ma (Figure 5.4).

With the increase in Antarctic ice, subtropical high-pressure systems moved from the south to dominate the continental interior (Figure 5.5). One outcome of this increased aridity was that some rivers draining to the south were unable to adjust their gradients to the post-Miocene tectonics, especially on a continental scale to the northeast, and became disorganised or moribund, with a number of swampy lakes adjacent to the tilt axis. These lakes contracted and disappeared, with marked faunal changes as the land dried up (Chapter 3).

The cooling of Earth's climate during the Quaternary glaciations, starting about 2.6 Ma, led to wetter and drier cycles related to glacial and interglacial periods, and concomitant fluctuation in the amount of water flowing into the interior. During the cold, dry glacial periods, aeolian landforms (dunes) spread out from central Australia and moved into surrounding areas, where they can still be seen today as more or less stable. Australia does not have as wide a variety of dune forms as the Sahara Desert, but the dunes can be dramatic; most are longitudinal, some tens of kilometres in length. The pattern of dune fields mirrors the anticlockwise pattern of the dominant winds in Australia (Figure 5.5).

WHY IS THE AUSTRALIAN OUTBACK RED? (BOX 5.1)

In a word, the answer is hematite (from the Greek *haimatites*, meaning bloodlike). This iron oxide (Fe_2O_3) is formed by weathering and is largely responsible for the pervasive red colour of rocks, soils, dust and sand that characterise central Australian landscapes. Every visitor knows how a combination of red dust and sweat produces stains on clothes that are almost impossible to remove. Hematite is also the predominant pigment in red ochre, one of the clays favoured by Aboriginal artists for their paintings.

The iconic Red Centre of Australia is broadly encompassed by the 500 mm mean annual rainfall isohyet, suggesting that redness is related to aridity—which is certainly true. Aridity came rather late to central Australia, geologically speaking, probably beginning in mid-Cenozoic times as the Australian Plate moved northwards to straddle the dry mid-latitudes, and progressively intensified during the last 2–3 Myr. However, much of the red colour in the rocks originated prior to the last 3 Myr, at times when the climate was significantly wetter than the present—for example, between 80 Ma and 60 Ma. Furthermore, the red colour is not just a surface feature; it commonly extends to depths of more than 50 m, as seen in openpit mine exposures throughout Australia (Figure 5.3b). On the other hand, the red sand dunes are much younger, mostly less than 1 Ma according to luminescence and cosmogenic nuclide dating. The sand dunes are composed mostly of white quartz grains, but a thin coating of hematitic clay gives the reddest dunes their distinctive colour. Thus the Red Centre has a complex history of redness.

Red sand dunes in the Sturt Desert, South Australia.

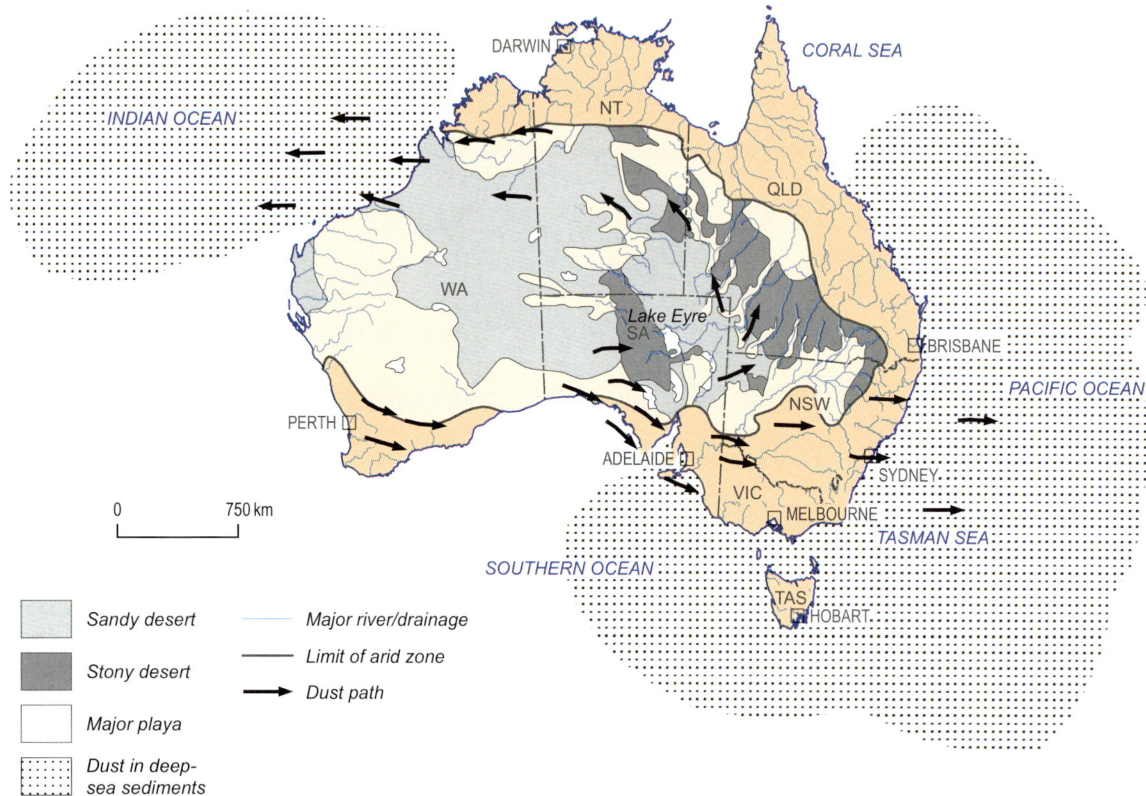

Figure 5.4: Major desert and dune systems of Australia. Note rivers and river catchments overlain by longitudinal dune systems and the distribution of stony and sandy deserts. (Sources: Bowler, 1976; Fujioka & Chappell, 2010)

The repeated rise and fall of lake systems tracked the changing climate and especially the availability of rainfall. Lakes came and went in both the east and the west of the continent, leading to the development of clay dunes, or lunettes. Early people settled by the lakes when they were full, leaving behind skeletons such as the 'Mungo Lady', and other evidence of their habitation at least as long ago as 40 ka and perhaps as early as 55 ka. The wet–dry cycles extended to the eastern highlands; Lake George, for instance, currently an intermittently dry lake basin on the Great Divide, was some 40 m deep during the last glaciation at 20 ka and overflowed into the headwaters of the Murrumbidgee River, and at the same time the lakes further west were full.

Winds also reach out from Australia from the northwest and southeast corners, carrying dust with them (Figure 5.4). Dust particles represent a link between the lithosphere, hydrosphere, atmosphere and biosphere (Figure 5.2a); they are therefore an important parameter in understanding climate. The analysis of dust can be used to better understand the El Niño/La Niña – Southern Oscillation (ENSO) climate system, as well as past shifts of the intertropical convergence zone on glacial–interglacial time scales (Chapter 1).

Because Australia does not have the nutrient-rich cold upwelling currents of other continents (Chapter 6), dust, and the iron carried with it from the continental interior, forms a vital part of the ocean's food web by partly regulating plankton growth. Today, around 4–7 Mt/year of Australian dust is deposited in the southwest Pacific Ocean, much via episodic dust-storm events. Australia contrasts with the other Southern Hemisphere dust sources (southern Africa: 60 Mt/year; Patagonia: 30 Mt/year) in that the dust amounts are less and they are very episodic. The reasons for these differences are undoubtedly related to the complex geology and regolith and the relatively higher amounts of vegetation in Australia's arid zone compared with other dust-source regions.

During the last glacial maximum, the continent was even more arid than today—it was also much windier, and dust was transported and deposited over vast distances. The dust may have enhanced marine productivity, drawing down atmospheric CO_2 and further lowering temperatures at that time. Accumulations of dust both in the eastern highlands and in marine cores off the east coast of Australia coincide with dry climate periods.

Recent work on the age of zircon grains from dust in eastern Australia shows that much of it was sourced from the Queensland highlands. Particles are transported down rivers into the interior, becoming finer as they go. Dust is then blown out of dry river and lake beds and makes its way in a step-like manner via winds back to the east (atmosphere). Because of the long history of particle recycling, the elemental composition of Australian dust tends to be different from that of dust from other continents. Australian dusts tend to be low in phosphorus and nitrogen, essential elements in ocean fertilisation. Interestingly, the iron in the Northern Hemisphere dust (Sahara Desert) interacts with that hemisphere's pollution (NOx, SOx), somewhat ironically making the iron more bioavailable for plankton growth in the Atlantic Ocean. The more pristine air masses (atmosphere) around Australia mean that the dust does not 'benefit' from this pollution, although workers suspect that the interaction with bushfire smoke may fulfil this role in part.

An obvious positive consequence of the dust is the accumulation of deep, red, loamy soils that are a valuable agricultural resource on the western slopes of New South Wales (Figure 5.6). Another obvious, but less positive, consequence is the large dust storms that sweep across southeastern Australia, particularly during times of drought. A dust storm in October 2002 eroded an estimated 96 Mt of top soil and entrained around 5 Mt of dust along a 2400 km-long front in eastern Australia. Similarly, in September 2009 a huge dust plume crossed New South Wales and southeast Queensland, turning Sydney into a surreal, sunless, red city (Figure 5.7).

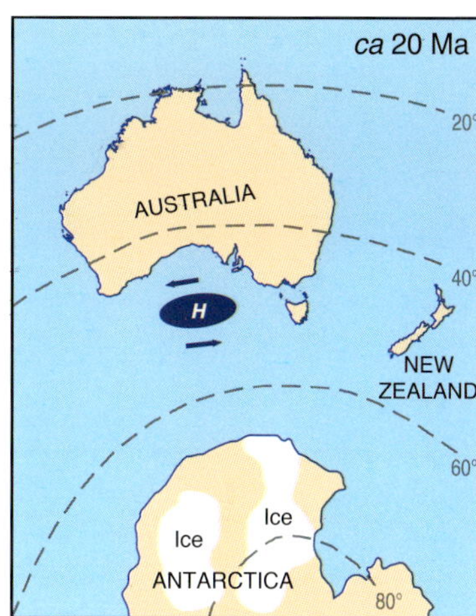

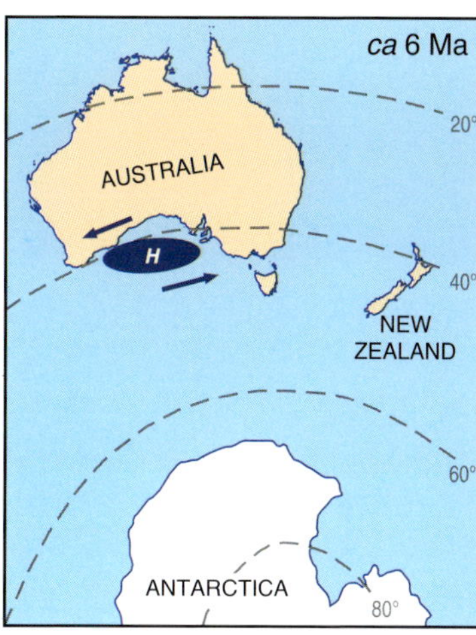

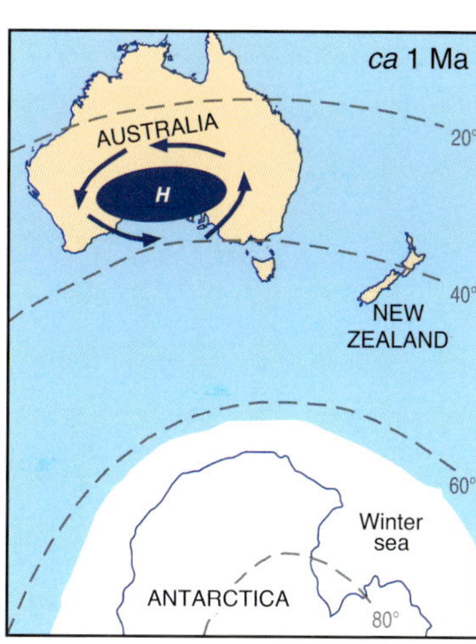

Figure 5.5: Factors influencing climatic changes of the last 20 Ma, 6 Ma and 1 Ma in the Australia–Antarctic region. Note the increase in ice cover in Antarctica, and the move of the high-pressure system into the centre of the continent. (Source: Fujioka & Chappell, 2010)

Figure 5.6: Deep loamy soils near Young, New South Wales. Young is the Australian capital of cherries, which are grown in these dust-derived soils. The region also prides itself on other stone fruits and fine wines. Originally named Lambing Flat, Young also featured in the early gold rushes (Chapter 8). A riot occurred there in 1861, with mainly European miners attacking Chinese miners, which resulted in curbs on immigration to the colony.

Why so flat?

One message of this book is that the Australian continent is very flat; it has relief never more than 1500 m and an estimated average slope of only around 1.4°. The elevation of Australia is also very low, with an average height of 325 m (Figure 5.8a), which contrasts with the global continental average of 870 m. Australia, however, has not always been so flat and low (or dry). Mountain ranges, of estimated Himalayan proportions, are envisaged to have formed the spine of central Australia in the Neoproterozoic. Although their elevations will remain uncertain, these mountains must have been high because the Petermann Orogeny (Chapter 2) exhumed deep crustal rocks (eclogites), and their load on the lithosphere formed foreland basins many kilometres thick (e.g. Amadeus Basin). The sediments that make up the Uluru monolith were laid down in a basin like the Siwaliks at the footwalls of the Himalaya in northern India. The reservoir rocks that host, or may host, Australia's hydrocarbon riches, such as the Triassic Mungaroo and Late Cretaceous Ceduna deltas (Chapter 4), were built from the eroded remnants of interior mountain ranges. The Flinders Ranges have been intermittently feeding detritus into adjacent basins for more than 55 Myr, with significant uplift in the past 5 Myr. Because topography drives fluid flow, an understanding of landscape evolution and its palaeotopography through time is very important for reconstructing ancient mineral systems (Figure 8.5).

Although Australia is flat, it is hardly dull. Intriguing subtlety exists in Australia's landscape and this is revealed by high-resolution elevation data such as that acquired by the space shuttle-derived radar

4WD track through claypan, Freeth Junction, Witjira National Park, South Australia.

Finally, Australia is distinguished from other dryland or arid regions of the world by the longevity of human occupation. The widespread use of fire for tens of thousands of years is thought to have resulted in alteration of the landscape through changes in the distribution of flora and fauna (Chapter 3), as well as in the hydroclimate.

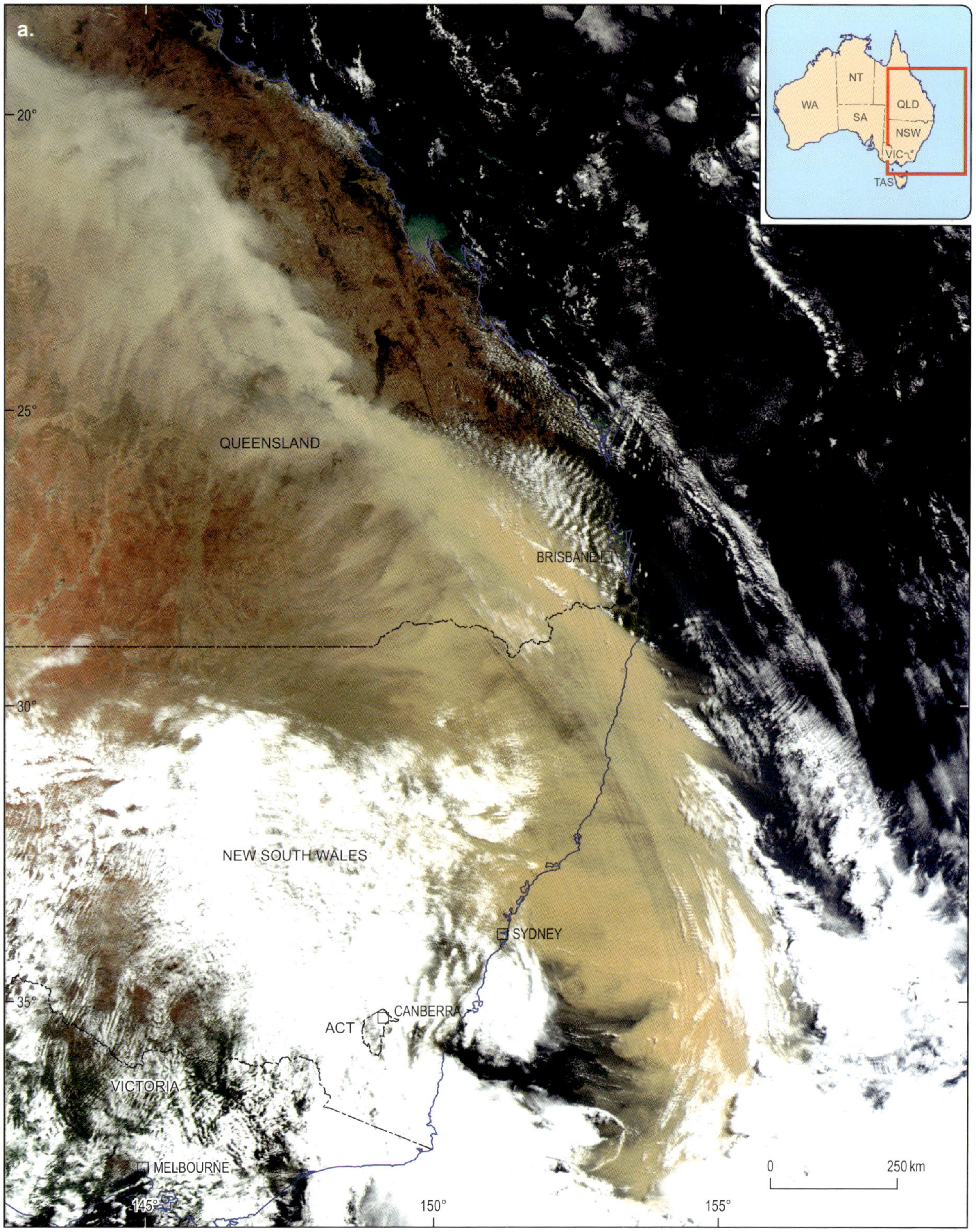

© Getty Images [J Reynolds]

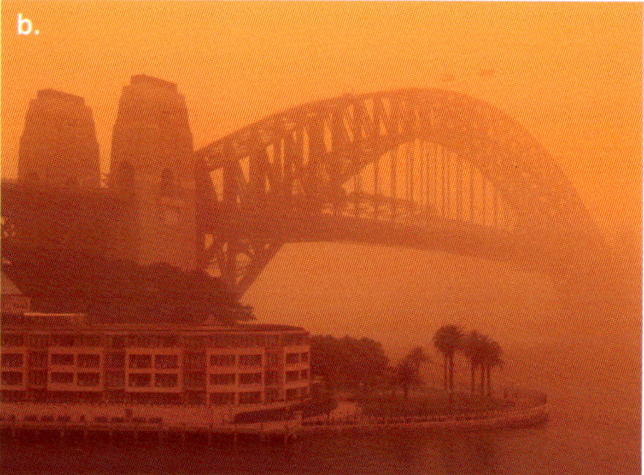

Figure 5.7: (a) A satellite view of the September 2009 dust storm across eastern Australia. (Source: NASA). (b) photograph of Sydney Harbour Bridge during the September 2009 dust storm.

systems. Such data highlight the complex mosaic of differing land surfaces in the Yilgarn Craton of southern Western Australia, which can be used to document landscape denudation processes (Figure 5.8b). At a more local scale, the use of very precise laser logging using light detection and ranging (LiDAR) has revolutionised the mapping of flat and apparently featureless landscapes. Standing on the banks of the Murray River and looking out over the flat floodplain, one would be forgiven for missing the minutiae of the landscape. The LiDAR images uncover an otherwise unrealised complexity, such as braided plains, scroll bars and filled billabongs, illustrating the dynamic nature and rich history of the erosion and deposition of these 'flat' landscapes (Figure 5.8c).

Why is Australia flat? As we shall see, the reasons are linked to the continent's deep and ancient geological roots, the tectonics since the breakup of Gondwana and the geomorphological processes that have operated in this framework. Let us now consider some of these controlling processes.

Tectonic controls of flatness

Australia became the island continent when it was finally released from its long connection with Antarctica. The breakup of Gondwana along the southern margin of Australia took around 100 Myr to complete. It was not until *ca* 34 Ma that final separation had occurred. Rifting also occurred on the eastern margin when the Tasman and Coral seas opened and India and other Asian fragments broke from the western margin (Chapter 4). Since the break, Australia has drifted more than 3000 km to the north-northeast, at a rate of 6–7 cm per year

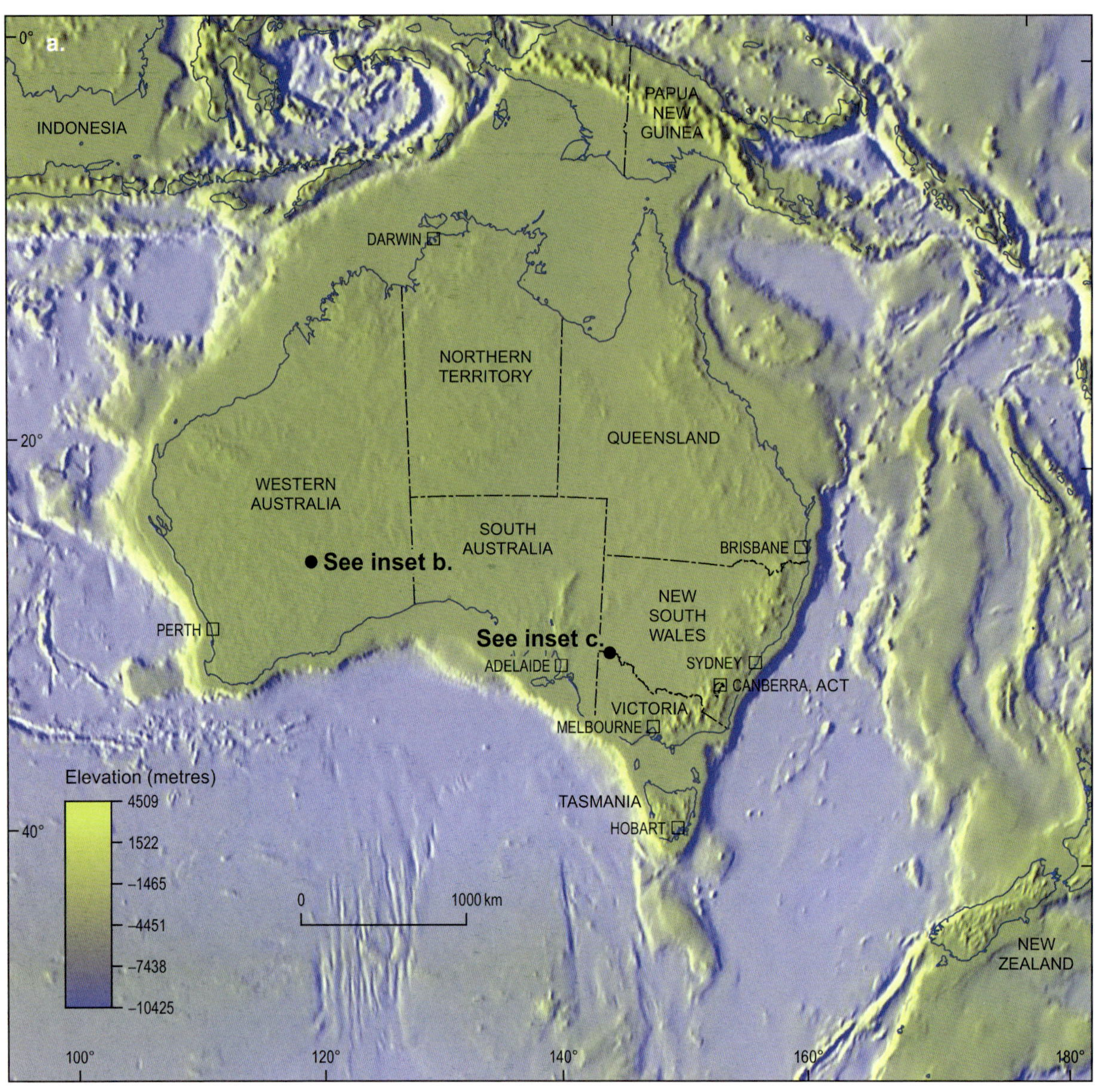

Figure 5.8: (a) Digital elevation model of Australia and its surroundings; seafloor on the west, south and east; and New Guinea to the north. The flatness of the Australian landmass stands out in contrast to the mountainous land to the north. (Source: Mike Sandiford, pers. comm.). (b) Digital elevation model from Shuttle Radar Topography Mission (SRTM) of southwest Western Australia, showing the small rises and scarps in an otherwise flat landscape. (c) Digital elevation model from LiDAR data over the Murray River corridor in South Australia, showing the subtle changes in elevation related to the evolution of the river.

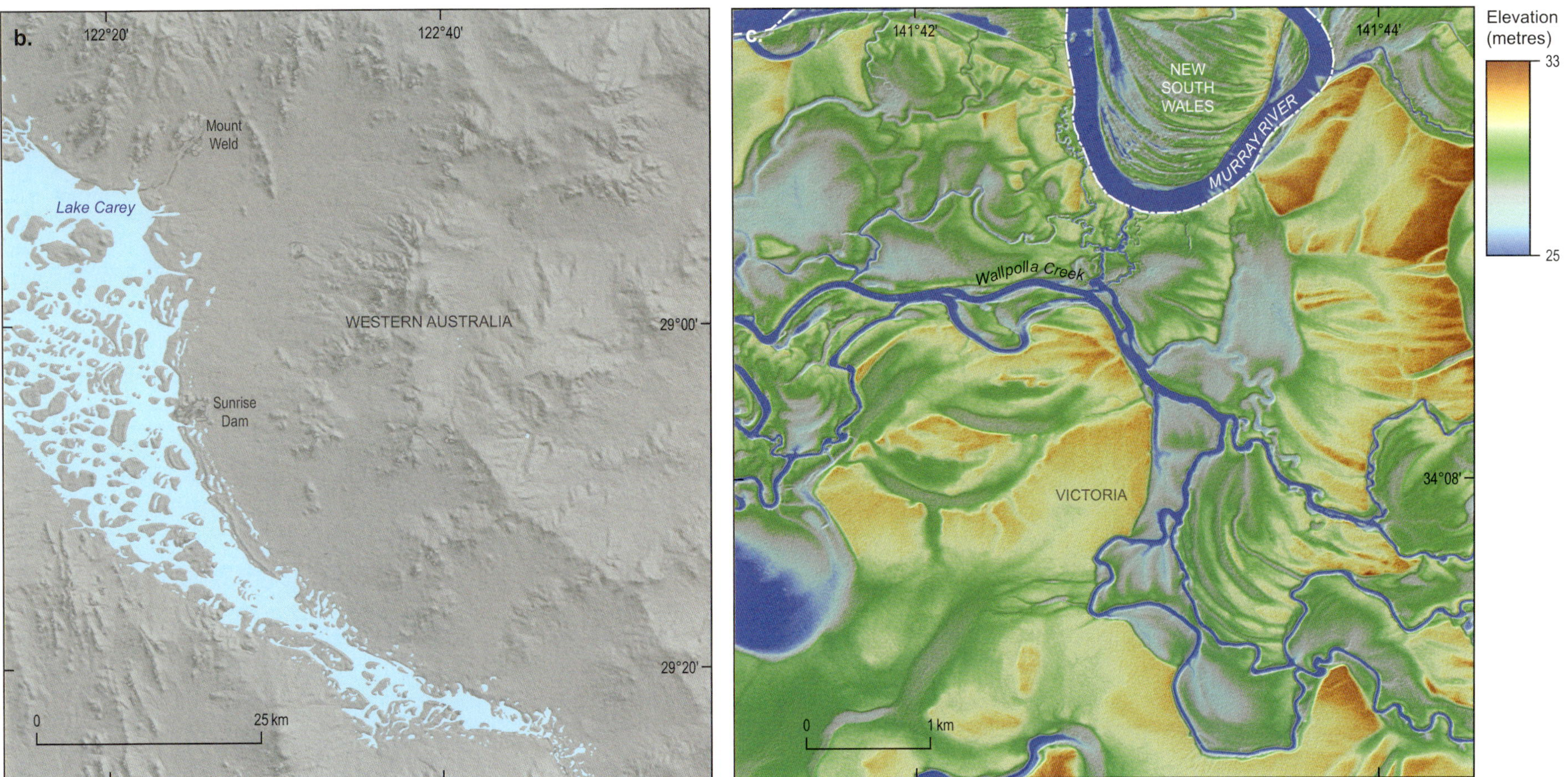

(Chapter 2). Impacts of the breakup, such as raised margins on the east and west, are still obvious in the landscape today. The drift north, in addition to moving Australia through increasingly warmer (and drier) climate belts, had the effect of tilting the continent to the north. This tilting may have been caused by movement of the Australian continent over irregularities in the geoid, rather than subduction pull of the oceanic lithosphere attached to the northern margin of the plate.

The Australian continent is located roughly in the middle of its plate and, except in the north, is far from active plate margins. Even there, the active margin is reflected in the mountains of New Guinea rather than on Australia. The distance from active margins has been an important tectonic factor in reducing the topographic consequences common in convergent margins, such as volcanoes and collisional mountain ranges associated with crustal thickening.

With regard to its flatness, Australia is different from other continents, but in degree rather than kind. Other Gondwana continents have high edges and low centres, and Africa and India are also surrounded on at least three sides by passive margins. Only South America is so different, in that, whereas it has a passive margin on the east, it has an active one on the west (Figure 5.9).

The strength of a continent is largely determined by the thickness of the lithosphere; in addition to the continent's intraplate position, Australia derives underlying strength from its very thick lithosphere. Continents with old (Archean) cratons tend to have thick lithosphere that is also cold and buoyant, making it strong and difficult to destroy (e.g. by subduction processes). A typical global continental average lithospheric thickness of around 90 km occurs in the younger east (Tasman Element). However, the centre and

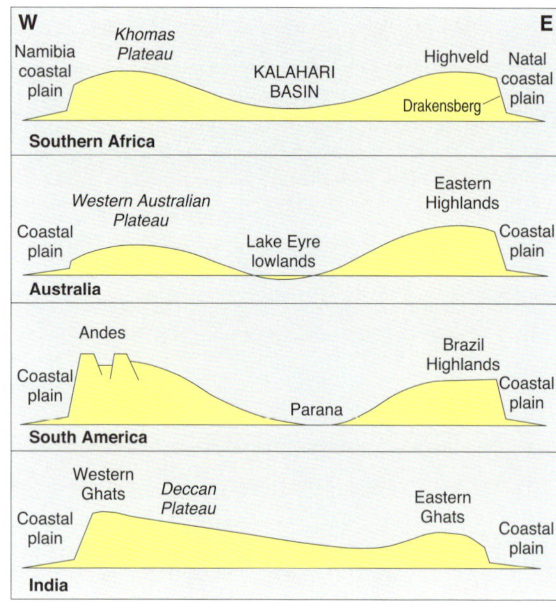

Figure 5.9: Schematic sections across four post-Gondwanan continents, showing the general saucer shape that all exhibit. (Source: Pain, 1985)

most of the old western cratons are underlain by lithosphere more than double the global average at more than 200 km thick (Chapter 2). Apart from a region within central Australia, the heat flow across the continent is low (Chapter 10), adding to the overall strength. This strength also means that the continent is difficult to deform and create uplifted mountain areas. The combination of intraplate position and lithospheric thickness is a fundamental reason for the relative stability of the Australian continent, and is why the ancient geological record has been so well preserved.

The last major orogenic event in Australia occurred around 250 Ma, and consisted of accretionary events along the eastern part of the Tasman Element, with addition of the New England Orogen. Since then, the eastern highlands have developed as a result of opening of the Tasman Sea, which began about 95 Ma in the south and spread north. The development of the eastern Australian passive margin led to the formation of the Great Divide along the axis of uplift, and the development, by erosion, of the Great Escarpment to the east. This escarpment in eastern Australia marks a major boundary between low rates of geomorphological processes on the inland side and much more active processes on the seaward side. Similar plateau edges, although not so dramatic, occur on other margins of Australia (Figure 5.10a).

The Pilbara, Western Australia, has been remarkably stable since the Archean. The region preserves evidence for many of Earth's early processes.

Image by Jim Mason

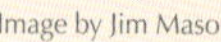

Figure 5.10: (a) Plateau edges surrounding the Australian continent, including the Great Escarpment of eastern Australia, as defined by Pain (1985), superimposed upon a digital elevation model of the continent.

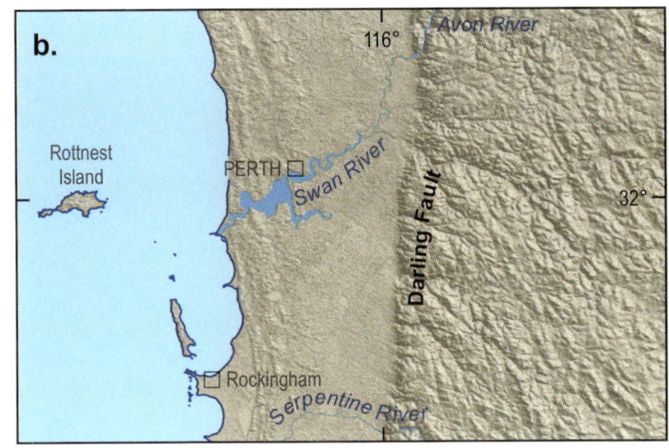

Figure 5.10 continued: (b) 30 m pixel SRTM image of the Darling Scarp east of Perth, Western Australia. Note the dissected Archean basement of the Yilgarn Craton to the east and the down-faulted Perth Basin to the west of the scarp. Aeolianite dune systems are parallel to the coast. (c) Photograph of the Wollomombi Falls flowing over the Great Escarpment in northeast New South Wales.

Neotectonic features

A range of geological evidence suggests that the current crustal stress regime was established during the Late Miocene (10–5 Ma), coinciding with significant changes at the Australian Plate margins, including the onset of major mountain building in New Zealand and New Guinea. At this time, intraplate uplift occurred in parts of Australia such as the Flinders Ranges (Figure 5.11).

Three distinct scales of neotectonic deformation have been described for Australia. At long wavelengths (thousands of kilometres), systematic variations in the height and inland extent of Neogene marine sediments indicate that the continent has been tilted down to the north at the same time as it moved north. This is well illustrated by the presence of Cenozoic marine sediments and shorelines along the edges of the Nullarbor Plain (including the spectacular high cliffs of the Great Australian Bight: Figure 5.12a), suggesting as much as 150 m uplift since the Miocene, but a distinct lack of similar sediments and shorelines in northern Australia. This tilting could also explain the difference in elevation of the Great Escarpment—up to 1500 m in the south but less than 50 m in far north Queensland.

At intermediate wavelengths (hundreds of kilometres), several undulations of about 100–200 m amplitude are evident on the 1–10 Ma time-scale, including formation of the Flinders Ranges and surrounding basins such as Lake Frome and Lake Eyre. The lowest point of Lake Eyre is some 15 m below sea-level, which initially seems odd—how can any part of a stable continent remain below sea-level without filling up with sediment? The answer may be that Lake Eyre has only been such a low point for a few million years and that such 'dynamic topography' is in response to flow instabilities in the upper mantle. The Bungunnia palaeolake (Figure 5.13) may also have been formed in the same way. Uplift along the southern margin has also been responsible for the preservation of beach ridges millions of years old in the Murray Basin (Figure 5.13). Part of the present coast consists of a beach ridge more than 125 km long, behind which is the Coorong lagoon. Inland from the Coorong is another inactive beach ridge, and then others further inland of increasing age, which can be traced for hundreds of kilometres into Victoria and New South Wales (Figure 5.13). Closest to the sea, the ridges are Quaternary, whereas

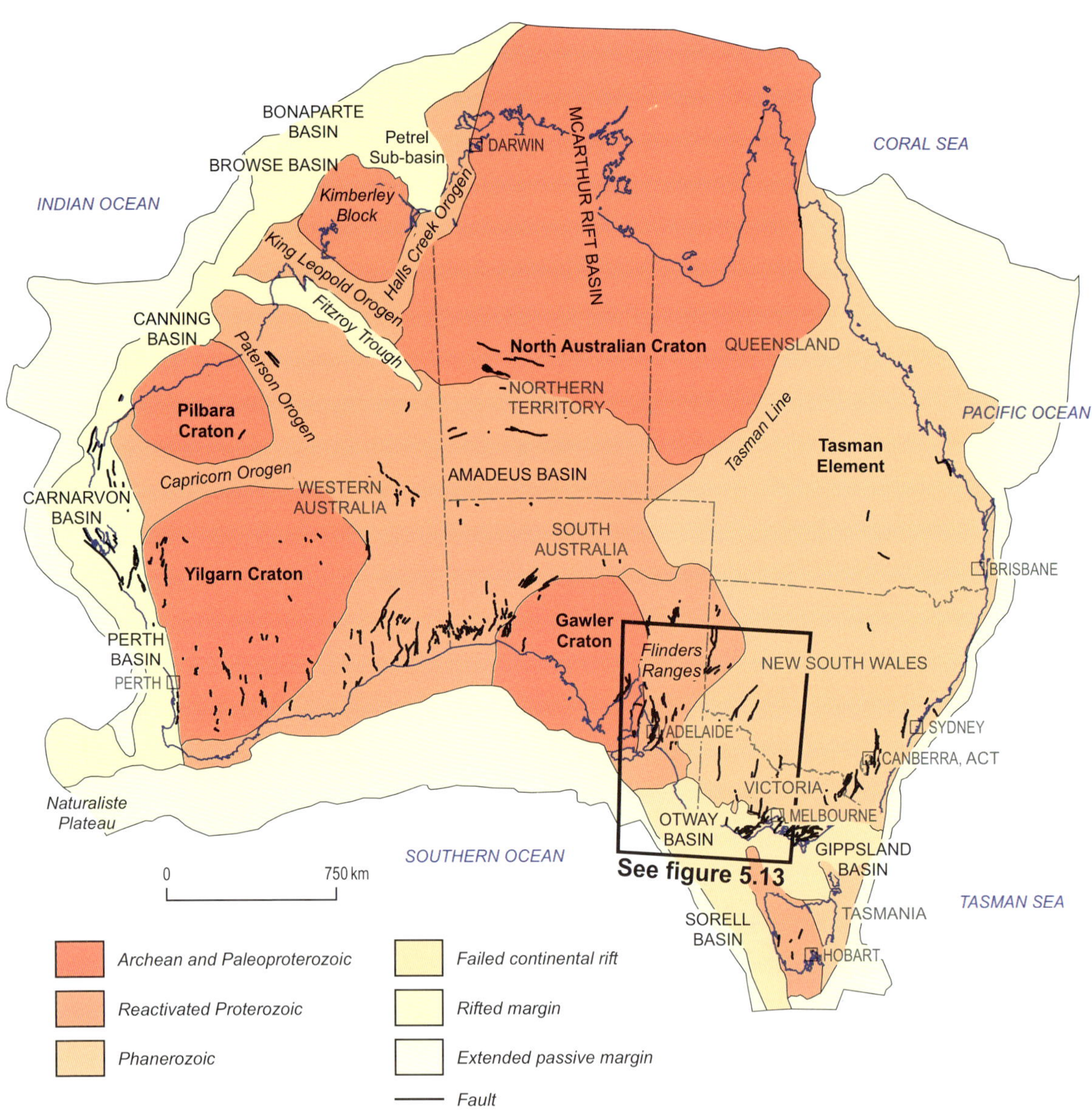

Figure 5:11: Distribution of faults in Australia that may have been active in the Late Cenozoic. (Source: Clark et al., 2011)

the oldest ridges inland are Miocene in age. Further beach ridges, still undated, are preserved west of the Otway Ranges, Victoria. In the early Quaternary, coastal erosion formed the Kanawinka Escarpment, a palaeo sea-cliff and one of the most prominent and laterally extensive geomorphological features in southeastern Australia (Figure 5.13).

At low wavelengths (tens of kilometres or less), brittle deformation associated with seismicity has produced numerous fault scarps (Figures 5.11 and 5.13). Only five fault scarps are known to have been associated with historic, earthquake-generated surface fault ruptures (Figures 5.11 and 5.13). Such faulting is typically associated with significant modifications to surface drainage, including ponding (e.g. Lake George) and diversion (e.g. of the Murray River by the Cadell Fault). On the Milendella Fault in the Mt Lofty Ranges, Proterozoic and/or Cambrian rocks have been thrust over Quaternary sediments (Figures 5.12b and 5.13).

Geomorphological controls on flatness

Australia is not only flat, but also lower in elevation than any other continent (Figure 5.8). We have seen that Australia has a deep and strong lithospheric root and is also located distant from the mountain-building effects of convergent plate margins. These tectonic controls are augmented by various geomorphological controls that helped to establish and maintain Australia's flatness.

The Lake Eyre and Murray basins are the lowest parts of the continent. Incidentally, these basins also coincide with some of the mainland's thinnest crust, at around 30 km. The highest areas are along the southern half of the eastern highlands, which also coincides with some of the thickest crust, at around 55 km. There is, however, a poor correlation between elevation and crustal thickness, and there is far more topography on the base than on the top of the crust (Figure 2.15). The western plateau of Western Australia, extending into South Australia and the Northern Territory, also has relatively high landscapes.

Early workers JT Jutson and WG Woolnough described ancient land surfaces on the Australian continent and linked them to deep and intense weathering profiles and duricrusts. Duricrusts are regolith materials (Figure 5.14) cemented by silica (silcrete), iron oxide (ferricrete) and/or carbonate (calcrete). The geologists also assumed an overriding simplicity, invoking a single, continent-wide surface, the 'Great Australian Peneplain' (or 'Australian Pediplain'), with an extensive duricrust sheet.

The complete story is much more complex. There are old surfaces covered with silcrete and ferricrete, but they developed at different times and in restricted areas. In central and eastern Australia, the old land surfaces are characterised by mesas and plateaus. These have thick and often siliceous weathering profiles on top, which contrast with the surrounding areas where weathering profiles are much thinner (and younger). In the west, there are similar features, usually with ferruginous duricrusts, and often expressed as one-sided, sloping plateaus or 'breakaways'. These spectacular landscapes are truly photogenic, especially in soft morning or evening light that makes the regolith colours 'glow'. These surfaces, and associated weathering profiles and duricrusts, cannot be correlated over the length and breadth of Australia. They are rarely the result of a simple period of weathering followed by erosion, but rather the result of complex interactions between weathering, denudation, erosion and deposition.

'Lazy' rivers

Rivers play a major role in shaping Australia's flatness and, as they are so slow moving, they could be described as 'lazy'.

Australia's general flatness means that its river systems have a low potential head, making them for the most part slow flowing, or even ephemeral. This slow-flowing nature is amplified by the mostly arid climate. Unlike most other continents, Australia does not have a transcontinental river system fed by major highlands with permanent snow. There is the Murray–Darling river system (Figure 5.10a), fed from modest mountains with a limited seasonal snowcap, but this is nowhere equivalent to the Mississippi of the United States (Box 5.2). This lack of major transcontinental rivers, in contrast to other continents, has hindered the development of the Australian interior and is another reason why most Australians live along the coastal fringe and not inland (Chapters 1 and 6). Although Australian rivers are meagre by world standards, they are still important sources of water (Chapter 7). They feature frequently in Aboriginal Dreamtime stories and also European art—think of 'Waltzing Matilda' and the 'jolly swagman camped by a billabong' (see *Did you know?* 5.1).

Image by Dan Clark

Figure 5.12: (a) High cliffs resulting from coastal erosion of uplifted Neogene marine sediments in the Great Australian Bight, South Australia. (b) The Milendella Fault thrusting Cambrian metamorphic rocks on top of Quaternary gravels in the Mt Lofty Ranges of South Australia. These gravels (in adjacent localities) record the 780 ka Matuyama–Brunhes palaeomagnetic reversal, making this a very young fault indeed. (Sources: Quigley et al., 2010; Clark et al., 2011)

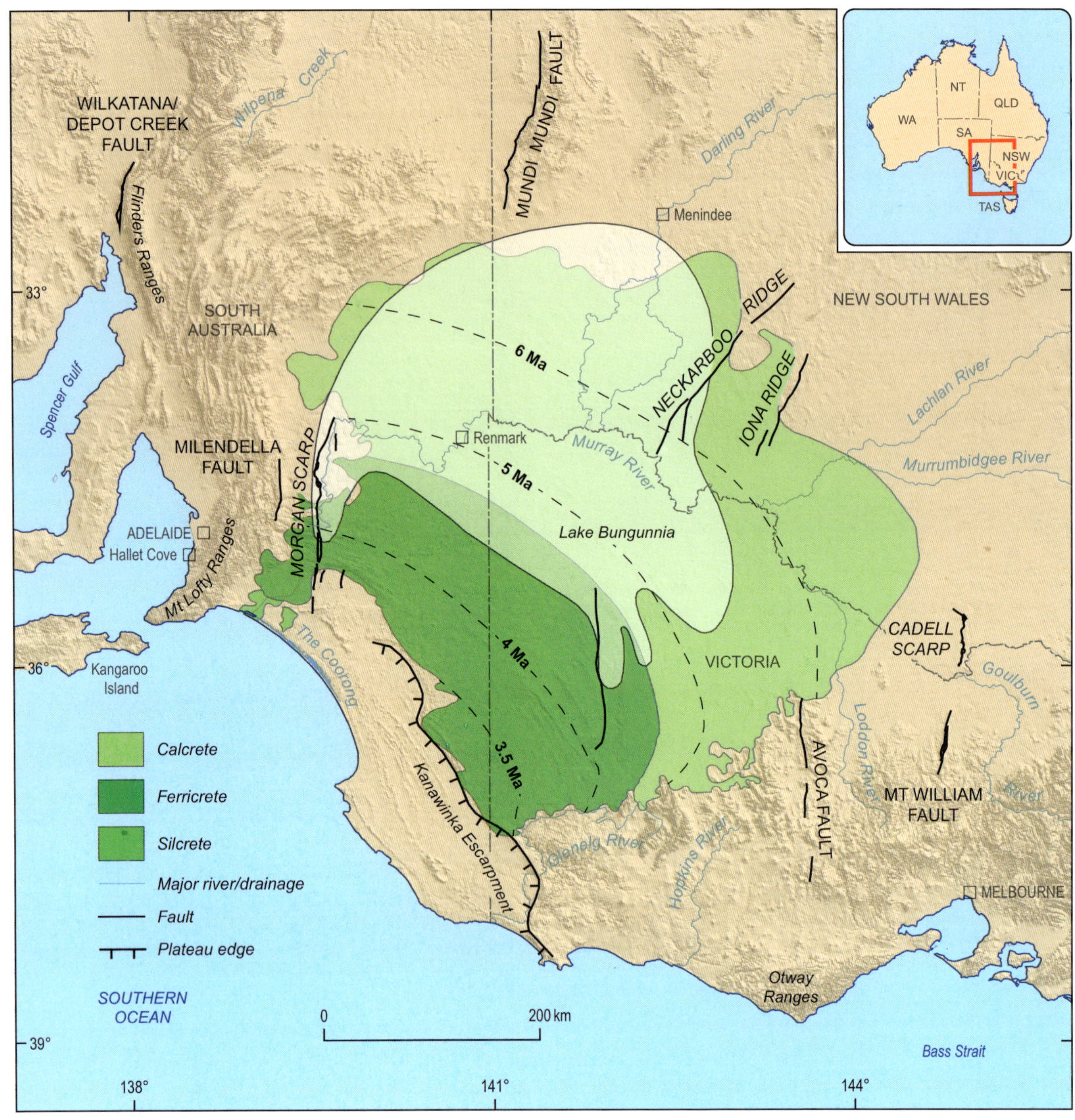

Figure 5.13: Beach ridges in the Murray Basin, southeastern Australia. Behind the modern arcuate beach ridge is a series of parallel ridges, extending back to the Miocene. It is not clear why there is a change from silcrete to ferricrete and then calcrete duricrusts dominant in the regolith as the ridges get younger, but this may be a consequence of age and weathering of beach materials.

Inland rivers in Australia have a very large flow variation, with mean annual discharges more than 1000 times more variable than those of most European and North American rivers (Table 5.1). A feature of flat landscapes, especially in arid climates, is that large areas have internal drainage. Most Australian rivers also show a consistent loss of flood discharge downstream. Some end in flood-outs, with water disappearing altogether, while others end in closed lake systems. Nearly half of Australia consists of areas that either drain internally or lack recognisable river systems (Figures 5.1 and 5.15). Thus, almost half of the Australian continent provides no runoff to surrounding oceans, in dramatic contrast to all other inhabited continents. This lack of runoff means that limited terrigenous sediment reaches the continental shelf, and carbonate sedimentation is predominant offshore (Chapter 6).

The Lake Eyre drainage basin is one of the largest internally draining areas in the world, covering about 1.2 M km^2, or some 16% of Australia (Box 1.5; Figures 5.1 and 5.15). Lake Eyre is a large salt lake that only intermittently fills with water. However, during the last interglacial, Lake Eyre was semi-permanent, fed by increased runoff from the north, probably caused by a stronger Australian monsoon. The ancestral Lake Eyre—Lake Dieri—was also much larger, perhaps 110 000 km^2 in area compared with 9500 km^2 today, incorporating other lakes such as Callabonna and Frome into one large lake (Figure 5.15). This mega-lake would have been filled largely by inflows from Cooper Creek and the Diamantina River, which must have been more permanent watercourses at that time.

The Breakaways near Coober Pedy, South Australia; silcrete-capped mesas.

Figure 5.14: Examples of 'left to right' nodular calcrete from Broken Hill, New South Wales. Massive silcrete with rainforest leaf fossils from Fowlers Gap, New South Wales, and groundwater ferricrete from the Jacinth heavy mineral sand mine, South Australia. These resistant lithologies cap many landscape features, such as the top of breakaways.

THE MURRAY IS NO MISSISSIPPI (BOX 5.2)

The Murray–Darling river system is Australia's longest at over 3670 km, with its headwaters in central Queensland and its mouth in South Australia. The town of Bourke, some 2300 km up the Murray–Darling, in northern New South Wales, is only 110 m asl. Such low-gradient rivers may seem unusual, but St Louis, some 2600 km up the Mississippi River in the centre of the United States, is only 142 m asl. Bourke was one of Australia's busiest ports in the late 19th century but, unlike the Mississippi, river flows were unreliable on the Darling and paddle steamers could be stranded upriver for years because of low water flows. Some comparative statistics are provided in Table 5.1.

The Fitzroy River in central Queensland, the largest river flowing east from the Great Divide, illustrates the difference in the size of rivers that flow directly to the sea. Like most Australian rivers, the Fitzroy flows are highly variable. This tropical river, complete with saltwater crocodiles, is situated in the cyclonic north (Chapter 6) and is prone to extreme flooding events associated with cyclones and tropical storms, especially in the summer months, like those of 2010–11.

Table 5.1: Statistics for Australian basins and rivers compared with the Mississippi–Missouri

Basin	Lake Eyre	Murray–Darling	Fitzroy	Mississippi–Missouri
Basin area (km^2)	1 200 000	1 072 000	142 665	3 202 230
Country area (km^2)	7 617 930	7 617 930	7 617 930	8 080 464* *(conterminous USA)
% of country	16	14	0.53	40
Longest river (km)	Cooper–Barcoo (1420)	Murray–Darling (3672)	Nogoa–Fitzroy (480)	Mississippi–Missouri (6275)
Discharge (m^3/sec)	Ephemeral	767	Highly variable	18 400

Pliocene and Pleistocene sands and clays are exposed at the Headings Cliffs on the Murray River near Renmark, South Australia.

Not only does topography impact river morphology, so does climate (Figure 5.15). The net amount of water available for a river is dependent on the amount of rainfall and the evaporation rate. In Queensland, where rainfall is more than 10% of the total amount of evaporation, there is enough surface runoff for rivers to be integrated into well-developed channel systems. However, where rainfall falls to less than 5% of evaporation, the rivers become disconnected and intermittent. This means that sediment moving down the rivers does not always reach Lake Eyre, but is deposited along the way. An important consequence of this pattern, especially in closed basins, is that erosion and sedimentation tend, over time, to reduce the relief of these areas, thus contributing to Australia's flatness.

Along the eastern highlands, and also along parts of western and northern Australia, rivers have two distinctive parts. Above scarps or knick points, they tend to have low-angle long profiles, and flow in broad valleys. In the east, many of them cross the Great Escarpment (Figure 5.10) as waterfalls and, near the coast, they are deeply incised and tend to have steeper long profiles than inland. It is important to remember that rivers above the Great Escarpment are controlled by a local base level. They do not 'know' about their elevation, so are not influenced by uplift along the eastern highlands, until knick points work their way upstream. Thus the inland reaches behave more like rivers draining towards the interior and, because of local deposition, are also evolving towards lower relief. For example, because of local deposition, the modern Shoalhaven River above the Great Escarpment south of Sydney flows at a higher elevation than it did in the Oligocene. In this, the Shoalhaven is similar to those rivers on the inland side of the Great Divide; only 10% of the Bland Basin (Figure 5.10), for instance, is erosional—the other 90% is covered in terrestrial sediments that reflect a decrease in relief.

This two-part river morphology is repeated around the edges of the Murray–Darling Basin, including the Darling River around Menindee, which has the appearance of an aeolian landscape drowning in fluvial sediments. The same pattern repeats at the Avon River in Western Australia,

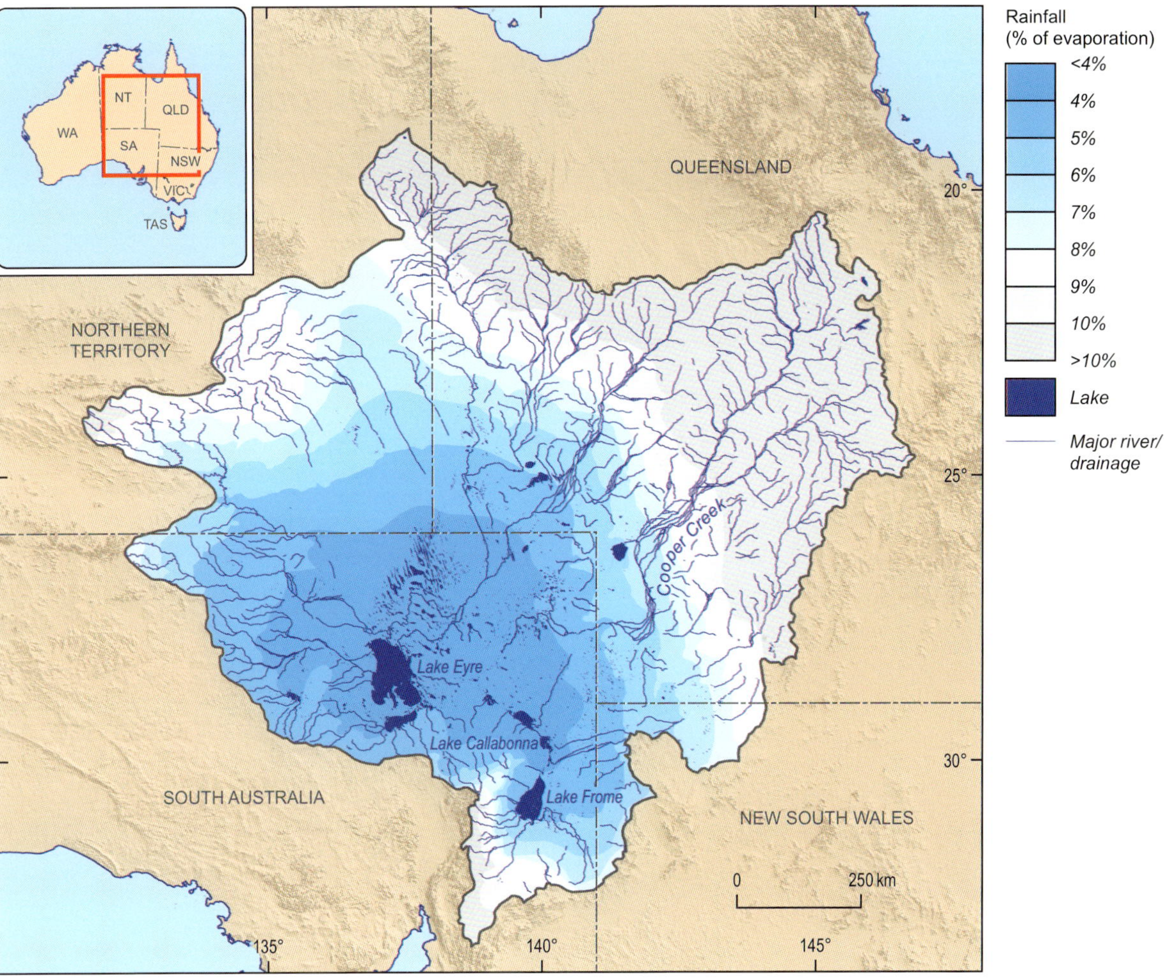

Figure 5.15: Water balance for the Lake Eyre drainage basin. Note how the channels change from a dendritic drainage pattern to more disconnected drainage pattern as evaporation increases and rainfall decreases (increasing aridity) further into the centre of the basin. (Source: Bureau of Rural Sciences)

Did you know?

5.1: Billabong—the resting place for a swagman

The word *billabong* was first recorded in Australian English in reference to the Bell River of southeast New South Wales. The explorer Thomas L Mitchell recorded in 1836:

> *The name this stream receives from the natives here, is Billibang. It is a Wiradjuri word of southern New South Wales and northern Victoria. In the extended and current sense – an arm of a river, made by water flowing from the main stream, usually only in time of flood, to form a backwater.* (Frederick Ludowyk, *The Australian modern Oxford dictionary*)

Billabongs are important refuges for wildlife—and people—in outback Australia as they are usually the last sources of surface water in dry conditions. They feature in Australian literature, such as AB 'Banjo' Paterson's iconic 1895 poem about a swagman 'Waltzing Matilda', carrying his swag or bedroll.

Aerial view of Outback Billabong, Queensland.

where low-gradient intermittent streams flow over the Darling Escarpment (Figure 5.10b) in incised channels, and further north, at the Fortescue and Robe rivers (Figure 5.1).

The biosphere is an important factor in shaping rivers and therefore the landscape and regolith (Figure 5.2a). Plants play their part in stabilising the landscape by mediating the impact of water scarcity. In the arid zone, the patchwork distribution of plants partitions water from rainfall into overland flow zones, limiting soil erosion and creating a landscape with local runoff sources, and run-on sinks. When the surface water flow does reach ephemeral streams, the channel-associated plants modify the flow conditions and velocity.

The effects of large plants are quite obvious; for example, river red gums (*Eucalyptus camaldulensis*) have a stabilising effect on river architecture, which in turn influences channel flow, erosion and deposition, to the extent that sediment ridges form. With time, these ridges lengthen and coalesce into anastomosing channel forms that are characteristic of inland rivers such as Cooper Creek (Figure 5.15). Other, less immediately obvious, examples of biosphere impacts are the plant litter from species like *Cassia* or spinifex grasses that accumulate and alter surface water flows by creating small 'dams' in what are otherwise very low-gradient landscapes. Soil crust-forming microphytic (and ancient) organisms such as cyanobacteria, mosses and lichens, as well as fungal hyphae, also play a role in stabilising the ground surface. These fragile crust-forming ecosystems are easily damaged by vehicles and overgrazing by hard-hoofed (imported) animals. Once the crust is broken, erosion can take hold, leading to serious land degradation.

Australia's palaeorivers

Not all of Australia's rivers are obvious at the surface today. 'Palaeorivers' can be elevated in the landscape as ridge lines meandering high above the present-day plains—these are examples of inverted relief. In other cases, the palaeorivers can be concealed by younger sediments, or even volcanic flows—these are 'palaeovalleys'. We will read more about the palaeovalleys and their importance for groundwater in Chapter 7.

Inversion of relief is a common feature in Australian landscapes. Relief inversion occurs when regolith materials that initially formed in lower parts of a landscape, usually valley floors, become cemented and therefore more resistant to weathering and erosion than the surrounding material, which is higher in the landscape. Weathering and erosion of the less resistant uncemented material result in the former (cemented and resistant) valley floor being preserved high in the landscape. A good example of relief inversion is preserved as the Miracking Conglomerate in South Australia, an alluvial silcrete that caps a series of parallel elongated mesas more than 200 km long, which are all that remains of a major drainage channel, complete with tributaries. Another example of inversion of relief caused by resistant ferricrete is the nodular iron ore of the Robe River deposit, Western Australia. These channel-iron deposits extend for tens of kilometres along the old valleys, and the modern valley follows this ancient path as a lateral stream (Figure 5.17). These palaeorivers are an important economic resource of iron (Chapter 9).

Palaeorivers are common on the Australian continent, and some are as old as early Neogene. Remnants of these drainage lines can be seen on

The Darling River at Bourke, New South Wales. By the 1890s, around 100 paddle steamers and barges worked the river.

satellite imagery (Figure 2.4), and are also visible on digital elevation models (Figure 5.10a). The Eucla Basin, for example, shows many signs of palaeorivers, some of which may have reversed their direction in the Neogene because the land surface, already very flat, was uplifted and tilted. Drainage reversals may also have occurred along the eastern highlands as the continental margin underwent passive-margin uplift.

Finally, with the end of the last glacial maximum, the sea-level rose, flooding older river systems, and shifting the mouths inland by tens to hundreds of kilometres in places. The Georges River and Sydney Harbour are rias—drowned remnants of deeper river systems. Deep submarine canyons cut the continental slope. Remnants of drainage channels can be seen on the submerged shelf inland of the Great Barrier Reef and snaking across the floor of the shallow Arafura Sea, while Pleistocene extensions of the Murray River can be traced across the southern shelf into the Murray Canyons. These changes in sea-level are just one of the impacts of long-term climate change on the Australian landscape (Chapter 6).

Figure 5.16: Aerial view of Cooper Creek, South Australia, in flood, January 2008. (Source: Lake Eyre Basin Intergovernmental Agreement)

Image courtesy of Cullen Resources Ltd

Figure 5.17: The channel-iron deposits in the Pilbara region, Western Australia, form cliffs and mesas. These cliffs and mesas were once a valley floor that became cemented (with iron oxyhydroides) and thus more resistant to weathering. Subsequent erosion has lowered the surrounding material, leaving the former valley floor high in the landscape. These deposits are highly prized for the quality of the iron ore resource (Chapter 9).

Regolith—everything between fresh rock and fresh air

Regolith can be broadly subdivided into in-situ and transported materials. In-situ regolith is weathered rock that has not undergone physical transport, although it may have undergone processes that have destroyed the primary fabric. In field descriptions, it may be characterised based on weathering intensity (slightly to highly weathered rock), or referred to in very broad terms as saprolith, which may then be subdivided into saprock (<20% of weatherable minerals altered), saprolite (>20% of weatherable minerals altered), or pedolith, where the primary rock fabric is totally destroyed (Figure 5.18). Transported regolith is material that has been eroded, transported and redeposited in the landscape, and includes alluvial, colluvial and aeolian sediments.

The distinction between in-situ and transported regolith is of critical importance to mineral explorers, since it allows more confident identification of possible sources of regolith materials, either from underlying or from distant source rocks. This is particularly the case when trying to explain the origin of surface geochemical anomalies, which might be linked to buried mineralisation (Chapter 11).

Weathering

Weathering involves chemical and mechanical processes associated with the breakdown or alteration of rocks and minerals in response to environmental conditions at or near Earth's surface. The weathered zone may grade imperceptibly into unweathered rock, especially on porous rocks, but sometimes there is a knife-sharp contact called the 'weathering front' between unweathered and deeply weathered rock (Figure 5.18). As rocks are subjected to chemical weathering, and their mineral constituents change to new, more stable, mineral assemblages, their elements are preserved in resistate minerals (e.g. zircon, monazite, rutile), are partly redistributed into new minerals, or are taken into solution. The materials lost in solution may be washed away entirely by rain and/or groundwater, or may be reprecipitated either lower in the weathering profile or downslope elsewhere in the landscape; this is the case with silcrete, ferricrete and calcrete duricrusts. Such elemental changes within the weathering profile may be large and, in some cases, result in economic mineral deposits (e.g. aluminium in bauxite, iron in iron ore). Weathering and regolith formation can dramatically alter the near-surface rock properties, especially density, porosity and permeability. These properties, along with their distribution, must be understood when applying geophysical techniques to probe beneath the regolith.

Evidence for very deep weathering comes from mining throughout Australia (Table 5.2) and logs of deep drillholes. Since groundwater has been found at depths greater than 2 km in places like the Great Artesian Basin, and is presumably interacting with enclosing rocks, weathering may extend to several kilometres. To accumulate such thicknesses of saprolite, weathering must have occurred at a faster rate than the weathering products were removed, which is usually associated with low relief in the landscape. The depth of weathering is not necessarily indicative of climatic conditions, because at depths of tens of metres the temperature is controlled as much by geothermal heat as by surface temperature, and the water necessary for chemical weathering can be present at shallow depths even in arid regions, as in the Great Artesian Basin (Chapter 7).

The degree or intensity of weathering can be estimated for the continent by combined analysis of gamma-ray spectrometric data and detailed elevation and slope information (Box 5.3). The map derived from this analysis depicts vast areas of intensely weathered material across the Australian landscape. The development of these large areas of intense weathering is due to the longevity of weathering and landscape preservation. Weathering intensity largely controls the degree to which primary minerals are altered to secondary components including clay minerals and oxides. As weathering intensity increases, there are changes in the hydrological, geochemical and geophysical characteristics of the regolith. Intense weathering is one of the factors in the low fertility of Australian soils (Box 5.3).

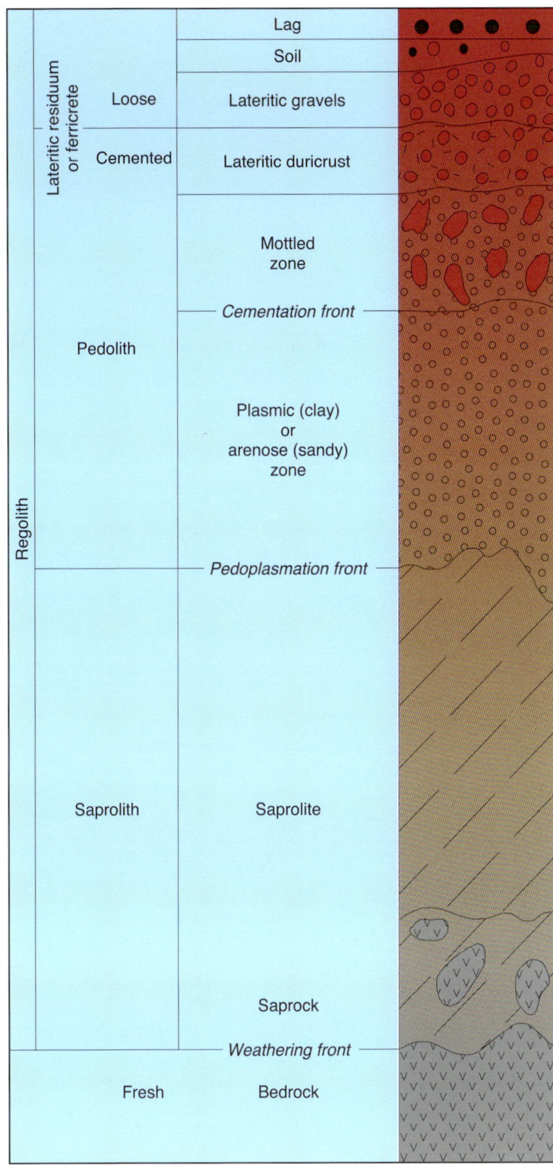

Figure 5.18: Regolith profile. (Source: modified from Eggleton, 2001)

A REGOLITH-COVERED CONTINENT. SO WHERE ARE THE FRESH ROCKS?

(BOX 5.3)

In Australia, the regolith profile can be up to several hundred metres thick. These deep profiles are partly a function of prolonged and deep weathering of the underlying bedrock and their preservation. Weathering involves chemical and mechanical processes that break down or alter rocks and minerals at or near Earth's surface, and is a prerequisite for material to be eroded. The interplay between weathering and erosional processes across different temporal and spatial scales is fundamental to understanding the nature and evolution of the landscape.

In Australia, it is possible to make a continent-wide map that estimates the intensity of weathering by using a combination of gamma-ray spectrometric, digital elevation and geological data. The weathering intensity map can be used to assess chemical and physical denudation processes, and the relative rates of regolith formation and removal over regional areas. Weathering has a major influence on the permeability and porosity characteristics of the soil and underlying saprolite, which in turn influence groundwater pathways and soil fertility. Deeply weathered landscapes often contain significant salt stores that can lead to soil salinity and saline discharge (Chapter 7). In terms of mineral exploration in regolith-dominated terrains, weathering can dilute or enrich pathfinder elements at the surface. So surface geochemical surveys can be interpreted in the context of variably weathered and leached landscapes. The weathering intensity map also has the potential to provide information on secondary enrichment processes within the regolith (e.g. palaeochannel uranium-bearing calcrete deposits).

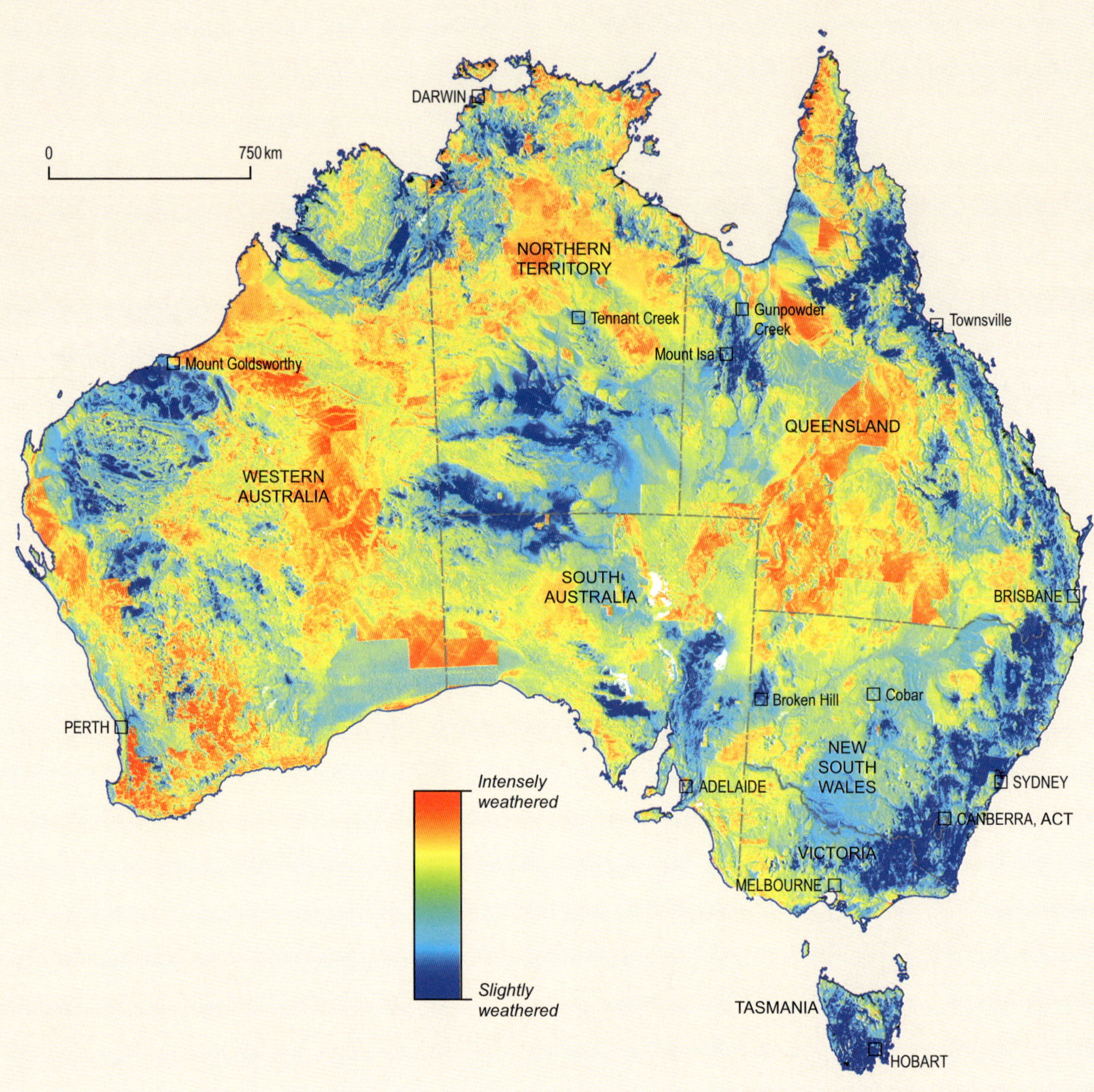

Figure B5.3: Weathering intensity map for Australia. The large areas of warm colours indicate the extent of intense weathering across the continent. Blocky parts of the image occur over gaps in the gamma-ray spectrometric data. (Source: Wilford, 2011)

Table 5.2: Examples of deep weathering in Australia

Weathering	Thickness	Location	Notes
Oxidation, leaching and secondary enrichment in sulfide ores	150–200 m	Cobar and Broken Hill (NSW), Gunpowder (Qld) and Tennant Creek (NT)	
Oxidation	350 m	Wingellina (SA)	
Oxidation and leaching of Cu-bearing carbonates	800 m	Mt Isa (Qld)	250 m bsl
Leaching and enrichment	250 m	Mt Goldsworthy (WA)	170 m bsl

Note: Localities are in Figure B5.3.

Erosion eventually starts to remove in-situ regolith materials, including the saprolite. Large blocks of unweathered rock (corestones) tend to be left behind, and make distinctive landscapes, such as granite tors (Figure 5.19a). At a larger scale, erosional remnants can be as big as the Uluru monolith (Figure 5.19b), which rises 348 m above the surrounding plain. In this dramatic landscape, it is interesting to wonder where the huge volume of eroded material was transported and redeposited.

Modern alluvium in southeast Australia is of mixed composition, and contains pebbles of basalt, shale, granite or whatever occurs in the catchment. Older alluvium is less diverse, and the oldest gravels, of Cretaceous to Eocene age, consist almost entirely of quartz. This is because of continent-wide deep weathering in the Mesozoic and early Cenozoic. At that time, there was no fresh rock exposed, and the only hard, unweatherable material available for bedload in rivers was quartz. Later saprolite was completely or partially stripped, exposing more and fresher rock. Over much of Australia where thick saprolite is still preserved, it may not be related to younger or present-day weathering. Some weathering profiles are related to palaeo-weathering, and may be preserved under younger sediments or basalt flows. In the case of old alluvium, some is buried under Cenozoic basalt as 'deep leads', including deposits of gold, diamonds, sapphires and other resistate minerals.

Soils and salt

The age and style of evolution of the Australian continent has meant that Australian soils are generally poor by world standards. They also formed very slowly and in most areas are very thin (Box 5.4). Even if there were enough water, most are not naturally fertile because plant nutrients have long since been removed by weathering. Australian agricultural soils, generally speaking, are deficient in a number of trace elements including copper (Cu), cobalt, manganese (Mn), zinc (Zn) selenium, iodine, molybdenum (Mo), boron, iron (Fe) and sulfur.

Australia's nutritionally poor soils mean that the Western Australian wheat belt, for example, has low average wheat yields (2 t/ha) compared with other wheat-producing areas of the world (e.g. United Kingdom at 7.78 t/ha, or China at

Figure 5.19: (a) Granite tors—the Devils Marbles of the Northern Territory. (b) Uluru (Ayers Rock), Northern Territory. Australia's iconic landscape feature near the centre of the continent.

HOW FAST DO SOILS FORM? (BOX 5.4)

Soil formation is broadly controlled by five factors: climate, organisms, parent material, topography and time. By studying soils on similar parent materials of different ages within a restricted geographical area, the first four factors can be held approximately constant and rates of soil formation can be calculated.

This has been done in north Queensland in the semi-arid region west of Townsville (*ca* 500 mm/year rainfall), where soil depths were measured on basalt lava flows ranging in age from 13 ka (essentially zero soil depth) to several million years, showing that net soil formation rates were *ca* 0.3 m/Ma. In comparison, soil formation rates on basalt lava flows in the humid region inland of Cairns (Figure 5.20) (4000 mm/year rainfall) are an order of magnitude higher.

Ultimately, soil formation represents the balance between the processes of soil creation (e.g. weathering, biological activity and aeolian additions) and erosion (e.g. surface erosion and leaching). In the case of soil formation on basalt lava flows, weathering of the fresh rock is a limiting factor. Soil formation is significantly faster on transported, pre-weathered parent materials such as alluvium, where soil depth is controlled much more by biological activity. Biological mixing (bioturbation) rates in soils in northern Australia, where termite activity predominates, have been measured in the range 0.02 mm/year to 0.4 mm/year, which is equivalent to the upper 1 m of soil being completely mixed in 50 kyr to 2500 years respectively.

Australia tends to have relatively thin and agriculturally poor soils. They are also fragile, making them prone to erosion and loss to fluvial and aeolian processes (as dust storms).

Other soils are old, leached and tired. A common phenomenon in Australia's wheat belt is that of sodic soils. These are old, clay-rich, often alluvial soils that have an excess of sodium (Na) bound to clay particles. These soils are known to be dispersive and hard setting, meaning that they slake (disperse) under rainfall or irrigation and are easily eroded, and set almost as hard as concrete when dry. They are also poorly wetting, meaning that they form a surface barrier to water penetration. These soils can also form a hard-setting barrier at the base of tillage, meaning that they can also become water saturated—the poor old farmers can't win. This means that sodic soils are difficult to maintain, are hard to till and are hard to crop. The most effective treatment for these soils is the addition of large amounts of gypsum to displace the Na ions with calcium ions, causing the soils to *flocculate*, forming *flocs*, or sand-sized clay aggregates, which better allow water penetration and improve soil structure.

Another problem with old, tired, clay-rich soils is that of excess tilling causing soil erosion, much like that which occurred in the dust bowl of the American central west in the early 20th century. Australia is a world leader in sustainable agricultural practices involving tracked vehicles (rather than wheels, to reduce soil compaction), minimum tillage (directly seeding using minimum till and seed-drill technologies, rather than turning over the entire topsoil), zero tillage (using seed-spreading technology) and stubble retention (not burning or ploughing-in stubble from previous crops). New technologies also allow the direct application of seed, fertilisers and pesticides by using satellite guidance to avoid over- or under-distribution and all help to retain soil structure and soil fertility as well as inhibit soil erosion by wind and water.

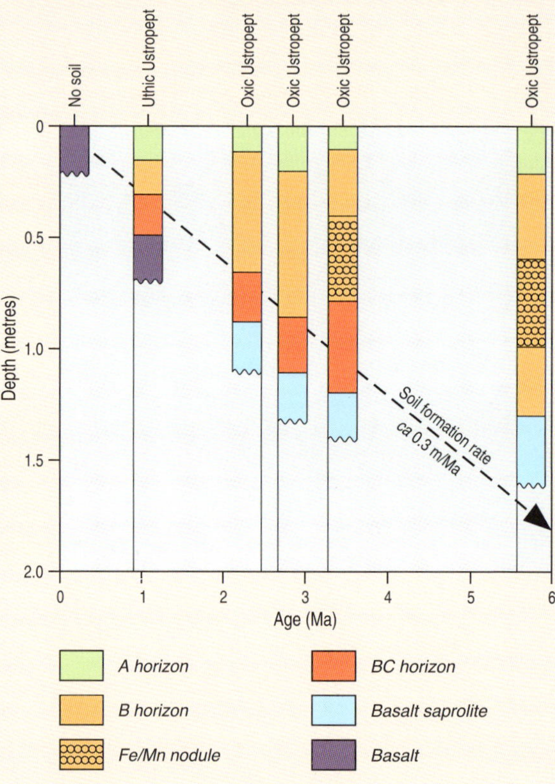

Figure B5.4: In the semi-arid region west of Townsville, Queensland, soil forms at an average rate of 0.3 m/Ma. Calculations based on soil depths developed on basalts of varying age. (Source: Pillans, 1997)

3.93 t/ha: 2003–04 figures). Even floodplains, naturally fertile in other parts of the world, tend not to be so in Australia because the primary sources of the alluvium are nutrient poor themselves. Another consequence of the Australian biophysical environment is that soils have low organic-matter contents. This is certainly the case in areas formerly covered with open woodland, and even the small areas of rainforest in northern Australia have much lower organic-matter content in the soil than elsewhere on Earth. The only naturally fertile soils tend to have formed on younger volcanic rocks, and this is why basalt soils of the Newer Volcanic Group in Victoria (Chapter 2) were sought after and settled on (Figure 5.20). European settlers in Australia thus had much less advantage in terms of soils than they did in other places, such as North America.

European settlement brought a dramatic increase in the human impact on Australia's soils. For example, many areas in the Sydney Basin have lost at least 50 cm of topsoil in the past 200 years. This is a large loss for an original soil that is only 1 m thick (Box 5.4). The spread of cropping and grazing using European techniques led to a phenomenon known as post-settlement alluvium. Most streams and rivers in the agricultural parts of Australia have a former floodplain surface covered with up to 1 m of modern alluvium that can be dated to the period following settlement and development of their catchments. The post-settlement alluvium often has European artefacts such as glass bottles and fence wire buried within it. In the Dundas Tablelands of Victoria, this is particularly marked, with bodies of alluvial sediment still making their way down rivers such as the Glenelg (Figure 5.21).

Figure 5.20: Soils formed on Cenozoic basalt of the Atherton Tablelands, north Queensland. These are highly productive soils suitable for agriculture, including dairying and tea-growing.

Australia is a very salty continent. Salt is continuously added to Australian agricultural soils from salt-laden rains—for example, in Canberra, the average salt addition is about 30 g/ha/year. A change in vegetation from open woodland with deep-rooted trees to shallow-rooted crops and grasses (Figure 1.10) led to a rise in the level of the natural water-table. Rising water-tables resulted in a reduction in the removal of salt from the surface to depth in many landscapes in both western and eastern Australia. The build-up of salt at the surface is called dryland salinity.

Figure 5.21: A body of sediment from post-settlement erosion moving down the Glenelg River in Victoria.

Ironically, this process is most active in good years; droughts in the 2000s led to a decrease in dryland salinity. However, the same droughts led to the accumulation of salts on floodplains such as that of the Murray River and, with a return to higher rainfall and river flows, this salt may well find its way into the Murray River and thence into Adelaide's water supply (Chapter 7).

Australia's salinity is exacerbated by the low-gradient rivers and internal drainage that are so characteristic of flat landscapes. The rivers are ineffective in removing salt from the continent, and thus there tends to be an accumulation of salt as well as sediments. This is compounded by the fact that the salt is redistributed back to the eastern highlands as aeolian dust carried by winds from the west. One of the perceptions about inland Australia is that it should be capable of supporting a flourishing agricultural industry if only there was enough water. There is water along the coastal strip and also at depth in the interior's groundwater (Chapter 7). This desire has led to many schemes to turn coastal rivers inland, and 'green the desert'. However, with the exception of the Snowy Mountains Hydroelectric Scheme, these plans have largely failed.

Rivers have, nevertheless, been tamed in other ways. The Murray River is one of the most regulated in the world, with four storage dams and 26 locks, weirs or barrages regulating river flow. Another example is the Balonne River in southern Queensland (Figure 5.1), where large amounts of water are diverted into storage in order to irrigate cotton. All these schemes are controversial, and all reflect a desire to try to overcome the geological impact of the long-term evolution of the Australian continent. Griffith Taylor (foundation Professor of Geography, Sydney University) famously concluded that a large part of central Australia was 'almost useless'.

How old is the Australian landscape?

Australia is often described as an ancient continent, a description that applies to many of its rocks and also its landscapes. The landscapes of today, however, are being shaped by modern weathering, erosional and depositional processes, so when we refer to ancient landscapes we really mean that they have ancient origins. In many parts of Australia, the relative tectonic stability, low relief, lack of Quaternary glaciation, low rainfall and resistant rocks are the combined factors that contribute to low erosion rates. As such, Australia's modern landscapes preserve much of their past history and origin, more so than in many other parts of the world.

The oldest known land surface and the oldest evidence for geomorphological (landscape forming) processes in the world is exposed near Strelley Pool in the Pilbara. At this location, there is angular unconformity between the older Panorama Formation and the Strelley Pool Formation. The unconformity represents a hiatus in volcanic activity between *ca* 3427 Ma and 3350 Ma. The unconformity cuts across volcanic and hydrothermal rocks of the Panorama Formation (Figure 2.22b). These rocks probably formed part of a rocky coastline, with the basal sands and gravels of the Strelley Pool Formation deposited later on a gravel beach. The beach deposit is then succeeded by organic stromatolitic carbonates, laid down by a transgressive sea.

Although the Strelley Pool exposure provides a fascinating insight into the nature of an Archean landscape, elements of that landscape have not persisted into the landscape of modern Australia. Palaeogeographic reconstructions indicate that parts of the Australian continent have been subaerially exposed for hundreds of millions of years (Table 5.3), and some areas in the west and north may have been exposed since the Precambrian, more than 500 Myr. So, are there relict landforms that have survived for hundreds of millions of years in these long-exposed landscapes? The answer is 'yes', as the following examples illustrate:

- The Kimberley Plateau in northwest Western Australia is an erosion surface cut across gently dipping sandstones. On softer rocks, valleys were incised below the main plateau level. The Precambrian Sturtian Glaciation occurred around 650 Ma, after the main plateau had been created, leaving telltale glacial scour marks on the sandstones and patches of glacial till in the valleys. Possibly, no Phanerozoic sediments ever covered the plateau, and the present landform may be traced to the Precambrian. However, apatite fission-track thermochronology suggests that this landform was once buried and subsequently exhumed (Box 5.5).
- The Jenolan Caves, near Sydney, contain volcanic sediments that were washed in by the sea during the Carboniferous Period, about 345 Ma, indicating that the caves were 'open' at that time.
- Relicts of Permian glaciation are widespread in every state of Australia, including well-preserved glacially striated pavements, glacial erratics, diamictites and ice-rafted dropstones. At Hallett Cove, a popular beach and geoheritage site in the southern suburbs of Adelaide, large Permian glacial erratics sit on the modern beach-face.
- Deeply weathered saprolite of Late Paleozoic and Mesozoic age has been identified in many parts of Australia, based on palaeomagnetic dating of oxidised samples and oxygen isotope analyses of kaolinite clays. Late Cretaceous to Oligocene sediments in the Gippsland Basin are dominated by quartz and kaolinite, consistent with stripping of a deeply weathered regolith from the adjacent granitic terrain. Higher atmospheric CO_2 levels, combined with higher rainfall and rainforest vegetation, may have promoted the deep weathering of deep granitic saprolites, which were then eroded to provide the sedimentary detritus.
- Uluru is probably the most famous landform in Australia, but its history is tricky to determine. This large erosional feature is composed of steeply dipping arkosic sandstone of Cambrian age and stands some 348 m above the surrounding plain. The rocks were tilted, folded and faulted during a succession of contractional events associated with the Alice Springs Orogeny (450–320 Ma), and the area has probably remained above sea-level since that time. Late Cretaceous pollen has been recovered from beneath the surrounding plain of Uluru, indicating that the monolith was high ground by that time (*ca* 70 Ma).

Did you know?

5.2: The Pilbara landscape is out of this world!

NASA and the Mars Society of Australia view the remote Pilbara of Western Australia as a possible analogue for the Martian landscape. The Pilbara's harsh, arid environment, stony outcrops with their Fe-rich patina, and fossil evidence for very early life (Chapter 3) are part of the attraction for budding astronauts to test their equipment for future missions to Mars.

Image courtesy of David Willson, Mars Society Australia

Testing models of potential Mars suits in Australia.

- The 'gibber' or stony deserts of central Australia are Australia's oldest arid landforms (Figure 5.4). They are characterised by a surface layer of siliceous pebble- to cobble-sized clasts (as the actual gibbers). Cosmogenic nuclide dating shows that gibber pavements west of Lake Eyre were formed between 2 Ma and 4 Ma. In contrast, other characteristic arid landforms, such as sand dunes and salt lakes, may be no older than about 1 Ma.
- Unlike Northern Hemisphere continents, Pleistocene glaciation was of limited extent in Australia, confined to a few square kilometres on and around Mt Kosciuszko, and more extensive in central and western Tasmania. Lake St Clair in Tasmania (Figure 5.1), which was scoured by ice during the Last Glaciation about 20 ka, is Australia's deepest lake (*ca* 190 m). The lack of extensive Pleistocene glaciation is one reason why the Australian regolith is so deep compared with the Northern Hemisphere, where large ice sheets scraped off and mixed much of the pre-Pleistocene regolith.

The ages of some well-known Australian landforms are collated in Table 5.3.

Table 5.3: Ages of some well-known Australian landforms

Landform	Bedrock age	Landform age	Dating method
Jenolan Caves	Jenolan limestone (Upper Silurian)	Carboniferous (340 Ma)	K/Ar Fission track
Glacial striations & erratics, Hallett Cove, Adelaide	Proterozoic	Permian	Biostratigraphy Regional geology
Gosses Bluff	Cambrian–Devonian	143 Ma	Ar/Ar
Great Escarpment	Paleozoic	Cretaceous (80–90 Ma)	Seafloor magnetic anomalies
Mt Kosciuszko	Paleozoic	Late Mesozoic?	Fission track Marginal basin history
Uluru	Cambrian sandstone	>70 Ma	Pollen
The Breakaways	Cretaceous	Cenozoic	Pollen Palaeomagnetism
Nullarbor Plain	Cenozoic limestone	Eocene	Marine fossils
Bungle Bungle Range	Devonian	Miocene?	Regional geology
The Three Sisters	Triassic sandstone	<15 Ma	K/Ar Regional geology
Lord Howe Island	Miocene	Miocene	K/Ar
Lake George	Ordovician–Silurian	Late Miocene–Pliocene	Pollen Palaeomagnetism
Simpson Desert gibbers	Cenozoic	Pliocene (2–4 Ma)	Cosmogenic nuclides
Simpson Desert linear dunes	Cenozoic	Pleistocene (1 Ma)	Luminescence Cosmogenic nuclides
Lake Eyre	Cenozoic	Pleistocene	Radiocarbon Luminescence
Lake Mungo	Pleistocene	Pleistocene	Radiocarbon Luminescence
Lake St Clair	Paleozoic–Mesozoic	Pleistocene	Radiocarbon Cosmogenic nuclides
Undara lava tubes	Pleistocene	Pleistocene (0.19 Ma)	K/Ar
Great Barrier Reef	Pleistocene	Pleistocene–Holocene	Sr isotopes Palaeomagnetism Radiocarbon
Mt Gambier	Pleistocene	Pleistocene–Holocene	Radiocarbon Luminescence
Twelve Apostles	Miocene limestone	Holocene	Global sea-level curve
Bondi Beach	Triassic	Holocene	Global sea-level curve

DATING LANDSCAPE EVOLUTION (BOX 5.5)

A large array of modern dating methods underpin our knowledge of landscape evolution in Australia, ranging from methods such as radiocarbon, which measures ages up to 60 kyr, to methods such as Ar/Ar that can measure ages up to and beyond the age of Earth (Figure B5.5). Each method uses different materials for dating, so no one method is applicable in all situations. Furthermore, many landforms are erosional features, so how do we date what is missing? Fortunately, no landscape is completely erosional, and there will always be materials and features that can be dated to at least partially reconstruct the landscape history.

Interestingly, two methods can actually be used to directly measure erosion rates—cosmogenic nuclides and apatite fission-track thermochronology (AFT). The former relies on the balance between nuclide production and erosional loss in surface rocks, while the latter relies on the relationship between fission-track production (from ^{238}U fission-age) and thermal annealing of tracks with increasing depth (temperature).

Several independent dating methods have been used to demonstrate that ancient (pre-Cenozoic) regolith and landforms are preserved in modern landscapes around Australia. These include palaeomagnetism (iron oxides), oxygen isotopes (clays), Ar–Ar (K-bearing Mn oxides) and biostratigraphy (pollen- and spore-bearing sediments). Other methods, such as (U–Th)/He and U–Pb, also show promise for dating pre-Cenozoic regolith, but have yet to be applied to suitable materials.

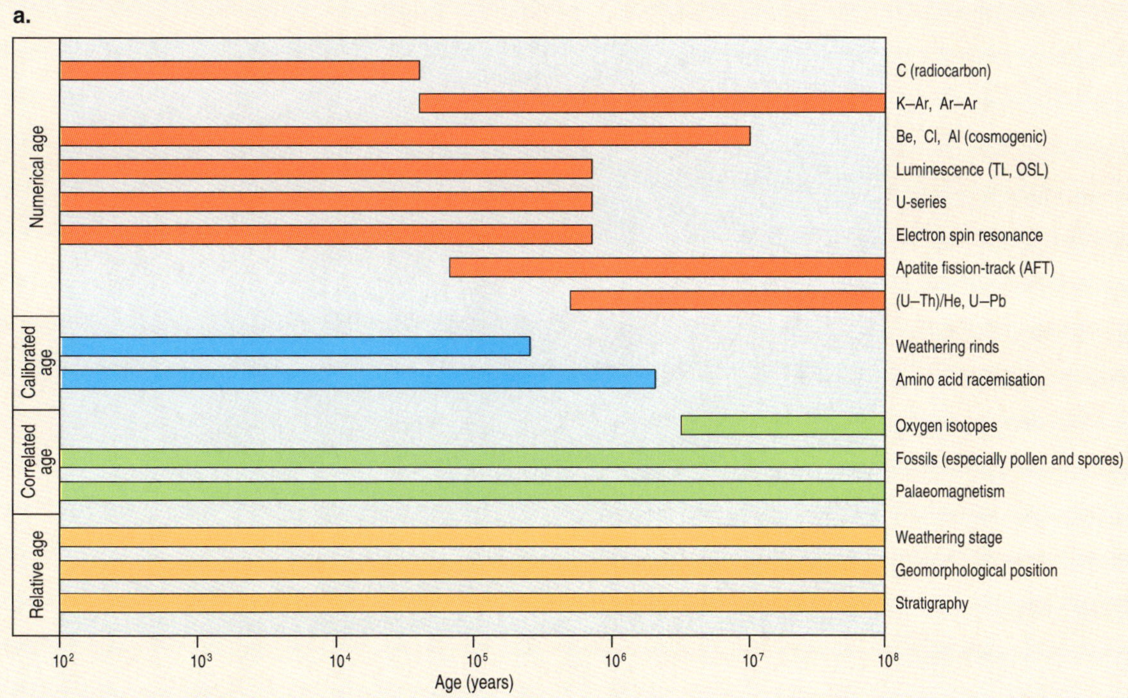

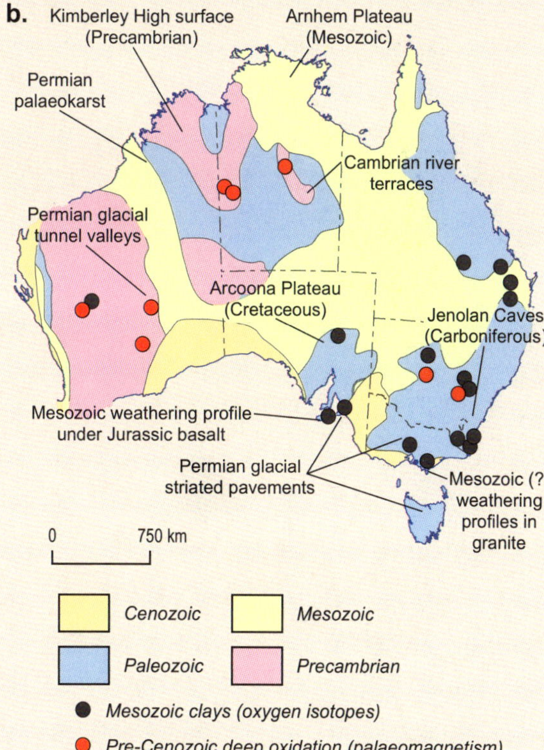

Figure B5.5: (a) Various techniques are available for dating regolith materials and landscapes. The techniques vary in their range, from hundreds to thousands of years to hundreds of millions of years. (b) Maximum duration of continuous subaerial exposure. (Source: Scott & Pain, 2008)

OSL = optically stimulated luminescence ; TL = thermoluminescence

Figure 5.22: At Porcupine Gorge, near Hughenden in north Queensland, Galah Creek has incised some 40 m since basaltic lava (K/Ar dated at 0.89 Ma) flowed down the palaeovalley floor of the creek. However, the incision is localised, and denudation rates on the surrounding plateau are much lower.

Denudation measures the overall lowering of landscapes by all erosional processes and should not be confused with erosion measured at specific points or lines in the landscape (e.g. scarps or river valleys). Long-term denudation rates can be derived from measuring the sediment volumes (and ages) in basins within or adjacent to the landscape. In eastern Australia, for example, Lake George and the Murray Basin both yield late Cenozoic denudation rates of around 3 m/Ma. More recently, denudation rates have been calculated using thermochronology, particularly apatite fission-track thermochronology (Box 5.5), which provides estimates of denudation over large areas, typically hundreds of square kilometres. At Porcupine Gorge in north Queensland, there is an example of the contrast between overall landscape denudation and erosion at specific points (Figure 5.22). The river at Porcupine Gorge has incised *ca* 40 m in the last 0.89 Ma, which is an average rate *ca* 44 m/Ma. The rate of denudation on the surrounding plateau, in contrast, is one to two orders of magnitude lower, based on cosmogenic nuclide measurements at several specific points. Thus, denudation rates and site-specific erosion rates are not strictly comparable—they measure different things.

For many geomorphologists, the survival of old landscape features is difficult to reconcile with results from techniques such as apatite fission-track thermochronology, which provide total denudation estimates of up to 4 km since the Permian over large areas of the Australian continent (Figure 5.23). The survival of pre-Cenozoic regolith and landforms in the modern landscape argues against denudation on the scale suggested. How could ancient regolith and landforms survive in the face of such massive denudation? There are probably two reasons: firstly, denudation is spatially and temporally variable and, secondly, many ancient landscape features may have been formed, buried and then re-exposed at the surface. For example, apatite fission-track thermochronology results from Yilgarn Craton suggest up to 4 km total denudation since the Permian, a figure that is broadly supported by vitrinite reflectance data from the infaulted Collie Basin indicating burial temperatures of up to *ca* 100°C. The Collie Basin, which contains some 1400 m of Permian sediment, including tillite and coal measures, may therefore represent a remnant of a much more widespread but largely eroded sedimentary cover across the Yilgarn Craton. The large volume of eroded material was presumably deposited in the adjacent sedimentary basins, which contain the calculated sediment volumes. Thus, three independent lines of evidence suggest that the survival of glacial landforms and deposits of Permian age in parts of the Yilgarn Craton can be explained by kilometre-scale burial and later exhumation.

Up till now, we have mostly been describing the landscape and regolith in terms that are not particularly favourable to sustaining civilisation. Using these measures, a lot of Australia could be considered 'almost useless'. However, it is not so bad; in fact, the regolith enriches Australia in many ways …

How regolith enriches Australia

Resources from the regolith have enriched the lives of Australians. Australia no longer rides on the sheep's back (also based on a regolith resource, when you think about it) but is now reliant on bulk commodities to fill out its balance sheet. Nevertheless, a glass of shiraz, grown in Australian regolith, is a lot more palatable than an orebody (Box 5.6).

The bulk commodities represent a substantial proportion of Australia's exports—more than 42% in 2008–09—and many of those are actually obtained fully or partially from the regolith (Chapter 9). They include bauxite, alumina and aluminium; ilmenite, rutile, leucoxene and zircon; iron ore and steel; Mn ore; nickel (Ni) laterite ore; magnesite; and, to some extent, gold (Au) and uranium (U). Coal could also be considered a palaeo-regolith resource. In fact, many ore deposits are affected by, or have been upgraded by, regolith processes in some way, through weathering, supergene enrichment, sorting and winnowing, but let's not stretch the analogy too far.

Since the first people occupied Australia more than 50 ka, regolith materials have been the focus of mining activities. Aboriginal people mined ochre and clay deposits for artistic and ceremonial purposes, and rock outcrops for materials to manufacture cutting, pounding and grinding tools. Mining is known to have occurred at hundreds of localities, but some of the most notable operations were at Koonalda in South Australia, and at Wilgie Mai, southeast of Geraldton in Western

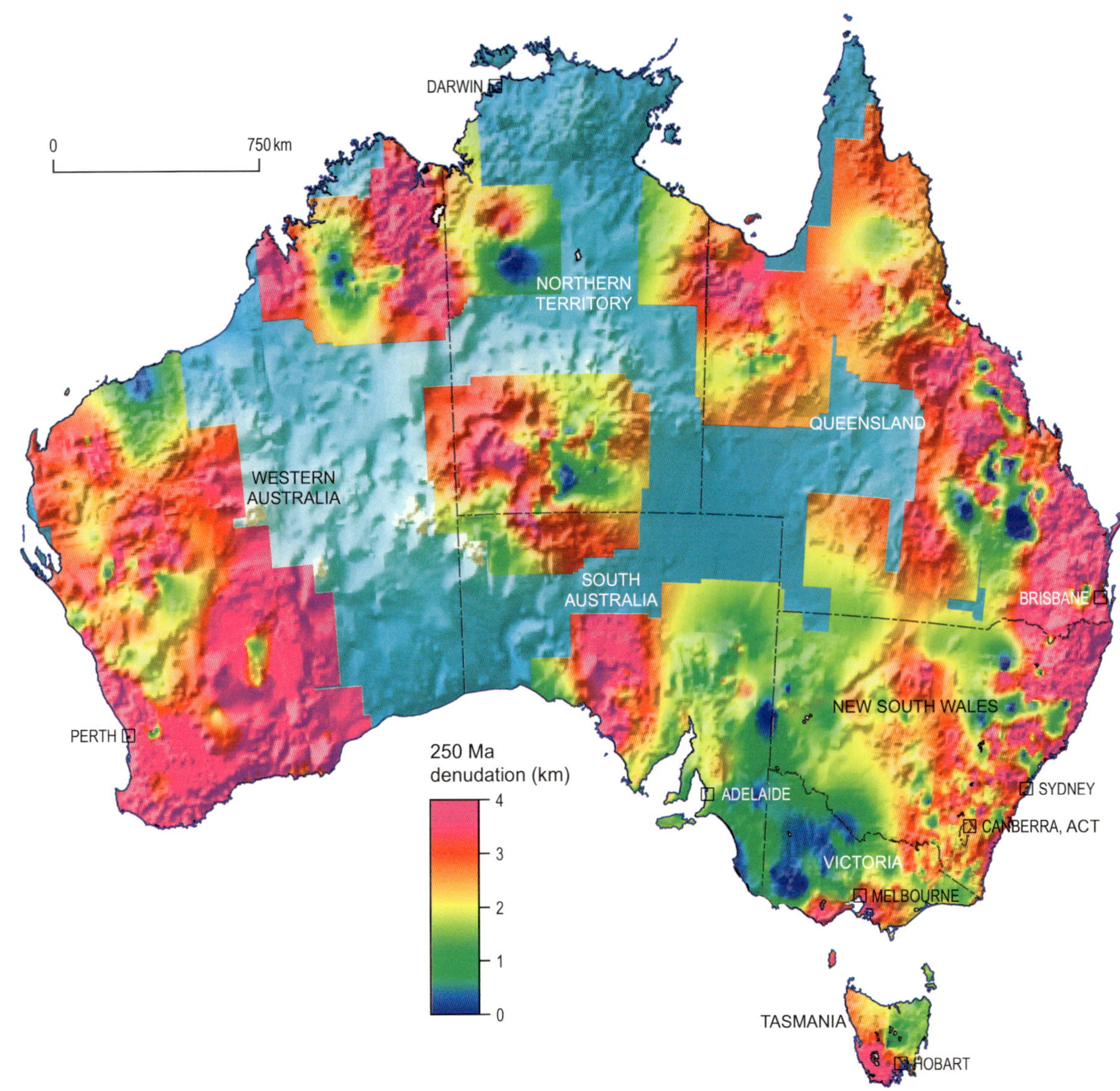

Figure 5.23: Denudation of Australia over the past 250 Myr, based on apatite fission-track thermochronology. Blue regions are areas of no data. These data suggest that large areas of the Yilgarn Craton in southwest Western Australia have been eroded by around 4 km. (Source: Kohn et al., 2002)

FLAVOURED WATER IN A FRUIT BAG (BOX 5.6)

Grapes are small bags of sweet-and-sour water. The water comes through the roots of the vine, whereas the flavour is influenced by the mineral and organic composition of the soil (regolith). *Terroir* is the French term for earth, but to a viticulturalist it means a combination of the vine variety, chemical and physical attributes of the soil and rock (regolith), slope, aspect, drainage and microclimate. Differing *terroir* leads to differing wine characteristics.

The world of wine is divided into Old World (traditional Europe) and New World (the rest). One characteristic of New World wines is that most grapes are grown in hotter climates than the more temperate Old World, which means shorter hang times. In the New World, grapes have longer growing seasons and tend to be riper, creating wines that are fuller and have more body and higher alcohol content. Australian wines are New World. They were once dismissed by traditionalists, but for the past few decades have been winning many top international wine competitions.

Australia's primary wine grapes are shiraz, chardonnay, and cabernet sauvignon. Australia sustains more than 1100 wineries in 26 different wine regions. Being such a large country with almost every climate and soil type, Australia is one of the few wine producers to make every one of the major wine styles. Wine exports amounted to more than $2.172 B in 2009–10.

One of the leading wine-growing areas is the McLaren Vale, located around 30 km south of Adelaide in South Australia. This wine region comprises two triangular-shaped basins, with the vines growing in seven distinct 'terranes' that range in bedrock age from the *ca* 750 Ma Burra Group to Holocene sands. The region boasts high-class full-bodied red wines. To the north of Adelaide, the Clare Valley's best shiraz and cabernet wines are grown from old vines on thick red-coloured terra rossa loams overlying dolomitic siltstone and slate.

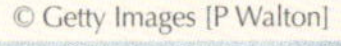

© Getty Images [P Walton]

The Mediterranean climate of the McLaren Vale, South Australia, supports some of Australia's finest vineyards.

Australia (Figure 5.1). At the Koonalda flint mine, a 300 m-long shaft was driven to a depth of 75 m from the surface. At the Wilgie Mia ochre mine, a mining face up to 30 m wide and 15 m deep was worked with the aid of heavy stone mallets, fire-hardened wooden wedges and wooden pole scaffolding to access the working face. Wilgie Mia may well be the longest continuously mined site in the world. Flint from Koonalda was traded throughout central Australia. Ochre from Wilgie Mia mine was particularly valuable, and may have been traded all the way from Western Australia to Queensland. Ochre from the Yarrakina mine at Parachilna, in the northern Flinders Ranges, was valued for its softness and lustrous qualities. Blocks were traded into the Northern Territory, western Queensland and eastern New South Wales, and are known to have been exchanged for shields and spear shafts.

Mining the regolith continued after the arrival of the First Fleet in January 1788. In March 1788, brick making began at what is now Darling Harbour in Sydney. By May 1788, the site had produced between 20 000 and 30 000 bricks, which were used to construct buildings, including the first Government House in 1789 and buildings in the Rocks area. The early brick-makers did not have access to a good source of lime to make their mortar, so oyster shells, probably derived from nearby Aboriginal middens, were used to make the mortar for the construction of Government House (Chapter 6).

In addition to the production of bricks and tiles, the clay at the brickfields was used to make pottery, including the then-ubiquitous terracotta pipe, commencing in 1790. As the settlement at Sydney Cove grew, brick-making activity expanded. The surrounding landscape was deforested, as trees were felled to fuel the kilns, and pock-marked by clay pits. The excavation of clay pits and brick-making continued on the perimeter of the settlement and migrated outwards as the settlement expanded; such pits produced some of the first dateable fossils (Chapter 3). The early settlers also experimented with glass making, using sand derived from the white dunes along what became South Head Road.

Figure 5.24: Spoil heaps from opal mining at Coober Pedy in South Australia. The harsh climate has meant that many residents of Coober Pedy live underground in the old workings.

As Australia developed economically, road-making activities expanded, and construction techniques began to change. The difficulty in obtaining and working with quality aggregate (hard, fresh rock) resulted in many local governments resorting to paving highways and roads with gravel and clay sealer; this is still a common practice. Clay-paved roads reduce dust and are self sealing when damp,

OPAL—AUSTRALIA'S NATIONAL GEMSTONE (BOX 5.7)

Opal is Australia's national gemstone—more than 95% of the world's opal comes from Australia, almost all of it from the southern margins of the Great Artesian Basin (Figure B5.7). Opal is chemically similar to quartz, but contains up to 20% water. The characteristic display of opal colours is produced by the diffraction of light by a regular array of silica spheres that make up the opal. The background colour may be white, grey, blue, black or colourless. A dark background produces 'black' opal, which is the most colourful and valuable type of opal, almost all of which comes from Lightning Ridge in northern New South Wales. Unlike most other mining sectors in Australia, opal mining is almost entirely done by individuals and family groups, rather than mining companies.

Some overseas opal is volcanic in origin and tends to have abundant inclusions of water, which make the gem brittle. Australian opal, in contrast, was deposited as a replacement mineral from silica-rich groundwater within 25 m of the surface, particularly in kaolinitic weathering profiles. Opal forms close to groundwater permeability barriers, such as within faults, cracks and joints, or replacing soluble minerals or organic materials (e.g. bone, wood or fossil). The famous pliosaur skeleton called 'Eric', from Lightning Ridge, is one such example. Flow structures in some opals indicate an initial aqueous phase, followed by dehydration of a silica gel. The discovery of heritage-listed Lark Quarry dinosaur footprints near Winton in outback Queensland (Chapter 3) was made by opal miners; there the Cretaceous plant rootlets are also opalised.

Opal has been regarded as a stone of good fortune, and it was believed that the gem possessed magical properties. Opal fell out of favour during the late 18th and 19th centuries, as it was associated with pestilence, famine and the fall of monarchies. The type, colour, size and soundness of precious opal are factors that determine the price, which is expressed per carat.

Opal is hosted in Cenozoic weathering profiles; it has been dated at *ca* 16 Ma (Miocene) in the Coober Pedy deposits, with several younger opal-forming events. Some opal is very young. Radiocarbon dating of organic material incorporated in the matrix of some black opal from Lightning Ridge has yielded Holocene ages, while partly opalised fence posts have been reported in some places.

The 'fire' in opal also features in Aboriginal Dreamtime stories. The Wangkumara people from around Cooper Creek believe that they were brought the gift of fire from opal. The story goes that the tribe sent a pelican away from their camp to find out about the country up north. The pelican felt ill and landed on a hill. The pelican discovered beautiful opal and started chipping it away to bring back to the people. A spark from the chipping caused the nearby dry grass to catch on fire, which spread slowly back to the camp. This fire was captured by the people to cook their meat (Connolly, 2012).

Opal miners are often 'rugged individuals', some of whom took up a claim in order to strike it rich, but could never leave their new homes because they'd sold all their possessions to cover their 'grubstake'. Other claims are owned and run by families and are handed down through generations; these claims are treated more as a holiday retreat and are only worked during school holidays to earn that 'little bit extra'. At White Cliffs (NSW) and Coober Pedy (SA), many of the citizens live entirely underground. Homes and entire hotels are constructed underground in disused opal mines to provide year-round comfort, escaping the searing summer temperatures in these remote communities. Australia's opal miners are modern-day troglodytes!

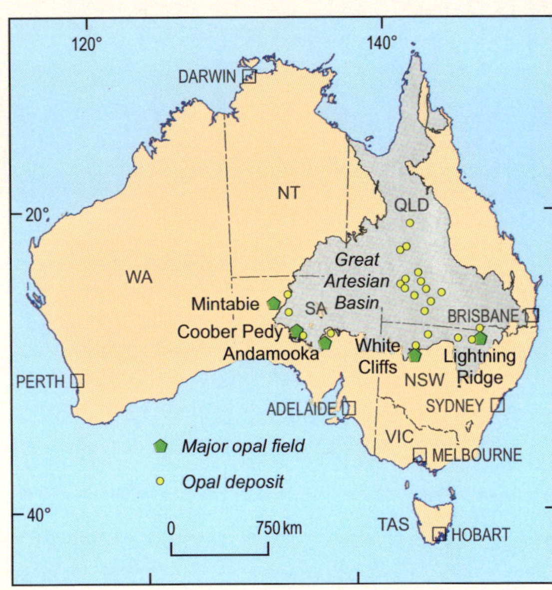

Figure B5.7: Location of the opal fields in Australia; outline of the Great Artesian Basin is also shown (Chapter 7).

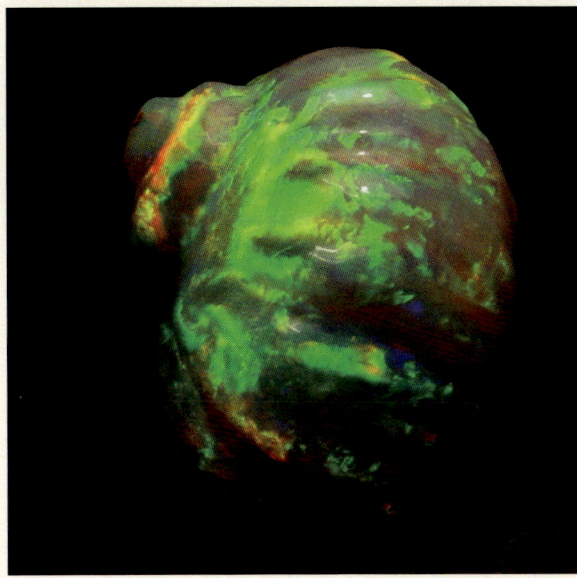

Image courtesy of DMITRE

Opalised snail shell.

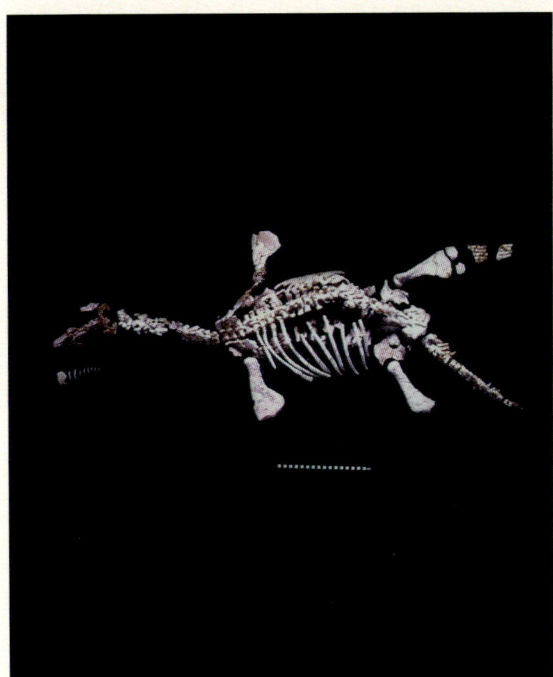

Image courtesy of Australian Museum

Eric the pliosaur, an opalised skeleton on display at the Australian Museum, Sydney.

© Getty Images [O Gerhard]

Noodling heaps at an opal mine in Coober Pedy, South Australia.

Figure 5.25: Junction Mine gossan, Broken Hill, New South Wales.

but become slippery when wet. It is very common for clay-paved roads in Australia to be closed for days during and after rain, and there are heavy fines for deliberately driving on closed roads.

The fines for driving on closed dirt roads are $1250 in South Australia, $13 300 in the Northern Territory and $1000 per wheel per kilometre in New South Wales; such is the value of these regolith resources!

Many indurated regolith materials (calcrete, silcrete, ferricrete and ferruginous nodules) were found to be suitable for use as road base, although their compressive strength is not always completely satisfactory. However, some regolith materials are not ideal; silcrete, for example, although hard, is also brittle and polishes very easily, resulting in a road surface that becomes dangerously slippery. The use of calcareous nodules as road base is very common in South Australia, and the use of pea-sized ferruginous nodules is very common in Western Australia, a direct reflection of the dominance and easy availability of those materials in the regolith of each state.

Silcrete and ferricrete have even been used as a building material for houses. Blocks and slabs were used by early settlers to make dwellings in South Australia and Victoria. In Western Australia, calcarenite, formed from cemented marine shells, was sawn and used to construct 'shell block' stone buildings.

Europeans very quickly worked out that regolith was useful not only for construction materials. Mineral prospecting in the new colony was given impetus by the discovery of alluvial gold—a regolith deposit—at Ophir in New South Wales (Chapter 8). This and other Au discoveries led to Australia's first population boom, but prospectors with high hopes were often defeated by the harsh realities of the Australian landscape, especially in the west where the long distances and lack of water, both for drinking and processing, proved to be particularly challenging. The problem still remains today (Chapter 7). Populations in the goldfields in eastern Australia declined as the relatively accessible deep leads were worked out and mining went further underground. Mining operations, including extractive technology, became more complex and more expensive. Towns like Coober Pedy (SA) and Lightning Ridge (NSW), where opal—another regolith deposit—is mined, are, however, still a haven for the rugged individualist (Figure 5.24; Box 5.7).

Regolith processes play an important role in upgrading subeconomic primary (in-situ) mineral deposits, or creating secondary mineral deposits, such as clay, Au, diamonds, tin, alumina (bauxite), Ni laterite, Mg (magnesite), iron ore (channel-iron deposits) and U. Many of the bulk commodity deposits mined today may not have been economic without the aid of secondary enrichment by regolith processes. The regolith-governed mineral systems differ from the hydrothermal mineral systems (Chapter 8) in their pressure and temperature, but the essential elements of the two types of system are much the same (Figure 8.4).

In-situ deposits may be enhanced by weathering processes, and it is not uncommon for gold miners to take a rich supergene oxide Au or Cu resource for a quick return on investment, but leave sulfides in the ground for later (if ever), because of the extra expense of treating the often-difficult sulfides. Examples come from all over Australia, but particularly the Yilgarn Craton, where supergene Au in Eocene palaeovalleys has helped many a company out of a tight squeeze when funds were short. Similarly, the giant Broken Hill Pb–Zn–Ag deposit had a thick gossan of secondarily enriched oxide, hydroxide and carbonate ore, some of which still exists at Browns Shaft, above the massive sulfide lodes (Figure 5.25). Broken Hill was often referred to as 'the hill of mullock' by early explorers, who misunderstood the value of the gossan. The gossan was quickly mined and smelted to obtain a fortune in Ag before the more difficult primary ores could be treated with metallurgy using froth flotation, the Australian invention developed at Broken Hill (Chapter 1). (See Figures 2.21a and 2.21b for localities of these deposits.)

Figure 5.26: Gove, Northern Territory, hosts one of Australia's main bauxite deposits.

Regolith processes are responsible for the in-situ upgrading of bauxite deposits at Gove (NT) (Figure 5.26) and the Darling Ranges (WA), but also the transport and deposition of bauxite pisoliths at Weipa (Qld). Because of this, Australia has the world's second largest economic demonstrated resources of bauxite, and the world's largest production levels (Chapter 9). Similarly, Pilbara iron ore was enriched in situ to form the direct-to-ship microplaty hematite ore of the Hamersley Basin. High-quality iron ore was formed by sedimentary and groundwater processes such as the channel-iron deposits at Pannawonnica and Robe River (Figure 5.1).

Australia also has the world's largest economic demonstrated resources and production levels of zircon, rutile and a number of other minerals that are found in heavy-mineral sand deposits, which formed on modern and fossil beaches as a consequence of coastal processes. Heavy-mineral sand deposits have long been known and exploited in Holocene coastal dunes and barrier islands in Queensland, New South Wales and Western Australia. An improved understanding of mineral sources and depositional palaeo-environments has resulted in many new finds in the Eucla and Murray–Darling basins (Figure 5.10). New understanding of palaeo-longshore drift processes in the Eucla Basin—a Cenozoic coastal barrier system—has resulted in the discovery of numerous rich heavy-mineral sand deposits, including Jacinth and Ambrosia, which alone are estimated to contain a combined 4.5 Mt of zircon, 2.6 Mt of ilmenite and 46 kt of rutile. Similarly, new models of coastal structure and sedimentation in the Loxton–Parilla Sands of the Murray Basin, developed since the 1980s, have brought to light many new heavy-mineral sand deposits; the basin had previously been thought to contain only large, sub-economic fine-grained deposits in Victoria. Advances in remote-sensing technology, particularly night-time thermal satellite imagery and detailed airborne geophysics, have been instrumental in uncovering palaeo-environments and these new deposits.

Figure 5.27: In-situ recovery injection and extraction wells and associated pipework at the Pepegoona uranium deposit, near the Beverley uranium mine, South Australia.

Australia now has four working U mines, which in order of size, are Olympic Dam (SA), Ranger (NT), Beverley and Honeymoon (SA). Olympic Dam is the world's largest single U deposit, containing more than 2.557 Mt of U_3O_8 in a hard rock deposit also containing a fabulous wealth of Cu, Au and Ag—this is BHP Billiton's crown jewel (Chapter 8). While it would be nice to claim this as a regolith deposit, it is not, nor is Ranger (108 000 t of U_3O_8). However, the Beverley mine (Figure 5.27) (21 000 t of U_3O_8), and Honeymoon mine (2900 t of U_3O_8) both in the Frome Embayment (SA), are sandstone-hosted U deposits formed in Cenozoic palaeovalleys by the concentration of U from groundwater. Numerous similar deposits are now recognised around Beverley, including Four Mile East and Four Mile West (32 000 t of U_3O_8 combined), Beverley North–Pepegoona (4000 t of U_3O_8), and the Yadglin and Pannikin deposits. The U mineralogy is dominated by coffinite (U silicate) and uraninite (U dioxide).

The realisation that the flat landscape of the Frome Embayment hides Paleogene and Neogene palaeovalleys that contain reductants suitable for fixing U from groundwater has led to further discoveries at Goulds Dam and Oban; exploration continues apace. Numerous other similar deposits have been found—for example, Angela–Pamela, Bigrlyi and Walbiri (NT) and Manyingee, Oobagooma and Mulga Rock (WA). Models for depositing U within the palaeovalleys include the traditional roll-front style, but also the accumulation of organic material in point bars of the palaeorivers, which act as chemical reductants causing U deposition. No matter the model, U is fixed and concentrated by a fundamental chemical process in the regolith that involves the precipitation of U from an oxidised fluid on contact with a reductant, whether that is anoxic groundwater, sulfide, hydrocarbon or organic material (*Did you know?* 5.3). Change in redox is a common process for the deposition of metals in many mineral systems (Chapter 8).

Another important regolith-related U deposit style is that of the calcrete-hosted surficial deposit. U mineralogy in these types of deposit is dominated by carnotite (U-vanadate), and these deposits occur in areas where calcrete occurs in proximity to source rocks of U and vanadium (V). Typically, the granite–greenstone terranes of the Yilgarn Craton are favoured because the granites leach the U (Chapter 10), and the greenstones (metamorphosed basalts) leach V, on weathering. A number of deposits or prospects are recognised, mostly in Western Australia, including Yeelirrie (52 500 t of U_3O_8), the Wiluna area (Lake Way: 5714 t of U_3O_8, and Centipede: 5355 t of U_3O_8) and Lake Maitland. The deposition of U in these deposits is not dependent on redox, but instead relies on a change in pH and an increase in Ca, K and V ion concentration in groundwater by evaporation. This occurs in juxtaposition with calcrete formation in the bottoms of long, wide drainage systems, resulting in carnotite precipitation within the calcrete bodies. Extraction is even easier as the carnotite can be dissolved using heap-leach techniques.

While occasional surprise gossans are still found sticking out of the ground, prospectors appear to have thoroughly worked over most of the rocky

Did you know?

5.3: Uranium deposition is reversible.

The element U is relatively cheaply and easily extracted using the in-situ recovery process, whereby the chemical process that precipitated the U minerals in the first place is reversed and the minerals are dissolved by re-oxidising the groundwater. This is achieved by using a weak acid–peroxide solution injected into the ground through a 4-, 5- or 6-injection well pattern and recovered through a central extraction well. Beverley was Australia's first in-situ recovery mine, with Honeymoon following.

Injection and extraction pipes at Beverley uranium mine, South Australia.

Figure 5.28: The Marnpi calcrete profile at Broken Hill, New South Wales. This is a nodular pedogenic (soil-formed) calcrete that can be used for mineral exploration.

terrain in Australia during the past 200 years, and most (if not all) of the easy-to-find mineral deposits have been discovered. Mineral exploration companies are now being forced to prospect in regolith-dominated terrains, where regolith blankets cover ore deposits at depths of less than 1 m to several kilometres around basin margins. If mineral exploration in regolith-dominated terrains is to be effective, then explorers need to develop a better understanding of regolith and landscape-forming processes (Chapter 11).

Seeing through the regolith

The greatest challenge facing mineral explorers in the 21st century is being able to 'see through' regolith. The work of the Cooperative Research Centre for Landscape Environments and Mineral Exploration (CRC LEME) showed that there are many regolith geological, geochemical and biogeochemical techniques that can be used to understand landscape evolution and 'see through' regolith (either by concentrating weak geochemical signals within the regolith, or by penetrating it) to discover what lies underneath. Some of the methods are well known but need to be applied to Australian conditions; some are anecdotal and need to be proven; and some have been recently developed. Here are a few examples.

Regolith-landform mapping

As geological mapping tells us about what is under the ground, regolith-landform mapping tells us about what is going on at Earth's surface. Regolith-landform mapping identifies regolith materials and the landscape in which they occur in a holistic approach, using geological, geomorphological, geophysical and botanical information to construct maps that describe the sources and sinks of regolith materials. It is now regarded as an important first step in mineral exploration, as important as accurate geological or structural mapping. Depending on scale, regolith-landform mapping divides a landscape up into facets (parts of landforms) or landforms and identifies in-situ regolith materials (fresh bedrock, saprolite or colluvium), transported regolith materials (alluvium, dust), and biota that have an affinity with certain regolith landforms where undisturbed. Such accurate mapping can help explorers understand the regolith, inform about possible geochemical sampling materials within a landscape including saprolite, soils, duricrusts or plants, and help target exploration strategies to learn whether a geochemical anomaly is in situ (i.e. the source deposit is directly below) or transported (i.e. the source deposit is laterally distant).

Calcrete sampling

Calcrete is an indurated regolith carbonate material that occurs commonly within the arid parts of Australia (Figure 5.28). In the 1990s, explorers realised that Au had an affinity with calcrete, and that calcrete could be a useful exploration tool because it is easily recognised, sampled and dissolved for analysis. Further research highlighted the use of calcrete to explore for other metals, including Cu, Zn and U. Mineral exploration companies and state governments have now invested large sums of money on widespread calcrete sampling in those areas where it is common. Of course, there are caveats with any exploration sampling

TERMITES, SOIL DEVELOPMENT AND MINERAL EXPLORATION (BOX 5.8)

Termites and other ground-dwelling animals construct their nests from the regolith materials around them. These constructions have been used the world over by mineral explorers as a first-pass method of exploring the regolith to see what lies underneath. Most termites and ants build low, unremarkable mounds, but explorers in southern Africa have long recognised the ability of invertebrates to excavate material from beneath the surface cover and sort grains of different sizes. Where diamonds exist, diamond indicator minerals like garnet and pyroxene are often found in nests—and the odd diamond too!

In Australia, a number of termite species build remarkable, naturally air-conditioned nests—'termitaria'—in areas of semi-arid and tropical/subtropical savanna. These constructions can range from mounds to 'cathedrals' several metres tall and are built from material excavated from below ground as a colony grows. Analysis of the termitaria reveals that the termites access different parts of the regolith depending on the seasonal depth to the water-table, and build their termitaria in multiple lobes, which often express these different depth sources. Where transported cover hides relatively shallow bedrock, termites will burrow down to the bedrock interface to depths of tens of metres, presumably in search of water, and bring samples of saprolite back to the surface.

Termites, and other burrowing animals such as worms, are responsible for large-scale soil turnover and development, especially in warmer climates. Because of their large taxonomic diversity, Australian termites are thought to create larger biological modifications to the soil than in other arid areas of the world. In northern Queensland, burrowing animals, principally termites, have been shown by thermoluminescent age dating to turn over shallow soils (about 1 m thick) above basalt flows at remarkable rates, between 11 kyr and 45 kyr. These ages indicate that it is possible for burrowing organisms to amalgamate and transport geochemical signals from a buried source, through a soil profile, in geologically short timeframes, certainly faster than ore pathfinder elements can be dissolved from primary minerals and removed by groundwater.

For mineral explorers, this is a boon, because the termites are doing the drilling for them. Termitaria in areas with shallow cover amalgamate the local geochemistry, including the bedrock, and may be sampled for geochemical analysis. The various particle size fractions contained within them express different Au and other mineral pathfinder elements and may be used to locate potential mineral resources under cover. In the Tanami Desert of northern Australia, termites have been shown to 'sample' Au-bearing saprolite from beneath 15 m of transported cover. In this area, the coarse, sandy fraction of the termitaria showed elevated levels of Mo and Au, and the silty-clay fraction showed elevated levels of Au, arsenic, Zn and Ni, brought directly to the surface from the saprolite underneath. Thus, termitaria sampling can be used as a low-impact, inexpensive, culturally sensitive method of conducting a first-pass geochemical assay of a mineral tenement.

© H Leue, Lonely Planet Images

4WD bushcamper and cathedral termite mounds, Kakadu National Park, Northern Territory.

© Getty Images [Vision and Imagination]

The Dog Fence at Cameron Corner, Sturt National Park, New South Wales. At 5614 km, it is one of the longest fences in the world. It was designed to keep dingoes and feral dogs out of southern grazing lands.

medium: calcrete commonly occurs over limestone as a weathering product (and therefore collected samples will be devoid of useful trace elements), and it is important to pay attention to the type of calcrete collected, and not mix groundwater (laminated) and pedogenic (nodular) calcrete samples, which is like comparing apples with oranges—they will have different background levels of metals.

Bioprospecting

Long recognised in northern Europe, Asia and North America, geobotany and biogeochemistry are now important exploration tools in Australia. Geobotany is the study of the relationship between plants and the geological conditions in which they occur, whereas biogeochemistry is the study of how plants express the chemistry of the geological substrate in which they grow.

A number of plants in Australia are now recognised as geobotanical indicators and grow only in chemically distinct (often metal-rich) soils: the widespread pearl bluebush (*Maireana sedifolia*) for regolith carbonate (calcrete); the copper weed (*Polycarpaea spirostylis*) in northern Queensland; the nickel bush (*Stackhousia tryonii*) in northern Queensland; the shrub violet (*Hybanthus floribundus*) in Western Australia for Ni and ultramafic rocks; and the thorny wattle (*Acacia continua*) for Zn at Broken Hill. Plant species suitable for biogeochemical sampling in search of minerals should be widespread and should accumulate or even hyperaccumulate (at orders of magnitude above background) ore metals or their pathfinder elements.

Animal activities can also provide a range of sampling media. Burrows provide soil samples from below the surface. These include the larger digging animals like wombats, but also the invertebrates like termites (Box 5.8). Sampling their diggings has proved to be a useful and inexpensive form of drilling. Herbivores eat plants, and their droppings can be analysed for the presence of anomalous metals. The scat of the western grey kangaroo, for example, has been successfully used to detect anomalous U mineralisation in the sand-dune country of northeast Western Australia.

Geophysical techniques

Geophysics, especially airborne electromagnetics, has been adopted and improved by Australian scientists for imaging the regolith and what lies immediately beneath. The technique measures relative conductivity of regolith materials and can be used to assess their thickness and distribution. This has proven useful for mineral exploration—for example, to find unconformities and conductors (potential reductants), which are elements of basin-hosted U mineral systems. The technique has also been adapted for research into regolith and groundwater salinity. In the latter case, with control provided by drilling data, airborne electromagnetics has been used to map the 3D distribution of aquifers and near-surface stratigraphic units. This, in turn, has allowed assessment of the likely impacts of irrigation and environmental water actions aimed at better use of water—for example, in the Murray River corridor through three states (Chapter 7).

Plainly speaking

In this chapter, we have explored the age and evolution of the landscape of the Australian continent and described how its age, and tectonic and climatic environments, have resulted in a flat landscape and a mainly deep regolith cover. The land has shaped the way people have settled and developed the continent, increasingly in the past 200 years. The mostly impoverished soils, a consequence of landscape age and stability, and the availability of water have restricted most settlement activity to the coastal belt. The nature of the regolith cover has been, and continues to be, of fundamental importance for water availability and for the successful discovery of minerals. People have also had an important impact on this ancient land, with soil loss, regulation of rivers, and irrigation with both surface water and groundwater. The first settlers, coming from wet tropical South-East Asia, had many thousands of years to adapt to the Australian environment; the last wave, coming from Europe, found the environment very different from the one they left behind. The Australian environment, so much a result of its underlying climate, geology, landforms and regolith, has presented a challenge to its settlers, one that by and large has been successfully met.

When you look at the map of Australia, you see the vast expanse of land, which is of course bounded by an extremely long coastline. We shall now show how the coast has shaped so many aspects of this nation.

Gibber plains near Chambers Gorge, Flinders Ranges, South Australia.

Bibliography and further reading

Regolith

Bishop P & Pillans B (eds) 2010. *Australian landscapes*, Geological Society, London, Special Publication 346.

Cathcart M 2009. *The water dreamers: the remarkable history of our dry continent*, Text Publishing, Melbourne.

Cunningham C 1996. *The Blue Mountains rediscovered: beyond the myths of early Australian exploration*, Kangaroo Press, Sydney.

Robin L 2007. *How a continent created a nation*, UNSW Press, Sydney.

Scott K & Pain CF (eds) 2008. *Regolith science*, CSIRO Publishing, Melbourne.

Taylor G 1911. *Australia in its physiographic and economic aspects*, Oxford University Press, London.

Thorley P 2009. *Desert tsunami: Australia's inland floods, from prehistory to present*, Central Queensland University Press, Rockhampton.

Cenozoic climate changes of the Australian landscape

Bowler JM 1976. Aridity in Australia: age, origins and expression in aeolian landforms and sediments. *Earth-Science Reviews* 12, 279–310.

Bowler JM & Magee JW 1978. Geomorphology of the Mallee region in semi-arid northern Victoria and western New South Wales. *Proceedings of the Royal Society of Victoria* 90, 5–25.

Bowler JM, Kotsonis A & Lawrence CR 2006. Environmental evolution of the Mallee region, western Murray Basin. *Proceedings of the Royal Society of Victoria* 118, 116–210.

Dunkerley D 2010. Ecogeomorphology in the Australian drylands and the role of biota in mediating the effects of climate change on landscape processes and evolution. In:*Australian landscapes*, Bishop P & Pillans B (eds), Geological Society, London, Special Publication 346, 87–120.

Fujioka T & Chappell J 2010. History of Australian aridity: chronology in the evolution of arid landscapes. In: *Australian landscapes*, Bishop P & Pillans B (eds), Geological Society, London, Special Publication 346, 121–139.

Jennings JN 1968. A revised map of the desert dunes of Australia. *The Australian Geographer* 10, 408–409.

McLaren S, Wallace MW, Gallagher SJ, Miranda JA, Holdgate GR, Gow LJ, Snowball I & Sandgren P 2011. Palaeogeographic, climatic and tectonic change in southeastern Australia: the Late Neogene evolution of the Murray Basin. *Quaternary Science Reviews* 30, 1086–1111.

Pillans B 2007. Pre-Quaternary landscape inheritance in Australia. *Journal of Quaternary Science* 22, 439–447.

Why so flat?

Belton DX, Brown RW, Kohn BP, Fink D & Farley KA 2004. Quantitative resolution of the debate over antiquity of the central Australian landscape: implications for the tectonic and geomorphic stability of cratonic interiors. *Earth and Planetary Science Letters* 219, 21–34.

Braun J, Burbidge DR, Gesto FN, Sandiford M, Gleadow AJW, Kohn BP & Cummins PR 2009. Constraints on the current rate of deformation and surface uplift of the Australian continent from a new seismic database and low-T thermochronological data. *Australian Journal of Earth Sciences* 56, 99–110.

Clark D, McPherson A & Collins C 2011. *Australia's seismogenic neotectonic record: a case for heterogeneous intraplate deformation*, Geoscience Australia Record 2011/11, Geoscience Australia, Canberra.

Gallagher K, Dumitru TA & Gleadow AJW 1994. Constraints on the vertical motion of eastern Australia during the Mesozoic. *Basin Research* 6, 77–94.

Hesse PP 1994. The record of continental dust from Australia in Tasman Sea sediments. *Quaternary Science Reviews* 13, 257–272.

Hesse PP, Humphreys GS, Smith BL, Campbell J & Peterson EK 2003. Age of loess deposits in the Central Tablelands of New South Wales. *Australian Journal of Soil Research* 41, 1115–1131.

Hill SM 1999. Mesozoic regolith and palaeo-landscape features in southeastern Australia: significance for interpretations of denudation and highland evolution. *Australian Journal of Earth Sciences* 46, 217–232.

Hou B, Frakes LA, Sandiford M, Worrall L, Keeling J & Alley NF 2008. Cenozoic Eucla Basin and associated palaeovalleys, southern Australia—climatic and tectonic influences on landscape evolution, sedimentation and heavy mineral accumulation. *Sedimentary Geology* 203, 112–130.

Jutson JT 1914. *An outline of the physiographical geology (physiography) of Western Australia*, Geological Survey of Western Australia Bulletin 61.

Knighton AD & Nanson GC 2000. Waterhole form and process in the anastomosing channel system of Cooper Creek, Australia. *Geomorphology* 35, 101–117.

Nott JF 1992. Long-term drainage evolution in the Shoalhaven catchment, southeast highlands, Australia. *Earth Surface Processes and Landforms* 17, 361–374.

Ollier CD 1982. The Great Escarpment of Eastern Australia: tectonic and geomorphic significance. *Journal of the Geological Society of Australia* 29, 13–23.

Ollier CD 2001. Evolution of the Australian landscape. *Marine and Freshwater Research* 52, 13–23.

Pain CF 1985. Morphotectonics of the continental margins of Australia. *Zeitschrift für Geomorphologie* supplement band 54, 23–35.

Quigley MC, Clark D & Sandiford M 2010. Tectonic geomorphology of Australia. In: *Australian landscapes*, Bishop P & Pillans B (eds), Geological Society, London, Special Publication 346, 243–265.

Sandiford M 2007. The tilting continent: a new constraint on the dynamic topographic field from Australia. *Earth and Planetary Science Letters* 261, 152–163.

Tooth S & Nanson GC 2000. The role of vegetation in the formation of anabranching channels in an ephemeral river, Northern plains, arid central Australia. *Hydrological Processes* 14, 3099–3117.

Twidale CR 1991. A model of landscape evolution involving increased and increasing relief amplitude. *Zeitschrift für Geomorphologie* 35, 85–109.

Twidale CR 1994. Gondwanan (Late Jurassic and Cretaceous) palaeosurfaces of the Australian Craton. *Palaeogeography, Palaeoclimatology, Palaeoecology* 112, 157–186.

Twidale CR 2000. Early Mesozoic (?Triassic) landscapes in Australia: evidence, argument, and implications. *Journal of Geology* 108, 537–552.

Woolnough WG 1927. Presidential address: part I—the chemical criteria of peneplanation, part II—the duricrust of Australia. *Journal and Proceedings of the Royal Society of NSW* 61, 1–53.

Regolith—everything between fresh rock and fresh air

Eggleton RA (ed.) 2001. *The regolith glossary: surficial geology, soils and landscape*, Cooperative Research Centre for Landscape Environments and Mineral Exploration, Canberra and Perth.

Ollier CD & Pain CF 1996. *Regolith, soils and landforms*, John Wiley and Sons Ltd, Chichester.

Taylor G & Eggleton RA 2001. *Regolith geology and geomorphology*, John Wiley and Sons, Chichester.

Taylor G, Eggleton RA, Foster LD & Morgan CM 2008. Landscapes and regolith of Weipa, northern Australia. *Australian Journal of Earth Sciences* 55(S1), S3–S16.

Wilford JR (in press). A weathering intensity index for the Australian continent using airborne gamma-ray spectrometry and digital terrain analysis. *Geoderma*.

Wilson MJ 2004. Weathering of the primary rock-forming minerals: processes, products and rates. *Clay Minerals* 39, 233–266.

How old is the Australian landscape?

Killick M 1998. Phanerozoic denudation of the Western Shield of Western Australia. *Geological Society of Australia Abstracts* 49, 248.

Kohn BP, Gleadow AJW, Brown RW, Gallagher K, O'Sullivan PB & Foster DA 2002. Shaping the Australian crust over the last 300 million years: insights from fission track thermotectonic imaging and denudation studies of key terranes. *Australian Journal of Earth Sciences* 49, 697–717.

Osborne RAL, Zwingmann H, Pogson RE & Colchester DM 2006. Carboniferous clay deposits from Jenolan Caves, New South Wales: implications for timing of speleogenesis and regional geology. *Australian Journal of Earth Sciences* 53, 377–405.

Pillans B 1997. Soil development at a snail's pace: evidence from a 6 Ma soil chronosequence on basalt in north Queensland, Australia. *Geoderma* 80, 117–128.

Stewart AJ, Blake DH & Ollier CD 1986. Cambrian river terraces and ridgetops in central Australia: oldest persisting landforms? *Science* 233, 758–761.

How regolith enriches Australia

Anand RR & Paine M 2002. Regolith geology of the Yilgarn Craton, Western Australia: implications for exploration. *Australian Journal of Earth Sciences* 49, 3–162.

Britt AF, Smith RE & Gray DJ 2001. Element mobilities and the Australian regolith—a mineral exploration perspective. *Marine and Freshwater Research* 52, 25–39.

Connolly MJ 2012. *Munda-gutta Kulliwari*, Dreamtime Kullilla-Art. www.dreamtime.auz.net

Eggleton RA, Taylor G, Le Gleuher M, Foster LD, Tilley DB & Morgan CM 2008. Regolith profile, mineralogy and geochemistry of the Weipa Bauxite, northern Australia. *Australian Journal of Earth Sciences* 55(S1), S17–S43.

Government of South Australia 2010. *Geology of the McLaren Vale wine region*, Primary Industries and Resource South Australia, Adelaide.

Kotsonis A 1999. Tertiary shorelines of the western Murray Basin: weathering, sedimentology and exploration potential. In: Extended abstracts of Murray Basin Mineral Sands Conference, Mildura, Victoria, 21–23 April, 57–63.

McNally GH 1995. Engineering characteristics and uses of duricrusts in Australia. *Australian Journal of Earth Sciences* 42, 535–547.

Petts AE, Hill SM & Worrall L 2009. Termite species variations and their importance for termitaria biogeochemistry: towards a robust media approach for mineral exploration. *Geochemistry: Exploration, Environment, Analysis* 9, 257–266.

Pillans B, Spooner D & Chappell J 2002. The dynamics of soils in north Queensland: rates of mixing by termites determined by single grain luminescence dating. In: *Regolith and landscapes in eastern Australia*, Roach IC (ed.), Cooperative Research Centre for Landscape Environments and Mineral Exploration, Canberra and Perth, 100–101.

Senior BR, McColl DH, Long BE & Whitely RJ 1977. The geology and magnetic characteristics of precious opal deposits, southwest Queensland. *BMR Journal of Australian Geology and Geophysics* 2, 241–251.

Taylor G & Eggleton RA 2008. Genesis of pisoliths and of the Weipa Bauxite deposit, northern Australia. *Australian Journal of Earth Sciences* 55(S1), S87–S103.

6

Living on the edge—waterfront views

Australia developed as a nation of coastal fringe dwellers even before the realisation in colonial times that there was no inland sea and the continent had a 'dead heart' of desert and scrub. The maritime character of the nation developed with the reliance on coastal seas for transport and trade during European settlement, through to the present day where the coast is the setting for most of Australia's population, industry, tourism and recreation. The geological history of the coast and its distinctive configuration, landforms and environmental regimes have produced a unique, highly diverse continental margin. In turn, the coast has profoundly influenced the pattern of settlement and development of Australia.

Brendan P Brooke, Peter T Harris, Scott L Nichol, Jane Sexton, William C Arthur, Ralf R Haese, Andrew D Heap, Martyn C Hazelwood and Lynda C Radke

Geoscience Australia

LIGNE
EQVINOC: TIALE
TERRA DOS PAPOS
NOVA GVINEA
Hoogh Landt
Groene Eylanden
Timore Landt
Banda
Droge bogt
hoge Landt
CAR:
PEN:
TARIA
Water plaets
Riuier Nassau
Staten Riuier
Van Diemens Riu.
Riuier Van Alphen
C. Van Diemen
R. Maet Suycker
Limmensbogt
Abel Tasmans Riu.
Vuyle Hoeck
Van Diemens Landt
HOLLANDIA
NOVA
detecta 1644.
TERRE AVSTRAL
découuerte l'an 1644.
TROPIQVE DE CAPRICORNE
G. F. de Wit landt
detecta 1628.
Landt
d'Eendraght
Houtmans Abrolhols
Turtel duyf
Landt uan P. Nuijts, opgedaan met het gulden zeepaerdt uan Middelburgh
16. Ianuary Anno 1627.
I. st Pierre.
I. st François
Terre de Diemens
découuerte le 24. nouembre
1642.
Wits Eylanden
Pedra blanca

Our coastal home

Snorkelling on the Great Barrier Reef over luxuriant coral reefs, sailing on the splendid waters of Sydney Harbour, with its iconic bridge and opera house, and swimming in the warm, rolling surf of Queensland's Gold Coast are popular images of coastal Australia. However, these images belie the wild and inhospitable character of most of the Australian coast, which posed great challenges to early European exploration and settlement of the continent. Indeed, the rugged form and prevailing oceanic and climatic regimes that characterise most of the Australian coast affected where Aboriginal people lived prior to European settlement and strongly influence where most Australians live today.

By the middle 1600s, Dutch traders and explorers managed to map in some detail most of the southern, western and northern coasts of the Australian continent (Figure 6.1). The Dutch, however, did not escape unscathed, with a number of their ships wrecked on the rocky reefs and cliffs of the west coast. However, no settlement was attempted, due to the inhospitable conditions encountered, particularly along the southern and southwest coasts, which are pounded by dangerous waves generated in the Southern Ocean and lack safe anchorage. The northwest and northern coasts were no less unwelcoming, with strong tidal currents, shifting shoals and vast archipelagos of reef-fringed islands. The Dutch sailors also found little freshwater or suitable soils or, most importantly, any potential for profitable trade.

It was more than a century later before there was competition between England and France to map and colonise the continent. Both began to map in detail the more benign southeast and east coasts of the mainland and Tasmania, and European settlement was considered feasible. It appears that these more watered and accessible sections of coast were likewise important for the Aboriginal people of Australia. Reports from early explorers such as Captain James Cook and Captain Matthew Flinders document the large number of Aboriginal encampments observed along these coasts, as well as the coast of northern Australia. It seems very likely that a large majority of the estimated (up to) 750 000 Aboriginal people who lived in Australia prior to European settlement resided on the coast or partially relied on coastal resources (Chapter 1). Soon after the arrival of the First Fleet in Sydney Harbour in 1788, communicable diseases that were brought by Europeans, especially smallpox, decimated the Aboriginal population along the east coast of Australia. The first Australians were then displaced with the expansion of the colony.

Shellfish is thought to have formed 30–50% of the protein consumed by many Aboriginal people, and the early European settlers. The extensive utilisation of coastal resources by Aboriginal people along the eastern seaboard was well recorded in shell middens that once lined many stretches of shoreline. Such was the extent and volume of these shell accumulations that they were mined for lime—the essential ingredient for mortar and cement (Chapter 5). As these deposits became depleted, oyster 'reefs' in bays and estuaries were 'mined' for lime. This exploitation of oyster reefs and middens formed the basis for an important coastal industry, supported by fleets of coastal vessels, and generated an important early

Figure 6.1 (opposite): 17th century Dutch map of the southern, western and northern coasts of Australia. Thevenot, M (Melchisedech), 1620?–1692.

Map courtesy of National Library of Australia, MAP NK 2785

(Chapter 5). These locations and the adjacent regions have remained the focus of population growth in Australia, in large part due to the attraction of the same coastal features that initially marked them as suitable for early European settlement. The cities of this region—Sydney, Melbourne, Hobart, Brisbane and Adelaide—make up more than 55% of the current national population (Figure 1.3), with another 8% located in Perth on the southwest coast; all these cities were established before the federation of Australia in 1901 (Figure 6.2). Over the past half-century, the residential areas of these cities and nearby coastal towns have rapidly expanded, and today more than 86% of Australia's population lives, within 50 km of the coast. In fact, 9 of the 10 largest urban centres in Australia (the exception is Canberra) were established in colonial times and are located on the southeast coast. The residents of these nine coastal cities make up more than 70% of the Australian population.

In this chapter, we examine the influence the coast has on society, as well as the impacts people have on coastal environmental systems. We outline the geological evolution of the coast and how Australians' experience and use of the coast vary markedly between coastal regions. Key features that make the Australian coast so distinctive are described, and we examine how these have helped shape Australian society. We also look at new geoscience knowledge of the coast and continental shelf that is increasing our understanding of its environmental character, value and hazards. This knowledge is better informing the management of Australia's essential coastal ribbon.

Aboriginal shell midden at Ordinance Point, Tarkine Wilderness, northwest Tasmania.

extension of European settlement out from the initial colonial outposts. Shell middens can still be found at many locations, especially on the less accessible and more remote stretches of coast, and are now protected by legislation.

The initial colonial settlements were located in southeastern Australia, at the few sites with safe anchorage, reliable freshwater supplies and soils that appeared to support European agriculture

Australia: a maritime nation

The establishment of coastal ports was essential to the economic development of the colonies and their integration into an independent nation. The early pattern of British settlement of the Australian coast was closely tied to the development of trade in whaling, mining and agricultural products. In 1968, Professor James Bird of Southampton University published *Seaport gateways of Australia*, with which he hoped 'to stimulate Australians to take an interest in their own ports, because Australia is a maritime nation and scarcely knows it'. Bird's book described the main considerations that influenced the colonial governments in their selection of seaport gateways in an island continent whose geology provided few suitable gulfs, long and navigable estuaries, or rivers of sufficient length and volume.

Australia's being an island for the past 10 ka ensured that people coming to or leaving from Australia have had to use a ship or boat. This only changed in the 20th century with the advent of aviation. For most of the post-World War II immigrants, however, even up to the 1960s, a ship of the sea was the vessel used. Being an island nation does extend a strategic benefit in terms of defence. The Royal Australian Navy, established in 1911, has a proud tradition in both surface and submarine vessels, and it is arguably the only 'blue water' fleet in the region.

Until the expansion of the wool trade, Australia's economy was dominated by maritime activities such as whaling and sealing. In Tasmania alone, there were 35 whaling stations by 1841. The whalers and sealers plied the Pacific Ocean and deep into the Southern Ocean, wreaking havoc on their prey. Their voyages led to the discovery of Macquarie Island (1810) and Heard and McDonald Islands (1853–54) (Chapter 1). Of more immediate impact to Australia was the establishment of bay whaling along the southern coast of Australia, with many of the settlements being beyond the reach of the colonial authorities of the day, often to the severe detriment of the Aboriginal people encountered.

By the middle 19th century, several ports had been established outside the major colonial cities to support trade in these commodities—for example, at Geraldton, Robe, Port Fairy, Bowen, Launceston and Newcastle. Today the whalers

Whaling was one of Australia's early industries. Sheltered bays, such as Twofold Bay at Eden, New South Wales, provided a haven for whales, from which they were harvested. Today, these same bays draw tourists to view whales, such as the humpback (*Megaptera novaeangliae*), as they migrate from Australia's warm northern waters to the Southern Ocean.

Image courtesy of Merimbula Marina, photographer Erich Schute

© M Gottschalk, Lonely Planet Images

Coffs Harbour jetty, Coffs Harbour, northern New South Wales.

and coastal trading ships have long gone, and port services have become more concentrated following the construction of rail and road networks. Former important port cities such as Rockhampton, Coffs Harbour, Warrnambool, Victor Harbor, Albany and Launceston no longer function as major trading ports but have developed into important regional centres that support agriculture, mining and tourism and provide key social services such as health care and education. Others, such as ports along the Western Australian coast (e.g. Geraldton) and the Queensland coast (e.g. Gladstone), are now important export hubs for the mining, petroleum and agricultural sectors.

In the past few decades, a number of new ports have been built far from the capital cities but close to key areas of agricultural and mineral production, to handle the rapid increase in the export of bulk commodities to Asia (Chapter 9). For example, Hay Point (coal) near Mackay, and Dampier and Port Hedland (iron ore) in the Pilbara and Kimberley regions are among the world's largest export hubs for these resources. The export of bulk commodities from ports now underpins Australia's economy. Australia exported more than $157 B in 2008–09 in raw materials such as iron ore, bauxite, coal and natural gas, and $32 B in agricultural products such as wheat, beef, wine and wool. In September 2011, one ship alone, the MV *Ocean Shearer*, exported 24 683 live cattle to Indonesia. These exports, and the export of a range of other raw materials from Australia's ports, represent an international trade worth approximately $249 B per year.

Figure 6.2: The Australian coast and shelf. Locations referred to in the text are shown. Also shown is an approximate render of the position of the coastline during the last glacial maximum, 21 ka, when sea-level was 125 m lower than today.

The pull of the temperate coast

The concentration of people on the southeast and southwest coasts of Australia has, at least in part, been driven by the amenity of the temperate and subtropical sandy bays, estuaries and open shores of these regions. The temperate coasts are attractive, such that the New South Wales Auditor General found a positive correlation in 2010 between workplace absence (sickies) and a worker's proximity to the coast. These coasts provide virtually year-round safe swimming, fishing and boating, along with stunning harbour and beach views and a cool summer sea breeze. Numerous sandy beaches in these areas are exposed to persistent ocean swells, and the resulting wave regime has led to the development of the Australian surfing culture and its associated clothing and surf-gear industries (e.g. 'global' brands such as Billabong™ and Rip Curl™). Surfing has become both a popular sport and an industry for many coastal cities and towns. For example, Sydney, with dozens of sandy beaches that have good surfing waves throughout the year, has been a 'surfing city' since the early 20th century. It was in Sydney in 1907 that Surf Life Saving Australia—the first of its type in the world—was established.

For many coastal towns, such as Noosa Heads, Tweed Heads and Ulladulla, Torquay and Margaret River, their proximity to good-quality surf breaks has been a significant driver of economic activity and population growth over the past 30 years. More recently, high population growth has extended to northern coastal cities such as Townsville, Darwin and Broome, largely driven by growth in tourism and the energy and mineral resources sectors (Chapter 9) in the adjacent hinterlands. Rapid urban expansion on the coast has been matched by a demographic shift away from many inland towns.

Swell generated in the Southern Ocean breaking on Cow Bombie, about 2 km offshore in the Margaret River region, Western Australia.

Image by Jamie Scott Images

NINGALOO REEF (BOX 6.1)

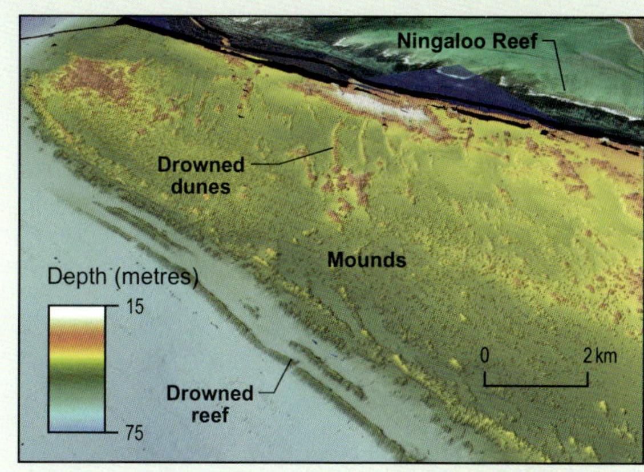

Figure B6.1: Perspective view of the seabed seaward of Ningaloo Reef at Point Cloates, Western Australia, showing drowned reefs and sand dunes. The bathymetry was produced from multibeam sonar soundings collected by Geoscience Australia on the Australian Institute of Marine Science's vessel RV *Solander* in 2008.

Ningaloo Reef extends almost 300 km along the coast of central Western Australia, forming a near-continuous fringing coral-reef system. Listed as part of the Ningaloo Coast World Heritage Area in June 2011, it is the largest fringing reef in Australia and the only example in the world of a fringing reef located along the west coast of a continent. The reef forms here because, unlike other continental west-coast currents, the Leeuwin Current feeds warm tropical waters southwards onto the narrow continental shelf. In places, the reef is only 200 m from the shore, which, along with the shallow lagoon, provides an accessible amenity for tourism (particularly ecotourism, including watching whale shark that come to feed at times of coral spawning) at localities such as Coral Bay.

The Ningaloo Coast has a long history of occupation by Aboriginal people, with shell middens preserved in coastal rock shelters dating back 35 ka. The archaeological record suggests that these sites were abandoned during the last glacial maximum when sea-level decreased but were reoccupied during the Holocene sea-level highstand. New seabed mapping of the continental shelf seaward of Ningaloo Reef has revealed in detail the drowned landscape that Aboriginal people might have utilised during the glacial period. This landscape includes an extensive system of relict reefs and dunes that would have formed during the Late Pleistocene to early Holocene. The drowned coast is widest offshore from Point Cloates, where it extends about 5 km west from Ningaloo Reef and includes hundreds of patch reefs up to 5 m high and ridges that are up to 20 m high and several kilometres in length.

Underwater photography shows that the mounds and ridges provide hardground habitat for dense coral and sponge communities. The distribution of these relict seabed features along the Ningaloo Coast has determined the location and abundance of modern reef communities on the continental shelf. Importantly, this geoscience information is being used to inform the design and management of areas for marine biodiversity conservation and sustainable use, including the design of a national system of marine protected areas.

The remote north

Despite some wistful thinking in faraway London, no major colonial settlements were established on the northwest and northern Australian coasts. Sir James Stephens, then Undersecretary of State for the Colonies, wrote in 1842:

> *To complete the greatness of the Australian nation, it will be necessary that they should have a greater variety than at present of climates, soils, products and exports. It is also necessary that they should secure harbours to the north and northwest. A belt of colonies drawn around the coast will give us absolute and undisputed possession of the interior … and for keeping to ourselves a source of commercial and maritime greatness which their tariffs, improvements etc., are daily rendering more important to us.*

These remote coasts have large tides and strong tidal currents, extensive cliffs and numerous rocky islands, coral reefs and sandbanks. Their shorelines are often borderd with extensive mangrove forests and mud and salt flats; they are frequently struck by cyclones; and they have dry hinterlands. The settlement of Victoria was established in Port Essington on the Coburg Peninsula (near Darwin) in the 1830s but lasted only a few years. Darwin and Broome are the only substantial enduring colonial settlements on the northern and northwest coasts. Broome was established in the 1880s to service the pearling industry, which has developed into a major export industry for the region. The success of this industry is strongly linked to the natural state of this isolated coast, the abundance of large protected bays, very low levels of river discharge

and large tidal ranges, which favour the growth of pearl shells. Pearling in its early days relied on Aboriginal and Japanese divers, and Broome is still strongly influenced by these people and their cultures. Overall, however, the northern coasts remain sparsely populated today. For example, on average along this more than 9000 km of coast, there is less than 1 person per square kilometre, and Darwin (193 000 people) and Broome (15 000) are the two largest cities in northern Australia (Figure 1.3).

In the past two decades, the same features that discouraged colonial settlement and subsequent urbanisation of the northern and northwest coasts have appealed to people wanting to experience a natural, 'wild' coast. Ecotourism has especially developed in the Kimberley region of northwestern Australia and on the World Heritage–listed Ningaloo Coast. Major attractions in these areas are the remoteness, dramatic landforms and pristine character of the marine environment, which includes archipelagos of hundreds of bedrock islands, fringing reefs and diverse marine ecosystems (Box 6.1).

Aerial photograph of the Alligator River Estuary in Kakadu National Park, Northern Territory. Water discharged from the catchment (upper right) is heavily laden with sediment. The estuarine water (upper left) is less cloudy. Also shown are the mangrove-lined shoreline and tidal creeks, and extensive intertidal mud flats.

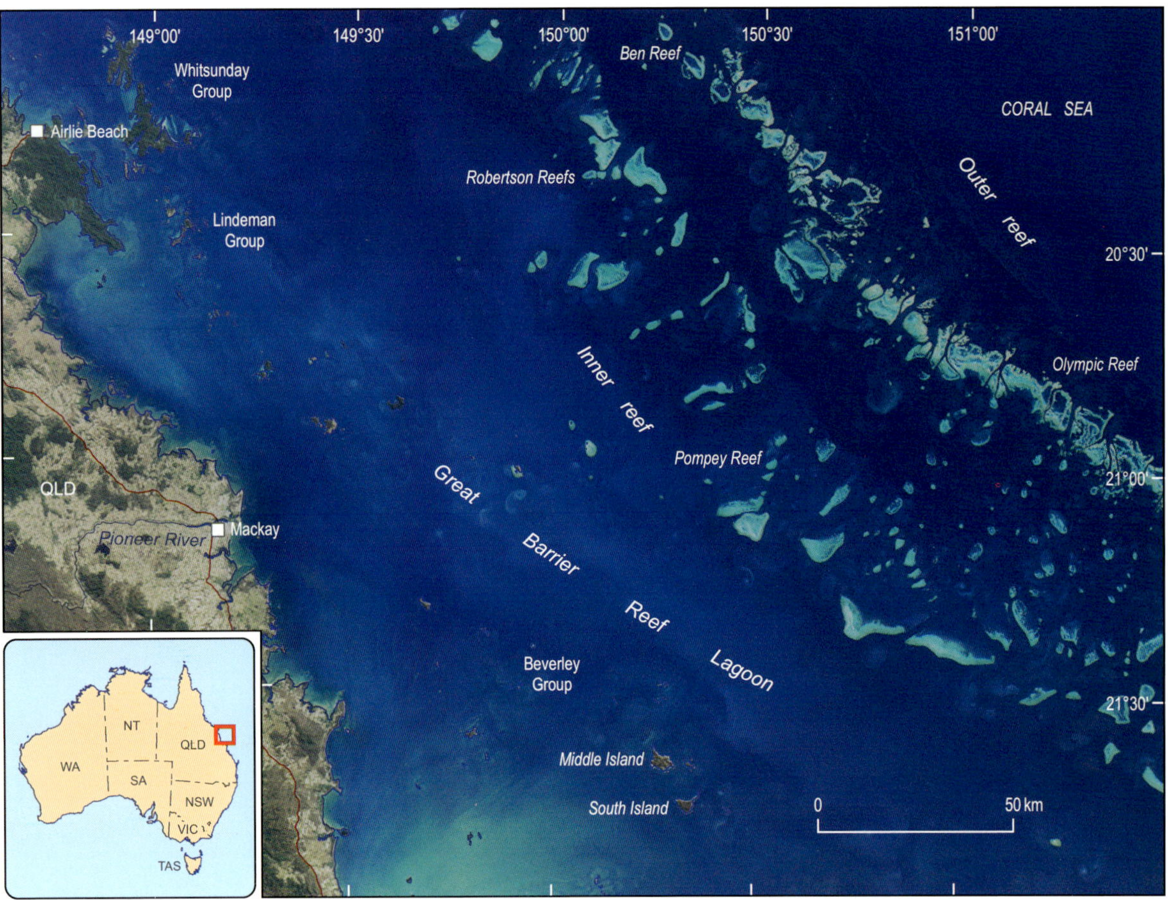

Figure 6.3: Multi-angle Imaging SpectroRadiometer (MISR) satellite image of the southern part of the Great Barrier Reef, highlighting numerous reefs and shoals and the lagoon used for shipping. (Source: NASA)

The Great Barrier Reef

The Great Barrier Reef of northeastern Queensland posed a major navigational hazard to early sailors, who feared running aground on the reef itself or on the numerous shoals inside the reef. Such was the fate of the renowned first English explorer, Captain James Cook, who in 1770 ran aground 146 km north of Cairns on what is now the Endeavour Reef, named after his ship. Although the Great Barrier Reef receives abundant swell from the Pacific Ocean, this wave energy is attenuated on the reef so that much smaller, locally generated waves are experienced on the mainland coast. Early colonial exploration and settlement were challenged, however, by a lack of deep anchorage, prevailing strong southeasterly onshore winds, extensive mangrove forests and broad tidal flats, especially in the meso- to macro-tidal reaches of the central and northern Great Barrier Reef. Nonetheless, shortly after settlement of the major colonial cities, a number of small colonial settlements sprang up along the mainland coast opposite the Great Barrier Reef to service a range of mining, agricultural and fishing industries—in particular, gold mining, grazing and sugar cane. Gladstone, Rockhampton, Bowen, Townsville and Port Douglas, for example, were all established between 1853 and 1870.

The reefs and shoals of the Great Barrier Reef remain a risk for shipping today, as demonstrated by the grounding of the huge coal carrier *Shen Neng 1* on Douglas Shoal in 2010. There are now, however, international shipping lanes that safely cross the reef and connect to a number of ports (e.g. Gladstone, Mackay, Townsville) to service thousands of vessels each year (Figure 6.3). The reef itself is also a major driver of economic activity. For example, during 2007, reef tourism generated $5 B, compared with $140 M by the long-established fishing industry. Although reef tourism is big business, the perception of the coast is still largely framed by its utility for recreational fishing. The clear-water outer reefs sit tens of kilometres offshore; there are no true surfing beaches; and, during the wet season, swimming is dangerous due to the presence of highly venomous jellyfish. Undoubtedly, the geomorphology of the Great Barrier Reef and lagoons, and the associated sea conditions have shaped very different perceptions and uses of this tropical reef coast compared with the temperate coast of southern Australia, where most Australians live.

Formation of the edge

Much of the configuration and broadscale landforms that comprise the Australian coast are a direct inheritance of Gondwanan break up in the Cretaceous and Paleogene (155 Ma and 34 Ma; Chapter 4). The Australian coast and continental shelf have changed little since the formation of the island continent because there has been no major tectonic activity to fracture or deform the continental margins. For example, thousands of kilometres of the present rocky coast of eastern and southwestern Australia largely retain the outline created more than 30 Ma.

Following the break up of Gondwana, the northward migration of Australia and spreading of seafloor around the new continent created the modern ocean basins (Figure 4.1). As a result, Australia's southern, eastern and western coasts became margins of the Southern, Pacific and Indian oceans, respectively, giving these coasts distinctive oceanographic and climatic regimes. Most notable are the continent's southern margin and the subsequent formation of the world's longest Southern Ocean coast. The record of marine inundation of the low-lying sections of southern Australia by the Southern Ocean during the Cenozoic, when the global sea-level was up to 100 m higher than at present (Box 2.6), is remarkably well preserved in the extensive marine and coastal deposits in the Eucla and Murray basins (Figure 5.7). Amazingly, in places, relict Cenozoic dune-capped barriers have been preserved, forming subtle elongate ridges in the modern landscape, especially in the Murray Basin (Chapter 5). The preservation of these beach and dune landforms over tens of millions of years is a product of the very low rates of sediment accumulation, coupled with minimal regional tectonic uplift and negligible landform erosion, the latter a strong feature of the flat, dry regions of southern Australia (Chapter 5).

In the Eucla Basin, Eocene and Miocene shoreline deposits have been uplifted by up to 150 m relative to global sea-level at that time (Chapter 2),

An aerial view of the 200 km unbroken line of coastal cliffs near 'Head of the Bight' in South Australia.

Main image © M Gottschalk, Lonely Planet Images, inset image © Getty Images [J Edwards]

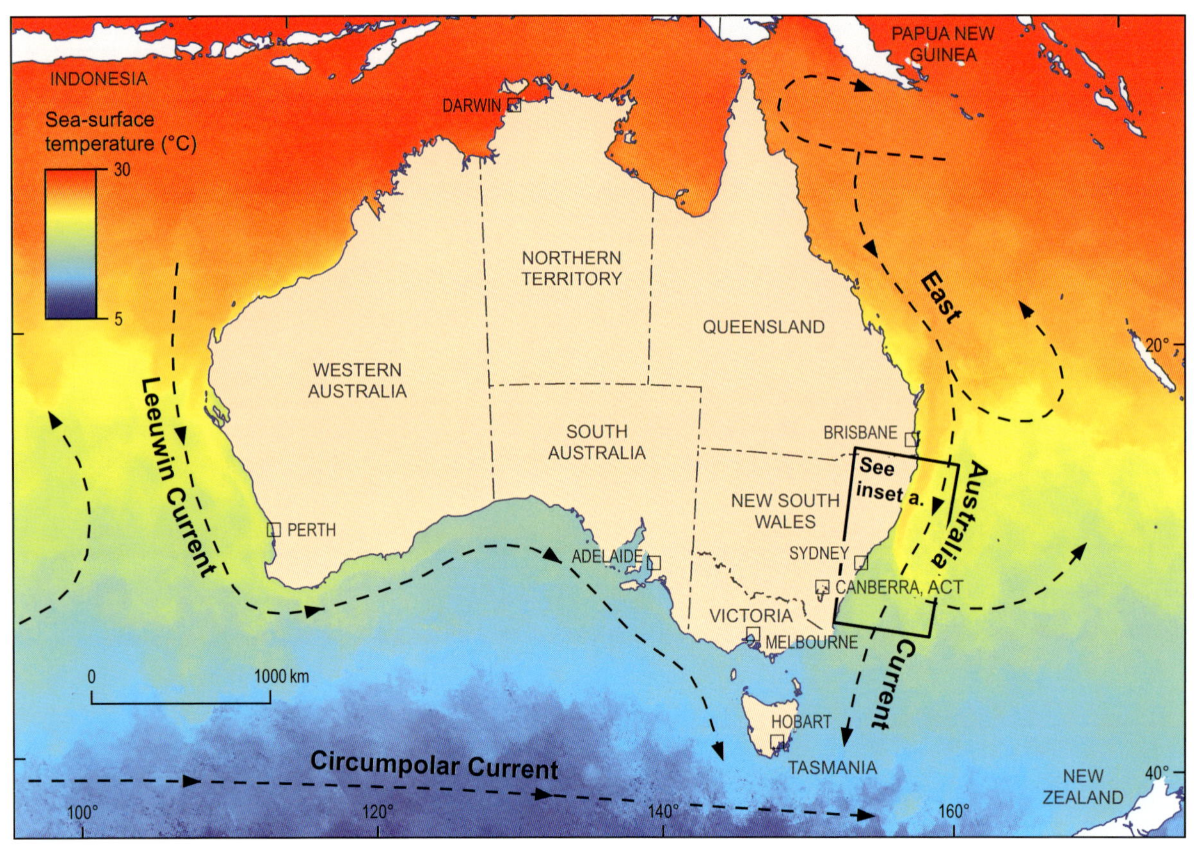

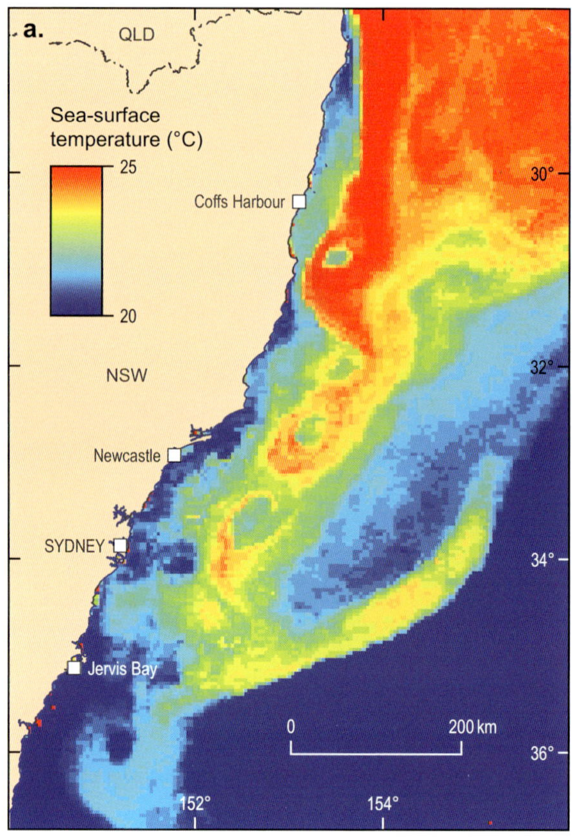

Figure 6.4: Satellite-derived sea-surface temperature, illustrating the warming influence of Australia's poleward-flowing boundary currents, the East Australia and Leeuwin currents. (a) Map shows the vortex nature of the East Australian Current as it flows southwards along the New South Wales coast. (Source: CSIRO Marine and Atmospheric Research)

revealing that parts of the southern margin have been gently uplifting at a rate of up to 20 m/Ma. These palaeoshoreline elevation data, coupled with measurements of small-scale deformation, suggest that the continent appears to be 'hinged' across the centre, with the northern margin gently subsiding towards the northern subduction zone, while the southern margin, or at least parts of it, has been uplifting (Chapter 5). Although these coastal movements are subtle compared with tectonic movements on coasts that are on or near active plate margins (e.g. western coasts of South and North America), they help to explain differences in the geological character of the southern and northern coasts of Australia. The southern coast has large stretches of coastal cliffs and strandplains composed of Cenozoic to Quaternary marine limestones, whereas much of the northern coast features drowned bedrock landforms, with no major exposures of Cenozoic marine deposits.

The collision of Australia with the Pacific and Eurasian plates began around 40 Ma and eventually created a barrier to the western-flowing warm equatorial ocean current. As a result, the current diverged, with one branch flowing down the east coast of Australia, forming the East Australia Current, and the other branch flowing around New Guinea through the Indonesian archipelago and southwards down the Western Australian coast. Australia is now unique among the continents in that it is bounded on both its eastern and western margins by warm-water, poleward-flowing boundary currents (Figure 6.4). The East Australia Current is a western boundary current of the Pacific Ocean, which transports about 15 M m^3/s (15 Sverdrups) of warm tropical waters along Australia's east coast. It is equivalent to the Gulf Stream and Kuroshio currents, which transport warm waters to the north along the eastern coasts of North America and Asia, respectively.

In contrast with the cool-water, equatorward-flowing boundary currents found on the eastern sides of all the other oceans, the eastern boundary current in the Indian Ocean is the warm, poleward-flowing Leeuwin Current. The Leeuwin Current transports about 6 Sverdrups of warm water southwards along Australia's west coast, exerting an important influence on the climate and marine ecosystem of Western Australia. In particular, it is responsible for the presence of tropical marine organisms on the southwest coast of the continent.

One important consequence of the Leeuwin Current is the absence of any significant fishery adjacent to Western Australia. In contrast, the Benguela Current System off southern Africa and the Humboldt Current System off South America are boundary currents that support rich coastal fisheries because of seasonal, wind-driven upwelling and highly productive continental shelf waters. Upwelling in Australia's shelf waters is generally not sufficient to sustain zones of high productivity, and these waters are described as oligotrophic (with little nutrient to sustain life). One western coastal area at Ningaloo, however, has a strong seasonal productivity, which brings whale sharks to feed and breed, and this has now been heritage listed as a marine park. The general lack of nutrient-rich water around Australia means that it ranks only about 60th in fish production, despite having the third largest exclusive economic zone of any nation (*ca* 14 M km^2).

Northern Australia entered tropical waters in the early Miocene (*ca* 24 Ma) as the continent steadily drifted northwards (Figure 5.5). These warm waters brought coral larvae from the north, providing the potential for the development of coral reefs in northern Australia. Reef growth was initiated adjacent to Torres Strait by around 20 Ma, eventually extending down the outer shelf of northeastern Australia to form the Great Barrier Reef, which today extends around

Whale shark (*Rincodon typus*) and golden trevally (*Gnathanodon speciosus*), Ningaloo Reef, Indian Ocean, Australia.

Image courtesy of Murray–Darling Basin Authority, photographer Michael Bell

Aerial view of mouth of the Murray River, South Australia. The Coorong Estuary sits inland of the Younghusband Peninsula, a large sand barrier.

2300 km from the strait to just east of Bundaberg. The reef limestone underlying the modern reef ranges from about 2 km in thickness in the north (at Ashmore Reef) but is typically less than 200 m thick along the outer shelf barrier reefs. Drilling of continental-shelf limestone deposits indicates that reef growth was initiated in the central Great Barrier Reef by around 500 ka. Older and thicker reef deposits are interpreted to occur on subsiding marginal plateaus of the Coral Sea offshore from the Great Barrier Reef (e.g. the Marion, Queensland and Kenn plateaus).

There has been an accumulation of marginal-marine deposits as Australia migrated northwards, especially in the shallow seas of the northwest margin (e.g. North West and Sunda shelves). Towards the end of the Neogene (*ca* 3 Ma), Earth cooled and, as the ice caps grew, sea-level dropped. Under these lower sea-level conditions, the old exposed rocks around the continental margin and the younger marine deposits were eroded by the sea. The modern continental shelf comprises these eroded areas of the continental margin (e.g. in New South Wales) in combination with submerged low-lying areas of Gondwanan rocks and a cap of Neogene and Quaternary marine sediment (e.g. North West Shelf, Great Australian Bight).

The shifting edge—Quaternary climate and sea-level

During the Quaternary Period, small ice sheets and glaciers were limited to western Tasmania and the southeast highlands of Australia. This is in contrast to the Northern Hemisphere where, during the last glacial period (*ca* 75–11.5 ka), the Fennoscandian (Europe) and Cordilleran (northern America) ice sheets spread across the continents and grew up to a few kilometres thick. In Australia, therefore, there has been no modification of coastal landforms by glacial erosion and sedimentation, and no isostatic movement of the land relative to the sea due to loading and unloading of the crust by ice, which commonly occurs with the accumulation and retreat of continental ice sheets. As a result, relict shoreline deposits in Australia provide important records of global changes in sea-level during the Quaternary (Box 2.6). For example, fossil coral reefs that grew during the last interglacial on the coast of Western Australia still sit at the same elevation (up to 3 m asl) as when they were growing, some 125 ka. This great stability of the Australian coast makes it an important location for assessing modern changes in sea-level related to global climate change.

A sediment-starved coast

Skeletal fragments of marine organisms such as algae, foraminifera, molluscs, bryozoans and coral are the main constituents of sediment on the majority of the Australian coast and shelf. Physical erosion (e.g. currents and waves) and bioerosion (e.g. fish and sea cucumbers) break down these skeletons. In tropical waters, bioerosion of sediment by sea cucumbers and coral by fish produces much of this sediment; for example, each individual of bumphead parrotfish (*Bolbometopon muricatum*) can remove about 5 t/year of reef. In southern and southwestern Australia, biogenically produced calcium carbonate gravel, sand and mud dominate the temperate continental shelf, covering around

2 M km^2 to form the world's largest carbonate province. The dominance of carbonate is a consequence of the very low sediment load of most of Australia's coastal rivers throughout the Quaternary (Chapter 5), with rivers virtually absent in some regions (e.g. Great Australian Bight).

Even the great Murray–Darling River system, which drains Australia's largest coastal catchment (>1 M km^2), struggles to supply much in the way of sediment to the coast. The long-term average sediment discharge is estimated at about 1 Mt/year. This represents a small fraction of the sediment discharge of most rivers with similar-sized catchments in the Northern Hemisphere, and is much less than the estimated 4–7 Mt/year of continental aeolian dust that is deposited in the Pacific Ocean (Chapter 5). As a result, the Murray River has not formed a delta at its terminus on the eastern coast of South Australia, and periodically its mouth is completely closed off by wave-transported sand. To maintain a connection with the Coorong estuary, dredging has been necessary in the river mouth.

Calcium carbonate–producing organisms have flourished in these sediment-starved shallow marine regions. Knowing the composition of shelf and coastal sediment is important for assessing how future changes in climate, sea-level and temperature may influence the coast. For example, the supply of sediment from the land may decline with a reduction in coastal rainfall, resulting in a negative beach-sediment budget and erosion of the shoreline, whereas the offshore biogenic production of carbonate sediment may be adversely affected by changes in water temperature or pH of the ocean or even overfishing.

During the Quaternary Period, carbonate sediment has accumulated on the shoreline and been blown landward as mobile dunes before becoming cemented. Today, remnants of these deposits form aeolianite (lithified dunes) cliffs, reefs and islands along vast stretches of the southern and southwestern coast. The immense volume of biogenically produced sediment locked up in aeolianite along the coast of Australia exceeds the volume of tropical carbonate that has accumulated in the Great Barrier Reef. For example, Quaternary aeolianite forms most of the coast between Perth and Shark Bay (*ca* 800 km), including the Zuytdorp Cliffs north of Kalbarri that rise up to 260 m above the shoreline and extend virtually unbroken along more than 200 km of coast. In 1836, Charles Darwin visited Albany, 400 km south of Perth, and examined the nearby aeolianite coast, concluding that these deposits provided unique insights into the past productivity of marine carbonate organisms whose skeletal remains are preserved in the deposits. Darwin also thought that since the dune sediments were deposited there had been significant changes in sea-level and the position of the shoreline.

Along most of the eastern and much of the northern margin, terrigenous sediment forms the major component of coastal deposits. Even along the more humid east coast, however, much of the mid- to outer shelf is carbonate dominated. These shelf carbonates include relict shoreline deposits that record the relatively greater contribution of temperate carbonate biota (e.g. molluscs, coralline algae) to the coast during glacial periods of low sea-level, when the supply of terrigenous sediment to the coast was lower than now because rainfall was much lower.

6.1: Submarine canyons

Submarine canyons were first found in Australia by geologist Reg Sprigg in pioneer dives off the South Australian coast in the 1940s. Deep-cutting erosional submarine canyons are structures that form important sites of high marine biodiversity and provide habitat and breeding sites for cetaceans and commercially important fish species, especially where the canyons extend onto the continental shelf.

Australia has relatively few, especially those that extend onto the shelf. This is probably because of the low rate of sediment supply to the shelf and the slow rate of associated processes of mass failure, which help form canyons. In turn, this is a product of the passive margin setting and the low rate of river sediment discharge to the Australian shelf.

An example of a 'shelf-incising' submarine canyon is the Perth Canyon, which is one of the few that were probably directly connected to coastal rivers, at least via sediment pathways. During Quaternary periods of low sea-level, the sea dropped up to 125 m, bringing the coast much closer to the head of the canyon.

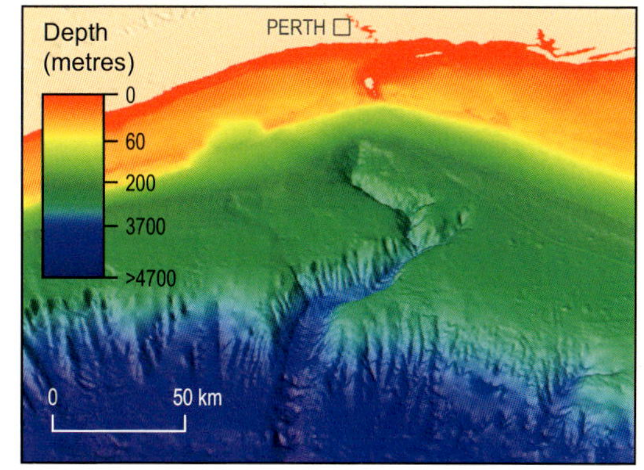

DISTINCTIVE FEATURES OF THE AUSTRALIAN COAST (BOX 6.2)

A range of characteristics of the Australian coast clearly distinguish it from other continents. These features are a consequence of the continent's geological evolution, which has set the configuration and aspect of the coast, in concert with a unique set of climatic and oceanographic regimes that determine wave and tide energy and the supply and distribution of sediment.

Key large-scale features include:

1. the extensive southern coast, the world's longest south-facing continental margin, which constantly feels the force of large waves from the Southern Ocean
2. thousands of coral reefs, including the vast barrier and fringing reefs around the margin of northern Australia
3. thousands of kilometres of coastal cliffs, especially along the south and southwestern margin, composed of Cenozoic marine deposits and Quaternary aeolianite
4. thousands of kilometres of coast with large gulfs, bays, beaches and headlands
5. the macrotidal coasts of northwestern and northern Australia, with vast networks of tidal creeks, mangroves and tidal flats, often juxtaposed with exposed bedrock cliffs and ridges
6. giant sand islands and coastal dunefields, especially on the southwestern, central-eastern and northeastern coasts.

Twelve Apostles, Great Ocean Road, Victoria.

© Getty Images [P Walton]

1. Cape Bauer, Eyre Peninsula, South Australia, on the continent's southern margin.

© Getty Images [P Folrev]

2. Large coral reef, offshore Airlie Beach, Queensland.

© R Lewis, Lonely Planet

3. Limestone cliffs, Pennington Bay, Kangaroo Island, South Australia.

© Getty Images [D Messent]

4. Bays and headlands.

© Getty Images [J Edwards]

5. Mangroves lining the Alligator River, Northern Territory (photo high tide).

Image courtesy of Tourism Queensland

6. Giant sand dunes, Fraser Island, Queensland.

The terrigenous fraction of modern coastal sediment in southeastern and southwestern Australia is dominated by highly resistant ancient quartz sand grains. These grains are extensively rounded and etched, indicating that the sand has been mobilised over long periods. Grains of zircon, another highly resistant mineral, are also found in trace quantities and have been dated using the SHRIMP (Chapter 2). The age of the zircon grains reveals the location of the original onshore bedrock source of the sediment and shows that sand is often transported hundreds, even thousands, of kilometres along the coast (e.g. Sydney to Brisbane). This longshore-drift transport is driven by coastal wave and wind regimes and, over longer time scales, by the regular sweeping of the shoreline back and forth across the continental shelf approximately every 100 ka (Box 2.6).

Changes in sea-level

Since the beginning of the Quaternary, the level of the sea has exhibited periodic excursions to around 125 m below its present position, during which the Australian continental shelf has been repeatedly submerged and exposed (Box 2.6). The cycles of Quaternary sea-level change are driven by the growth and retreat of ice sheets and glaciers during 20 or more ice ages. During the peak of the most recent glacial period at 21 ka, when ice sheets were at their maximum in the Northern Hemisphere, sea-level was 125 m lower than it is today because of the great volume of water locked up in ice. At this time, the coast lay several to hundreds of kilometres seaward of its present position, depending on the

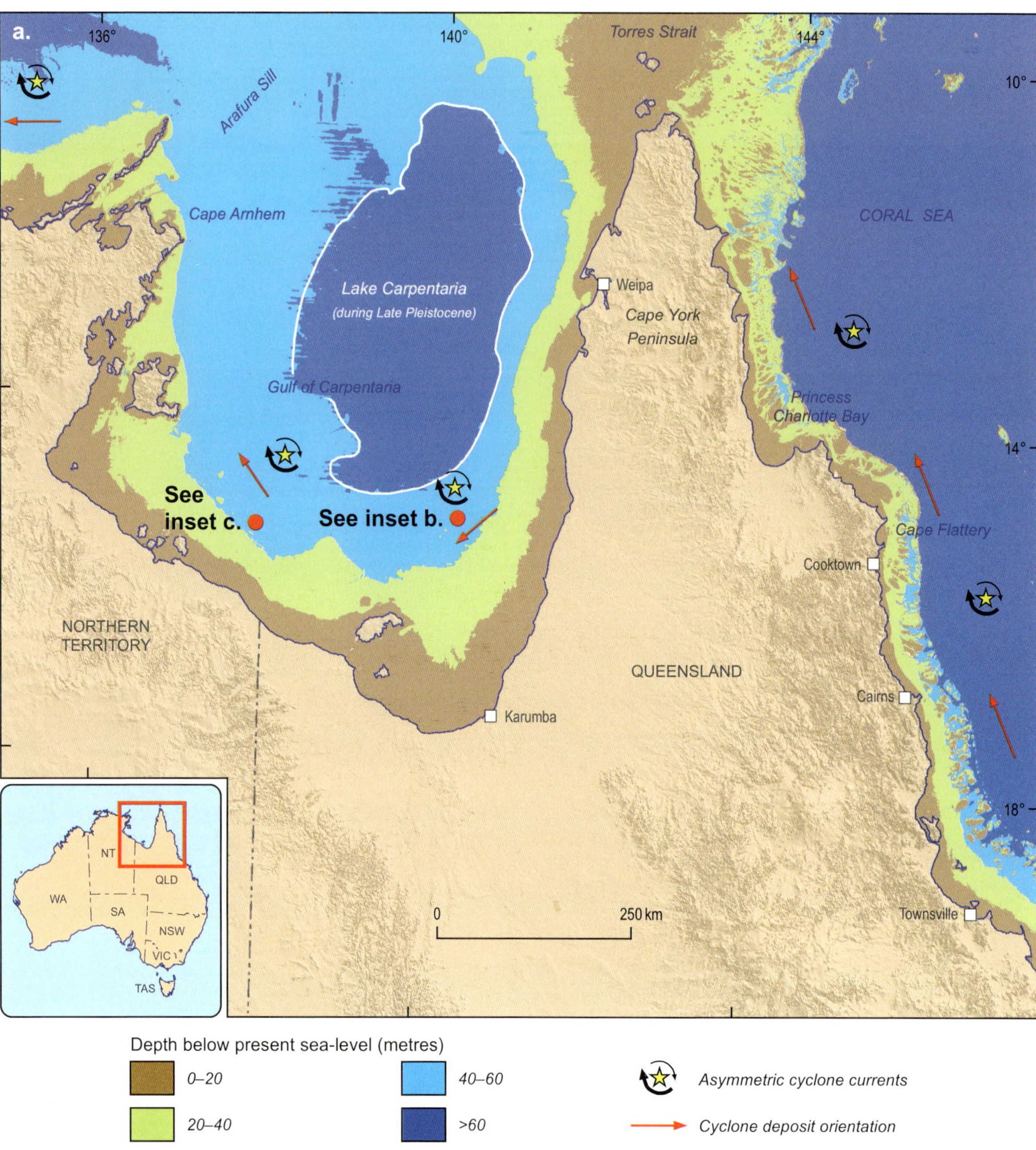

Figure 6.5: The submerged lake and reefs in the Gulf of Carpentaria. (a) Depth contours illustrate the approximate locations of shorelines when sea-level was 20 m below present (brown), 40 m below present (pale green) and 60 m below present (light blue). The dark blue area outlined in the centre of the gulf illustrates the maximum extent of Lake Carpentaria during the Late Pleistocene, around 21 ka. (b) and (c) Multibeam sonar bathymetry maps of submerged 'give-up' coral reefs (locations shown in a). These reefs are part of a reef province that extends around the southern margin of the gulf. (Source: Harris et al., 2008)

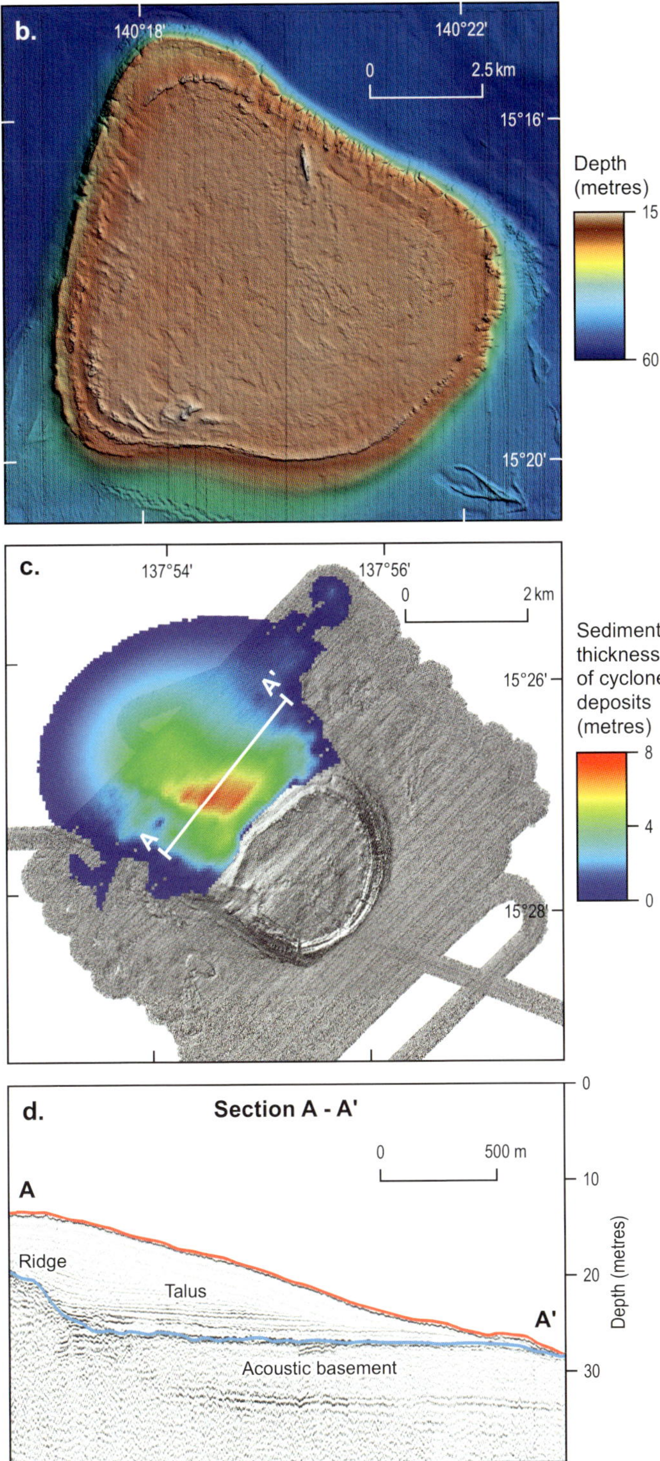

gradient of the shelf, exposing some 2 M km^2 of the Australian continental shelf. Vast coastal plains, the Sahul, existed on the North West Shelf, the Gulf of Carpentaria and the Great Barrier Reef lagoon.

During these sea-level lowstands, Australia's rivers flowed across the continental shelf, and a few connected to the heads of large submarine canyons, such as the Perth Canyon (Swan River, WA) on the southwest margin (see *Did you know?* 6.1) and the Sprigg Canyon (Murray River, SA) in the southeast. During this period, land bridges formed between mainland Australia and adjacent offshore islands including Kangaroo Island (SA) and Tasmania, and with Papua New Guinea, enabling Aboriginal people to move between these areas. At the same time vast, shallow, brackish lagoons and lakes (epicontinental lakes) were formed in the low-lying areas of Bass Strait, the Gulf of Carpentaria and Bonaparte Gulf. Rising sea-level around 12 ka flooded the land bridge between Tasmania and the rest of Australia, and the Tasmanian Aboriginal people were cut off from contact with the mainland. Their cultural development continued in isolation from mainland tribes for thousands of years, until European contact occurred in the late 1700s.

The formation of epicontinental lakes has occurred repeatedly during the Quaternary. For example, the shallow basin within the Gulf of Carpentaria has been submerged and exposed (Figure 6.5); it was transformed into a lake during times when sea-level was at least 53 m lower than at present—below the depth of the Arafura Sill. During the last glacial maximum, Lake Carpentaria may have been as much as 500 km long, 250 km wide and 15 m deep. This lake phase was short lived, however, perhaps only around 10 kyr duration. When the Arafura Sill was breached by rising global sea-level, the gulf became a brackish to marine lagoonal environment. Recent studies of the Late Quaternary history of the gulf have shown that repeated exposure and submergence are recorded in basin-wide alternating marine and lacustrine sequences. During the most recent post-glacial sea-level rise, the Torres Strait Sill was breached around 10 ka and the Coral Sea was reconnected to the Gulf of Carpentaria.

In glacial periods, the land bridge between Cape York and Papua New Guinea formed at the location of present-day Torres Strait. This exposed land provided a migration route for Aboriginal and Torres Strait Islander people, which was open for most of the past 125 ka (Figure 6.5). During this period, sea-level

Figure 6.5 continued
Cyclone deposits adjacent to a submerged reef (c) in the Gulf of Carpentaria (reef surface is 26 m below sea-level). An acoustic sub-bottom profile over the deposit adjacent to the reef (d) shows the base of the deposit (blue) and the seabed reflector (red) that are used to derive sediment thickness (c). The orientation of observed cyclone deposits is controlled by the dominant net cyclone-induced current direction (Hearn & Holloway, 1990) on the coast of northeast Australia. Heavy arrows (a) indicate likely cyclone-induced stronger inshore currents, and light arrows indicate weaker offshore currents. These cyclone-induced transport pathways likely extend along most parts of northern Australia's shallow continental shelf. (Source: Harris & Heap, 2009)

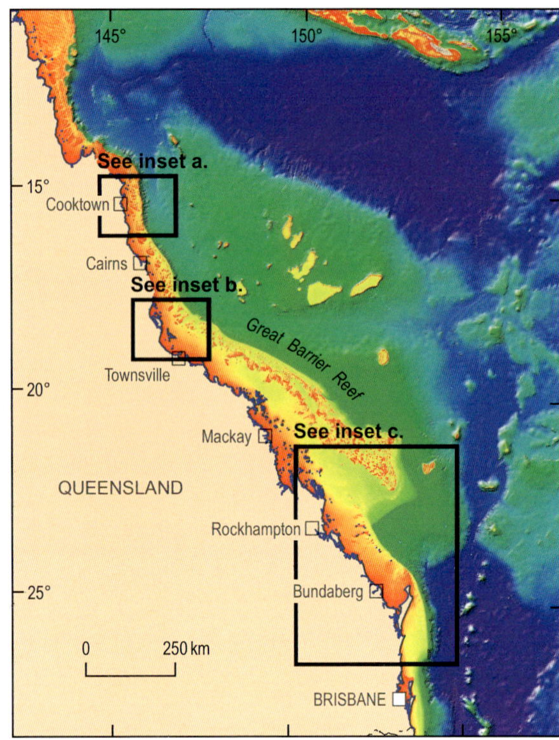

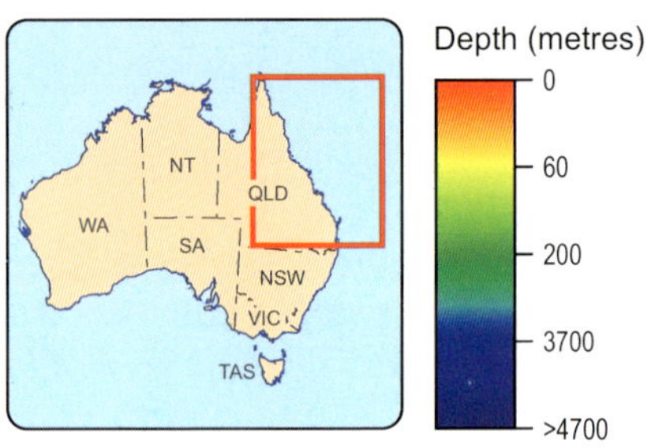

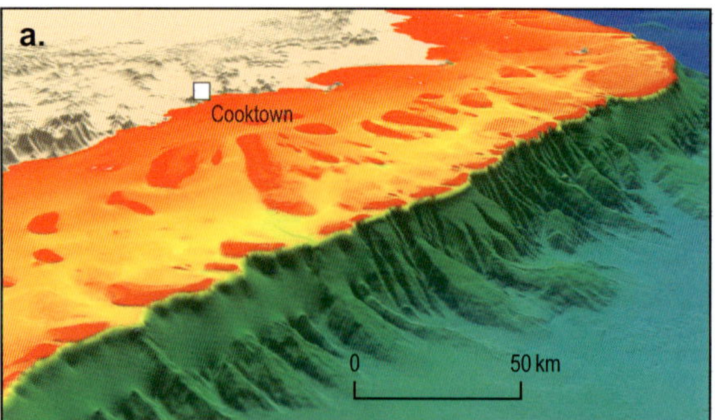

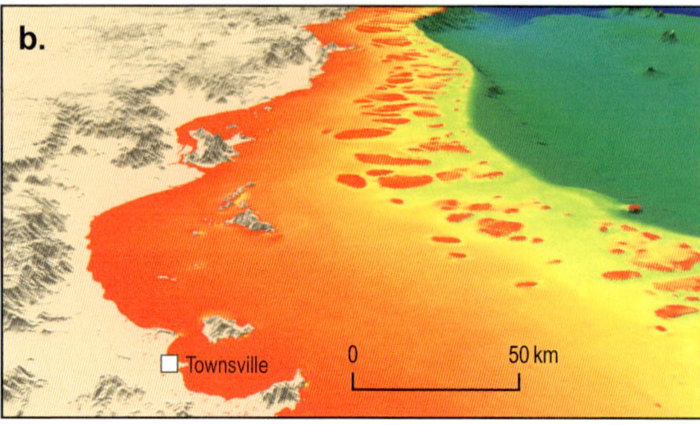

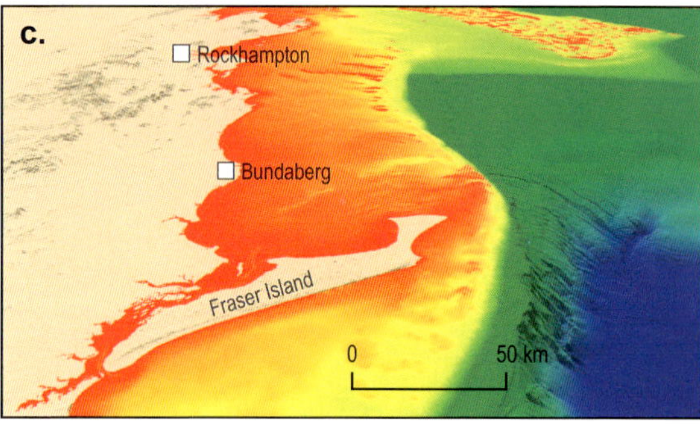

Figure 6.6: Bathymetry map for northeast Australia, showing the Great Barrier Reef. Locations of perspective views (a, b, c) are also shown. (a) Perspective view of the northern Great Barrier Reef in the Cooktown region, showing a chain of fossil reefs along the shelf edge and submarine canyons that incise the steep continental slope (land is grey). (b) Perspective view of the Great Barrier Reef in the Townsville region, showing the reef structures on the outer shelf and broad lagoon on the mid and inner shelf. (c) Perspective view of the southern end of the Great Barrier Reef in the Bundaberg region, showing the southern end of the reef, with Fraser Island, the world's largest sand island, in the foreground.

was at its present high position for only around 13% of the time, and most often (>87% of the time) it was more than 10 m below its present position (Box 2.6). The extended coastal plains were undoubtedly important sites of occupation by Aboriginal and Torres Strait Islander people during these times, but today lie submerged. An example of this comes from around Perth, where Aboriginal artefacts made from Eocene-aged chert have no obvious exposed source area. Curious too is the fact that the size of the chert artefacts diminished with time, the people clearly recycling older tools. It was not until offshore oil drilling intersected the Eocene stratigraphy that the source of the chert was found. The low-stand sea-level must have exposed reefs of chert that were used, but, once covered by the sea around 6.5 ka, only the previously mined resources were available—for recycling.

Sea-level rose from a lowstand position of around 125 m below present at 21 ka to reach its present level around 6.5 ka, rising at an average rate of around 11 mm/year, rapidly flooding the continental shelf. Adjustment of the continental shelf to the greater load of water produced only very minor movements (<2 m) along some sections of the Australian shoreline during this sea-level highstand.

The living edge—coral reef growth and sea-level

Knowledge of how reefs responded to past changes in sea-level and climate provides important insights into how they may respond to predicted future climate change. During the Quaternary, coral reefs were established on the continental shelf of Australia when rising sea-levels flooded the shelf, and then died when sea-level receded. At lowstands of the sea, coral reefs formed on the edge of the shelf in northern Australia and along the upper continental slope of the Great Barrier Reef, but coral growth ceased on these reefs when subsequent sea-level rise stranded them in deep water. Conversely, reefs that grew during periods of high sea-level were exposed during the subsequent glacial periods of low sea-level. The exposed reefs formed limestone hills that stood tens of metres above the coastal plain and were eroded by rivers and rainfall. The Great Barrier Reef we see today is therefore the product of cycles of reef growth and erosion, in concert with the numerous Pleistocene sea-level cycles (Figure 6.6; Box 2.6).

With rising sea-levels, coral reefs are usually re-established upon limestone foundations formed by the relict reefs that grew during a previous cycle. Only a portion of the available relict limestone platforms are, however, recolonised by active reef-forming coral colonies during rising sea-level. In the Great Barrier Reef, most platforms may support local, sparse patches of living reef-forming corals and diverse sponge and algal assemblages, but their upper surfaces largely comprise relict reef limestone and Holocene corals, forming what are known as 'give-up' coral reefs—reefs that did not grow fast enough to keep pace with a rising sea-level.

In the Gulf of Carpentaria, a recently discovered reef province is composed almost entirely of give-up coral reefs (Figure 6.5). Uranium-thorium ages of coral taken from drill-core samples show that reef growth here commenced shortly after the limestone pedestals were submerged by rising sea-level around 10.5 ka, making them the oldest Holocene reefs known in Australia. Reef growth persisted for around 2 ka, but had ceased at most locations by approximately 8 ka. Measurements of reef growth rates (0.95–4 m/ka) indicate that the reefs were unable to keep pace with contemporaneous rapid sea-level rise (>10 m/ka), which is consistent with a 'give-up' reef growth history. Similar relict Holocene reefs have been reported from locations in northern Australia in the Arafura and Timor seas, and surrounding Lord Howe Island, 700 km northeast of Sydney, which sits at the present southern limit of coral reef growth (Figure 6.7). The relict reef around Lord Howe Island is around 20 times larger than the modern reef.

Australia has become greatly endowed with coral reefs during the Quaternary, ranging from large offshore barrier-reef complexes to fringing reefs attached to the mainland coast and islands. These structures comprise a framework of cemented dead coral and other reef sediments that have accumulated over the past several thousand years of reef growth, with a cap of living coral, coralline algae and associated reef organisms. Other

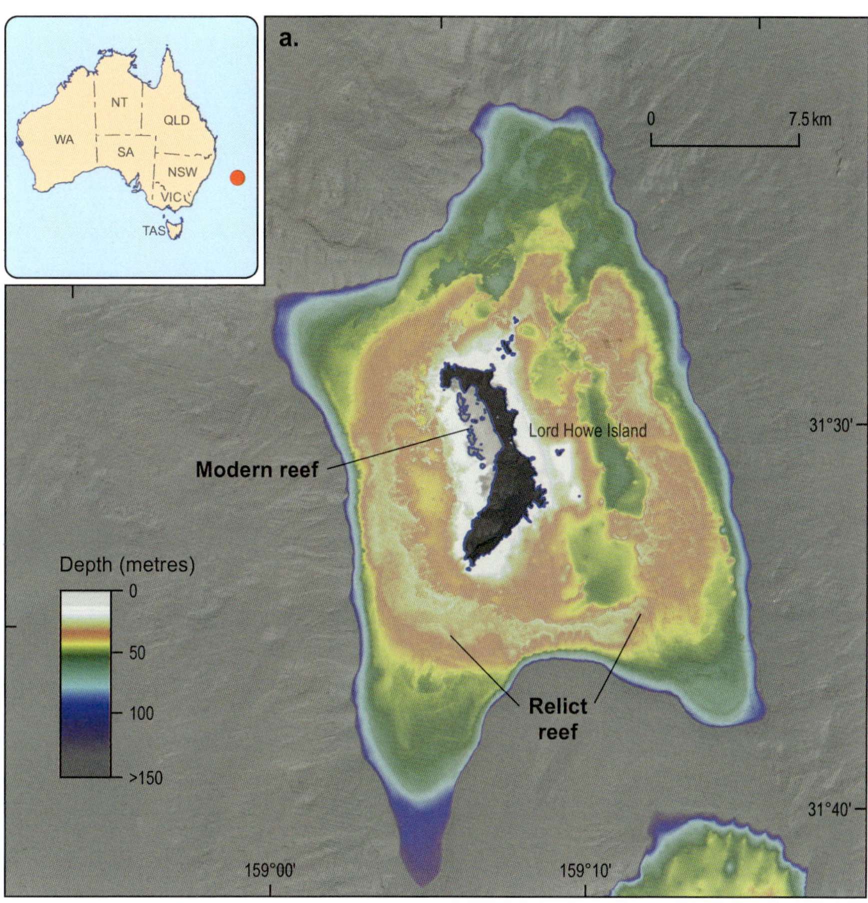

Image courtesy of Ian Hutton

Figure 6.7: Holocene and modern coral reefs at Lord Howe Island. (a) Digital relief model of the submarine platform that surrounds Lord Howe Island (black), showing the modern and much larger relict reef. (b) Aerial view of Lord Howe Island, showing the modern reef and shallow lagoon (pale blue) that fringes the western coast of the island. (Source: Ian Hutton)

Image by Neville Coleman OAM

Zuytdorp Cliffs, Western Australia.

large, biogenically formed seabed structures include carbonate sandbanks. These deposits predominantly comprise skeletal fragments of carbonate-producing biota, such as red and green algae (*Halimeda*) and foraminifera. Very large carbonate banks occur on the North West Shelf and Queensland Plateau and are recognised for their high biodiversity and pristine condition.

Reefs are well developed on the tropical and dry-tropical northeastern (Great Barrier Reef), northern (e.g. Wessel Islands) and northwestern (e.g. Ningaloo Reef; Box 6.1) coasts of Australia. These localities are suitable for coral colonisation and reef growth because of a firm substrate, as well as low rates of sediment and nutrient input from the land. Australia forms the base of the so-called 'Coral Triangle', a region that includes Indonesia, East Timor, the Philippines and Papua New Guinea. This region has the greatest diversity of marine life known to exist anywhere on Earth. However, the luxuriant coral reef growth that is a dominant feature of the Great Barrier Reef and Ningaloo Reef continental shelves does not occur along Australia's northern margin, which borders the Coral Triangle. There are fringing and patch reefs on the inner shelf along the Kimberley coast (e.g. Montgomery Reef) and isolated atolls and banks further offshore (e.g. Ashmore Reef, Big Bank Shoals, Scott Reef), but the shelf edge along the northern margin is otherwise devoid of coral reefs living at sea-level. This is despite the widespread availability of submerged limestone platforms that were apparently suitable for coral colonisation. Identifying the reasons for this pattern of reef development remains a challenge for marine scientists.

Beautiful one day, perfect the next: the modern Australian coast

The modern coast of Australia takes many forms, but it is the sand, surf and sun that are the well-known characteristics of the Australian coast, which today totals more than 36 000 km in length. Sandy shorelines make up around half this length, but equally stunning are the vast stretches of cliffs and rocky ground that comprise one-third of the shoreline. Muddy shorelines are a major feature of the less well-known northern coast (north of the Tropic of Capricorn), where mud flats form the seaward margin of immense coastal lowlands, as well as the soft seaward margin of rocky areas. In general terms, most rocky coast occurs in southern and northwestern Australia, while large stretches of sandy coast occur right around the continent but predominantly in the south (Box 6.2). Here we examine the geological, geographical, oceanographic and climatic processes that combine to produce the much-loved and internationally renowned Aussie beaches, bays and coastal lakes, as well as the lesser known, but very distinctive, rocky and muddy coasts.

The hard edge

Australia has a unique range of rocky coasts, from rounded outcrops of granite in southwestern and southern Australia to the towering cliffs of marine limestone that form the sharp southern edge of the Nullarbor Plain (Figure 6.8). The rocky southern coast also presents great juxtapositions in the geomorphology and age of the coast. Here, huge, smooth domes of Proterozoic granite sit adjacent to, or under, the sharp, ragged cliffs of Pleistocene aeolianite, the unconformity between these rocks marking the passage of 1330 Myr. Somewhat counterintuitively, in Australia it is the soft, younger rocks that form most of the cliffed and rocky shoreline, such as the Quaternary aeolianite cliffs and reefs on the southwestern (e.g. Zuytdorp Cliffs) and southern (e.g. Eyre and Yorke peninsulas) coasts, and the Cenozoic marine limestone cliffs along the Nullarbor Plain. However, the old, hard Gondwanan rocks do form extensive, dramatic rocky shorelines, such as on the eastern (e.g. Sydney and Otway Ranges), southern (e.g. Kangaroo Island, Tasman Peninsula) and northern (e.g. Kimberley) coasts.

Most rocky shorelines have been cut back by waves during the Holocene and past periods of high sea-level. The rate of this, however, varies with lithology and exposure to wave energy. The hard sandstone coast around Sydney, which is exposed to moderate wave energy, has a long-term cliff retreat rate of around 1 mm/year, whereas the much softer Cenozoic limestone coast of the Great Ocean Road (southeast SA and western Vic.)—the 'Twelve Apostles' coast (Box 6.2)—exposed to very high wave energy, has an average cliff retreat rate of around 25 mm/year. Nevertheless, even the high wave-energy regime of southern and southwestern Australia has had little discernible impact on sections of Proterozoic-aged granite coast.

Beach near Fraser's Reef in Bundjalung National Park, New South Wales.

Shaping the soft coast

Australia's sandy beaches and coastal dunes were deposited over the last few thousand years, during the present period of high sea-level. Estuaries were

Figure 6.8: Coastal localities of Australia, including notable cliffs, beaches and dunes.

also formed where streams and rivers were dammed or deflected as shoreline deposits accumulated, and with the flooding of valleys and lowlands by the rising sea. A key feature of many sections of the Australian coast is that these Holocene and modern deposits commonly overlie or sit against beach and estuarine landforms and sediments that date back to the Last Interglacial (125 ka), which was the previous period of warm climate and high sea-level (Box 2.6). The great stability of much of the coast, together with low rates of erosion, has resulted in the sea now being almost at the same level as the last highstand.

There are three major coastal sedimentary provinces in Australia. On the east coast, sediments are predominantly quartz rich and terrigenous; the south and west coasts are marine-sourced carbonate dominated; and the northwest is a mix of quartz and carbonate sediment. In the northwest, north and northeast, most terrigenous sediment has been delivered to the coast by tropical rivers. Large-scale examples of sandy beach deposits composed of quartz sand occur in northeastern and southeastern Queensland (e.g. Cape Flattery and the Gold Coast, respectively), northern and central New South Wales (e.g. Byron Bay and Forster), eastern Victoria (e.g. Ninety Mile Beach) and northern and eastern Tasmania (e.g. Bakers Beach, Marion Beach) (Figure 6.8). Coasts composed of carbonate or mixed carbonate/quartz sand occur in the south (e.g. the Coorong and Eyre Peninsula coasts) and along the southwestern coast. These areas are exposed to the strong wind and wave regimes generated in the Southern and Pacific oceans and are therefore also referred to as wave-dominated coasts.

Australia's northern coast is a mix of muddy, rocky and sandy environments. Much of this coast has a large tidal range, low wave energy and extensive muddy lowlands. There are numerous river deltas; broad, sandy beaches; and large barriers. In the Kimberley region, these depositional environments are commonly located within large bedrock-fringed embayments (e.g. Walcott Inlet). Tides in the north range up to 12 m, generated by the interactions between the regional tidal circulation system, the broad continental shelf, the highly indented shoreline and the orientation of the coast. Currents generated by these tides are capable of constantly moving sandy sediment along the seabed. Sediment mobility can be a major concern for shipping. For example, the Ports Corporation of Queensland maintains a major shipping channel through Torres Strait plus several port facilities on the remote islands in the strait where islander communities rely on sea transport for their livelihood and communication with the rest of Australia.

Australia's beaches and dunes

Beaches, dunes and estuaries can be classified into different types, based on their major geomorphological, sedimentary and oceanographic characteristics (Box 6.3). These classifications highlight well how particular regions have distinctive coastal processes and landforms. For example, Australia has more than 10 700 beaches that occupy approximately 50% of the total coastline. They fall into four broad morphodynamic beach types: wave dominated, tide dominated, tide modified, and beaches with reef or rock outcrops (Box 6.3). Each type has a particular shape and sediment texture, produced by the varying relative influence of wave and tide power on the composition of beach sediment.

Knowledge of the morphology and dynamics of beaches is important for understanding how beaches change over seasonal cycles, and how they may react to storms, as well as to changes in waves and sea-level that may result from any future climate change. Another important application of this knowledge is the prediction of the amenity

Inglis Creek in the remote Kimberley region of northwest Australia, showing bedrock ridges, mangrove forests and broad muddy tidal flats that are characteristic of this macrotidal coast.

and safety of the beach. This information informs coastal land-use planning and Surf Life Saving Australia, whose members patrol more than 250 of Australia's most visited beaches.

Beaches in southern Australia

Wave-dominated beaches, synonymous with the coast of southern Australia, have clearly been the most favoured backdrop for European settlement and urban expansion. These beaches have a broad low-gradient beach face made of well-sorted sand, and a wide surf zone with well-developed sandbar and troughs. They are often set between prominent headlands, comprise clean, white quartz or carbonate sand, with clear, clean water, and with a wave climate that produces regular year-round surf. A defining feature of much of the southern and southwestern coasts is the negligible discharge of terrigenous sediment by ephemeral rivers compared with the volume of bioclastic carbonate sediment produced on the shelf and moved onshore to form beaches by the strong wave regime (e.g. Dog Fence Beach, SA).

Beaches in northern Australia

Northern Australian beaches are largely shaped or significantly influenced by tides. They generally have a narrow, steep beach face of relatively coarse shelly sand, fronted by an extensive low-gradient sand or mud flat. At low tide, the shoreline may be hundreds of metres to a few kilometres (on low-gradient coasts) from the sandy beach face, the width increasing with tidal range. These beaches usually lack good surfing waves as a consequence of

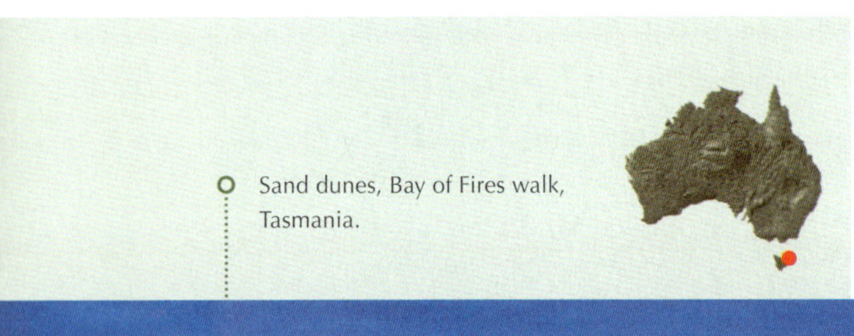

Sand dunes, Bay of Fires walk, Tasmania.

the large tides, beach morphology and generally low wave energy. They occur where there are macrotidal regimes or where tidal energy is high relative to wave energy (e.g. Cable Beach near Broome, Carmila Beach near Broad Sound). There are also large stretches of the northern and northeastern coast with sandy beach systems formed under much smaller tidal regimes (2–3 m range), in places backed by large sand barriers and dunefields similar to those found in southern Australia (e.g. Cape Arnhem, Cape Flattery and Corio Bay).

Coastal dunes

Approximately 85% of the sandy beaches in Australia are backed by coastal dunes. These aeolian landforms comprise sand that was transported to the beach by waves from the continental shelf (offshore) or from rivers (alongshore), then eroded from the beach by strong onshore winds and deposited inland. Coastal dunes range from shore-parallel deposits a few hundred metres to a few kilometres wide, that rise just a few metres above the beach, to vast sand seas that extend tens of kilometres inland, with dunes that rise tens to hundreds of metres. Dune barriers lie parallel to the beach and protect adjacent estuaries, lakes and coastal lowlands from the wave energy of the open ocean. They are common on the east, southeast and southwest coasts and at numerous sites in Tasmania.

Some of Australia's low coastal dunes have been readily transformed into urban areas—for example, much of the Gold Coast, parts of eastern Sydney and suburbs that line Warnbro Sound near Perth sit on coastal dunes and beach-ridge plains. In contrast, vast sand seas and large dunefields are conspicuous features of the wild southern and southwestern Australian coasts (e.g. Kaniaal and Warren beaches) and remain largely undeveloped due to their lack of freshwater, poor and unstable soil, and remoteness. These near-pristine dune landscapes are also characterised by high levels of terrestrial biodiversity, and several now sit within national parks and state reserves (e.g. Canunda National Park, D'Entrecasteaux National Park).

Most coastal dunes in Australia were emplaced during the Holocene (typically 11–2 ka), which includes the later stage of the last marine transgression (11–6.5 ka; Box 2.6). Dunes continued to be deposited during the subsequent period of stable sea-level (last *ca* 6.5 ka), but often at a lower rate. Today, the rate at which sand is supplied to beaches is even lower than during the late Holocene. In rare cases, there is a negative sediment budget, with the beach undergoing long-term erosion and the shoreline moving landwards (e.g. Dee Why beach in Sydney, Fraser Island). Eroding shorelines such as these can create major problems for coastal communities, as discussed later.

Another common feature of many of Australia's large coastal dunes is that they overlie or abut relict coastal dunes. For example, the giant sand dunes of southeastern Queensland (e.g. Fraser, Moreton and Stradbroke islands, and Cooloola) and the central coast of New South Wales (e.g. Stockton Bight) are where Holocene and recent sand dunes cap large Pleistocene dunes. Pleistocene dune barriers sit landward of modern barriers in numerous locations along the southern and southwestern coasts (e.g. Ninety Mile Beach; Ainslie Sand, northeastern Tasmania; and the Coorong Coast).

Did you know?

6.2: Mapping the unknown edge—the benthic environment

Developing a comprehensive knowledge of seabed environments is a major challenge in Australia because of the vast size (*ca* 8.2 M km^2) and diversity of the marine estate. Indeed, the biodiversity of the seabed is one of the nation's least known key natural resources. In recent years, the true geomorphic detail of the seafloor has been revealed by high-resolution multibeam sonar mapping down to spatial scales of 2–3 m (see Box 6.1). This approach is proving highly effective as it provides near-continuous coverage of large areas (hundreds of square kilometres) of seabed compared with site-specific information provided by scuba surveys, for example. In coastal waters and the adjacent shallow continental shelf, it is particularly important to map seabed morphology in this way because it can, to varying degrees, represent the diversity of seabed habitat types that exist over large areas at fine spatial scales and enable observations of seabed biota to be accurately related to the geomorphology of the site at which it was recorded (see Appendix 6).

Image by Rachel Przeslawski

AUSTRALIA'S BEACHES (BOX 6.3)

Why are Australia's beaches so good for surfing? Well, not all of them are. Australia's beaches can be classified into four major types: wave dominated, tide modified, tide dominated, and beaches fronted by rock or reef flats (Figure B6.3). In Australia, wave- and tide-dominated and tide-modified beaches represent nearly 90% of all beaches. They are typically bound by headlands and are usually free from offshore rocks and reefs. Wave-dominated beaches occur principally around the southern half of Australia; tide-influenced and tide-dominated beaches occur mainly along the northern half; and beaches with rock and reef flats are slightly more abundant in the north (Figure B6.3).

In reality, each beach can be classified by a 'modal state' that is aligned with the most common wave conditions to which it is subject. The morphology of a beach changes in response to variations in wave energy. Thus, wave-dominated beaches experience persistent high wave energy throughout the year and develop a wide, low gradient profile. This is in contrast to Northern Hemisphere low-gradient beaches that often experience greater seasonality, with a distinctive low–wave energy season.

The morphology of wave-dominated beaches results in their being particularly good for surfing, but they are also relatively hazardous for swimming due to the plunging breakers and well-developed rip currents. For beaches fronted by rocks and reef flats, waves breaking on the steep upper beachface at high tide produce very dangerous surf conditions. These beaches are therefore some of the most hazardous, particularly along the southern Australian coast. At low tide, the rock or reef platform is exposed and waves break far offshore.

Bondi Beach, New South Wales.

© Getty Images [D Messent]

© R I'Anson, Lonely Planet Images

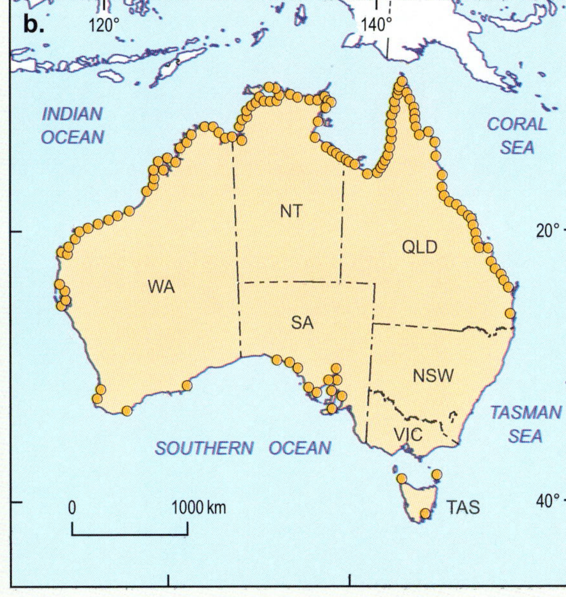

Top image is of Manly Beach in Sydney, New South Wales—a typical wave-dominated beach with a wide surf zone, shore parallel bars and rip currents generated by relatively high wave energy. The image below shows the very broad low-tide sand flat at Cable Beach, near Broome, Western Australia, a tide-dominated beach.

Figure B6.3: Distribution of beach types around the Australian coast. (Source: Short & Woodroffe, 2009)

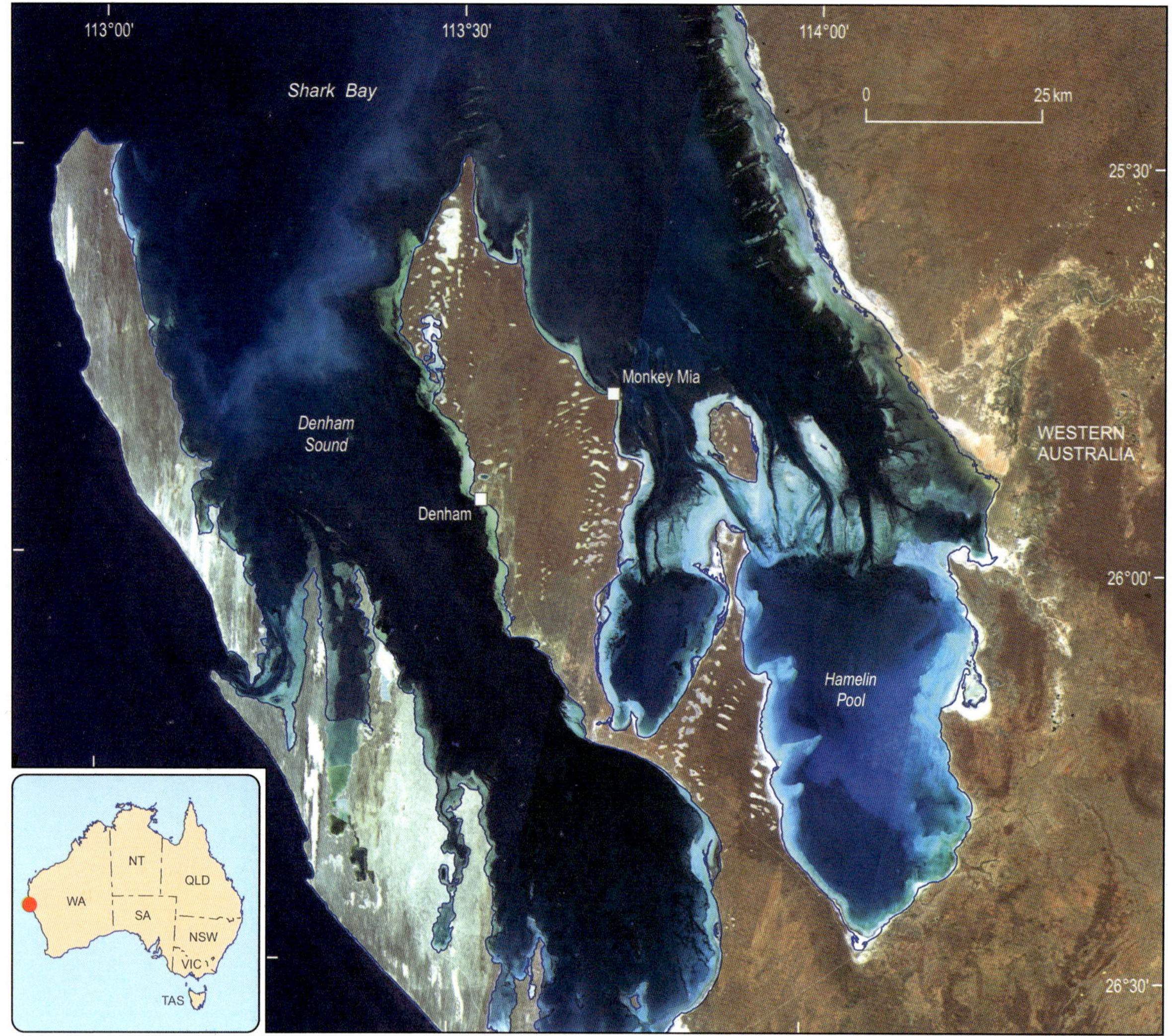

Figure 6.9: Satellite image of Shark Bay, Western Australia.

Both the Holocene and Pleistocene dunes can provide valuable groundwater resources that are used by many coastal communities (e.g. Moreton Bay, Perth and Rottnest Island), and often host groundwater-dependent ecosystems (Chapter 7). While most large coastal dunes remain undisturbed, some systems located close to transport routes and industry have been used to extract valuable heavy minerals (e.g. rutile, zircon, ilmenite) that occur in sufficient concentration to warrant mining (Chapter 5). Examples of mineral-sand extraction from coastal dunes include North Stradbroke Island off the coast of Brisbane, Stockton dunes adjacent to the port of Newcastle, and Coburn, north of the port of Geraldton.

There are significant differences between the coastal dunes of northern and southern Australia. In southern Australia, dunes are more extensive and contain a far larger volume of sand than those in the north. A few exceptions are large coastal dunefields in Arnhem Land, and Cape St Lambert. Overall, these differences are a product of the higher wind- and wave-energy regimes of beaches in the south, where the coast feels the impact of strong swell and onshore wind, especially from the Southern Ocean, and a supply of sand from offshore (shelf) or alongshore (river). In contrast, the coast of northern Australia experiences much lower levels of wave energy and therefore lower rates of sediment delivery to beaches, as well as less frequent strong onshore wind.

Australia's coastal waterways and estuaries

Gulfs and large embayments

Australia has several large, shallow-water gulfs and bays in a wide range of geographical settings. Some of the larger examples are Shoalwater, Harvey and Moreton bays; Jervis and Twofold bays; Port Phillip Bay; Great Oyster Bay and Macquarie Harbour; Spencer Gulf and Gulf St Vincent; Shark Bay, Exmouth and Admiralty gulfs; and Joseph Bonaparte Gulf (Figures 6.2 and 6.8). Most are protected to varying degrees from ocean swell, and their size, shape and shallow depth often produce unique wave and tidal regimes. Likewise, many have unique and highly diverse marine ecosystems that provide some of the more productive coastal fisheries and important tourism sites. A very large tropical example is the Gulf of Carpentaria (300 000 km^2). In early 1606, the Dutch explorer Willem Janszoon, aboard the *Duyfken*, landed at the Pennefather River on the eastern coast of the gulf, near Weipa (Chapter 9). This coast remains sparsely populated (Figure 1.3) and has changed little since Janszoon's visit, preserving internationally significant feeding and breeding areas for turtles, dugongs, dolphins and seabirds; several Aboriginal communities maintain their traditional fishing and hunting areas here. The gulf coast also contains nationally important commercial fishing grounds, especially for prawns and a range of fish (e.g. barramundi, mackerel), and some of the world's largest bauxite mines and export facilities (e.g. Weipa and Gove; Chapter 9).

Shark Bay is one of the world's unique and pristine coastal seas and is on the Australian and World Heritage lists. With a coastline of 1500 km and a water area of 10 000 km^2, this is a key tourism location for Western Australia. The bay supported three Aboriginal tribes when the first European, Dirk Hartog, visited in 1616, followed by a series of Dutch, French and British explorers in the 17th to early 19th centuries. Today, the population living on the shores of Shark Bay is less than 1000. The bay contains the world's largest seagrass meadows (4000 km^2), which support thousands of dugongs, and provides habitat for dolphins and whales. In the bay's shallow, very saline reaches grow the world's most diverse and abundant examples of stromatolites (Chapter 3).

Estuaries

Estuaries are transition zones between marine and terrestrial environments. Frequent changes in the relative influence of land and marine processes result in estuaries being highly dynamic environments. We rely on estuaries for essential ecosystem services, such as shoreline protection and disturbance regulation (e.g. mangrove mitigation of cyclones), nutrient cycling, habitat diversity, food production and recreation. There are more than 1000 estuaries and other coastal waterways around Australia (Figure 6.10), and these have been

Mangrove forest lining the Embley River estuary, Queensland.

Figure 6.10: The distribution of gulfs and different estuary types around the Australian coast. There is a north–south divide, with wave-dominated waterways in the south and tide-dominated systems in the north. Large tracts of the southern and southwestern coasts have no estuaries due to the lack of rivers; few people live along these coasts (Figure 1.3). There is a greater variety of estuary types in northeastern Australia.

classified into different types based on sediment supply and the relative influence of wave, tide and river processes. Here, the term 'coastal waterway' is used collectively to include estuaries, deltas, lagoons and tidal creeks, which are abundant on Australia's wave- and tide-dominated coasts.

From a geological perspective, the extent to which the 'accommodation space' has filled with sediment is indicative of its degree of maturity; estuary 'basins' that are completely filled with sediment and are prograding seawards are 'deltas'. Estuaries and deltas with a relatively strong wave influence at their entrances are termed 'wave dominated', whereas those that have a relatively high tidal influence are 'tide dominated'. Each type has a distinctive arrangement of geomorphological features. For example, sandy barriers usually sit across the mouth of wave-dominated systems (e.g. Coorong Estuary), while tide-dominated estuaries have wide, funnel-shaped mouths (e.g. South Alligator River), and deltas may protrude seawards from the line of the coast (e.g. Burdekin River delta). Each waterway type has a distinctive arrangement of sedimentary environments or geomorphological habitats (e.g. fluvial deltas, saltmarsh, mangroves, tidal sandbanks and intertidal flats) and may differ in terms of the degree of vulnerability to anthropogenic disturbances.

Shaped by waves

Several of Australia's state capitals and largest population centres are located around large wave-dominated estuaries in southeastern and southwestern Australia (Figure 6.10). Australia's major port facilities have been established in these waterways—for example, Sydney Harbour, Fremantle Harbour (Perth) and Port Phillip Bay (Melbourne). The port of Brisbane is located on the Brisbane River Delta that projects into Moreton Bay, a very large (1523 km^2), partially filled, wave-dominated estuary. Moreton Bay is flanked by two giant sand islands, Moreton and North Stradbroke islands, which form a protective barrier from the Coral Sea, sheltering the central basin of the bay itself.

Most wave-dominated systems experience microtidal conditions (spring tidal range <2 m) and have sandy beaches around part or all of the estuary shoreline. These estuarine or 'bay beaches' are often the focus for urban development, given their protected setting. In some regions, however (e.g. southeast Tasmania), the bay beaches can be particularly vulnerable to erosion from rises in sea-level and changes in wind and wave regimes. This represents a significant management challenge where buildings have been sited close to the shoreline, which is the case in many large cities (e.g. Sydney Harbour; Moreton Bay, Brisbane) and coastal towns.

Wave-dominated estuaries and lagoons, by definition, have received a relatively small amount of terrigenous sediment input during the Holocene. These systems are geologically 'immature' as they have only partially filled with sediment and often have a relatively deep, central, muddy basin. These estuaries, therefore, have the capacity to gradually accumulate and retain sediment from the river catchment as they slowly infill, which has significant implications for the health of an estuary, as we will see later.

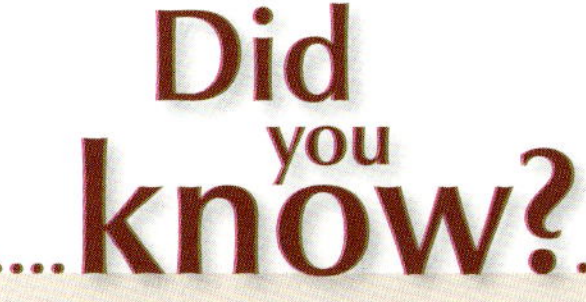

6.3: Australia's near-pristine estuaries

Near-pristine systems are assumed to have amounts of nutrient and sediment delivered in streams or groundwater at levels close to that of pre-European settlement. By implication, Australia's near-pristine estuaries are usually located in the most remote and inaccessible parts of the coastline. In fact, about three-quarters are situated to the north of the Tropic of Capricorn, where population and development are relatively sparse. As part of the 2001 national assessment of waterways, 20% of these northern waterways were considered largely unmodified, and only 7% were classified as either modified or extensively modified. In comparison, more than 50% of estuaries situated in the more populated areas south of the tropic were classified as modified or severely modified; approximately 25% were largely unmodified; and only 19% were in near-pristine condition. These near-pristine systems are largely restricted to conservation parks and reserves, where catchments have been little impacted by human activity and coastal development is restricted.

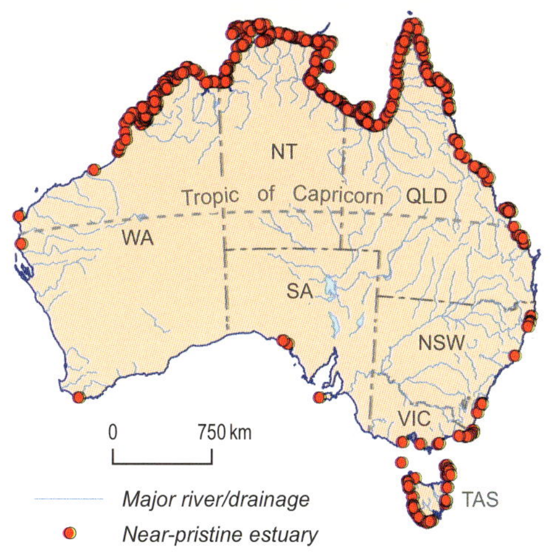

© Getty Images [P Harrison]

The Entrance, New South Wales.

Shaped by tides and rivers

Tide-dominated estuaries and deltas, together with tidal flats, are defining features of the coast of northern Australia (Figure 6.10), where macrotidal regimes (spring tidal range >4 m) are prevalent. Unlike the southern coasts, many of these waterways sit in broad, low-gradient coastal plains. In these systems, modern and relict marine, coastal and terrigenous sediments are reworked by strong tidal currents into the estuary and along the coast, creating highly turbid waterways and muddy shorelines. Macrotidal coasts are now being considered for their potential as a renewable source of energy; no tidal-energy infrastructure, however, is currently in place in Australia (Chapter 10).

In the southern Gulf of Carpentaria and on the northeast coast of Queensland, many waterways are interlinked by smaller tidal channels. These 'connected' coastal waterways are recognised for their crucial role as fish nurseries and habitat for fish such as Australia's iconic barramundi, as well as for their recreational and biodiversity conservation values. They also support the rapidly expanding aquaculture and tourist industries. Another regionally unique estuary type is found on the highly embayed Kimberley coast. Here, estuaries occupy the dramatic 'drowned' joint-aligned bedrock valleys that can extend in a straight line several kilometres inland. These waterways have provided access to the rugged Kimberley plateau, which is largely inaccessible by road.

The port of Darwin is located on a tide-dominated estuary with a macrotidal (6 m) range, low wave energy (wave height predominantly <0.6 m) and little input of terrestrial sediment. There are a number of large tide-dominated estuaries in this region, which have formed in the mouths of tropical rivers and receive significant river discharge in the wet season, produced by the arrival of the southern monsoon. The remote South Alligator and Daly rivers, for example, have the classic funnel-shaped river mouth and tidal reaches that meander tens of kilometres inland. This channel morphology reflects the dominance of tidal energy over the highly seasonal river discharge in transporting sediment; a large proportion of sediment discharged from the estuary in the wet season is subsequently reworked by tidal currents back into the estuarine reaches of the river. These estuaries can have vast networks of tidal inlets and creeks lined with mangrove forests, and often connect to extensive saline flats that are only inundated during high tides, including spring high tides. Another defining feature of these estuaries is the large population of saltwater crocodiles. Colloquially know as 'salties' (in contrast to the much smaller freshwater species), these large (up to 6 m long) protected reptiles add a significant element of danger to living in northern Australia (Chapter 3); at the same time, they are an important part of these tropical ecosystems and a key attraction for the tourism industry. Extensive freshwater wetlands occur adjacent to the tidally influenced environments in many parts of northern Australia, especially during the wet season (e.g. Mary River).

Approximately 20% of coastal waterways in Australia are wave- and tide-dominated deltas (Figure 6.10). They occur predominantly in wetter regions of northern and eastern Australia and on the Pilbara coast. Deltaic coasts often include floodplains or coastal lowlands that comprise alluvial and marine sediments, which were deposited as the coast advanced seawards. Coastal plains provide important groundwater resources and valuable arable land, which is increasingly being lost to extensive urban developments. For example, large-scale urban expansion has occurred on the floodplains formed by the Brisbane River, the Swan River, the Yarra River, and the Tweed, Clarence and Shoalhaven rivers.

Australia's near-pristine estuaries

Approximately half of Australia's estuaries were classified as near pristine during the first (2001) national assessment of coastal waterway condition. Near-pristine estuaries have not been significantly impacted by human activities, and are considered to exist in an *almost* natural state. These estuaries are valuable natural assets, with many cultural heritage values. They support nearshore fisheries, and are important as Aboriginal lands, as undisturbed habitat for biodiversity conservation and for ecotourism. Moreover, near-pristine estuaries are important because they can provide benchmark information for monitoring change in modified systems in terms of water quality, ecology and geomorphology. Most countries do not have this important opportunity for conservation of biodiversity and scientific inquiry because they lack estuaries in such good condition.

Australia's near-pristine estuaries are characterised by limited and sustainable fishing, no aquaculture, water movements through fringing wetlands and estuary inlets unaltered by structures (e.g. weirs or tidal embankments), and catchments that retain more than 90% of their natural vegetation cover. However, in reality, it is unlikely that any estuary is in absolutely pristine condition because human activities have altered global biogeochemical cycles over the past century.

Muddying the water: environmental pressures on the coast

The coast, where most Australians live (Chapter 1), is also the focus of a range of broadscale primary industries such as agriculture, forestry, fishing and aquaculture. It is the location of rapidly expanding facilities for the minerals and energy sectors, and is the setting for fast-growing urban areas. At the same time, the coast's natural attractions underpin most of the nation's tourism.

To better inform the management of these important environments, it is essential to have a comprehensive understanding of coastal sedimentary systems, especially of the processes that control the transport and deposition of sediments. These processes determine the morphology of estuaries, coastlines and the continental shelf, as well as the pathways of nutrients and contaminants.

A decline in the quality of water in coastal waterways is common in Australia's large urban centres, with some notable recent improvements following reductions in urban contaminant discharge (Box 6.4). But water quality is also a major issue for estuaries in rural regions, especially those connected to catchments with intensive agriculture. Water-quality problems have also been reported for the vast shallow shelf inside the Great Barrier Reef. As we shall see later, water quality is of major concern for the health of Australia's coral reefs, as well as people.

ENVIRONMENTAL ISSUES IN URBAN COASTAL WATERWAYS (BOX 6.4)

Moreton Bay (Brisbane, Qld)

- Nutrients (N and P) and sediment from the catchment causing west to east gradient in water quality—with poor quality in the western reaches and good quality in the eastern reaches of the bay.
- Seagrass loss, especially in the western reaches of the bay.
- Episodic nature of rainfall on a catchment that has been significantly cleared—resulting in strong pulse of sediment loads.
- Around 50% of the 48 000 km stream vegetation in poor condition.
- Decline in aquatic species diversity in some streams and areas in the bay.

Downtown Melbourne on the shores of Port Phillip Bay, Victoria.

Derwent Estuary (Hobart, Tas.)

- Heavy metal contamination of water, sediment and biota.
- Nutrient and organic enrichment from sewage treatment plants and paper production.
- Poor recreational water quality at beaches.
- Loss of habitats and species.
- Marine pests and coastal weeds.
- Artificial environmental flows from catchments.
- Barriers to fish migration.

Port Jackson (Sydney, NSW)

- Poor recreational water quality following heavy rainfall due to stormwater outflows of nutrients, bacteria and litter.
- Metals and persistent organic chemicals in sediments as a legacy from past industry and due to stormwater input.

Port Phillip Bay (Melbourne, Vic.)

- Reductions in freshwater flow, causing increased salinity.
- Loss of habitats (e.g. seagrass areas) and species (e.g. sand flathead).
- Increased turbidity due to dredging of shipping channels.

Swan–Canning Estuary (Perth, WA)

- Seasonal bottom-water anoxia in the upper estuary.
- Occasional toxic algal blooms.

Reduced river discharge and ecosystem health: the Coorong Estuary

The potential impact on estuaries of large-scale intensive agriculture in their catchments is well demonstrated by the Coorong Estuary at the mouth of the Murray River in South Australia. The estuary has experienced major declines in water quality and ecosystem health since the development of large-scale irrigation in the Murray River's catchment, the Murray–Darling Basin—Australia's largest river catchment. This decline accelerated during a decade-long drought in the 2000s. A key issue has been the extensive diversion of water for the irrigation of crops (e.g. rice, cotton and grapes) in the Murray–Darling Basin, which has led to a greatly reduced inflow of freshwater to the Coorong. As a consequence, salinity in the southern estuary has been up to eight times the salinity of normal seawater. In turn, there has been a system-wide die-back of seagrass and major reduction in habitat for fish and birds.

Land clearing and increased sedimentation

Despite several estuaries achieving near-pristine status, most Australian catchments have undergone varying degrees of large-scale land clearance and subsequent use of land for agriculture, forestry and urban development. Sediment cores from estuaries provide valuable records of the impact of these changes on estuarine sedimentary processes and ecology, which are usually otherwise not well documented. Sediment cores from estuaries, especially wave-dominated ones, typically reveal very low rates of sediment accumulation before the removal of native vegetation (usually much less than 0.5 cm/year) and an abrupt increase in sedimentation in the years immediately following the removal (up to several cm/year). Pollen and other microfossils (e.g. diatoms, spores, charcoal)

Aerial view of the Coorong river and lake system, South Australia.

Image courtesy of Tourism Queensland

Aerial view of sugar cane fields, Bundaberg, Queensland.

preserved in the sediment provide evidence of the decline in native flora, the appearance of exotic 'weed' taxa (e.g. *Pinus*), an increase in large-scale burning and changes in the estuarine aquatic taxa. In some cases, such as in Macquarie Harbour (Tas.), the sediment record actually shows a post-European reduction in charcoal and an increase in shrub pollen, evidencing the disruption to the regular Aboriginal burning of the catchment.

The estuarine sediment record also provides evidence of changes in water quality. In particular, an increased occurrence of algae can be indicative of increased nutrient concentrations, as shown for the Gippsland Lakes in eastern Victoria, which experienced phosphate loading in the mid-20th century. Similar but longer term impacts on estuarine systems related to major changes in catchment land use have been recorded in estuaries in the Northern Hemisphere. Land clearance in Australia, however, has been largely in the period of mechanisation (post-World War I), with large areas cleared in the past 50 years. Rates of change evident in Australian estuaries, therefore, can be more rapid than occurred thousands of years or centuries earlier in European and North American estuaries.

Acid sulfate soils

Low-lying plains around Australia's coastal waterways have soils and underlying sediments that often contain abundant quantities of iron sulfides (principally pyrite). When these soils are exposed to air through physical disturbance related to urban and agricultural development, the sulfides oxidise to produce sulfuric acid and heavy metals.

When released into coastal waterways, these contaminants can have devastating environmental and economic impacts, including fish kills, groundwater contamination and corrosion of infrastructure (Chapter 7).

The term acid sulfate soils (ASS) describes soils that: (i) may produce sulfuric acid (potential ASS), (ii) are currently producing sulfuric acid (active ASS) or (iii) have produced sulfuric acid (post-active ASS) in amounts that have long-term effects on soil. Based on a recent estimate, Australia has 58 000 km^2 of coastal ASS. Most of these soils developed during the Late Pleistocene (125 ka) and early–mid Holocene (*ca* 125–6.5 ka), when sea-level was similar to or slightly higher than today, and shallow coastal depositional environments extended into low-lying plains that are currently less than 5 m asl. The iron sulfides were produced through a process called sulfate reduction, which requires anoxic conditions and a source of sulfate, iron and reactive carbon, which are abundant in coastal waterways. Indeed, ASS continue to form today in subtidal and intertidal environments such as salt marshes, tidal flats and mangroves.

The disturbance of ASS is generally caused by human activities, such as the removal of native vegetation, agriculture and engineering works (e.g. drainage). There are a wide range of case studies (Chapter 7) that describe impacts and remediation options for coastal ASS (e.g. Lake Alexandrina, Gulf St Vincent, Shoalhaven River coastal plain). The financial costs of ASS impacts on Australian primary industries and infrastructure were estimated at $10 B in 2000.

Water quality collapse: the Peel–Harvey Estuary lesson

Peel Inlet and Harvey Estuary south of Perth form a shallow (mostly <2 m) wave-dominated barrier estuary. The estuary covers 136 km^2 and opens to the Indian Ocean. Before the introduction of farming in the 1830s, the estuary was characterised by low nutrient loads, clear water and extensive areas of seagrass. The Group Settlement Scheme established after World War I to settle former soldiers saw widespread land clearing of the catchment and drainage of the swampy coastal plain for farming. Soon after World War II, artificial fertilisers were introduced, allowing the area to be used for beef and dairy grazing and piggeries. The annual application of fertiliser across the Peel–Harvey catchment then soared, with superphosphate use rising from 5 t in 1945 to 25 000 t by 1975. The Peel–Harvey was then eutrophic and poorly flushed, with excessive nutrient loads derived from the fertiliser and intensive production of livestock. Adverse consequences included 50 000 t (dry weight) of macroalgal weed in Peel Inlet in 1979, and large toxic algal blooms in the surface waters.

In order to improve the health of the estuary, superphosphate application was reduced to 9500 t in 1987. Up to 20 000 m^3 of weed was removed each year, and in 1994 a channel was cut through the sandy barrier to the ocean. The new channel increased tidal flushing, thereby diluting the load of dissolved nutrients, improving light penetration to the estuary bed and reducing the potential for the growth of weed and toxic algae. By 1996, the volume of weed harvested had decreased to 1300 m^3. Harmful algal blooms are now greatly reduced and restricted to the upper reaches of the estuary, and the distribution and extent of seagrass has significantly increased.

The Great Barrier Reef lagoon

The Great Barrier Reef is internationally recognised as one of the world's best-managed coral reef ecosystems. Nonetheless, the lagoon has water-quality problems, especially the inshore zone (<20 m depth, <20 km offshore), exhibiting signs of environmental decline—in particular, from an increase in the sediment loads of rivers that drain the coastal catchments. The long residence time of fine sediments in the inshore zone leads to higher water turbidity and light attenuation. The latter reduces the depth range of phototrophic biota, including corals. In contrast to the impacts on reefs of recently enhanced sediment loads and turbidity, there are some coral reefs that do survive well in naturally turbid waters (e.g. Paluma Shoals), predominantly in areas where strong tidal currents generate persistent high turbidity.

The population along the mainland coast of the Great Barrier Reef is relatively low at 700 000 people along its 2000 km length. Despite the low population density, the load of sediment, total nitrogen and total phosphorus from rivers has increased from two-fold to more than four-fold over the past 150 years, based on the analysis of sediment cores from the lagoon, river-discharge data and sediment-discharge modelling. Approximately 80% of the catchment draining into the Great Barrier Reef lagoon supports some form of agricultural production. Cattle grazing covers the largest area,

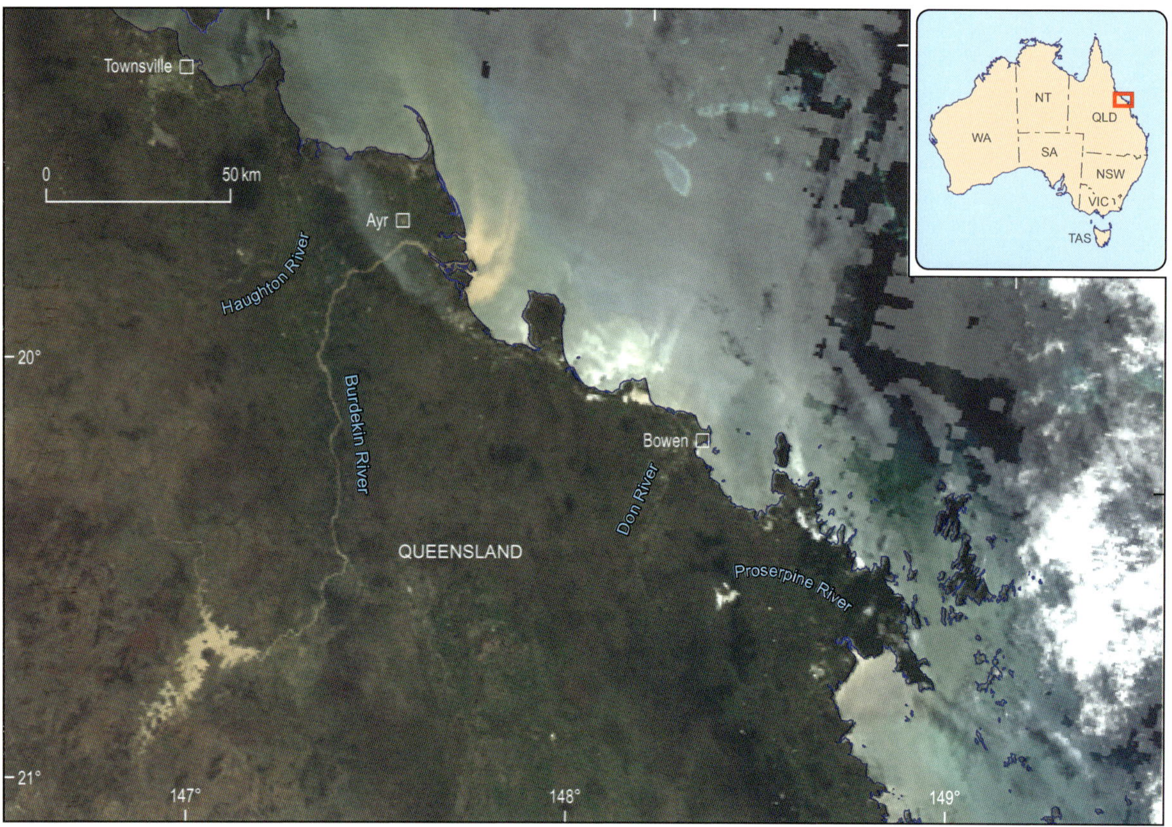

Figure 6.11: Sediment outflow from the Burdekin River, Queensland, due to flooding following the passage of ex-tropical cyclone Nelson in February 2007. The Burdekin River mouth is a good example of a tide-dominated delta. Note the long-shore drift to the north of the sediment discharge and the narrow point—Cape Bowling Green. This long-shore drift illustrates the patterns predicted in the conceptual model shown in Figure 6.4.

followed by cropping (mainly sugar cane) and minor aquaculture. Records of the trace element (e.g. Ba, Ca and Y) concentrations of seawater preserved in the aragonitic skeletons of dated fossil coral indicate that the significant increases in suspended sediment in the lagoon are linked to these catchment land uses, particularly livestock grazing.

An additional potential threat to the health of the Great Barrier Reef is posed by pesticide and herbicide runoff from agricultural lands. Contamination of nearshore coastal waters with a variety of pesticides used on sugar cane and banana crops (e.g. diuron, atrazine, hexazinone) and herbicides used in the grazing industry has been detected at the mouths of the Russell and Mulgrave rivers, south of Cairns. Although at low levels (nanograms per litre) that do not exceed the Australian and New Zealand Guidelines for Fresh and Marine Water Quality, pesticides in the marine environment are known to exacerbate the effect of thermal stress on some hard corals, making them more susceptible to bleaching and mortality, especially during episodes of raised sea-surface temperature. Discharge of pesticides onto the Great Barrier Reef occurs year round and has the potential to have widespread impact; low levels of pesticides have even been detected in the reef waters well removed from intensive agricultural catchments to the south, such as Princess Charlotte Bay.

River flow, and the associated impacts of sediment and nutrient discharge into the Great Barrier Reef lagoon, is temporally and spatially highly variable (Figure 6.11). The level of impact depends on the flood volume of the coastal river, the shoreline and seabed geomorphology and the prevailing oceanographic conditions. Dissolved nutrients delivered to the lagoon during flood events help macroalgae to dominate seabed habitats, which explains their proliferation in impacted inshore areas. Significantly, once a section of reef is invaded by macroalgae, it may become unavailable for the settlement of coral larvae as, for example, after a coral-bleaching event.

Crown-of-thorns 'starfish' (*Acanthaster planci*) predate on corals, and their population outbreaks are one of the biggest short-term threats to corals and coral reefs in the Great Barrier Reef. Major outbreaks of the sea stars have been recorded since the 1960s and may be linked to catchment runoff.

Increased nutrient loads from rivers enhance the food available for juvenile sea stars, increasing their survival and the frequency of population outbreaks. Enhanced sedimentation also significantly lowers the rate of survival of young corals and thus the capacity to recover from disturbances, such as sea star outbreaks and coral bleaching. Crown-of-thorns 'starfish' outbreaks may not be a new phenomenon, their remains having been identified in drill cores at least 5 ka, although this interpretation is contested.

Clearly the health of the Great Barrier Reef is a concern to many Queensland coastal communities and businesses that rely on a healthy reef for their livelihood. This issue is also of national importance because of the reef's role in Australia's environmental heritage and economy. Addressing the causes for the decline in water quality in the Great Barrier Reef is a priority of the Queensland and Australian governments, which have committed hundreds of millions of dollars to this endeavour via the 2009 Reef Water Quality Protection Plan. This comprehensive plan aims to reverse the decline in water quality entering the Great Barrier Reef from the land by 2013; one key goal is to ensure that by 2020 the quality of this water will have no detrimental impact on the health of the reef. The health of the reef is also of international importance because of the pressing need globally to conserve coral reef ecosystems.

Hazards of living on the edge

Natural disasters on the coast have helped shape Australia's history, especially through the way communities, regions and the nation have pulled together to respond to emergencies and prepare for these events. Notable examples include the deadly impacts of tropical cyclone (TC) Mahina (1899) on several small communities along the east coast of Cape York Peninsula; TC Tracy (1974) on the city of Darwin; the very destructive Brisbane and Sydney hailstorms (1986, 1999); and widespread coastal floods in New South Wales (1955) and southeast Queensland (1974, 2011). While some of the impacts of natural disasters can be mitigated, the risk cannot be completely eliminated. For example, although the severe impacts of TC Larry (2006) and TC Yasi (2011) were restricted to local communities, the hinterlands suffered broadscale flooding and wind damage that significantly reduced agricultural production, damaged transport infrastructure and interrupted production at mines for several months (Chapter 9). The consequences of these impacts were then felt across the country as the growth rate of the national economy slowed. Natural disasters also cause enormous intangible effects following deaths and dislocation, and loss of property, heritage and jobs.

Tsunamis, tropical cyclones and coastal storms are hazards that regularly have major impacts on the Australian coastal zone. Oral history, Aboriginal Dreaming stories and coastal sedimentary records also provide evidence of the occurrence and impact of natural hazards in Australia prior to European settlement. Written records of these events begin with the diary entries of the early European settlers, newspaper articles and anecdotal accounts. Today, a relatively large and increasing proportion of people, buildings and infrastructure is exposed to coastal natural hazards because most of the population is concentrated in the narrow coastal

Banana crops destroyed by tropical cyclone Larry, Innisfail, Queensland (March 2006).

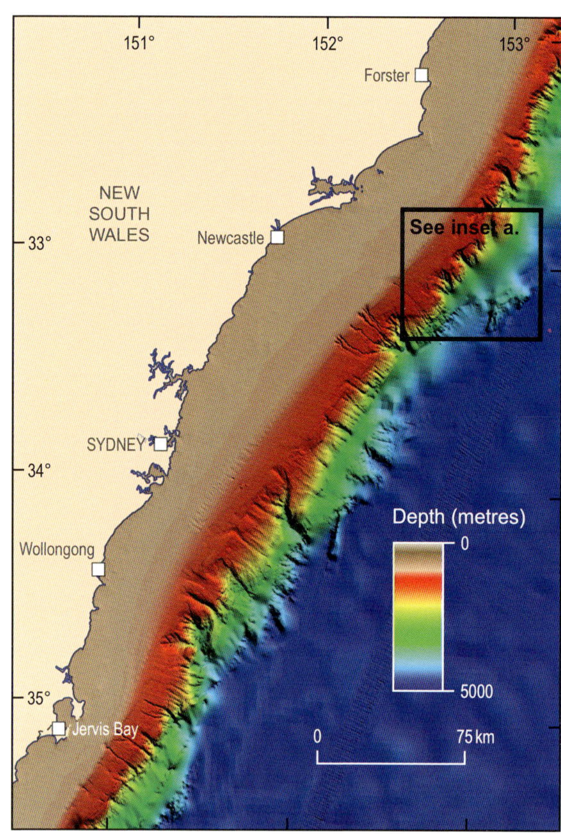

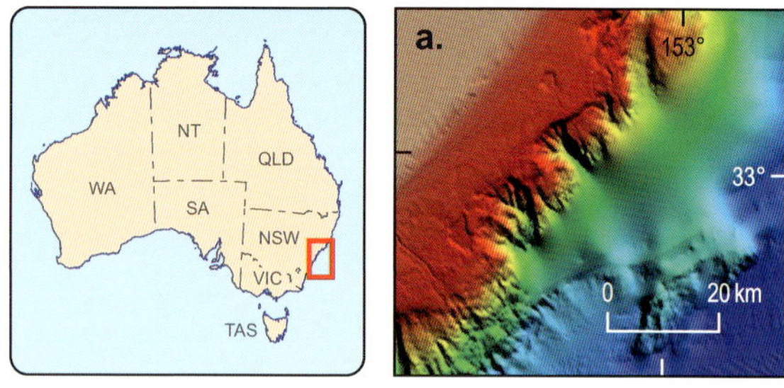

Figure 6:12: Slope failures (slumps) on the edge of the continental shelf adjacent to the New South Wales coast. (Source: Glenn et al., 2008)

zone. Here we examine the processes that generate tsunamis, tropical cyclones and shoreline erosion on the Australian coast and how these geohazards have influenced where and how Australians live on the coast.

Tsunami

Modern Australia has never experienced the full catastrophic impact of a large tsunami such as the 2004 Indian Ocean tsunami or the tsunami that struck northern Japan in 2011. However, tsunamis have struck the Australian coast during historic and prehistoric times, and the continent is surrounded to the north and east by some 8000 km of active plate boundaries capable of generating tsunamis that would reach the coast within two to four hours (Chapter 2). Furthermore, exposure to this hazard is significant because 50% of Australians live within 7 km of the shoreline, and even more spend a significant amount of their leisure time at or near the beach. The attraction of the Australian beach for international tourists adds to the number of people potentially exposed to this hazard.

Potential sources of tsunamis include submarine earthquakes, volcanic eruptions, submarine landslides, and catastrophic slope failure associated with oversteep and unstable volcanic cones. Due to the relative frequency of earthquakes, our understanding of this source of tsunami generation is more advanced than for other sources, but there is still significant uncertainty regarding the frequency and maximum magnitude of earthquake-generated tsunami. Other potential sources of tsunamis (e.g. meteorite impact) are less well documented and are only beginning to be understood.

An example of the uncertainty in the estimate of maximum earthquake magnitudes relates to the 2011 earthquake off the northeast coast of Japan. This event demonstrated the possibility of very large seabed movement on a relatively short (500 km) segment of a subduction zone giving rise to a far larger tsunami than had previously been anticipated. The impact of this event was catastrophic, with waves exceeding 12 m in height arriving at the Japanese coastline and then amplified by the configuration of the coast. In Australia, the tsunamis that have historically created the largest run-ups have been generated by earthquakes off the south coast of Indonesia and have inundated parts of the coast of Western Australia. The largest recorded run-up in Australia was 8 m by the 2006 Java tsunami; in 1994, another Javan tsunami inundated the Exmouth Peninsula, with tsunami deposits found on the nearby coast aligned with gaps in Ningaloo Reef. There are also Dreaming stories of prehistoric large inundation events in Exmouth Gulf that led Aboriginal people to leave and never return to that area. The 2004 Indian Ocean tsunami was observed along the length of the Western Australian coast but created only minor damage and no loss of life: moored boats were damaged at Geraldton, and strong rips and currents swept several people out to sea at Rockingham, south of Perth.

The subduction zones considered to be the main sources of tsunami hazard for eastern Australia are those stretching from Papua New Guinea through the Solomon Islands and New Hebrides to New Zealand (Figure 2.1). In historical times, earthquakes along this zone have generated several tsunamis that were recorded on the east coast but with only minor inundations. The Puysegur Trench is also considered to be a significant source of tsunami hazard for southeastern Australia, particularly Tasmania. Tsunamis can also be generated by foci sourced further away, such as the large earthquakes that have occurred in the eastern Pacific, off South America. The tsunami generated by the magnitude 9.5 Chile earthquake in 1960 did not propagate towards Australia. However, although this event caused no major coastal inundation, it did create currents strong enough to tear boats from their moorings in several harbours along the east coast. The 2011 Japanese tsunami also did not produce inundation, but it did alter the patterns of the East Australian Current for a few days (tide gauges observed 30–50 cm wave heights).

Although earthquakes are seen as the most likely source in the region, there are at least five active volcanic source regions capable of generating tsunamis that could affect Australia. These volcanic sources could be underwater caldera collapse, submarine landslides or large-volume pyroclastic flows into the sea. The only documented eruption to affect Australia's continent, the caldera collapse of Krakatau on 26–27 August 1883, generated a tsunami that, according to eyewitness reports, reached several locations along the coast of Western Australia.

Little is known about the magnitude and frequency of tsunamis generated by submarine landslides (slope failure), in part because they are harder to observe and because they are often coincident with an earthquake. Distinguishing between a purely earthquake-sourced tsunami and a co-sourced slope-failure/earthquake-generated one requires detailed bathymetric studies that are not routinely conducted in Australia. A recent seabed survey along the New South Wales continental slope identified that slope failures have occurred in the past and that there are several potential sources for future ones (Figure 6.12). The historical scars represent volumes up to about 20 km^3 and reside in water depths that may have generated an initial sea-surface displacement of several metres. The nature of the mass movement, however, is a critical factor in determining whether a tsunami may be generated. Better understanding this potential

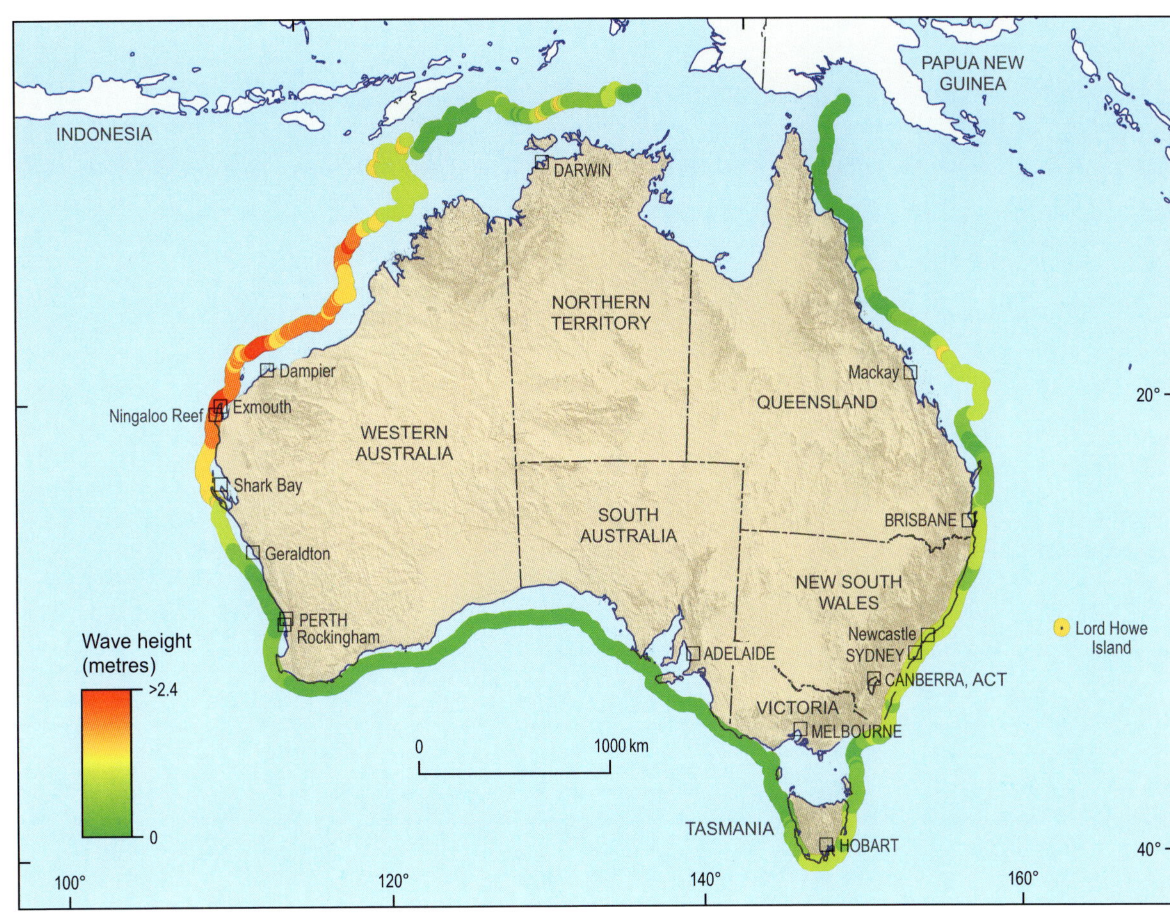

Figure 6.13: An example from the national probabilistic offshore tsunami hazard assessment. This assessment is only for tsunamis generated by subduction-zone earthquakes and shows the maximum wave height at the 100 m depth contour expected for a given return period. The most vulnerable areas to tsunamis are located along the northwest coast of Western Australia, which has high-value energy and minerals infrastructure facing the subduction zone beneath Indonesia.

Did you know?

6.4: Tropical cyclone Tracy—a chance to build better

On Christmas morning 1974, TC Tracy crossed the coast at Darwin, bringing winds of more than 270 km/h and inflicting widespread destruction on the city. To this day, Tracy remains one of the most memorable natural disasters in Australia's history. More than 60 people were killed, many as a result of injuries suffered as residential buildings were ripped apart when the cyclone slowly passed over the city. In total, around 60% of the houses in Darwin were completely destroyed and less than 5% remained habitable. The impact of Tracy resulted in the largest evacuation in Australian history, with 35 000 of Darwin's 48 000 residents evacuated in a massive airlift to the southern capitals.

The impacts of Tracy brought about significant changes in building codes for tropical cyclone–affected regions of the country, as well as fundamental changes in catastrophe risk insurance in Australia.

hazard is important, as this part of the New South Wales coast is densely populated and has a range of critical infrastructure. Computational modelling of tsunami generation, propagation and inundation around Australia has been used to estimate the national offshore tsunami hazard (Figure 6.12) and onshore tsunami hazard in selected locations.

The impact of a tsunami on communities along any affected section of the coastline is highly dependent on the population and infrastructure at that location. The 2006 Java tsunami that destroyed a family's campsite on the Shark Bay coastline would have had a much different outcome if the same event had impacted a highly populated location. Inundation is not the only concern from tsunami impact, as strong rips and currents have been observed after tsunami events around the world, including Sydney and Perth. This is a significant hazard for the public in leisure activities on the beach, and the nearshore through to commercial activities in ports and harbours. Furthermore, these rips and currents are known to persist for 24–48 hours after the initial earthquake that generated the tsunami, and this has implications for Australia's ports in the northwest (e.g. Dampier) and on the east coasts (e.g. Newcastle and Mackay) that operate on a 24-hour basis. The Joint Australian Tsunami Warning Centre issues warnings to the public before a tsunami reaches Australian shores.

Cyclonic events

Tropical cyclones

As noted above, cyclones generate very destructive wind, torrential rain and storm surge that can cause major damage to communities. The sparse population along most of the northern and western coasts of Australia means that many tropical cyclones have little direct impact on people. Conversely, regions with a high frequency of cyclones have been viewed as being far less favourable for European settlement. On occasion though, tropical cyclones pass over larger coastal communities or track further south and impact on large cities and more densely settled regions (Figure 6.14). Tropical cyclones that impact the Australian coast form in the Coral Sea, the Gulf of Carpentaria (*Did you know?* 6.4) or the eastern Indian Ocean. On average, 15 tropical cyclones develop over these waters each season (November–April), of which about six will make landfall, with the northwest coast between Broome and North West Cape most frequently impacted.

Tropical cyclone activity in the Australian region is significantly modulated by the El Niño–Southern Oscillation (Chapter 1). During El Niño years, there is a decrease in cyclone frequency over the Coral Sea, with the converse during La Niña years. This is reflected in the frequency of cyclone landfall, with landfall in El Niño years around 50% that in La Niña years along the Queensland coast. There is also a weak influence on the intensity of tropical cyclones, with a slight decline in intensity during El Niño years. Since the mid-1970s, there has been a significant shift towards more frequent El Niño events, and this has resulted in a reduction in tropical cyclone activity along the east coast. Coinciding with this relatively quiescent period, there has been major growth in communities along the Queensland coast. A shift back towards a more energetic Coral Sea would see these communities exposed to more frequent tropical cyclones.

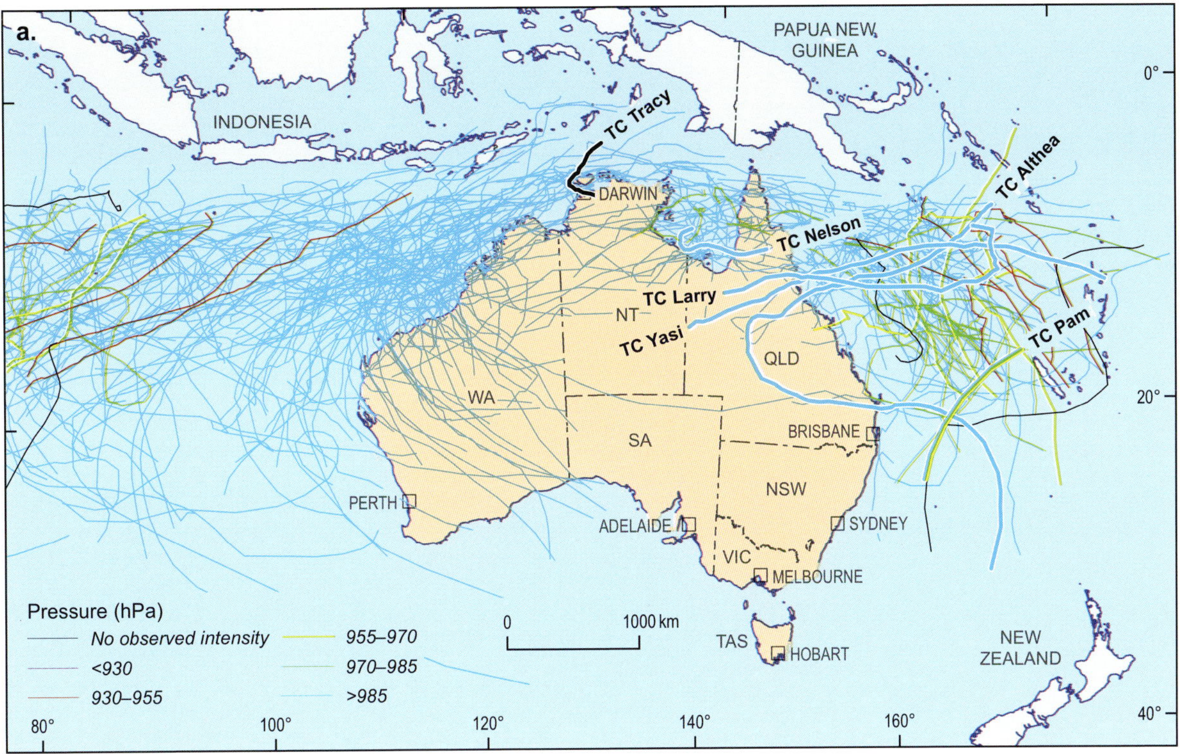

Image courtesy of NOAA

Figure 6.14: (a) Map showing historical tropical cyclones in the Australian region, 1981–2011. The majority of cyclones make landfall along the northwest coastline, which is sparsely populated but hosts significant infrastructure and resources. Cyclones recurving to the southeast across southern Western Australia bring destructive winds, heavy rain and storm surge to southern parts of the continent. (Source: Geoscience Australia). (b) Tropical cyclone Yasi from space. This storm had an eye more than 100 km in diameter and maximum wind speeds of 110 knots over a sustained 10-minute interval. TC Yasi struck land only 10 km from the point where TC Larry made landfall in 2006. (Source: NOAA)

The impact of a tropical cyclone on the Australian coastal zone is manifold. Severe winds destroy structures (Figure 6.15) and critical transport networks. Coastal vegetation such as mangroves that protect the coast from wave and wind energy can also be destroyed, increasing local erosion due to wave action for some time following the passage of the cyclone. Heavy rain accompanying tropical cyclones often leads to flooding in coastal rivers and increased transport of sediments, nutrients and pollutants from land areas into estuaries and coastal waters. The storm surge produced by cyclones can have a very destructive impact on the shoreline (Figure 6.16). The low atmospheric pressure in the centre of a cyclone causes sea level to rise into a large dome by several tens of centimetres. As a cyclone approaches the coast, the extreme winds drive water onto the coast, increasing water levels dramatically. When combined with astronomical tides and breaking waves, water levels can exceed the highest astronomical tides by several metres. This recently occurred with TC Yasi in northeastern Queensland, with a surge of up to 3 m.

The size and intensity of a cyclone are the most readily observed factors affecting the magnitude of a storm surge as it approaches the coast. Local bathymetry, coastline orientation and astronomical tide are the other major influences. Areas of coastline where the edge of the continental shelf is close to shore (and deep waters are relatively close to coastal population) are less likely to experience significant storm surge. However, deep offshore water allows large waves to be generated more

Figure 6.15: Damage to residential buildings in Innisfail, Queensland, following tropical cyclone Larry (2006).

Figure 6.16: Storm surge damage and sand deposition from tropical cyclone Yasi (2011) at Tully Heads, Queensland.

readily, increasing coastal impacts such as beach erosion. For example, large waves generated by TC Pam in 1974 caused significant damage to Sydney's beaches, despite the storm passing close to Lord Howe Island more than 700 km to the northeast. Conversely, coastlines with shallow offshore waters are more susceptible to large storm surges, such as the northwestern coast of Western Australia. The large astronomical tidal range around Australia's northern coastline also plays a major role in determining the outcome of a cyclone—Townsville was spared a devastating storm surge in 1971 when TC Althea made landfall at low tide.

An important source of information on cyclone magnitude and frequency over long time-scales is provided by Holocene shoreline deposits that were emplaced during cyclone-generated storm surges. These geological records suggest that the recurrence interval of high-magnitude cyclones in northeastern Australia may be considerably shorter than has been predicted based on the historical record of cyclones. This information is important for planning for the potential impact of cyclones on coastal communities and infrastructure.

Current projections of tropical cyclone activity have large uncertainties at all scales, but the scientific consensus is that there will be a global decline in tropical cyclone frequency with continued global warming. The intensity of tropical cyclones is, however, likely to increase slightly as the climate warms. Several studies in the Australian region indicate similar trends to the global projections, with a decline in frequency and an increase in the proportion of intense cyclones.

Impact of cyclones on the seabed

Cyclone-generated currents in northern Australia can radically reshape the seabed. The potential impacts of this erosion on cables or pipelines include the removal of support through scouring of the underlying sediment and burial by mobilised sediment. Potentially, this sedimentation could result in a disruption to gas and oil production, port activities and communications. Infrastructure damage can also be produced by the stress induced by the cyclone-generated currents. Currents associated with tropical cyclones can mobilise as much as the upper 1–2 m of seabed sediment, resulting in the formation of distinctively structured storm beds ('tempestites').

Cyclone-generated currents also appear to profoundly influence the overall distribution of sediment along the coast. In the Great Barrier Reef and the Gulf of Carpentaria, widely distributed accumulations of reef sediment are attributed to sediment mobilisation under currents generated by tropical cyclones (Figure 6.5). The orientation of these deposits is indicative of a consistent, along-coast transport pathway. An explanation for this pattern is that currents generated by the passage of a cyclone are asymmetric in plan view, such that stronger flows are generated between the eye of the cyclone and the coast, giving rise to sediment transport along hundreds of kilometres of coast. The result of the passage of many cyclones over geological time-scales is a consistent force for the net along-coast sediment transport on the inner to mid-shelf, possibly extending throughout northern Australia.

Mid-latitude storms

Intense low-pressure systems that traverse the Southern Ocean can hit the southern coast of Australia. However, the Southern Ocean storms rarely hit populated areas. An exception is the storm that crossed the coast at Esperance in 2007, which caused severe flooding and widespread infrastructure damage. Intense lows also form over the warm waters of the East Australian Current, between southern Queensland and northern Tasmania. These storms often impact on the southeastern coast, the most developed and densely populated section of the Australian coast (Figure 1.3). In recent decades, the most powerful east-coast low occurred in May 1974, generating extreme winds, large swells and storm surges that eroded the beaches along hundreds of kilometres of the New South Wales and Queensland coast (*Did you know?* 6.5).

East-coast lows, although generally short lived, are capable of generating winds and heavy rain comparable to tropical cyclones—the *Sygna* storm of May 1974 generated wind gusts of 160 km/h but lasted less than 24 hours. The Great Divide of eastern Australia (Figure 5.6) amplifies the temperature gradient between the relatively warm East Australia Current and the cooler land, providing conditions conducive to rapid intensification of these storms (in a matter of hours) close to the coastline. Similar coastlines around the world, with warm currents adjacent to mountain ranges, such as along the eastern seaboard of the United States, also experience these types of explosive storms.

Often the east coast events are remembered for the ships lost during the storm (e.g. the 1974 Sygna storm, the *Pasha Bulker* storm in June 2007), but they affect more than just shipping. The storm responsible for grounding the *Pasha Bulker* in Newcastle caused flooding in the adjacent Hunter Valley, with an insurance bill estimated at $1.35 B. Large swells generated by the Maitland Gale (May 1898) resulted in Stradbroke Island being cut in two, but only after repeated storms in the preceding years had narrowed the shore at Jumpinpin. Periods with an increased frequency of east-coast lows (e.g. through the 1970s) result in greater shoreline erosion because beach systems do not have sufficient time to recover between storm events. For example, the New South Wales sandy coastline was extensively eroded during a series of storms in 1974 and again in 1978, resulting in widespread reduction of beach width and property damage.

Climate change impacts

Some climate change scenarios predict up to a 1.1 m sea-level rise for the Australian coast by 2100. Rises of this magnitude have been experienced in the recent geological past (Chapter 2). The greater number of people and their associated infrastructure today, however, mean that the impacts of sea-level rise will be greater. The 2010 national assessment of Australia's coastal vulnerability to the climate change impacts of a rise in sea level and related erosion of sandy shorelines found that more than $220 B of infrastructure (replacement value in 2008 dollars) is potentially at risk from such a rise in sea-level. Queensland and New South Wales

6.5: The changing face of the beach

Over a timescale of years to decades, sandy beaches go through repeated cycles of erosion and accretion in tune with changes in wave energy. The wave-dominated beaches of southeastern Australia are particularly responsive to these changes. Periods of increased storm wave activity can result in a temporary loss of sand and, in some cases, erosion of the foredune and loss of property. Records of natural beach change began during a period of storm activity and beach erosion in southeast Australia from 1974 to 1976, with gradual recovery of these beaches by the 1990s. At Moruya Beach (NSW), the 1974 storms initiated a period of beach erosion that resulted in the landward displacement of the beach by 40–50 m. Sand was not lost from the system, but was stored temporarily in the nearshore and was gradually redeposited to form a new foredune.

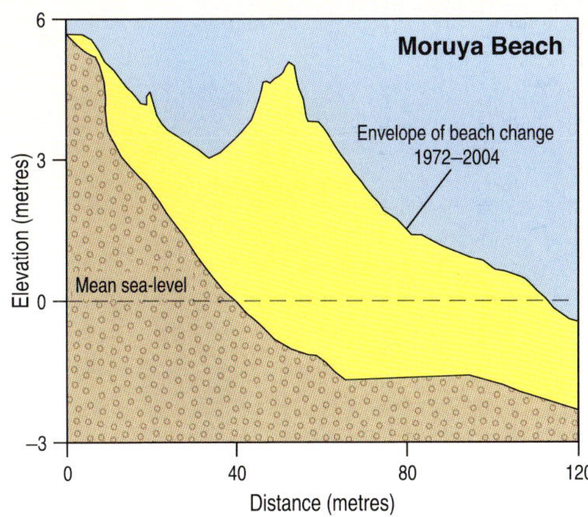

Figure DYK 6.5: Envelope of beach profile change at Moruya Beach, New South Wales, between 1972 and 2004. (Source: McLean & Shen, 2006)

THE GOLD COAST (BOX 6.5)

On the Gold Coast, in southeast Queensland, a range of environmental issues are a consequence of extensive and rapid urban expansion. This narrow coastal strip extends 32 km from Coolangatta in the south, on the New South Wales border, to Southport at the mouth of the Nerang River. This strip of coast consists of a series of four sandy embayments, each bounded by headlands: the southern beaches between Snapper Rocks and Currumbin, Palm Beach, Burleigh Beach, and the 17 km stretch of continuous beach from Miami to the Nerang River entrance. European settlement began in the 1840s and mainly involved grazing and forestry. However, after World War II there was a moderate increase in the rate of settlement that continued until the development boom of the 1960s. Development was initially focused on the sand-dune system immediately behind the beach but then moved to the swampy area behind the dunes. The population has increased from around 50 000 in the late 1960s to more than 527 000 in 2010. Population growth was more than 3% per year through the past decade, with projections by Queensland Treasury of around 750 000 people in 2031.

Severe beach erosion was recorded as early as the 1890s but did not become a significant issue for communities until the mid-1960s after a series of tropical cyclones resulted in extensive erosion of the shoreline into the developed land. A seawall was constructed, and the beach was replenished with imported sand along the length of the Gold Coast to protect the built areas from future erosion.

Since the 1980s, a beach nourishment program has been operating, initially pumping sand from the Tweed River mouth in New South Wales, near the southern end of the Gold Coast, then dredging sand from directly offshore from the beaches (approximately 200 000 m^3/year). Currently, more than 500 000 m^3 of sand is pumped each year from Letitia Spit in New South Wales to maintain beaches along the Gold Coast. Despite these efforts, sections of the beaches are still susceptible to periodic erosion, and an assessment of the vulnerability of the Gold Coast to predicted climate change impacts by 2100, such as shoreline erosion and marine inundation, suggested that $5–7 B of residential and commercial infrastructure was potentially at risk.

Aerial photo of Surfers Paradise, the Gold Coast, 1952.
Image courtesy of John Oxley Library, State Library of Queensland, Neg: 8035-0001-0007

Aerial photo of Surfers Paradise, the Gold Coast, 2003.
Image courtesy of John Oxley Library, State Library of Queensland, Neg: 10026-0001-0145, image creator Daryl Jones.

Image courtesy of Tourism Queensland

were found to have the greatest vulnerability when considering both the value and number of buildings affected (Box 6.5). A 2009 study by Sydney University attempted to quantify some of the costs associated with sea-level rise. The study found that those beachfront properties at higher risk of erosion were 40% discounted in market value (2007 figures) compared with equivalent lower risk properties along the same beach. The same study also estimated that Manly beach, with its annual 4 million visitors, generated almost $260 M of the state's economy.

There are many uncertainties involved in estimating the potential impacts of rising sea-level hazards and in developing adaptation strategies. These include the rate of sea-level rise, the associated extent and rate of coastal erosion, and the complex interactions of waves and currents at the shoreline. Responses could range from building shoreline protection structures to the planned retreat of people, buildings and infrastructure out of the zone of potential erosion.

All of Australia's major capital cities (except Canberra) and a significant number of other densely populated coastal areas will be potentially affected by inundation induced by sea-level rise, higher tides and more frequent or intense storms. For example, Perth, Adelaide, Melbourne, Sydney, Cairns, Brisbane and the Gold Coast all have extensive low-lying coastal areas that include critical infrastructure and valuable commercial and residential precincts. The impact of inundation in these settings is potentially on a national scale. For example, floods produced by coastal storms that impacted southeast Queensland in December 2010 and January 2011 had direct (flooding of houses and businesses) and indirect (reduced production and exports, health problems) impacts that cost the Queensland and Australian economy $5–6 B. Storms and floods are not the only impact of rising sea-levels. Inundation of coastal groundwater aquifers by salt water will damage these resources, especially where excessive pumping of the aquifers has occurred.

Beach sediment nourishment techniques are currently being utilised to offset sand losses at a number of locations, such as on the Gold Coast and the metropolitan beaches of Adelaide. These projects are costly, and appropriate sources for beach sand are becoming difficult to locate. With an increased appreciation of the natural protective function of coastal dunes, more holistic solutions are starting to be used to protect and enhance the amenity of beaches, such as revegetating dunes that back the shoreline.

Final say—better valuing the edge

Two common adages of real estate are 'you can't beat a waterfront view' and 'location, location, location'. The allure of a coastal home in temperate Australia has been a major factor in the recent rapid expansion of the large coastal cities of Australia, driven by a growing population and prolonged strong economic growth generated by the other 'boom' in the export of commodities. This expansion of coastal cities has been an important driver of the domestic economy and has further

Waterfront living at Sandy Bay, Hobart, Tasmania.

increased the proportion of the population in these cities. In the more remote and sparsely populated sections of coast are the ports that play a key role in the expansion of the commodities trade and the economic prosperity of Australia.

There is much more to the 'clinging to the coast' by Australians than the desire for a waterfront view. Much of the Australian coast is dry, wild and remote and represents the antithesis of a 'good location' for European settlement and development. As a consequence, modern Australian society is concentrated in the more benign humid, temperate coastal regions. The spatial distribution of the Australian population, therefore, reflects the strong influence of the continent's geological evolution on the geomorphological and oceanographic character of the coast. This interaction of geology, ocean and climate has produced the distinctive, as well as the less well-known, coastal environments of Australia.

Point Quobba near Carnarvon, Western Australia.

© Getty Images [P Walton]

Australians are now much more aware of the need to better manage coastal systems, both in urban areas and beyond, because of the key role these environments play in Australians' quality of life and the national economy. There is also a growing awareness and understanding of the natural hazards that coastal communities are exposed to—how the level of exposure will increase as the coastal population continues to grow and the intensity of destructive events threatens to increase in the future. Australians have a long history of attempting to shape the land for their own purposes, with sometimes disastrous consequences. However, it is really only with an informed understanding of coastal processes and of how the coast has shaped Australian society, and a more widespread appreciation of this knowledge, that a more sustainable relationship can develop between Australians and this unique, valuable and often fragile edge.

Having walked along the shore, we will now dip our feet into the water and consider how fresh and salt water influence Australia.

Image by Jim Mason

Phillip Island, Victoria.

Bibliography and further reading

Living on the edge

Achterstraat P 2010. *Sick leave: Department of Premier and Cabinet*, Audit Office of New South Wales, Sydney.

Bourman RP, Murray-Wallace CV, Belperio AP & Harvey N 2000. Rapid coastal geomorphic change in the River Murray estuary of Australia. *Marine Geology* 170, 141–168.

Brown CM & Stephenson AE 1991. Geology of the Murray Basin, southeastern Australia. *Bulletin of the Bureau of Mineral Resources Australia* 235.

Colhoun EA, Hannan D & Kiernan K 1996. Late Wisconsin glaciation of Tasmania. *Papers and Proceedings of the Royal Society of Tasmania* 130, 33–45.

Cresswell GR & Golding TJ 1980. Observations of a south flowing current in the southeastern Indian Ocean. *Deep-Sea Research* 27A, 449–466.

Davies PJ, Symonds PA, Feary DA & Pigram CJ 1989. The evolution of the carbonate platforms of northeast Australia. In: *Controls on carbonate platform and basin development*, Crevello P, Wilson JL, Sarg JF & Read JF (eds), Society of Economic Paleontologists and Mineralogists Special Publication 44, 233–258.

Earl K, Holm O & Powell T 2002. *Outline of geological studies in the Great Barrier Reef and adjacent regions: industry, Commonwealth Government and international*, Geoscience Australia Record 2002/22, Geoscience Australia, Canberra.

Fairbridge RW 1995. Eolianites and eustasy: early concepts on Darwin's voyage of HMS Beagle. *Carbonates and Evaporites* 10, 92–101.

Gregory JW 1906. *Dead heart of Australia: a journey around Lake Eyre*, John Murray, London.

Harris PT & Baker EK (eds) 2011. *Seafloor geomorphology as benthic habitat: GeoHAB atlas of seafloor geomorphic features and benthic habitats*, Elsevier, London.

Harris PT, Heap AD, Marshall JF & McCulloch M 2008. A new coral reef province in the Gulf of Carpentaria, Australia: colonisation, growth and submergence during the early Holocene. *Marine Geology* 251, 85–97.

Hill KL, Rintoul SR, Coleman R & Ridgway K 2008. Wind forced low frequency variability of the East Australia Current. *Geophysical Research Letters* 35 (L08602), 1–5.

Hopley D (ed.) 2011. *Encyclopedia of modern coral reefs: structure, form and process*, Encyclopedia of Earth Sciences Series, Springer, Berlin.

Hopley D, Smithers SG & Parnell KE 2007. *The geomorphology of the Great Barrier Reef—development, diversity and change*, Cambridge University Press, Cambridge.

Hou B, Frakes LA, Alley NF & Heithersay P 2003. Evolution of beach placer shorelines and heavy-mineral deposition in the eastern Eucla Basin, South Australia. *Australian Journal of Earth Sciences* 50, 955–965.

House of Representatives Standing Committee on Climate Change, Water, Environment and the Arts 2009. *Managing our coastal zone in a changing climate—the time to act is now*, Parliament of the Commonwealth of Australia, Canberra.

Murray-Wallace CV, Bourman RP, Prescott JR, Williams FM, Price DM & Belperio AP 2010. Aminostratigraphy and thermoluminescence dating of coastal aeolianites and the later Quaternary history of a failed delta: the River Murray mouth region, South Australia. *Quaternary Geochronology* 5, 28–49.

Murray-Wallace CV, Brooke BP, Cann JH, Belperio AP & Bourman RP 2001. Whole-rock aminostratigraphy of the Coorong Coastal Plain, South Australia: towards a million year record of sea-level highstands. *Journal of the Geological Society of London* 158, 111–124.

Pearson M 1990. The lime industry in Australia—an overview. *Australian Historical Archaeology* 8, 28–35.

Pigram CJ, Davies PJ, Feary DA & Symonds PA 1989. Tectonic controls on carbonate platform evolution in southern Papua New Guinea: passive margin to foreland basin. *Geology* 17, 199–202.

Playford PE 1996. *Carpet of silver: the wreck of the* Zuytdorp, University of Western Australia Press, Nedlands.

Short AD 2010. Sediment transport around Australia—sources, mechanisms, rates and barrier forms. *Journal of Coastal Research* 26, 395–402.

Short AD & Woodroffe CD 2009. *The coast of Australia*, Cambridge University Press, Cambridge.

Siddall M, Chappell J & Potter E-K 2006. Eustatic sea level during past interglacials. In: *The climate of past interglacials*, Sirocko F, Litt T, Claussen M & Sanchez-Goni M-F (eds), Elsevier, Amsterdam, 74–92.

Veron JEN 2008. *A reef in time—the Great Barrier Reef from beginning to end*, The Belknap Press of Harvard University Press, Cambridge, Massachusetts.

Wasson RJ, Olive LJ & Rosewell CJ 1996. Rates of erosion and sediment transport in Australia. *IAHS Publication* 236, 139–148.

The modern Australian coast

Boyd R, Dalrymple R & Zaitlin BA 1992. Classification of clastic coastal depositional environments. *Sedimentary Geology* 80, 139–150.

Brearley A 2005. *Ernest Hodgkin's Swanland: estuaries and coastal lagoons of southwestern Australia*, University of Western Australia Press, Perth.

Dalrymple RW, Zaitlin BA & Boyd R 1992. Estuarine facies models: conceptual basis and stratigraphic implications. *Journal of Sedimentary Research* 62, 1130–1146.

Dodson JR & Mooney SD 2002. An assessment of historic human impact on south-eastern Australian environmental systems, using late Holocene rates of environmental change. *Australian Journal of Botany* 50, 455–464.

Duller GAT & Augustinus P 2006. Reassessment of the record of linear dune activity in Tasmania using optical dating. *Quaternary Science Reviews* 25, 2608–2618.

Galloway RW 1981. An inventory of Australia's coastal lands. *Australian Geographical Studies* 19, 107–116.

Harris PT, Heap AD, Bryce SM, Porter-Smith R, Ryan DA & Heggie DT 2002. Classification of Australian clastic coastal depositional environments based upon a quantitative analysis of wave, tidal, and river power. *Journal of Sedimentary Research* 72, 858–870.

Heap AD, Bryce S & Ryan DA 2004. Facies evolution of Holocene estuaries and deltas: a large-sample statistical study from Australia. *Sedimentary Geology* 168, 1–17.

Heap AD, Bryce SM, Ryan DA, Radke LC, Smith CS, Smith R, Harris P & Heggie DT 2001. *Australian estuaries and coastal waterways: a geoscience perspective for improved and integrated resource management*, AGSO Record 2001/07, Australian Geological Survey Organisation, Canberra.

Murray E, Radke L, Brooke B, Ryan D, Moss A, Murphy R, Robb M & Rissik D 2006. *Australia's near-pristine estuaries: current knowledge and management*, Cooperative Research Centre for Coastal Zone, Estuary and Waterway Management, Brisbane.

Roy PS & Thom BG 1991. Cainozoic shelf sedimentation model for the Tasman Sea margin of southeastern Australia. In: *The Cainozoic in Australia: a re-appraisal of the evidence*, Williams MAJ, de Deckker P & Kershaw AP (eds), Geological Society of Australia Special Publication 18, 119–136.

Roy PS, Williams RJ, Jones AR, Yassini R, Gibbs PJ, Coates B, West RJ, Scanes PR, Hudson JP & Nichol S 2001. Structure and function of south-east Australian estuaries. *Estuarine, Coastal and Shelf Science* 53, 351–384.

Ryan DA, Heap AD, Radke LC & Heggie DT 2003. *Conceptual models of Australia's estuaries and coastal waterways: applications for coastal resource management*, Geoscience Australia Record 2003/09, Geoscience Australia, Canberra.

Saunders KM, Hodgson DA, Harrison JJ & McMinn A 2008. Palaeoecological tools for improving the management of coastal ecosystems: a case study from Lake King (Gippsland Lakes) Australia. *Journal of Paleolimnology* 40, 33–47.

Short AD 2005. *Beaches of the Western Australian coast: Eucla to Roebuck Bay*, Sydney University Press, Sydney.

Short AD 2006. Australian beach systems—nature and distribution. *Journal of Coastal Research* 22, 11–27.

Sircombe KN 1999. Tracing provenance through the isotope ages of littoral and sedimentary detrital zircon, eastern Australia. *Sedimentary Geology* 124, 47–67.

Turner L, Tracey D, Tilden J & Dennison WC 2004. *Where river meets the sea: exploring Australia's estuaries*, CSIRO Publishing, Melbourne.

Woodroffe CD 2002. *Coasts: form, processes and evolution*, Cambridge University Press, Cambridge.

Muddying the water: environmental pressures on the coast

Brearley A 2005. *Ernest Hodgkin's Swanland: estuaries and coastal lagoons of south-western Australia*, University of Western Australia Press, Perth.

Brooke B, Nichol S, Hughes M, McArthur M, Anderson T, Przeslawski R, Siwabessy J, Heyward A, Battershill C, Colquhoun J & Doherty P 2009. *Carnarvon Shelf survey: post survey report*, Geoscience Australia Record 2009/02, Geoscience Australia, Canberra.

Eyre BD & Maher DT 2010. Structure and function of warm temperate east Australian coastal lagoons: implications for natural and anthropogenic change. In: *Coastal lagoons: critical habitats of environmental change*, Kennish M & Paerl H (eds), CRC Press, Boca Raton.

Fabricius K 2007. *Conceptual model of the effects of terrestrial runoff on the ecology of corals and coral reefs of the GBR*, report to the Australian Government's Marine and Tropical Sciences Research Facility, Townsville.

Fitzpatrick RW, Marvanek S, Powell B & Grealish GJ 2010. Atlas of Australian acid sulfate soils: recent developments and future priorities. In: *Proceedings of the 19th World Congress of Soil Science*, Gilkes RJ & Prakongkep N (eds), International Union of Soil Sciences, Brisbane, 24–27.

Fitzpatrick RW, Powell B & Marvanek S 2008. Atlas of Australian acid sulfate soils. In: *Inland acid sulfate soil systems across Australia*, Fitzpatrick RW & Shand P (eds), CRC LEME Open File Report 249, 75–89.

Fletcher M & Thomas I 2010. A Holocene record of sea-level, vegetation, people and fire from western Tasmania, Australia. *The Holocene* 20, 351–361.

McCulloch M, Fallon S, Wyndham T, Hendy E, Lough J & Barnes D 2003. Coral record of increased sediment flux to the inner Great Barrier Reef since European settlement. *Nature* 421, 727–730.

Morse K 1993. Who can see the sea? Prehistoric Aboriginal occupation of the Cape Range Peninsula. In: *The biogeography of the Cape Range*, Humphreys WF (ed.), Records of the Western Australian Museum Supplement 45, 227–242.

National Working Party on Acid Sulfate Soil 2000. *National strategy for the management of coastal acid sulfate soils*, NSW Agriculture, Wollongbar Agricultural Institute, Wollongbar. www.mincos.gov.au/__data/assets/pdf_file/0003/316065/natass.pdf

Negri AP, Flores F, Röthig T & Uthicke S 2011. Herbicides increase the vulnerability of corals to rising sea surface temperature. *Limnology and Oceanography* 56, 471–485.

Queensland Government 2009. *Reef water quality protection plan 2009 for the World Heritage Great Barrier Reef and adjacent catchments*, Queensland Government, Brisbane.

Shaw M, Furnas MJ, Fabricius K, Haynes D, Carter S, Eaglesham G & Mueller JF 2010. Monitoring pesticides in the Great Barrier Reef. *Marine Pollution Bulletin* 60, 113–122.

Hazards of living on the edge

Anning D, Dominey-Howes D & Withycombe G 2009. What price climate change?—valuing Sydney beaches to inform coastal management decisions: a research outline. *Management of Environmental Quality* 20, 408–421.

Arthur WC, Schofield A & Cechet B 2008. Assessing the impacts of tropical cyclones. *Australian Journal of Emergency Management* 23, 14–20.

Bureau of Meteorology 2010. *About tropical cyclones*. www.bom.gov.au/cyclone/about

Fountain L, Sexton J, Habili N, Hazelwood M & Anderson H 2010. Stormsurge modelling for Bunbury, Western Australia. Geoscience Australia Professional Opinion 2010/04, Geoscience Australia, Canberra.Glenn KC, Post A, Keene J, Boyd R, Fountain L, Potter A, Osuchowski M, Dando N & Party S 2008. *NSW continental slope survey: post-cruise report*, Geoscience Australia Record 2008/14, Geoscience Australia, Canberra.

Harris PT & Heap A 2009. Cyclone-induced net sediment transport pathway on the continental shelf of tropical Australia inferred from reef talus deposits. *Continental Shelf Research* 29, 2011–2019.

Hearn CJ & Holloway PE 1990. A three-dimensional barotropic model of the response of the Australian North West Shelf to tropical cyclones. *Journal of Physical Oceanography* 20, 60–80.

McLean R & Shen J 2006. From foreshore to foredune: foredune development over the last 30 years at Moruya Beach, New South Wales, Australia. *Journal of Coastal Research* 22, 28–36.

Nicholas T & Hazelwood M 2011. Coastal geomorphic mapping—detailed infill mapping of the coastal geomorphology between Newcastle and Wollongong, New South Wales. Geoscience Australia Professional Opinion 2011/01, Geoscience Australia, Canberra.

Nott J 2007. The importance of Quaternary records in reducing risk from tropical cyclones. *Palaeogeography, Palaeoclimatology, Palaeoecology* 251, 137–149.

Queensland Treasury 2011. *Population and housing profile Gold Coast 2011*, Office of Economical and Statistical Research, Brisbane.

Short AD & Trembanis AC 2004. Decadal scale patterns in beach oscillation and rotation Narrabeen Beach, Australia—time series, PCA and wavelet analysis. *Journal of Coastal Research* 20, 523–532.

7

Groundwater—lifeblood of the continent

This chapter explores the journey of the development and use of Australia's groundwater resources, from the first Australians, to the frontier settlers, to the present day. Over this time, some 50 ka, the changes to groundwater systems in Australia have been profound. Water availability has shaped life and society in many ways, with aridity shaping the landscape and soils and determining where we live, grow our crops, raise animals and build our cities in this, the driest inhabited continent. Recognising the connections that groundwater can have with rivers, wetlands and ecosystems means understanding groundwater processes, and this, underpinned by the knowledge of geology, is fundamental to resolving many of Australia's natural resource management problems. Our understanding of these connections has improved, and emerging technologies have helped with management of Australia's groundwater for people, now and in the future.

Ross S Brodie,[1] Ken C Lawrie[1] and D Philip Commander[2]

[1]Geoscience Australia; [2]University of Western Australia

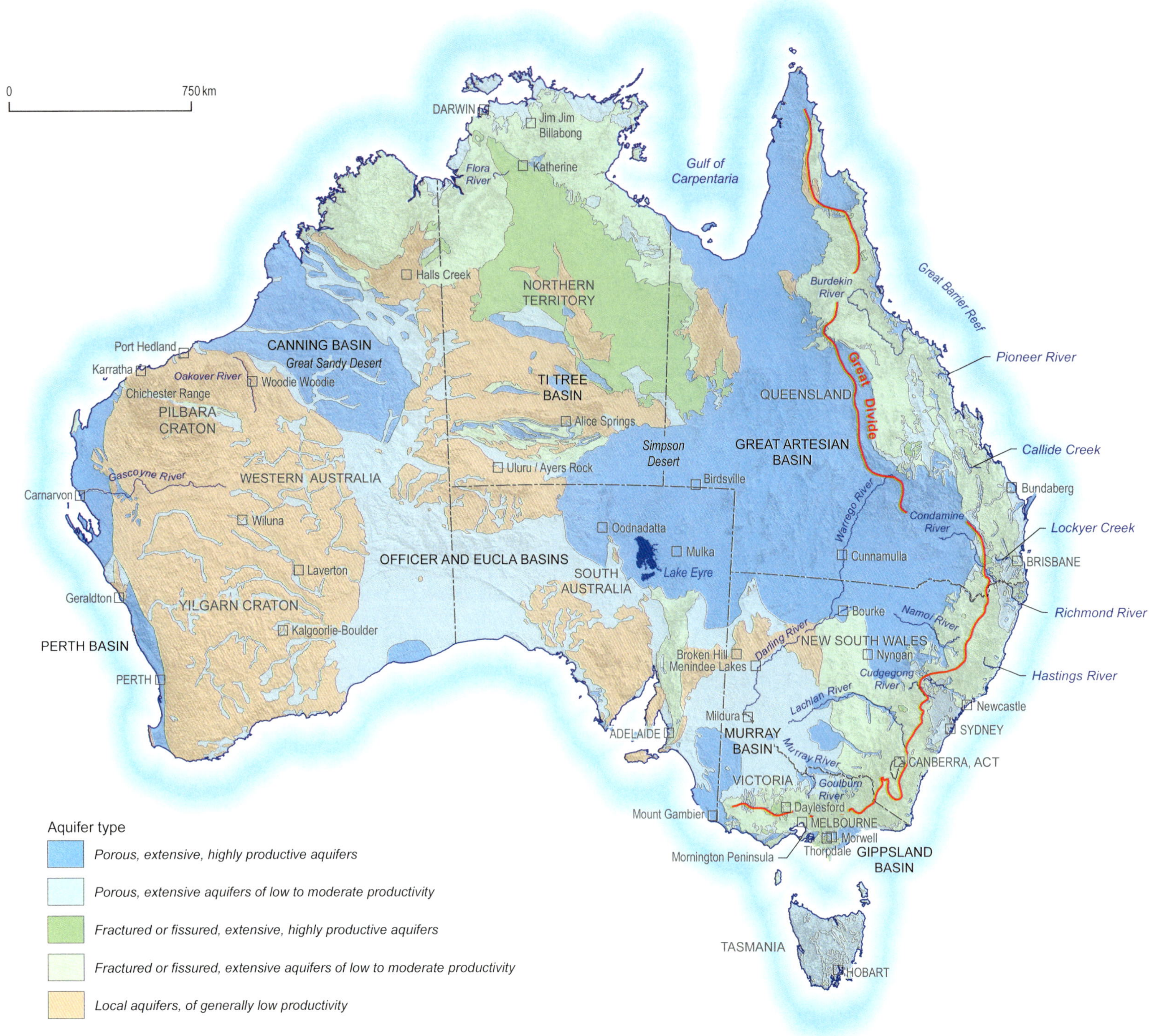
0
750 km
Aquifer type
Porous, extensive, highly productive aquifers
Porous, extensive aquifers of low to moderate productivity
Fractured or fissured, extensive, highly productive aquifers
Fractured or fissured, extensive aquifers of low to moderate productivity
Local aquifers, of generally low productivity
DARWIN
Jim Jim Billabong
Katherine
Flora River
Gulf of Carpentaria
Halls Creek
NORTHERN TERRITORY
CANNING BASIN
Great Sandy Desert
Port Hedland
Karratha
Oakover River
Woodie Woodie
Chichester Range
PILBARA CRATON
TI TREE BASIN
Alice Springs
QUEENSLAND
Great Divide
Burdekin River
Great Barrier Reef
Pioneer River
Callide Creek
Bundaberg
Simpson Desert
GREAT ARTESIAN BASIN
Uluru / Ayers Rock
Birdsville
Gascoyne River
WESTERN AUSTRALIA
Carnarvon
Wiluna
Oodnadatta
Mulka
Lake Eyre
Warrego River
Condamine River
Lockyer Creek
Cunnamulla
BRISBANE
Laverton
OFFICER AND EUCLA BASINS
SOUTH AUSTRALIA
Geraldton
YILGARN CRATON
Kalgoorlie-Boulder
Bourke
Namoi River
Richmond River
PERTH BASIN
PERTH
Broken Hill
Menindee Lakes
Darling River
NEW SOUTH WALES
Nyngan
Cudgegong River
Hastings River
Lachlan River
Mildura
Newcastle
SYDNEY
ADELAIDE
MURRAY BASIN
Murray River
CANBERRA, ACT
VICTORIA
Goulburn River
Daylesford
Mount Gambier
MELBOURNE
Morwell
Thorpdale
Mornington Peninsula
GIPPSLAND BASIN
TASMANIA
HOBART

Real dry

In the driest inhabited continent, Australia's environment and people face the persistent challenge of securing water. Although groundwater may be less than one-fifth of the total accessible freshwater resource, it provides nearly a third of the water that is consumed. With a growing population and growing economy, this amount is increasing. Groundwater is the only source of water for many regional towns and remote Indigenous communities, and for many of the mining operations in the country's interior that underpin the national economy. Groundwater also provides almost a third of the high-security water that is vital for agriculture and towns during droughts. It could be said that groundwater is the lifeblood that sustains communities, industries and habitats, particularly during the tough times.

Geology, landscape and climate dictate the nature and extent of Australia's groundwater resources. The geological history (Chapter 2) provides the underlying architecture of aquifers and aquitards that define groundwater reservoirs and seals. The national hydrogeology map highlights the intrinsic relationship between the groundwater resource and the geological fabric (Figure 7.1).

What is groundwater? Groundwater is water that is found underground in cracks and pore spaces in soil and rock. There is a finite supply of water on Earth that is continually recycled naturally via the hydrological cycle (Figure 7.2). The hydrological cycle describes the constant movement of water above, on and below (groundwater) Earth's surface. Groundwater travels underground by percolation driven by gravity and pressure until the water-table intersects the ground surface. Water can then be discharged at springs or any other body of surface water (such as rivers, lakes and oceans). Once returned to the surface, this water can be used by plants, stored on the surface or evaporated. The percolation or flow of groundwater occurs through aquifers, which are bodies of permeable rock that can store and transmit significant quantities of water. Aquitards are bodies of less permeable rock that seal or contain groundwater resources or minimise the flow from an aquifer.

The ancient cratons of the Yilgarn and Pilbara (WA) tend to yield only limited groundwater from fractures and the weathered profile (Figure 7.1). Instead, the alluvial sands, gravels and calcretes associated with Cenozoic palaeodrainage tracking across the landscape provide the water supply for people, stock and mines. The sedimentary basins that border these cratons, such as the Canning, Officer and Eucla basins to the east, and the Perth and Carnarvon basins to the west, represent significant groundwater stores. In particular, the breakup of Gondwana and initiation of the Perth Basin in the Permian has left a legacy of extensive aquifers in a sedimentary pile up to 15 km thick in which freshwater can extend to depths of 2 km—a fortunate resource for the state capital.

In central Australia, the consequences of the Ordovician Larapintine Seaway and subsequent deformation during the Devonian–Carboniferous Alice Springs Orogeny are reflected in the complex configuration of the Neoproterozoic to Devonian Amadeus, Georgina, Wiso, Daly and Ngalia basins (Figure 2.9). This further Gondwanan legacy

Figure 7.1: (opposite): National Hydrogeological Map of Australia, showing the type and productivity of the principal aquifer and the linkage with regional geology. The map also shows the key localities referred to in this chapter. The large dark-blue areas highlight significant groundwater resources in sedimentary basins, such as the Great Artesian Basin and the Canning Basin. Fractured rock aquifers, in the Yilgarn and Pilbara cratons in the west and along the Great Divide paralleling the eastern seaboard, tend to be local scale and of low–moderate productivity. Significant groundwater resources are hosted in largely Cenozoic alluvial deposits. (Source: Jacobson & Lau, 1987)

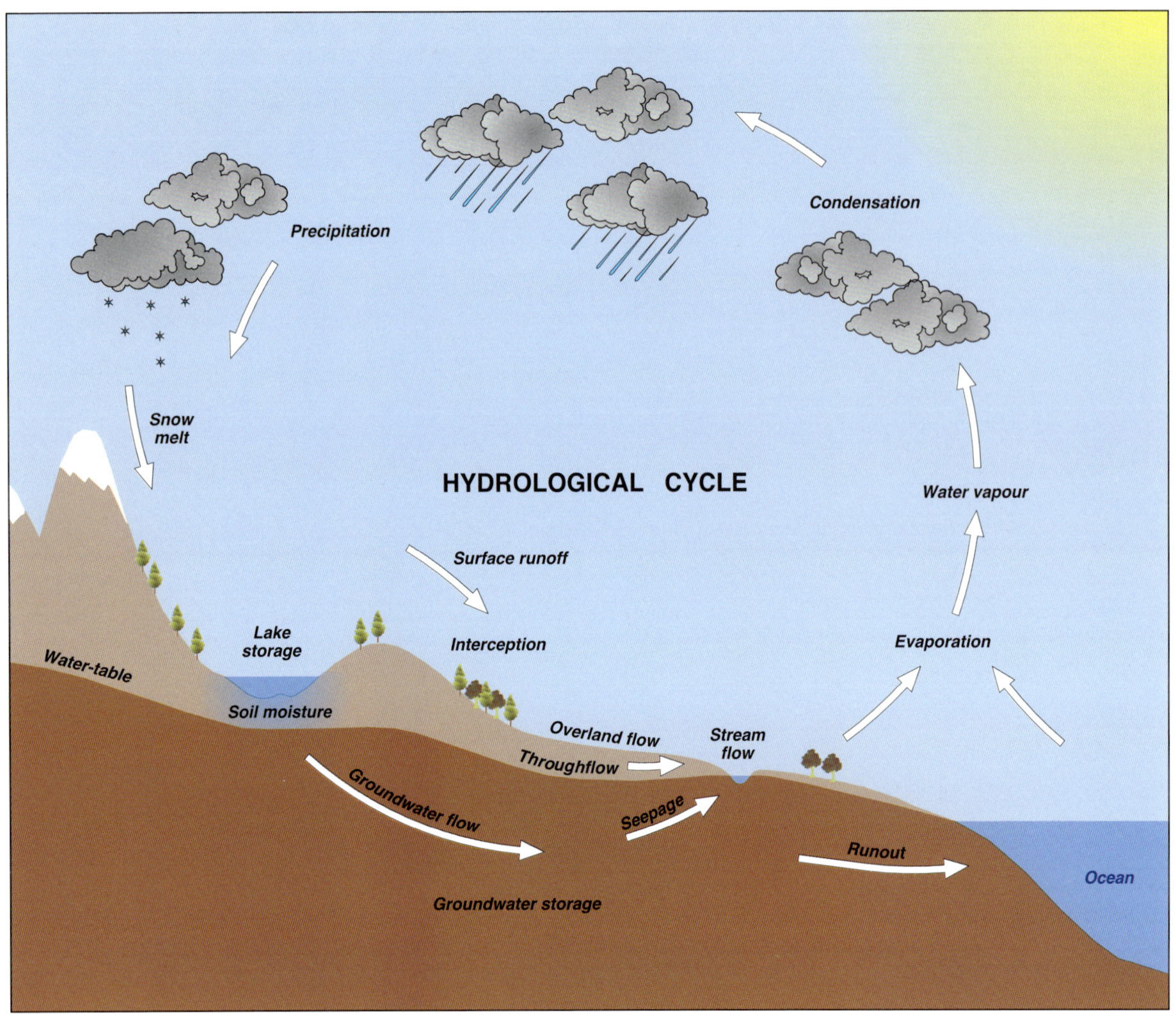

Figure 7.2: Hydrological cycle, showing the key role groundwater plays in the cycle. (Source: University of Texas)

includes the Amadeus Basin, hosting the Mereenie Sandstone aquifer tapped by the Alice Springs borefield, and the principal karstic and fractured carbonate aquifers of the Daly Basin, the Oolloo Limestone and Tindall Limestone, which are used for stock, irrigation and town water supply, and provide important baseflow to the Daly River. In South Australia, the most productive aquifers are limestone found in Cenozoic basins such as the Otway and Murray.

In eastern Australia, the sheer size of the Great Artesian Basin dominates the National Hydrogeological Map (Figure 7.1). During the Jurassic, quartzose and volcanic detritus was deposited by major river systems, followed by marine sediments from the Early Cretaceous shallow sea (Box 4.3). The end product is a vast geological construct of extensive aquifers that crop out and receive recharge along the elevated basin margins, partnered with bounding aquitards that enable pressurised conditions and the blessing of flowing artesian water. In stark contrast, the patchwork of Paleozoic metamorphic, sedimentary and igneous rocks along the Great Divide tends to host relatively local-scale and low-yielding fractured rock aquifers (Chapters 2 and 5).

Dotted along the Great Divide is the chain of the Cenozoic mafic Newer Volcanic Group (Chapter 2), a legacy of continental drift over a mantle hotspot and valued for its combination of productive soils and aquifers. However, the true significance of the eastern highlands from a groundwater perspective is the rivers that drain from it. These include the Cenozoic alluvial aquifers of westerly flowing rivers, such as the Condamine, Namoi and Lachlan, and easterly flowing rivers, such as the Burdekin, Pioneer and Lockyer (Figure 7.1). Coastal sedimentary basin aquifers are also important in the east. For example, in the aquifers of the Gippsland Basin east of Melbourne, the freshwater wedge can extend up to 40 km offshore. Sand dune deposits along the coast can also be regionally important; for example, the Tomago Sandbeds, formed during a Pleistocene sea highstand, provides water security for the coastal city of Newcastle.

Groundwater: a critical resource

The first Australian use of water

Aboriginal Australians have inhabited this land for millennia, navigating the changes in climate as it has cycled from wet to dry periods through at least 50 or 60 kyr. How has the longest continuous surviving human culture prevailed throughout much of the land, particularly when the interior was so arid? The answer may lie in three enduring strategies of traditional knowledge, oral instruction and stylised mapping.

Traditional knowledge of the location and nature of water supplies has accumulated by keen observation over thousands of years. Precise classification of water sources in terms of location, volume, quality and duration of supply provided a critical survival strategy and shaped daily life. Aboriginal languages distinguish various water sources: *gnammas* or rockholes holding temporary supplies after rain; intermittent claypan depressions; riverine waterholes; or soaks and perennial springs supplied by groundwater. The Karijini people of the West Canning Basin, for instance, have a concept of 'top' water and 'bottom' water, roughly corresponding with rain-fed and groundwater-fed features. Aboriginal terminology can also extend to differentiating between groundwater provinces. In the Jawoyn language (NT), *maminga* is water carbonated or high in lime, igilarrang is a similar 'light water' hosted in sandstone, and *yiman wiyan* is 'heavy water' hosted in basalt.

After rain in desert areas, Aboriginal groups could venture out to the country around the temporary water points, secure in the knowledge that they had a perennial source to which to retreat. Hence, permanent habitation of the desert was really only made possible by the more reliable groundwater supplies that acted as critical refuges. These regions were also refuges for the biota (Chapter 3) that provided the food supply for the Indigenous people. The sand dunes of Australian deserts can conceal an extensive palaeodrainage system representing buried remnants of Eocene rivers, and they are a good example of groundwater refuge. Many of these palaeovalleys are occupied now by salt lakes, which have fresh springs discharging around the edges. In places, caves and excavated wells in calcretes have also provided access to shallow groundwater. In the Simpson Desert (SA and NT), the Wongkamala and Wangkangurru people were able to survive by digging wells known as *mikari*. These are located in interdunal depressions, and are dug into gypcrete to depths of 4–6 m in a landscape where there are no surface clues as to the presence of groundwater. Geology has shaped tribal boundaries; in the Geraldton area, for instance, distinction is made between the 'rockhole' people of the inland fractured rock terrain, and the 'well' people of the Perth Basin.

Oral instruction, through storytelling and ceremonies, allows the knowledge base on water supplies to be passed on from generation to generation (Box 7.1). Such stories relate to the Dreaming, the time when the Ancestral Beings traversed the land, creating life and significant landscape features along the way. The Dreaming,

Artwork by Michael J Connolly (Munda-gutta Kulliwari). Dreamtime Kullilla-Art. www.dreamtime.auz.net

The Rainbow Serpent is a highly significant cultural entity of the Dreaming, connecting many Aboriginal tribal groups across Australia and always associated with water.

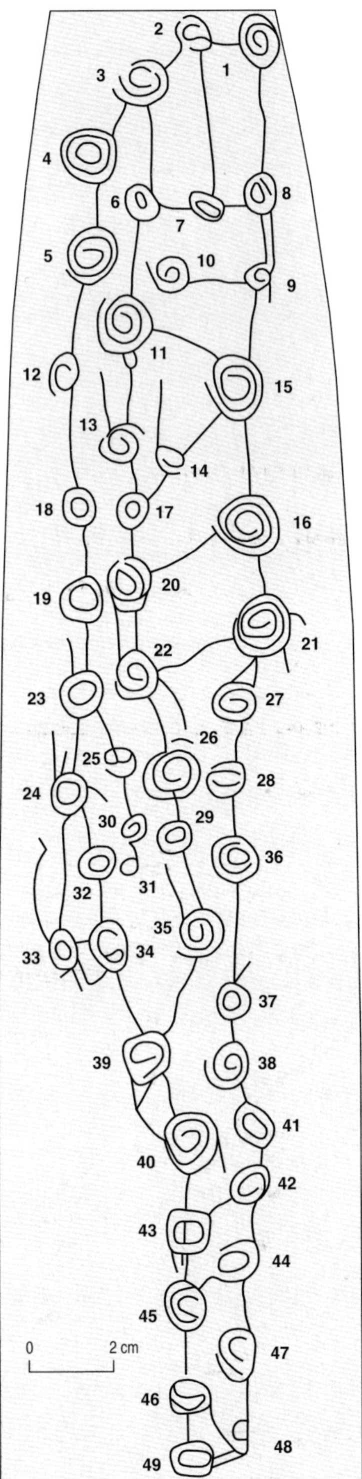

1. *Labbi-labbi*
2. *Tananga*
3. *Liuwiringa*
4. *Kunnamannera*
5. *Maiyada-maiyada*
6. *Wirra-wirra*
7. *Kirindji*
8. *Kanandibaroo*
9. *Markodarindja*
10. *Kampanbarro*
11. *Wirrkaldjarra*
12. *Pinna*
13. *Luwano*
14. *Kira*
15. *Tjul'tjun'waridji*
16. *Dandju*
17. *Tildi*
18. *Wakilbi*
19. *Kuna*
20. *Pintinba*
21. *Yinindi*
22. *Yalbirrimanno*
23. *Tanda*
24. *Kurandal*
25. *Palta*
26. *Kura*
27. *Binbiyan*
28. *Tjipallalla*
29. *Yirabanda*
30. *Dangalli*
31. *Yappadarra*
32. *Timbabiddi*
33. *Yuldumallo*
34. *Kunagarri*
35. *Mukubanda*
36. *Mari-mari*
37. *Karruwildji*
38. *Wallabarrarba*
39. *Kiribarro*
40. *Yanna*
41. *Wangadjarro*
42. *Wornba*
43. *Tjimarri*
44. *Kunananno*
45. *Wirrarigulong*
46. *Danneriyono*
47. *Miltji-miltji*
48. *Papulba*
49. *Lola*

Figure 7.3: A stylised map of the regional water resources of the Bindibu (Pintupi) people of the Great Sandy Desert, Western Australia, carved into the back of a spear thrower. (Source: drawn by Bayly (1999) from a photograph of Thomson (1962))

or *Tjukurrpa* in the Arrernte language around Alice Springs, also means 'to see and understand the law', emphasising the role of Dreaming stories in conveying traditional knowledge and cultural boundaries. The Rainbow Serpent is a highly significant cultural entity of the Dreaming, connecting many Aboriginal tribal groups across Australia and always associated with water. The Rainbow Serpent can journey across the sky, on land and underground. The water cycle is neatly encapsulated in this powerful and not entirely benign creator ancestor. Water, as represented by this mythological serpent, is the dimension that connects the Earth and sky, with groundwater the underground home and tracks of these beings. In the Wardaman language (NT), *barlba* is the term used to describe these underground travels and pathways to the surface. Groundwater-fed springs are so entwined with the Rainbow Serpent that, if one dries up, this is attributed to the Rainbow Serpent being disturbed and moving on.

Storytelling also takes on an important spatial dimension. For example, the journeys of Dreaming ancestors such as the Rainbow Serpent may be recounted as a strict sequence of short songs, each relating to a particular place. An oral map of ancestral travels is built up, allowing Aboriginal people to navigate between strategic water supplies using the paths defined in this way. Such 'songlines' or 'dreaming tracks' can create a network many thousands of kilometres long and connecting many tribal groups. Geology can play a role—for example, in the Perth Basin, the songlines tend to follow the sedimentary geological features, which give rise to springs and soaks.

Stylised mapping can be used to supplement the oral tradition of communicating water knowledge. These maps were portrayed in many forms, including decorated weapons or domestic implements, rock engravings, body paintings or sand drawings. Concentric circles typically represent the water supply, and these are connected with lines, perhaps in the appropriate compass direction, but not to scale in terms of distance. Most importantly, these maps are placed into context by detailed narrative, with each water source named and described. This allowed the transfer of detailed water resource knowledge from generation to generation. The anthropologist Donald Thomson gives an account of how this was managed by the Bindibu of the Great Sandy Desert. Thomson suddenly realised that, through such oral instruction, a carved spear thrower (Figure 7.3) was in fact a highly conventionalised map of the regional water resources.

British colonies and pastoral expansion

The early coastal fingerholds of British settlement struggled to secure a perennial water supply in a land very different from the homeland (see *Did you know?* 7.1). Fortunately, sand dunes can contain fresh groundwater lenses, so this was a ready supply for many coastal settlements. Also, groundwater discharge to wetlands and streams allowed water supply to persist during dry times.

GROUNDWATER DREAMING (BOX 7.1)

Hydrogeology and groundwater processes can be seen embedded in certain Dreaming stories, reflecting Aboriginal affinity with the Australian landscape. Here are two examples from very different geological settings.

The first is set in the Mt Gambier volcanic complex and Australian Geopark (SA):

> *Craitbul, the giant ancestor of the Boandik people, and his family had only one tool, a wooden digging stick, and their bare hands to dig out underground tubers. This was their daily food, which they cooked in an earth oven. They happily established their first camps at Mt Muirhead and Mt Schank until the bird spirit Bullin warned them of the evil spirit Tennateona. They fled to safer refuge at Mt Gambier where they again dug their earth oven. One day, water bubbled up into the bottom of the oven and put out the fire. They dug other ovens, but each time the water rose and extinguished the fire. This occurred four times. Disgruntled, Craitbul and his family finally settled in a cave on the side of a nearby peak.* (Roberts & Mountford, 1979)

These volcanoes are about 4.5 ka; the youngest in mainland Australia. The eruptions would have been a compelling experience for the Boandik people. The dry crater floor of Mt Schank is above the regional water-table. In contrast, there are four groundwater-fed volcanic craters at Mt Gambier: Valley Lake, Blue Lake, Browne's Lake and Leg of Mutton Lake. An interesting feature of the Mt Gambier volcanism is the occurrence of remnant blowholes caused by steam discharge from superheated groundwater, which would fit with the steam hissing as the oven fire was put out in the story.

The second example is a Wardaman Rainbow Serpent creation story for the Flora River, about 135 km southwest of Katherine (NT):

> *The black-headed python Walujapi created the Bulkbulkbaya spring in the Flora River by thrusting a digging stick into the ground. Walujapi pushed the digging stick through until it reached the spring water near Mataranka and the water flowed back, creating Bulkbulkbaya.* (Cooper & Jackson, 2008)

Modern hydrogeological mapping has confirmed that baseflow in the Flora River is from the Tindall Limestone Aquifer, with regional groundwater flow in a westerly direction from the Mataranka area.

Figure B7.1: (a) Mt Schank, South Australia, is around 4.5 ka; the dry crater floor sits above the water-table. (b) The spectacular Blue Lake near the city of Mt Gambier in South Australia. It is one of the groundwater-fed volcanic crater lakes mentioned in the Craitbul Dreaming story of the Boandik people. (c) Tufa dams on the Flora River, Northern Territory. These are formed during the dry season when river flow is mostly groundwater from springs discharging upstream. The water is saturated with $CaCO_3$ that precipitates with evaporation to form the dams. Local Aboriginal Dreaming stories acknowledge the provenance of the groundwater baseflow.

Did you know?

7.1: Groundwater and early European settlement

- Sydney's water supply for the first 40 years, the iconic but now buried Tank Stream, was fed from springs near what is now King and Spring Streets. The Tank Stream was replaced as a water supply by Busby's Bore, which was actually a tunnel to gravity feed water from the Lachlan Swamps (now Centennial Park); the swamps were fed by groundwater from the Botany Sands aquifer.
- The initial attempt to settle the Port Phillip area was not at Melbourne, but at Sullivan's Bay, 60 km to the south. A shallow well, constructed from six wooden barrels sunk into the sand, was the only water source for the 467 convicts, marines and free settlers.
- Captain James Stirling reported on the prospects in 1827 of founding a city on the Swan River: 'supply of fresh water from springs and lagoons is abundant ... it may be confidently assumed that water is plentiful all over this territory'. This may have been an optimistic assessment, but Perth's early groundwater supply is still commemorated by Spring Street in the city.

Image courtesy of State Library of New South Wales. Image no. SSV1/1852?/2

Old Tank Stream Sydney, 1852; watercolour by John Black Henderson.

European 'taming' of the land fanned inland from these coastal enclaves. This was encouraged by reports of favourable grazing lands in the journals of early explorers and surveyors who finally managed to cross the Great Divide west of Sydney (Chapter 5). Pastoralists and farmers followed the rivers, but utilisation of the land away from permanent watercourses involved the sinking of wells. Some of the first New South Wales government expenditure on rural water conservation works was the allocation of £2900 (around $300 K today) in 1866 to sink wells along a stock route from the Darling to the Lachlan and Warrego rivers.

Grazing in many areas is totally dependent on the ability to obtain groundwater. Throughout the extensive pastoral country of Western Australia's Precambrian bedrock, limited supplies of water can be obtained from the overlying surficial deposits, or the weathered bedrock itself. Such is the widespread occurrence of groundwater that station paddocks are laid out in a grid pattern with regular spacing of water points at paddock corners. In this way, European order could prevail over an ancient Australian landscape. The ubiquitous windmill, or more correctly windpump, was transformed from its North American origins in 1854 into a symbol of Australia.

The Canning Stock Route

The vast ranges and plains of Australia ensured, as was the case in the Americas, that 'cowboys' played a role in the development of the country. One theme of the bush is that of the drover, wandering from one waterhole to the next with his 'mob' of cattle. By the turn of the 19th century, stock routes were needed to drive the herds of cattle from vast outback stations to markets at the coast. These stock routes became colloquially known as the 'long paddock' (long field). One of the most famous was the Canning Stock Route in Western Australia.

Established in 1910 as a 1850 km-long route to drive cattle from Halls Creek in the Kimberley to Wiluna in the south, the Canning Stock Route traverses four deserts and is the longest historic example of its kind in the world. Of paramount importance to Alfred Canning's initial 1906 expedition was to locate water supplies at regular intervals along the route. The desert terrain is mostly underlain by Permian to Jurassic sandstones, and the water-table is generally deep. The surveyed route utilises Cenozoic palaeovalley aquifers in which the groundwater is shallow, and where groundwater discharges to salt lakes (Figure 7.4). Good groundwater supplies could also be obtained from surficial calcretes.

Canning did not use geological knowledge to choose the route. Instead, he captured local desert people and forced them to lead him to traditional watering points. Of the 54 water sources staged along the route, 48 are constructed wells of which 37 are on or near Aboriginal *jumu* (soakwaters) or *jila* (springs). The crude and illegitimate methods Canning used to find water were the subject of a Royal Commission. The enmity between Aboriginal inhabitants and the Europeans was reflected in the fatal spearing in 1911 of three drovers at Well 37, and subsequent reprisal shootings. By 1917, nearly half of the wells had been damaged or destroyed and, by 1931, only eight mobs of cattle had used the route. Although the wells were reconditioned, changing markets and a shipping alternative eventually resulted in the demise of the stock route, with the last cattle run in 1959.

Today, the route is busier than it has ever been, used not by drovers but by 4WD travellers keen to experience the landscape's beauty and remoteness and learn more of the art and culture of the traditional Aboriginal owners.

The Great Artesian Basin: Australia's largest groundwater resource

The Great Artesian Basin is Australia's largest groundwater resource and underlies one-fifth of the nation's land surface (Figures 7.1 and 7.5). European pastoral settlement of the Australian interior would not have been complete without this water. With pumping equipment being expensive and difficult to maintain, the tapping of free-flowing groundwater in the basin would have been manna from heaven for the early pastoralists (Figure 7.6). The first flowing bore was unintentionally drilled in 1878 when a shallow well was reconditioned and deepened to 53 m, on remote Kallara station, southwest of Bourke (NSW). This bore was adjacent to the Wee Wattah mud springs, prompting the drilling of similar shallow bores with small flows near other springs across the basin.

A number of scientists independently predicted the potential for artesian bores in different parts of the basin. In 1879, pioneering South Australian geologist Ralph Tate inferred this from the mound springs near Lake Eyre. In 1880, the first New South Wales Government Geologist, Charles Wilkinson, used the Kallara flowing bore to come to the same conclusion for the vast surrounding plains. His counterpart in Queensland, R Logan Jack, also made favourable reports for artesian conditions in the western interior of that state.

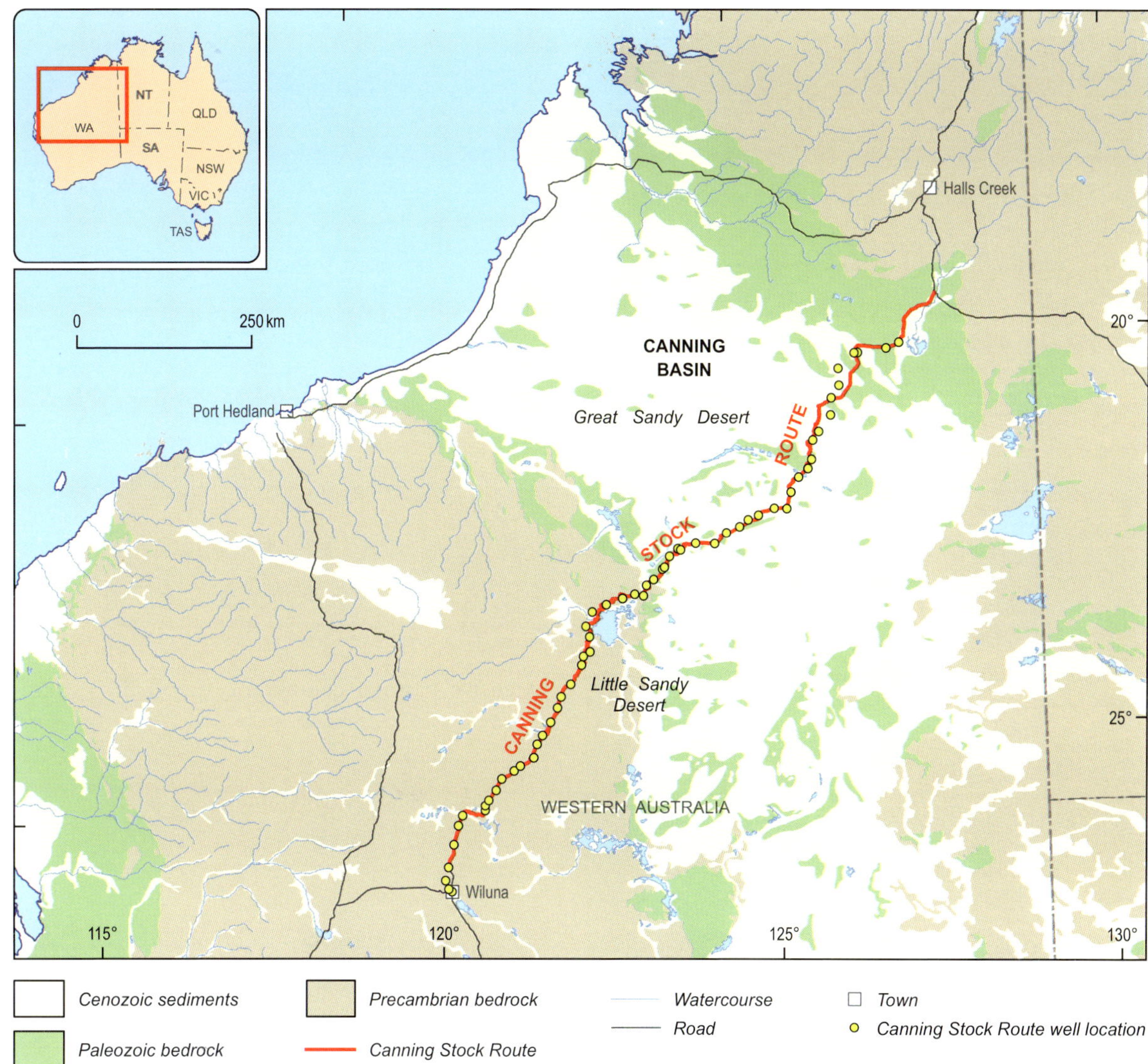

Figure 7.4: The Canning Stock Route between Halls Creek and Wiluna, Western Australia. The water points along the route took advantage of shallow groundwater resources hosted in calcretes and palaeochannel alluvials.

Incidentally, on the other side of the continent, Western Australian Government Geologist ET Hardman had pronounced the artesian prospects in the Perth Basin as poor, owing to the lack of upturned sediments, although he was later proved wrong. The very first known artesian bore in Australia, albeit brackish, was a 51 m-deep coal exploration hole in the Perth Basin in 1871 (See *Did you know?* 7.2).

The advent of the Canadian pole-tool technology suddenly brought drilling capability to depths exceeding 300 m. Application resulted in spectacular successes in 1887 with the Thurralgoona bore near Cunnamulla (Qld) and the Kerribree bore near Bourke (NSW). By 1899, a total of 524 flowing artesian bores had been drilled in Queensland alone. On the eve of the 20th century and Australian federation, artesian water had become part of the Australian psyche, with the belief that free-flowing groundwater could be obtained if only one drilled deep enough. An assured water supply distributed by thousands of kilometres of bore drains brought manifest change to the people and ecology across the vast inland plains, none more so than the Great Artesian Basin.

The Great Artesian Basin is an exemplar for continental-scale confined groundwater systems, being one of the world's largest. It contains sandstone aquifers spanning across the Triassic, Jurassic and Cretaceous with combined thicknesses

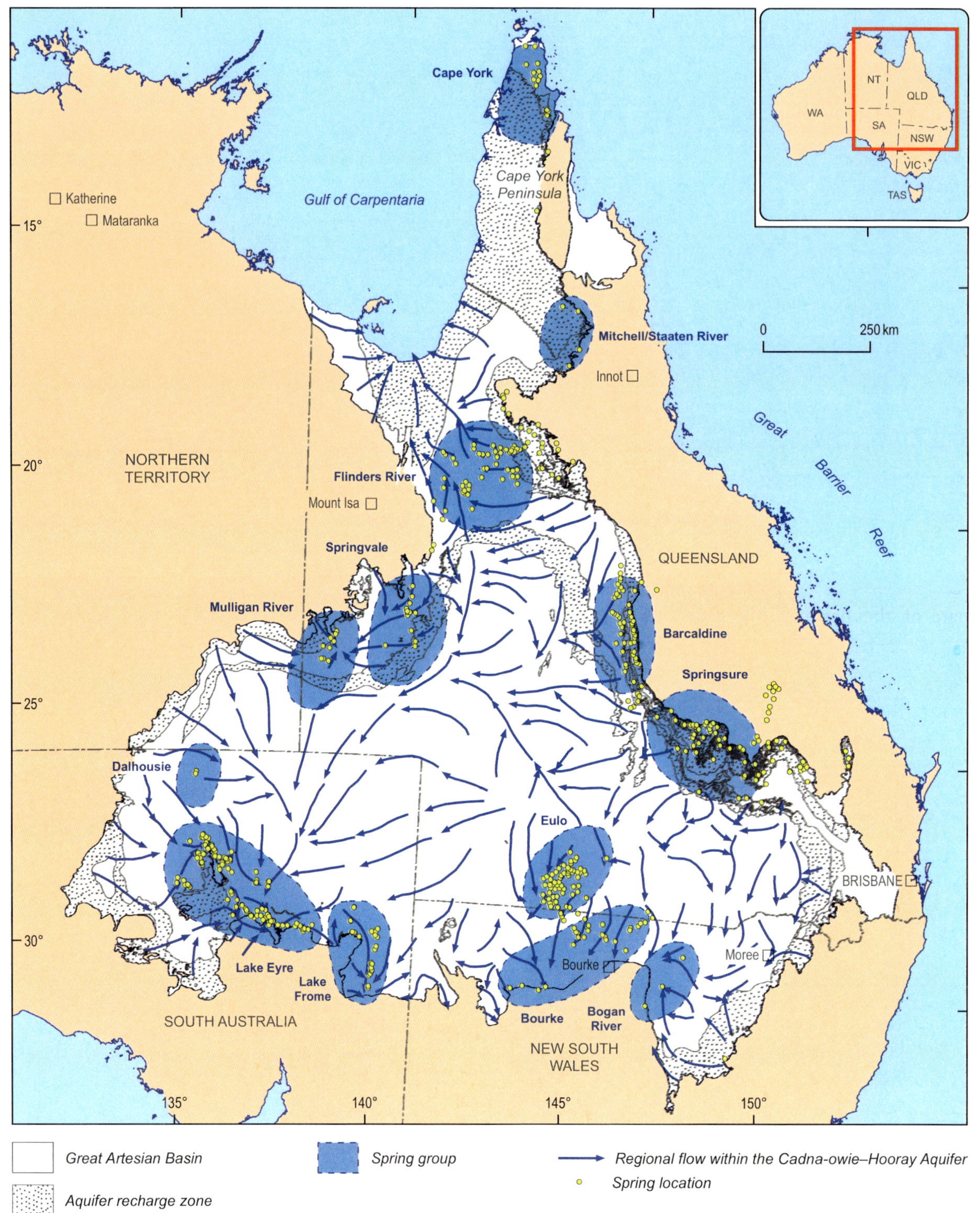

Figure 7.5: Map of the Great Artesian Basin and localities. Also shown is the extent of the groundwater flow paths, the springs and the spring groups. (Source: Radke et al., 2000)

between 100 m and more than 3000 m. The Great Artesian Basin is also an aggregate of sedimentary basins, most notably the Eromanga, Surat and Carpentaria (Figure 2.8).

Potentiometric contours are lines of equal hydraulic head, and these, together with hydrochemistry, reveal that the Great Artesian Basin aquifers are recharged by rainfall and river leakage through exposed intake beds along the basin margins, particularly to the east (Figure 7.5). The relatively high topographic elevation of these recharge beds is the key hydraulic driver of the basin. Conditions become pressurised down-gradient in the basin, where the aquifers become confined by overlying marine siltstones and mudstones. Isotope analyses (^{36}Cl and ^{14}C) show a progression in groundwater age, some exceeding 2 Ma in the southwest part of the basin. Groundwater velocities have been estimated at 1–5 m/year. These rates may appear to be slow, but over time they translate to potential flowpaths exceeding 2000 km in 2 Myr. Groundwater from the commonly exploited confined aquifers is typically a sodium-bicarbonate type with salinities of 500–1000 mg/L.

Groundwater storage in the basin is estimated to be about 65 000 km^3. The basin, however, is not managed in terms of extraction volumes but in terms of aquifer pressure. More than 4700 flowing artesian bores have been drilled in the basin since the 1870s. Basin outflow peaked at more than 2 GL/day in 1918, with aquifer pressures declining ever since. Such a decline has meant that groundwater flow in many natural springs and artesian bores has decreased or even ceased. Also, water use has been inefficient due to uncontrolled flows and the leaky open-earth drains commonly used to distribute water (Figure 7.6). Secondary problems have been caused, such as salinisation,

Song of the Artesian Water

As the drill is plugging downward at a thousand feet of level,
If the Lord won't send us water, oh, we'll get it from the devil;
Yes, we'll get it from the devil deeper down.

But it's hark! the whistle's blowing with a wild, exultant blast,
And the boys are madly cheering, for they've struck the flow at last;
And it's rushing up the tubing from four thousand feet below,
Till it spouts above the casing in a million-gallon flow.
And it's down, deeper down —
Oh, it comes from deeper down;
It is flowing, ever flowing, in a free, unstinted measure
From the silent hidden places where the old earth hides her treasure —
Where the old earth hides her treasures deeper down.

AB Paterson (1896)

Image courtesy of John Oxley Library, State Library of Queensland, Neg: 50456

Did you know?

7.2: First attempt

Drilling in the Great Artesian Basin was not the first attempt to find 'artesian' water in Australia. In 1850, the New South Wales Governor established the Artesian Well Board 'for the purpose of considering and reporting as to the best means of conducting an undertaking for endeavouring to obtain an abundant supply of pure water for the City of Sydney, by boring on the Artesian principle within the walls of Darlinghurst Gaol'.

The gaol was selected because its elevated position allowed water to be distributed by gravity, and hard-labour convicts could be used on a treadmill to power the boring plant. The chairman of the gaol board, geologist Reverend WB Clarke was not overly confident of the venture, estimating that it would take three years to bore to 500 ft (150 m). His pessimism proved justified, as the hole was abandoned at 75 ft, not because of any geological reason, but because (unpaid) convicts sabotaged the equipment.

Image courtesy of State Library of NSW. Image no. SV1/GAO/Dar/2

Watercolour of Darlinghurst Gaol by HL Bertrand (1891).

Image courtesy of John Oxley Library, State Library of Queensland, Neg:109158

Figure 7.6: Historical photograph of a flowing artesian bore in the Queensland part of the Great Artesian Basin. (Source: John Oxley Library, Great Artesian Basin Coordinating Committee)

erosion and the spread of weeds and feral pests. These problems are being addressed in part by a basin-wide program of capping or reconditioning of bores and replacement of open drains with reticulated pipelines.

The decline in aquifer pressures that followed the initial exploitation of the Great Artesian Basin stirred scientific enquiry into its groundwater origins. Even before 1900, it was widely recognised that groundwater had a meteoric origin—namely, sourced by recharge from rainfall. The main protagonist for this majority view for the Great Artesian Basin was the New South Wales Government Geologist, Edward Pittman. However, a protracted scientific debate was sparked when JW Gregory, Professor of Geology at Melbourne University, advocated an alternative plutonic origin—groundwater sourced from the cooling of deep-seated igneous rocks. His views were first revealed in a lecture in 1901 and later expanded upon in his book *The dead heart of Australia* (1906), following his travels to the mound springs near Lake Eyre.

One of the many lines of evidence used by Pittman for a meteoric source came from a 19th century equivalent of a catchment water balance. Government Astronomer HC Russell estimated that less than 1.5% of the rain falling in the upper Darling River catchment flowed past Bourke, and concluded that a significant proportion of rainfall neither evaporated nor became surface drainage. The civil engineer TE Rawlinson had appealed for an enquiry in 1878:

> *into the cause of the disappearance of the vast bodies of river water which collect on the inner water-shed of the coast ranges of Australia*

and considered that the interior of Australia would

> *ultimately be proved to be the storage reservoir where are conserved the rain and river waters which other theories fail to account for.*

Pittman also cited overseas geologists, including TC Chamberlin of the University of Chicago, who had developed a list of the necessary geological prerequisites for artesian conditions. These included surface exposure of permeable intake beds, inclined strata, and both underlying and overlying impervious beds, features that were mirrored in the emerging understanding of Great Artesian Basin geology.

Gregory, although recognising the existence of meteoric water, asserted that the Great Artesian Basin waters were mostly ancient, plutonic, deep-seated crustal water. He claimed that relatively high temperatures and enrichment of sodium (Na), potassium (K) and zinc (Zn) in certain artesian bores were characteristic of such a source. The corrosion of bore casing he thought was due to elevated concentrations of boric acid, which was claimed as another indicator of plutonism. This interpretation was refuted by Pittman, who demonstrated that high Zn concentrations could simply be attributed to corrosion of the Zn-galvanised bore casing pipes.

Although Pittman and Gregory had an amiable professional relationship, this did not prevent them from engaging in vigorous public debate. It was evident that the whole matter required careful investigation and review, and to this end the New South Wales Government invited other states to form a consultative board and the first, of what ultimately was to be five, Interstate Conferences on Artesian Water. The official statement from this 1912 conference in Sydney concluded that the artesian water 'is almost, if not entirely, derived from rainfall' (Figure 7.7).

Water and mining: the thirsty work of industry

The year 1851 marked the transformation of a nation. Gold was found by a waterhole near Bathurst, about 200 km inland from Sydney, quickly followed by Victorian discoveries in Ballarat and Bendigo. Over the next two decades, the population of Australia increased four-fold (Chapters 1 and 8). Reported finds across Australia were inevitably followed by a rush, involving thousands of ill-prepared would-be miners. The location of hastily erected inland settlements could not necessarily be chosen by the presence of permanent water. Some fields had to be evacuated by the authorities when water supplies ran out, and many miners died of thirst and waterborne disease.

The Eastern Goldfields (WA), discovered in the 1890s (Chapter 8), present a classic study of the trials and tribulations of securing water. According to the young Herbert Hoover (later United States president), the area around Leonora was 'red dust, black flies and white heat'. Reliable sources of local water were difficult to secure. Wood-fired condensers were initially used to desalinate the local groundwater, but these were responsible for deforesting the surrounding region. The distilled water was also outrageously expensive, with the price for less than 15 litres being the equivalent of an unskilled worker's weekly wage. The government even wasted a large sum of money

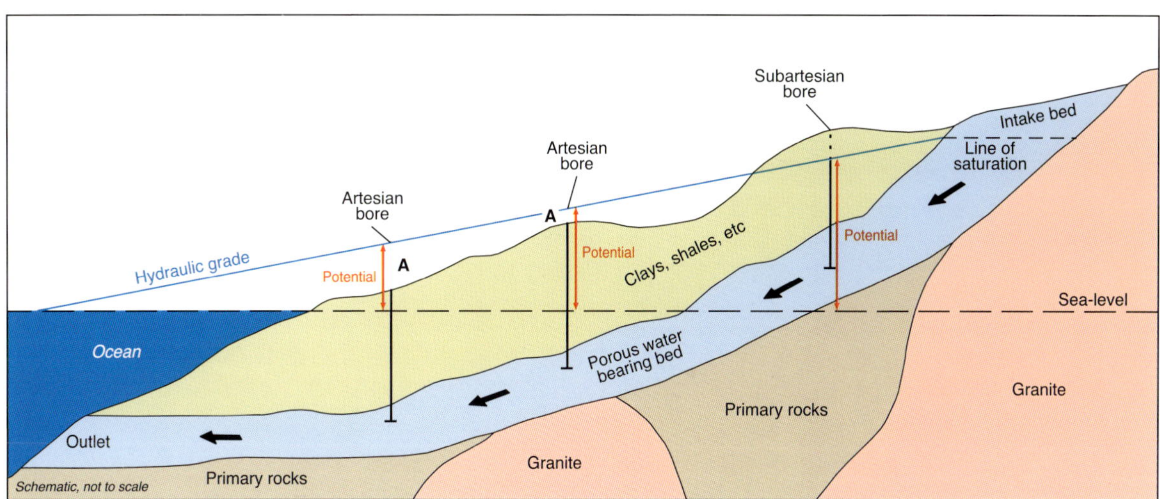

Figure 7.7: Redrawn historical depiction of an artesian basin with an outlet to the ocean, based on the Artesian Water Report of the Second Interstate Conference on Artesian Water, Brisbane, 1914. By the start of the 20th century, the geological and hydrological principles driving artesian groundwater conditions in confined sedimentary aquifers had been firmly established. (Source: modified from Faggion, 1995)

Image courtesy of Resource Equipment Ltd

Figure 7.8: Mines that enter below the water-table need to be dewatered. Pumping water is energy intensive, and the water needs to be disposed of in an environmentally sustainable way.

in drilling to 900 m for 'artesian water' in solid granite, ignoring the Government Geologist's adverse advice. It was only with the construction of engineer CY O'Connor's epic 600 km pipeline from Mundaring Weir near Perth that a viable water supply was finally delivered to the goldfields in 1903. In 1904, the daily water consumption on the goldfields was more than 5.5 M litres. Eighty years later, there was a resurgence of interest in the saline groundwater, with the introduction of carbon-in-pulp gold processing that could utilise the resource (Chapter 8). Drilling discovered a network of palaeochannels in basal Eocene sands that proved to be excellent aquifers. Bores typically spaced 1 km apart along the palaeochannels now access the saline groundwater for ore processing.

Some goldfields encountered the opposite problem of too much (ground) water. The Victorian deep-lead gold (Au) deposits are hosted in Cenozoic alluvials, commonly buried by basalt flows, which are also important aquifers. Mine dewatering proved to be a major challenge. For example, it was reported that in its heyday 170 horse-drawn winches (called whims) were used to dewater the gravel-hosted alluvial Au at the Majorca field in Victoria.

Mine dewatering has continued to be a significant issue across Australia. Since the 1990s, most of the major iron ore mines in the Pilbara have extended their operations to depths below the water-table (Figure 7.8). Mining is on a truly gigantic scale in the Pilbara. Some 400 Mt of iron ore are exported annually (Chapter 9), and this involves pumping millions of tonnes of groundwater to allow the ore to be mined from open pits. Total allocation for mining extraction in the region is 268 GL/year, of which about half is for dewatering. Much of the groundwater is hosted in Miocene-aged channel-iron deposits, which are porous and permeable pisolitic iron accumulations in the bottom of valleys in the Hamersley Range. Today they form major freshwater aquifers, recharged periodically by rapid runoff caused by cyclonic rain. The mining involves removing the whole aquifer, and water pumped to dewater the pits is either discharged to normally dry creek beds, creating perennial flow and a new ecosystem that will collapse when dewatering ceases, or is reinjected in the aquifer further downstream.

Groundwater is also a major consideration in mining the Chichester Range deposits of Marra Mamba Iron Formation, adjacent to the Fortescue Marsh in the central Pilbara, where the underlying groundwater is saline. This presents a challenge in dealing both with fresh groundwater and with hypersaline groundwater, which cannot be discharged to the surface or used for processing. At one mine, 24 GL/year needs to be pumped, with much of this water reinjected into the aquifer.

In the East Pilbara, on the edge of the Great Sandy Desert, 16 GL/year of fresh groundwater is discharged to the Oakover River from the Woodie Woodie manganese (Mn) mine. Proposals to use the dewatering to supply the water-deficient coastal towns of Karratha and Port Hedland have been considered, but the distances of hundreds of kilometres, and the transient nature of dewatering, militate against finding a use for this mining by-product.

Food for the nation and beyond: Australian irrigation

Making the desert bloom was a deeply held yearning for European settlers to Australia, and it would only be a matter of time before the free-flowing Great Artesian Basin bores were investigated for irrigation development. In the 1890s, the New South Wales Government subdivided about 700 acres (280 ha) around the Pera Bore near Bourke for this purpose. However, the Great Artesian Basin groundwater is enriched in Na relative to calcuim (Ca) and magnesium (Mg), even though overall salinity is low. This meant that it was not long before difficulties were experienced, with local soil structure degrading after irrigation. It took 14 years until the irrigation scheme finally had to be abandoned.

Most irrigation schemes using groundwater in Australia were established following World War II. Governments actively promoted the development of groundwater resources by providing technical advice, drilling rigs and financial assistance to farmers. By 1950, the New South Wales Government was operating 20 drill rigs throughout the state. Attention was drawn to the inland alluvial deposits of the Murray–Darling drainage system, with early drilling investigations focusing on the Lachlan and Namoi valleys. Shallow groundwater had been used for stock and domestic purposes, but, with the discovery of deeper and more productive aquifers, graziers became irrigators.

This groundwater development was replicated to varying degrees in the other states. Major irrigation of sugar cane and horticultural crops now occurs along the Queensland coast, such as in the Burdekin and Bundaberg districts, which use groundwater accessed from alluvial and deltaic aquifers. Near Alice Springs in central Australia, early-season table grapes are grown using groundwater from the Cenozoic Ti Tree Basin. Alluvial aquifers of the Gascoyne River near Carnarvon are used to irrigate fruit and vegetable crops for the Perth market, 1000 km to the south. In South Australia, groundwater is turned into wine in grape-growing regions such as McLaren Vale, Langhorne Creek and the Barossa Valley (Chapter 5).

With an average annual rainfall of more than 1020 mm, Thorpedale, Victoria, produces 17% of the nation's potatoes and 30% of the seed stock. Dams and creeks are used for irrigation, particularly in the hotter months.

WHAT ARE GROUNDWATER MANAGEMENT UNITS (GMUs)? (BOX 7.2)

In Australia, groundwater resources and their management are the responsibility of the states and territories. From the early 1900s, the main groundwater management focus was on the regulation of artesian water, because it was realised very early on that uncontrolled flows would lead to pressure reduction and essentially resource depletion (Figure B7.2). The Interstate Conferences on Artesian Water, held between 1912 and 1924, were the first national attempts to discuss groundwater management. Conference proceedings also collated the first bore databases, albeit only for deep 'artesian' bores.

From the 1950s, a boom in groundwater use, mainly for irrigation, was followed by concerns that certain aquifers were becoming stressed, as indicated by falling groundwater levels. This prompted state governments to define management boundaries around areas of interest so that groundwater use could be controlled. For example, in Western Australia, the first groundwater management areas were proclaimed in the 1970s, and virtually the whole state was proclaimed within two decades. Different states use different terminology for these management areas, but nationally they are referred to as groundwater management units (GMUs).

State agencies undertake activities such as exploratory drilling, groundwater assessments and maintenance of monitoring networks as part of their management responsibilities. Intergovernmental coordination is facilitated through the Council of Australian Governments (COAG). This enables development of reforms of national significance, such as the National Water Initiative and the Murray–Darling Basin Plan. It also provides for national overviews, such as Australian Water

Outback water storage tank and windpump near Alice Springs, Northern Territory.

Resources 2005, which indicated the following for Australia:

- There are 367 groundwater management units, for which 33% have a draft or final management plan in place.
- About 5% of groundwater management units are overallocated, meaning that the total groundwater volume that can be extracted exceeds the environmentally sustainable level that has been defined (Figure B7.2). Although only a small proportion of the total groundwater management units are overallocated, these do include some of Australia's major groundwater resources. A further 23% of groundwater management units are classified as highly developed.
- There are areas of particular groundwater development, such as near-urban areas in the nation's west—Alice Springs and Darwin (NT), Perth (WA); irrigated alluvial valleys in the east, such as the Namoi and Lachlan (NSW); and key areas with limited or episodic surface water, such as the Great Artesian Basin.

A new injection of funding from the Australian Government was initiated in 2008, principally under the National Water Commission's $82 M Groundwater Action Plan. Also, under the Commonwealth *Water Act 2007*, the Bureau of Meteorology is charged with the national compilation and delivery of water information, and is currently building capacity to store national statistics on groundwater use from metered bores and groundwater levels from monitoring bores.

Groundwater managers are facing increasing complexity in fully allocated systems, with the advent of trading, integration of surface-water and groundwater management, extension of metering and consideration of water resource management charges to be levied on users. The need for complex numerical models to assist in determining sustainable yields, and tools to assess trading of groundwater allocations means increased demand for hydrogeologists and groundwater managers. This need was recognised by the funding, from 2009, of a new National Centre for Groundwater Research and Training, based in Adelaide. The challenge for the next generation of hydrogeologists is to improve understanding of the geology of aquifers, aided by 3D geological models, to build better numerical representations reflecting heterogeneity, and to better understand recharge processes and climate change impacts.

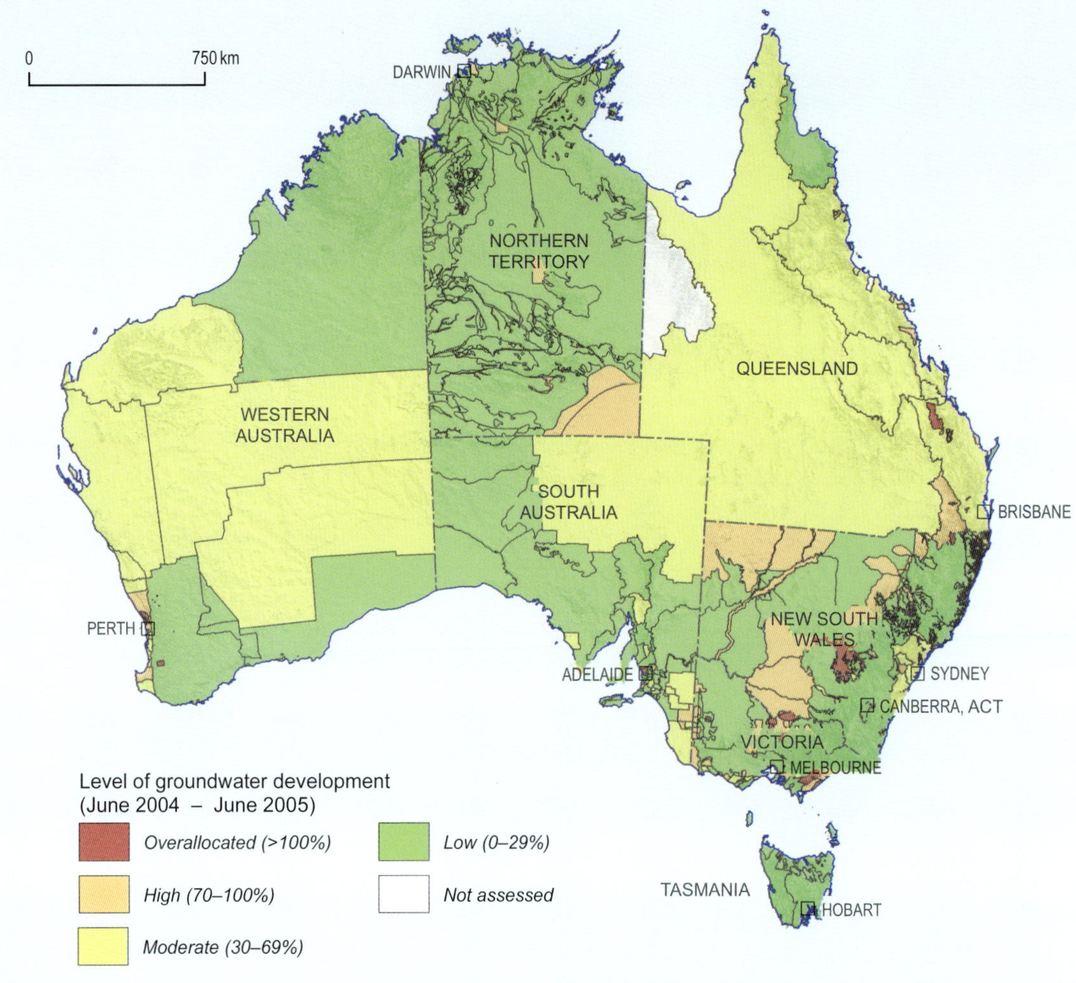

Figure B7.2: National map of level of groundwater development, as sourced from the Australian Water Resources 2005 national assessment. The level of development within groundwater management units is classified by comparing entitlements with the defined sustainable extraction limits. Dark-orange areas are overallocated groundwater management units, and light-orange areas are highly developed groundwater management units. (Sources: National Water Commission, 2006; Geoscience Australia)

Did you know?

7.3: Aussies like to drink!

Australians are large users of water, consuming 920 kL of water per capita in 2004–05. This figure includes water used in agriculture, by other industries and by households. On a per-person basis, the 2004–05 water consumption was 18% lower than in 2000–01 (1100 kL per capita). Agricultural water consumption per person also decreased over this time, from 772 kL per capita in 2000–01 to 598 kL per capita in 2004–05.

Average domestic (household) water use also declined from 117 kL per capita in 2000–01 to 103 kL per capita in 2004–05. In 2008–09, the Water Services Association of Australia calculated domestic water use in the capital cities at 43.6 kL per capita. The lowering of per-capita (and total) water use coincided with prolonged drought, which resulted in the introduction of tough water restrictions and reduced availability of water for agricultural purposes.

	2003–04	2004–05	2005–06	2006–07	2007–08	2008–09	2026[a]	2056[a]
Adelaide	105	101	101	102	84	83	85	71
Brisbane	102	104	73	60	51	53	84	84
Canberra	97	94	104	95	76	79	93	78
Melbourne	77	74	75	68	60	57	63	59
Perth	111	107	105	110	104	106	87	76
Sydney	83	78	75	74	68	74	70	63

Residential water use per capita (kL)
[a] Projected water use

(Sources: Australia Bureau of Statistics 2010; Water Services Association of Australia, 2010)

Groundwater use, particularly for irrigation, has rapidly expanded over the past three decades. This reflects the prolonged periods of drought over this time, combined with the transfer of demand due to capping of surface-water licences. In the Murray–Darling Basin, groundwater extraction has practically doubled since the 1990s. In 2009, nearly 2500 GL of groundwater was used across Australia to irrigate crops and pastures, which represents a third of the total use. The regulation of water in Australia is linked to the country's federation. Water is a state or territory responsibility, but the flow of water of course does not know of these political boundaries (see Box 7.2).

The people's groundwater: supply for cities, towns and communities

Many rural and remote communities across inland Australia rely on groundwater for their water supply. The discovery of the Great Artesian Basin was a boon for inland town water supplies, with artesian bores serving 24 towns in Queensland by 1900. Alice Springs is strategically located in the geographical centre of Australia and is an important Indigenous, pastoral, mining, tourist, transport and communications hub. Most of the town water supply is from the Roe Creek borefield, accessing the Devonian Mereenie Formation in the Amadeus Basin. More than 250 GL of groundwater have been pumped since the establishment of the borefield in 1964, to service an average consumption of 1000 L/person/day—a considerable amount for a region with average annual rainfall only 286 mm. Groundwater has also played a long-established role for communities closer to the coast. For example, by 1910, some 40 artesian bores had been drilled for public or institutional supply for the Western Australian capital, Perth.

Extraction from Roe Creek borefield now exceeds modern recharge, as the groundwater levels in the borefield continually drop by more than 1 m/year. This is because the groundwater supply is essentially 'fossil water' or palaeowater, being derived from a previous wet period, and not now significantly being recharged. Such aquifers can be found in most arid regions and may not have been replenished for thousands or even millions of years, depending on the palaeoclimatic record. Australia is not alone in its reliance on groundwater. Depletion of fossil water is effectively mining the resource, with notable examples in the United States, the Middle East, North Africa and parts of Asia. It is acknowledged that long-term dewatering will occur in the Mereenie aquifer, and management is focused on reducing operating costs and delaying major investments such as a new borefield.

For most of the 20th century, however, there has been a surface-water focus to service the needs of the 80% of Australians who live on the coastal fringe (Chapter 6; *Did you know?* 7.3). This was through the construction of major dams in the hinterland catchments around the Australian capitals. The 2009 National Water Accounts estimated that surface water made up 96%, or nearly 9000 GL, of the water supply for Australians. Long drought conditions through the 1990s–2000s and rising average temperatures have provided a renewed impetus to bring groundwater back into the provision of urban water supplies. This impetus came early for the southwest corner of Western

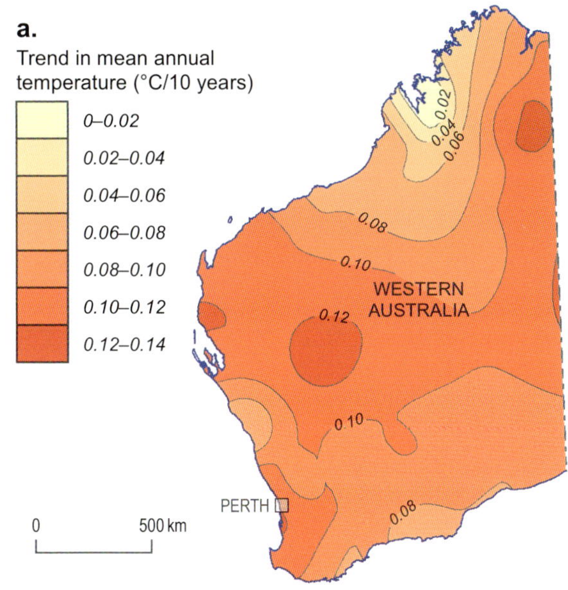

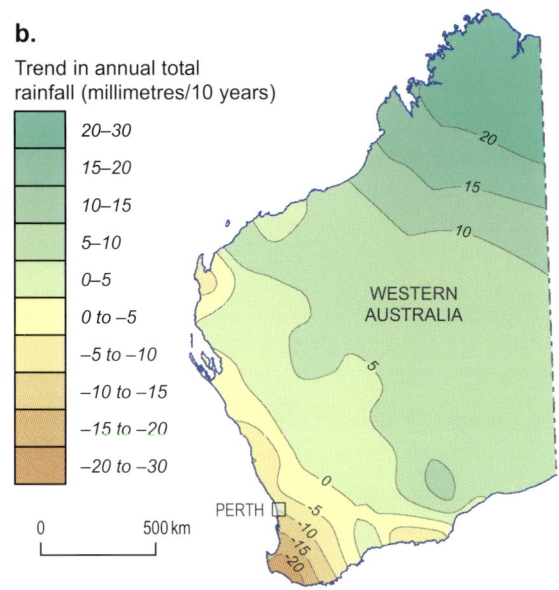

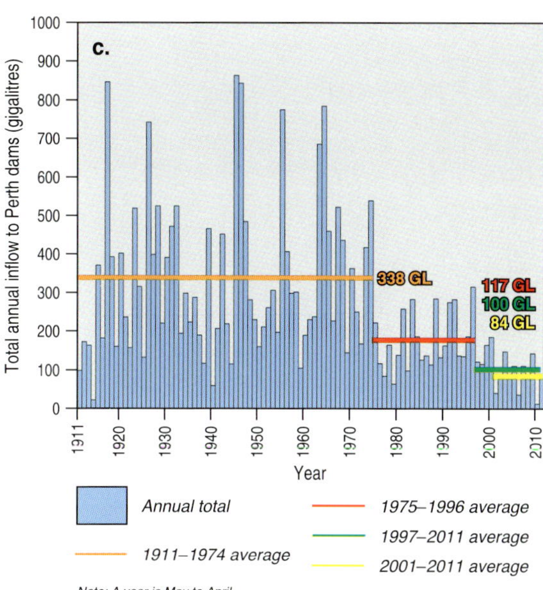

Figure 7.9: Changing climate in Western Australia. The most dramatic changes are occurring in the southwest of the state, where most of the population lives (Figure 1.3). (a) Increasing trends in mean annual temperatures, Western Australia. (b) Downward trend in mean annual rainfall, Western Australia. (c) Declining annual streamflow into Perth dams. (Sources: Western Australia State Water Plan, 2007; Bureau of Meteorology, 2011)

Australia (Figure 7.9a and b). Since 1975, there has been a 15% decline in winter rainfall, reflected in runoff into Perth metropolitan dams being reduced by 75% (Figure 7.9c). To make things even worse, climate models predict further rainfall declines, in the order of 20% by 2030.

A more diversified and integrated approach to urban water supply has been the response. Surface-water reservoirs are still vital, but 35–50% of urban supply is derived from groundwater and 15–20% from seawater desalination. Perth is blessed with a favourable distribution of coastal plain sandy sediments that, combined with the structure and sedimentology of underlying Mesozoic rocks (Figure 7.10), provides a number of significant fresh groundwater resources, including:

- the Gnangara Mound, a 70-m thick water-table mound within Quaternary sands north of Perth, which is the single biggest contributor to the city's urban water supply
- the Jandakot Mound, a similar but smaller shallow unconfined groundwater resource to the south of Perth
- the Leederville Aquifer, formed within Early Cretaceous marine sandstones in the Perth Basin and about 500 m thick
- the Yarragadee Aquifer, a significant Jurassic confined freshwater aquifer in the Perth Basin and underlying the Leederville Aquifer. Groundwater in the Yarragadee Aquifer has been dated at >40 ka, and fresh groundwater extends well offshore, reflecting the period when sea-level was lower and the continental shelf was emergent land (Chapter 1). Current annual extraction for urban water supply is about 45 GL/year.

As Perth is built on sandy Quaternary coastal plain sediments, this shallow unconfined aquifer has long been used by householders for garden watering during the hot, dry summers. It is

© DW Stock Picture Library

Artesian bore, Simpson Desert, South Australia.

estimated that one in four households has a bore, usually consisting of well liners to the water-table and a spear point connected to a centrifugal pump, although modern bores are usually drilled and equipped with an electric submersible pump. Many people have installed bores themselves, so there is widespread community awareness of the existence of groundwater. The taste of iron is prevalent in much of Perth's bore water, but this is simple to precipitate and filter. The garden bores around the city betray their iron content through staining on paths, buildings and fences.

Domestic use of the shallow groundwater is encouraged because it offsets the demand on treated public-scheme water, such as the surface-water reservoirs, deeper groundwater or seawater desalination. It also recognises that the water-table has risen due to land clearing and urbanisation. The phrase 'underground rainwater tank' is now being used to communicate graphically the idea of reusing the recharged rainwater. The Western Australian Government provides hydrogeological maps and an online groundwater atlas to help guide householders on the shallow groundwater resource under their properties.

In eastern Australia, urgent interest in groundwater for urban water supplies was reignited during the droughts of the first decade of the 21st century. In Sydney, the Triassic Hawkesbury Sandstone, which had defeated the early attempts of Darlinghurst gaolers, was again a focus of investigative drilling. Potential borefield sites with the capacity to provide up to 45 GL/year as an emergency drought supply have been identified. Brisbane City Council also undertook a program of drilling aimed at

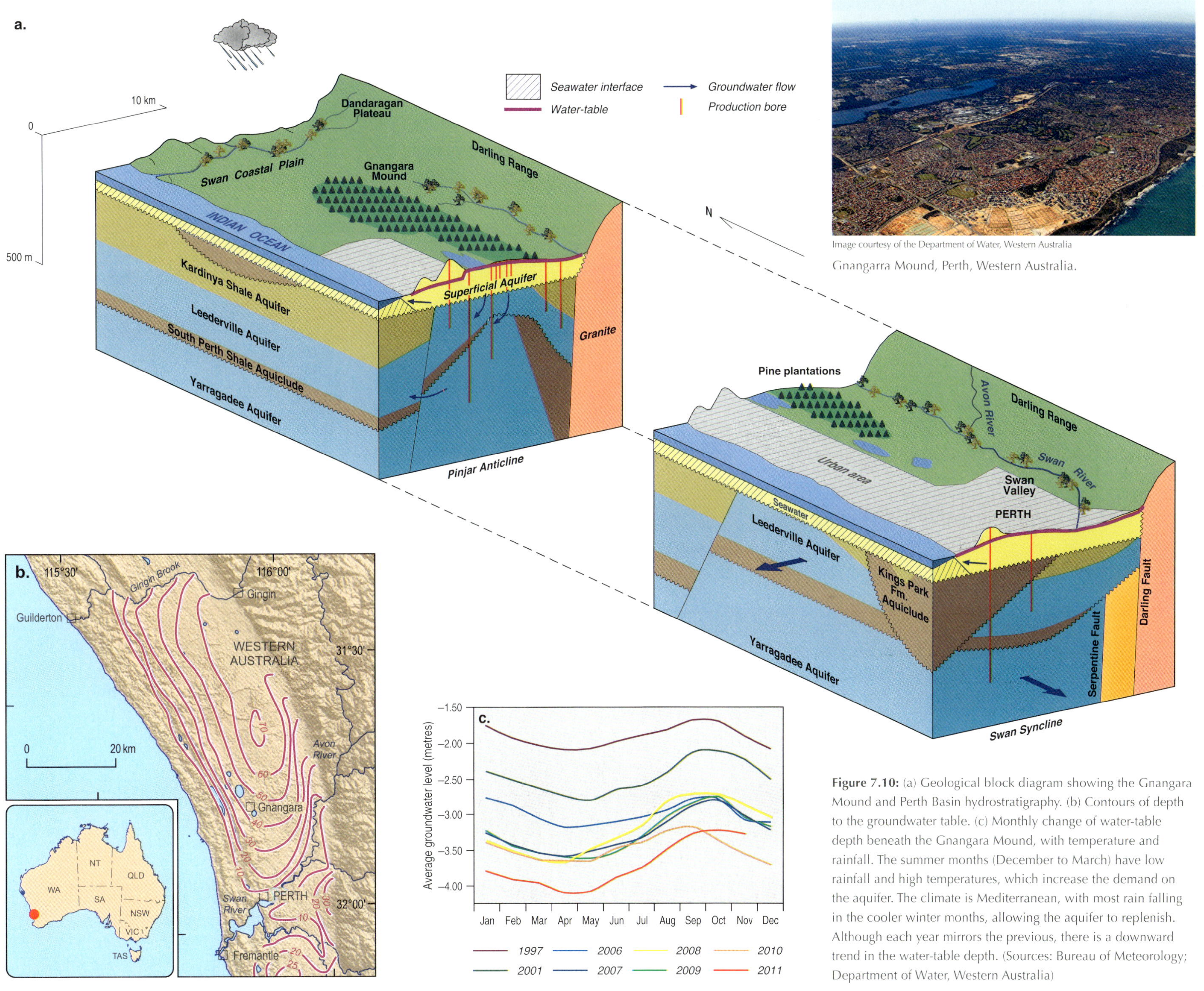

Image courtesy of the Department of Water, Western Australia

Gnangarra Mound, Perth, Western Australia.

Figure 7.10: (a) Geological block diagram showing the Gnangara Mound and Perth Basin hydrostratigraphy. (b) Contours of depth to the groundwater table. (c) Monthly change of water-table depth beneath the Gnangara Mound, with temperature and rainfall. The summer months (December to March) have low rainfall and high temperatures, which increase the demand on the aquifer. The climate is Mediterranean, with most rain falling in the cooler winter months, allowing the aquifer to replenish. Although each year mirrors the previous, there is a downward trend in the water-table depth. (Sources: Bureau of Meteorology; Department of Water, Western Australia)

Did you know?

7.4: Rain doesn't necessarily follow the plough

One story—just one phrase—can acquire exceptional prominence and power as a description of a place, if not a determinant of its future. 'Rain follows the plough' is one example from the late 19th century. At that time, there was intense debate in Australia over whether clearing of native forest (trees, 'bush') for pastoralism or agriculture increased or reduced rainfall. One side had a potent phrase; the other did not. The result of this linguistic imbalance, almost certainly, was more clearing, as 'rain follows the plough' became a mantra that encouraged often unsustainable agriculture and enduring environmental cost.

Popular enthusiasm for irrigation as a basis for agriculture was intense, with the poet Henry Lawson (Chapter 8) one of its keenest advocates. Now, as irrigation assumes proportions Lawson never imagined, its impact is arousing increasing concern. 'When the river runs backwards' is a phrase still gaining life—a powerful warning that natural processes are being overturned, putting the environment in jeopardy.

Image courtesy of Giovanni Marinelli

Clearing trees in the 1950s, southwest Western Australia.

substituting 20 ML/day of its surface-water supply, locating a number of small to moderate groundwater supplies. About 100 km inland, authorities in Toowoomba drilled bores into the Jurassic Helidon Sandstone of the Great Artesian Basin as the water situation became critical.

Bore water, mineral water: it's all groundwater

Groundwater can be many things to many people. Mention 'bore water' to Australians and the response may be somewhat negative. Memories can be triggered of trying to shower or launder using hard bore water. The smell of hydrogen sulfide or taste of iron may return. These concerns can be legitimate as groundwater contains a range of dissolved constituents. Salinity is the major water-quality factor, usually reflected in high NaCl contents, but there are other natural constituents that may preclude certain uses, in spite of the overall salinity. For example, throughout inland Australia, nitrate, even in fresh groundwater, frequently exceeds the drinking water guideline of 50 mg/L. This imbalance is a health concern due to the possibility of methemoglobinaemia (blue baby syndrome). The elevated nitrate levels are not due to agricultural contamination, but reflect bacterial activity in the arid soils.

Compare this with the response when 'mineral water' is mentioned to Australians. A smorgasbord of images may come to mind—pure and natural, exotic and sophisticated, health and vigour—depending on the demographic being targeted by the bottling companies. Drinking mineral water was not a British heritage, but began with brands from many European countries being imported for migrant communities and then promoted widely through restaurants. Bottled water, for both bulk and individual consumption, started to become popular in the 1980s. By 2010, it had evolved into a $560 M industry serving 19% of Australians. However, public concerns are emerging about the environmental impact of producing, transporting, using and disposing of plastic water bottles. In 2009, the town of Bundanoon, 150 km inland of Sydney, became the first Australian town to ban the sale of bottled water. This was followed by the University of Canberra in the nation's capital, which became the first university to place such a ban.

The other traditional use of groundwater is for baths (see Box 7.3).

Groundwater: a geological agent of change

European settlement has clearly brought profound change to the Australian landscape. In the previous section, we described some of the history and implications of groundwater resource development across Australia. However, groundwater is part of the hydrological cycle (Figure 7.2); it is intrinsically connected to the land, rivers, ecosystems and, therefore, people. Modifications to groundwater systems can have broad and sometimes unanticipated implications and, as such, groundwater can be viewed as a geological agent of change.

TAKING THE WATERS (BOX 7.3)

The Daylesford–Hepburn Springs area located in central Victoria is known as Spa Country, famous for its carbonated water springs. In 1864, its citizens petitioned the Victorian Government for the mineral springs to be protected from the impacts of mining. Many of the town residents were from European countries with a spa heritage, including Switzerland, Italy, Germany and England. The residents appreciated the water's intrinsic value, above that of gold, which led to the creation of the Hepburn Mineral Spring Reserve. The area was a popular destination for Melburnians to 'take the waters' after the arrival of the railway in 1880, but patronage of the spas declined from the 1920s. There has been a resurgence in recent decades, with facilities upgraded to accommodate a renewed interest in 'wellness': natural healing, hydrotherapy and spa relaxation.

There are more than 100 documented springs in the area, typically located in river valleys and streambeds. The spring waters are naturally effervescent and mildly acidic, and have low chloride (<80 mg/L) and high bicarbonate (>1400 mg/L). Their ambient temperature suggests a shallow water circulation in the order of hundreds of metres, supported by the silica geothermometry. The groundwater is sourced from local rainfall recharge, but carbon isotope data link the CO_2 to volcanic activity. The area has the highest concentration of eruption points in the Newer Volcanic Group.

The Victorian Spa Country is not the only place in Australia to visit spas. The Japanese-themed Peninsula Hot Springs geothermal complex on the Victorian Mornington Peninsula uses groundwater at 50ºC from a 637 m deep artesian bore (Figure B7.3). Further afield, artesian baths dotted across the vast expanse of the Great Artesian Basin provide a truly unique outback spa experience. Hot mineralised artesian groundwater is used in public baths and swimming pools in some of the country towns—the Moree Hot Artesian Pool Complex (NSW) uses artesian groundwater at 40ºC from an 868 m deep bore to create this town's major tourist attraction and the focus for its annual Healing Waters Festival. In other parts of Australia, there are examples of hot springs providing pleasure for tourists and locals alike—the Mataranka, Katherine and Tjuwaliyn springs (NT), Hastings Cave (Tas.), Zebedee Springs in the Kimberley (WA) and Innot hot springs (Qld) (Figure B7.3).

Image courtesy of Peninsula Hot Springs

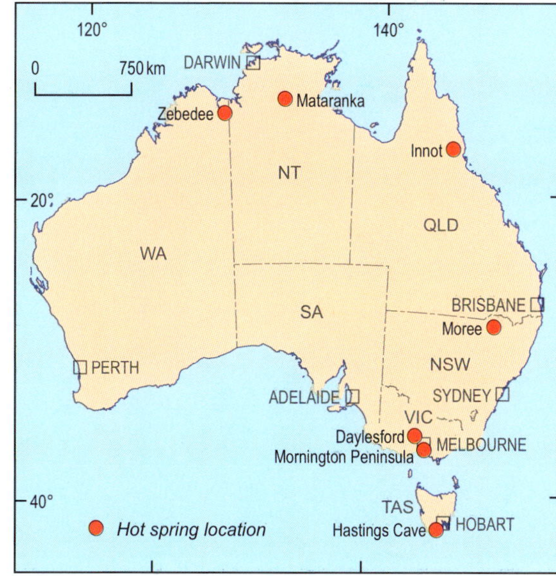

Figure B7.3: Location of some Australian hot springs.

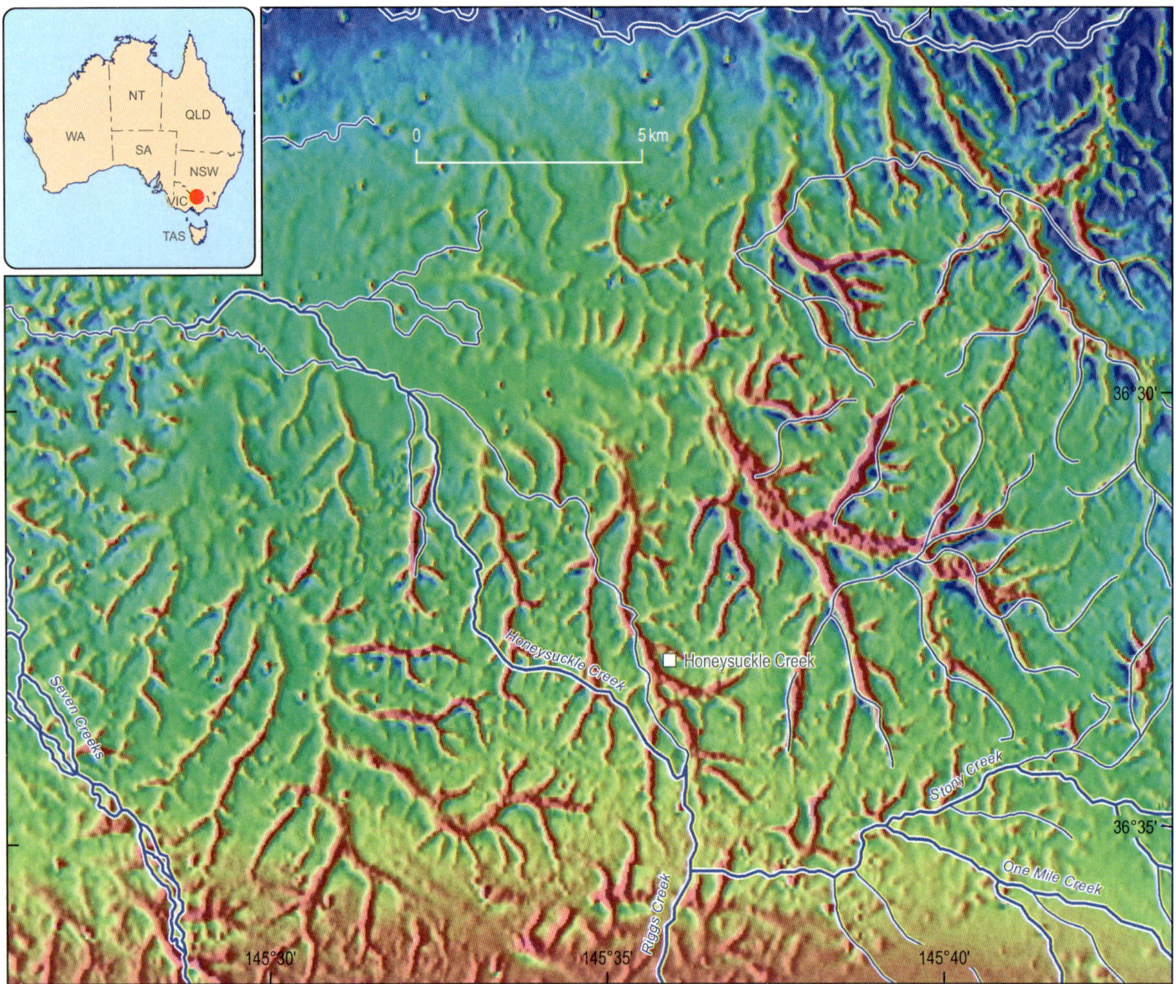

Figure 7.11: Total magnetic intensity (1st vertical derivative) of the Honeysuckle Creek area of the southern Murray Basin, near Shepparton (Vic.). The red colours are magnetic highs; blue colours are magnetic lows. The dendritic pattern mapped by the magnetic data is caused by clasts containing maghemite that were deposited in a Plio–Pleistocene palaeodrainage network. These palaeodrainage networks are the 'deep leads' that may contain rich accumulations of alluvial Au (Chapters 5 and 8), as well as groundwater. They are now covered by young alluvium. The present-day drainage follows some of these old paths, but deviates markedly in many places. (Source: Gibson & Wilford, 2002)

Changing rivers and aquifers

Rivers are located in catchments with underlying geology, and should be perceived as not just the river channels and floodplains but also the aquifers within which they reside. Because of the wide array of geological settings, there is variability in the hydraulic connection between a river and a given aquifer. In its upper reaches, a river may receive groundwater discharge from surrounding highly dissected fractured rock aquifers. This persistent baseflow can be critical to maintaining river flow and aquatic ecosystems during dry periods. Further downstream, the river may actually lose water to the underlying alluvial aquifer; this should not necessarily be viewed as a 'loss' as it can replenish a significant groundwater resource.

A geological understanding of the catchment, therefore, is a precursor to understanding the interactions between groundwater and surface water. Mapping the distribution of restrictive clay layers (aquicludes or seals) or, alternatively, palaeochannels (aquifers) that act as preferential pathways is important (Figure 7.1). Also important is identifying geological structures such as faults or basement highs that control the geometry and hydraulic properties of aquifers (Figure 7.10a). For example, in the Cudgegong Valley, 250 km inland of Sydney, bedrock highs or faulting constrict groundwater throughflow in the alluvial aquifer, resulting in shallower water-tables and gaining conditions in the river. Away from these geological features, the regulated Cudgegong River is largely a losing stream.

Surprisingly, the importance of river–aquifer connectivity to water allocation has only recently been recognised in Australia (Figure 7.12). Water management in the Murray–Darling Basin is symptomatic of this. In 1997, a cap was placed on the volume of surface water diverted from the river system; however, basin groundwater use was not constrained. Combined with drought conditions, groundwater use increased steeply with migration of demand. Groundwater now accounts for 16% of total water use in the basin, but a far greater proportion during droughts. By 2030, groundwater could make up more than 25% of total average basin water use. Of concern is how current and projected increases in groundwater extraction will decrease baseflows to rivers, particularly during low-flow conditions, so impacting on the integrity of the surface-water cap. It is also estimated that 25% of current groundwater use will eventually be sourced from river 'leakage' induced by the water-table declines from groundwater pumping. As recently as 2005, double accounting, where the available resource is counted as both a surface-water and a groundwater stock, was occurring in the national audit because of a failure to understand the extent of the connectivity. A conjunctive approach is required, in which groundwater and surface-water resources are managed in a coordinated way.

Change and the groundwater–wetlands connection

Colonial settlers in Australia typically considered wetlands as unsanitary and of little value, and as a consequence wetlands were drained, progressively filled in, or used as refuse tips. Fortunately, public perception has changed with increasing environmental consciousness, and the signing of the 1971 intergovernmental UNESCO Ramsar Convention for the 'conservation and wise use of wetlands and their resources'. Public interest has spurred the need to better understand wetland hydrology, including groundwater linkages.

The wetlands of the Gnangara Mound, north of Perth, are a classic example of groundwater and surface-water interaction (Figure 7.10). Here, an elevated dune plain is the surface expression of

Jim Jim Billabong in Kakadu National Park, a World Heritage Area in the Northern Territory.

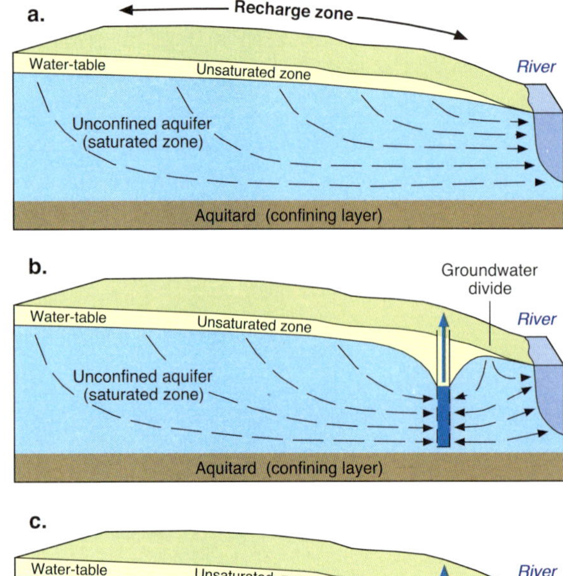

Figure 7.12: Impact of shallow groundwater extraction near rivers. (a) Pre-extraction state of groundwater and river water, where groundwater flows towards the river. (b) Drilling of a bore adjacent to a river will deflect the groundwater flow towards the bore. (c) Ongoing extraction may eventually draw water from the river into the bore. Note the lowering of the water-table around the bore—this is called a cone of depression. (Source: Winter et al., 1998)

Quaternary coastal sands hosting a groundwater mound, some 70 m thick. Wetlands occur in the interdune depressions and are windows to the shallow water-table. As such, the wetland hydrology is defined by the groundwater trends, which reflect medium-term rainfall cycles. The rainfall record, expressed as the cumulative departure from the mean, shows that until 1914 there was a relatively dry period, followed by a wet period from 1915 to 1975, before another dry period (Figure 7.9c). Groundwater trends are shown by the change from swamps to lakes around 1920, and the reversion to swamps in the late 1970s. Also, the water balance of the Gnangara Mound is controlled by the wetlands. At high water levels, evapotranspiration is increased, and this is diminished when groundwater levels fall and wetlands dry up.

The offshore realm: change and submarine groundwater discharge

Aquifers can extend offshore, allowing the opportunity for submarine groundwater discharge. This flow can influence the distribution of freshes (low-salinity zones), nutrients and contaminants in the marine environment. The existence of submarine discharge of groundwater has long been recognised in Australia. In the local language of the Guggu Yimithirr of eastern Cape York Peninsula, freshwater is *buiur*, and *buiur bindi* is a spring in the ocean, referring to groundwater discharge in the intertidal zone. In 1896, Queensland Government Geologist RL Jack, when examining the water budget for the Great Artesian Basin, pondered that:

> *The bores and springs combined do not give out the tithe of the water absorbed every wet season, and, as the water does not escape by land, and can hardly be supposed to find a hiding-place in the centre of the Earth, the only alternative is that it must escape where the absorbing strata crop out beneath the floor of the ocean. I have been informed by a thoroughly reliable shipmaster that there is a spot far out in the Gulf of Carpentaria where almost fresh water can be taken up in a bucket.*

Direct evidence of submarine groundwater discharge can be found along the lagoon of the Great Barrier Reef (Chapter 6). Fresh groundwater emanates from seafloor depressions, called 'wonky holes' by the local fisherman because they are a hazard to trawling equipment. These features are circular, 10–30 m across, and found in about 20 m of water depth within 10 km of the Queensland coastline (Figure 7.13). They are interpreted to be associated with palaeochannels incised into the shelf during sea-level lowstand that are filled with coarse sediment and capped by finer terrestrial material during later Holocene sea-level rise. Sea life is relatively prolific near these holes, but there are concerns that these springs are also conduits for agrochemicals, excessive nutrients or acid waters from the land, which could have a negative impact on the World Heritage coral reef (Chapter 6).

Changing groundwater impacts on Australia's unique ecosystems

There is increasing awareness of the groundwater dependency of many of Australia's unique ecosystems. As aquifers can contribute to rivers, wetlands and estuaries, many aquatic ecosystems can have at least a partial reliance on groundwater. Groundwater can provide critical baseflow to rivers during dry periods and thus sustain life. The platypus needs perennial stream flow or refuge pools, making it a signature species in this regard. Discrete springs within rivers can provide a stable environment for unique and specialised biota. This is also the case for the hyporheic (mixing) zone between surface water and groundwater found in the permeable riverbed sediments. This mixing encourages biological and chemical activity, creating a greater diversity of microfauna than in the river itself.

More broadly, many Australian vegetation communities also rely on shallow groundwater sources, particularly in periods of prolonged drought. The river red gum forests of the Murray–Darling Basin and the *Banksia* woodland on the Gnangara Mound of Perth are just two examples. Root access to the water-table via the capillary zone can be permanently, seasonally or opportunistically important. The depths to which trees send down their roots in search of water can be surprising. Rooting depths of 40 m have been observed for jarrah trees in Western Australia, and studies of bloodwoods in central Australia showed groundwater uptake from 20 m depth. The true extent of the groundwater dependency of Australia's flora has yet to be discovered.

As the links between groundwater and ecology are many and varied, there is a veritable groundwater menagerie. Here are just some examples of Australian fauna that have a dependency on groundwater:

- Native animals of the semi-arid interior have a groundwater reliance because springs and soaks are critical refuges, particularly during droughts. This is the case also for the diversity of birdlife, so much so that Aboriginal people sighted and tracked the more sedentary species, like zebra finches, pardalotes and various pigeons, using them as desert water diviners.
- Investigations at the Mon Repos beach turtle rookery near Bundaberg (Qld) suggest that groundwater conditions can influence the breeding cycle of loggerhead turtles. Eggs are laid in chambers about 0.5 m into the moist sand, within the capillary zone above the shallow water-table. This provides ideal conditions for incubation, including high humidity, low salinity and adequate ventilation. Too wet and the eggs become inundated; too dry and the chamber collapses while being dug.
- Groundwater can have an influence on dugongs, the 'cows of the sea' that graze offshore seagrass beds. Submarine groundwater discharge can be an important contributor of nutrients to such offshore habitats, particularly where marine upwelling or river discharge is limited. For example, studies of seagrass and kelp beds north of Perth estimated that half of the nitrogen required for growth came from groundwater discharge.
- Groundwater conditions can obviously have an overriding influence on cave ecosystems. On the southwest tip of Western Australia, eucalypt tree roots access groundwater within the caves of the Leeuwin–Naturaliste Ridge. In turn,

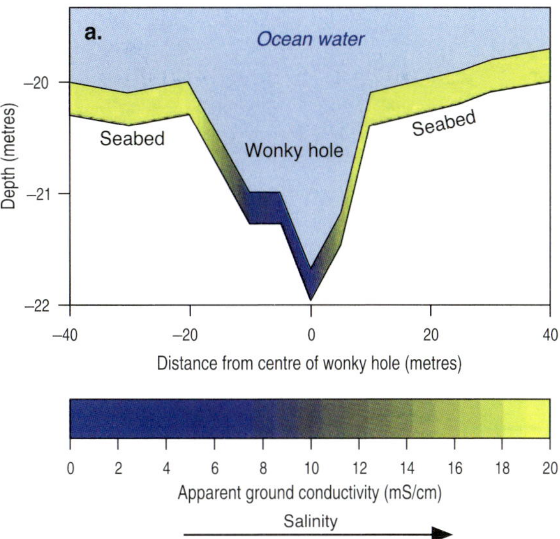

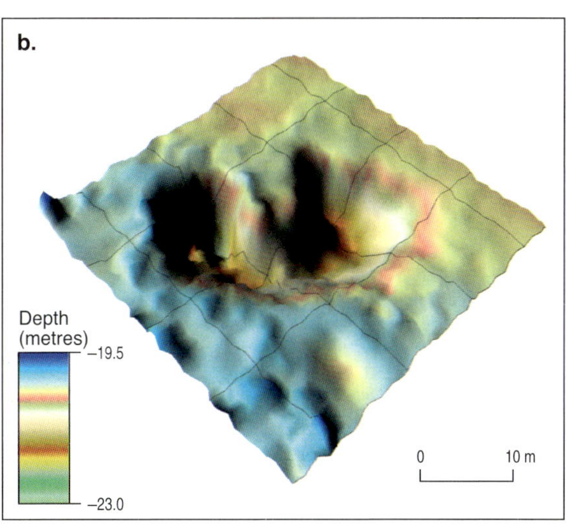

Figure 7.13: Example of a 'wonky hole' associated with submarine groundwater discharge in the lagoon of the Great Barrier Reef. (a) Apparent ground conductivity transect, showing significantly reduced conductivity due to the freshwater from the 'wonky hole' depression. (b) Mapping of the seafloor depression by acoustic multibeam imagery. (Source: Dr Thomas Stieglitz, School of Engineering and Physical Sciences, James Cook University, Townsville, 2005)

© D Wall, Lonely Planet Images

Often the only source of surface water in the driest regions of Australia are mound springs, such as the Bubbler Mound Spring, Oodnadatta Track, South Australia (see Figure 7.14).

the root-mat and fungal growths support a complete food web and one of the richest cave fauna communities in the world. This ecology is threatened by declining groundwater levels, thought to be a consequence of lower rainfall (Figure 7.9b). A similar situation prevails for root-mat ecosystems found in caves near Yanchep, about 50 km north of Perth, which rely on groundwater of the Gnangara Mound. Sprinkler systems have been installed to maintain moisture for the root mats.

- Animals have been found to be living in the void spaces within aquifers. Such 'stygofauna' are mostly crustaceans, but also include insects, worms, gastropods and mites. These animals display the evolutionary features of a subterranean existence, including no or limited eyes or pigment, enhanced non-optic senses and a worm-like body form. A diverse array of endemic species has been found within shallow calcrete aquifers associated with Cenozoic palaeovalleys in Western Australia.

The mound springs of arid Australia: a unique ecosystem

Near Lake Eyre in the arid northern part of South Australia, you will find that the most reliable water supply is not the lake but on top of the hills (Figure 7.14). Groundwater discharge from the Great Artesian Basin has precipitated carbonates and other minerals, which, when combined with windborne sediments, have created conical mounds that dot the flat, arid landscape. The active mound springs typically range up to 10 m in height, but there are ancient relict spring deposits that rise

40 m above the surrounding plains. About 60% of the estimated total spring discharge across the basin of 1500 L/s occurs in South Australia.

The mound springs have great heritage significance (see *Did you know?* 7.5). They represented reliable drought refuges for local Aboriginal people and were an important part of a continent-wide Indigenous communication and trading route. The Witjira–Dalhousie Mound Springs, 350 km north of Lake Eyre (Figure 7.14), for example, are renowned for the extent and volume of Aboriginal artefacts. These springs feature strongly in the Wangkangurru people's song-lines and story-lines.

The mound springs were also important for the Europeans. The springs were used by explorer John McDouall Stuart, the first European to cross the continent from south to north in 1862. He defined the route for the Overland Telegraph between Adelaide and Darwin, which, when completed in 1872, revolutionised communication with Britain and the rest of the world.

Mound springs are emblematic for groundwater-dependent ecosystems. They represent island habitats in a desert 'sea', and contain many rare and endemic species of plants, fish, snails, isopods, amphipods and ostracodes. Living in the Wiljira–Dalhousie Mound Springs are four endemic fish species (Dalhousie catfish, Dalhousie hardyhead, Dalhousie mogurnda and Dalhousie goby), as well as unique snails and crustaceans. In 2001, all Great Artesian Basin discharge springs were collectively listed under the *Commonwealth Environment Protection and Biodiversity Conservation Act 1999* as an endangered ecological community.

The complex of mound springs extends over an area of approximately 400 m by 500 m. This complex has groundwater flow that deposits Ca and other salts from the mineral-rich waters, and these deposits, combined with windblown sand, mud and accumulated plant debris, settle around the spring outflow, forming mounds that resemble small volcanoes. Elizabeth Springs was added to the National Heritage List in August 2009 as it is a significant refuge for a number of plants and animals that, due to the springs' isolation, have evolved into distinct species not found anywhere else in the world, including a freshwater hydrobiid snail and the threatened Elizabeth Springs goby.

Change and the geological inheritance of salinity

In southern Australia, groundwater salinity can be high as a result of interior drainage and high salt storage in the unsaturated zone. Both these factors have contributed to the salinisation of farmland and surface waters. The link between land clearing and salinisation was first documented in 1924 by railway engineer Walter E Wood in Western Australia. In the steam age, the quality of water used in boilers was critical; Wood noticed that the water in the Mundaring reservoir was becoming saltier, and he correctly identified the cause as the ringbarking of trees to provide more surface-water yield from the catchment. The subsequent rise in the water-table and the mobilisation of salts stored in the unsaturated zone ultimately resulted in saltier runoff. Although this recognition led to the replacement of the trees in the Mundaring catchment, the expansion of agriculture across

Did you know?

7.5: Heritage springs and endemic species

Artesian springs are the natural outlets of the extensive artesian aquifer from which the groundwater of the Great Artesian Basin flows to the surface. The first systematic survey of the basin's springs was by JA Griffiths (1896–98). Today, only 20 of the 107 spring groups identified more than 100 years ago remain active. Lowering of aquifer pressure caused by groundwater extraction from the thousands of bores is thought to be responsible for this decline. The environmental impact is enormous, with the loss of springs leading to the extinction of endemic biota.

Springs can range in size from only a few metres across to large clusters of freshwater pools known as 'supergroups'. Elizabeth Springs, situated approximately 300 km south-southeast of Mt Isa in western Queensland, forms part of the Springvale supergroup of springs, which that are largely extinct with the exception of this last one (Figure 7.5). The Elizabeth Springs goby is endemic to these waters.

Image courtesy of Peter Unmack

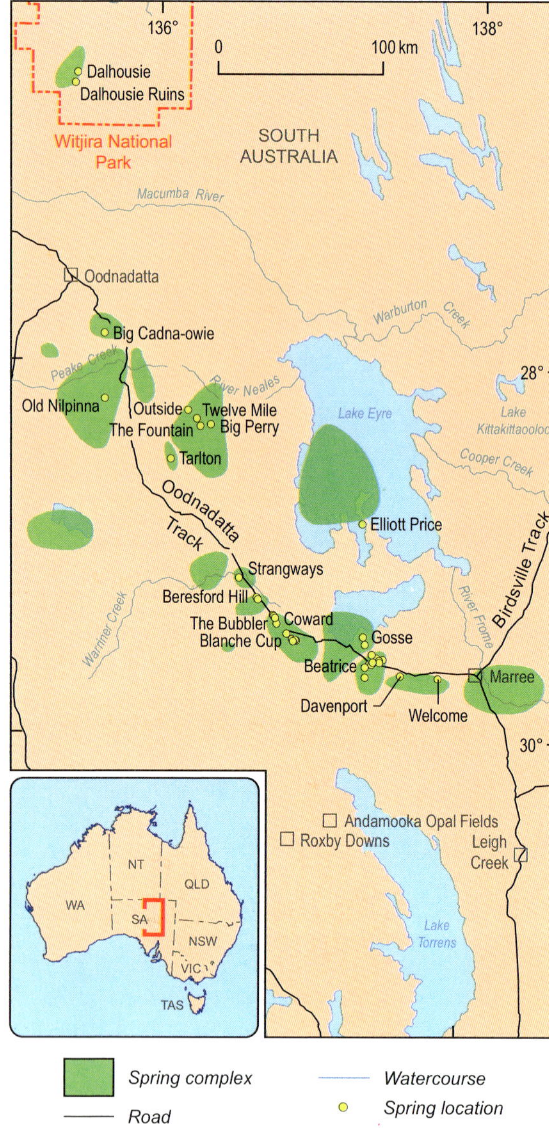

Figure 7.14: Mound springs of south and west Lake Eyre in South Australia. The Oodnadatta Track follows the route taken by the explorer McDouall Stuart, who in 1862 followed the ancient Aboriginal path that linked this arid region's mound springs. One of the largest mound springs is The Bubbler, which formed by groundwater discharge from the Great Artesian Basin. Such wetlands are unique groundwater-dependent ecosystems in an otherwise arid landscape.

Australia has largely ignored the dangers of salinisation (Figure 1.10). In the Western Australian wheatbelt, wells that were once fresh became salty and unusable.

Land and stream salinisation, commonly known as 'salinity', began to be recognised as a serious problem only in the 1960s. The 1990s was the Australian Decade of Landcare, and many farming communities were involved in regenerating the land, to address issues such as salinity. Much effort was directed to reducing groundwater recharge by tree planting, embraced enthusiastically by the younger generation, but not necessarily by their grandfathers who had spent much backbreaking effort on clearing the land in the first place.

Salt is part of the Australian landscape, and there are many sources (Chapter 5). Ocean or 'cyclic' salt is delivered inland with rainfall or as dryfall, as much as 300 kg/ha/year near the coast to less than 15 kg/ha/year hundreds of kilometres into the interior. Flushing is limited in the extensive areas of old, weathered soils, low rainfall and subdued topography, and salt accumulates over time in the soil and groundwater. Also, a geological history involving the advance and retreat of the sea (Chapter 2) has led to deposition of marine sediments with their (connate) salts, which may eventually become remobilised. Along the coastal margin, ingress of 'marine' salt can occur by intrusion of seawater into aquifers, as well as surface inundation during rare extreme tidal or tsunami events (Chapter 6). Rock weathering also releases 'mineral' salts. 'Aeolian' salt can be imported into a catchment with dust deposited by wind from the arid interior. Windborne sediment and salt deposition can be significant (Chapter 5). A dust storm in December 1987 was estimated to have eroded about 6 Mt of topsoil with subsequent deposition rates of 2 t/km^2/day. On the Southern Alps of New Zealand, the same dust tinged the snow red.

Widespread removal of native vegetation has led to a fundamental change in the water balance within cleared catchments. Soil loss, combined with the removal of deep-rooted water-efficient native vegetation and replacement with European cropping systems, has fundamentally increased groundwater recharge rates. This has been magnified in irrigation areas, where the water balance has been even more dramatically altered. Significant water-table rises and the mobilisation of salts that were stored in the subsurface is manifested in land and water salinisation.

Indeed, national assessments published in 2000 presented a grim picture of the extent of dryland salinity, with 4.7 M ha estimated at risk, projected to rise to 6.4 M ha by 2050. However, this forecast is now considered to be overly pessimistic, partly because subsequent work has revealed that hydraulic equilibrium has already been reached for certain cleared catchments across Australia. This means that water-table depths and the extent of saline scalds largely vary with trends in rainfall. For example, in upland New South Wales catchments, groundwater levels (and incidence of salt scalds) rose following the transition to a wetter phase after 1947 and declined with the return to relatively dry conditions in 2000.

KEEPING SALT OUT OF THE MURRAY (BOX 7.4)

The Murray River drains a semi-arid landscape containing significant stores of salt. In certain reaches, saline groundwater discharges into the river, a process exacerbated by water-table rises caused by enhanced recharge from extensive clearing of native vegetation and irrigation development. In response, various land management and engineering solutions have been implemented, with the goal of maintaining river salinity at the downstream town of Morgan (SA) below 800 μS/cm (electrical conductivity—EC—units) for 95% of the time (Figure B7.4a).

Salt interception works, involving the large-scale pumping and disposal of saline groundwater away from the river, are a major plank of the strategy. Collectively, these schemes reduce the salt load to the river by about 450 000 t/year, by pumping about 55 GL/year of saline groundwater to reduce the hydraulic gradient towards the river. This groundwater is usually piped to an evaporative disposal basin (Figure B7.4b) where crystallised salt can be harvested.

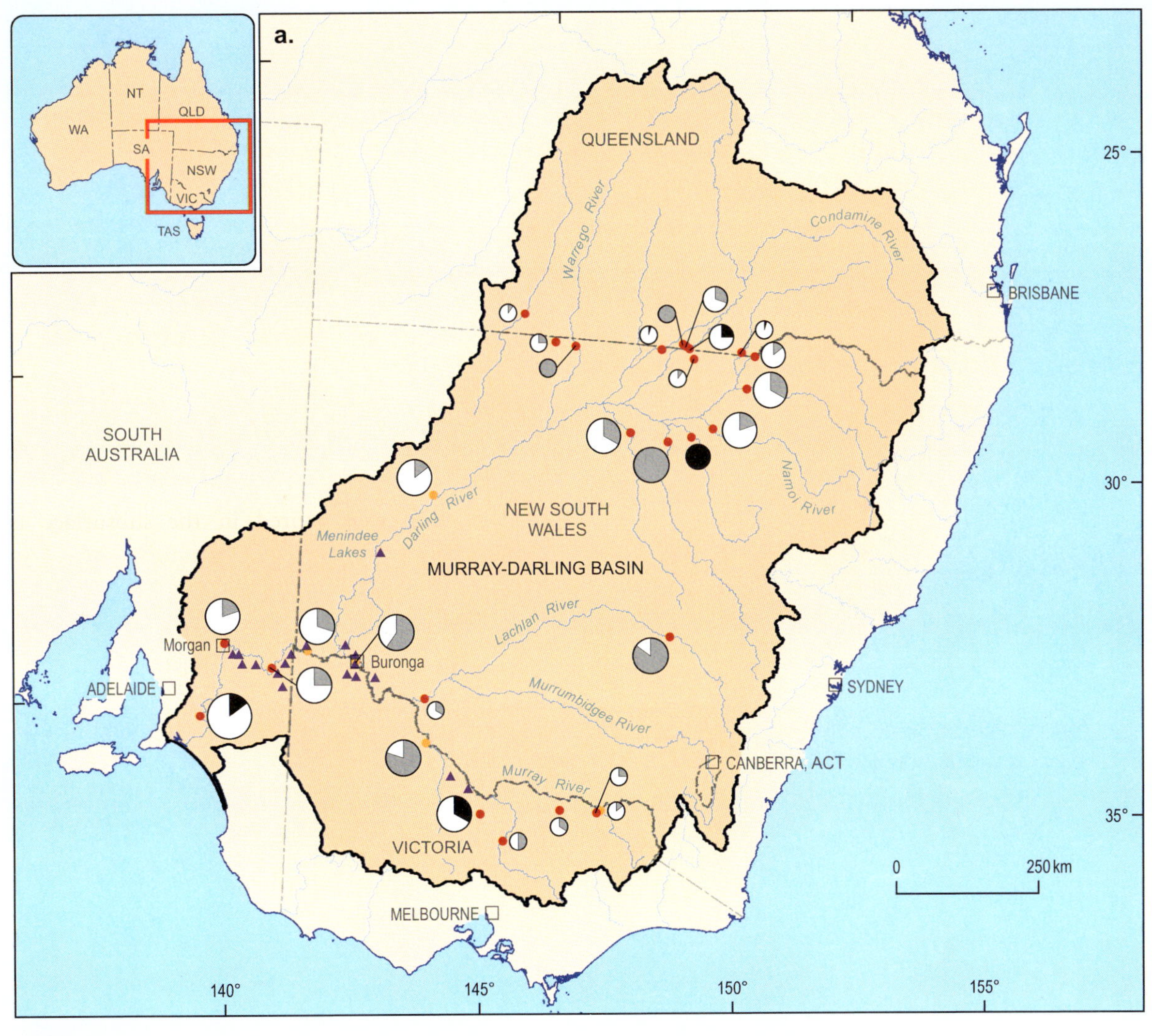

Size of baseline
(year 2000 baseline EC)
- 50–200
- 200–400
- 400–600
- 600–1400

- % above baseline
- % below baseline
- Interpretation site
- Salt interception scheme
- Australian Wetlands and River Centre site

Figure B7.4a: Baseline river salinity measured in 2008–09 for catchments as part of a coordinated salinity management strategy for the Murray–Darling Basin. Salt interception schemes, (borefields) are a significant engineering solution as part of the basin strategy. (Source: Murray–Darling Basin Authority)

Image courtesy of Murray–Darling Basin Authority, photographer Arthur Mostead

Figure B7.4b: Discharge of saline (ca 45 000 EC) groundwater from the Buronga Salt Interception Scheme at the Mourquong evaporative disposal basin, South Australia. About 23 000 t of commercial-grade salt is harvested each year from the disposal basin. (Source: Murray–Darling Basin Authority)

© Getty Images (Universal Images Group)

Artesian bore flows to supply open bore drains for stock water, southern Gulf of Carpenteria region, Queensland.

It is important to note that not all land salinisation is groundwater related. 'Transient' salinity is the term coined for the build-up of salt in the soil profile due to dense clay subsoil horizons limiting downward leaching. The near-surface salt store fluctuates with rainfall and evapotranspiration, and is not associated with a shallow water-table. Such subsoil horizons tend to be alkaline, to have high levels of Na and to be deficient in nutrients, with potential plant toxicities due to elevated boron, carbonate and aluminate. These deficiencies are a significant resource management issue in their own right, estimated to cost the farming economy about $1.3 B each year (Box 7.4).

Changing land use: a trigger for acid waters

Salinisation was an unanticipated consequence of clearing of inland catchments. Similarly, the clearing and drainage of coastal catchments has brought about unintended negative impacts (Chapter 6). Floodplain and backswamp drainage works and tidal barrages were established, ostensibly as a flood mitigation measure, but also to 'reclaim' the land for agricultural purposes. However, the negative consequence was not salinisation but the release of acid water or oxygen-depleted 'black' water into waterways and estuaries. Public concern mounted with huge fish kills, particularly after drought-breaking rains. So what was happening?

To answer this, we have to go back to 6.5–8 ka when the rapid mid-Holocene sea-level rise that created the Torres and Bass straits also deposited estuarine clays as palaeovalley infill and thin veneers over coastal sands. This also laid the foundations for acid sulfate soils that contain pyrite. These iron sulfides were formed by bacteria combining iron from the organic-rich waterlogged sediments with sulfate in the brackish tidal waters. It has been estimated that there is about 1 Bt of pyrite within these sediments along the Australian coast. When maintained in a natural waterlogged state, the sediments, termed potential acid sulfate soils, are stable. However, activities such as road construction, drainage, cropping, urban development and mining can expose these sediments to air. This triggers oxidation of the pyrite, which in turn generates sulfuric acid, forming an acidified profile of actual acid sulfate soils. This can lead to denuded acid scalds. Following rain, the store of acid can migrate into drains, estuaries and wetlands (Figure 7.15). The acid water can have a pH less than 2 and dissolves iron (Fe), aluminium (Al) and other metals, resulting in a toxic concoction for aquatic life (Chapter 6). Following a major flood in 1994, more than 1000 t of sulfuric acid, 450 t of Al and 300 t of Fe were released into the lower reach of the Richmond River (northern NSW). Dissolved silica can also initiate algal blooms, and precipitates of Fe and Al can smother aquatic plants. Secondary oxidation of Fe also consumes oxygen, contributing to the 'black' water events.

Understanding the complex interactions between groundwater and surface water is fundamental to managing acid sulfate soils. Backswamp drainage lowers the shallow water-table, exposing the soil pyrite to air and initiating acidification. Rising water-tables can bring the acid products to the land surface. Seepage of acid groundwater occurs when the shallow water-table is higher than the drain water level. Acid export can result, with tidal drawdowns causing low drain levels and a high hydraulic gradient towards the drain. This is exacerbated if the acid sulfate soil has high hydraulic conductivity, mainly due to cracks and macropores caused by soil shrinkage or biological activity.

The economic impact of acid sulfate soils along the coast is significant, considering that this is where most Australians live (Chapters 1 and 6). The cost of works to avoid or prevent acid drainage associated with coastal urban development is estimated to exceed $100 M every year in Queensland alone. To make things worse, it has recently emerged that acid sulfate soils are not just limited to the coastal zone. Sulfidic muds can be found in inland wetlands, particularly ones associated with long-term inundation and impacted by salinity. Recent droughts have led to these wetlands drying out, initiating sulfide oxidation and acid release. In 2002, the alarm was raised at Bottle Bend Lagoon on the Murray River, when the pH dropped to less than 3, killing the fish and also the surrounding trees (Figure 7.16).

Acid groundwater containing toxic dissolved metals has also been identified in the Western Australian wheatbelt, particularly in the extensive valley floors. These groundwaters typically have a pH of 3–4, with the acidity attributed to oxygenated recharge water reacting with shallow Fe-rich groundwater. The latter is thought to have initially formed in topsoil horizons during prior waterlogged conditions, which then percolated downwards to the water-table, leaving behind residual alkalinity in the form of carbonate

Image courtesy of Assoc. Prof. Scott Johnston, Southern Cross GeoScience

Figure 7.15: Discharge of an acid water plume into the estuary of the Hastings River, mid-north coast of New South Wales. (Source: Johnston et al., 2004)

Image courtesy of Murray Darling Basin Authority, photographer Arthur Mostead

Figure 7.16: Bottle Bend Lagoon, near Mildura, Victoria, on the Murray River, affected by both salinity and acid sulfate soils. Salinity has been measured at 140 000 electrical conductivity units (EC), and pH has been measured at <3. For comparison, the salinity of seawater is *ca* 55 000 EC.

Image courtesy of Department of Sustainability and Environment, Victoria. Photo by Richard Teychenne

Figure 7.17: Bloom of the cyanobacterium (blue-green alga) *Synechococcus* in the Gippsland Lakes, Victoria, in January 2008. Algal densities were above water-quality trigger levels for about five months. Groundwater discharge as well as surface runoff can transport nutrients into waterways to provide conditions for the triggering of toxic algal blooms. (Source: Department of Sustainability and Environment, Victoria)

soil nodules. As the groundwater still contains dissolved Fe, it has the potential to become even more acidic if exposed to the air. There is concern that the shallow groundwater pumping and the network of over 5000 km of deep drains used to control shallow saline groundwater may also release acid and toxic metals to aquatic ecosystems. The Al, Fe and Mn levels in the acid groundwaters can exceed 10 mg/L, with high concentrations (sometimes >1 mg/L) of lead, copper, nickel, zinc and uranium.

Change and development: a legacy of nutrients and contaminants

A legacy of waste disposal practices in Australia was uncovered in the 1970s when attention was directed to point-source pollution, mainly from industry, but also from municipal waste disposal and leaking petrol stations. Shallow sands such as the Perth coastal plain and the Botany Sands in Sydney were the focuses of investigation. For a long time before regulation, waste products that could soak away at the back of a business were an externality: 'out of sight, out of mind'. The problem for the community is that groundwater has a 'long memory' (*Did you know?* 7.6), and consequently some 600 likely pollution point sources were identified in the Perth area alone.

Groundwater can also be a vector for the transport of dissolved nutrients into waterways. The discharge of groundwater with elevated levels of nitrogen and phosphorus brought about by agricultural or urban activities can cause eutrophication in rivers and wetlands. This is the excessive growth of aquatic plants that prevents the penetration of sunlight, and upon decay depletes dissolved oxygen levels (Figure 7.17). Cyanobacteria (blue-green algae) blooms kill fish and can also be toxic to humans and livestock. In 1991, such an outbreak occurred over a 1000 km reach of the Darling River, resulting in the New South Wales Government declaring a state of emergency.

Surface changes: subsidence and lowering of the land

Land subsidence is the lowering of the land surface due to changes that are taking place underground, including the overextraction of groundwater. This subsidence can be reversible in confined aquifers that are subject to elastic compression or expansion as groundwater pressure changes, but can be irreversible where fine-grained sediments such as clay or peat compact on dewatering.

Confined aquifers hold water in elastic storage; hence withdrawal of groundwater leads to a compaction of the aquifer and to subsidence at the land surface. Recovery of water levels or the addition of water can similarly lead to a rise in the land surface. Generally, this is too small to be measured by conventional means and has no significant impacts; however, one of Geoscience Australia's high-resolution GPS stations in the Perth Basin is ideally situated to measure these changes. The GPS station at the Hillarys Marina tide gauge in Perth's northern coastal suburbs is almost coincident with the centre of the drawdown cone in the Jurassic Yarragadee Aquifer, one of two confined aquifers used for Perth's water supply. Since the mid 1990s, extraction from the Yarragadee has substantially increased with population growth, resulting in a maximum drawdown of some 50 m at Hillarys, together with around 10 m of drawdown in the overlying Cretaceous Leederville Aquifer. The Hillarys GPS station showed a 50 mm subsidence of the land surface between 1997 and 2011, coincident with these potentiometric head changes (Figure 7.18).

The best-known Australian case of land subsidence associated with groundwater dewatering is that of the Latrobe Valley in the Gippsland Basin. Here, mining at Morwell has removed some 1500 Mt of coal for Victorian power stations and 850 GL of groundwater for mine stability. Around the mining operations, this has led to groundwater levels dropping by up to 125 m and the land subsiding by more than 2 m in places. There are concerns that land subsidence is a much broader issue for the Latrobe Aquifer, which is extensive, and is used for irrigation and industry, as well as being dewatered offshore with oil and gas production in Bass Strait (Chapter 4). Monitoring along the Gippsland coastline shows a consistent decline in groundwater levels of about 1 m/year. Modelling has inferred that subsidence in the most affected coastal area would be in the order of 0.5–1.2 m by 2056.

A groundwater future for the driest inhabited continent

We have seen a rapid expansion in the use of groundwater resources in Australia over the past 40 years. This 'boom' in groundwater development, particularly of shallow alluvial systems in floodplain and coastal landscapes, has been driven by government limits placed on surface-water licences, combined with rainfall declines and prolonged droughts, particularly in southern and western Australia. During this period, scientific investigation and monitoring have just not kept pace with the development. We are also only just starting to recognise the connectivity and influence that groundwater can have. A positive influence is providing for river baseflow and ecosystem health. A negative influence is the mobilisation of salts, acid and contaminants. Sustainable management of Australia's groundwater systems is therefore critical to protecting our unique ecosystems, heritage and cultural values, and long-term socioeconomic advancement.

Australia also has a global role to play in helping to feed the burgeoning world population, particularly amid declining land and water resources. Despite all this, there are significant data and knowledge gaps in the understanding of how our groundwater systems actually work and how they interact with the rest of Australia's landscape.

So what are the issues and technologies that are on the horizon that will influence how we use and manage our groundwater systems in the future?

Groundwater and a changing climate

The vagaries of the Australian climate have long been recognised by farmers and scientists alike. However, it has really only been over the past few decades that the impacts of longer term climate trends have been realised. This was bought home in the southwest of Western Australia, with the mean

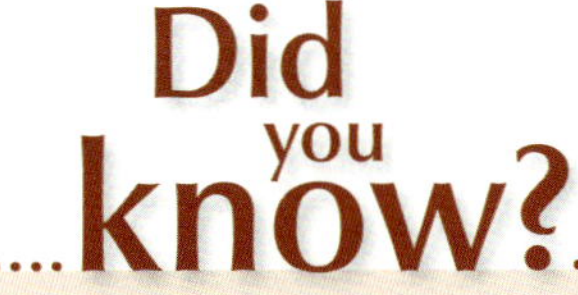

7.6: Fremantle's 'lost' water

As in the case of Darlinghurst, gaols had a ready supply of labour to dig wells. Between 1888 and 1894, inside the gaol at Fremantle, the port city in Western Australia, wells were sunk with connecting tunnels dug along the water-table in the underlying Quaternary coastal limestone. This network serviced Fremantle's water supply, but the addition of two more steam engines led to overpumping and a consequent increase in salinity.

In 1910, the gaol and Fremantle were connected to the metropolitan water supply scheme, and knowledge of the tunnels was lost. In 1989, a bore in the prison began pumping oily water, and analysis of the oil traced it to a nearby fuel pipeline supplying ships in the port. It was found that fuel oil had leaked adjacent to the prison and accumulated in the rediscovered tunnels.

The contamination has now been remediated and, as the gaol is not used as a prison any more, visitors can inspect the tunnels by boat.

Image courtesy of Fremantle Prison

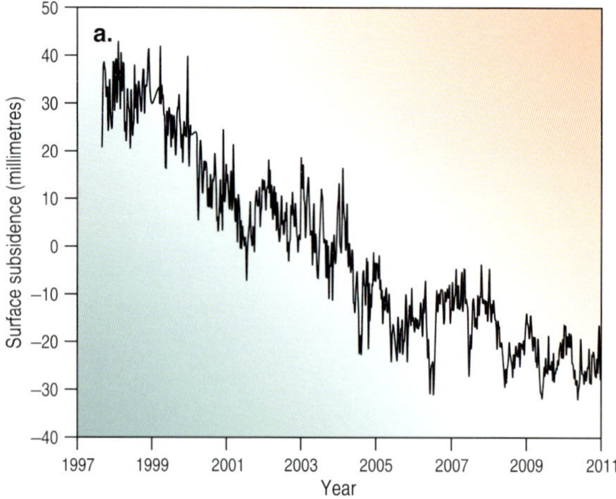

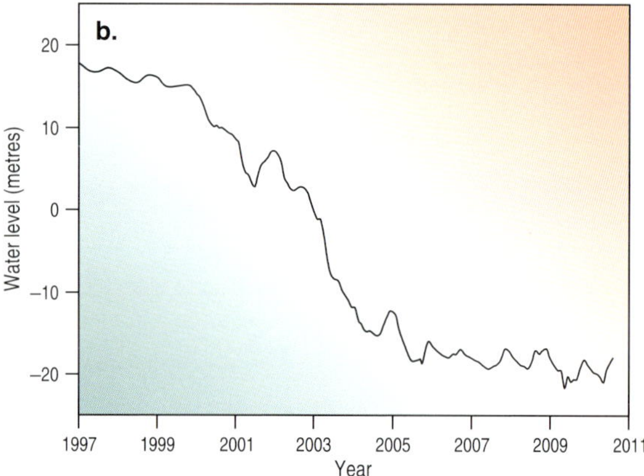

Figure 7.18: (a) Land subsidence recorded by high-resolution GPS associated with (b) declines in groundwater levels due to extraction from the confined Yarragadee and Leederville aquifers, Hillarys Marina, Perth, Western Australia. (Source: Department of Water, Western Australia; Geoscience Australia)

annual rainfall decline since 1975 (Figure 7.9). This, and the water crisis in the Murray–Darling Basin, spurred (among other Australian research) a series of CSIRO Sustainable Yield projects involving assessment of regional water resources in the context of future climate change.

The analysis is constrained by the current high level of uncertainty embedded in the global warming projections and the sensitivity of the global climate to responding to greenhouse gases. Some conclusions regarding groundwater in Australia are as follows:

- In the Murray–Darling Basin, the climate change impact on groundwater levels through decreased rainfall recharge was considered minor in comparison with the impact from current and future predicted growth in groundwater use related to population growth.
- In northern Australia, rainfall is highly seasonal, and modelling suggests that the nature of the rainfall (such as duration and intensity) is critical to defining recharge outcomes. A lower recharge may not necessarily be the result of a lowering of total rainfall.
- In southwest Western Australia, the impact of a future hotter and drier climate on groundwater levels will vary, depending on catchment factors such as vegetation cover, soils, water-table depth and groundwater extraction. Groundwater systems in the sandy coastal plain where native vegetation has been cleared are anticipated to be fairly resilient to all but extreme climate change. This is because reduced drainage discharge and evapotranspiration losses will largely offset reduced recharge.
- In Tasmania, the impact of climate change on groundwater levels is anticipated to be very minor.

These assessments highlight the complexity of understanding the relationship between rainfall and recharge. They also suggest that groundwater systems are more resilient to climate change than surface water, so these resources will likely take on an even larger role in providing water security into the future (Chapter 11).

'Water banking' options for managed aquifer recharge

Dams have been the primary solution to the problem of providing water security in Australia. In the Murray–Darling Basin, public-managed reservoirs have a total capacity of about 35 km^3, whereas the average basin runoff is 21 km^3/year. However, evaporative losses can be huge, particularly from storages in the semi-arid interior. The Menindee Lakes in western New South Wales have been engineered to retain floodwater from the Darling River. The maximum capacity of these lakes is about 2000 GL, but on average more than 20% (430 GL) is lost through evaporation each year. To put this into context, this annual loss represents more than 10% of the 3000–4000 GL/year of water savings proposed under the guide to the Murray–Darling Basin Plan.

Storing water underground in aquifers is one way of beating evaporation. Managed aquifer recharge is the term used for a portfolio of approaches to replenish aquifers purposefully for subsequent recovery or environmental benefit (Figure 7.19). The application of managed aquifer recharge is in its infancy in Australia, contributing about 7 GL/year to urban water supplies. Assessments undertaken for Perth, Adelaide and Melbourne suggest that this could be scaled up to at least 200 GL/year in these three cities alone. Perth already has a passive system where drainage sumps, and sometimes wetlands, in the dunal depressions receive road runoff and allow infiltration into the shallow aquifer. Runoff from roofs and paved driveways also infiltrates through soak wells in the sandy soil. The recharge from winter storms is then extracted during summer from more than 100 000 garden bores. In water-short Adelaide, urban stormwater is being seen as a resource to collect and recharge into both fractured rock and sedimentary aquifers, to be recovered during peak summer demand. In many Australian cities, the annual volume of urban stormwater runoff can be more than the water being reticulated. However, less than 3% of urban stormwater runoff is currently being harvested, with water storage being a major impediment.

Managed aquifer recharge operations in Australia were actually pioneered not in the cities but in irrigation areas. Infiltration pits and sand dams built from the 1960s in the Burdekin Delta now recharge up to 45 GL/year to offset groundwater pumping and prevent seawater intrusion (Figure 7.20). Also in Queensland, weirs were built across the Callide and Lockyer creeks with the express purpose of replenishing the shallow alluvial aquifers. In the Angus–Bremer irrigation area of South Australia, drawdowns and rising groundwater salinity in the exploited aquifer were mitigated through injection of surface water.

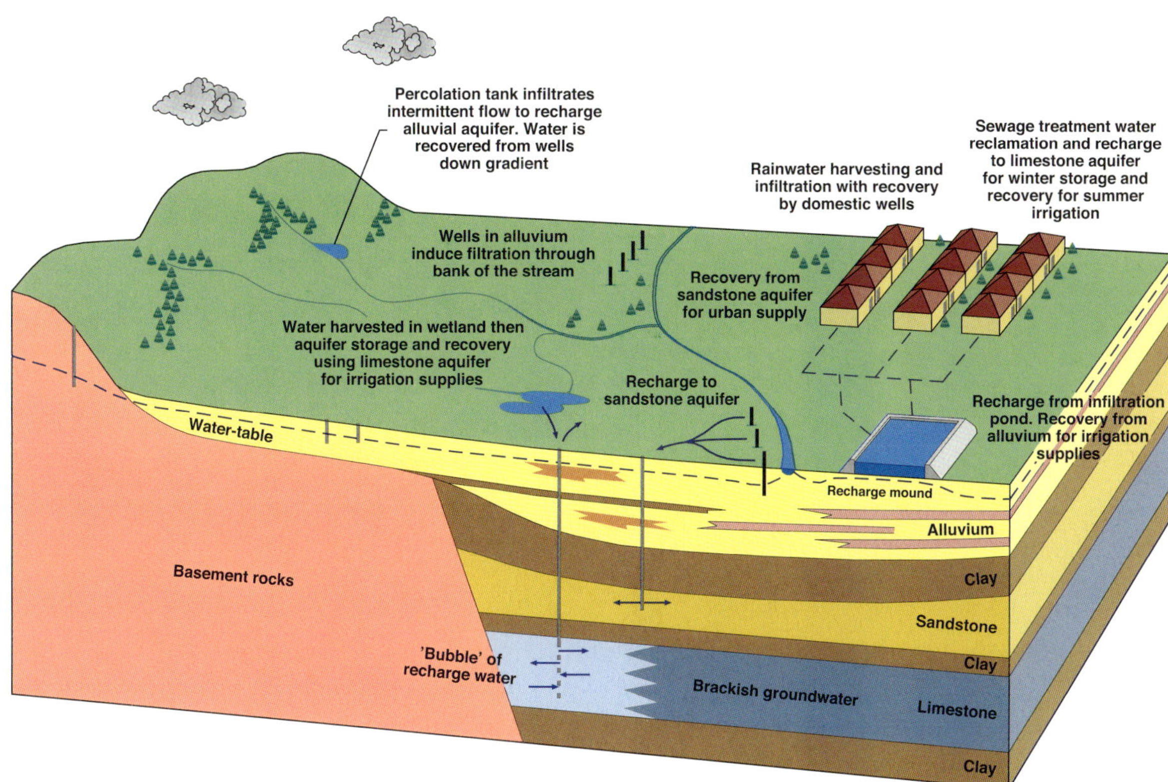

Figure 7.19: Some examples of managed aquifer recharge approaches to enhanced aquifer storage and recovery, including well injection, infiltration basins and percolation tanks. (Source: Dillon et al., 2009)

The full potential of actively using aquifers as underground water storages for urban, rural and remote water use across Australia has yet to be realised.

Large sedimentary basins: a groundwater El Dorado?

In May 2000, the CEO of Anaconda Mining called a press conference to reveal details of a plan to pipe newly discovered groundwater in the Officer Basin (Figure 7.1) to Kalgoorlie and Perth. This followed initial drilling to locate a contingency supply of process water for the Murrin Murrin nickel laterite operation near Laverton in the Eastern Goldfields (WA). An incredulous press

Image courtesy of Burdekin Water Boards and CSIRO

Figure 7.20: Managed aquifer recharge operations in the Burdekin Delta, Queensland, using sand dams to facilitate river infiltration. (Source: Dillon et al., 2009)

GROUNDWATER AND DROUGHTS: INSIGHTS FROM SPACE (BOX 7.5)

© Getty Images [M Hartley]

The Murray–Darling Basin has only recently emerged from a multiyear drought. The impact on the rivers and wetlands, and the consequence for ecology, communities and industries of the basin, have been devastating. So how did the basin's groundwater resources fare during this severe event?

Satellite-based Gravity Recovery and Climate Experiment (GRACE) data have been combined with hydrological monitoring to provide insights into the basin-scale water balance. The GRACE space mission accurately measures the distance separating two tandem low-orbiting satellites to derive spatial and temporal changes to Earth's gravity field. The satellites are 220 km apart and can detect changes smaller than a micrometre per second in relative velocity. When there is an increase in gravity ahead of the pair, the front satellite speeds up. The distance between the pair therefore increases. The opposite occurs when there is decreased gravity ahead of, or between, the satellite pair. Researchers use these data to estimate changes in the terrestrial water storage, which is the combination of surface water, soil moisture and groundwater at the regional scale.

The GRACE time-series data provide a potted history of the drought. In the first two years (2001–02), about 80 km^3 of soil moisture and 12 km^3 of surface water were rapidly lost—low levels basically continued from there on. By 2003, only 13% of total surface-water storage capacity remained. The impact on groundwater was more persistent, with water-tables still declining six years into the drought, amounting to a deficit of 104 km^3. Of the total water lost between 2002 and 2006, the vast majority (*ca* 83%) was groundwater. The average annual groundwater loss over this time was about 17 km^3, or about seven times the allowable extraction limit placed on groundwater users across the basin. This confirms that aquifers often represent the largest water store in semi-arid regions and that groundwater droughts have the longest time lag.

Lake Eildon on the Goulburn River of southern Victoria holds 3300 GL, or more than six times Sydney Harbour. The drought years of the 2000s had a major impact on the lake water levels.

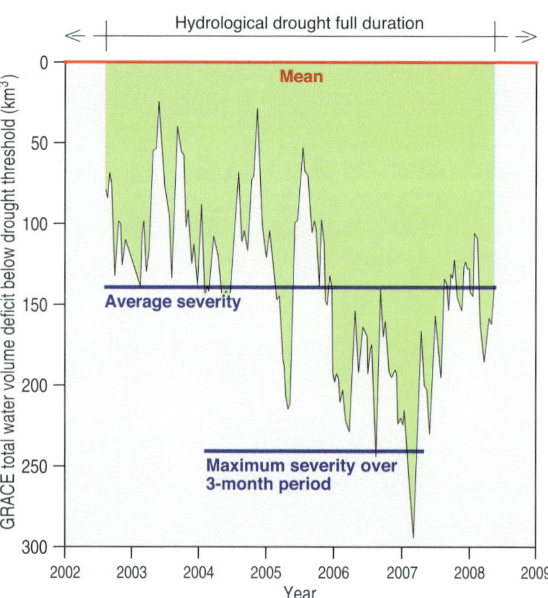

Figure B7.5: Time series of total water storage, integrated over the Murray–Darling Basin, as estimated from GRACE. (Sources: Leblanc et al., 2009; Tregoning et al., 2009)

ran stories on the 'inland sea' that could supply Perth for 4000 years (cf. Box 1.5). One headline claimed 'mining company trips over basin the size of England'. In fact, the Officer Basin had been mapped and described previously, with the main aquifers being the 400 m thick sequences of the Permian Paterson Formation and Devonian Lennis Sandstone. However, its groundwater potential was largely unknown outside a few unpublished government reports. The Anaconda drilling subsequently revealed that the groundwater was, as predicted, brackish. The potential of the Officer Basin, however, is still largely unknown, and the existence of significant fresh groundwater cannot be discounted.

To the north, the Canning Basin underlying the Great Sandy Desert is Australia's second largest artesian basin (Figure 7.1). The basin covers 530 000 km^2, about the size of mainland France, and contains extensive Permian, Jurassic and Cretaceous sandstone aquifers. This basin certainly does contain large volumes of low-salinity water, which is only used along the coast. Carbon dating of the groundwater in the West Canning Basin suggests that the groundwater is replenished regularly, and is not a fossil remnant of a previous wetter period (Chapter 5).

These basins are commonly viewed as large untapped groundwater resources. However, dreams of piping the groundwater to Perth for urban use have been thwarted by the tyranny of distance, as preliminary costings confirm that the energy, greenhouse emissions and overall supply cost would be significantly greater than for seawater desalination. The other factor is the sustainability of extraction, as recharge rates are largely unknown. Groundwater development of these large sedimentary basins in the future is more likely to service local or regional needs.

We have seen earlier that the Great Artesian Basin is Australia's largest groundwater resource. The Murray–Darling Basin, comprising the southern part of this large basin, was so stressed by the prolonged droughts of the 1990s–2000s that the impacts could be measured from space (Box 7.5).

Untapped resources: brackish and saline aquifers

The Australian map of groundwater salinity (Figure 7.21) shows the extent to which only brackish or saline groundwater is available for large tracts of the continent. This is particularly the case for the southern and inland areas, where surface water availability can also be severely limited. Here, a flat, ancient landscape and the lack of flushing rains and rivers have conspired in the prolonged accumulation of salt (Chapter 5). However, the advancement of desalination technologies may yet unlock this groundwater resource.

Although Australia is the driest inhabited continent, only about 1% of the global installed desalination capacity is found here. The bulk of desalination in Australia focuses on using seawater for our coastal capitals; such capacity is projected to increase 10-fold from 45 GL/year in 2006 to 450 GL/year by 2013 (Figure 7.22). Desalination of local brackish groundwater is already occurring for key remote communities in the arid interior,

Shuttle Radar Topography Mission (SRTM) image of sand dunes of the Great Sandy Desert, Western Australia, which cover Australia's second largest artesian basin.

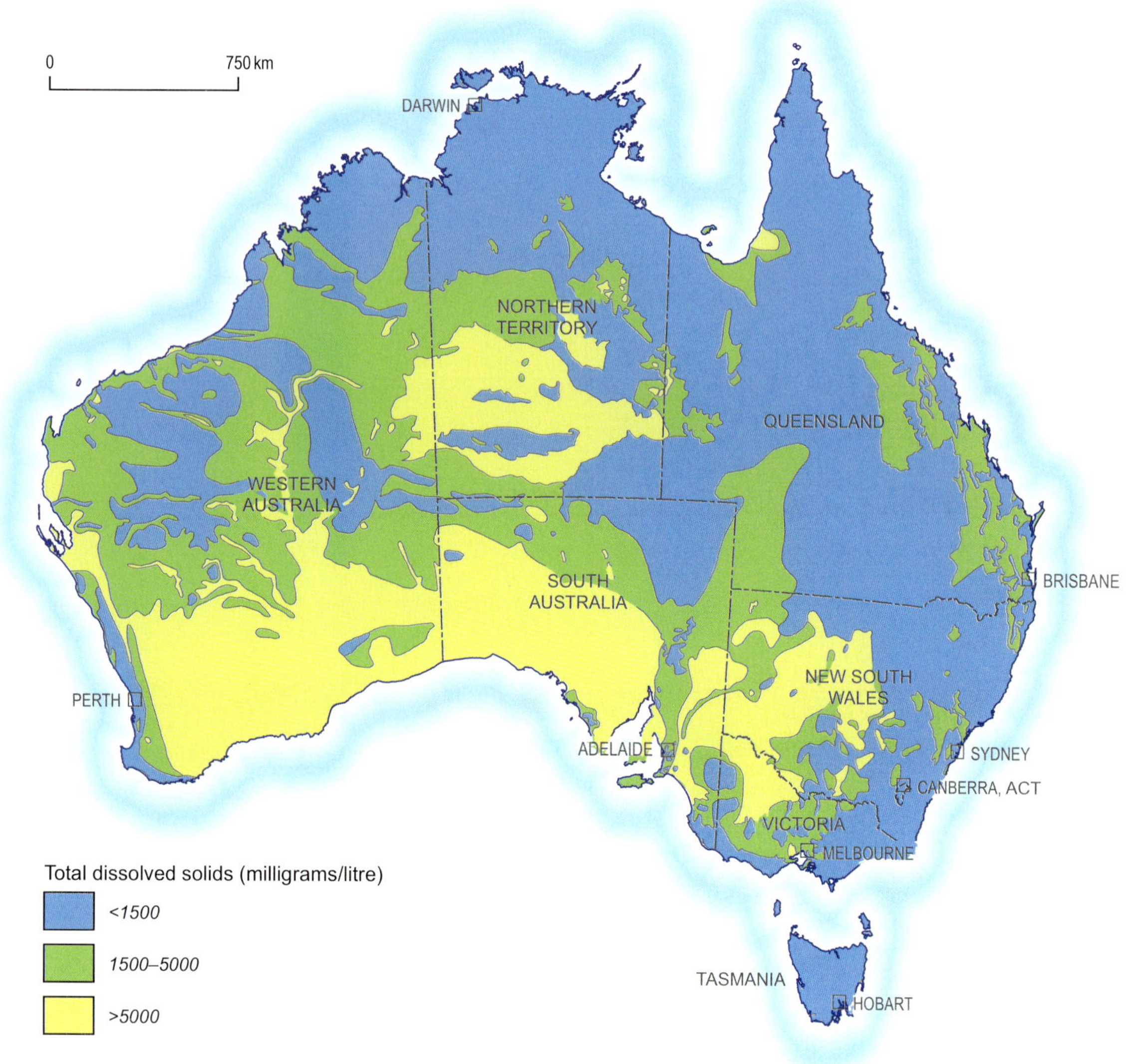

Figure 7.21: Groundwater salinity within the main aquifers across Australia. (Source: Lau et al., 1987)

such as Yulara, near Uluru. The lowering of capital, operating, maintenance and energy costs as desalination technologies advance could provide impetus for improved water supply and economic development across large areas of Australia. Over recent decades, improvements in pumps have halved the energy requirements of reverse osmosis plants, with new energy-recovery equipment halving this again. The linking of desalination with renewable energy sources such as solar and geothermal is also showing promise (Chapter 10). It is important to note that for much of Australia the available groundwater is brackish (say, <5000 mg/L) rather than of ocean salinities (>30 000 mg/L). This is important because the energy requirements for reverse osmosis of brackish water (0.7–1 kWh/m^3) are significantly less than for oceanic equivalents (3–3.7 kWh/m^3). Much of the desalination research agenda, such as forward osmosis techniques, new-generation membranes and nanotechnologies, aims to reduce the current seawater energy benchmark of 3.5 kWh/m^3 to the theoretical minimum of 0.8 kWh/m^3.

Groundwater and new energies

The picture many people have of artesian water is the steam coming off. Artesian water from deep confined aquifers is by its nature relatively hot, depending on the aquifer depth and geothermal gradient. Typical geothermal gradients in sedimentary basins range from 2°C/100 m in sandstone to 5°C/100 m in shales, so that groundwater temperatures commonly range from 30°C to 70°C. Local extremely high gradients have been measured—for instance,

1°C/5 m across brown coal seams in the Morwell open cut in Victoria. This raises the potential of using the geothermal energy captured within hot sedimentary aquifers (Chapter 10).

In the Great Artesian Basin, hot bore water, up to 100°C, has historically been discharged into bore drains to allow it to cool sufficiently for stock to drink. Cooling towers equipped with electric fans are also used, resulting in the irony of using fossil-fuel electric power to remove the thermal energy. There are examples across the basin of the direct use of the hot bore water, typically for swimming pools and therapeutic spas (Box 7.3). There are also examples of using open-loop groundwater-source heat pumps to generate electricity; examples of these small Rankine cycle plants have operated successfully in Mulka (SA) and Birdsville (Qld) (Figure 7.23). At the Perth zoo, the bore drilled into the Perth Basin in 1905 was reputedly used to heat the reptile house. It was only in the 1990s that hot Perth Basin groundwater began to be used for heating swimming pools. The North Perth Basin is also an exploration target for hot geothermal groundwater suitable for electricity generation. The hunt for renewable carbon-neutral energy sources has led to interest in the geothermal resource potential of hot sedimentary aquifers and fractured rocks across Australia (Chapter 10).

One aspect of the future exploitation of energy resources in Australia is the potential impact on groundwater resources. This is coming to the fore with the rapid development of coal-seam gas reserves, notably from the Jurassic Walloon Coal Measures of the Bowen and Surat basins (Qld)

Figure 7.22: Outlook to the year 2013 for desalination sites across Australia. This shows the significant investment in seawater desalination to secure water supply for coastal capital cities, as well as smaller scale operations using brackish groundwater to support remote communities and mines. (Source: Hoang et al., 2009)

Image courtesy of Ergon Energy

Figure 7.23: At Birdsville in western Queensland, an organic Rankine cycle heat exchanger is used to produce electricity from 98°C groundwater from the Great Artesian Basin. The groundwater is supplied at a rate of 27 L/s from a 1280 m deep bore, and the plant has a net output of 80 kW or about 30% of the town's power. (Source: Primary Industries and Resources South Australia)

(Chapter 9). The extraction process involves wells drilled into the coal seam to initially extract groundwater to depressurise the seam so that the gas is desorbed and released. Of concern is the appropriate disposal, reuse or reinjection of the coal-seam water, as well as the potential for leakage from the adjacent Great Artesian Basin and overlying alluvial aquifers (Chapter 11).

A geoscience systems approach to future hydrogeology

By its very nature, an understanding of groundwater systems needs a multidisciplinary systems-based approach, requiring a cohort of geologists, hydrologists, ecologists, modellers, engineers, social scientists and economists, among others. At the core, however, lies the need for a geological understanding of how aquifers and aquitards work.

Geoscience Australia's Broken Hill Managed Aquifer Recharge (BHMAR) project can be used as a case study to highlight the importance of the need for a geoscience foundation to groundwater studies. This project was tasked to discover and evaluate groundwater resource and managed aquifer recharge targets in a data-poor region of the Darling River floodplain in western New South Wales. As part of an initiative funded by the Australian Government Department of Sustainability, Environment, Water, Population and Communities, the overarching objective was to investigate options for enhancing the efficiency of the Menindee Lakes Scheme, a series of large, shallow, water storages in a semi-arid landscape. The project found that, by integrating the full range of geoscientific tools, datasets and skillsets, it was possible to uncover more effectively the extent, distribution and nature of good-quality groundwater resources and aquifer storage.

A regional helicopter-based electromagnetic survey mapped fresh groundwater associated with river leakage in unconsolidated sands and gravels, as well as mapping the clay aquitards that are important as seals in managed aquifer recharge borefields (Figure 7.24a). The airborne electromagnetic data provided the 3D framework of the region in terms of conductivity, which relates closely to salinity. A series of monitoring bores were constructed during an extensive sonic drilling program (Figure 7.24b). The bores provided a 1D sample of the aquifer's geometry and characteristics, while the airborne electromagnetic data provided the spatial context to map out and interpolate, in 3D, the aquifer's geometry across the region.

The excellent core recovery from the bores enabled detailed sediment and pore fluid analysis, including grain-size distribution, various X-ray determinations (XRD/XRF), moisture content and fluid salinity (Figure 7.24c). These data on the vertical hydrogeological profile were supplemented by downhole geophysical logging, including gamma, conductivity and nuclear magnetic resonance of the project bores (Figure 7.24d). These rock property data were used in the processing of the airborne electromagnetic data, ensuring confidence in what the remote-sensed signal was measuring.

Extensive sampling and chemical analysis of the groundwater, as well as of the rivers, lakes and rainfall, were evaluated from a water quality perspective, and also provided an understanding of key hydrological processes, such as river leakage, hydraulic connections between aquifers and potential water–rock interactions during managed aquifer recharge operations.

An airborne LiDAR survey was used to generate a high-resolution digital elevation model (DEM) of the region (see also Chapter 5). This dataset revolutionised geomorphological understanding. For example, it was used to derive the flood inundation history, leading to an improved knowledge of the recharge dynamics of the system. The LiDAR survey surface DEM map was supported by more detailed bathymetric surveys of the bed of the Darling River (Figure 7.24e). Together, these high-resolution models and maps highlighted the localised nature of river leakage. The LiDAR survey also helped to determine the geomorphological and stratigraphic framework of the region. The age of the sediments was determined by biostratigraphic dating using palynology (spores and pollen: Chapter 3). The landscape age was determined by optically stimulated luminescence techniques.

An understanding of the geological history and depositional environments can in turn be used in the interpretation of the geometry and hydraulic properties of key aquifers and aquitards. Routine pump testing of project bores was used to derive estimates of aquifer hydraulic properties in target areas (Figure 7.24d).

Image by Heike Apps

The Eucla Basin, Western Australia, contains an unconfined limestone aquifer—famous for its cave systems and a confined sandstone aquifer at the base. The groundwater is deep and generally high in salinity.

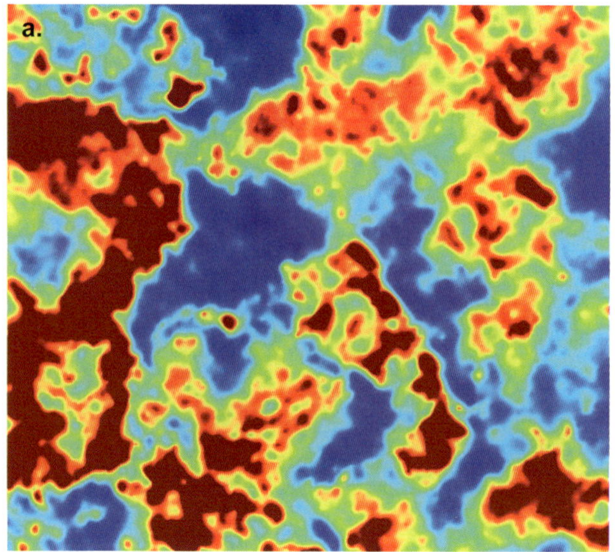

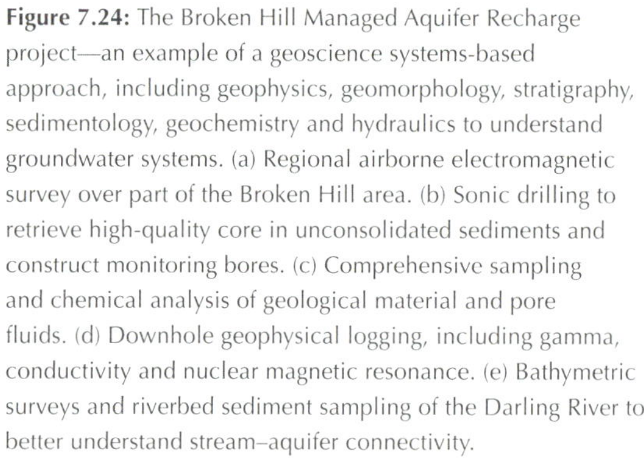

Figure 7.24: The Broken Hill Managed Aquifer Recharge project—an example of a geoscience systems-based approach, including geophysics, geomorphology, stratigraphy, sedimentology, geochemistry and hydraulics to understand groundwater systems. (a) Regional airborne electromagnetic survey over part of the Broken Hill area. (b) Sonic drilling to retrieve high-quality core in unconsolidated sediments and construct monitoring bores. (c) Comprehensive sampling and chemical analysis of geological material and pore fluids. (d) Downhole geophysical logging, including gamma, conductivity and nuclear magnetic resonance. (e) Bathymetric surveys and riverbed sediment sampling of the Darling River to better understand stream–aquifer connectivity.

Each of the various tools that were applied provided a particular new insight into understanding the groundwater and aquifer resource for Broken Hill. Their true value and power are maximised when the tools are fully integrated into a geoscience systems-based approach in space and time. This approach shows us that geoscience innovations have a vital place in understanding groundwater processes. Such a geoscience systems-based approach that integrates the latest advances in areas such as geological characterisation, landscape evolution, geophysics, hydraulics and hydrogeochemistry can make important inroads into unravelling the complexities of groundwater systems and contribute to critical land and water management issues.

© Getty Images [G Tedder]

Aerial view of cattle around a shrinking waterhole in drought near the town of Nyngan, western New South Wales.

Going with the flow

Australia is the driest continent (other than Antarctica). Water is the lifeblood that has sustained life here. Water, whether from surface rivers, lakes, groundwater, rainfall or underground aquifers, is also a key resource for future security. Here we have looked at the role and science of groundwater and how Australia is coping with this dwindling resource, with key innovations being developed here. All humans have to find ways to deal with the lack or overabundance of water at times, and the last decade-long drought has severely tested all aspects of the hydrogeology, as well as the people. The changing water cycle and increasing aridity have shaped the landscape and soils, and determined where we built our cities.

Now we shall look in more detail at the other resources that are vital for Australia—minerals and energy.

© Getty Images [A Campanile]

Desiccation cracks in the dry bed of Lake Eyre, South Australia.

Bibliography and further reading

Groundwater: a critical resource

Australian Bureau of Statistics 2010. *Water account Australia 2008–09*, release 4610.0, ABS, Canberra.

Australian Bureau of Statistics 2010. *Water use on Australian farms 2008–09*, release 4618.0, ABS, Canberra.

Bayly IAE 1999. Review of how indigenous people managed for water in desert regions of Australia. *Journal of the Royal Society of Western Australia* 82, 17–25.

Box JB, Duguid A, Read RE, Kimber RG, Knapton A, Davis J & Bowland AE 2008. Central Australian waterbodies: the importance of permanence in a desert landscape. *Journal of Arid Environments* 72, 1395–1413.

Brodie RS 2002. Putting groundwater on the map: a status report on hydrogeological mappingin Australia. In: *Balancing the groundwater budget*, proceedings of the International Association of Hydrogeologists conference, Darwin, 12–17 May 2002. www.connectedwater.gov.au/documents/IAH02_HydroMap.pdf

Bureau of Meteorology 2011. Map of average conditions. www.bom.gov.au/climate/averages/maps.shtml

Cooper D & Jackson S 2008. *Preliminary study on indigenous water values and interests in the Katherine Region of the Northern Territory*, CSIRO Sustainable Ecosystems, Darwin.

CSIRO 2008. *Water availability in the Murray–Darling Basin. A report to the Australian Government from the CSIRO Murray–Darling Basin Sustainable Yields Project*, CSIRO, Australia.

Faggion AE 1995. The Australian groundwater controversy 1870/1910. *Historical Records of Australian Science* 10, 337–348.

Gibson DL & Wilford J 2002. Aspects of regolith and landscape of the Strathbogie–Caniambo–Dookie area: the need for interpretation of detailed geophysical datasets in the light of regional data and models. In: *Victoria Undercover Benalla 2002 conference proceedings and fieldguide*, Phillips GN & Ely KS (eds), CSIRO Publishing, Victoria, 235–247.

Jacobson G & Lau JE 1987. *Hydrogeology of Australia*, 1:5,000,000 scale map, Bureau of Mineral Resources, Geology and Geophysics, Canberra.

Lau JE, Commander DP & Jacobson G 1987. *Hydrogeology of Australia*, Bulletin 227, Bureau of Mineral Resources, Geology and Geophysics, Canberra.

Murray–Darling Basin Authority 2010. *Guide to the proposed Basin Plan: overview*, Murray–Darling Basin Authority, Canberra. http://thebasinplan.mdba.gov.au

National Water Commission 2006. *Australian water resources 2005*, National Water Commission, Canberra. www.water.gov.au

Radke BM, Ferguson J, Cresswell RG, Ransley TR & Habermehl MA 2000. *Hydrochemistry and implied hydrodynamics of the Cadna-owie Hooray Aquifer, Great Artesian Basin*, Bureau of Rural Sciences, Canberra.

Report of the Second Interstate Conference on Artesian Water, Brisbane, 1914, AJ Cumming, Government Printer.

Roberts A & Mountford C 1979. *The dreamtime book*, Rigby, Adelaide, 166–167.

Smith DI 1998. *Water in Australia: resources and management*, Oxford University Press, Melbourne.

Thomson DF 1962. The Bindibu Expedition III. The Bindibu. *Geographical Journal* 128, 262–278.

Weaver TR, Cartwright I, Tweed SO, Ahearne D, Cooper M, Czapnik K & Tranter J 2006. Controls on chemistry during fracture-hosted flow of cold CO_2-bearing mineral waters, Daylesford, Victoria, Australia: implications for resource protection. *Applied Geochemistry* 21, 289–304.

Groundwater: a geological agent of change

Boulton AJ & Hancock PJ 2006. Rivers as groundwater-dependent ecosystems: a review of degree of dependency, riverine processes and management implications. *Australian Journal of Botany* 54, 133–144.

Bureau of Rural Sciences. *Connected water—managing the linkages between surface water and groundwater*, Australian Government, Canberra. www.connectedwater.gov.au

Fensham RJ & Fairfax RJ 2002. In the Footsteps of J Alfred Griffiths: a Cataclysmic History of the Great Artesian Basin Springs in Queensland. Australian Geographical Studies: 40, 210–230.

Fitzpatrick R & Shand P 2008. Inland acid sulphate soils: overview and conceptual models. In: *Inland acid sulphate soil systems across Australia*, Cooperative Research Centre for Landscape Environments and Mineral Exploration Open File Report 249, 6–74.

Freij-Ayoub R, Underschultz J, Fangjun L, Trefry C, Henning A, Otto C & McInnes K 2007. *Simulation of coastal subsidence and storm wave inundation risk in the Gippsland Basin*, CSIRO Petroleum Report 07-003, CSIRO Wealth from Oceans Flagship.

Hamilton S 2004. River–groundwater interactions in the Cudgegong Valley, Mudgee NSW. In: *Conference proceedings*, 9th Murray Darling Basin Groundwater Workshop, Bendigo, 17–19 February 2004.

Hatton T & Evans R 1998. *Dependence of ecosystems on groundwater and its significance to Australia*, occasional paper 12/98, Land and Water Resources Research and Development Corporation, Canberra.

Humphreys WF 2006. Aquifers: the ultimate groundwater-dependent ecosystems. *Australian Journal of Botany* 54, 115–132.

Johannes RE & Hearn CJ 1985. The effect of submarine groundwater discharge on nutrient and salinity regimes in a coastal lagoon off Perth, Western Australia. *Estuarine, Coastal and Shelf Science* 21, 789–800.

Johnston SG, Slavich P & Hirst P 2004. The acid flux dynamics of two artificial drains in acid sulfate soil backswamps on the Clarence River floodplain, Australia. *Australian Journal of Soil Research* 42, 623–637.

National Land and Water Resources Audit 2001. *Australian dryland salinity assessment 2000: extent, impacts, processes, monitoring and management options*, Australian Government, Canberra.

New South Wales Department of Environment and Climate Change 2009. *Salinity audit—upland catchments of the New South Wales Murray–Darling Basin*, NSW Department of Environment and Climate Change, Sydney.

O'Grady A, Carter J & Holland K 2010. *Review of groundwater discharge studies of terrestrial systems*, CSIRO Water for a Healthy Country National Research Flagship.

Rengasamy P 2006. World salinization with emphasis on Australia. *Journal of Experimental Botany* 57, 1017–1023.

Sammut J & Lines-Kelly R 1996. *An introduction to acid sulfate soils*, Australian Government Department of Environment, Sport and Territories & Australian Seafood Industry Council.

Sinclair Knight Merz 2001. *Environmental water requirements of groundwater dependent ecosystems*, Environmental Flows Initiative technical report 2, Commonwealth of Australia, Canberra.

Stieglitz T 2005. Submarine groundwater discharge into the near-shore zone of the Great Barrier Reef, Australia. *Marine Pollution Bulletin* 51, 51–59.

Tomlinson M & Boulton A 2008. *Subsurface groundwater ecosystems: a review of their biodiversity, ecological processes and ecosystem services*, Waterlines Occasional Paper no. 8, National Water Commission, Canberra.

Tregoning P, Leblanc M, Ramillien G, Tweed S & Fakes A 2009. *Monitoring groundwater variations in the Murray–Darling Basin using space gravity measurements*, Research School of Earth Sciences, Australian National University, Canberra.

Winter TC, Harvey JW, Franke OL & Alley WM 1998. *Groundwater and surface water a single resource*, circular 1139, US Geological Survey, Denver.

A groundwater future

CSIRO 2009. *Groundwater yield in south-west Western Australia*, summary of a report to the Australian Government from the CSIRO South-West Western Australia Sustainable Yields Project, CSIRO, Australia.

CSIRO 2009. *Water availability for Tasmania*, report one of seven to the Australian Government from the CSIRO Tasmania Sustainable Yields Project, CSIRO, Australia.

CSIRO 2009. *Water in northern Australia*, summary of reports to the Australian Government from the CSIRO Northern Australia Sustainable Yields Project, CSIRO, Australia.

Dillon P, Pavelic P, Page D, Beringen H & Ward J 2009. *Managed aquifer recharge: an introduction*, Waterlines Report Series no. 13, National Water Commission, Canberra.

El Saliby I, Okour Y, Shon HK, Kandasamy J & Kim S 2009. Desalination plants in Australia, review and facts. *Desalination* 247, 1–14.

Geoscience Australia 2007. *Direct-use of geothermal energy: opportunities for Australia*, Geoscience Australia, Canberra.

Hoang M, Bolto B, Haskard C, Barron O, Gray S & Leslie G 2009. *Desalination in Australia*, CSIRO Water for a Healthy Country National Research Flagship.

Leblanc MJ, Tregoning P, Guillaume R, Tweed SO & Fakes A 2009. Basin-scale, integrated observations of the early 21st century multiyear drought in southeast Australia. *Water Resources Research* 45, W04408.

Service RF 2006. Desalination freshens up. *Science* 313, 1088–1090.

UNESCO Centre for Membrane Science and Technology 2008. *Emerging trends in desalination: a review*, Waterlines Report Series no. 9, National Water Commission, Canberra.

8

Foundations of wealth—Australia's major mineral provinces

Since the 1850s, Australians have prospered not only from the wool off the sheep's back but also from the mineral wealth bequeathed by Australia's geological history. Our mining history has contributed significantly to our national identity. Much of the country was explored and settled because of its vast mineral wealth. Today, Australia's economy is highly dependent upon minerals that provide most of our export income. Major mineral provinces in Australia—the Victorian goldfields, the Eastern Goldfields in Western Australia, Broken Hill, Mt Isa and Olympic Dam—although linked to different tectonic events, share many commonalities. These include a spatial association with major fault zones, a temporal association with major thermal events, an association with changes in tectonics, and a broad association with major crustal boundaries. This mineral wealth was created as a consequence of the large-scale tectonic processes that built Australia from its disparate elements.

David L Huston,[1] Richard S Blewett,[1] Roger Skirrow,[1] Anthea McQueen,[1] Jiao Wang,[1] Lynton Jaques,[1,2] and Denis Waters[1] (with a contribution from Frank Bierlein[3])

[1]Geoscience Australia; [2]Visiting Fellow, Research School of Earth Sciences, Australian National University; [3]Areva NC Australia

Image by Jim Mason

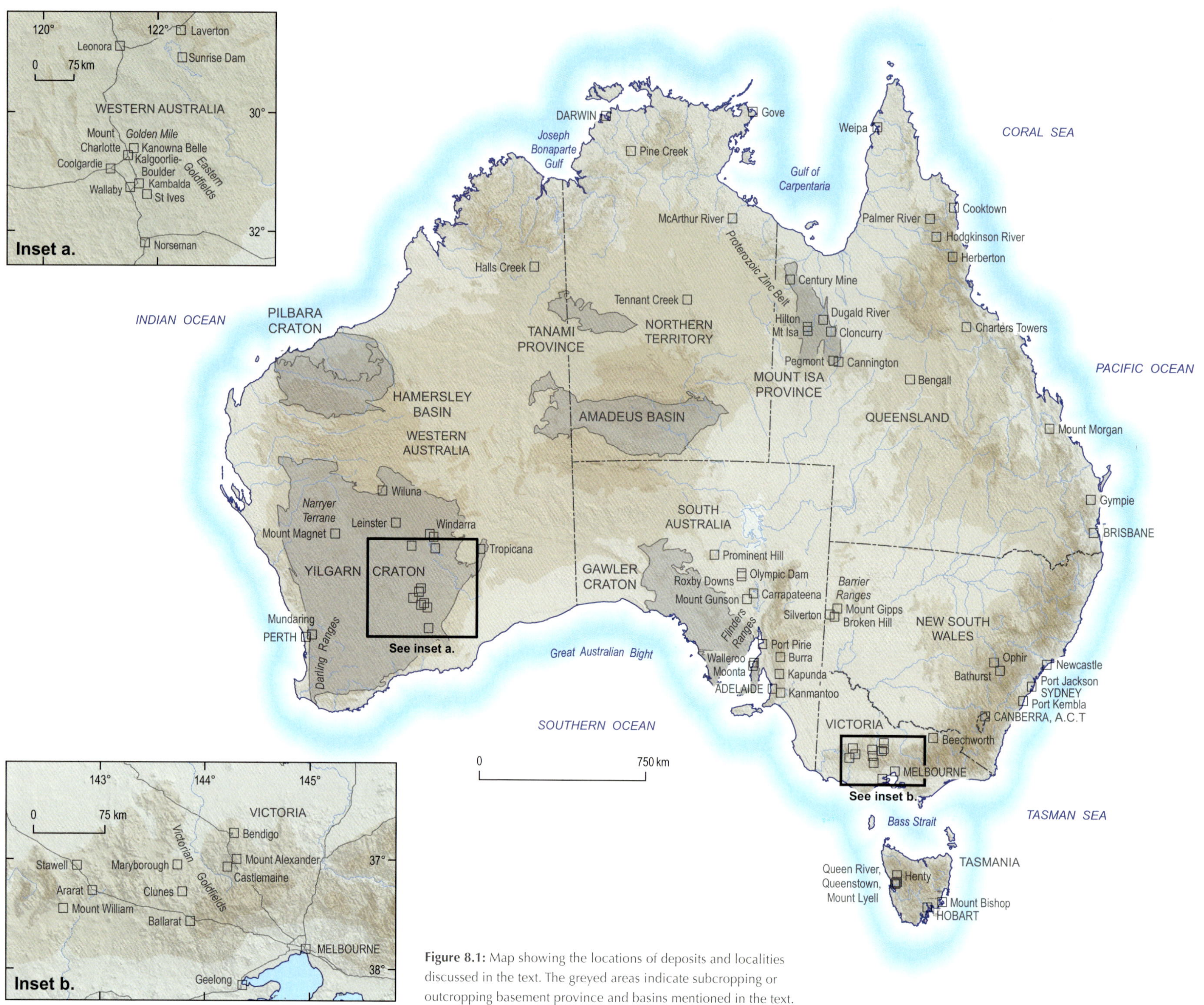

Figure 8.1: Map showing the locations of deposits and localities discussed in the text. The greyed areas indicate subcropping or outcropping basement province and basins mentioned in the text.

Sources of wealth

Australia, considered by some as the 'Lucky Country', has vast agricultural lands, unique flora and fauna, a sunny climate, diverse people and, importantly, abundant mineral resources (Chapter 1). Early last century, Australia was, on a per-capita basis, one of the two or three wealthiest countries in the world. The country is still in the top 20, and in the top 10 'most liveable' by some reckonings. In the past, much of this wealth was based on agriculture (wool and wheat), although mining played a very important role, especially after the nation-changing discovery of gold (Au).

Today, the bulk commodities of iron ore, coal, hydrocarbons and bauxite are generating most of Australia's mineral wealth (Chapter 9), whereas the early mineral wealth was generated by metallic elements such as Au, copper (Cu), silver (Ag), lead (Pb) and zinc (Zn). The 1851 discovery of Au ushered in changes and events—mass immigration, early multiculturalism, the opening up of the outback (Figure 8.1), the Eureka Stockade, unionism—that still resonate today. Gold made many Australians, as well as the country, rich.

The benefits of Australia's mines go beyond wealth. The first mass migration of non-Anglo–Celtic people to Australia occurred in response to the discovery of Au. People came from all over the world, but one of the largest waves of immigrants came from China. Many of the early goldfields and mining districts were highly multicultural. Resentment shown towards the Chinese resulted in violence in several goldfields, and these sentiments eventually led in part to the development of the 'White Australia Policy', a contingency that selected Australian immigrants based upon race that was not abolished until the 1960s. Descendants of the Chinese miners still live in the Victorian (and other) goldfields, and the Chinese Spring Festival forms an integral part of Bendigo's annual Easter Fair.

Historically, mining also has had its downside, including environmental, societal and even economic effects (Chapter 9). For individual miners, the greatest downside is the hazardous nature of mining, with fire, collapses, large plant and machinery, and chronic disease being of particular importance. In the modern age, these

An Australian gold diggings. Painting of life on the Victorian goldfields around 1855 by Edwin Stocqueler (1829–1900).

Figure 8.2: Photographs of the effects of mining on the environment in western Tasmania. (a) 'Moonscape' around Queenstown. (b) The Queen River. (c) The nearby Henty mine site. These photographs illustrate differences between historical and modern mine sites: the Mount Lyell mineral field, first discovered in 1883, has been mined for more than a century. The need for timber for the mines led to the destruction of rainforest, and pyritic smelting techniques produced acid rain, which prevented regrowth, resulting in massive soil erosion in this area; this combination produced the 'moonscape' environment that has only recently, after the closure of the smelter, begun to regrow. In contrast, the Henty gold mine, which opened in 1996, has a small footprint with minimal environmental impact.

hazards have been significantly reduced, largely due to improvements in worker training, health and safety regulations and mining methods.

Like virtually every other human activity, mining can have significant effects on the environment (Figures 8.2a and b). Modern mines generally have a much smaller impact on the environment than those in the past (Figure 8.2c). Today, modern rehabilitation programs can successfully restore the landscape and vegetation after mining. Although improvements have been made, one of the largest challenges to the mining industry is to minimise the future effects of its operations on the environment and maintain support in the public arena—its social licence to operate (Chapter 9).

In this chapter, we present an abbreviated history of Australian mining and its importance to shaping the Australian people. We introduce the concept of a mineral system and use this concept to describe four important metallogenic provinces: the Victorian goldfields, the Eastern Goldfields of Western Australia, the Proterozoic zinc belt in central Australia, and the Olympic Dam Iron-Oxide-Copper-Gold (IOCG) Province in South Australia. Each of these discoveries had, and continues to have, a shaping influence on the development of Australia. We bring these major metallogenic provinces together and show that they are not isolated phenomena; rather, their individual geneses (mineral systems) have much in common.

A short history of discovery and mining in Australia

Many social historians attribute wool as the main commercial factor that influenced the development of Australia. The equally great contribution of mining has been often overlooked. For example, from 1851 until the early 1870s, Au was the predominant Australian export. Through much of the rest of the 19th century, the mineral industry was very close to wool as the predominant export industry. Since the late 1940s and especially now into the 21st century, mining and related industries have accounted for the bulk of Australia's export earnings, eclipsing the agricultural sector many times over (Chapter 9).

Early mining in Australia

The history of Australian mining began with Aboriginal mining of various rocks and ochres for stone tools and pigments (Chapter 5). Trade of these commodities enabled members from different language groups to share aspects of 'The Dreaming'. Specific cultural knowledge and practices were shared and reinforced during meetings. Trade occurred between tribes, with particularly valued commodities being distributed across the country (see *Did you know?* 8.1).

The utilisation of stone continued with the arrival of the First Fleet in 1788, when convicts were assigned to cut sandstone blocks from the shores of Port Jackson for the Governor's residence, warehouses, military barracks, prisons and other buildings. Within 10 years of the arrival of the First Fleet, coal was discovered in 1791 by convicts near Newcastle (NSW) and later to the south and west of the settlement (Chapter 9). The find led to the establishment of the Coal River penal settlement in 1801. These areas provided fuel for heating and cooking, and later steam locomotion in the young colony of New South Wales. Australia's first truly commercial mining venture was at Newcastle in 1799, when coal was exported to Bengal (India). From those humble beginnings, Newcastle has developed into a major metropolitan centre and is now the world's largest coal export port.

The first metallic mine in Australia was the Glen Osmond Ag–Pb mine, which was opened near Adelaide in 1841. Mining saved the near-bankrupt economy of South Australia, with the discovery of Cu, first at Kapunda in 1843, followed by Burra in 1844. Up to the 1850s, Burra was the world's richest Cu mine, and for the first 20 years of operation, it paid dividends of 300% per annum, making it a significant economic driver in the early days of the colony. More than 1000 miners were employed in the Burra Mine in 1849–50. Before these Cu finds, South Australia's population in 1844 (Appendix 8.1.1) was around 17 000; it more than quadrupled in 10 years, to more than 85 000 people. By the 1870s, South Australia had replaced Cornwall as the largest Cu-producing region in the British Empire. Many of the skilled Cornish mine workers came to South Australia, both pulled by the allure of rich deposits and pushed by the potato blight and the declining mining industry in Cornwall.

Like Australia's coal resources, close proximity to the sea was an important factor in the early discovery and mining of South Australian Cu. Ore was initially brought by bullock wagon to Wallaroo for shipment to the smelters of South Wales. Later, ports were used (Chapter 6), and coal was shipped from Newcastle to South Australian smelters, with some lower grade ore back-loaded for smelting at Newcastle. Funding for these early mining ventures and the development of the Victorian goldfields (see below) was sourced mainly from overseas (London), as the first Australian stock exchange did not open until 1861 in Melbourne.

Gold, silver, lead and copper rushes of the mid- to late 19th century

Really large mining-driven changes began with the gold rushes. The winter of 1851 saw the first gold rush at Ophir near Bathurst in New South

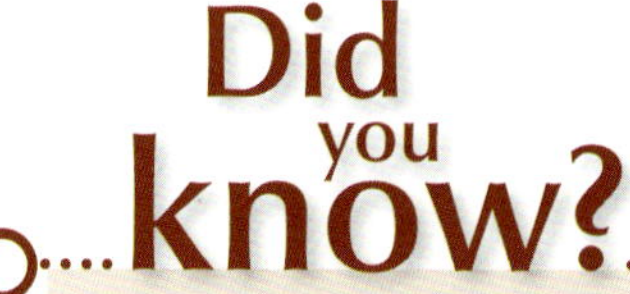

8.1: The first miners and traders

Aboriginal people were undoubtedly the first miners and traders, and possibly also geologists, in Australia. Some stone varieties were prized as traded ornaments; archaeologists refer to these as manuports. Trade was widespread: pearl and baler shells from northern Australia have been found on the Great Australian Bight on the southern coast; opal from inland was used for tools on the southeastern Queensland coast; bryozoan chert was well known for implements in Western Australia (Chapter 6) and Victoria. Highly specialised stone axe heads sourced from greywackes at Herberton (Qld) were prized by the Kalkadoon tribes of the Mt Isa area, some 1000 km to the west, and greenstone axe heads from Mount William (Vic.) were traded throughout southeast Australia. Even fossils such as ammonites and one Diprotodon skull found in the Ord Basin near Kununurra (WA) were prized, although for what reason is now lost. Giant marsupial teeth were used as charms.

Image by Chris Fitzgerald

Aboriginal stone tool from Queensland that is part of a message stick indicating a site of importance.

Wales. Although controversies remain about who first discovered Au, Edward Hargraves, using his California experiences, promoted the new field and started the rush. A further rush followed shortly after in Victoria. In response to these rushes, the New South Wales and Victorian colonial governments initiated a licensing system for prospecting, which was partly designed to encourage unsuccessful miners to return to their jobs in the cities and on farms. Development of this licensing system, which replaced the Crown's exclusive mineral rights, was one of two important drivers on the development of the mining industry in the 19th century. The other was economic depression, which drove city people to the bush and resulted in discoveries like Broken Hill (Zn–Pb) in New South Wales, and Coolgardie (Au) in Western Australia (Figure 8.1). The combination led to the first expression of 'revolution' and demonstration of the Australian spirit at the Eureka Stockade, where miners protested against licensing and raised their own flag, now preserved at the Ballarat Museum.

Burra, South Australia, (1876) just before the mine changed over to open cut. The criss-cross structure (centre) supports an aquaduct, which carried water to the water wheel in the crushing house (left). Concentrating sheds can be seen in the foreground.

Image courtesy of State Library of South Australia. Image no. B1450

Other than the arrival of the First Fleet, the discovery of Au shaped the make-up of the Australian people more than any other event. The first people on the goldfields were from Sydney and Melbourne and the surrounding sheep properties. When news reached England, the mass influx of Au seekers from overseas began. Australia's population soared dramatically, trebling in 10 years, with, as noted above, the first large numbers of Asian immigrants. By 1861, there were more than 24 000 Chinese immigrants on the Victorian goldfields of Ararat, Ballarat, Beechworth, Bendigo, Castlemaine and Maryborough. There were more than 11 000 Chinese on the New South Wales goldfields. In the 1870s, there was an influx of Chinese miners to Queensland after the discovery of Au in the Palmer and Hodgkinson rivers and in Cooktown. Chinese miners worked not only Au but also other metals such as tin (Sn), tungsten and Cu. The influx of people from around the world captured the imagination of poet Henry Lawson.

The patterns of discovery were similar at many of the new goldfields. The alluvial resources ('placer' deposits) were usually the first to be discovered, but they were commonly depleted in a few years, with most diggers moving on to the next new find. Where deeper vein or hard-rock Au resources were discovered, fields lasted much longer (for more than 120 years in some cases). These resources required capital and company structures, and provided employment for many of the original diggers. There was, however, a strong appeal in digging for yourself. Although many diggers made great fortunes, others lost everything, including their lives.

The gold rushes began in the southeast corner of the continent, and then worked their way anti-clockwise around Australia (Appendix 8.1.1). In the 1860s, the first big goldfields were found in Queensland, first at Gympie, and then at Charters Towers in the early 1870s. These Au discoveries saved the almost-bankrupt colony's economy. Gold was then discovered at Pine Creek, south of Darwin, in the 1870s, and in the 1880s gold rushes reached the northwest of Western Australia with a rush to Halls Creek. By the end of the 1880s, prospectors were finding Au some 200 km east of Perth, and then, in 1893, at the richest of all Australian goldfields, the Golden Mile at Kalgoorlie. This period of Au prospecting also yielded tales of lost Au, the most famous being Lasseter's reef. In 1897, an Australian-born prospector from the Victorian goldfields, Lewis 'Harry' Lasseter, supposedly found a rich vein of Au somewhere near the border between Western Australia and the Northern Territory. The myth persists to this day, and fortune hunters still roam the Australian outback in search of Lasseter's and other lost Au reefs.

As prospectors went inland, most looking for Au, other commodities were found (Appendix 8.1.1). Sn was discovered at Mount Bishop in northwest Tasmania in the early 1870s, and in 1883 Ag and Pb were detected at Broken Hill. CU was discovered at Mount Lyell in western Tasmania and at Mount Morgan and Cloncurry in Queensland by the 1880s. At Mount Morgan, oxidised ores rich in Au graded downwards into Au–Cu ores at depth. By 1900, Australia had the beginnings of a diverse mining industry, with coal at Newcastle and Port Kembla, Au in virtually every part of Australia, Cu in many places, and Sn in western Tasmania. In the 1870s, Australia was probably the largest Sn producer in the world.

Extract of the third verse of 'Eureka'

I hear the broken English from the mouth at least of one
From every state and nation that is known beneath the sun
The homely tongue of Scotland and the brogue of Ireland blend
With the dialects of England, from Berwick to Land's End;
And to the busy concourse here the West has sent a part,
The land of gulches that has been immortalised by Harte;
The land where long from mining-camps the blue smoke upward curled;
The land that gave that "Partner" true and "Mliss" unto the world;
The men from all the nations in the New World and the Old,
All side by side, like brethren here, are delving after gold.

Henry Lawson (1889)

Image courtesy of National Library of Australia. Image no. nla.pic.vn-4699151

New expansion and contraction: 1900–39

Although Au mining had dominated the second half of the 19th century, by the beginning of the 20th century, prospectors had found most of the easily won sources. By about 1914, the prospectors had found virtually all the deposits that were payable using the technology at the time. The value of Au declined, the cost of mining increased, and

Image courtesy of National Museum of Australia

Harvest of Endurance scroll, 'Chinese Miners', Australia China Friendship Society.

World War I began. The importance of Au as an economic and therefore social driver declined. Mining continued, but the wealth it generated did not rival the earlier influence. Rather, the only major finds of the first half of the 20th century were Pb, Zn and Cu deposits at Mt Isa in Queensland, but their full potential was not realised until the 1950s (see below).

The Great Depression of the late 1920s and 1930s had a mixed effect on the Australian mining industry. The industry was mostly depressed because of low commodity prices, but a rise in the price of Au and a government bonus led to a surge in Au prospecting and production. Consequently, Au mining in Western Australia had a boom, and a new goldfield was discovered at Tennant Creek in the Northern Territory. At the same time, Newcastle and other coal-producing areas had major strikes as that industry and its labour supply adjusted to economic depression.

The post-war boom: 1945–75

Aside from Au in the 1930s, the mining industry went into near hibernation until the late 1940s, when a boom began that was to last two decades. This boom differed from the earlier ones in that it involved a different mix of commodities and was driven in part by changes in government policies. The development of military 'Cold War' and then civilian nuclear programs, particularly in the United States and Great Britain, required uranium (U), a commodity that previously had no use. Other commodities that were found at this time included bauxite (an ore of aluminium–Al) at Gove and later Weipa and the Darling Range, and iron ore, firstly at Koolanooka and then in the Pilbara region (Chapter 9). Although this boom also saw the discovery of important base-metal deposits, Au exploration was virtually non-existent, only resuming in the 1980s after the Au price had been floated relative to the US dollar and technological innovation allowed mining of low-grade deposits (see below).

This mid-20th century mining boom also saw the development of a new style of discovery. Previously, all discoveries in Australia had been made by prospectors, semi- or self-educated men who roamed the outback looking for the telltale signs of mineralisation, Au nuggets in streams or veins, or the green bloom of malachite on surface outcrops. Exploration now became increasingly capital and technology dependent, and geoscientists—geologists, geochemists and, particularly, geophysicists—became important players in finding deposits. As the costs of exploration increased, mineral exploration was carried out by junior exploration companies and the exploration divisions of major mining houses. Many of these discoveries were made using newly developed geophysical techniques, giving geophysics the aura of a 'magic bullet' for success. Exploration was also facilitated by government surveys (Chapter 2). The Bureau of Mineral Resources, Geology and Geophysics, better known as BMR—now Geoscience Australia—was established in 1946, with the aim of mapping the continent to assist industry to uncover new mineral and energy resources.

Changes in government policy also drove this boom. As late as 1945, Australia was believed to be deficient in iron ore, and its export was banned. Removal of these bans resulted in the discovery of huge iron ore deposits in the Hamersley Basin of Western Australia in the 1960s, which has become one of the great iron ore provinces of the world (Appendix 8.1.1; Chapter 9). Decreases in shipping costs and the development of open-cut mining methods made these resources payable, and very profitable. Giant bauxite deposits and manganese (Mn) deposits at Gove in the Northern Territory were discovered at this time. The coupling of easily won coal using open-cut mining methods and high-quality alumina meant the synergistic development of Al refineries adjacent to some of Queensland's ports (Chapter 9).

The nickel boom: 1960s to 1970s

While prospecting for Au in 1947, George Cowcill collected green-stained ferruginous rock specimens in the Kambalda area of Western Australia. Later, in 1954, thinking the samples might be U-bearing, he submitted them for chemical analysis, only to find that they contained nickel (Ni). A decade later, Western Mining Corporation geologists explored for Au and base metals in the same area, but, as happens commonly in exploration, they discovered something quite different, the great Kambalda Ni deposits (Appendix 8.1.1). Discovery of the Lunnon shoot (initial intersection of 8.3% Ni over 2.7 m) was announced in 1966, sparking a boom that continued until 1972. The host magnesium (Mg)-rich ultramafic rocks, called komatiites, were difficult to identify in the weathered rocks of the Yilgarn Craton. These komatiites hosted a style of Ni deposit previously unrecognised anywhere else in the world. In some places, the komatiites weather to silica-rich cap rocks, bearing little chemical relationship to their parent rock, and only preserving their characteristic spinifex rock texture (Chapter 2). Fortunately, these rocks are magnetic, so the use of aeromagnetic data rapidly assisted explorers in identifying prospective ground. The Ni boom was intense, with competition leading to the pegging of very speculative leases with a low prospect of discovery.

Nevertheless, more than 70% of the mined Ni deposits in Western Australia were discovered in these early years. In the late 1960s, demand for Ni soared to meet armament requirements for the Vietnam War and as a result of industrial relations problems at the Canadian company Inco, then the largest Ni producer in the world. Many new companies were listed, some without leases, let alone mines, to take advantage of the high prices.

Gold booms of the late 20th century

From the end of World War II, the US dollar had been pegged to the gold standard. By the early 1970s, Au had truly lost its lustre. Towns synonymous with the Australian gold rushes were nearly dead; Kalgoorlie, for instance, had only around 100 working miners in 1976. However, when the convertibility to Au by the US dollar was terminated in 1971, the price of Au began to rise steadily, peaking in 1980 at more than US$850 (or more than US$1800 per ounce in 2011 dollars). Opportunities materialised to acquire large shares of the most prospective ground at rock-bottom prices. From this, the Golden Mile was reborn, this time as a 'super pit' but with grades less than a fifth of those of the early mines.

Technology also played its part in the late-20th century Au resurgence. The development of the carbon-in-pulp process and refinements in open-cut mining methods in the 1980s permitted mining of the large low-grade halos of previously mined, high-grade deposits. By 1993, Kalgoorlie was once again Australia's largest Au producer, a century after its founding; the Super Pit has a projected mine life to 2017. Other old fields have also been re-energised, including Kanowna, Kambalda and Coolgardie (WA), Charters Towers (Qld), and Stawell (Vic.).

Interest in Au continued apace, and by 1990, around 10% of all listed stocks on the Australian Stock Exchange were partly or wholly involved

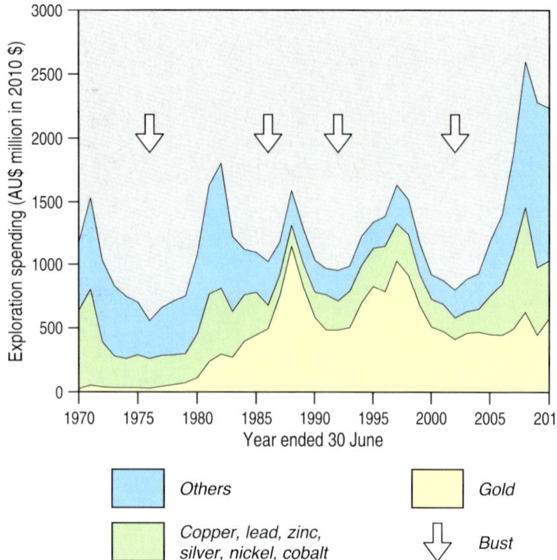

Figure 8.3: Australian mineral exploration expenditure in constant 2007–2008 dollars. (Based on Australian Bureau of Statistics data deflated by the consumer price index). Note the four busts following the booms in expenditure. These low points were not good times for the employment of exploration geologists.

in Au mining. Companies synonymous with other commodities also chased the allure of Au. For example, North Flinders Mines' foray into the Granites–Tanami Province of the Northern Territory was a far cry from their original target of base metals in the Flinders Ranges of South Australia. Mineral fields not previously known for Au, such as the bauxite fields southeast of Perth, were also tested, leading to the discovery of the giant Boddington Cu–Au deposit.

Boom time from 2005 onwards

The latest boom began in about 2005 and continues, albeit with a short hiccup associated with the 2007–08 global financial crisis. This boom differs in important ways from previous ones in that it crosses virtually all commodity sectors—Au, Ni, base metals, U, iron ore, coal and bauxite—with major price increases in most commodities. In addition, there has been increasing interest in new commodities such as rare-earth elements, Mn and lithium (Li), which are important inputs into high-technology and other strategic industries (Appendix 8.1.1). Because the bulk commodities of coal, iron ore, natural gas and aluminium dominate the economics of this boom, they are discussed in more detail in Chapter 9.

Mining busts of the late 20th century

Busts have just as large an impact on the industry and society as the booms they follow (Figure 8.3). The mining industry is highly susceptible to economic downturns; as commodity prices fall, mines close, jobs are lost and mineral exploration declines. Since 1970, there have been four busts with periods ranging from four to eight years (Figure 8.3). The 'deepest' bust, which accompanied the mid-1970s oil crisis, involved a 65% decrease in real exploration expenditure relative to the peak of the Ni boom. The most recent bust, which started just before the turn of the 21st century, differed from those before in that many junior exploration companies did not go bankrupt but diversified into other business activities, including information technology ventures that collapsed later during the 'dotcom' bust of the early 2000s. Another important consequence of this bust was the consolidation of the industry, with the disappearance of virtually all mid-tier Australian producers (e.g. Mt Isa Mines Ltd, Western Mining Corporation Ltd) into large multinational conglomerates. This was accompanied by the dismantling of industry-based research groups, with the research niche taken up by consultants, academia and government.

Australia's giant mineral systems

Australia is among the top five countries, in terms of both production and resources, for many commodities, including Au, Cu, U, Zn, Pb, Ag and others discussed in Chapters 4 and 9. Australia is a big country, with the full spectrum of geological time recorded, making this resource endowment widely distributed in time and space. Almost all Australian states and territories have world-class deposits containing at least one of these commodities, which are hosted by Archean, Proterozoic and Phanerozoic rocks (Figure 2.21).

It takes a lot of energy and a lot of fluid to make a giant mineral deposit. A mineral deposit is, however, only a symptom of a much larger system—a mineral system (Figure 8.4). For a more complete explanation of this approach, see Appendix 8.3.1. In essence, a mineral system works like this:

1. Fluids from various sources with favourable physical and chemical properties dissolve metals from a large volume of dilute metal-rich source rock.
2. The metal-rich fluids follow pathways that focus and concentrate them through a much smaller volume of rock.
3. Most of the rock volume has very low permeability. For fluid flow to occur, the rocks must be fractured or faulted and linked with a connected architecture.
4. The architecture must also be able to throttle or focus the large volumes of metal-rich fluid through a smaller volume of rock.
5. In this smaller rock volume, the metal-rich fluid encounters a change in the physical and/or chemical conditions.
6. The properties of the metal-rich fluid at this critical time become unfavourable for holding the dissolved metals in solution.
7. The metals are deposited, and the remaining fluid is expelled and dispersed.
8. The system is preserved.
9. Geodynamic processes (energy, kinematics and dynamics) drive the mineral system.
10. Mineral systems operate at scales from the global to the microscopic.

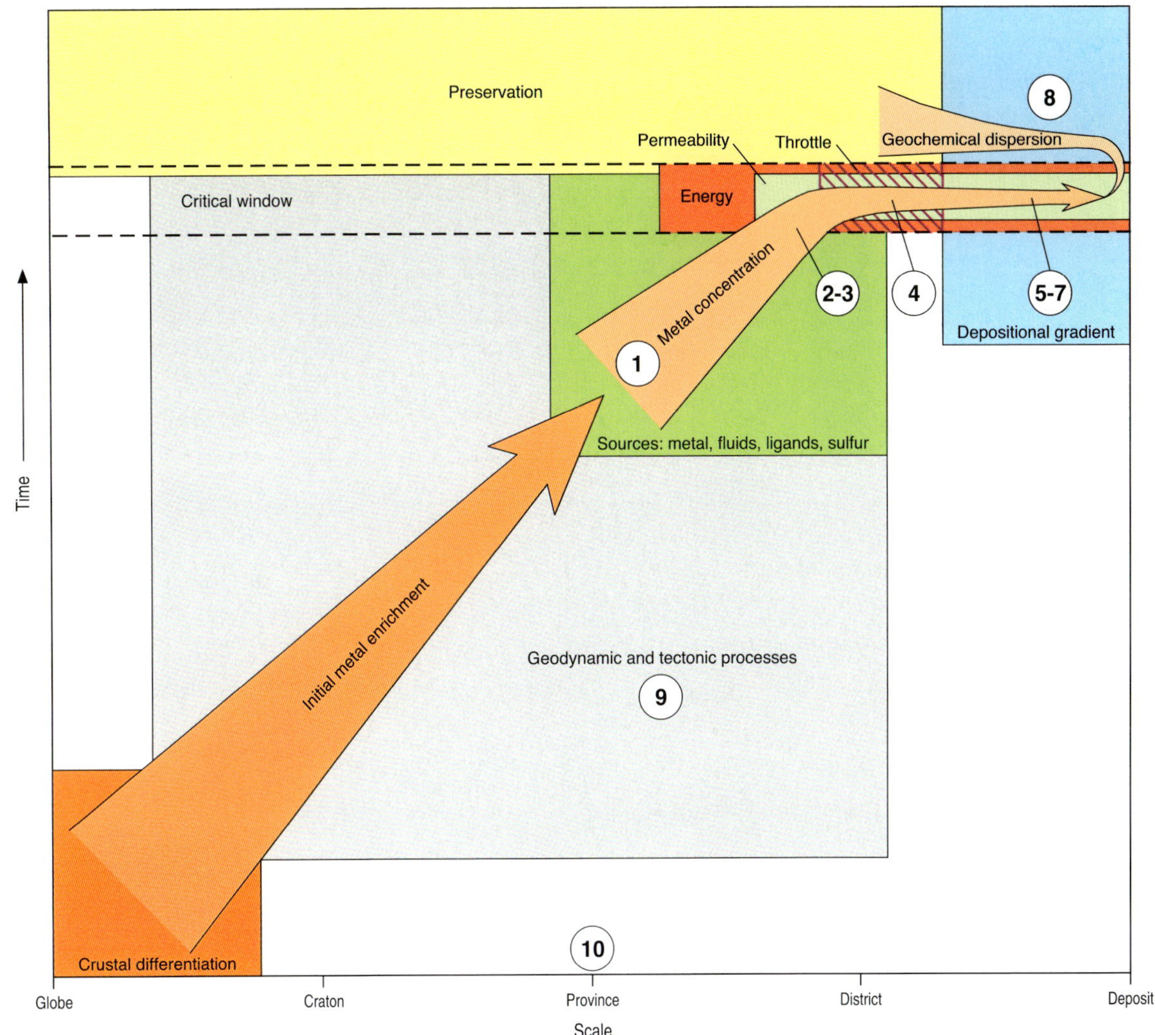

Figure 8.4: A space and time illustration of a mineral system. In this system diagram, a mineral deposit is formed during a critical window of time when there is a conjunction of sufficient energy, permeability and depositional gradients to focus metal and fluid for ore concentration. The time period before formation ensures sufficient preparation and initial concentration of metal sources. The time period after formation ensures that the deposit is preserved, and not eroded. The numbers on the diagram refer to the parts of a mineral system listed in the text. In contrast to a petroleum system, fluids are not trapped in a mineral system; they are focused and exit the depositional site, leaving a geochemical dispersion footprint that may be preserved at scales many times larger than the deposit itself.

© Getty Images [R Woldendorp]

Aerial view of open-cut gold mine, Mount Magnet, Western Australia. Note the brown-yellow colours deep into the pit, showing the penetration of oxidation and weathering (Chapter 5).

The benefit of a mineral system is that its scale is many orders of magnitude larger than the mineral deposit itself, which is just a favourable symptom of an effective system. This means that explorers can use the much larger scale indicators of the ore system in their search for the deposit itself. For example, a deposit only 500 m wide may have a fluid outflow zone many tens of kilometres wide, such as in the Eastern Goldfields. Similarly, the zone of depletion of the metal-rich source rock may be many tens of kilometres wide, such as in Broken Hill. These wide 'alteration' footprints then become amenable to remote sensing or geophysical detection, and thus can be used to vector more effectively and cheaply to ore.

Given the size and diversity of Australia's mineral endowment, we have selected four key hydrothermal mineral provinces: the Victorian goldfields; the Eastern Goldfields in Western Australia; the Proterozoic zinc belt that spans New South Wales, Queensland and the Northern Territory; and the Olympic Dam Iron-Oxide-Copper-Gold (IOCG) Province in South Australia. These four giants are interesting not only for their history, but also because they illustrate the concept of a mineral system.

Victorian goldfields—the gold rush that changed a nation

The discovery of Au in Victoria was the first of the major minerals-related economic booms in Australia. In addition to the wealth created in Bendigo and Ballarat and then transferred to Melbourne, this gold rush began to transform

Australia from a continent populated by the relatively few Aboriginal people and European (largely British and Irish) immigrants to the multicultural society of today. Victoria's immigrant population increased from about 10 000 in 1840 to 500 000 in 1860 (Box 1.2). The mining history of the Victorian goldfields has influenced the Australian character in other ways. For instance, the Eureka Stockade rebellion was, at least on the surface, a dispute about mining licences, and the first wave of non-European immigration to Australia was the Chinese miners who came to work the Au and stayed to create businesses.

When, in early 1849, a young shepherd, Thomas Chapman, unearthed 38 oz near Amherst in what was to become the Victorian goldfields, a nascent gold rush was nipped in the bud by troopers ostensibly sent to enforce trespass laws, although the real reason may have been to prevent mass desertions of workers from rural farms and the growing city of Melbourne.

According to official records, payable Au was first found in early 1851 at Clunes (Figure 8.5), about 130 km northwest of Melbourne. However, possibly because of its remoteness, this goldfield did not experience the mad rush seen at Ophir in New South Wales. When first visited by troopers in late July, only 50 men were working the goldfields, although Clunes for a short time became the largest inland town in Victoria. The first major rush in Victoria was to Ballarat, after Au was discovered by John Dunlop and James Regan as they returned to Melbourne from Clunes. By the end of September 1851, nearly 1000 men were seeking Au, and the Governor of Victoria was compelled to adopt the New South Wales licensing system. Within a year, around 20 000 diggers were trying their luck in the Ballarat goldfield (Box 8.1). The attraction was strong, as, in the year 1856 alone, more than 20% of Tasmania's convicts tried to abscond, many drawn to the Au riches of Victoria.

The discovery of Ballarat opened a floodgate; within a year, Au had been discovered at Mount Alexander, near Castlemaine, and then Bendigo (Figure 8.5). At first, most Au was mined from alluvial workings, although this was mostly exhausted by 1900. The earliest years were also the richest; for most of the 1850s, Victoria produced 70–90 t/year of Au. After 1860, production declined steadily until around the turn of the century, which saw a small revitalisation of production (to 25–30 t/year). By 1930, production had virtually ceased, with less than 1 t/year. Minor revivals (to 5 t/year) occurred during the Great Depression in the mid- to late 1930s and in the recent Au boom from the late 1980s to now. The Victorian goldfields did not, however, experience the renaissance seen in the Eastern Goldfields of Western Australia, possibly because the nature of the Victorian deposits (nuggety) made them less amenable to the advancements in metallurgical (carbon-in-pulp) and bulk-mining methods that have driven the most recent Au boom. The location of many deposits within or near town centres in the Victorian goldfields, along with local opposition, served as major barriers to open-cut mining.

Victoria's golden riches: gifts from the Benambran Orogeny

The formation of Victoria's goldfields is intimately linked to the evolution of the Lachlan Orogen. This orogen, which makes up most of the Tasman Element of eastern Australia, formed as a convergent margin that was active through the mid- to late Paleozoic assembly of Gondwana and Pangaea between 490 Ma and 230 Ma (Chapter 2). Although eastern Australia was mineralised throughout this assembly, the most important metal-forming period was the 490–435 Ma Benambran cycle, which produced a series of island and continental arcs and backarc basins that extended from northern Queensland to western Victoria and were cratonised during the 440–435 Ma Benambran Orogeny (Figure 2.33).

The most significant deposits of the Benambran cycle formed near its conclusion. In the central Lachlan Orogen, porphyry Cu–Au, epithermal Cu–Au and Au-only deposits mostly formed at 440–435 Ma, largely associated with emplacement of potassium-rich (shoshonitic) volcanic centres during late extension of the oceanic Macquarie Arc. In the Victorian goldfields of the western Lachlan Orogen, contraction associated with the Benambran Orogeny formed most lode-Au deposits (Figure 8.6, Box 8.2). The closeness in time and space of these two contrasting mineral systems presents a conundrum in understanding the Benambran convergent margin.

Figures 8.6a and 8.6b–c illustrate two possible geodynamic scenarios to account for this conundrum. The first involves the attempted subduction of a seamount or microcontinental block, resulting in the locking up of the arc at *ca* 440 Ma (Figure 8.8a). Rollback and slab tearing associated with this lock-up resulted in extension and alkaline magmatism with associated porphyry Cu–Au deposits in the Macquarie Arc. Compression

THE ECONOMIC IMPACT OF THE VICTORIAN GOLD RUSH (BOX 8.1)

The discovery of vast quantities of gold (Au) in Victoria in the early 1850s had a major impact on Victorian economic development and on that of the rest of Australia. Gold was Australia's most valuable export commodity from 1851 to 1870, and at its peak in 1852 the mining sector comprised more than 35% of Australian gross domestic product.

A boom in one sector or region requires an adjustment to resource allocation in the economy as a whole and is almost always associated with changes to income redistribution. The overall impact of the gold rush was overwhelmingly positive, although there were some negative impacts in some sectors of the economy that were forced to adjust in the short and medium term. Wages for gold miners rose rapidly, increasing five-fold from 1851 to 1852 (Figure B8.1a), leading to increases in wages in other sectors and temporary shortages of labour. Attracted by high wages for miners, labour flowed into Victoria, and the population increased by more than 450% from 1851 to 1860 (Figure B8.1b, Box 1.2). Although some of this inflow came from other states (particularly South Australia and Tasmania), most of the increase was accounted for by immigration. This large increase in the supply of labour mitigated the upward pressure on wages.

The rapid expansion of the mining sector placed considerable inflationary pressure on goods and services (Figure B8.1a). Although there were dramatic increases in prices for domestically produced and consumed goods and services, prices for traded commodities were relatively stable because Australian importers and exporters were (and are) generally price takers on world markets and the exchange rate was fixed. Higher wages increased the supply costs for businesses, putting further upwards pressure on prices. Overall, consumers received higher incomes but faced increased prices, while producers of traded goods (particularly wool) experienced increased costs but received unchanged output prices.

There was considerable investment in public infrastructure in Victoria as a result of the gold rushes. An increased demand for transport and telegraph services, particularly in regional areas, was combined with increased revenue and an improved credit standing. This growth in regional infrastructure benefited the agricultural sector when the mining sector declined, enabling growth to remain strong for a decade after the boom peaked.

Similar pressures are present in today's economy as a result of the current mining boom, which is centred largely in Western Australia and Queensland. Demand for Australia's mineral exports has driven up the value of the Australian dollar, causing ongoing adjustments in other sectors, primarily the manufacturing and tourism industries, as they absorb the impact of the high Australian dollar (Chapter 9).

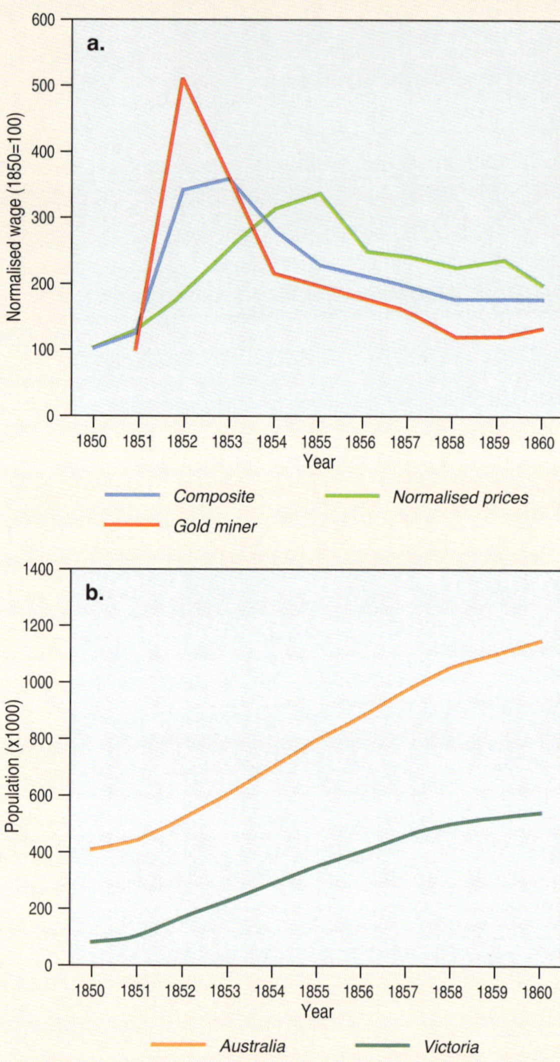

Figure B8.1: Changes in economic indicators (a) and population (b) associated with the discovery of Au in 1851 and the subsequent Victorian gold rush. Graph shows changes in normalised miners' wages and changes in normalised prices in Victoria between 1850 and 1860. (Source: Maddock & McLean, 1984)

inboard of the arc deformed the Western Lachlan, resulting in lode-Au mineralisation. Alternatively, the West and Central Lachlan were separated by a transform, the Baragwanath Transform, which accommodated extension, and porphyry Cu–Au deposits, to the northeast, but contraction, and lode-Au deposits, to the southwest (Figure 8.6c). A third option, not shown in the figure, is that the Macquarie Arc was translated southwards from its original position during oroclinal folding associated with the later (*ca* 410 Ma) Bindian Orogeny.

The early Paleozoic rocks of the Bendigo and Stawell zones lie in a wedge-shaped extensional basin, the 'Castlemaine basin', between Proterozoic blocks to the west and east. At the surface, the Bendigo Zone is dominated by Early Ordovician turbidites with lesser Cambrian basaltic rocks (Figure 8.5). Cambrian-aged mafic volcanic rocks in the Stawell Zone are associated with the fault that marks the western boundary of the adjoining Bendigo Zone. Interpretation of the deep seismic reflection data suggests that mafic volcanic rocks form a thick wedge beneath the Ordovician turbidites (Figure 8.7). The basalt within these zones has both tholeiitic (iron—Fe-rich) and boninitic (Mg-rich) associations. The basalt geochemistry has been interpreted as indicative of an extensional backarc basin setting for the Bendigo and Stawell zones. Moreover, these basalts are also enriched in Au, and they may have been one of the main sources of metal (Figure 8.4).

The Cambro-Ordovician fill of the Castlemaine basin is transected by a series of, mostly, moderately to shallowly west-dipping faults,

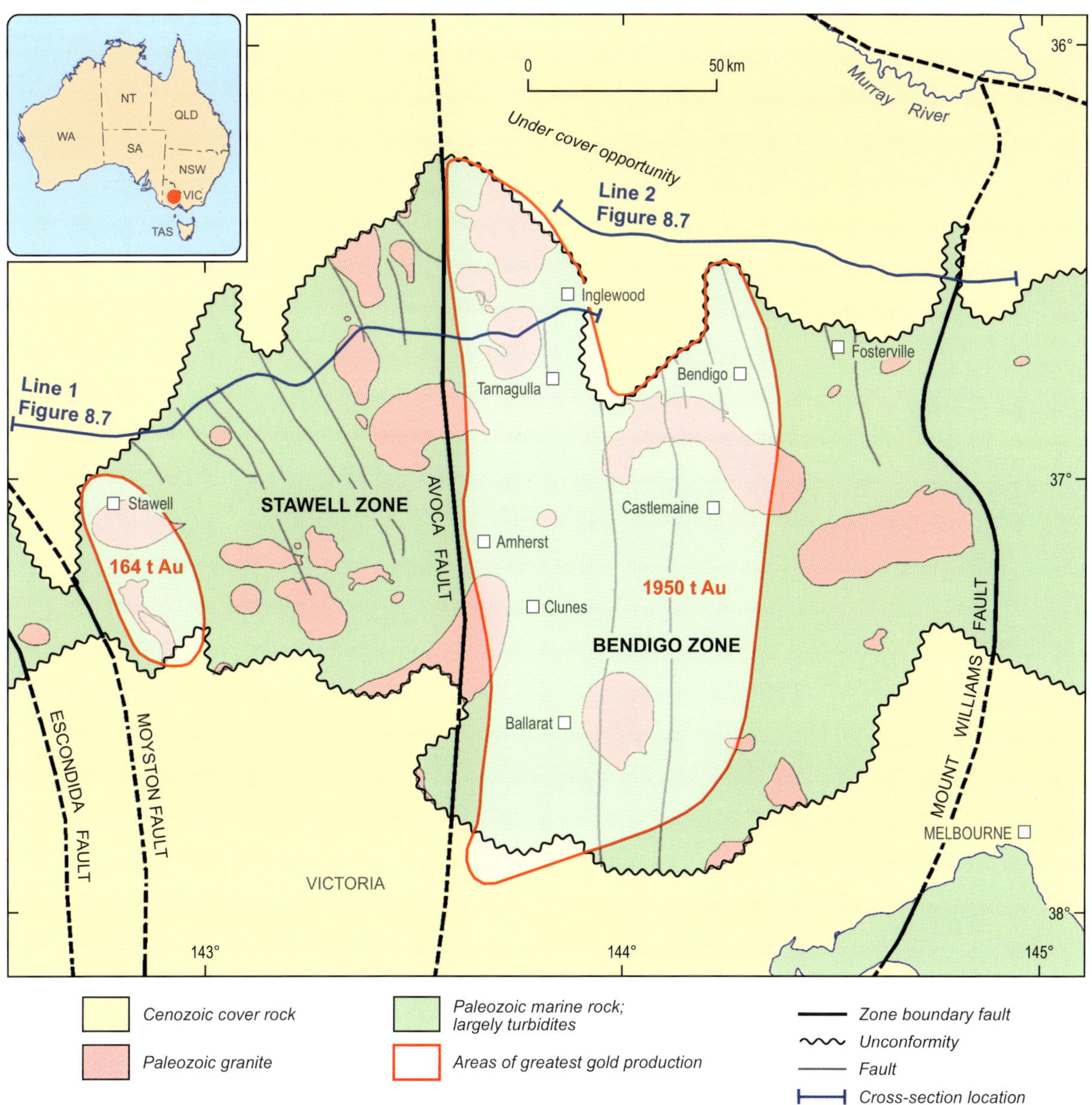

Figure 8.5: Simplified geology of the western Victorian goldfields, showing the location of Au deposits and major faults. Note that the Paleozoic Au-bearing geology extends to the limits of Cenozoic cover. Opportunities exist for Au mineralisation, especially to the north under the Murray Basin. The blue lines indicate the locations of the seismic traverses shown as an interpretation in Figure 8.7. (Source: Willman et al., 2010)

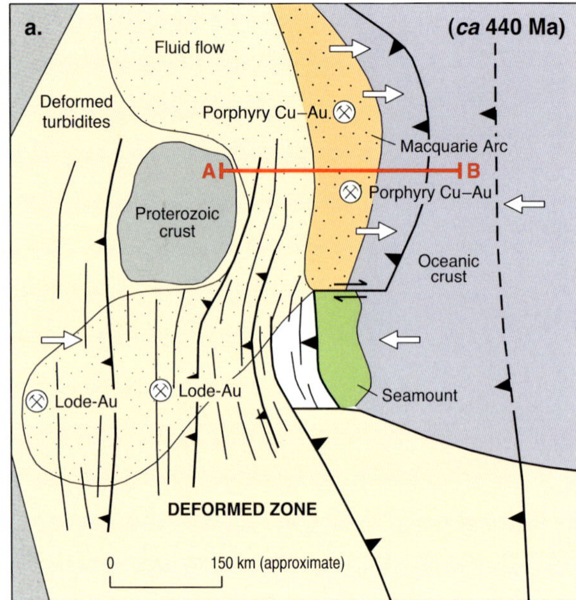

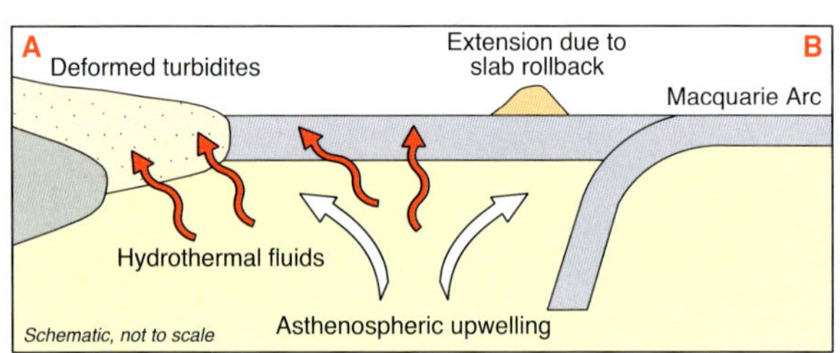

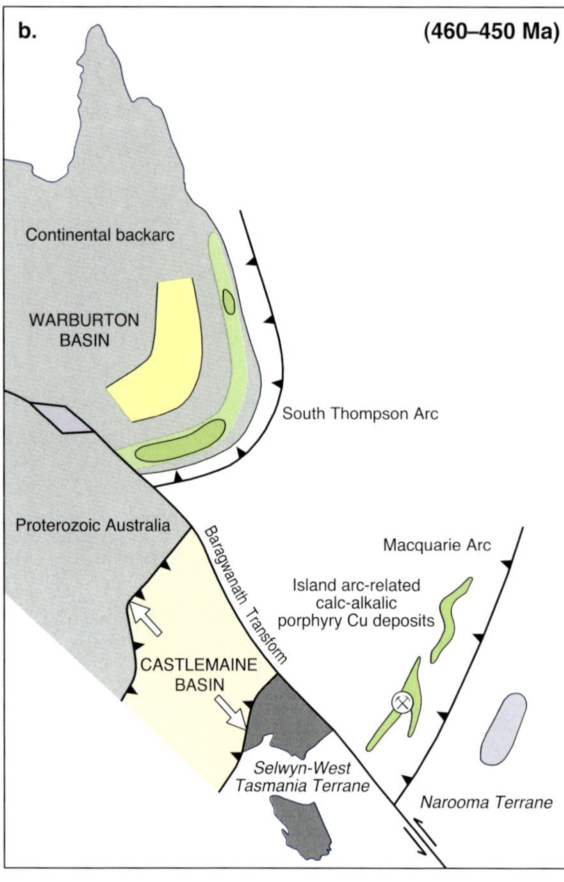

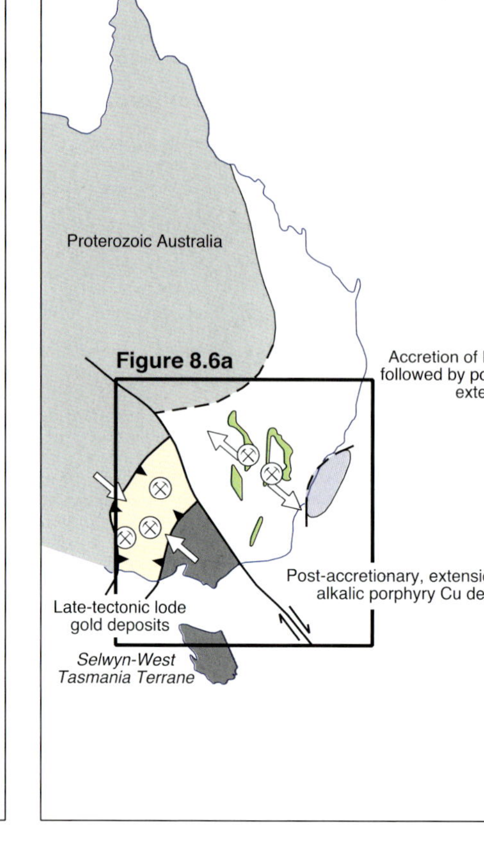

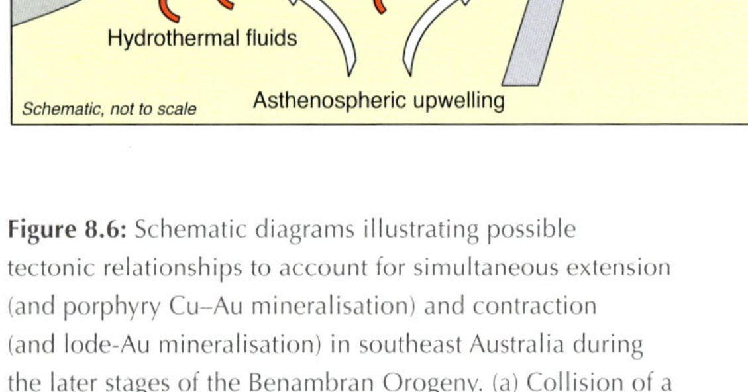
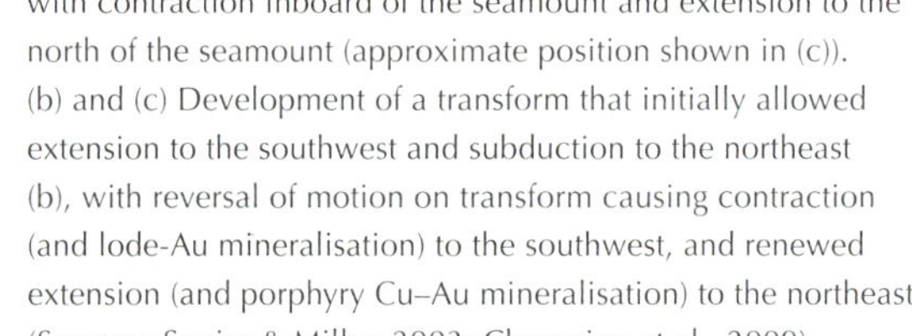

Figure 8.6: Schematic diagrams illustrating possible tectonic relationships to account for simultaneous extension (and porphyry Cu–Au mineralisation) and contraction (and lode-Au mineralisation) in southeast Australia during the later stages of the Benambran Orogeny. (a) Collision of a seamount or micro-continent causing slab lock-up and tearing, with contraction inboard of the seamount and extension to the north of the seamount (approximate position shown in (c)). (b) and (c) Development of a transform that initially allowed extension to the southwest and subduction to the northeast (b), with reversal of motion on transform causing contraction (and lode-Au mineralisation) to the southwest, and renewed extension (and porphyry Cu–Au mineralisation) to the northeast. (Sources: Squire & Miller, 2003; Champion et al., 2009)

some of which have apparent reverse motions (Figure 8.7). The margins of the basin are defined by inward-dipping faults that juxtapose basin fill against Proterozoic basement rocks and are most likely the original basin-bounding faults. The faults within the basin also probably initiated as normal faults related to basin formation, but were inverted into reverse faults during the *ca* 440 Ma Benambran and later orogenies. This early basin architecture forms a fundamental control on the richness of the Victorian goldfields, particularly in the Bendigo Zone, as it provided not only Au-enriched source rocks but also pathways for later fluid flow during reactivation of the early structures.

Isotopic dating of the Au deposits (Box 8.2) shows that the Victorian goldfields experienced three separate Au mineralising events, the first at 440–435 Ma, associated with the Benambran Orogeny; the second at 420–410 Ma, associated with the Bindian Orogeny; and the last at 380–365 Ma, associated with early stages of the Kanimblan cycle (Box 8.2). Of these, the earliest (Benambran Orogeny) was most prolific, producing more than half of the total Au from the Victorian goldfields. This event differs from the later events in the Victorian goldfields, and Archean Au events in the Eastern Goldfields Province (see below), in that it was not synchronous with granite magmatism. The oldest granites appear to have been emplaced 25 Ma after the Benambran-aged Au was deposited. These granites overlap the Bindian Orogeny, and many of the early Kanimblan deposits have a close spatial and/or temporal association with magmatism.

Most Au in the Stawell and Bendigo zones is spatially located within 5 km of the major first-order faults, many of which were initiated during basin development. However, these faults are generally unmineralised. The most significant goldfields in the Stawell Zone are in the hangingwall of the Moyston Fault, and most Au in the Bendigo Zone is associated with west-dipping back thrusts (Figure 8.7). In detail, the Au deposits are associated with second- and third-order faults and associated folds (Figure 8.8). The first-order faults acted as the main fluid conduits that linked Au-enriched mafic volcanic rocks deeper in the basin with the second-order faults, which provided a throttle for the metal-rich fluid, focusing it into favourable structural and/or stratigraphic sites for deposition (Figure 8.8).

In the Bendigo and Stawell zones, the dominant Au depositional sites are structural. In most cases, uniformity in Ordovician siliciclastic basin-fill did not create sufficient chemical gradients to deposit Au. Rather, the Au is hosted in saddle reefs along anticlinal closures, in cross-cutting and bedding-parallel veins, and in tensional veins in fold closures (Figure 8.8). The auriferous veins are commonly associated with sericitic (white mica) alteration halos and hosted mainly in sandstone units, which were able to fracture in response to the tectonic stress in a brittle manner and thus create dilatant zones (space). When the metal-rich fluids encountered this space, the pressure of the fluid dropped and the metal-rich fluids expanded, causing phase separation or boiling, and hence deposition of Au (and quartz).

Having produced 80% of the Au from the Victorian goldfields, the Bendigo Zone dwarfs the adjacent Stawell and Melbourne zones in terms of endowment, whether measured on an acreage basis or in total Au produced. The Bendigo Zone forms the deepest part of the Castlemaine basin and is characterised by a greater abundance of mafic volcanic rocks at depth (Figure 8.7). In contrast, the Stawell Zone, which produced only 8% of Victorian Au, has a lower abundance of mafic volcanic rocks, and the Melbourne Zone, which has produced little Au, particularly during the Benambran event, is apparently not underlain by mafic volcanic rocks (Figure 8.7). The apparent correlation between the abundance of Au-enriched mafic volcanic rocks and Au productivity highlights the importance of an enriched source for Au metallogeny (Figure 8.4).

Key factors in forming the Victorian goldfields

The Victorian goldfields illustrate some features that may be fundamental to forming giant mineral systems. These goldfields formed during the assembly of Gondwana–Pangaea and are associated with a major stress switch from extension (basin formation) to compression, associated with the Benambran Orogeny. As shown above, this orogeny was the main energy driver in the mineral system and was likely caused by the accretion

Alluvial gold specimen with rosy quartz, Western Australia.

DATING ORE DEPOSITS—THE VICTORIAN EXAMPLE (BOX 8.2)

Geochronology is an essential tool for understanding the formation of ore deposits. Knowing when a mineral deposit formed is also of vital importance from an economic viewpoint as it has implications for how and where new resources might be found. In the vast majority of cases, the timing of ore formation is tied to the evolution of the geological domain in which deposits are hosted. As mineral deposits are generally the result of tectonic processes that operate in well-defined geological environments, constraining the timing of gold (Au) mineralisation in the Victorian goldfields was critical not only to the understanding of how, why and where these enormously rich ore systems developed, but also to deciphering the processes that dominated the tectonic evolution of southeastern Australia at 490–350 Ma.

Prior to the mid-1990s, the geochronological and tectonic framework of the Au-bearing rock sequences in Victoria was only very broadly constrained by general structural interpretations, the age of the sedimentary rocks, and the approximate timing of deformational events and the intrusion of magmatic rocks. There was considerable debate about whether mineralisation occurred during one 'climactic' event or in stages. The cause for the Au endowment was also contentious, with some scientists favouring granites, which account for more than 30% of the outcropping rocks in Victoria, as the ultimate driver.

As Au cannot be directly dated, determining its age has to rely on the indirect dating of 'proxy' minerals such as hydrothermal mica, feldspar or arsenopyrite, which are commonly associated with, and assumed to have formed concurrently

West-dipping reverse fault zone and associated fault fill and extension veins from Wattle Gulley gold mine, Victoria.

Image courtesy of Stephen Cox, Australian National University with, the Au. Indirect age determinations can also be obtained from the dating of intrusive rocks that truncate or are truncated by an Au-bearing quartz vein. In all instances, formation ages can be determined by means of analytical techniques that use radiogenic isotope systems such as Ar–Ar, K–Ar, Pb–Pb, U–Pb, Re–Os and Sm–Nd; ratios between the parent and daughter pairs provide precise age constraints.

In the past 15 years, numerous studies have been undertaken with the aim of resolving the age of mineralisation, and its relationship to magmatism, in the Victorian goldfields. These data have greatly advanced our understanding of the geotectonic setting in Victoria and processes that formed one of the world's richest Au provinces. Most importantly, Au mineralisation across Victoria was not the result of a province-wide mineralising event. Instead, the deposits formed episodically and in response to a diachronous west–east progression of deformation, metamorphism and exhumation typical of a developing mountain range. Major Au deposits in the west (e.g. Stawell) and central parts (e.g. Ballarat, Bendigo) formed at *ca* 440 Ma and are significantly older than smaller deposits in the east (e.g. Woods Point) that formed at 380–370 Ma. The first phase of Au, at 420–400 Ma, was younger than the first phase of gold mineralisation, but the second period of magmatism was coeval with the second period of Au mineralisation at 380–370 Ma. All of these processes were intimately linked to an extensive period of terrane collision in the Paleozoic along the Palaeopacific margin of proto-Australia (Chapter 2; Figure 2.26).

Frank Bierlein

Historic Mining Exchange building in Lydiard Street, Ballarat, Victoria.

Sketch of Sturt Street, Ballarat, in the late 1800s.

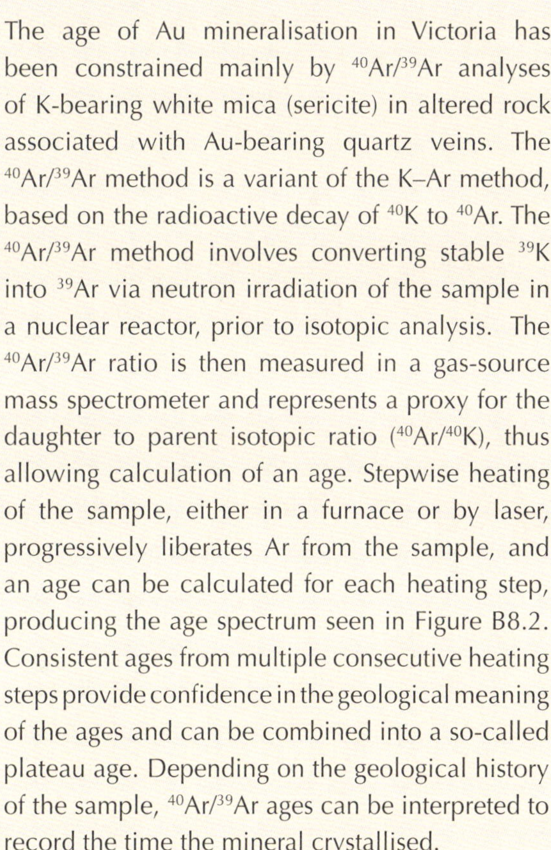

The age of Au mineralisation in Victoria has been constrained mainly by $^{40}Ar/^{39}Ar$ analyses of K-bearing white mica (sericite) in altered rock associated with Au-bearing quartz veins. The $^{40}Ar/^{39}Ar$ method is a variant of the K–Ar method, based on the radioactive decay of ^{40}K to ^{40}Ar. The $^{40}Ar/^{39}Ar$ method involves converting stable ^{39}K into ^{39}Ar via neutron irradiation of the sample in a nuclear reactor, prior to isotopic analysis. The $^{40}Ar/^{39}Ar$ ratio is then measured in a gas-source mass spectrometer and represents a proxy for the daughter to parent isotopic ratio ($^{40}Ar/^{40}K$), thus allowing calculation of an age. Stepwise heating of the sample, either in a furnace or by laser, progressively liberates Ar from the sample, and an age can be calculated for each heating step, producing the age spectrum seen in Figure B8.2. Consistent ages from multiple consecutive heating steps provide confidence in the geological meaning of the ages and can be combined into a so-called plateau age. Depending on the geological history of the sample, $^{40}Ar/^{39}Ar$ ages can be interpreted to record the time the mineral crystallised.

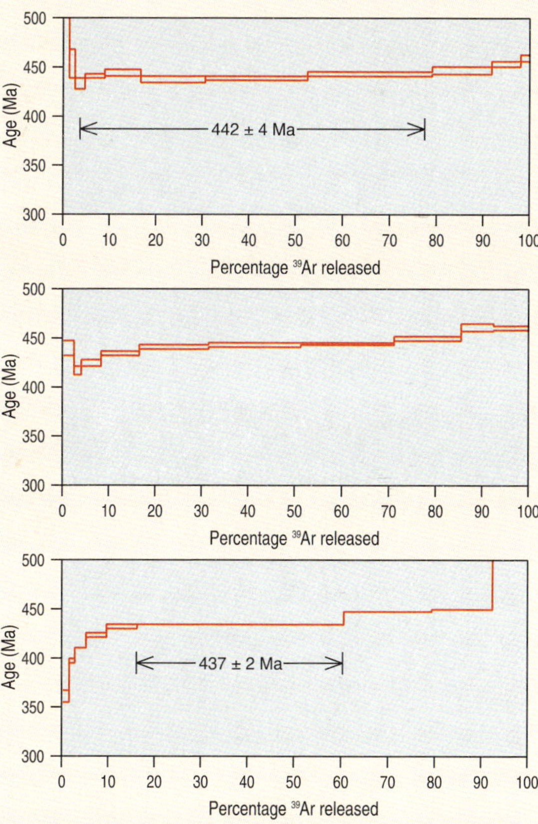

Figure B8.2: $^{39}Ar/^{40}Ar$ spectra from Au-related white mica, Central Deborah deposit, Bendigo. (Source: Bierlein et al., 2001)

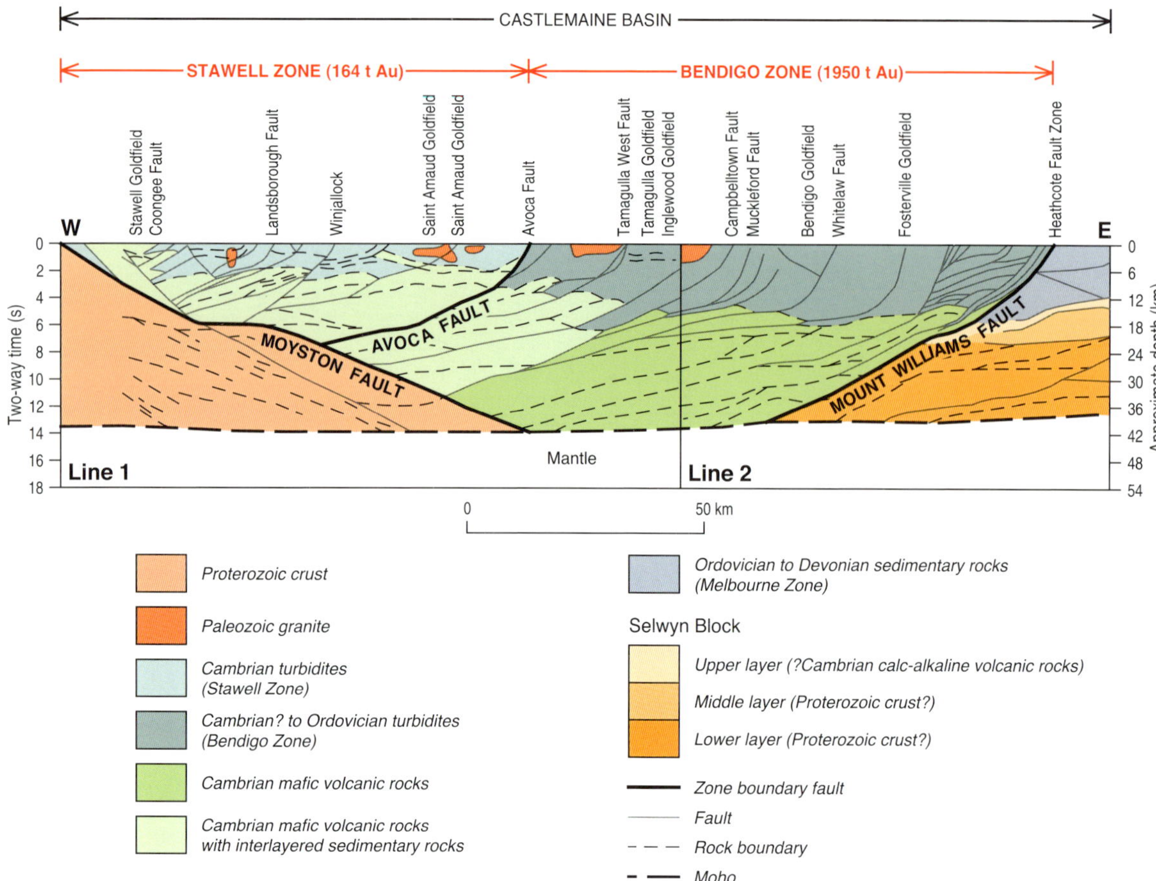

Figure 8.7: Composite seismic section across the Victorian goldfields, showing the relationship of Au deposits to structural and lithological architecture. Note the fault-bound, wedge-shaped Stawell and Bendigo zones, which host most of the Au. Note also the Cambrian mafic rocks throughout most of the crustal section. The location of individual seismic traverses is shown in Figure 8.6. (Source: Willman et al., 2010)

of crustal blocks from the present-day east. The deep-penetrating faults of the Castlemaine basin formed deep pathways connecting Au-enriched and mantle-derived mafic volcanic rock source regions, located deep in the basin, to higher structural levels. Under compression during the orogeny, these deep faults were reactivated as thrusts, and the deep metal-rich and hot fluids were driven up the interconnected pathway to the structural sites, where the Au was deposited. The Victorian goldfields share elements with other world-class lode-Au mineral systems, including the Eastern Goldfields, such as a pre-existing structural architecture formed during basin development, a mantle-derived Au-enriched source, and a major episode of orogeny.

The Eastern Goldfields—Australia's Fort Knox

The Eastern Goldfields (WA) is Australia's premier Au province, with around 4200 t of Au recovered in a little over a century from mines stretching more than 600 km from Norseman in the south to Wiluna in the north (Figure 8.9a). The jewel in the crown is the Golden Mile at Kalgoorlie, with more than 1600 t of Au mined. The region is also renowned for its Ni, ranging from the world-class deposits at Kambalda and Leinster to more infamous examples, such as Poseidon's Windarra deposit near Laverton (Chapter 9).

The Maduwangka people, whose traditional land encompasses Kalgoorlie in Western Australia, had their lives changed in 1893 when three Irish Au prospectors—Paddy Hannan, Dan Shea and Tom Flanagan, discovered Au near Mount Charlotte. Within days of their discovery of 8 lbs of Au nuggets, more than 700 Au diggers pegged their claims in and around Kalgoorlie. The area was originally known as Hannan's Find, and later changed to Kalgoorlie, meaning 'silky pear bush' in the Maduwangka language.

This Au discovery coincided with the 'Panic of 1893' that prompted bank crashes in Melbourne, and encouraged 40 000–50 000 Victorians to head west, helping to quadruple Western Australia's population during the 1890s. The timing was also favourable for cashed-up British bankers, who provided needed capital in the hope of winning a share of the 'Westralian' bonanza.

For more than a century since then, the fortunes of the region have waxed and waned, largely driven by metal prices, as well as available technology. By 1903, the population of Kalgoorlie was 30 000, many of whom had migrated from the eastern colonies. These 'T'othersiders', as they were known, overwhelmingly voted for federation in 1901, ensuring that Western Australia joined the Commonwealth of Australia. By early 1976, the population had declined to just over 19 000. The combination of high inflation and a stagnant economy threatened the existence of Kalgoorlie; today, the town, twinned with Boulder, is a thriving regional city of more than 33 000 permanent residents, plus a large fly-in/fly-out (FIFO) workforce based in Perth. As the past has shown, however, favourable economics and technological innovation will determine whether this jewel remains bright.

Being located in semi-desert, Kalgoorlie has been, and continues to be, constrained by its water supply. As elsewhere in arid Australia, early prospectors had little water and had to resort to dry blowing techniques to 'pan' for Au. Water was so precious that liquor was reportedly used to clean wounds in the local hospitals. In 1903, the first reliable water supply was delivered via a 557 km-long pipeline from Mundaring in the hills to the east of Perth. This was a major engineering feat in its time (Chapter 7).

Although some of the finds opened up in the late 19th century are still being mined, many new Au deposits have been discovered in the past two decades. Noteworthy examples are the Wallaby and Sunrise Dam, in rock types not traditionally considered prospective for Au. These deposits were

Image by Mining Photo

The Super Pit at Kalgoorlie (WA) is Australia's largest producing gold mine (850 000 oz per year). More than 10 Moz of Au has been produced from the mine, which is now 3.5 km long, 1.5 km wide and 410 m deep.

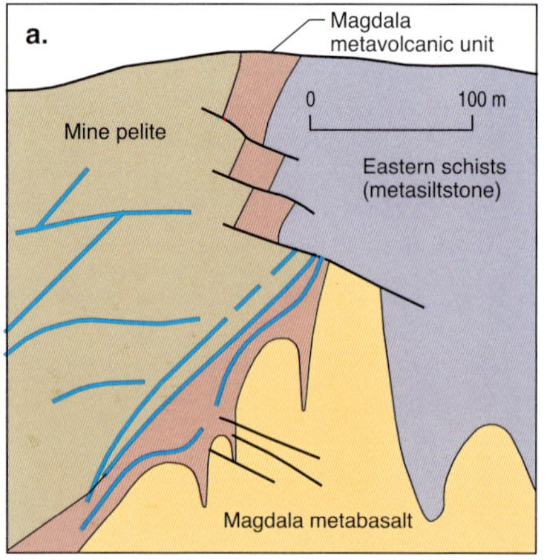

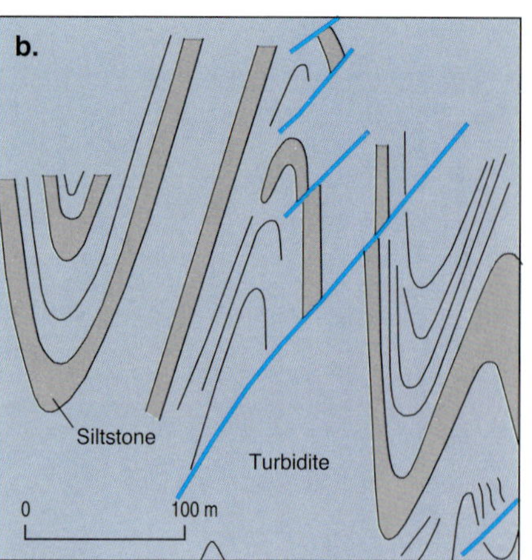

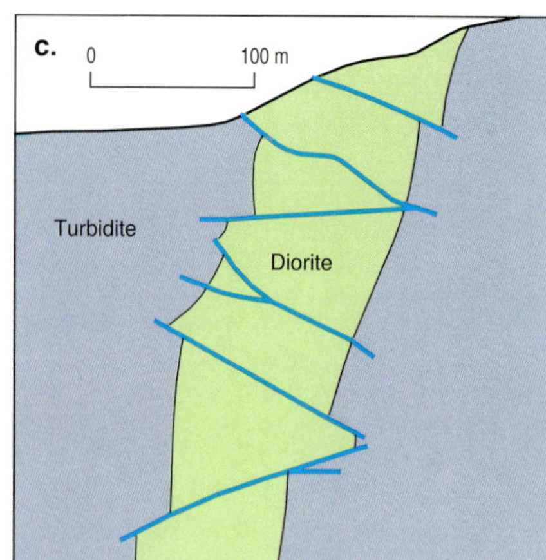

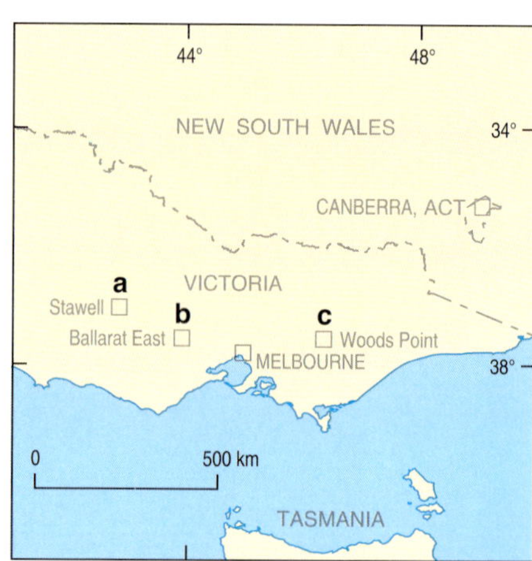

Figure 8.8: Geological cross-sections, showing the geological settings of Au zones (blue) in the Victorian goldfields. (a) Quartz flats and verticals from Scotchmans United, Stawell. (b) 'Leatherjackets' from First Chance and Last Chance, Ballarat East. (c) Floor reefs from Mornington Star, Woods Point. (Source: Phillips et al., 2003)

discovered by sophisticated geological, geophysical and geochemical exploration techniques, many developed or refined by Australia's innovative exploration and mining industry (Box 8.3).

Eastern Goldfields: a favourable convergence in time and space

The important features of the giant Au mineral systems of the Eastern Goldfields (Figure 8.9) include the rapid formation of new crust and basins on the rifted edge of an old cratonic block, followed by the development of deep-crustal faults and the intrusion of magmas from metasomatised mantle sources. A series of tectonic stress switches, giant energy systems that produced major crustal melting, an effective structural focus for fluids and metal, and a suitable physico-chemical complexity all facilitate metal deposition.

Most significant lode-Au provinces develop in an accretionary tectonic setting on the extended margin of an old cratonic block. Major crustal age subdivisions and structures are identified in a neodymium (Nd) isotope model-age map (Figure 8.9b). This map is a proxy for the age of the crust; it shows that the Eastern Goldfields are located on the extended margin of an older continental block. These accretionary margins are associated with long-lived retreat of the subduction zone by the rolling back of the hinge zone, driving extension behind the subduction zone and ultimately crust formation. This extension can be punctuated by phases of contraction (compression), which are driven by hinge advance, slab shallowing or collisions. The terminal phase, and commonly the Au mineralisation phase, is associated with collision, culminating in cratonisation and, hence, preservation.

The Yilgarn Craton records a long geological history, with the earliest elements older than 3000 Ma. Towards the end of the Archean, the craton underwent a rapid crustal growth phase, doubling its size between 2750 Ma and 2630 Ma, particularly along the eastern margin of the proto-craton to form the Eastern Goldfields Superterrane (Chapter 2). Basins formed in the upper crust to more than 10 km thickness. They were not filled with turbidites as in the Victorian goldfields, but with thick accumulations of mostly mafic volcanic rocks (greenstones). The mid- and lower crusts were built with thick granite batholiths, with a combined total thickness exceeding 30–40 km. This is an example of a granite–greenstone terrane, a characteristic rock association of the Archean.

Deep seismic reflection profiles across the Eastern Goldfields reveal important deep faults, some of which transect the entire crust (Figure 8.10). As was the case in the Victorian goldfields, these faults played a key role in greenstone basin formation. Later, they provided pathways for metal-rich fluids to flow from the deep crust and/or mantle during mineralisation. Most of the major Au districts in the region are located in the hangingwall of these faults.

Structural controls may even extend into the mantle, which can be partly mapped by seismic tomography (Chapter 2). Records across the Yilgarn Craton of the variations in seismic wavespeed from distant earthquakes provide a velocity structure for the upper mantle below. A high-velocity layer is imaged at around 120 km beneath the Eastern Goldfields, and this has been interpreted as the delaminated remnants of dense lower crust. The Golden Mile at Kalgoorlie, the home of the Super Pit, is located approximately 120 km above the edge of this high-velocity layer. This edge has been interpreted as a boundary (structure), and it is interesting to note similar 'structures' beneath other major deposits of the Eastern Goldfields (Figure 8.11). If these features are associated with mineralisation, then they illustrate the truly remarkable scale of a mineral system.

Giant mineral systems require vast amounts of energy. The crust hosting the Golden Mile deposit is believed to have developed in an energetic tectonic environment, possibly a backarc position, during a period of vigorous crustal growth. Inferred subduction beneath the Eastern Goldfields would have done several important things: it would have taken oceanic crust deep enough to melt to form voluminous granite; provided an extensional driving force during periods of slab retreat to form basins; hydrated and fertilised the upper mantle with Au and volatiles; and brought hot mantle to the base of the crust, providing a thermal driver for magmatism and fluid flow.

Other signals of the energy in the system are recorded by the komatiites, which require extreme temperatures (>1300°C) to form. The region around Kalgoorlie hosts the greatest volume of these rocks anywhere on Earth. They were erupted in a single geological event (2715–2705 Ma) from an anomalously hot upwelling mantle that was a rich source of Au. These ultramafic rocks, together with other greenstone rocks, were deposited into basins that were controlled by active growth faults during extension. It is no coincidence that the largest Au (and Ni) deposits occur adjacent to these basin-controlling faults.

When an imposed stress exceeds the yield strength of the crust, the crust breaks. Such breaking is manifest as faults, which are the main pathways for fluid flow in a hydrothermal mineral system because intact rock (like granite) is essentially impermeable. Stress also provides a source of energy to drive the fluids (Figure 8.4). Changes in stress, such as direction, mode (extension or contraction) or magnitude, can also be important triggers and drivers in a mineral

Dry blowing in the Eastern Goldfields. Due to the lack of water in the Eastern Goldfields, gold diggers had to develop different ways of prospecting and mining. Dry blowing uses air flow to separate gold from dust and soil.

system. The Eastern Goldfields underwent several periods in which the prevailing stress changed. Many of these changes accompanied the deposition of Au, especially when the stress directions were at an angle to the prevailing structural grain of the rock packages. During extension, fluids tend to be drawn down into the crust; during compression, fluids tend to be squeezed and driven upwards in the crust.

Granite makes up around 65% of the area (Figure 8.9a) and forms the major component of the crust, both at the surface and at depth. Most granite is of the tonalite-trondhjemite-granodiorite series, which forms from the melting of hydrous mafic rocks, including oceanic crust, at high pressure (Chapter 10). Major crustal growth by granite emplacement occurred between 2680 and 2665 Ma, mostly beneath the overlying greenstone basins. The crust was thickened by the emplacement of enormous volumes of felsic magma into the mid-crust. A gravitational instability was thus developed, with dense greenstones on top of weak, hot and less dense granites. The density instability was relieved by extension, possibly also driven by delamination of the lower crust, the remnants of which are imaged in the seismic tomography of the upper mantle (Figure 8.11). Extension of the lithosphere had a number of important consequences for the Au mineral system:

- It created a series of detachment faults that seismic reflection data show as cutting deep into the crust (Figure 8.10), producing fluid pathways connecting the mantle. These pathways allowed the drawdown of basinal fluids into the crust, where they interacted with rocks or mixed with upward-moving magmatic-hydrothermal, metamorphic or mantle-derived fluids.

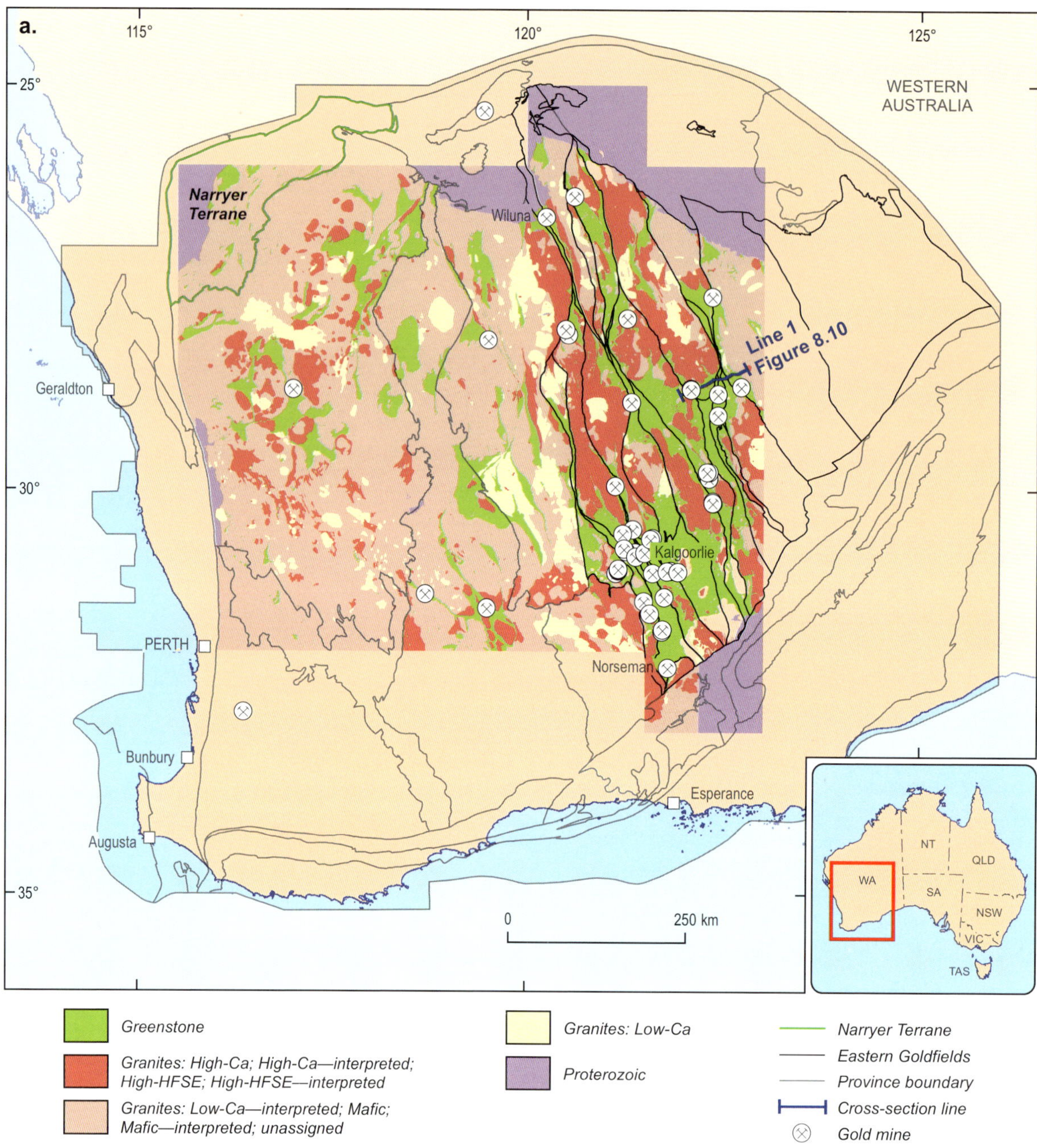

Figure 8.9: (a) Simplified solid geology map of the Yilgarn Craton, showing Au deposits, b) variations in neodymium (Nd) isotope model-ages as determined from 3000–2630 Ma granites (modified from Cassidy & Champion, 2004). The model-ages are best interpreted as the age when the crust separated from the mantle. The Eastern Goldfields—in blue and green colours in (b)—was built onto the extended margin of the old western half of the craton (in red colours). The major changes in crustal age (colours) correspond to major boundaries in the crust and mantle lithosphere. The light grey lines in (a) are the main geological boundaries of the Eastern Goldfields Superterrane.
HFSE = high field strength elements

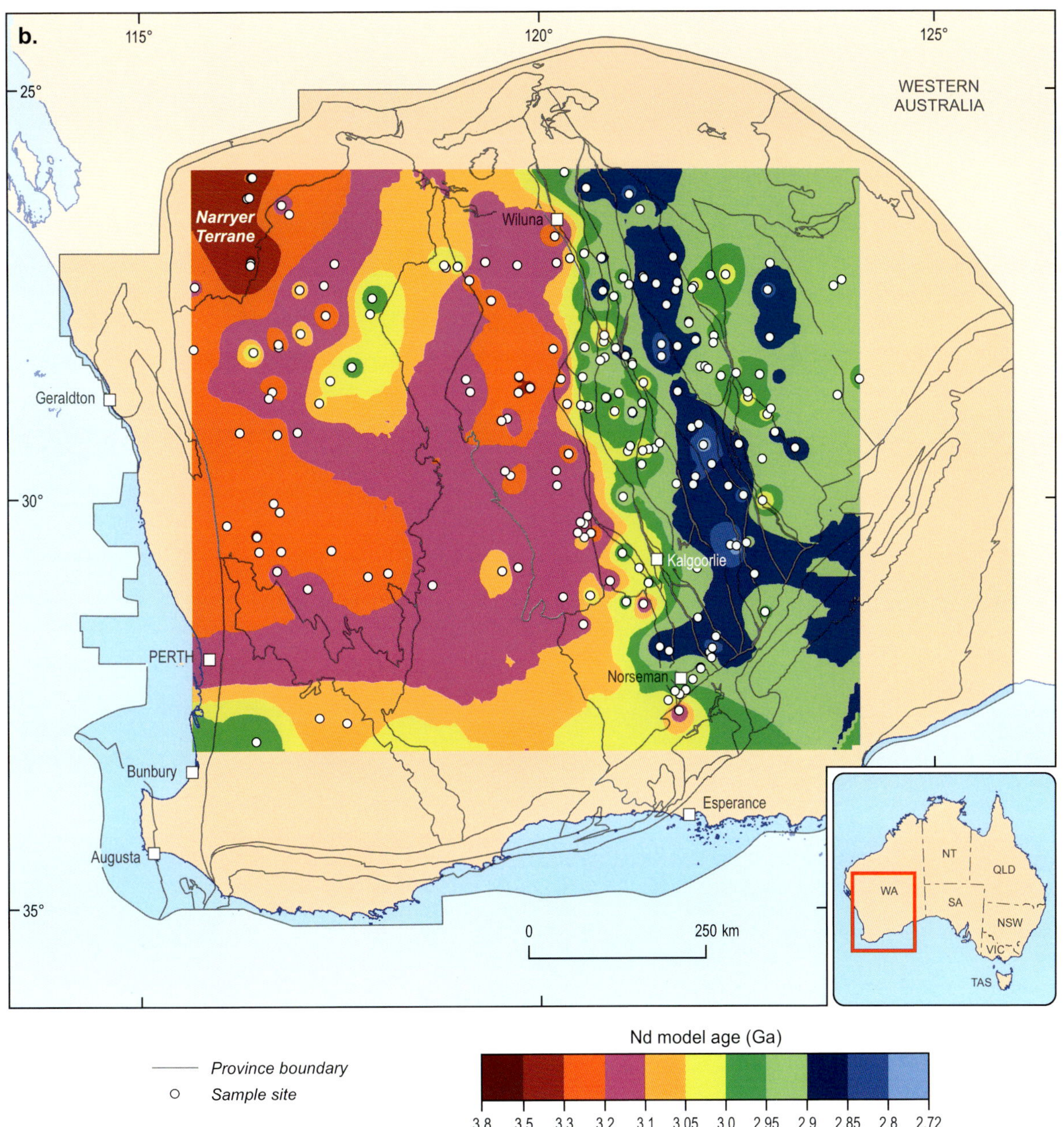

- These faults also localised the emplacement of magmatic rocks, including syenite, lamprophyre and Mg-rich granites that are spatially associated with many Au deposits (Figure 8.12a) and are the products of melting fertilised mantle.
- In the mid- to upper crust, extension resulted in a high geothermal gradient and associated metamorphism of the greenstones, generating fluids.

Extension also 'dropped' the thickest greenstone basins, which have the capacity to host most of the Au, deeper into the crust, thereby preserving them from later erosion.

Domes and regional anticlines are the most favourable structures for hosting large Au deposits, and they occur over the domain to deposit scale in the Eastern Goldfields. In some locations, domes stacked one above the other are linked with the deep-penetrating faults described above (Figures 8.10 and 8.13). This architecture is highly effective in focusing fluids/magmas and energy from deep in the crust or mantle into the upper crust. Computer models show that these fault-bound domes focus fluids upwards into the domal crests, especially where the dome underlies a basin containing a fluid seal (such as a shale).

The final tectonic stages in the evolution of the Eastern Goldfields saw a change to compression, with folding and strike-slip faulting, and accompanying high-temperature metamorphism and the emplacement of a new type of granite (Figure 8.12b). These low-Ca granites were emplaced rapidly (2655–2630 Ma) across the

EASTERN GOLDFIELDS: RESOURCES AND ECONOMIC IMPACT (BOX 8.3)

Much of Australia's recent Au production has come from the Eastern Goldfields of Western Australia (Figure B8.3a). The increase in mining activity in the region from the early 1980s provides a good example of how changes to prices and costs drive production decisions in the mining industry. The price of Au had soared by the end of the 1970s, which was followed by a steep rise in Au production in the 1980s (Figure B8.3b). Higher prices increased the profitability of existing projects, and encouraged investment in further Au exploration, production and processing, and in new technologies at each stage of the Au supply process. Similarly, the reverse has been true. When prices fall in Australian dollar terms, production tends to decrease and exploration activity slows or stops.

The adoption of new technologies also contributed to the major increase in production in the 1980s (Figure B8.3b). Carbon-in-pulp and carbon-in-leach processes that enable Au to be recovered from low-grade (as little as 1–2 g/t Au) ore deposits were introduced over the period 1983–1990. These production technologies increased the availability of 'economically feasible' Au resources. They also improved the likelihood of successful exploration activity by increasing the chance that extraction of an identified resource would be considered economically viable. New exploration technologies, such as bulk-cyanide-leach Au analyses, were also developed and adopted in the 1980s.

Mining production decisions tend to involve long time lags, as exploration activity and capital investment decisions are made several years in advance of production commencing. The Australian Au supply can be modelled with relationships between production, capital investment and exploration expenditure. The price of Au accounts for around 40% of the variability in exploration expenditure, particularly for less advanced or 'greenfields' projects. Changes in exploration activity (expenditure) are the major driver for production changes. Gold exploration expenditure peaked in 1987–88 as a result of the higher prices and adoption of new technologies of the 1970s and 1980s (Figure B8.3b).

Au exploration expenditure in Western Australia rose by 26% to $452.5 M in 2010–11—almost 70% of the national expenditure. The Eastern Goldfields is the dominant province, with exciting new opportunities on the margins. The Tropicana Au deposit (Figure 8.1; Appendix 8.1.1), for example, is hosted in high-grade gneisses at the boundary of the Yilgarn Craton and the Albany–Fraser Orogen. This is a significant new Au play in a remote part of Western Australia. The company expects 300–400 people to be employed at the mine, producing 200 000–450 000 oz per year over 10 years.

Discovery of the Tropicana deposit in 2005, one of very few recent greenfields discoveries in Australia, has led to an exploration rush to the Yilgarn–Albany-Fraser boundary. Virtually all of this region has been taken up for exploration, and significant additional discoveries have been made, which will lead to further economic development of this remote region.

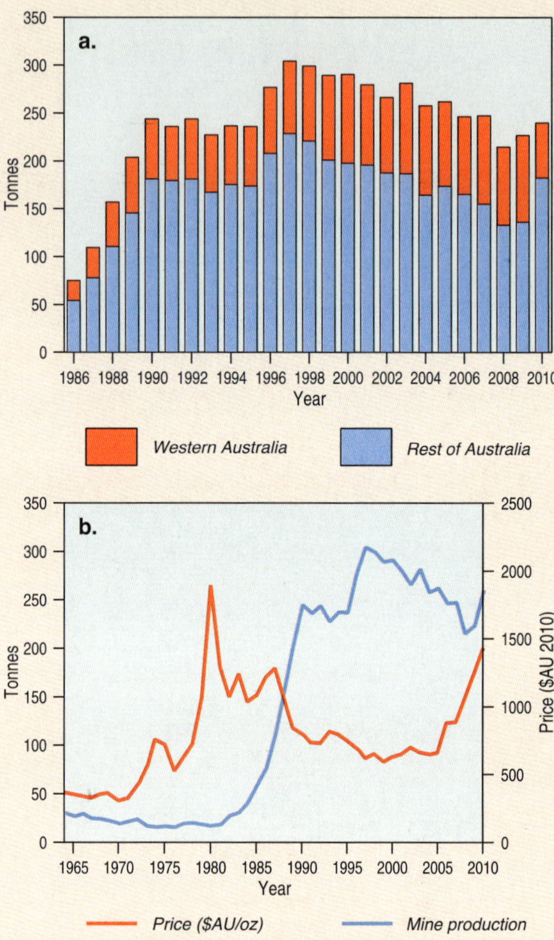

Figure B8.3: (a) Gold production in Australia (1986–2010). Note how Western Australia has consistently produced around 75% of Australia's Au. (Source: Australian Bureau of Agricultural and Resource Economics). (b) Gold production and prices between 1964 and 2009, showing how production lags the price. Note the steady decline in production until 1980, after which production rapidly increased as a consequence of the price increase, uptake of new technology (carbon-in-pulp) and the availability of investment capital. (Sources: Australian Bureau of Agricultural and Resource Economics & Australian Bureau of Statistics)

entire Yilgarn Craton (Figure 8.9a). They were formed by the melting of older granites in the mid- to lower crust (Chapter 10), and were emplaced into the cores of the upper-crustal domes, providing heat that drove fluid flow and, possibly, fluids themselves, during the final and most significant Au event (2647–2630 Ma). The timing of their emplacement coincided with the final stage of major heat loss from the crust, which stabilised the lithosphere and preserved the Yilgarn Craton as an entity for the next >2.5 Ga.

The mineral-system concept predicts that broad outflow zones should emanate from the depositional site, altering a large region around the deposit (Figure 8.4). The surface expression of alteration associated with Au mineralisation can be large, locally with dimensions of tens of square kilometres. The general pattern of alteration is asymmetric in the Eastern Goldfields (Figure 8.14). Its signal can be mapped as variations in the composition of white mica, using hand-held instruments such as PIMA (Portable Infrared Mineral Analyser) or air- or satellite-borne multispectral scanners. Most of the Au was deposited along the gradient in white mica composition between phengite and muscovite. These gradients can reflect changes in redox, temperature or pH of the alterating fluid(s). Stable-isotope studies of C, S and O, together with pathfinder elements such as bismuth (Bi), molybdenum (Mo), W, Li, arsenic (As) and antimony (Sb), as well as the distribution of alteration minerals themselves, suggest that Au mineralising fluids came from a number of sources: (1) reduced basinal and metamorphic fluids; (2) oxidised felsic magmas, especially those from a metasomatised mantle source; and (3) potentially reduced mantle fluids.

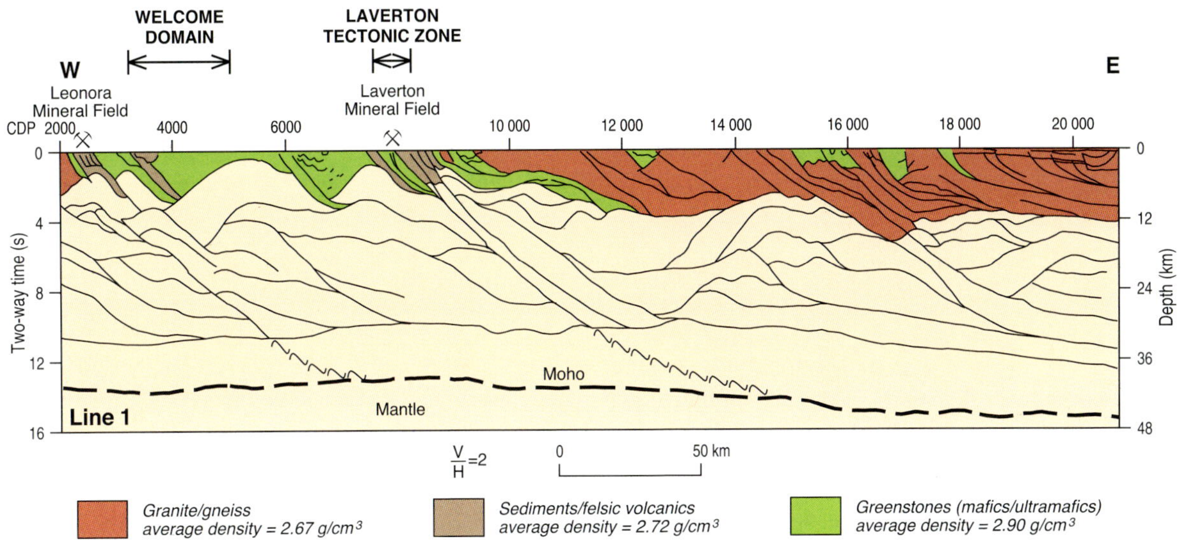

Figure 8.10: Seismic section across the Eastern Goldfields, showing the structure of the crust. Note the association of mining districts with major structures that transect the crust and the common antiformal geometry to many interpreted reflections, many of which are stacked one upon the other. (Source: Goleby et al., 2004)

The depth expression of alteration associated with the Au mineralisation is also large. The magnetotelluric technique (MT) data measure the electrical properties at depth in Earth and can be used to infer the presence of deep structures, fluids and alteration zones. Beneath Kalgoorlie, there is a large zone of anomalous conductivity in the upper crust and mantle, which may be a result of alteration on a lithospheric scale. Large-scale alteration has also been inferred from seismic reflection data, where 'bland zones' of seismic reflectivity near the base of the crust may mark an alteration zone.

Gold deposition (Figure 8.12c) can occur by a variety of mechanisms, including fluid mixing; fluid-rock reaction, especially in iron-rich lithologies; phase separation; and vapour condensation. The effectiveness of these processes is one of the major determinants of deposit size and grade.

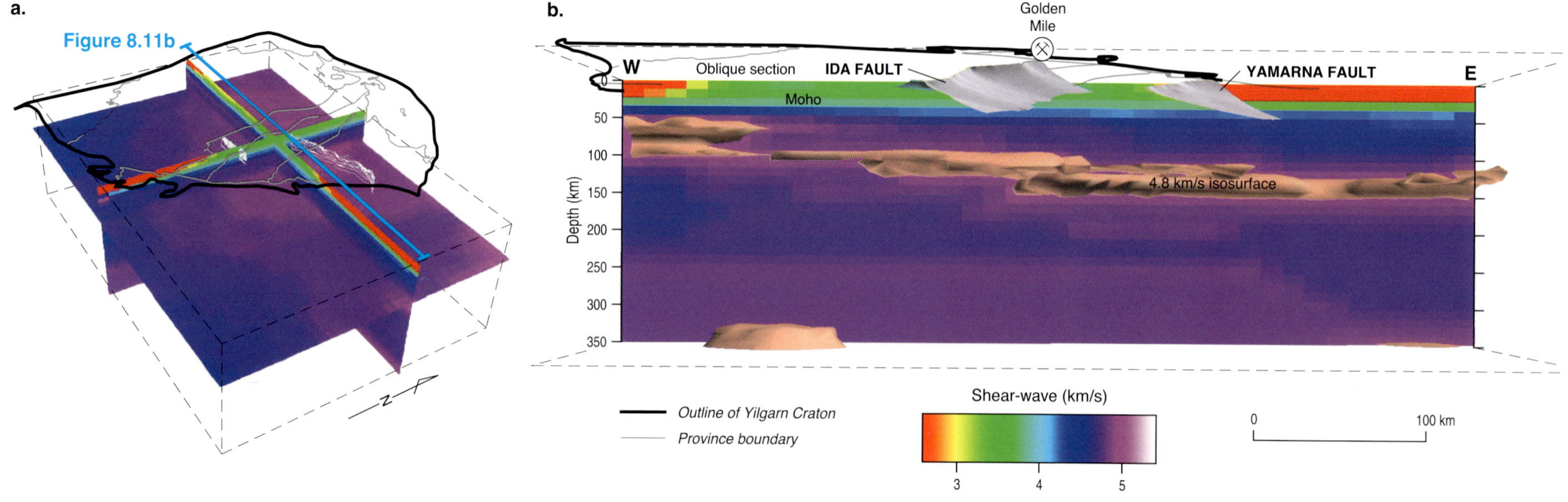

Figure 8.11: Seismic tomographic image of the shear-wave velocity structure of the Yilgarn Craton. (a) Oblique view, from above towards the northwest, of the shear-wave velocity tomographic volume of the Yilgarn Craton. Cross-sections are east–west and north–south to a depth of 350 km, and the horizontal slice is at around 130 km. Outline of the craton and the interpreted seismic reflection sections are also shown. V:H = 1. (b) Detailed view looking north of the >4.8 km/s shear-wave layer (pink isosurface) in the upper mantle lithosphere. The bright purple colours shown in the horizontal slice (in a) depict a section through the data, which this pink isosurface was constructed from. This isosurface marks a high-velocity layer around 20 km thick, which may be the dense remnant of lower delaminated crust. The isosurface has steps or boundaries, with an example beneath the Golden Mile. The Moho and major faults through the crust are labelled. (Source: Korsch & Blewett, 2010)

Key factors in forming the Eastern Goldfields

Why are the Eastern Goldfields so rich? Like the Victorian goldfields, the Eastern Goldfields formed during a major period of continent assembly—in this case, possibly associated with assembly of the Kenorland supercraton (Chapter 2). The Abitibi Subprovince in Canada, the other major Archean lode-Au province in the world, formed by similar processes over a similar time period, and was possibly part of Kenorland. These two provinces formed rapidly, with most crustal growth occurring within a period of 100 Myr or less. This contrasts with the older granite-greenstones of the Pilbara Craton, which also formed during the Archean but over a billion years (Chapter 2). The Pilbara Craton has an Au endowment orders of magnitude less than the Eastern Goldfields, highlighting the importance of rapid crustal growth in Au metallogeny. This rapid growth promoted high Au endowment in three ways: (1) the initial development of deep crustal-penetrating faults, followed by their reactivation during stress switches; (2) the emplacement of Au-enriched source rock derived from the mantle; and (3) a vast amount of concentrated energy that drove fluid movement together with stress switches. Although many of these features are also apparent in the Victorian goldfields, the data from the Eastern Goldfields emphasise the importance of deep structures linking domes with enriched mantle, stress switches and high heat flow for Au endowment.

Proterozoic Zn–Pb deposits: Broken Hill and Mt Isa

By the mid-1860s, pastoralists began dividing up the Barrier Ranges in southwestern New South Wales into a series of sheep stations. As in Victoria, many shepherds also worked as prospectors, initially for Au, and later for Cu and Ag. Discovery of Ag gave rise to the town of Silverton in 1882

Figure 8.12: Photographs of rocks from the Yilgarn Craton. (a) Syenite dykes (red rocks) cutting basalt at the Jupiter Au deposit near Laverton. Height of pit is around 50 m. (b) Low-Ca granite pavements from southeast of Leonora. Rainwater remains in these natural rock or 'gnamma' holes, providing a source of drinking water in this semi-arid desert region (Chapter 7). (c) Quartz vein with visible Au from underground in the Victory–Defiance mine at St Ives.

(Figure 8.15a), but remoteness inhibited its development. In September 1883, Charles Rasp climbed to the top of a long, narrow ridge locally named 'the broken hill' and sampled ironstone at the top, hoping to find tin. He found, however, Ag and Pb, although in quantities that could not be profitably mined, shipped and processed at the time. Rasp formed a syndicate with six others from the Mount Gipps sheep station to take out a mining lease and sink a shaft. Public floating of the syndicate in 1885 created the Broken Hill Proprietary Company Limited, which eventually became BHP Billiton Ltd, currently Australia's largest and the world's sixth largest company.

Sinking of the shaft did not initially produce the rich ore expected by the syndicate. Although Ag and Pb were present in the rocks, they were not sufficiently rich. Rather, ore, rich in silver chloride, was being tipped onto the mullock heap. Only the vigilance of one syndicate member discovered this blunder, and Broken Hill was on its way to becoming recognised as the largest accumulation of Zn and Pb on Earth.

The arrival of the railway from Port Pirie (SA) in 1888 vastly helped the economics of the deposit, because 'ore' from the early shafts suddenly could be mined at a profit, and the future of the Broken Hill deposit seemed assured. However, in the early 1890s, Broken Hill was hit by a triad of calamities: the price of Ag dropped sharply during the 'Panic of 1893'; the rich oxide ores began to be exhausted, leaving lower grade and metallurgically more difficult sulfide ores; and the mines experienced their first major strike, the Eighteen Weeks strike of 1892. The first two problems were partly solved by the third problem. After the strike was broken, wages were decreased by 10%, and the working week increased from 46 hours to 48 hours. A solution to the problem of sulfide ores was froth flotation, developed in Melbourne in 1903 by Charles Potter, G D Delprat and Auguste de Bavay. This technology produced a much cleaner separation of ore metals and allowed processing of 5.7 Mt of low-grade 'waste', valued at the time at £30 M. This process was a revolution in the mining industry as it allowed exploitation

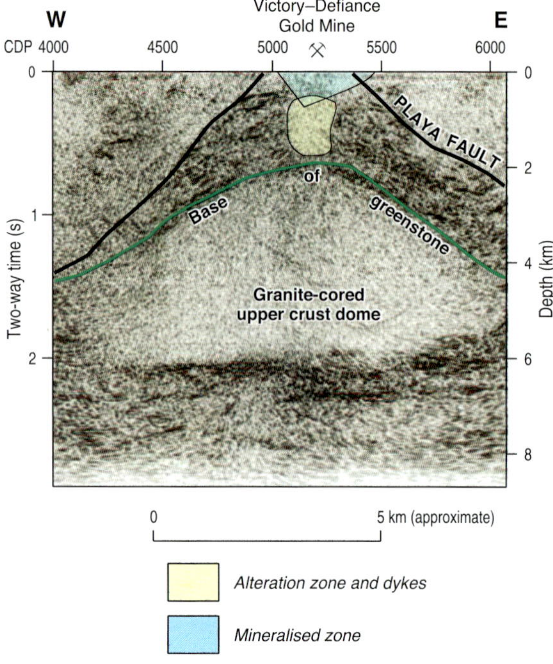

Figure 8.13: Domes are a common structure of Eastern Goldfields Au deposits. This seismic image (courtesy of Gold Fields St Ives Ltd) shows a granite-cored dome beneath the Victory–Defiance Au mine at St Ives. The granite is seismically transparent and is overlain by the reflections from the greenstone stratigraphy (base outlined). The pale washed-out zone above the apex of the dome and beneath the deposit is interpreted as an alteration zone from mineralising fluids and abundant granitic/porphyry dykes, which scatter the seismic signal.

of complex sulfide ores in Australia and overseas (Chapter 1). Two of the three inventors of froth flotation were not chemists, but brewers!

Although many technical problems at Broken Hill were solved early in the 20th century, the history of Broken Hill in the following half-century was turbulent, tied not only to fluctuations in Pb and Zn prices, but also to a rising union militancy fuelled by health and safety concerns, the prosperity of some (not all) mines, and the conservatism of most mining companies. Up to the 1920s, Broken Hill had one of the highest levels of industrial unrest in Australia. This period also saw a consolidation of the field, with the number of companies decreasing through closure and merger from 11 in 1920 to 3 in 1940, when BHP left the field. Now only one company is actively mining the Broken Hill lode, and the city of 19 500 people is diversifying its economy with tourism and the arts.

Mt Isa in northwestern Queensland has a very different history from that of Broken Hill. The Mt Isa Province, although hosting a number of major Zn–Pb deposits, began life as a Cu province (Appendix 8.1.1). The first mineral discovery of malachite and native Cu was made near present-day Cloncurry by prospector Ernest Henry in 1867; this became the Great Australia mine. In the early 1880s, Henry and other prospectors made Cu discoveries through much of the Mt Isa Province. The first Zn discovery was the still-undeveloped Dugald River deposit in 1881. However, it was not until 1923 that John Miles discovered Pb- and Ag-rich gossan at what would become the Mt Isa mine.

At first, the Mt Isa deposit was exploited in a chessboard of small leases worked by individual miners, but this began to change with the incorporation of Mt Isa Mines Ltd in 1924. By November 1925, Mt Isa Mines was the only large player, although some small mines continued to operate for 20 years. At the time, Mt Isa was the most distant mine from any port in Australia. This tyranny of distance and the relatively low grade of the Mt Isa ores discouraged investment, and it was not until 1927 that exploration began to define a large sulfide ore body. Metallurgical test work demonstrated that concentrates could be extracted from the ores using a suite of processes, including flotation, which was first developed for Broken Hill. However, the Great Depression and capital demands for development of the Mt Isa mine intervened. By 1930, funds could not be raised, and it looked as though the mine would close before the first ores had been smelted.

The saviour was American Smelting and Refining Company (ASARCO), which made significant investments and became a major shareholder in Mt Isa Mines. In 1931, the first Pb was smelted. Unlike Cu, which is present in numerous small deposits, prospects and showings in the Mt Isa Province Zn–Pb deposits are uncommon but can be extremely large. Including Broken Hill, 5 of the world's 10 largest Zn–Pb deposits are present in the Australian Proterozoic zinc belt, but discovery of these deposits has been spasmodic. After Mt Isa in 1923, the next deposit in this belt, Hilton, was discovered in 1948, followed by Here's Your Chance (HYC, also known as McArthur River) in 1955 and Century in 1990.

These and the smaller Lady Loretta (1969) and Broken Hill-type Cannington (1990) deposits were discovered through systematic exploration by major companies. The discoveries became the impetus for developing towns and settlements in outback Australia and spawned some of Australia's major mining houses.

The Australian Proterozoic zinc belt: riches from the breakup of Nuna

The central third of Australia is underpinned by rocks that were mostly deposited during the Proterozoic (Chapter 2). The eastern margin of this Proterozoic block, which extends from the Curnamona Province in the south to the Georgetown Inlier in the north (Figure 2.8), formed as a series of basins during the breakup of Nuna, Earth's first supercontinent (Box 2.5). Nuna amalgamated between 2100 Ma and 1750 Ma, and began to breakup at *ca* 1700 Ma. With breakup came the formation of vast sedimentary basins, which continued to develop until 1600–1590 Ma, when contractional deformation terminated basin formation in most areas. Although the Mt Isa and Curnamona provinces are currently separated by younger rocks, including the Neoproterozoic and Phanerozoic Officer and Amadeus basins (Figure 2.9), the provinces were likely contiguous between 1700 Ma and 1590 Ma (or even younger). This correlation is based upon similarities in their geological history and metallogenesis, and on continuities in geophysical datasets when later basins are removed.

The first hints of Nuna's breakup might have begun as early as *ca* 1780 Ma, when an extensional basin dominated by continental basalt and fluvial to shallow-marine sedimentary rocks developed on pre-existing basement in the Mt Isa Province. However, extension leading to Nuna's breakup did not begin in earnest until *ca* 1690 Ma, when turbiditic siliciclastic rocks, accompanied by tholeiitic basalt, were deposited in the Curnamona Province (Willyama Supergroup) and along the eastern margin of the Mt Isa Province (Soldiers Gap Group) (Figures 8.15, 8.16 and 8.17). Sedimentation continued with a mixture of shallow- to moderate-water depth siliciclastic and carbonate rocks, which were deposited as thick sequences between *ca* 1660 Ma and *ca* 1590 Ma.

The pre-1660 Ma basins were most likely the product of the separation of Proterozoic Australia from Laurentia to the (present-day) east as Nuna

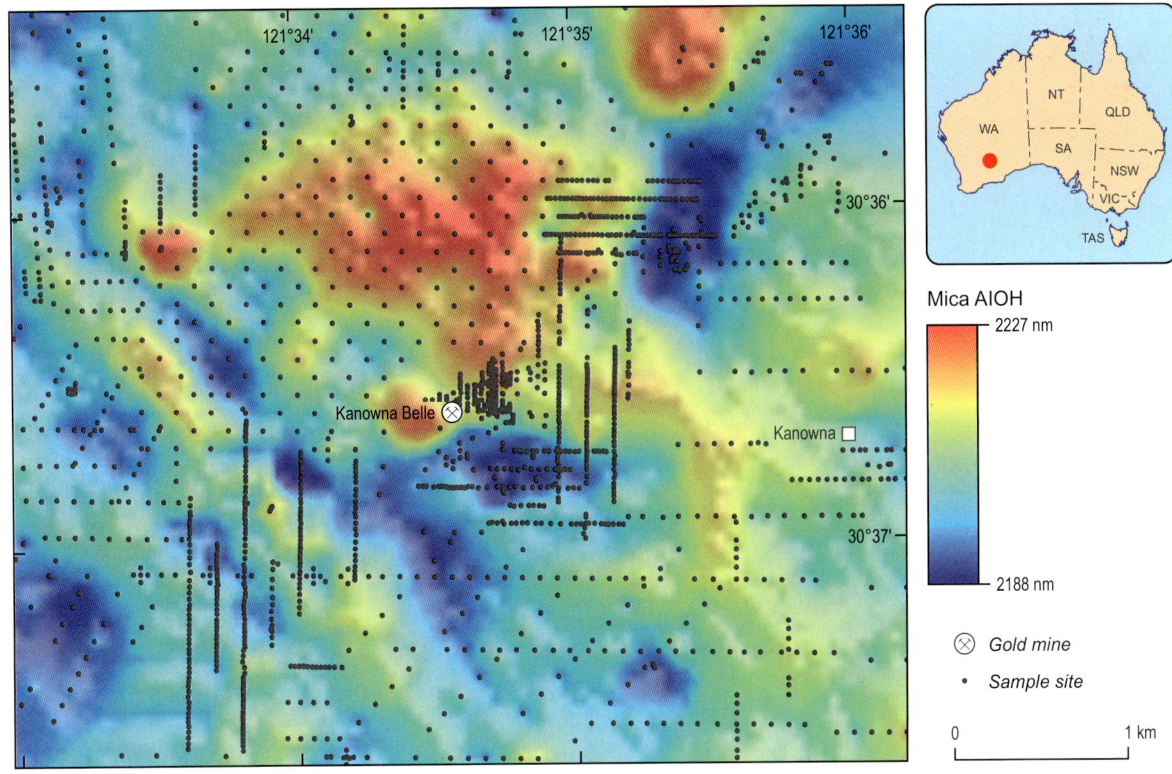

Figure 8.14: Variations in the composition of white mica from the Kanowna Belle area northeast of Kalgoorlie, Western Australia, as determined from PIMA (portable infrared mineral analyser) data. The map shows variations in the wavelength of the Al-OH absorption feature, which is dependent on the phengite content of white mica. In the image, red represents phengitic mica and blue represents muscovitic mica. The Kanowna Belle deposit produces around 250 000 oz of Au per year and lies on a major gradient in white mica composition. Gradients in white mica composition can relate to changes in redox, temperature or pH of the alteration fluid. (Source: Predictive Mineral Discovery Centre, 2008)

© Getty Images [C Claver]

Browns Shaft, Line of Lode mine, Broken Hill, New South Wales.

broke up (Figure 2.30). Extension associated with this initial rifting produced deepwater basins that were extensively intruded by high-level mafic sills (Figure 8.17a). These rift-related mafic rocks provide a number of ingredients to this mineral system. The intrusion of mafic sills into a sedimentary basin alters the temperature gradients, facilitating the convection of seawater through the basin. The mafic rocks are Fe rich, so, in combination with the high heat flow, the Fe^{2+}-rich rocks converted the convecting seawater into a relatively high-temperature (*ca* 250°C), reduced, H_2S-rich fluid (Figure 8.17a). A fluid with these chemical characteristics can carry significant Zn (and Pb), but little Cu (Appendix 8.3.1).

Convection of hot fluid through the basin was an efficient way of extracting the Zn and Pb from the sediments in the basin and concentrating the metals in the ore fluid. This process also altered the rocks, producing impermeable layers and zones (seals) that focused the fluids (see *Did you know?* 8.2). The hot fluids became less dense and rose up along active fault systems that breached the seals, towards the seafloor. These metal-bearing hot fluids cooled rapidly when they mixed with cold seawater at or just beneath the seafloor (Figure 8.17a). This large change in the physical conditions of the ore fluid resulted in deposition of Zn and Pb as a series of semi-massive to massive sulfide lenses (e.g. Broken Hill, Cannington and Pegmont).

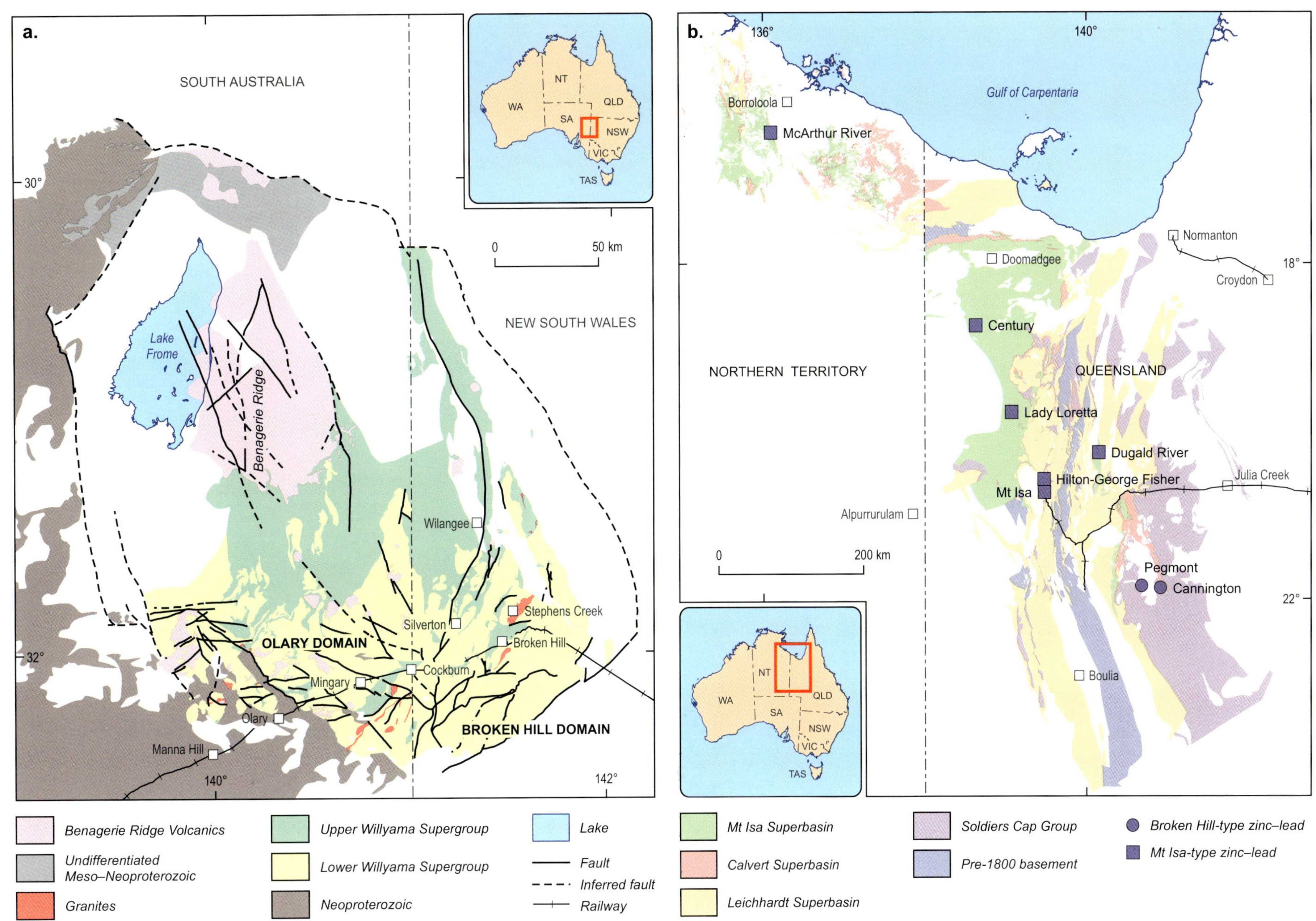

Figure 8.15: The Proterozoic Australian zinc belt, showing geology of (a) the Curnamona Province, and (b) Mt Isa Province. The locations of deposits discussed in the text are shown.

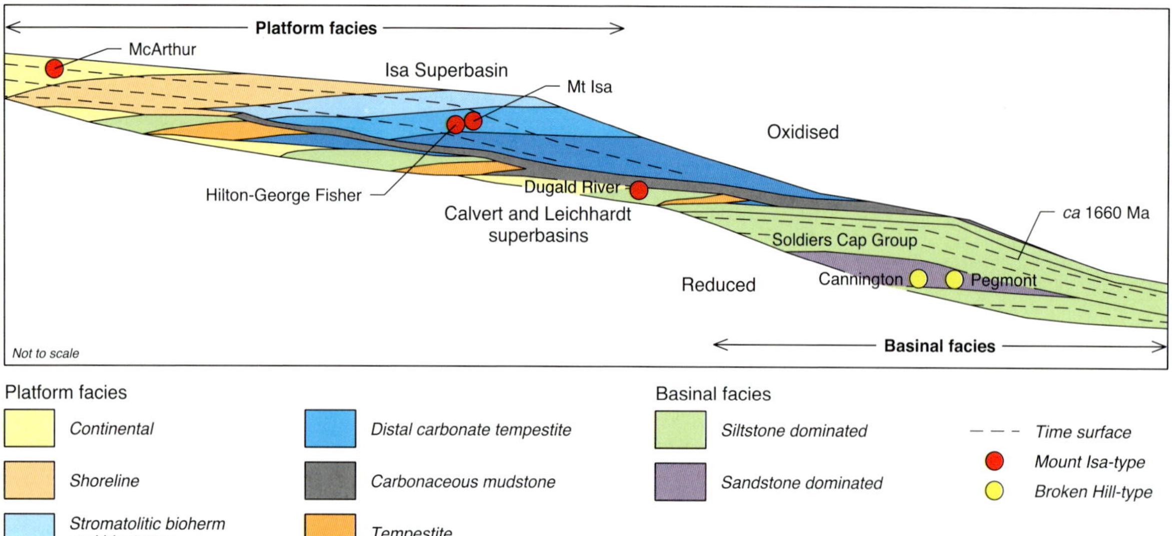

Figure 8.16: Schematic stratigraphic relationship between the stratigraphy in the Mt Isa Province and the locations of major Zn–Pb–Ag deposits. Broken Hill-type deposits were formed during initial extension and formation of a passive margin, whereas the Mt Isa-type deposits formed on the platform developed on the resulting passive margin. The dashed black lines indicate time surfaces; colours and patterns indicate sedimentary facies assemblages. (Source: adapted from Southgate et al., in review)

As the mineralisation at Broken Hill and in the east of Mt Isa have been affected by subsequent medium- to high-grade metamorphism, the grain size (Figure 8.18a) and, possibly, the tenor (value) of the ores have been upgraded, enhancing the economics of mining and metallurgy of the Broken Hill and Cannington mines.

As breakup of Nuna continued, the extensional basin widened, and there was a concomitant decrease in heat flow and development of the overlying passive margin or sag basin beginning at *ca* 1660 Ma. Unlike the older underlying basins, the younger basins did not develop coeval mafic volcanic and intrusive rocks. They were filled with siliciclastic and carbonate rocks (Figures 8.16 and 8.17b). Without the heat engine of the earlier mafic intrusions, fortunately for us, there was another way to drive seawater deep into a basin. Cauliflower chert (Figure 2.29) and halite pseudomorphs in rocks of the clastic-carbonate sequences are indicators of evaporites and highly saline conditions. Seawater interacting with these evaporite-bearing rocks would have generated basin fluids that are highly saline and dense. Highly saline fluids are a very effective solvent for dissolving metal, particularly when oxidised, from the rocks through which it passes, even at low temperature. The Mesoproterozoic oceans and atmosphere were now oxygenated (Chapter 2); these saline oxidised fluids became dense and flowed downwards, increasing in temperature to around 200°C. They stripped Zn and Pb from the pre-existing volcanic rocks deep within the basin, and produced regional-scale semi-conformable alteration zones dominated by hematite-bearing assemblages. These alteration zones, with a scale of tens of kilometres, are much larger than the deposit (Figure 8.17), making this element of the mineral system a bigger target for exploration (Figure 8.4). The metal-bearing fluids became hotter due to the geothermal gradient deep in the basin, making them buoyant and prone to rising up tectonically active faults that focused fluids into local sub-basins higher in the crust (Figure 8.17c).

Distant forces—for example, at plate boundaries—can influence the internal dynamics of basins. We see this today with the Flinders Ranges uplift in South Australia being driven from the plate boundary in New Zealand (Chapter 2). Apparent polar-wander paths are a fossil record of past plate movements. Sharp bends in the apparent polar-wander path are generally interpreted as evidence for collisions and plate reorganisation. The apparent polar-wander path for northern Australia has several bends that coincide with the periods of known mineralisation (Figure 8.19). Therefore, fluid flow was likely initiated by these far-field tectonic events (Figure 8.17b). The *ca* 1640 Ma McArthur River deposit was likely triggered by the Liebig Orogeny, which is found 800 km (present-day) to the south and occurred when the Warumpi Province was accreted onto the North Australian Element (Chapter 2). As these basins were distant from plate margins, the evidence for these deformational events may not be obvious within the deposits themselves. This is another example of the large scale in which a mineral system operates (Figure 8.4).

Although distal events triggered hydrothermal fluid flow, localisation of ores was controlled by local factors such as faults and basin facies. All major Mt Isa-type deposits in the Australian zinc belt are associated with faults that were active at the time of mineralisation. In many cases, they were initiated as normal faults, some of which may have been inherited from earlier basin systems, as was the case in the Victorian goldfields and Eastern Goldfields. At the McArthur River deposit, however, ore deposition was possibly localised in a pull-apart basin associated with the Emu strike-slip fault system (Figure 8.17b).

Once the metal-rich fluid flowed into the sub-basin, it migrated either to the seafloor or laterally below the seafloor (Figure 8.17b), where changes in temperature, pressure and perhaps chemistry drove the metal-rich fluids to deposit their metals. The Zn and Pb were deposited as the metal-rich fluids mixed with an H_2S-bearing reduced fluid. This reduced fluid could have been sourced from anoxic seawater or a reduced diagenetic fluid from the basin. H_2S can also be produced by the

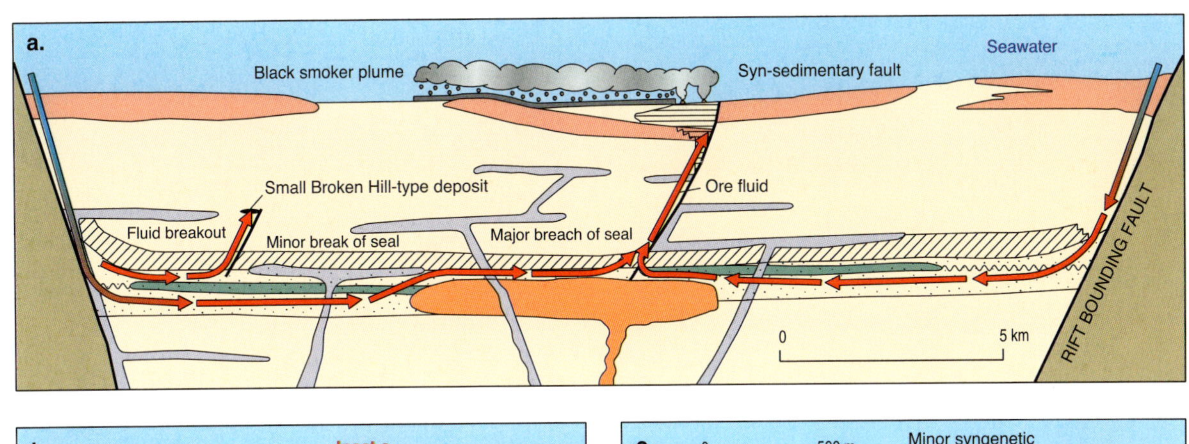

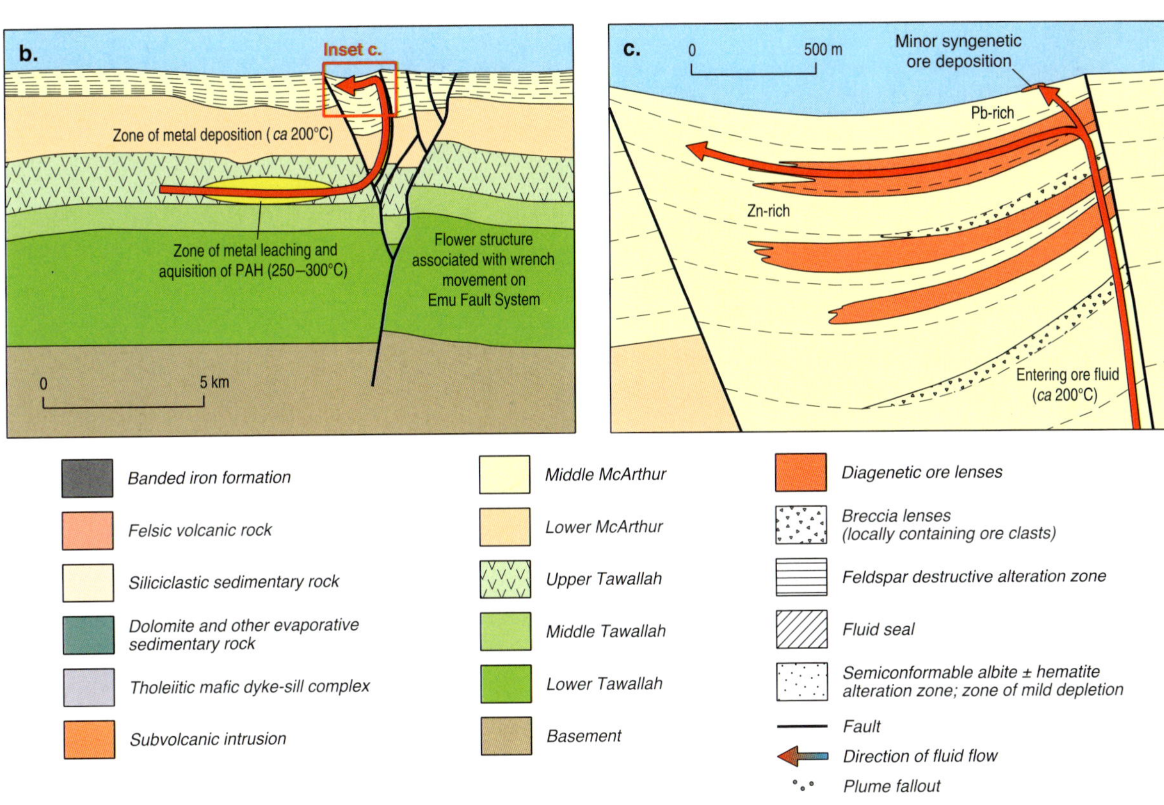

Figure 8.17: Schematic models for (a) Broken Hill-type mineral system, (b and c) Mt Isa-type mineral system from McArthur River. In the Broken Hill-type mineral system, extensional faults most likely controlled seawater drawdown and the upflow of the metal-bearing fluid. Seawater evolved into ore fluid along stratiform aquifers, resulting in the development of extensive stratiform albitic alteration zones that were oxidised and leached of Zn and Pb. Metal deposition occurs as the fluids are quenched when mixed with seawater at or just below the seafloor. In the Mt Isa-type mineral system, dense evaporative brines were either drawn or deep basinal brines were released by distal tectonic drivers to move upwards along basin-controlling faults. Base metals are leached during regional fluid-rock interaction. Base metals are deposited either at the seafloor upon interaction with H2S-bearing seawater or at depth in the sedimentary pile, where H2S is provided either by mixing with H2S-bearing fluid or by organic reduction of sulfate in the ore-forming fluid.
BHT = Broken Hill-type; PAH = polyaromatic hydrocarbon

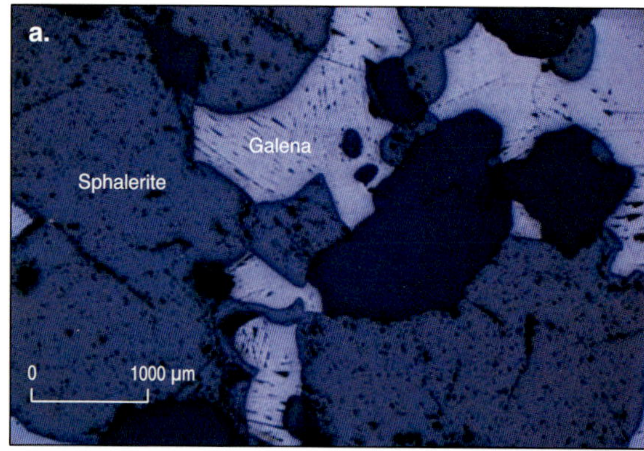

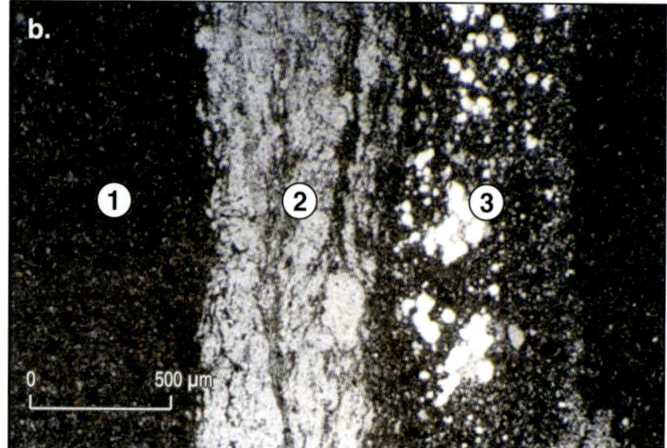

Images courtesy of Ross Large, University of Tasmania

Figure 8.18: Photographs showing contrasting grain size of ore. (a) Coarse-grained metamorphosed ore from Broken Hill. (b) Microscopic textures of very fine-grained, unmetamorphosed ores from the McArthur River deposit. In (b), the numbers indicate differing compositional bonds that characterise the ore.

reduction of sulfate by microbes, which derive their energy from this process and, interestingly, generate oxygen as a by-product. Sulfide- and organic carbon-rich dolomitic siltstones favour these reduction processes, and typically host Mt Isa-type Zn–Pb deposits. These depositional horizons are also conductive to electricity and therefore amenable to mapping using geophysical techniques. They also correspond to the maximum flooding surfaces when local sea-level was its highest, making them amenable to mapping using sequence-stratigraphic techniques (Figure 8.16).

The very fine-grained ores formed by these processes can present challenges to ore processing. At the Mt Isa and nearby Hilton–George Fisher deposits, recrystallisation associated with greenschist facies metamorphism has coarsened the ores sufficiently to facilitate metallurgical processing. In contrast, the fine-grained McArthur River deposit (Figure 8.18b) has not been metamorphosed and so requires more complex and costly processing; it was not mined until 40 years after its discovery in 1955, when metallurgical advances allowed exploitation.

Key factors in forming the Australian Proterozoic zinc belt

Like other major mineral systems in Australia, Zn–Pb systems in the Proterozoic zinc belt are responses to changes in the geodynamic setting. As we explained above, these deposits are temporally and genetically associated with the breakup of Nuna. Although the associated mineral systems operated in an overall extensional environment, the exact timing seems to have been controlled by adjustments in plate motion, which changed local stress fields, triggering flow of seawater or basinal brines that evolved into ore fluids at depth within basins. In some cases, structures controlling fluid flow appear to be inherited from pre-existing basins. Whereas Broken Hill-type deposits are associated with thermal anomalies related to rifting and may have mantle input for some metals, isotopic and fluid inclusion data suggest a crustal source of metals and a normal geothermal gradient for Mt Isa-type deposits. This differentiates Mt Isa-type mineral systems from other giant Australian mineral systems, in that mantle involvement was not present, possibly because most metals involved in these systems, particularly Pb, are enriched in the crust relative to the mantle and bulk Earth (Appendix 8.3.1).

Olympic Dam IOCG Province: new era of discovery

The discovery of the Olympic Dam deposit in northern South Australia is legendary in the history of Australian mineral exploration. This story begins in 1861 in the Moonta area (Figure 8.20), when Irish shepherd Patrick Ryan discovered Cu minerals in a rabbit burrow. The Moonta mines subsequently became major contributors to the South Australian Cu industry for the rest of the 19th century and into the 20th century. However, at the end of World War II, production ceased for more than two decades, only restarting in 1969 with the reopening of mines at Burra, Kanmantoo and Mt Gunson (Figure 8.20). The new interest also spurred a new round of exploration, but in a manner that differed substantially from the early discoveries. This new exploration was based on scientific exploration models funded by major mining companies.

As an example of this new way of doing exploration, Western Mining Corporation Ltd (WMC) undertook an Australia-wide search for Proterozoic Cu based on the concept that altered basalt acted as a source of Cu in sediment-hosted Cu deposits. This search quickly focused on South Australia, where altered basalts and a number of Proterozoic Cu deposits were known, including those at Moonta. Based on a mineral system model developed in 1972, a multidisciplinary team was set up by WMC to search for Cu. The model envisaged Cu being leached by oxidised fluids from basalts located in the lower parts of a basin and being transported upwards along faults, where Cu deposition would occur through chemical reduction of the fluid on encountering reduced parts of the sedimentary package. A critical step in the discovery of Olympic Dam was the release in 1974 by the Bureau of Mineral Resources of aeromagnetic and gravity data over the Stuart Shelf of northern South Australia. The WMC team recognised the potential of several coincident magnetic and gravity anomalies for sediment-hosted Cu deposits of the type predicted in their model. After a lineament analysis, five anomalies were highlighted, and a drilling program commenced in 1975 near Roxby Downs Station. This was no ordinary program, however, because it was predicted from the geophysical modelling that the target depths would be very deep, to hundreds of metres, and in a remote, true 'greenfields' region. Drillhole RD1, sited near the Olympic stock water

Photo courtesy of MARUM, University of Bremen

Black smoker near Fenway vent field eastern Manus Basin, Papua New Guinea. (Source: PacManus; cruise Sonne SO216, ROV dive # 311; station 41ROV)

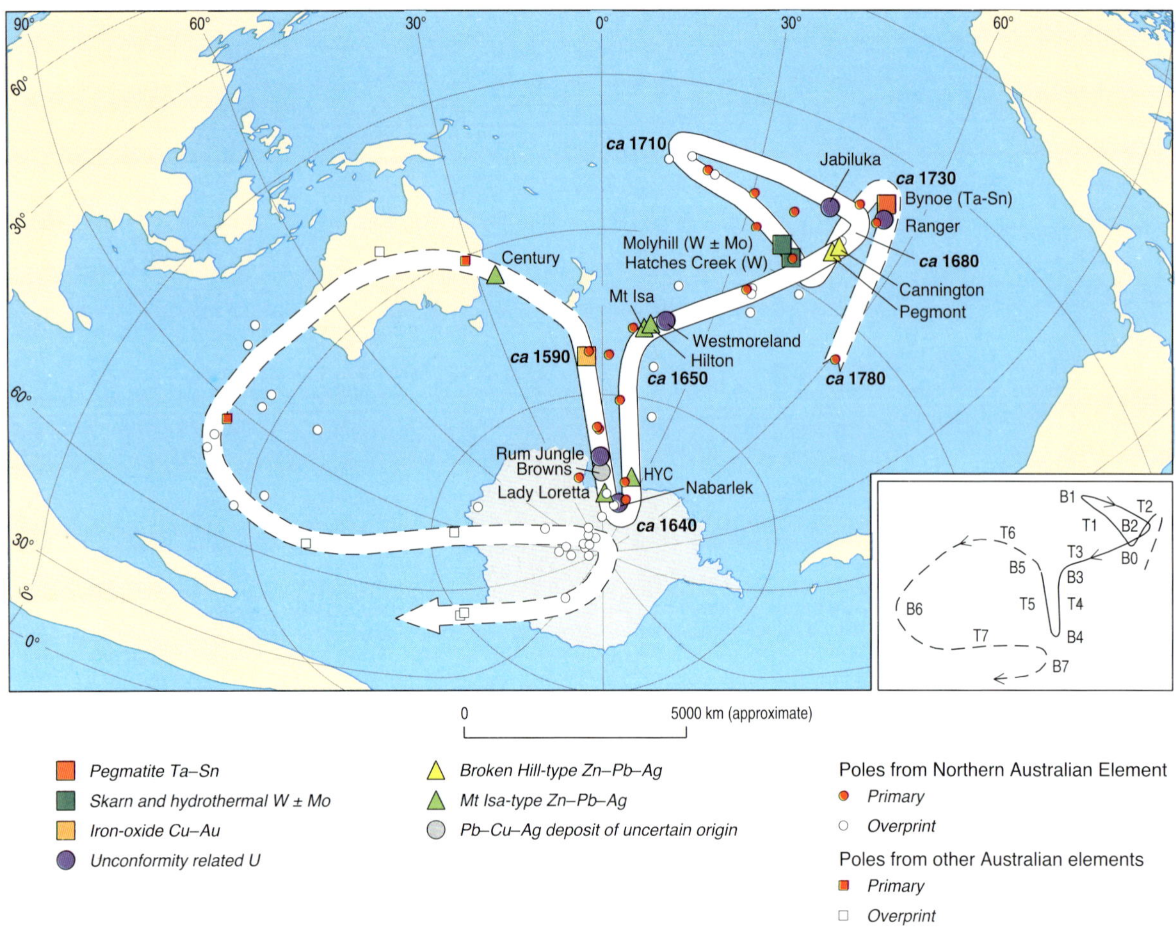

Figure 8.19: Apparent polar-wander path for the North Australia Element in the late Paleoproterozoic to early Mesoproterozoic. These paths are thought to track the plate's movements, with bends related to plate reorganisations from changing far-field subduction dynamics and/or collisions with other continents. The timing of many of the major Zn–Pb deposits was coincident with bends in the path, suggesting that far-field tectonic forces were a factor in the mineral systems. The inset shows a simplified apparent polar-wander path, highlighting bends (B0 to B7) and quasi-linear tracks (T1 to T7). The dashed parts of the wander path have greater uncertainty. (Source: Idnurm, 2000)

dam, named in honour of the 1956 Melbourne Olympic Games, intersected long intervals of 0.5% to 1.5% Cu. Surprisingly, this stunning discovery was made not in sedimentary strata or in altered basalts but in unusual hematite-rich breccias. Persistence, diligence, bold leadership and a willingness to think more broadly than the original target concept were some of the qualities that led the WMC team to continue exploration until drill hole RD10 indicated the magnitude of the discovery, when it intersected 200 m of high-grade Cu in 1976 (Box 8.3).

Formation of a supergiant: the Olympic Dam IOCG + U deposit

When the Olympic Dam deposit was discovered in 1975, the mineralised hematite-rich breccias were unfamiliar in terms of known mineral deposit types, and certainly differed from the style of sedimentary-hosted Cu deposit expected by the WMC explorers. The discovery of Olympic Dam was thus a major factor in the later recognition of a completely new class of mineral deposit, known as iron-oxide-copper-gold (IOCG) deposits.

The slow-changing, nearly flat landscape of outback South Australia today (Chapter 5) could not have been much different at 1590 Ma. A cataclysmic event spanning large parts of the Proterozoic continental blocks now found in South Australia, the Northern Territory and Queensland resulted in one of Earth's great outpourings of magma, known as the Hiltaba event. This event was focused in the Gawler Craton of South Australia, an Archean to Proterozoic cratonic block largely covered by younger sedimentary rocks (Chapter 2). The ultimate cause of melting that formed the extensive and voluminous Gawler Range Volcanics and Hiltaba Suite granitic intrusions is still not well understood.

Whatever the cause, the magmas resulting from crustal melting were superheated and enriched in 'high-field-strength-elements' such as thorium (Th), rare-earth elements and U, some of which were concentrated in the Olympic Dam deposit. Mantle melting also occurred, which resulted in production of mafic volcanic and intrusive rocks, particularly along the eastern boundary of the Archean cratonic core. This boundary was imaged by a seismic reflection traverse over the Olympic Dam deposit (Figure 8.21). The seismic data indicated that the boundary was a northeast-dipping suture juxtaposing Archean and Paleoproterozoic domains. As was the case in the Eastern Goldfields, this mineral system also records a large conductor in the magnetotelluric (MT) data at depth beneath the Olympic Dam region (Figure 8.21). This suture may have terminated a period of convergence that involved northeast- to north-dipping subduction below the Olympic Dam IOCG Province. In this scenario, the mantle below the province would have been fertilised during subduction, facilitating generation of high-temperature melts (A-type) during extension associated with the magmatic event. As we observed in the previous mineral-system examples, the crustal 'architecture' set up prior to mineralisation at 1590 Ma was crucial for focusing mantle and crustal melts, and also hydrothermal fluids implicated in the formation of the Olympic Dam deposit.

The Gawler Craton is covered by extensive regolith (Chapter 5) and by many sedimentary basins, making geophysical methods one of the few cost-effective exploration tools. The vast amounts of Fe deposited in these systems mean that potential fields have been very useful—in particular, the 'inversion modelling' of gravity and magnetic data in 3D. Such modelling at Olympic Dam suggests that magnetite-rich alteration zones extend to depths of around 20 km (Figures 8.21 and 8.22).

Copper–Au-rich hematite alteration, on the other hand, is modelled in the uppermost 5 km or so (Figure 8.21). These data, together with the seismic reflection and magnetotelluric data, show that the Olympic Dam deposit, gigantic as it is, represents just one part of a crustal-scale ore-forming system that involved magma generation, major faults tapping the mantle, and hydrothermal fluid flow on a regional scale in the upper and middle crust.

Over the past decade, broadscale studies of the Gawler Craton and discoveries of new iron oxide copper–gold (IOCG) deposits have shown that Olympic Dam is not alone as once thought but is part of a metallogenic belt, the Olympic Dam IOCG Province, extending more than 500 km along the eastern margin of the Gawler Craton (Figure 8.20). Earlier studies of the Olympic Dam deposit recognised a minor component of magnetite in the deposit, with Cu, Au and U overwhelmingly associated with hematite-rich breccia bodies. We now know that the IOCG hydrothermal systems are systematically zoned, with shallow hematite–sericite–chlorite alteration assemblages associated in places with Cu–Au ± U mineralisation, and deeper magnetite-rich assemblages accompanied by either biotite- or potassium-feldspar-rich assemblages, but with generally less Cu, at least in this part of Australia. Also at deep levels, and more regionally widespread, are zones rich in albite, actinolite or clinopyroxene (indicating Na–Ca metasomatism) (Figures 8.23 and 8.24).

Differential uplift and erosion within the Olympic Dam IOCG Province has resulted in 'exposure' (although still below young cover sediments and regolith) of different crustal levels (Figures 8.20 and 8.22). For example, in the Moonta district and Mount Woods Inlier, mainly deeper levels are now in the near surface

Did you know?

8.2: More than 600 'SydHarbs' of fluid made Broken Hill

- Broken Hill Pb–Zn–Ag deposit = Pb 34 × 10^6 tonnes or 34 × 10^{12} grams
- Typical ore fluid Pb concentrations are 100 ppm Pb or 1 tonne fluid to 100 grams
- 34 x 10^{12}g x 1t/100g = 34 × 10^{10} tonnes (fluid) or 34 × 10^{13} litres = 340 km^3
- (1 × 10^{12} litres = 1 km^3)
- (1 SydHarb = 562 × 10^9 litres = 0.562 km^3)
- So 340 km^3 of fluid = Broken Hill Pb = 605 SydHarbs

Image courtesy of Tourism Australia

OLYMPIC DAM RESOURCES AND ECONOMIC IMPACT (BOX 8.3)

Image courtesy of BHP Billiton

It was not until 1988 that the first Cu was produced from what has emerged as one of the world's greatest ore deposits. Olympic Dam ranks as the world's largest single deposit of U, fourth largest Cu deposit, and fifth largest Au deposit. Total resources (including past production, as of June 2009) stood at 9.23 Gt @ 0.86% Cu, 0.33 g/t Au, 0.027% U_3O_8, and 1.5 g/t Ag.

Production totalled 197 kt of Cu, 3.62 t of Au, 30.1 t of Ag and 3987 t of U in 2011, a year when government approved a proposal for a staggering six-fold increase in production. This proposal involves excavating an open pit 4 km long and 1 km deep. Water requirements are one of the critical issues for the proposed expansion, and a coastal desalination plant (more than 320 km distant) is the preferred option (Chapter 7).

Long-life mines are a prize sought not only by mining companies but by states and nations. The impacts of Olympic Dam on the South Australian economy and regional development have been huge. The Roxby Downs township created by the mine has grown to a community of 4500 inhabitants. Olympic Dam has provided an important boost to employment in South Australia and across the Australian mining industry. Annual royalties of more than $30 M are paid to the South Australian Government. Present production contributes $2.0 B to the gross state product, and directly employs around 4150 people. The proposed expansion is estimated to add $18.7 B (in 2008 dollars) to gross state product over 30 years in net present value terms compared with a 'business as usual' scenario.

and dominated by magnetite–biotite and albite–actinolite alteration assemblages. Because Cu–Au mineralisation is generally only weakly developed in the magnetite-rich zones in the Olympic Dam IOCG Province, these regions are thought to have lower potential for large hematite-rich IOCG deposits such as Olympic Dam.

Conversely, the shallow crustal levels that existed at 1590 Ma in the Olympic Dam region appear to have been preserved, and it is here that two major deposits have been discovered in the past decade: Prominent Hill and Carrapateena (Figure 8.20). Although each deposit appears to be at least an order of magnitude smaller than Olympic Dam, both show many of the same features as their big neighbour, such as hematitic alteration assemblages and brecciated zones; they are clearly part of the same mineral system (*Did you know?* 8.3).

Although reconnaissance in nature, radiogenic isotopic studies suggest that a substantial proportion of the Cu in the Olympic Dam deposit was sourced from mantle-derived rocks or magmas. These magmas are represented here by mafic/ultramafic dykes of Hiltaba Suite age and/or by mafic volcanic rocks within the co-magmatic Gawler Range Volcanics. Stable isotope and geochemical studies point to the presence of two,

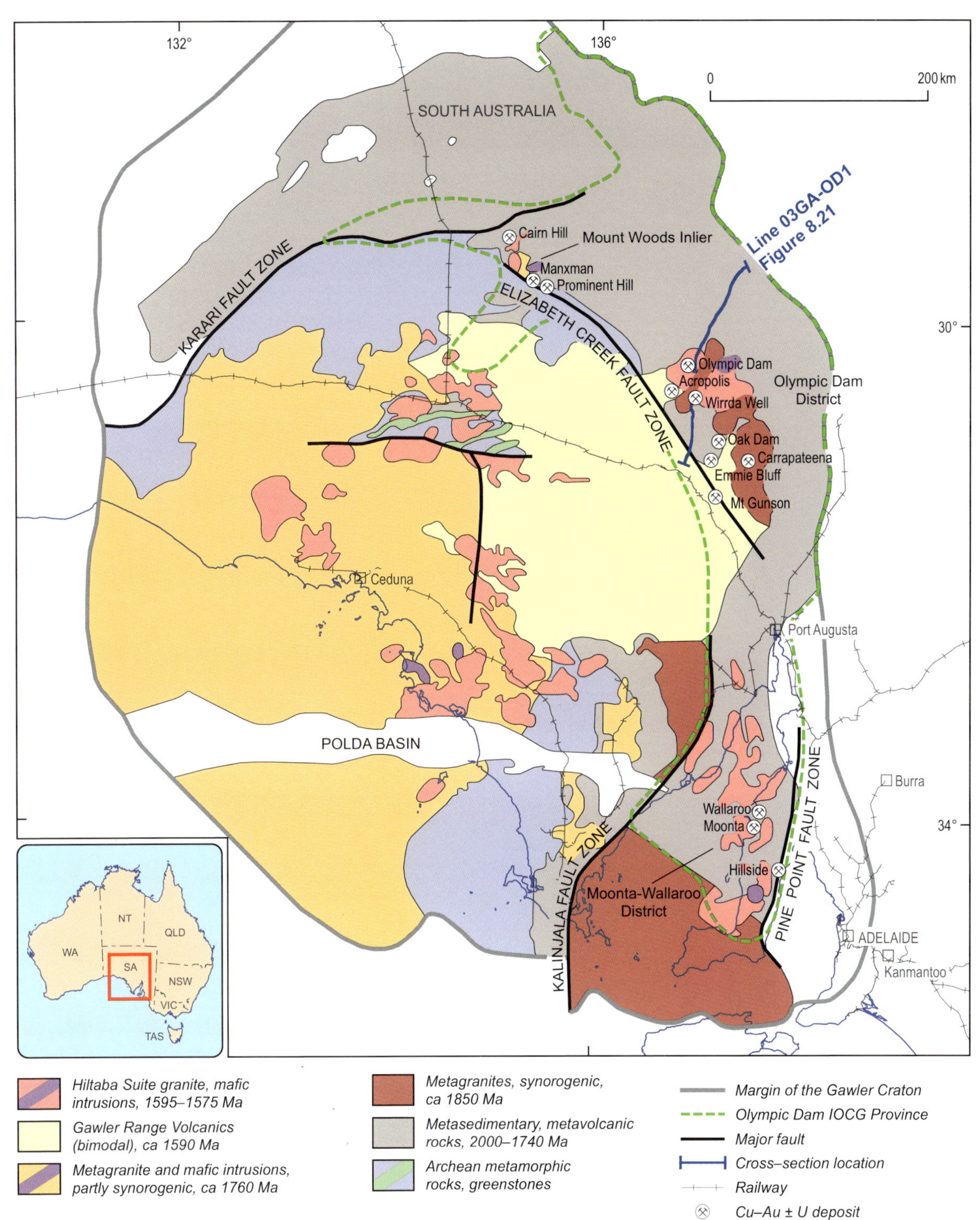

Figure 8.20: Location of the Olympic Dam copper–uranium–gold deposit, with simplified solid geology of the Gawler Craton (post-1500 Ma rock removed). Note the large area of Gawler Range Volcanics and Hiltaba Suite granites. The Olympic Dam IOCG Province is also shown along the margin between the Archean and Proterozoic components of the Gawler Craton. (Source: Hayward & Skirrow, 2010)

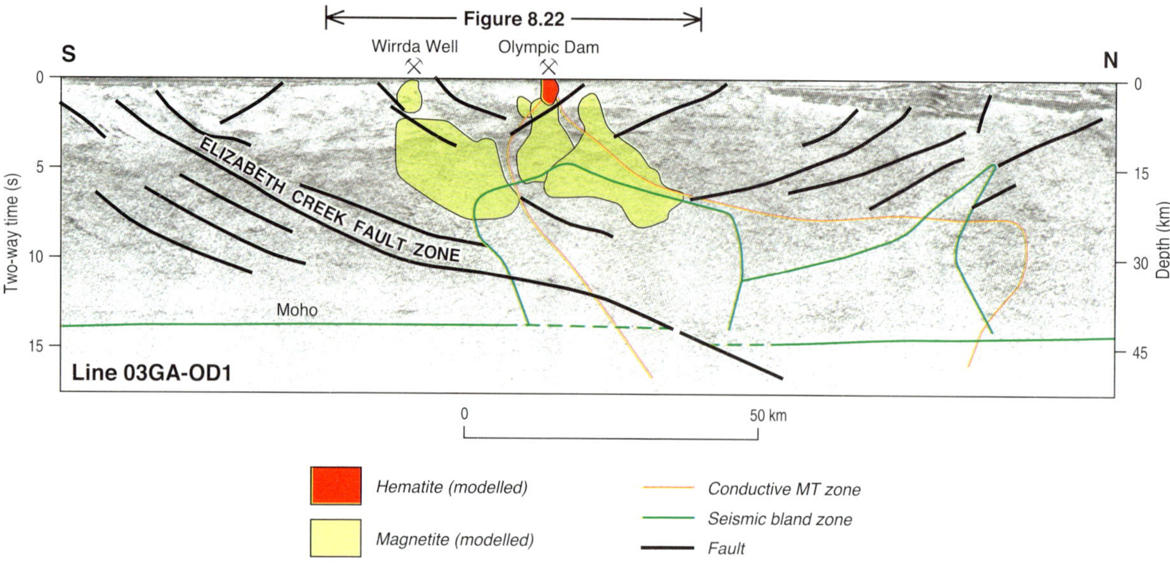

Figure 8.21: Seismic section passing through the Olympic Dam deposit. The enormous scale of the mineral system is well illustrated in this image of seismic reflection data and modelled potential field data. Interpreted structure shows that Olympic Dam is located in the hangingwall of the Elizabeth Creek Fault Zone that offsets the Moho. It is also coincident with the apex of an anomalous highly conductive zone in the magnetotelluric (MT) data that extends the full depth of the crust, and bland zones in the seismic data, which probably represent alteration and/or intrusive complexes in the lower crust. The magnetite roots of Olympic Dam are deep, with the inversion isosurfaces of inferred magnetite distribution extending to nearly 20 km. Hematite is modelled in the upper 5 km, and only in proximity to the Olympic Dam deposit. (Sources: Lyons & Gobby, 2005; Drummond et al., 2006; Heinson et al., 2006; Hayward & Skirrow, 2010)

or possibly even three, different fluid sources: 1) hypersaline high-temperature brines; 2) highly oxidised, sulfate-rich lake water or groundwater; and 3) CO_2-rich magmatic fluid.

The role of magmatic fluids is controversial, as is the origin of the breccias, which are even more extensive within the host Roxby Downs granite than the 9 Bt of hematite-rich ores. Phreatic, phreato-magmatic, hydrothermal and tectonic processes have all been proposed as contributing to breccia formation. More recently, the concept of deep CO_2 effervescence from magmas, perhaps triggered by mixing or mingling of mafic and felsic melts, has been proposed as a driver for brecciation. Geophysical datasets point to large-scale geodynamic processes operating through the entire crust. More work is required to assess this model, and test whether a low-density magmatic fluid also carried other volatiles such as fluorine (F), phosphorus (P), sulfur (S), and even some Cu and Au into the Olympic Dam deposit.

Key factors in forming the Olympic Dam IOCG Province

All these observations can be integrated into a mineral-system model (Figure 8.24). The model illustrates the crustal scale of the hydrothermal system, its deep controlling architecture, the roles of multiple fluids, and the importance of both felsic and mafic magmatism in the system. Unlike the other major mineral systems discussed here, the Olympic Dam IOCG Province is the product of two separate geodynamic systems: an early convergent system that fertilised the mantle and set up a crustal architecture, and a later extensional system that produced high-temperature crustal- and mantle-derived magmatism and associated mineralisation. These two different events occurred more than 100 Myr apart. These crustal and mantle processes have produced a large-scale, zoned mineral system, the effects of which can be mapped at the crustal scale using seismic, magnetotelluric, gravity and magnetic data. The system involved Cu-bearing oxidised fluids that reacted with rock to produce the hematitic alteration assemblages that characterise the Olympic Dam deposit. As this deposit formed in the very uppermost crust, a key factor in this mineral system is preservation (Figure 8.4).

Giant Australian mineral systems

Despite forming at various times and in quite different geodynamic settings, Australia's major mineral provinces and associated mineral systems have many common factors (see also Chapter 9).

The most consistent commonalities of these factors are an association with changes in regional stress patterns, links to global- and/or continental-scale geodynamic processes, an association with a thermal event, and links with major faults and shear zones. In addition, many of the deposits formed at or near margins of major contemporary crustal blocks, a relationship also noted from many petroleum systems (Chapter 4).

Energy is a critical driving factor in all mineral systems. Mineralisation accompanied thermal events, as indicated by coeval magmatism or regional metamorphic events. This is best expressed in the Eastern Goldfields Province, where the major period of lode-Au mineralisation overlapped granite emplacement, contractional deformation and regional metamorphism. The Olympic Dam IOCG Province was accompanied by voluminous high-temperature magmatism, and Broken Hill-type deposits in the Proterozoic Zn belt are associated with coeval mafic magmatism. These magmato-thermal events drove fluid flow, and, in some cases, these magmas may have provided metal and/or fluids. Conversely, Mt Isa-type mineral systems and the upgrading of banded iron-formation to iron ore (Chapter 9) involved lower temperature, oxidised fluids that were driven by processes other than magmatism, such as basinal fluid flow driven by density differences and accompanying far-field tectonic forces.

In all major mineral provinces, mineralisation is associated with a switch in the stress regime at the province to continental scales. This can involve a change from extension to contraction (or compression to transpression), as in lode-Au

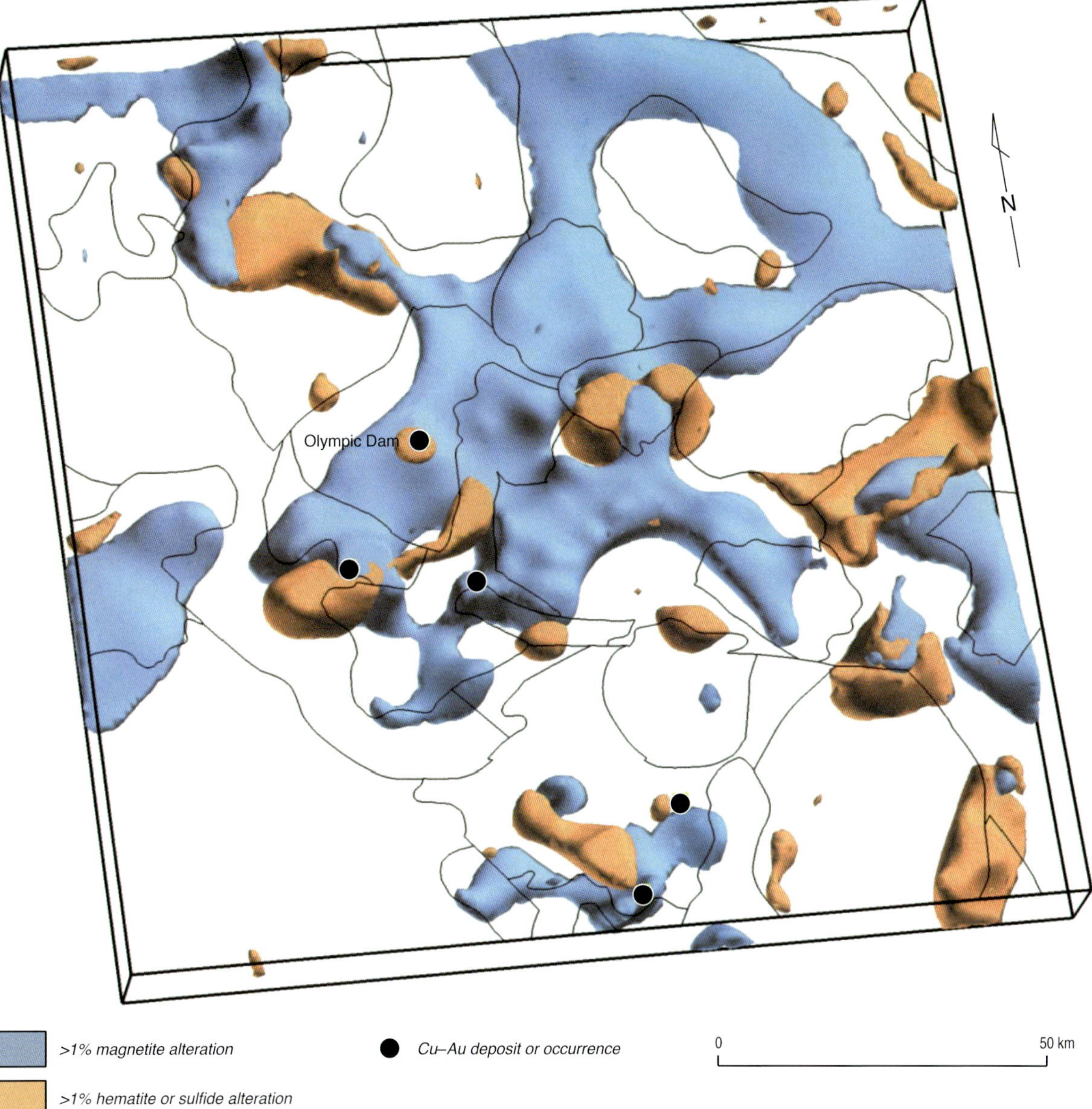

Figure 8.22: 3D perspective view showing volumes of hematite- and magnetite-dominated alteration assemblages in the Olympic Dam region. The image is based on geologically constrained inversions of regional gravity and magnetic data.

Images courtesy of DIMITRE

Figure 8.23: Images of hydrothermal alteration in drill core from the Olympic IOCG Province, Gawler Craton (core diameter 4–5 cm).

provinces in the Victorian goldfields and Eastern Goldfields. Alternatively, mineralisation can be associated with a switch from contraction to extension, as in the Olympic Dam IOCG Province, the Proterozoic Zn belt and the Hamersley iron ore province (Chapter 9). In at least one case, mineralisation in the Proterozoic Zn belt seems to be associated with the initiation of strike-slip faulting. These stress switches are commonly controlled by global- and continental-scale processes or cycles, such as the supercontinent cycle (Figure 2.20).

Because most of the rocks present at the crustal level where these hydrothermal mineral systems operate are essentially impermeable, all of Australia's major mineral provinces are associated with faults and/or shear zones. These structures provide the plumbing system for fluid flow. The association between ore and structure occurs at many scales. For example, in lode-Au provinces, entire districts, not individual deposits, are associated with major crust-penetrating structures; the deposits generally are associated with second- or third-order structures. Major crust-penetrating structures provide fluid pathways that tap fluid and metal sources, but the amount of fluid flowing along these pathways can overwhelm the reactive capacity of wall rocks. Reaction occurs along lower order structures because smaller fluid fluxes do not overwhelm reactants, so ore deposition can effectively occur along these structures.

Another factor that characterises some major mineral provinces is a direct or indirect link to the mantle. In three of the five major Australian mineral provinces considered, a direct or indirect mantle contribution of metal (and possibly fluids) is likely. Seismic reflection data suggest direct tapping of the mantle by crust-penetrating faults in the Victorian goldfields and the Eastern Goldfields. In the Olympic Dam Province, a likely source of Cu and possibly Au is mantle-derived mafic rocks. In the Proterozoic Australian Zn belt and in the Hamersley iron ore province, a direct mantle link is less clear.

Factors that characterise some, but not all, major Australian mineral deposits include a relationship to the Great Oxygenation Event (Chapter 3), direct or indirect sourcing of metal from the mantle, and a spatial association to the edges of major crustal blocks. Prior to the Great Oxygenation Event, which began *ca* 2460 Ma, most surficial fluids on Earth, including seawater and meteoric fluids, were reduced. The oxidation of the atmosphere produced oxidised and sulfate-rich surficial fluids, which under some circumstances can be very potent low-temperature ore fluids for dissolving and transporting U, Zn and Cu (Appendix 8.3.1). However, Fe is highly insoluble in such fluids. Hence, after the Great Oxygenation Event, the availability of low-temperature oxidised fluids allowed formation of Mt Isa-type Zn–Pb deposits, as well as other Cu and U deposits. When these fluids passed through banded iron-formation, they would have dissolved silica, carbonate and phosphate, which are gangue minerals, but left the Fe. The result was the relative upgrade of Fe content of these rocks to form the prized high-grade hematitic iron ore (Chapter 9).

We must also consider the metal source in a mineral system (Figure 8.4). The original differentiation of Earth and subsequent tectonic processes led to certain elements being concentrated in either the mantle or the crust (Appendix 8.3.1). Siderophile metals, such as Ni, platinum (Pt) and Au, are enriched in the mantle relative to the crust, so the mantle or mantle-derived rocks are excellent sources of these metals relative to crustal rocks. In contrast, elements such as Pb and U, and, to a lesser extent, Zn, are concentrated in the crust relative to the mantle, suggesting that crustal sources are best for lithophile and some chalcophile elements. These relationships are consistent with the patterns observed in major Australian mineral provinces.

Mineral systems that have a component of metal and/or fluid sourced from the mantle are closely located with respect to major crustal boundaries, either convergent or divergent boundaries. For example, lode-Au deposits formed near convergent margins in both the Victorian goldfields and the Eastern Goldfields, most likely as a consequence of the accretion of blocks, some of which may have been exotic. IOCG deposits in the Olympic Dam and other Australian provinces are spatially associated with major boundaries between crustal blocks, as indicated by both seismic reflection, magnetotelluric, magnetic, gravity and radiogenic (Nd and Pb) isotope data. Broken Hill-type deposits are also associated with these boundaries.

In all of Australia's major mineral provinces, many or all of these factors have converged during a critical time window to make the mineral system operate. Additionally, most of these provinces are located in crustal blocks that have been preserved from later loss through erosion or tectonism, possibly because the blocks were cratonised quickly after mineralisation (Figure 8.4). These factors may be the principal geological reasons why the provinces discussed here have provided much of Australia's mineral wealth and have been important drivers of the nation's growth and development. This tectonic stability is one of the reasons that the continent is old, flat and red (Chapter 5).

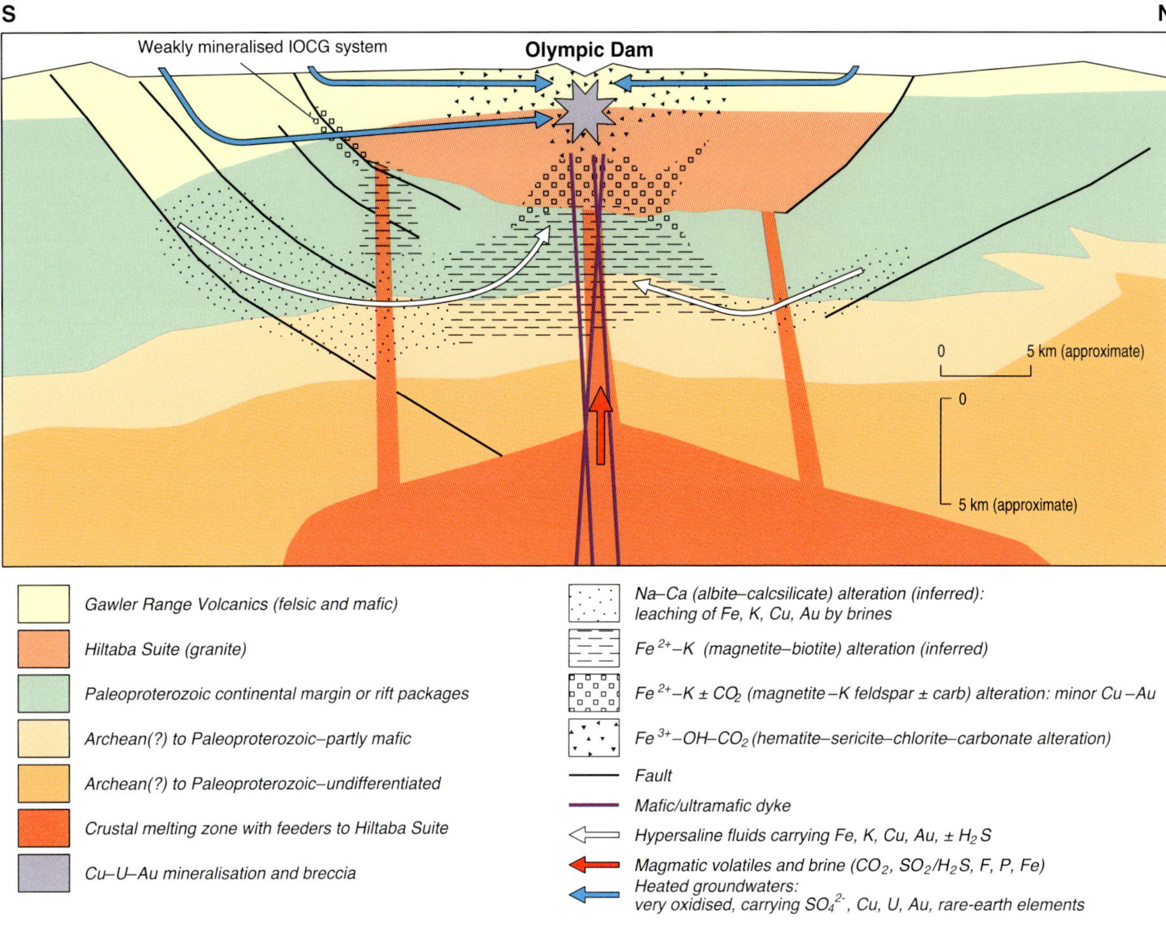

Figure 8.24: Model geological cross-section of the Olympic Dam IOCG+U mineral system, illustrating possible fluid flow, zonation of alteration, and elemental fluxes.

Metallic exploration and mining: future challenges

In the next half-century, the Australian metallic mining industry faces a number of challenges, both in meeting ever-increasing expectations that businesses operate in a manner responsible to environment and society, and in maintaining

Did you know?

8.3: Why is hematite so important?

Hematitic alteration represents the effects of oxidised hydrothermal fluids acting in the presence of abundant Fe; these fluids are considered to have carried some, if not most, of the Cu, U and Au to form Olympic Dam and the other IOCG deposits in the province.

The reaction of these Cu-bearing oxidised fluids with chemical reductants appears to have been a critical process in the formation of the Olympic Dam deposit. Whether the reductants—for example, magnetite, pyrite, or Fe^{2+}-bearing silicate minerals—were already present in the host rocks or were present in a second, reduced fluid that mixed with the oxidised fluid remains uncertain. The process of fluid mixing is believed to have resulted in Cu, U and Au deposition.

Image by Chris Fitzgerald

and increasing the resource base. Over the past decade, businesses have increasingly used the 'triple bottom line'—planet, people and profit—to determine best practice. The greatest impact of this new approach has been enhanced environmental practices (compare 'old' with 'modern'; Figure 8.2). Given societal expectations, the trend to higher environmental standards will increase. However, this trend conflicts with the current move to mine larger, lower grade deposits that have bigger environmental footprints, presenting a major challenge to industry.

A second environmental challenge to the mineral industry is the appetite of mines for water and energy, inputs that can be difficult to provide in remote locations. Finding a source of water (Chapter 7), particularly in competition with other industries, can be just as important as defining ore reserves. The cost of diesel-generated electricity has led some companies to consider solar and other alternative energies as a source of power for remote operations.

Industry also faces challenges in maintaining its skill base in the face of changing lifestyles, and the reluctance to work (and live) in remote, generally arid locations. A related challenge is building infrastructure for a large work force. The preferred fly-in/fly-out (FIFO) option (Chapter 9), which became widespread after the introduction of the fringe-benefits tax, can cause strain on staff and their families. Making a stable career in the industry can also be difficult. The boom and bust of the resources cycle (Figure 8.3) has historically led to imbalances in skilled staff; during booms, there are insufficient personnel to support the industry, whereas, during busts, job security is poor and many skilled workers lose their jobs and leave the industry, often for good.

The other major societal issue facing the industry is ensuring that traditional owners gain from mining. Many mining operations in Australia are remote and located on land owned or claimed by traditional owners, who historically have not shared in the wealth and other benefits of mining. Legislation now requires greater consultation and negotiation with traditional owners when deposits are discovered, leading to agreements between industry and land councils to ensure that the wealth is shared through royalties and jobs. Clashes between traditional culture and the industry still occur.

Of at least equal importance to meeting societal expectations is the challenge of maintaining and increasing the resource base. Australia owes its position as a major mining country to successes of past prospecting and exploration. This position is underpinned by a relatively small number of major to world-class mines, and some of these are nearing the end of their mining lives (Appendix 8.1.1). Although Australia's Au and base-metal resources have grown strongly over the past 30 years, most of this growth has been through the addition of resources at existing mines, and not the discovery of new deposits. Despite large expenditures, only a handful of new major or world-class deposits have been discovered in the past 20–30 years, leading to the perception that Australia is mature in terms of mineral exploration (Chapter 11).

The early discoveries were mostly exposed at the surface or could be detected at the surface using geological, geochemical or geophysical methods. In the 1960s, this began to change, and discoveries began to be made under significant thicknesses of cover (Chapter 11). The most significant discovery was Olympic Dam in 1975, under 300 m of cover.

Discoveries since have been made under cover, with a consequent increase in discovery costs, as the testing of deep targets and concepts is inherently more expensive.

The higher cost of deep exploration means that one of the major challenges to exploration is the development of new methods. Using a mineral-systems scientific framework, the methods should include development of robust 4-D predictive geological models; cover maps and new detection methods; and concepts such as the role of the deep crust, the mantle and major crustal boundaries and lineaments in localising mineralisation. These methods will focus exploration prior to expensive drilling, and if successful they will shift the negative perception that Australia is a 'mature' exploration destination. Many of Australia's known mineral provinces extend under cover (e.g. Figure 8.5), making Australia highly prospective for further discoveries.

Gold mining and exploration, Western Australia.

Image by Jim Mason

The Pilbara, Western Australia.

Bibliography and further reading

Short history of mining in Australia

Blainey G 1993. *The rush that never ended: a history of Australian mining*, 4th edn, Melbourne University Press, Melbourne.

Butlin NG 1985. *Australian national accounts 1788–1983*, Source Papers in Economic History no. 6, Australian National University, Canberra.

Close SE 2004. *The great gold renaissance—the untold story of the modern Australian gold boom 1982–2002*, Surbiton Associates, Melbourne.

Doran CR 1984. An historical perspective on mining and economic change. In: *The minerals sector and the Australian economy*, Cook LH & Porter MG (eds), George Allen & Unwin Australia, Sydney.

Hill D 2001. *Gold!: the fever that forever changed Australia*, Random House, Australia.

Idriess IL 1931. *Lasseter's last ride*, Angus & Robertson First Edition, Sydney.

Simon J 2003. Three Australian asset price bubbles. In: *Asset prices and monetary policy*, Richards A & Robinson T (eds), proceedings of the Reserve Bank of Australia conference, Sydney, 18–19 August 2003, Reserve Bank of Australia, Canberra.

Sydenham S & Thomas R 2000. *Chinese at the Australian goldfields*; *Women on the Australian goldfields*; *Rebellion: the Eureka Stockade*. www.kidcyber.com.au

Wilkinson R 1996. *Rocks to riches: the story of Australia's national geological survey*, Allen & Unwin, St Leonards, New South Wales.

Australia's mineral systems

Jaques AL, Jaireth S & Walshe JL 2002. Mineral systems of Australia: an overview of resources, settings and processe. *Australian Journal of Earth Sciences* 49, 623–660.

Woodall R 1985. Limited vision: a personal experience of mining geology and scientific mineral exploration. *Australian Journal of Earth Sciences* 32, 231–237.

Wyborn LAI, Heinrich CA & Jaques AL 1994. Australian Proterozoic mineral systems: essential ingredients and mappable criteria. *Australasian Institute of Mining and Metallurgy Publications Series* 5/94, 109–115.

The Victorian goldfields

Battellino R 2010. Mining booms and the Australian economy. *Australian Bureau of Statistics Bulletin* March 2010, 163–69.

Bierlein FP, Arne DC, Foster DA & Reynolds P 2001. A geochronological framework for orogenic gold mineralisation in central Victoria, Australia. *Mineralium Deposita* 36, 741–767.

Champion DC, Kositcin N, Huston DL, Mathews E & Brown C 2009. *Geodynamic synthesis of the Phanerozoic of eastern Australia and implications for metallogeny*, Geoscience Australia Record 2009/18, Geoscience Australia, Canberra.

Maddock R & McLean I 1984. Supply-side shocks: the case of Australian gold. *Journal of Economic History* 44, 1047–1067.

Phillips GN, Hughes MJ, Arne DC, Bierlein FP, Carey SP, Jackson T & Willman CE 2003. Gold—historical wealth, future potential. In: *Geology of Victoria*, Birch WD (ed.), Geological Society of Australia Special Publication 23, 376–433.

Squire RJ & Miller JM 2003. Synchronous compression and extension in East Gondwana: tectonic controls on world-class gold deposits at 440 Ma. *Geology* 31, 1073–1076.

Willman CE, Korsch RJ, Moore DH, Cayley RA, Lisitsin VA, Rawling TJ, Morand VJ & O'Shea PJ 2010. Crustal-scale fluid pathways and source rocks in the Victorian Gold Province, Australia: insights from deep seismic reflection profiles. *Economic Geology* 105, 895–915.

The Eastern Goldfields lode-gold province

Australian Bureau of Agricultural and Resource Economics 2009. *Australian commodity statistics 2009*, ABARE, Canberra.

Australian Bureau of Agricultural and Resource Economics 2010. *Australian mineral statistics 2010*, March quarter 2010, ABARE, Canberra.

Australian Bureau of Statistics 2010. *Consumer price index*, March 2010, ABS, Canberra.

Blewett RS, Czarnota K & Henson PA 2010. Structural-event framework for the eastern Yilgarn Craton, Western Australia, and its implications for orogenic gold. *Precambrian Research* 183, 203–229.

Cassidy KF & Champion DC 2004. Crustal evolution of the Yilgarn Craton from Nd isotopes and granite geochronology: implications for metallogeny. *The University of Western Australia Publication* 33, 317–320.

Goleby BR, Blewett RS, Korsch RJ, Champion DC, Cassidy KF, Jones LEA, Groenewald PB & Henson P 2004. Deep seismic reflection profiling in the Archaean northeastern Yilgarn Craton, Western Australia: implications for crustal architecture and mineral potential. *Tectonophysics* 388, 119–133.

Hagemann SG & Cassidy KF 2000. Archaean orogenic lode-gold deposits. Society of *Economic Geologists* 13, 9–68.

Hogan L 2004. *Research and development in exploration and mining: implications for Australia's gold industry*, Australian Bureau of Agricultural and Resource Economics eReport 04.3, ABARE, Canberra.

Hogan L, Harman J, Maritz A, Thorpe S, Simms A, Berry P & Copeland A 2002. *Mineral exploration in Australia: trends, economic impacts and policy issues*, Australian Bureau of Agricultural and Resource Economics eReport 02.1, ABARE, Canberra.

Korsch RJ & Blewett RS 2010. Geodynamics and architecture of a world class mineral province: the Archean eastern Yilgarn Craton, Western Australia. *Precambrian Research* Special Issue 183.

Predictive Mineral Discovery Cooperative Research Centre 2008. *Concepts to targets: a scale-integrated mineral systems study of the Eastern Yilgarn Craton*, Project Y4 final report, parts I–V, January 2005 – July 2008.

Robert F, Poulsen KH, Cassidy KF & Hodgson CJ 2005. Gold metallogeny of the Superior and Yilgarn cratons. Economic Geology 100th Anniversary Volume, 1001–1033.

Western Australian Department of Treasury and Finance 2004. *An economic history of Western Australia since colonial settlement*, Department of Treasury and Finance, Perth.

Proterozoic zinc–lead deposits

Cooke DR, Bull SW, Large RR & McGoldrick PJ 2000. The importance of oxidized brines for the formation of Australian Proterozoic stratiform sediment–hosted Pb–Zn (SEDEX) deposits. *Economic Geology* 95, 1–19.

Gustafson LB & Williams N 1981. Sediment–hosted stratiform deposits of copper, lead, and zinc. *Economic Geology* 75th Anniversary Volume, 139–178.

Idnurm M 2000. Toward a high resolution Late Palaeoproterozoic-earliest Mesoproterozoic apparent polar wander path for northern Australia. *Australian Journal of Earth Sciences* 47, 405–430.

Large RR, Bull SW, McGoldrick PJ, Walters S, Derrick GM & Carr GR 2005. Stratiform and strata-bound Zn–Pb–Ag deposits in Proterozoic sedimentary basins, northern Australia. *Economic Geology* 100th Anniversary Volume, 931–964.

Parr JM & Plimer IR 1993. Models for Broken Hill-type lead–zinc–silver deposits. Geological Association of Canada Special Paper 40, 253–288.

Southgate PN (ed.) 2000. A thematic issue on Carpentaria–Mt Isa zinc belt: basement framework, chronostratigraphy and geodynamic evolution of Proterozoic successions. *Australian Journal of Earth Sciences* 47, 337–657.

Southgate PN, Kyser TK & Large RR (eds) 2006. A special issue devoted to Australian Zn–Pb–Ag deposits: a basin system and fluid flow analysis. *Economic Geology* 101, 1103–1312.

Southgate P, Neumann N, & Gibson G (in review). Depositional systems in the Mt Isa Inlier between 1800 and 1640 Ma: implications for Zn-Pb-Ag mineralisation. *Australian Journal of Earth Sciences*.

The Ochre Pits of the West Macdonnell Ranges, Northern Territory, are the one the important sources of ochre for the Western Arrente people in their ceremony and artwork. The pits are one of the most visited sites by tourists in the West Macdonnell National Park.

Image courtesy of Toby Hudson, Wikimedia Commons

Walters S & Bailey A 1998. Geology and mineralization of the Cannington Ag–Pb–Zn deposit: an example of Broken Hill-type mineralization in the Eastern Succession, Mount Isa Inlier, Australia. *Economic Geology* 93, 1307–1329.

Williams N 1978. Studies of base metal sulfide deposits at McArthur River, Northern Territory, Australia: the Cooley and Ridge deposits. *Economic Geology* 73, 1005–1035.

Williams PJ 1998. A special issue on the metallogeny of the McArthur River–Mt Isa–Cloncurry minerals province. *Economic Geology* 93, 1119–1488.

Olympic Dam

BHP Billiton 2009. Olympic Dam expansion draft environmental impact statement—executive summary.

Drummond BA, Lyons P, Goleby B & Jones L 2006. Constraining models of the tectonic setting of the giant Olympic Dam iron oxide–copper–gold deposit, South Australia, using deep seismic reflection data. *Tectonophysics* 420, 91–103.

Haynes DW 2006. The Olympic Dam ore deposit discovery—a personal view. *Society of Economic Geologists Newsletter* 66, 1, 8–15.

Haynes DW, Cross KC, Bills RT & Reed MH 1995. Olympic Dam ore genesis: a fluid mixing model. *Economic Geology* 90, 281–307.

Hayward N & Skirrow RG 2010. *Geodynamic setting and controls on iron oxide Cu–Au (U) ore in the Gawler Craton, South Australia. Advances in the understanding of IOCG deposits*, PGC Publishing, Adelaide, 119–146.

Heinson GS, Direen NG & Gill RM 2006. Magnetotelluric evidence for a deep crustal mineralizing system beneath the Olympic Dam iron oxide copper-gold deposit, southern Australia. *Geology* 34, 573–576.

McPhie J, Kamenetsky VS, Chambefort I, Ehrig K & Green N 2011. Origin of the supergiant Olympic Dam Cu–U–Au–Ag deposit, South Australia: was a sedimentary basin involved? *Geology* 39, 795–798.

Reeve JS, Cross KC, Smith RN & Oreskes N 1990. Olympic Dam copper–uranium–gold–silver deposit. In: *Geology of the mineral deposits of Australia and Papua New Guinea*, Hughes FE (ed.), Australasian Institute of Mining and Metallurgy Monograph 14, 1009–1035.

Skirrow RG & Davidson GJ 2007. A special issue devoted to Proterozoic iron oxide Cu–Au–(U) and gold mineral systems of the Gawler Craton: preface. *Economic Geology* 102, 1373–1375.

Upton D 2010. *The Olympic Dam story: how Western Mining defied the odds to discover and develop the world's largest mineral deposit*, Upton Financial (self published).

Williams NC & Skirrow R 2004. The lowdown on Gawler copper and gold. *AusGeo News* 74, Geoscience Australia, Canberra.

Williams NC & Dipple GM 2007. Mapping subsurface alteration using gravity and magnetic inversion models. In: *Proceedings of Exploration 07, Advances in Geophysical Inversion and Modelling* 29, 461–472.

Williams PJ, Barton MD, Johnson DA, Fontboté L, De Haller A, Mark G, Oliver NHS & Marschik R 2005. Iron oxide copper–gold deposits: geology, space-time distribution, and possible modes of origin. *Economic Geology* 100th Anniversary Volume, 371–405.

Challenges

Australian Academy of Science 2010. *Searching the deep Earth: the future of Australian resource discovery and utilisation*, proceedings of the Theo Murphy Think Tank 2010, Canberra, 19–20 August 2010, Australian Academy of Science, Canberra.

426-24

9

Sustaining Australia's wealth—economic growth from a stable base

Despite continuing global supercontinent cycles, the Precambrian core of the Australian continent has endured. Most of the continent is now remote from active plate boundaries, and this, together with its core strength, has ensured relative geological stability over the last 200 million years (Myr). Much of the continent is deeply weathered. The relative stability, however, has ensured that the vast mineral and energy resources of Australia have been preserved and, indeed, created. These resources include the bulk commodities of iron ore, bauxite, coal and natural gas, which now generate much of the wealth for the Australian people. The resources sector, to which these bulk commodities contribute the majority of the value, is worth over $206 billion per year to Australia in export earnings. Increasing demand from our Asian neighbours is driving further growth in the sector. Demand for Australia's natural resources, coupled with sound financial management and governance, ensured that Australia largely avoided the ravages of the 2008 global financial crisis (GFC).

Paul J Kay, Richard S Blewett and David L Huston

Geoscience Australia

Image by Eric Taylor

Resources from a stable base

Australia is one of the most stable and strongest economies in the world. A major factor in this has been the unprecedented boom in demand for Australia's high-quality bulk resources—in particular, from the rapidly developing Asian economies (Figure 9.1). In 2011, Australia was the 11th largest economy in gross domestic product (GDP) per head and 18th largest in total GDP. According to the International Monetary Fund, Australia's GDP at purchasing power parity was US$919 billion (B) or US$40 836 per person in that year.

Stemming from its size and long geological history, Australia has a rich endowment of geological resources. Australia is a top-five producer of the world's key resource commodities, including:

- world's leading producer of bauxite, alumina, rutile, zircon and tantalum
- second largest producer of lead, ilmenite and lithium
- third largest producer of iron ore, uranium and zinc
- fourth largest supplier of LNG (liquefied natural gas)
- fourth largest producer of black coal, gold, manganese and nickel
- fifth largest producer of aluminium, brown coal, diamonds, silver and copper.

This impressive list of resource endowment for Australia is dominated economically by the main bulk commodities. These are high-volume/tonnage resources that require massive infrastructure and capital to develop. The Bureau of Resources and Energy Economics documented the key resources exports in 2011 as iron ore ($58.4 B), coal ($47 B), gold ($14.6 B), LNG ($11.1 B) and bauxite/aluminium products ($9.3 B), forming a pivotal (44%) share of the nation's export trade. Each commodity attracts around $10 B or more in export income annually, and collectively they have had a unique influence on Australian society. The four bulk commodities of coal, iron ore, LNG and bauxite products make up 40% of the nation's exports. They provide raw materials supporting the industrial development of modern economies in Asia, while delivering substantial economic and social benefits to the Australian people.

The favourable geological endowment of these resources is complemented by Australia's stable democratic political system and a transparent regulatory environment. These factors have enabled commercial entities to invest the necessary billions of dollars in project development. This chapter presents a short history of the Australian economy and the major role of resources in shaping it. We discuss the main drivers of demand and the benefits in jobs, the development and economic achievement that rides on the back of the resources industry. We also consider some of the costs to society and the environment and the balance reached by sound governance. A summary of the geology of the main bulk commodities—iron ore, bauxite, coal and gas—is given, and the unique conditions that have led to this resource endowment are explained.

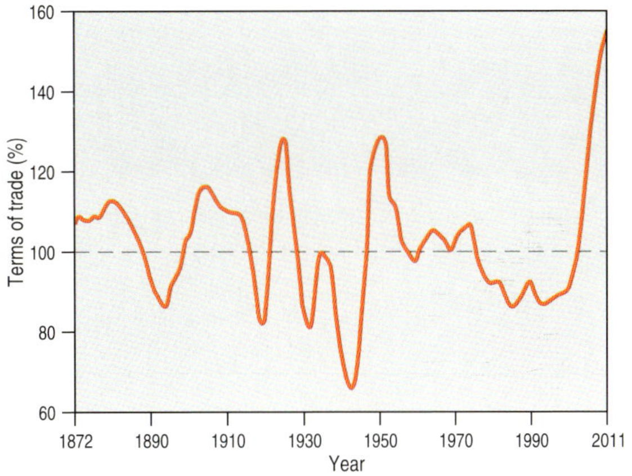

Figure 9.1: Australia's terms of trade, 1900–2000 average = 100. Following the highs of the 1950s (led by wool), Australia's terms of trade was on a downward trend. In the early 2000s, this trend changed to strongly positive, led by the demand from Asia for Australian resources. (Sources: Reserve Bank of Australia; *The Economist*, 2011)

Opposite: Iron ore reclaimer, Cape Lambert Port, Pilbara, Western Australia.

Photo courtesy of Rio Tinto

Figure 9.2: Locality map including place names mentioned in the text.

A brief economic history of Australia

Resources and the evolution of Australia's economy

For thousands of years, the different regions of Australia had an economy based on the trade of goods rather than money. Aboriginal and Torres Strait Islander people exchanged prized stone implements, shells, ochres and pigments, along trade routes extending across many parts of the country (Chapter 1). Following British settlement of New South Wales in 1788, an Australian economy began to evolve. This evolution can be viewed in seven periods of time.

The early years: colonial expansion (1788–1820)

The economic development of Australia during the colonial expansion period depended upon the influx of convict labour, which was financed by capital from Britain. The early financial history of the colony of New South Wales was marked by a shortage of money. The Bank of New South Wales (today known as Westpac) was established in 1817 to bring some order to the finances of the colony and to act as a bank of issue. Competitor banks soon followed and, with capital structures in place, the colony developed rapidly once large areas of arable land were identified.

Wealth from the sheep's back (1821–50)

In the first decades, the arable land to the west of the Blue Mountains of New South Wales (Figure 9.2) was an unknown. The first crossing of the mountains in 1813 led to a number of access routes to the arable grasslands, and the official seal of approval from Governor Macquarie for expansion was secured in 1820. A rapidly evolving worldwide wool industry was pivotal in the expansion. Wool was cemented as a key primary product of Australia at that time and, by 1820, Australia had a GDP per capita of $1528, the second highest in the world after the Netherlands.

Putting Australia on the map: the golden years (1851–90)

The public announcement of gold discoveries created a nexus with past development, particularly in the remarkable goldfields of central Victoria.

Alluvial gold washing, Mount Alexander goldfields, Victoria, *ca* 1852.

Image courtesy of National Library of Australia, image no. VN 4496154

No previous mineral discovery in Australia was comparable with the new alluvial gold finds. Gold attracted labour from overseas and drew Australian workers from their jobs (Chapters 1 and 8). For the 10 years from 1851 to 1861, Victoria produced 750 tonnes of gold, some 40% of the world's output in that period. As the gold rushes ended, miners were absorbed into the Australian economy, resulting in increased availability of casual labour, much of which went into further developing the agricultural sector and other activities such as railroad building.

Despite these changes in the rural parts of Australia, the country remained predominantly urban. Only in Victoria (due to the rush to the goldfields) did the urban population fall relative to that in rural areas. Later mineral discoveries of the period, such as Mount Lyell (1881), Mount Morgan (1882) and Broken Hill (1883) led to new development, but this could not entirely offset the decline of the Victorian gold industry.

Arrested expansion (1891–1900)

An intense economic depression during the early 1890s led to a precipitous trough in activity by 1893, with 12 banks defaulting, taking two-thirds of Australia's bank assets with them. Falling export prices, ongoing drought, a cessation of capital inflow and land boom collapses led to a general decline in each of the Australian colonies. In contrast with most of Australian history since 1788, emigration from Australia occurred during this period. The discovery of gold in the supremely rich Eastern Goldfields (Chapter 8) saw a flood of easterners move to the west, and the expansion of the economy and population of the colony of Western Australia. These people took with them the values of a united eastern and western Australia, and it was their yes-vote in the Federation Referendum that ensured that Western Australia joined the Commonwealth.

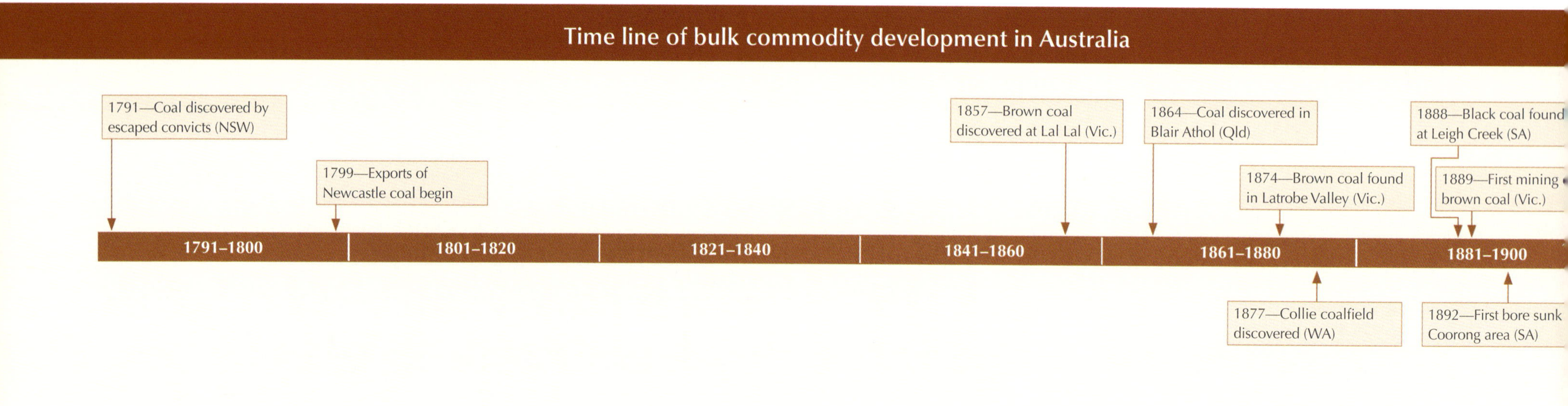

New expansion and contraction (1901–39)

The strength of Australian balance of payments in the early 20th century made it possible to liquidate most of the commitments made to overseas creditors generated during the down-time of the 1890s. Australia also once again became an attractive destination for migrants, many of whom were subsidised by governments. At federation in 1901, Australia's GDP included agriculture (20%) and mining (10%), together accounting for 95% of exports and more than 30% of the labour force. The GDP per capita at this time was $4299, second only to New Zealand in world standards.

The impact of World War I (1914–18) was widespread, with a decline in the flow of capital for long-term investment. However, one earlier theme continued strongly, that of increased metropolitan concentration based economically on rural development, a theme that continues today. In 1910, some 40% of the Australian population resided in the six capital cities. The Commonwealth Bank of Australia was founded in 1911, issuing notes backed by the nation's mineral resources. Following a pause after the 1929 depression and three bank failures in 1931, the growth in output of Australia continued.

The beginnings of the bulk commodity industry (1940–75)

By the end of the 1930s and into the 1940s, Australia had developed a rising manufacturing output, much of it initially to serve the war effort. From the end of the 1940s to the late 1960s, this manufacturing growth continued, representing a break with the largely agricultural or at least rural past. Australia's prosperity was improving, and simultaneously mineral exploration companies increased activity. In 1950, Australia's GDP per capita was $7218, or fifth in the world, led now by the powerful United States of America.

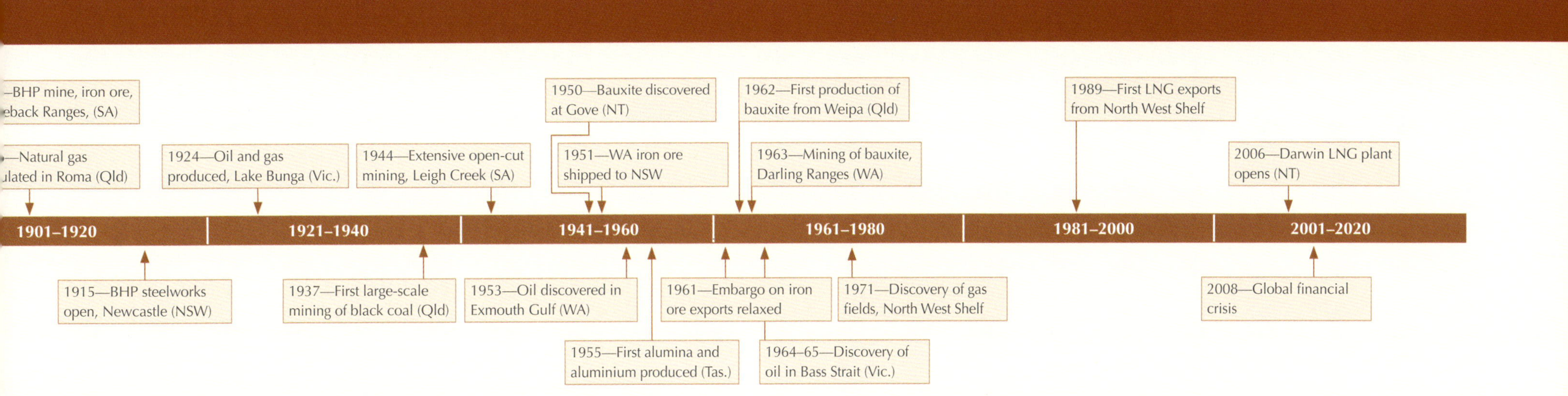

Did you know?

9.1: Fallout from the Poseidon crash

A small nickel (Ni) deposit at Windarra, Western Australia, was the subject of one of the most spectacular stock market 'bubbles' in Australian history: the Poseidon crash of 1970. Speculators drove the 80-cent stock to $280 before its inevitable crash. Many investors suffered losses and these, together with the negative media coverage, soured investment in the mining industry.

The government established the Rae Committee (1974), which recommended a number of changes to stock market regulation. The Poseidon bubble also led to the development of codes for the reporting of ore reserves, which were the forerunners of the present-day Joint Ore Reserves Committee (JORC) code, which has (along with the Canadian NI 43–101) become an international standard for reporting ore reserves.

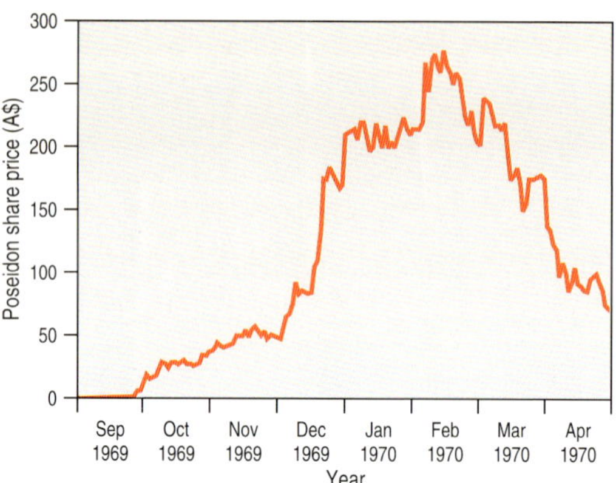

Figure DYK 9.1: The share price for Poseidon from late 1969 to 1970. The nickel boom saw an 80-cent stock rise to $280 in a matter of months. Some made fortunes, but many investors lost their money from the inevitable crash from these dizzy heights. (Source: Simon, 2003)

Following exploration successes, a highly capital-intensive mining sector developed, but it did not add to the total work force in the same way as manufacturing or agriculture. The emergence of large markets in Japan, in particular, together with rapidly improving geological knowledge acquired by government geological surveys and measures taken to improve the efficiency of mining, fuelled this resurgence. Apart from Japan's economic growth, the general expansion of the world economy in the 1950s and 1960s meant an ever-increasing demand for minerals.

Australia, with its established industry, had the framework in place that enabled the development of new deposits to meet this demand. A major expansion of Australia's aluminium industry followed the decision to mine bauxite in the Darling Ranges, Western Australia, in the early 1960s. The perception that high-grade iron ore resources were limited in Australia was turned around with the discovery of vast resources in the Pilbara in the 1960s. The economics of working some previously uneconomic deposits changed remarkably because of technological advances, which lowered the cost of mining and transporting huge quantities of material. These advances would not have been decisive without the emergence of Japan as a major buyer of coal, iron ore, bauxite and, later, LNG. Around 1964, a pivotal crossover occurred, with Australia's growing exports to Japan exceeded the falling exports to the United Kingdom. In 1973, the GDP per capita in Australia was $12 485, or 10th in the world.

The Mineral Securities Australia Ltd (Minsec) collapse of 1971 sparked the most serious money panic in Australia since 1893. A poor call on the value of Poseidon shares during the nickel boom (*Did you know?* 9.1) culminated in the failure of some financial institutions. Exacerbating the Australian situation was the 1974 international oil shock and associated impacts on the global economy.

Shaping the economy and the people: bulk commodities (1975 to present)

The brief history detailed above highlights the uncertainty of economic security. Australia has moved, or by some measures lurched, from one means of sustaining itself to another. By the late 1970s, the rate of growth of the mining industry in Australia, which had been maintained for more than 15 years, began to slow. New mines had been developed around the world to meet a forecast demand for minerals, which turned out to be overly optimistic. The Australian industry's costs had increased but mineral prices generally had not. The industry was largely dependent on exports and had to compete for sales with an increasing number of mines in other countries; some of these mines were less affected by cost increases, or were assisted in various ways by their governments. New coal mines were established in Australia after the second oil shock in 1979, but world demand slowed, leaving Australia with surplus capacity for a time.

Between 1983 and 1985, the then Treasurer Paul Keating deregulated the banking system by floating the Australian dollar (1983), granting 40 new

© Getty Images (W Krecichwost)

The resources boom epicentre, a view of the Perth city skyline from Kings Park, Western Australia.

foreign exchange licences (1984), and granting banking licences to 16 foreign banks (1985). Despite slow global growth, exports to Japan continued to expand. In the late 20th century, massive new industrialising markets emerged in China and, to a lesser extent, India. In 2000, GDP per capita was US$20 866 or 23rd in the world. By 2006, however, it was US$37 433 and 15th in world, the increase a function of the high prices paid for Australia's quality resources. The trend has continued to 2011 at US$40 836 or 18th in the world.

One aspect of the ongoing economic development of Australia is the impact on the quality of life of Australian people. Australia's population is increasing at 1.4% per year, which is leading to increasing impacts on the resident population. These effects are not limited to economic matters, but include environmental, employment and social costs (including housing shortages and rising house prices). The prevailing circumstances mean that some of these costs can be ameliorated by financial means, such as higher wages and lower taxes, which are largely funded through 'dividends' from the resources export industry.

The most important difference from previous booms is the driver, the industrialisation of the BRICK group of countries—Brazil, Russia, India, South Korea and, particularly, China. Other mining booms can be linked to the industrialisation of economies—Japan in the 1960s and 1970s, for instance—but the size of the population in the BRICK group has made the current boom larger and longer than previous booms. Over the past two decades, real growth in China has averaged 9.4%. By 2025, China's cities are expected to grow by more than 200 million people, which equates to creating three Sydneys

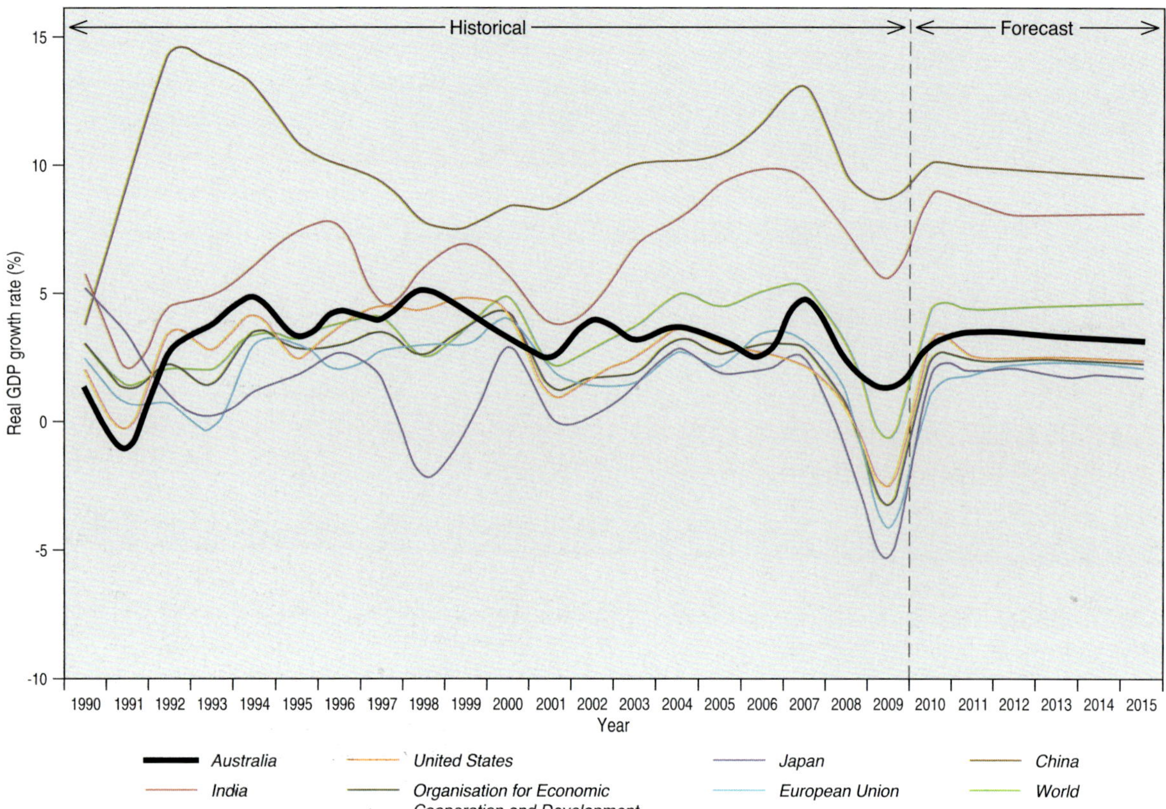

Figure 9.3: Growth of GDP in selected countries. Note that Australia endured a much smaller drop in growth rate following the global financial crisis than did equivalent developed economies (such as the European Union and Organisation for Economic Cooperation and Development). Australia's growth mirrored more closely that of India and China, as a result of their ongoing demand for Australia's bulk commodities. (Source: World Economic Outlook Database from IMF)

per year. As a consequence of this growth, China has become a net importer of many commodities, thereby driving up their global price, and China is now the largest destination for Australian exports. One of the reasons that Australia did not feel the full effects of the GFC is that Australian exports, particularly of minerals, were tied to China, which continued to grow (Figure 9.3).

The demands of China and India for raw materials have also affected, and will continue to affect, the ownership of Australia's mining industry. Traditionally, the funding and resulting equity in Australian mining houses came from Britain, Europe and North America, but in the current boom much of the capital invested in Australian mining has come from China. This reality will continue to affect government policy on overseas investment.

A final difference between this and previous booms may be its longevity (Figure 9.1). Many analysts argue that the present boom looks set to continue at breakneck speed for the foreseeable future—although, no doubt, the same was said for each of the previous booms. The industrialisation of the BRICK group of countries will eventually bring 2 billion people to developed-world living standards. The resources required to achieve this may continue to fuel Australia's current mining boom for years to come, provided that investment continues for projects and Australia maintains a competitive edge in minerals geoscience.

Living in contemporary Australia

Most Australians in the early 21st century have an enviable quality of life. Although not without its societal and environmental problems and issues, Australia is fortunate today to have one of the best-performing economies in the developed world. Australia is a middle-sized power, but exerts an influence on the world beyond what one would expect from the 22.7 million people who live 'down under'. The contemporary economic dynamic that has positioned Australia is underpinned by geology, especially by the big four bulk commodities of coal, iron ore, bauxite products and LNG.

These bulk commodity drivers are feeding the demand of our neighbours in Asia, and this has led to a post-2000 world where the economic reality is no longer predictable, as seen in Figure 9.1. China is Australia's largest merchandise export market, accounting for $46.5 B in 2009–10. This is followed in order by Japan ($37.1 B), South Korea ($16.5 B), India ($16.2 B) and the United States ($9.5 B).

Strong domestic demand can readily constrain a country's capacity to export. For example, the United States, with around 27% of global coal resources compared with Australia's 9%, has not developed its export industry for coal to the same degree (Figure 9.4; Table 9.1). This is a direct result of the need for the United States to first satisfy its domestic market and only export the surplus. There is a contrast between the two nations in domestic and export consumption; for Australia, domestic demand is less than 70% of production. Limited domestic demand has facilitated Australia's role as the world's largest exporter of coal and iron ore.

Australia does not have a 'top 10' LNG resource, nor the largest known resources of bauxite, but it is the world leader in bauxite production and a prominent and growing producer of LNG (Table 9.1). Whether resources are developed and

Table 9.1: Australia's world rank for production and resource base

World rank	Resources	Production
Bauxite	2	1
Coal	4	4
Iron	4	3
LNG	14	4
Rutile[a]	1	1
Zircon[a]	1	1
Ilmenite[a]	2	2
Uranium	1	3

(Sources: Geoscience Australia, 2011; US Energy Information Administration, 2011; DRET, 2011)
[a]Rutile, zircon and ilmenite are derived from mineral-sands deposits.

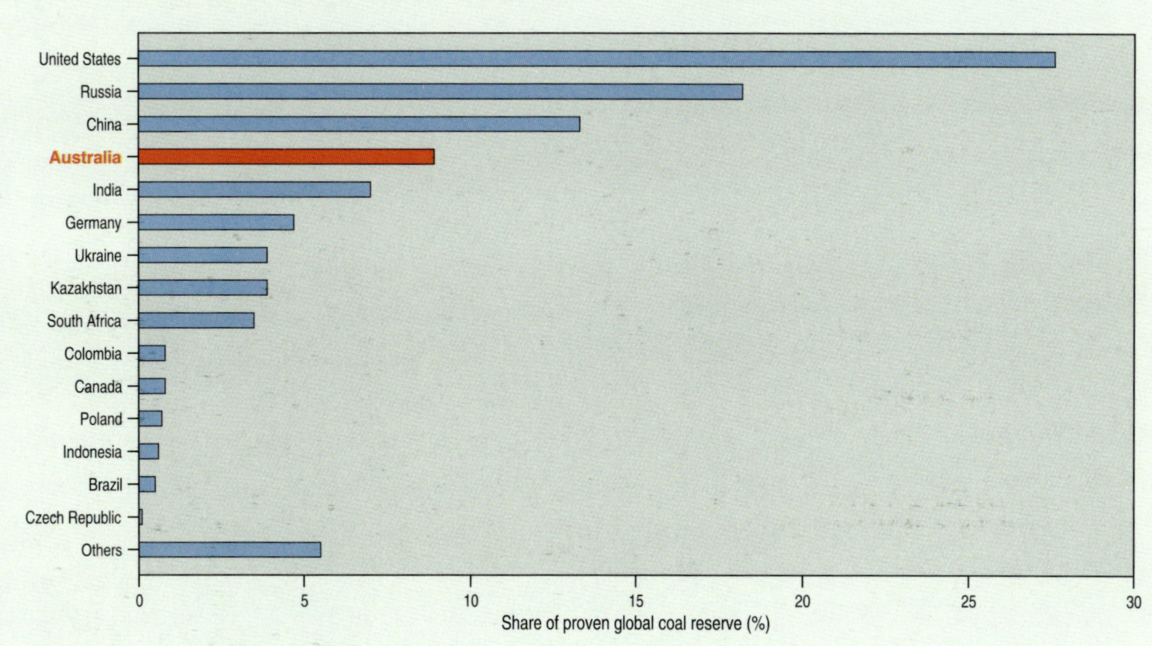

Figure 9.4: Share of proven global coal reserve (2010). Australia holds a relatively small coal resource compared with its dominance of the global export market. (Source: BP Statistical Review of World Energy, 2011)

Image by Mining Photo

The Gladstone alumina refinery, a key technology investment in central Queensland.

exported depends on a myriad factors, including the domestic market, location of resources, political stability, educated workforce and established legal structures. Although Australia does have a strong resource base, it is these external factors that have spearheaded development and led to an extensive export industry. Access to land and funding for exploration remain key factors for the industry's future because new deposits provide the opportunity to expand resources.

Australia has earned a well-deserved reputation as a reliable and competitive supplier of a wide range of high-quality mineral and energy products (Figures 9.5 and 9.6). Prevailing conditions in the global marketplace constrain Australia to the role of 'price taker', rather than setting prices according to the cost of production. The peaks and troughs of demand reflect economic cycles (Figure 9.3) and further reinforce Australia's role as a price taker.

Historically, industrialisation and growing incomes have led to shifts in the composition of economic activity to encompass higher shares of consumption and services, at the expense of other sectors such as investment, construction, manufacturing and mining commodity industries. Although China and India are likely to follow this path, the vast size of their population means that this transition could be expected to take years, possibly decades. Given this, it is likely that growth in their consumption of raw materials will remain robust for some time yet.

The rapid industrialisation of China, in particular, has led to an unprecedented market for Australia. China is the most rapidly mobile population in history, with hundreds of millions of people leaving rural areas for the eastern cities in a matter of decades. Cities require resources, such as steel, concrete, energy, water and food. This process of industrialisation, and the requirement particularly for coking coal and iron ore, repeat the industrial 'revolution' that occurred in the west some 200 years ago.

Contrary to some expectations, there does not appear to be a long-run tendency for the prices of mined commoditities to fall relative to the price of manufactured goods. Mineral prices have remained very strong in recent years. Most notably, increased demand from steel producers, especially in China, has driven up the prices of iron ore and coal. Australian mining houses are trying to ramp up supply in response, as are producers elsewhere in the world. The efforts of mining companies to increase production can be hampered by large fixed costs, the capital-intensive nature of

Photo courtesy of DW Stock Picture Library, Sydney

Australians enjoying the sun at a cafe near the Sydney Opera House, New South Wales.

modern mining, the lack of critical infrastructure and competing sectors for investment return. Consequently, new production may lag the initial investment decision by a number of years. A result of this lag is that prices run up excessively, and this can be followed by periods of weakening as global supply expands, with prices falling in absolute terms during recessions. Past experience suggests that the projected increase in demand will underpin current prices for the medium term, but beyond that prospects are less certain.

The depletion of higher quality resources could, in time, make commodities more expensive to mine. Furthermore, declining grades and the requirement for more complex levels of processing could lead to increased environmental impacts and higher production costs. Hence, there is an ongoing imperative to discover and delineate new resources to sustain future production.

Societal benefits

Jobs and services for Australians

The early 21st century has seen Australia's capital cities thrive. Restaurants and cafes work to capacity, traffic is frustratingly busy and employment is strong, all indications that the economy is providing positive outcomes. The minerals sector directly employs only a small proportion of the Australian workforce, around 130 000, representing only about 1% of the total labour force. The intense activity in the bulk commoditites sector is characterised by materials handling and technical/professional support based at largely static locations, thus having some parallels with the manufacturing sector.

The social effects of the cycles in the minerals and energy sectors can be profound. During boom times, workers are drawn to remote areas for high wages. In 2010, the Australian Bureau of Statistics calculated average mineral industry wages at around $103 000 compared with hospitality at $46 000 or construction at $65 000. Often the communities these workers are drawn from find the remaining labour pool inadequate to fill necessary jobs. At the same time, the remote towns rapidly increase in population, straining physical and social infrastructure. Underwriting the development of regional infrastructure, including railways and roads, manufacturing and information technology, the Australian minerals industry also contributes to

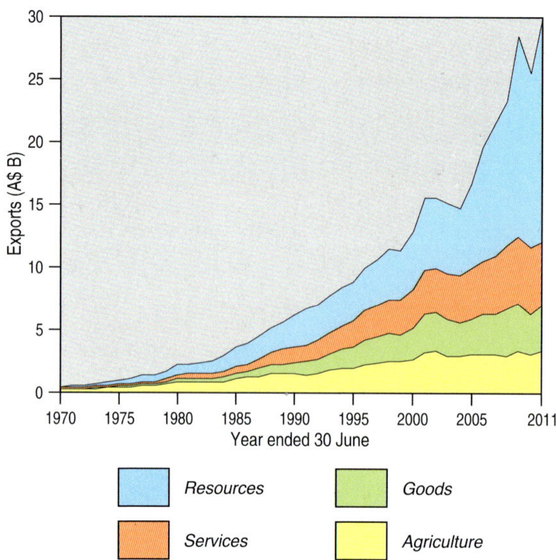

Figure 9.5: Australia's exports in dollars of the day by financial year ended 30 June. Note the dramatic rise in resources growth from 2005. (Source: Bureau of Resources and Energy Economics, 2012)

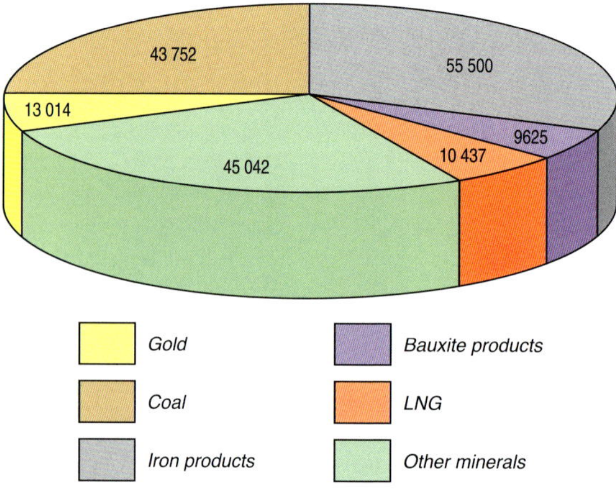

Figure 9.6: Australia's mineral export value in millions of Australian dollars to financial year ended 30 June 2011. Note how coal and iron products dominate the mix. (Source: Bureau of Resources and Energy Economics, 2012)

the economy in multiplier effect. There is a saying: 'for every job in the resources sector, another four jobs are created'. While the precise ratio is debatable, it is significant that the reverse situation also occurs after the booms, with attendant social implications.

The benefits of the minerals industry to the Australian community extend far beyond direct labour and exports (Figures 9.5 and 9.6). The Australian minerals industry provides the greatest value adding of any sector to the Australian GDP per person employed. The wealth generated by the bulk commodities industry has contributed substantially to the support of community services such as social equity payments, hospitals, police, infrastructure and other government services. Australia is a developed country with a highly educated workforce and is generally an early adopter of new technology. In effect, the bulk commodities, along with other resources exports, allow Australia to fit the mould of a developed first-world country, without a large manufacturing sector that traditionally characterises such economies.

Regional infrastructure and economic development of regional Australia

Large areas of Australia have been opened up by developments associated with the bulk commodity export industry, despite the limited area actually required for the mining operations. Towns, community facilities, transport and communications infrastructure, including roads, ports, airports and railways, have been built almost entirely by resource development, funded by private capital.

The Pilbara and surrounding areas in Western Australia provide a contemporary illustration of the development impact of resources. Prior to the iron ore discovery, there was a small population and limited infrastructure in this remote region. Today, there are significant communities at Newman, Port Hedland and Karratha, towns largely based in mineral extraction and shipping, with purpose-built railways and ports, along with supporting infrastructure such as airports. The heavy-duty railways of the Pilbara region have pushed the limits of wheel-to-rail interface technology, adding useful knowledge to railways worldwide. Rio Tinto plans to further expand iron ore production capacity on the railway line linking its mines to the port at Dampier. Without the development of the Pilbara's iron ore and the North West Shelf gas, the region would probably have remained largely underdeveloped.

Late 19th and early 20th century mineral sector development saw new mines planned as large-scale operations, with the construction of adjoining residential areas from the very beginning. Large workforces needed to be housed, and provided with community services and transport facilities developed to handle millions of tonnes of product each year. Rather than provide these facilities, governments in Australia made it a condition of many new mining leases that the companies did so, with major financial investments not only in the mining operation but in social infrastructure, such as streets, houses, schools, hospitals and recreation facilities.

National and regional economic benefits

The minerals sector contributed 9% of Australia's GDP in 2011, being the second largest sector after finance and insurance, according to the Reserve Bank of Australia. Only in economies such as Canada, Norway and the Middle Eastern gulf states does the production of minerals and energy play as significant a role. In most developed economies, mining represents less than 2% of GDP.

The benefits of the minerals industry to the Australian community extend far beyond export profits, with economy-wide impacts made through the multiplier effect. Put simply, the demands of the resources sector support a myriad other activities, such as services, manufacturing and technical and intellectual development. Despite the remote location of the resources, the cities around Australia are dependent upon the resources industry to varying degrees. Perth is the support and supply base for numerous Western Australian petroleum and mineral operations. The city's economy strongly reflects this dependence on prevailing circumstances in the resources sector. The Western Australian Government alone has a 'Royalties for Regions' program that aims to spend around 25% of the royalty revenue in the regional areas that created the wealth. In 2009–10, an additional $619 million to the regular Budget programs was provided for regional communities. Projects are wide ranging and include health, education and infrastructure, as well as significant spending ($80 million over five years) on acquiring new geological knowledge of the state.

Societal impacts

The success and opportunities brought to Australia by the bulk commodity trade have not been without issues for society. Around 0.02% of the land area of Australia is physically affected by mining and resource development. This is about the same sized area as the citizens' residential driveways. Although this area of mining disturbance is small compared with that caused by other activities such as real estate development, farming and grazing, road construction and urban growth, the minerals industry must responsibly manage the environment in which it operates.

The risks of higher inflation, acute skill shortages and potentially some symptoms of what economists know as 'Dutch Disease' have been at least partial consequences. Dutch Disease was first identified in the natural gas export boom years of the Netherlands. One effect of Dutch Disease is that a currency is strengthened in its price parity with other currencies. It can have a negative effect on the competitiveness of other industries and activities in those regions not benefiting from the resource trade. A current example in Australia is the high price of the Australian dollar (above parity with the US dollar through much of 2011), which makes Australia an expensive destination for inbound tourists and overseas destinations cheap for Australian travellers. Australia had been able to avoid more severe impacts, but the phenomena associated with a 'two-speed' economy have influenced the policy considerations of government and other authorities. To avoid negative impacts of inflation the Reserve Bank of Australia was raising interest rates, while comparable economies were lowering theirs.

Photos courtesy of Sibelco Australia

Land use prior to mining influences whether rehabilitation might be a return to grazing land or, in this sand-mining case on Queensland's Stradbroke Island, facilitating the growth of native species to create a self-sustaining native ecosystem.

The resources sector is cyclical in nature. Rapidly increasing production and export levels of the previous five years stalled during the 2008 GFC, following significant falls in metal prices. This downturn brought reductions in capital spending, exploration and workforce in Australia and internationally. Viewed in hindsight, this may appear as a small negative deviation; however, the effect on industry and individuals at the time was dramatic.

The rise of fly-in/fly-out (FIFO) mining, whereby personnel are flown to and from capital cities on charter flights direct to the mines, may also present societal challenges. Rosters are variable, but nine on-days days and five off-days are common. Life at the mine is based around a generally well-equipped camp, with a canteen, wet mess (bar), internet access, gym and other recreational facilities. Although the majority of the workforce is male, women are well represented in some areas, particularly as truck drivers and geologists. FIFO has not been a major part of the bulk commodities workforce to date, due to the existence of well-established residential towns. However, it has seen significant growth from a small base. People are drawn by higher potential wages from the trades and professional jobs in the southern cities of Australia to work in remote regions at the mines and other support services to the minerals industry. Many employees, however, view the FIFO lifestyle as a transient phase. Once they have achieved a financial or career goal, many workers move onto a more traditional job environment, perhaps in the company's urban headquarters or with another employer.

The environment and social licence to operate

We have stressed that many bulk commodity operations are located in areas remote from major metropolitan centres. Even so, the industry increasingly finds that it is required to justify its activities in competition with other potential land users. From the 1960s onwards, the mining industry became increasingly influenced by public concern for the quality of the environment—leading to what is known as a social licence to operate. With the rising awareness that preservation of natural features such as scenery, plant and animal habitats, cultural and even geoheritage had a value to society, governments increased controls, for instance, on discharge of potentially polluting emissions, such as water containing sediments or chemicals, and noxious gases.

Initially based on the more densely populated coastal strip, particularly ocean beaches, the mineral sands industry was early to face difficulties in their social licence to operate. Although Australia

The Pilbara iron ore operations use some of the world's longest trains. This train on the Tom Price Railway is destined for Dampier, Western Australia, where the ore will be unloaded prior to export.

Image by RSW Images

7097

Image by Mining Photo

Mining directly employs few people relative to the export income it derives; however, the economic multiplier effect leads to jobs in many other sectors.

continues to play a major role in the global output of commodities mined from mineral sands, especially heavy minerals such as rutile, zircon and ilmenite (Table 9.1), some proposed mines have been mothballed and are under threat of closure due to community objections. From a revenue base of $1.15 B in 2005–06, mineral sands fell to $1.08 B in 2010–11. This is despite Australia's prominent position as a supplier and the rapidly increasing global demand for commodities. The media profile and community awareness of the mineral sands industry reflects the proximity of the resource to areas recognised as high value to the Australian population for tourism and conservation purposes, as opposed to the experience of the coal, iron ore, aluminium and natural gas industries, in general. Significant new issues have arisen in recent years that may lead to consequences for the other bulk commodity industries that have been faced by minerals sands project proponents for many years. The suite of conflicts associated with competing land use, such as at Liverpool Plains in New South Wales for the Caroona coal mine proposal, proposed coal-seam gas projects in southeast Queensland or conventional gas at James Price Point in the Kimberley region of Western Australia, will be dealt with on a case-by-case basis.

Recent coal-seam gas proposals in the eastern states have experienced objections paralleling those that arose earlier over sand-mining proposals. The physical location of the wells and associated infrastructure can impact upon a significant area of arable land (e.g. Bowen and Surat basins in Queensland) (Figure 2.9). One company even has plans for drilling for coal-seam gas in an inner Sydney suburb. The recovery process—called 'fracking'—uses a chemical mixture injected under pressure to improve the permeability structure of the gas-bearing coal seams. There are community concerns about the safety of the groundwater due to pollution from these chemicals, some of which may be carcinogens.

The dewatering of coal seams may impact on existing groundwater users (Chapter 7). Brackish water is also produced during dewatering. Disposal of this saline production water can present an environmental issue. Utilising a series of evaporation ponds to reduce the water volume takes up land area, and the resulting highly saline brine must be disposed of sensitively. Recent approvals have required the implementation of other technologies to treat production water and dispose of associated waste streams. Coal-seam gas operators are required to treat and dispose of the wastewater to maintain groundwater pressure as a first priority, and in some cases to allow other beneficial use of the extracted and treated water. Reinjection of treated water to repressurise artesian aquifers and the use of saline water for local coal-wash plants are options being actively explored.

Prior to the advent of commercial coal-seam gas operations, natural gas accumulations were usually vented to the atmosphere. As methane is a highly intensive greenhouse gas, with a global warming potential significantly higher than carbon dioxide (CO_2), this represented not only a wasted resource, but also an emission source for greenhouse gases. Exploration and mining leases are now required by law to be rehabilitated; some rehabilitation involves reshaping and revegetating the surface, often resulting in a significant increase in land quality. Most interesting is that resource companies are the largest sector employing Australian environmental scientists and, increasingly, Indigenous workers. Companies are often able to draw upon the knowledge of local Indigenous people to ensure the best results for rehabilitation.

Environmental and social concerns in relation to the mining industry have also become global. Widespread uptake of new communication technologies occurred around the same time as the Australian mining industry grew from a large national sector into a world player. This increasing globalisation has led to an increased consciousness regarding the environmental consequences of exports once they have left Australia's shores. For example, there have been concerns over CO_2 emissions resulting from the burning of exported Australian coal, although many studies have shown that mixing the coal with lower quality local resources can result in large environmental benefits. The low sulfur and ash content of Australian coal has led to marketing efforts to 'package' it with local coal for more efficient use in developing nations. Gujarat NRE Coking Coal Ltd exports Australian coal to India, where it uses the ash by-product in the company's cement works, thereby maximising resource utilisation. The low sulfur content of Australia's coal resources presents a number of environmental and engineering advantages globally over many other coal resources.

The market recognises this relative advantage with regard to sulfur content, and Australian coal commands a market premium as a consequence (Figure 9.7). About 40% of the world's electricity is generated using coal, and this proportion is steadily increasing. By comparison, nearly 80% of electricity in Australia is generated by coal-fired power stations. In capital-intensive industries, such as power generation, it is technologically difficult to achieve rapid change. Therefore, if there were to be a global move to restrict the use of coal because of climate change concerns, such a transition would likely take decades.

Production of bulk commodities inevitably results in emissions of CO_2; however, significant advances have been made in reducing the impact through the incorporation of newer technologies.

Australian industries have long argued that they operate in a world market and that, if similar environmental operating conditions are not placed on all producers, they will be at a competitive

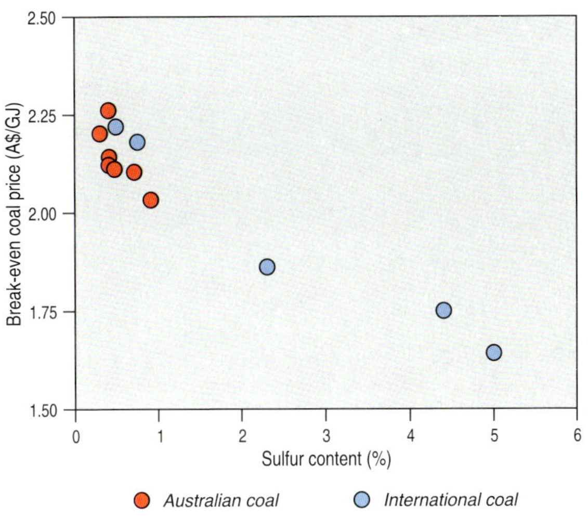

Figure 9.7: Impact of sulfur content on the break-even coal price expressed in gigajoules (GJ). The graph shows that the low sulfur coals (<1%) command the highest prices. (Source: Australian Cooperative Research Centre for Coal in Sustainable Development)

© Getty Images [P Hendrie]

Modular accommodation buildings for workers at the Mount Whaleback mine, Newman, Western Australia.

disadvantage. Australian and international law on climate change remains in a state of flux, and pressure is mounting in the domestic and international arenas. It is conceivable that developments in international law could force changes in the way that Australia's greenhouse emissions and trade impacts on global emissions are calculated.

Technology and innovation: keeping ahead of the game

Contrary to the perception in some parts of society that the resources sector is a 'sunset' industry, the sector has consistently exhibited a high degree of innovation and technical advancement. Contemporary examples are the development and use of software, remote-controlled operations and advanced geoscience expertise; historical examples include initiatives such as mineral flotation and electrolytic smelting. Australian companies are major developers, users and exporters of computer software used to improve and increase the efficiency of mining operations. To lessen its exposure to external forces, Australian industry operators have long been at the forefront of implementing technologies of scale in both coal and iron ore operations, driving down costs and improving productivity. These are important attributes to keep Australian mining operations price competitive and thus major suppliers of the world's mineral commodities.

A more recent phenomenon is the increasing number of mine personnel who are able to operate equipment remotely from the capital cities. For example, Rio Tinto has satellite links with its Australian mines that allow workers some 1200 km away to operate drill rigs, load cargo, use robots to place explosives and perform production blasts. Robotic machines monitored by electronic eyes in the Pilbara region transmit images and data to the Perth operations centre, dig out the broken ore, move it to automated conveyor belts and activate water sprays to remove fines and reduce dust. These remote-controlled operations not only reduce direct financial costs, but also importantly reduce exposure of workers to potentially hazardous situations. The West Angelas iron ore mine in the southern Pilbara is an example of technological development, with an autonomous haul-system fleet accessing the Marra Mamba iron ore deposit, using a FIFO roster for on-site staff. The company's Perth data centre employs 300 personnel; engineers are able to monitor large screens to set movement commands for the remotely automated machines. Ore haulage using driverless trucks has been trialled at Rio Tinto's West Angelas iron ore mine in the Pilbara. A fleet of 160 automated trucks, which will be run from the company's Perth operations centre, is scheduled for 2015.

New technology can compensate for declining quality of ore, such as the carbon-in-pulp technique for gold extraction (Chapter 8). The development of new technology has largely been driven by necessity as easy and more accessible mineral resources have been exploited, leaving remaining resources in more remote locations that are more difficult to develop. The shift to more technologically advanced operations, through the use of remote operators, has the potential to help resource companies attract and retain skilled and valuable employees who wish to minimise the time they spend in remote locations.

Governance: getting the balance right

One romantic view of mining in the 19th century gold rushes is the individual prospector who pegged his claim and worked the ground by hand. In contrast, the mining of bulk commodities is underpinned by massive economies of scale, organised labour, complex technologies and substantial infrastructure investment. Commercial entities developing these projects take substantial financial risks, which governments can mitigate by providing high degrees of regulatory certainty. A stable society, stable political system and well-established legal environment with an adherence to the 'rule of law' have provided the necessary framework for investment that encouraged developers to proceed with large-scale export projects in Australia.

Australia's federal system of government is enshrined in its constitution of 1901. The Australian people live in a country that is divided into six states and three territories (including Norfolk Island), further subdivided into around

Flinders Ranges, South Australia, are the result of Cenozoic uplift of these Neoproterozoic rocks.

Image by Jim Mason

Did you know?

9.2: Community ties in Weipa

Weipa in western Cape York is a small coastal community of about 2000 people, with a unique Aboriginal heritage. Local traditional owner groups include the Alngith and a number of others, including the Aboriginal communities of Napranum, Aurukun, Mapoon, Injinoo and New Mapoon.

Weipa's history is predominantly linked with Rio Tinto–Comalco Aluminium's bauxite mine, which has been operating since the early 1960s. Mining predominates in Weipa, although tourism, especially fishing and ecotourism, is growing rapidly.

The traditional owners, Comalco Aluminium Limited and the Queensland Government, signed on the 14th March 2001, the Western Cape Communities Co-Existence Agreement (WCCCA), making a commitment to focus on Aboriginal economy, culture, education, employment and training. Already there have been improved educational attendance, academic performance and employment outcomes for the local people. Traineeships for local Aboriginal people have increased, and the company has worked with the community to preserve sites of cultural significance and supported cultural activities.

Rocky Point settlement, Weipa, Queensland.

700 shire or city councils (local government). Broad economic policy settings are a federal responsibility and, with the exception of the offshore regime, the regulation of the resources sector is a state or territory responsibility. The monitoring of foreign investment and, where necessary in the national interest, its control is an Australian Government responsibility. Recognising the need for capital inflows to sustain regional growth, Australia's foreign investment policy is designed to encourage investment consistent with the interests of the community. One formal mechanism to ensure consistency is the assessment of foreign investment proposals, with approval subject to their not being considered contrary to the national interest.

Governments in Australia do not seek to participate directly in resource developments; however, there is a historical legacy of some state government involvement in the mining of coal for power generation. State governments, by virtue of their wide powers to regulate matters within their own boundaries, are more directly involved in the day-to-day administration and regulation of mining operations. The granting of exploration rights and mining leases, the approval of mining operations, and the levying of royalties and other like charges is a state responsibility. State governments also consider environmental protection and rehabilitation aspects of development proposals. The cost for all of the environmental works associated with a project is borne by the proponent. State and Australian governments carry out precompetitive 'exploration' and geoscientific mapping in onshore and offshore Australia (Chapter 2). However, the Australian Government has a responsibility in terms of investment attraction, derived through its national economic management role and maintains overriding responsibility for petroleum activities in Commonwealth waters (Chapter 4).

The majority of Australia is arid, and subterranean groundwater in many areas is a scarce, valuable resource (Chapter 7). However, once regulatory requirements are met, groundwater supply can usually be secured where surface water is not available, but this comes at a cost. The large distances between significant sites and settlements in Australia are an overriding aspect. Government support for the resources sector has assisted in the development of rail, road and port infrastructure in some of the more populated regions. Private investment has usually facilitated infrastructure development in the more remote parts of the country, such as the Hamersley iron ore province and the bauxite mines in the tropical north.

A wide range of issues concerning development and sustainability face the minerals sector. Companies need to sustain a high level of community acceptance, and this has been recognised publicly through the triple-bottom-line reporting of social, economic and environmental performance under the Code for Environmental Management developed by the Australian minerals industry. Almost without exception, the majority of Australian mining and petroleum companies prepare and issue sustainability reports. The finite nature of mineral resources and the strong potential for rapid decline in demand represent other challenges. Resources companies and Australian governments work to ensure that these issues are minimised; however, declining grades do raise the potential for increased environmental impacts.

Native title—Indigenous Australians gain a right

The High Court of Australia's landmark decision in 1992 to extinguish the colonial concept of *terra nullius* and recognise native title of the traditional owners was an important step in Australia's social development. By developing regulatory systems that take into account the requirements of government and the capacity of industry, and engaging affected people, especially Indigenous people, in the process, a community licence to mine is sought and often achieved.

The 1992 *Mabo v Queensland* (No. 2) High Court decision finally recognised that Australia's Aboriginal and Torres Strait Islander people held legally responsible rights to land. The High Court proclaimed that 'the common law of this country recognises a form of Native Title which, in the cases where it has not been extinguished, reflects the entitlements of indigenous inhabitants, in accordance with the laws and customs, to their traditional lands'.

The *Native Title Act 1993* commenced on 1 January 1994. In 1998, the Federal Parliament passed a comprehensive package of amendments, which commenced on 30 September 1998. Under the Act (or approved state/territory legislation), applicants for onshore mining or petroleum titles are required to undertake formal negotiations or consultations with native title holders or registered native title claimants who have registered a claim over the area prior to the grant of the titles.

The recognition of native title has significantly enhanced the position of many Indigenous communities in rural and regional Australia. These

Image by Mining Photo

Loaded coal conveyor at wash plant of mine, Clermont Mine, Emerald, Queensland.

Photo courtesy of Newmont Asia Pacific

Aboriginal involvement in mine rehabilitation.

communities have become integral stakeholders in mineral and petroleum resource development. The use of Indigenous land use agreements between Indigenous groups and mining and exploration companies has proven a win–win in a number of cases. Indigenous people have secured cultural support, training and work, while companies have been able to more readily attract workers already living in remote areas such as Cape York or the Pilbara (*Did you know?* 9.2).

Partnership programs reflect the aspirations of Indigenous communities for training and employment and for establishing commercial enterprises. Commercial arrangements are also expanding, under which traditional owners take on ownership of mining ventures and support organisations. For example, the Alice Springs-based CDE Group, which provides mining and civil works services to industry, is 51% owned by Lhere Artepea and 33% owned by the Rusca family, making its Indigenous ownership 84%. The success of the CDE Group was recognised formally when the company won the Contract Miner of the Year award at the 2008 Australian Mining Prospect awards.

The carbon wealth of Australia: coal and the Australian people

The early discovery of coal in Australia stands apart from the more recent, technical or geological discoveries of bauxite, iron ore and gas (LNG). A group of escaped convicts, led by William and Mary Bryant in March 1791, found coal when they beached their vessel south of Newcastle near the entrance to Lake Macquarie. From the ship's record 'we picked with an axe as good coals as any in England … they burned exceedingly well'. For the Newcastle area, this discovery initiated a sequence of pivotal developments that continue to influence the social and cultural fabric of the Hunter region today.

In the years following the Bryant discovery, coal occurrences were reported at many other centres to the north and south of Sydney. The coal industry began in 1798 when ship owners gathered surface coal at Newcastle and brought it to Sydney for sale. Export of Newcastle coal began in 1799 with a shipment to Bengal in India. Coal was discovered in Tasmania by the French near Recherche Bay in 1793; after settlement, coal was mined there from 1803 onwards on a small scale.

Coal is now mined in all Australian states (Figure 9.8). The location and high quality of Australia's coal resources have led to a heavy domestic reliance on this fuel source for power generation. Queensland and New South Wales have the largest black coal resources and production. Victoria hosts the largest resources and only production of brown coal. The coal industry is one of the mineral industry's largest employers, Queensland having more than 88 225 and New South Wales more than 19 000 direct employees at the start of 2011. In that year, Queenslanders produced more than 9500 tonnes of coal per employee from 52 operating mines.

Geological setting of Australia's coal

Australian coal of economic significance occurs in Permian, Triassic, Jurassic, Cretaceous and Cenozoic sedimentary basins. The most economically significant coal accumulations are in the Permian and Cenozoic; however, minor but important measures exist in the Mesozoic. Generally, the Permian coals are hard, with a coal rank between high-volatile bituminous and anthracite. The Mesozoic coals are high-volatile bituminous, relatively hydrogen-rich and carbon-poor (perhydrous) coals. The Cenozoic coals are low-rank lignite.

The Carboniferous was the main coal-forming period for the Northern Hemisphere. In contrast, Australia's Carboniferous coals are poorly developed, with only thin coals known from the Coen region in Cape York Peninsula of Queensland, and also along the northern margin of the Sydney Basin in New South Wales. Australia lay in close proximity to the South Pole during the Carboniferous (Chapter 2). Although climatic conditions for plant growth were suitable, as evidenced by the abundant plant remains in the fossil record, it was relatively dry (Chapter 3). Ice sheets and local valley glaciers extended across the Australian land mass during the Late Carboniferous to Permian (Chapter 2). Formation of peat, and hence coal, was limited, with some minor occurrences in the Cooper and Arafura basins in central and northern Australia.

Permian coal is widely distributed throughout Australia, from the Collie Basin in Western Australia, the Arckaringa Basin in central Australia, the Sydney Basin in New South Wales and the Bowen, Cooper and Galilee basins in Queensland (Figures 2.9 and 9.8). The main Permian coal-producing basins in eastern Australia were formed in the foreland of the convergent Gondwana margin, to the west of a volcanic arc system. The foreland basins and associated arc developed on the early Paleozoic Lachlan Orogen of the Tasman Element (Chapter 2).

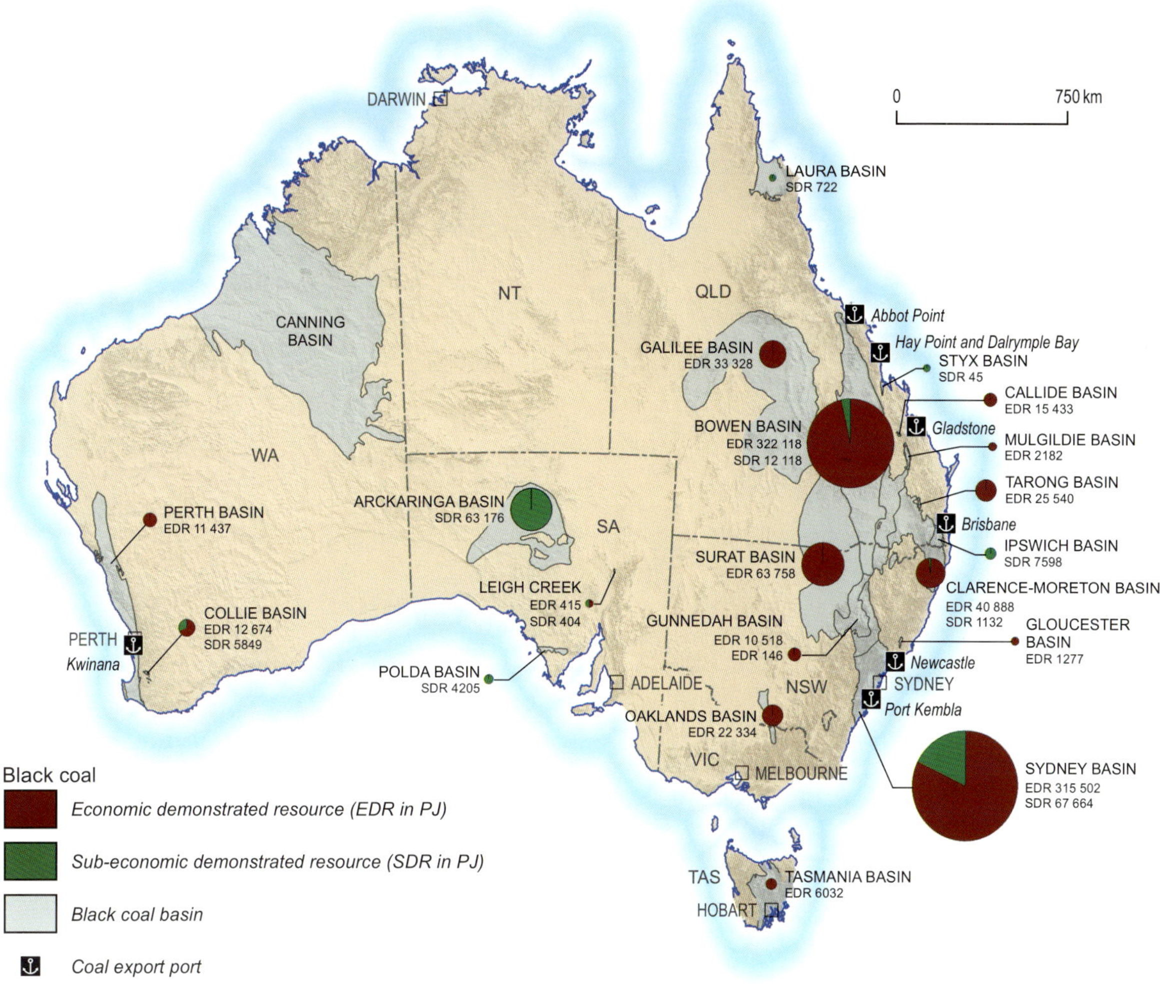

Figure 9.8: Map of the main black coal resources in Australia. Note the dominance of the resources near the east coast, adjacent to the main population centres of the country. Not surprisingly these resources have shaped the energy choice of the Australian people. Nearly 80% of Australia's electricity is generated from burning coal. (Source: Geoscience Australia & Australian Bureau of Agricultural and Resource Economics, 2010) EDR = economic demonstrated resources; PJ = petajoule; SDR = sub-economic demonstrated resources

© Getty Images (P Harrison)

Thick coal seams located close to the surface make open-cut mining very profitable. Hunter Valley, New South Wales.

Why do Australia's Permian coals command such a high price?

Australia's Permian coals are prized around the world; they command high prices on markets and are known for their exceptional thickness, quality and rank. For both coking (metallurgical) and power generation (thermal), these coals have outstanding qualities.

Australia does have other coals, but the Late Permian Sydney and Bowen basins form the bulk of the export coal from Australia. These basins host abundant, relatively easily mined, vitrinite-rich coals, with a sulfur content of less than 1% and a low ash content, generally less than 10% (Figure 9.7).

The main controls on the formation of the Permian coal appear to be a combination of palaeoclimate and palaeotectonic setting. To form thick coal seams, a steady and ongoing creation of suitable accommodation space is needed to allow the initial peat deposits to accumulate. Each of the 30 m (or more) thick seams of coal at Blair Athol in Queensland would have originated from a layer of peat that was more than 400 m thick.

The foreland setting of the basins tended to remain at sea-level, ensuring that the accommodation space was balanced by sediment supply—an ideal situation for peat accumulation. The high-latitude palaeogeography was moist and temperate in the Permian, which resulted in limited decomposition of the accumulating organic matter. This played a key role in peat accumulation and preservation, and is analogous to the modern peatland environment of northern Canada.

Volcanism to the east of the present deposits contributed volcanic detritus, now represented by tuff horizons. Locally, tuffs are interlayered with the coals, such as in the Wollombi Coal Measures from the Newcastle area. An eruption analogous to that at Mount St Helens in 1980 flattened a peat forest in a west-southwest direction from a laterally directed volcanic blast. A thick blanket of volcanic ash from that event overwhelmed the peat swamps in which trees were still growing.

The virtual absence of pyrite (FeS_2) and other sulfur-bearing minerals in most Australian coal resources (Figure 9.7) reflects the fact that these terrestrial successions were flushed by freshwater with low sulfate content. Higher sulfur levels accompanied by high boron levels occur locally in the central eastern area of the basins, reflecting Late Permian marine incursions. Hydrogen sulfide (H_2S), produced by bacterial reduction of the sulfate and combined with dissolved iron in the seawater, only permeated through the peat following these marine transgressions. Pyrite is most abundant in the Greta seam of the Sydney Basin and the Garrick seam of the Bowen Basin. In both cases, the coal seams are overlain by a marine succession.

Coal is a complex organic-rich rock produced by the low-temperature metamorphism of peat, the effects of which are known as rank. The carbon content increases and the water content decreases with increasing rank. The coal rank in the Sydney–Gunnedah–Bowen basins decreases markedly from east to west and appears to be controlled mainly through burial depth, although magmatic intrusions do locally increase coal rank. The highest coal rank occurs in the northeast Bowen Basin. High-rank coals emit their volatiles at high temperatures; low-rank coals tend to produce a crumbly, powdery residue. Medium-rank coals emit volatiles at the same temperature at which they become fluid, causing the coal to froth. The weakly resistive liquid coal material fills with bubbles to form an expanded, carbon-rich material with a large surface area and strong resilient framework, known as 'coke', which is used predominantly for the smelting of iron.

The Sydney Basin is geologically contemporaneous with the Bowen Basin and covers an area of some 35 000 km^2 (Figure 9.8). The Sydney Basin coal successions are covered by a thick and, relative to the Bowen Basin, continuously preserved cover of Triassic sediments. As a consequence, development of coal resources has concentrated on deposits near the basin margins where the cover is thinner. The Sydney Basin passes into the Gunnedah Basin to the northwest, which is a source of both metallurgical and thermal coal and a region of coal-seam gas (CSG) exploration.

The process of coalification in the Bowen Basin took place largely during the Triassic under a relatively high-heat-flow regime, when the highest temperatures developed in the northern and eastern parts of the basin. Low-volatile bituminous or anthracite coals were formed in the north and east as a consequence. Whereas the coals from across the basin are suitable for thermal purposes, only in the northeast was the temperature optimal for forming the moderately volatile coals suitable for coking—an extremely high-value coal resource.

Image by Mining Photo

Dust supression at coal reclaimers and loaders, Dalrymple Bay coal terminal, south of Mackay, Queensland.

High-quality thermal coal can be derived from much of the Sydney–Gunnedah–Bowen basins, with the high quality due to a generally low ash and sulfur content. This coal is ideal for the production of electricity or other uses that require a high heat with few environmental effects, except the generation of CO_2. Metallurgical or coking coal has a set of characteristics that make it of high value and greatly in demand for industrial development requiring large quantities of iron and steel. Coking coals are found in other basins, but only in the Sydney–Gunnedah–Bowen basins where the coal seams have been uplifted and preserved at shallow depths are they amenable for economic shallow-underground and open-cut mining techniques.

Another favourable factor in the preservation of eastern Australia's Permian coal measures is the limited amount of later deformation due to the 350–220 Ma Hunter Bowen Orogeny (Chapter 2). Many of Australia's coal beds are gently dipping, making them relatively easy to mine, whether using open-cut or underground mining techniques. In contrast, the Hercynian and Alleghanian orogenies deformed the European and North American coal measures, producing bituminous anthracites, such as in the coal fields of South Wales and the Appalachians.

Black sunlight: power and industrialisation of a nation

Mining of coal occurs in every state of Australia, but quality varies. The eastern states of New South Wales and Queensland produce more than 96% of all black coal (Figure 9.8), and Victoria produces most of Australia's brown coal. Domestic coal is used to generate nearly 80% of the total electricity, which amounted to 266 TWh in 2008–09. Around 75% of coal mined is exported, mainly to Asia. Brown coal (lignite) is younger and less matured, with a low heating value. Used widely for power generation, it is also made into briquettes and can be converted to liquid or gaseous fuels. The extensive Cenozoic brown coal deposits equate to hundreds of years of reserves in the southeast of Victoria, in the Gippsland Basin and especially in the Latrobe Valley, with the coal used in adjacent power stations. Australian brown coal is not exported. The onshore coal deposits correlate to the coaly sediments offshore that are the source rocks for oil and gas—for example, in the Gippsland Basin (Chapter 4).

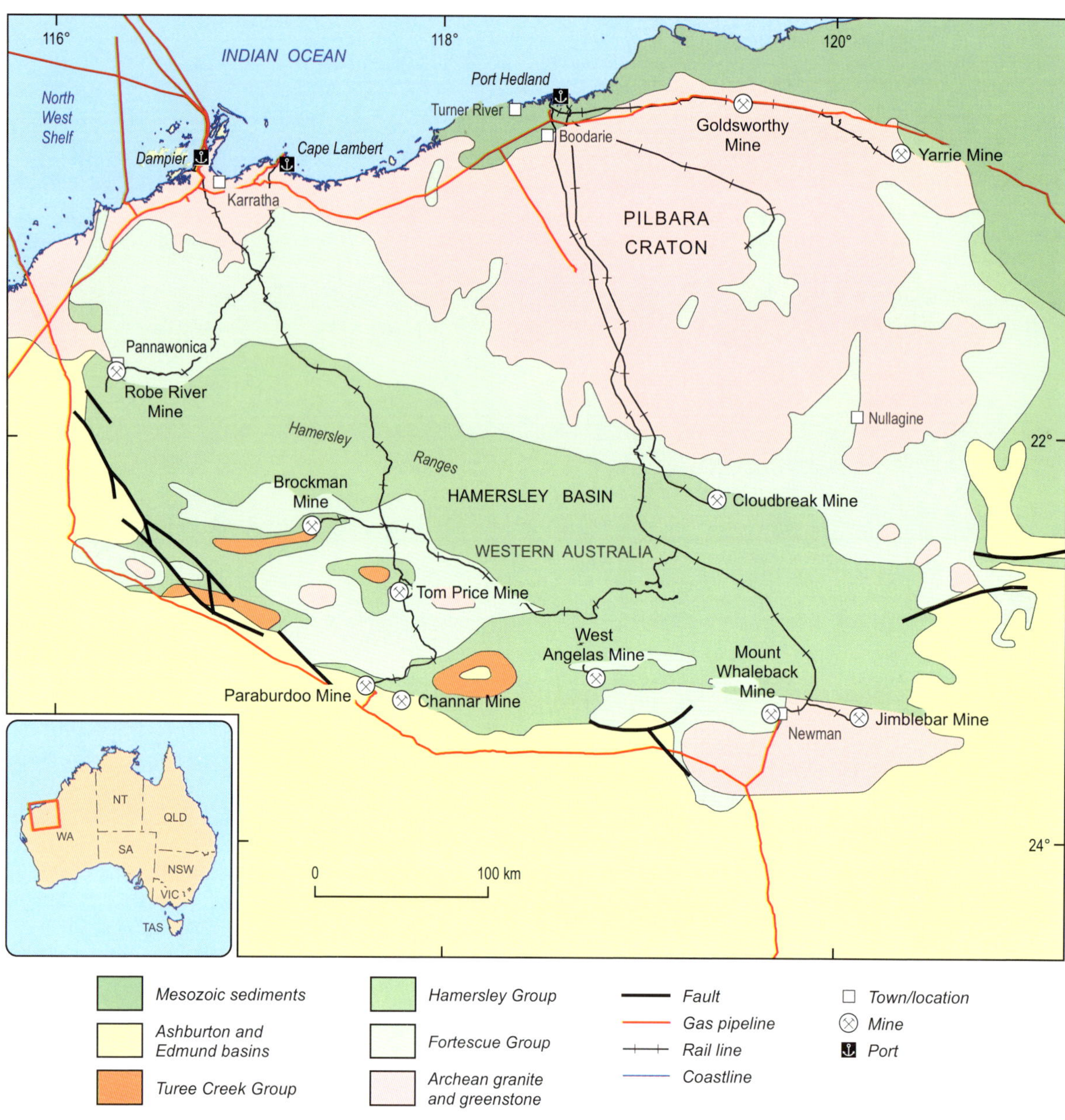

Figure 9.9a: Geology of the Hamersley iron ore province, showing the locations of major deposits and infrastructure. Note the proximity of the North West Shelf gas resources and the iron ore of the Hamersley Basin. The promise of marrying iron ore with nearby energy resources provided the genesis of investment to make value-added hot-briquette iron for export.

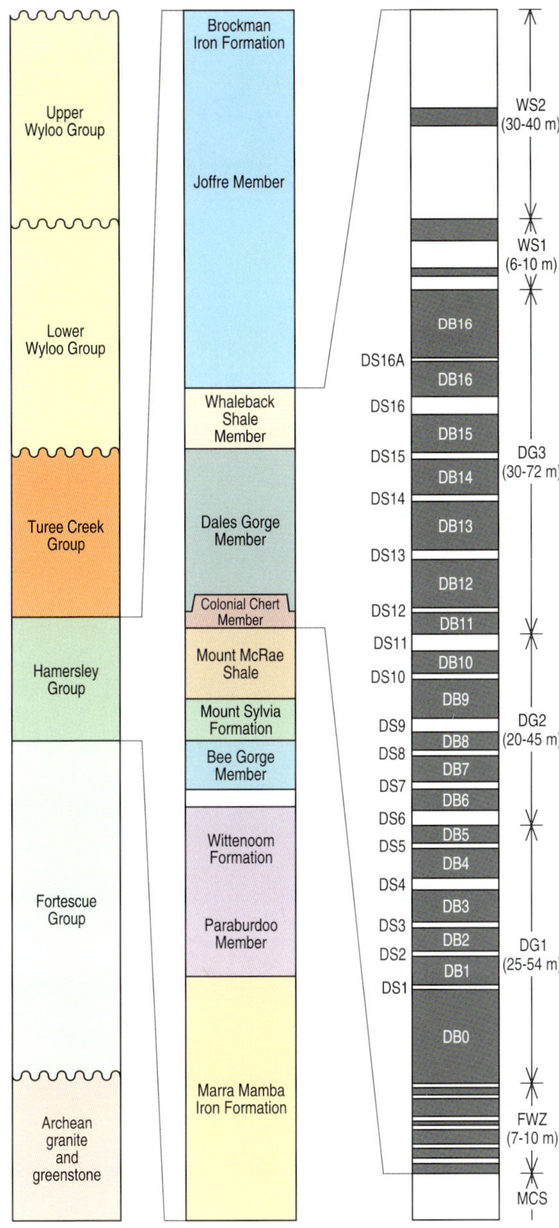

Figure 9.9b: Geological column of the Hamersley Basin, showing details of the lower Brockman Iron Formation. The right column shows the detailed stratigraphy of the Mount Tom Price mine, which consists of alternating iron formation (DB series) and shale (DS series). (Source: modified from Taylor et al., 2001)

From the red centre to greenbacks: Australia's iron ore and steel

Red, the iconic colour of Australia's outback, reflects a landscape rich in iron oxides (Chapter 5). To Australians, this richness of iron extends from the past, with the red ochre, much valued by Aboriginal people, being hematite. The Pilbara region of northwest Western Australia is prized for its hematite, with a world-class export industry feeding the steel mills of a growing Asia.

Iron ore deposits were first discovered in Tasmania in 1822. By the late 19th century, iron ore had been identified in all of the colonies. Low-grade iron ore provided feed for Australia's first blast furnace at Mittagong, New South Wales, in 1848. High-grade iron ore mining commenced in South Australia's Middleback Ranges in 1900. Prior to a government ban on exports (see later), specialist resources such as the magnetite of Savage River in Tasmania and the pisolitic iron ore of the Cobar region of New South Wales provided limited-scale iron exports.

Although prospector Lang Hancock is generally credited for discovering the Hamersley iron ore province in 1952, the potential of this area had been recognised more than six decades earlier by Western Australian Government geologist Harvey Page Woodward. Hancock, however, did set in train political changes necessary to develop the province (Box 9.1). Since the 1960s, Australia's iron ore industry has grown remarkably. Australian exports of iron ore rose from virtually zero in 1965 to 80 Mtpa in 1974 and further expanding to more than 293 Mtpa in 2009. In 2011, $59 B worth of exports were generated, making it the largest export industry in Australia after coal ($47 B). In terms of people, the Pilbara region has grown from a sleepy backwater to having a resident population of more than 51 000 people in 2011, although many workers are transient, preferring a FIFO roster. Growth of iron ore production in the Pilbara region has not been confined to the larger global corporations. Fortescue Metals Group, for example, started to mine the Cloudbreak, Christmas Creek and Marra Mamba Iron Formation deposits in the Fortescue Valley between Newman and Nullagine. Furthermore, development of the Christmas Creek to Port Hedland railway line has added significantly to the region's infrastructure.

Hancock's prize: the Hamersley iron ore province

High-grade iron ore deposits, particularly microplaty hematite ores, generally require multiple stages for their formation. The first stage involves deposition of banded iron-formation, and the second (and later) stage(s) involves upgrading of this iron-bearing deposit by dissolution of deleterious material during moderate-temperature hydrothermal fluid flow and supergene processes.

Banded iron-formation, particularly the type that dominates the Hamersley iron ore province, formed on Earth before the *ca* 2.46 Ga Great Oxidation Event. This early Proterozoic event saw a rapid and dramatic increase in atmospheric oxygen, driven by an increase in oxygen-producing life forms (Chapter 3) and/or the escape of hydrogen into space. During the late Archean, the deposition of sedimentary rocks in the Hamersley Basin occurred with little free atmospheric oxygen. The oceans were overall reduced, although they were probably stratified with a more oxidised upper layer. Interaction between the deep, reduced oceanic waters and mafic crust resulted in bottom waters that were highly enriched (to hundreds of parts per million) in reduced iron.

The Hamersley iron ore province is hosted by rocks of the Fortescue and Hamersley groups (Figure 9.9), or their weathered/eroded derivatives. They formed between 2.78 Ga and 2.41 Ga, and consist of the basal mafic–volcanic–dominated Fortescue Group and the conformably overlying sediment-dominated Hamersley Group. The latter contains four banded iron-formations, the *ca* 2.60 Ga Marra Mamba Iron Formation, the *ca* 2.52 Ga Mount Sylvia Formation, the *ca* 2.46 Ga Brockman Iron Formation and the *ca* 2.45 Ga Boolgeeda Iron Formation. The Marra Mamba and Brockman iron-formations contain all major primary iron ore deposits currently being mined in the Hamersley Province (Figure 9.9).

The Hamersley Group was deposited upon a relatively shallow-water platform formed by the Fortescue Group and the older granite–greenstone

Exposure

(5) Erosion and weathering stage
Deep weathering of ore bodies leads to removal of apatite

Microplaty hematite–martite ore
Modification stages
Structural modifications of ore bodies (not for Mount Tom Price)

Burial stage
Burial of ore bodies under thick sequences of the Wyloo Group

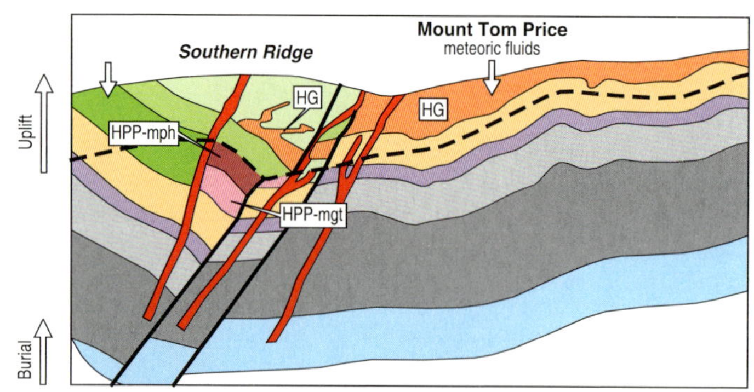

Erosion stage
Continued uplift across major faults leading to erosion of the tops of mineralised systems.
Deposition of hematite conglomerates

(4) Leaching of carbonate and remaining silica
Microplaty hematite–martite–apatite mineralisation

(3) Deep meteoric circulation stage
Deep circulation of oxidised, low-salinity meteoric fluids
Overprint by microplaty hematite
Microplaty hematite–martite–apatite-ankerite mineralisation

Continued uplift
'Cessation' of upward flow
Downward flow takes over

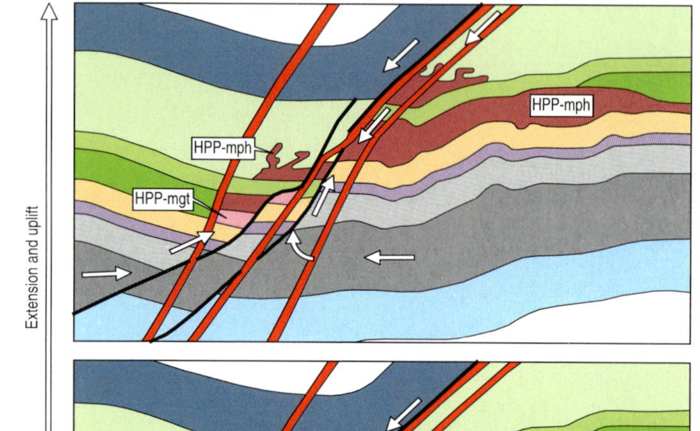

(2) Hypogene stage
Upwards migration of basinal brines
Magnetite–carbonate–apatite mineralisation
Intrusion of dolerite dykes

Start of uplift
Fluid overpressure in sealed Paraburdoo Member

Panhandle fold event
Localised WNW trending fold corridors

Extension
WNW to NW trending normal faults (Southern Batter Fault)

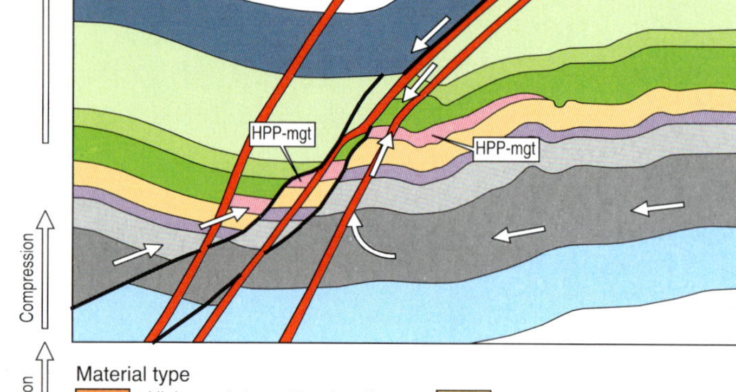

Ophthalmia fold and thrust belt
Development of a foreland basin, followed by regional deformation (folding and thrusting)

Burial stage

(1) Deposition stage
Deposition of banded iron-formations in a tectonically inactive basin

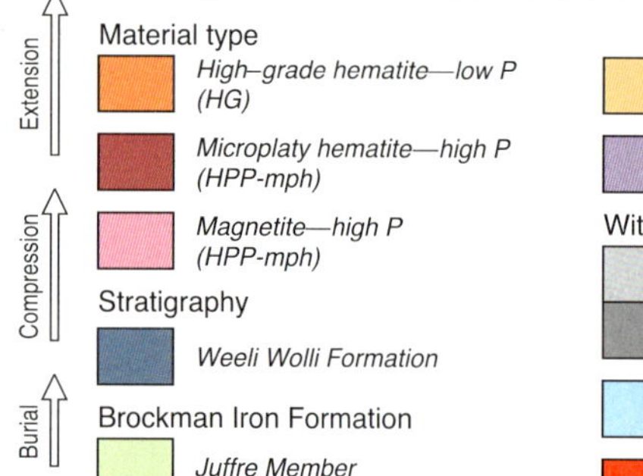

Figure 9.10: Model for the genesis of iron ore deposits, based largely on the geology of Mount Tom Price deposit. (1) Deposition of banded iron-formation proto-ore at *ca* 2.46 Ga. (2) Initial dissolution of silica associated with upward migration of basinal brines along active normal faults at *ca* 2.01 Ga. (3 & 4) Dissolution of silica and carbonate and oxidation of magnetite to hematite and martite associated with downward migration of deeply circulating, oxidised meteoric fluids at *ca* 2.01 Ga. (5) Dissolution of apatite during Cenozoic weathering. (Source: modified from Taylor et al., 2001)

THE PILBARA IRON ORE: DISCOVERY OF A NEW MINERAL PROVINCE (BOX 9.1)

Large iron ore resources were known as early as 1890 from the Pilbara region, with research by the Western Australian Government geologist Harvey Page Woodward. He wrote that:

> *this is essentially an iron ore country. There is enough iron ore to supply the whole world, should the present sources be worked out.*

Despite this earlier report, the Australian Government was under the impression that the country had only limited resources of iron and, with war looming in the late 1930s, an export embargo on iron ore was enacted in 1938. In support of the national government, the Government of Western Australia also banned the pegging of claims for iron ore prospects.

In the midst of these constraints on development, prospector Lang Hancock in 1952 'rediscovered' the iron ore in the Pilbara. He lobbied for a decade to get the bans lifted, which they finally were in 1961; he then revealed his discovery and staked his claim. Hancock and his partner Ken McCamey entered into a deal with the Rio Tinto Group to develop the find.

Hancock was born in 1909 in Perth to one of Western Australia's oldest land-owning families, and the story of his iron ore discovery in the Pilbara has entered Australian mining folklore.

The initial discovery was made on 16 November 1952 while Hancock was flying south to Perth with his second wife, when they were forced by bad weather to fly low. He described his discovery as follows:

> *I was flying down south with my wife Hope, and we left a bit later than usual and by the time we got over the Hamersley Ranges, the clouds had formed and the ceiling got lower and lower. I got into the Turner River, knowing full well if I followed it through, I would come out into the Ashburton. On going through a gorge in the Turner River, I noticed that the walls looked to me to be solid iron and was particularly alerted by the rusty looking colour of it, it showed to me to be oxidised iron.*

Once government reluctance with the mining of iron ore was overcome, development of the Pilbara iron ore region began. The industry grew with gathering momentum; ports, railways, mines and support towns were constructed in what was once one of the most remote parts of the world. Aided by information from the Bureau of Mineral Resources (now Geoscience Australia) and the Geological Survey of Western Australia, the pace of exploration was stepped up. Australia became a major raw materials exporter, especially to Japan and Europe. The exploration boom led to mining operations at Mount Tom Price and at Paraburdoo in the following decade (Figure 9.9a). From a zero base in 1961, the industry has grown so that it is now the largest export industry in dollar terms in Australia. There has been a 10-fold expansion in exploration expenditure for iron ore since 1994. This has led to the identification of more than 60 new iron ore deposits, mostly in Western Australia. Despite massive increases in production in recent years (e.g. 488 Mt in 2011), the resource life for iron ore in Australia has been maintained.

The red rocks of the Pilbara attracted Hancock to the region.

Photo courtesy of Rio Tinto. Copyright © 2009 Rio Tinto

Iron ore carriers awaiting dispatch to North Asia. Parker Point, Pilbara, Western Australia.

basement of the Pilbara Craton (Figure 9.9). Deposition of iron-formation occurred during the upwelling of reduced, iron-rich waters from the adjacent deep oceanic basins onto the shallow, and relatively oxidised, Hamersley Basin shelf. Reduced Fe^{2+} was oxidised during upwelling, becoming insoluble Fe^{3+} and depositing as ferric hydroxides. Silica was also precipitated during this upwelling, along with phosphate and carbonate. Background sedimentation introduced some aluminium, resulting in an iron- and silica-rich rock with significant levels of phosphate and carbonate, and with variable amounts of aluminium. Such rocks have been mined elsewhere in the world as iron ore, but are less economic than hematite-rich ores, which are formed through hydrothermal and supergene upgrading of banded iron-formation, as discussed below.

The intensity of deformation increases from north to south across the Hamersley Basin, which was affected by five contractional deformation events, including the *ca* 2.14 Ga Opthalmia and the *ca* 1.77 Ma Capricorn orogenies, the latter the consequence of collision between the Pilbara and Yilgarn cratons (Chapter 2). In the northern part of the basin, the strata are generally flat lying and undeformed, but in the south the rocks were strongly folded during these orogenies about west-northwest-trending axes. Along the southern margin, fold vergence and reverse fault movements suggest tectonic transport to the north. Most major deposits are localised in the more deformed southern part of the basin (Figure 9.9a).

In addition to the contractional events, the Hamersley Basin was also affected by events that formed younger extensional basins, at *ca* 2.03 Ga and also at *ca* 2.00 Ga. The basal unit of the older basin contains clasts only of banded iron-formation, whereas the basal unit of the younger basin contains clasts of hematite, suggesting that upgrading of the ores took place between 2.03 Ga and 2.00 Ga. The best constraint on the timing of the upgrade comes from an altered mafic dyke that intruded along iron ore–related normal faults at the Paraburdoo deposit. This dyke gives an age of *ca* 2.01 Ga, which is the most likely age of mineralisation.

In detail, the Hamersley Basin iron ore deposits are lithologically and mineralogically complex, the result of a series of both hydrothermal and supergene events (Figure 9.10). Three discrete mineral assemblages are present: (1) magnetite–carbonate–apatite, (2) hematite–martite–apatite, and (3) hematite–martite, with each stage adding value to the iron ore.

- The earliest assemblage to overprint the banded iron-formation was a magnetite–carbonate-apatite assemblage, which formed by the reaction of moderate-temperature (150–250°C) reduced fluids that altered the pre-existing iron-formation adjacent to normal faults. These fluids are interpreted to be basinal brines that were expelled upwards from deeper in the Hamersley Basin and migrated along normal faults during extensional uplift. This event was associated with the intrusion of the *ca* 2.01 Ga mafic dykes and might have been related to early extension associated with the *ca* 2.00 Ga basin (Figure 9.10). Interaction of fluids and iron-formation resulted in the removal of silica and the addition of carbonate.
- Continued erosion associated with extensional uplift allowed ingress of deeply circulating meteoric fluids; upward movement of basinal brines stopped (Figure 9.10). This resulted not only in a change in the direction of fluid flow from upward to downward but also in a change in the characteristics of the mineralising fluids. Meteoric fluids after the Great Oxidation Event were oxidised, had low salinity and were relatively low temperature (*ca* 100°C). These fluids altered the early magnetite–carbonate–apatite assemblage to a microplaty hematite–martite–apatite–(ankerite) assemblage and then to a hematite–martite–apatite assemblage. Not only did meteoric fluids oxidise ferric to ferrous iron, but they also continued to dissolve the remaining silica and carbonate, thus removing these deleterious components from the ores. Hematite-rich clasts present at the base of the younger extensional basins indicate that these processes were likely complete by *ca* 1.95 Ga, the most likely age of deposition.
- The best-quality iron ore produced in the Hamersley Basin comes from a low-phosphorus hematite–martite assemblage. This assemblage only overprints the earlier assemblages above the present-day base of weathering. This suggests that the final upgrading of ores through dissolution of apatite occurred as a consequence of relatively recent, probably Cenozoic, weathering (Figure 9.10).

Much younger Cenozoic landscape evolution was superimposed on the older Hamersley Basin landscape, and was controlled by basement lithology, structure and landscape position.

Before the Cenozoic, the region was mantled by a deeply weathered regolith (Chapter 5). Erosion of this material occurred in the upland areas, with deposition into lowland and valley areas. This eroded and weathered regolith is in itself part of the mineral system, with the channel-iron deposits such as Robe River palaeovalley and the alluvial fans forming significant iron ore resources (Figure 9.9a).

The Hamersley iron ore province: keys to geological beneficiation

The process of ore formation in the Hamersley iron ore province is not a typical process of metal concentration but, in effect, a process of contaminant removal and geological ore beneficiation. Even so, contaminant removal has surprising commonalities with other mineral-systems (Chapter 8). These include a relationship with extensional faults, and an association with a thermal anomaly, as indicated by the presence of syn-ore mafic dykes. Extensional faults were fluid conduits for both early upward-migrating and later deeply circulating meteoric fluids.

Another important control on mineralisation, as noted above, was the relationship to the Great Oxidation Event at *ca* 2.46 Ga (Chapter 3). Because of the prevailing reducing conditions of seawater prior to this event, high concentrations of iron were able to be deposited as banded iron-formation. The subsequent upgrading, when iron became, in effect, insoluble in oxidised surficial fluids, prevented iron mobility but beneficially allowed for removal of silica, carbonate and apatite. These contaminants reduce iron ore quality, with apatite contributing undesirable phosphorus (Figure 9.10).

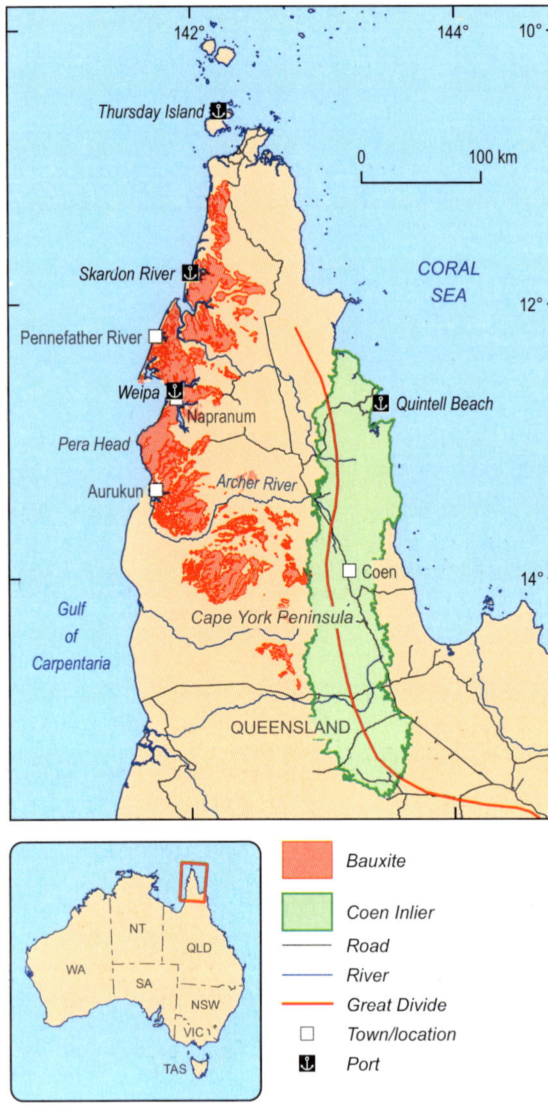

Figure 9.11: Map of the distribution of bauxite. Note its vast areal extent along the west coast of Cape York Peninsula. This is a tropical region with highly seasonal rainfall (monsoon) patterns. The ore body is located down gradient of the highest elevations in the Coen Inlier. (Source: after Taylor et al., 2008)

Further benefit came from the stripping of silica from the high-grade hematite ore, with concomitant removal of asbestos and asbestiform minerals, which are known for their negative health impacts, especially cancer. Some magnetite ores, even in the Pilbara region, do contain the asbestiform amphibole riebeckite, a low- to moderate-grade metamorphic mineral, which consists of hydrated sodium iron–magnesium silicate.

Bauxite: a legacy of Australia's deeply weathered past

Aluminium and hence bauxite are a vital resource for modern life. In just over 150 years since its first commercial production, aluminium has become the world's second most used metal, after steel. Its strength and low density make it ideal for transport and packaging applications. Aluminium is a unique metal: strong, durable, flexible, impermeable, lightweight, corrosion resistant, readily recyclable and a good conductor of electricity. A unit of aluminium cable carries twice as much electricity as a unit of copper, and most overhead and many underground transmission lines are now made of aluminium.

Bauxite is not just a tropical weathering material. In Australia, bauxite is known from the tip of Cape York Peninsula in Queensland to Tasmania and is mined in three localities: Weipa (Queensland), Gove (Northern Territory) and Pinjarra (Western Australia) (Figure 9.2). Bauxite forms by weathering in well-drained areas over extended time periods, with the original parent rock playing a critical role. Most weathering profiles in Australia have lost their major elements in the following order: Ca, Mg, Na, K, Fe^{3+}, Si, Fe^{2+} and finally Al. This process is why much of the Australian regolith is dominated by silcrete, ferricrete and, in places, bauxite, and why it is red coloured (Chapter 5).

Why is Weipa the world's largest bauxite deposit?

Weipa, on the western side of Cape York Peninsula is the largest bauxite deposit in the world (Box 9.2; Figure 9.11). The ore resource covers an area of 11 000 km^2, of which 520 km^2 is ore grade. Why is it so large and located where it is? Using a modification of the mineral-systems approach (Chapter 8), we can help to answer this question. Modification to the approach is needed because bauxite is formed by successive removal of other minerals and material to form bauxite ore, rather than the concentration of a desired element into an ore resource.

The bauxite must form where the regolith is well drained and permeable in nature, to allow the dissolving fluid and solutes to be removed, leaving the bauxite as a residue. Following the mid-Cretaceous marine regression from the western part of Cape York Peninsula, the region has been subjected to weathering and erosion, seasonally high rainfall, a fluctuating watertable, and efficient leaching. The relative tectonic stability of western Cape York Peninsula has helped to protect the resource from the erosion that would occur in uplifted areas. The original source of aluminium is the extensive successions of arkosic sandstone that crop out along western Cape York Peninsula. The feldspars, micas and hornblende in these parent rocks weather into kaolinite, goethite and

Bauxite mine worker next to equipment, Weipa, Queensland.

hematite, with silicic acid being mobilised. Quartz in the sediments eventually dissolves, further concentrating residual aluminium.

The Weipa bauxite formed on rocks of the Carpentaria and Karumba basins (Figure 2.8)—in particular, highly siliceous Cretaceous marine volcanolithic sediments and Paleogene quartzose terrestrial gravels, sands and muds. The bauxite ranges in thickness from 3 m to 12 m at elevations ranging from sea-level to about 80 m asl. The weathering of the source rock results in a varied loss of thickness and, at Andoom, just north of Weipa, it appears that a 40 m thick profile of the Cretaceous Rolling Downs Group has been reduced to 25 m, which now contains a 3–5 m thick layer of high-grade bauxite. Alluvial fans and rivers filled with quartzose sediments incise the plateau, and no marine sediments have been identified since the Cretaceous. The bauxite is underlain by a lateritic profile. In areas of low topography, 'red soil' sourced from local sediments overlies the resource, while the whole sequence is overlain by soils.

There is a high degree of mineralogical variability in the various mining areas around Weipa, particularly with regard to quartz, hematite, kaolinite and some aspects of trace element chemistry. From this chemistry, it appears that the typical bauxite pisoliths (pea-sized concretions) were derived from weathered source rock of essentially the same composition as the substrate under their current location. The isotopic chemistry (oxygen and

BAUXITE IDENTIFICATION IN AUSTRALIA (BOX 9.2)

Willem Janszoon, captain of the *Duyfken*, made the first recorded sighting of the Australian coast at the Pennefather River, some 40 km northeast of Weipa, in 1606. Captain Matthew Flinders sailed the *Investigator* into Albatross Bay and named Duyfken Point in 1802, noting in the ship's log the sighting of 'some reddish cliffs' south of the bay at Pera Head.

Remoteness from markets, poor soils, a monsoonal tropical climate and the rough terrain of the northwest Cape York Peninsula discouraged overlanders and settlers for most of the 19th century, although attempts to settle commenced as early the 1840s. Weipa was later established as a mission and CFV Jackson, Assistant Queensland Government Geologist, referred to the presence of 'brown pisolitic ironstone' between Mission and Embley rivers, but follow-up investigations were not carried out. Jackson wrote at the time:

> *I think it probable that if these deposits were systematically examined and sampled, they would be found to include masses of higher-grade ores; and, if such were the case, they might ultimately prove of some value, especially on account of their ready accessibility from the sea.*

In 1947, FW Whitehouse collected samples of bauxite at the mouth of the Archer River, to the south of Weipa. As a result, a field survey was considered, but the samples sent were low grade.

It was not until 1955 that the significance of Weipa's reddish cliffs was recognised by geologist Harry J Evans. Leading a field party on a reconnaissance of possible oil-bearing structures, Evans skirted the west coast of the Cape towards the end of his expedition 'just to see what the country looked like'. He discovered a large outcrop of bauxite near Weipa on 16 July and continued

A truck loaded with bauxite travels back to the crusher at Huntly mine, Western Australia.

Image courtesy of Alcoa

Image courtesy of Ian Oswald-Jacobs (ian@ioj.com.au)

Red-coloured bauxite exposed in cliffs near Weipa, Queensland. The bauxite is a residue of the deeply weathered pale-coloured bedrock below.

© O Strewe, Lonely Planet Images

Pisolitic bauxite: in places, this bauxite forms a surface layer of small balls of bauxite.

Published with the kind permission of Rio Tinto

Crossing Mission River to collect ore from north of Weipa at Andoom, Cape York, Queensland.

to collect ore samples over a large area. When analysis revealed the ore to be high grade, Evans returned to continue exploration, guided by a local Aboriginal elder, Matthew. He examined 84 km of coast in a 3 m dinghy. Realising the significance of the deposit he noted in his diary:

> *As the journey down the coast revealed miles of bauxite cliffs, I kept thinking that, if all this was bauxite, then there must be something the matter with it, otherwise it would have been discovered and appreciated long ago.*

A subsidiary of the American company Consolidated Zinc, Enterprise Exploration Pty Ltd commenced proving operations in June 1956, using the coastal vessel *Wewak*. A camp was developed east of today's Napranum township on the banks of the Embley River. By 1957, Consolidated Zinc had formed a new company called Commonwealth Aluminium Corporation Ltd to develop the Weipa deposits. The acronym Comalco is derived from the initial letters of the company.

Exploration and planning for the Weipa deposit progressed well, with the first trial shipment of 9849 tonnes of bauxite being sent to Japan in April 1961. A new town was built at Rocky Point, and bauxite processing and shipping facilities were constructed at Lorim Point. The first shipment of 10 000 tonnes of bauxite to the Gladstone alumina refinery in December 1966 marked the second significant step towards establishing a fully integrated aluminium industry in Queensland. The development of the readily accessible high-quality coal resources of the Bowen Basin provided an important catalyst for establishing the energy-hungry smelting component of an integrated industry.

Comalco and partners commenced aluminium production at the Gladstone smelter in 1984. Operations at Weipa include the mining, crushing and washing of bauxite, and ore handling through port facilities for transport to alumina refineries. Weipa has become one of the largest bauxite mining sites in the world and is a major contributor to Rio Tinto's bottom line. Development of the south of the Embley Project, which will replace production from the depleted East Weipa reserves, is expected to extend the life of the operation by 40 years and lift annual production.

The location of further deposits at Gove in the Northern Territory and the Darling Ranges east of Perth has led to the identification in Australia of more than 6200 Mt in economically demonstrated bauxite resources. The bauxite deposits generally occur as pisolitic residuals of Cenozoic laterite, developed on several rock types including sedimentary rocks, basalt and schist.

In 2011, the Bureau of Resources and Energy Economics reported that the export value of bauxite was $300 M. For the same period, the export value of alumina was nearly $5.5 B and that of aluminium ingots was $4 B. The transformation of ore (bauxite) to more refined products (alumina and aluminium) clearly adds significant value to Australia's bulk resources.

deuterium) of the present-day groundwater is the same as the pisoliths, which are thought to have started to form in the mid-Cenozoic (*ca* 50 Ma). This suggests that monsoonal conditions encountered today were prevalent when the pisoliths formed, but that this likely occurred when Weipa was at latitudes 1000–2000 km further south than today (Chapter 2). In contrast, the bauxites of Brazil have differing isotopes, suggesting that conditions for bauxite formation there were different from the trade wind–dominated rainfall encountered today. In this way, bauxites are a measure of ancient climatic conditions.

The possibility of a biological role in the formation of bauxite has been raised. In some cases, it appears that beetles might have influenced the formation of the marble-sized pisoliths. Other possible bio-agents are microbes such as *Bacillus*. The bauxite at Weipa is almost unique in the world, in that it is made up of free-running pisoliths with limited matrix material. These have a consistent size distribution, sphericity, roundness, density, total Al_2O_3 and major element chemistry, despite their varied complex interiors. One possibility is that the pisoliths formed further up slope than their current position, and were transported and deposited down to their present location.

The science case for reworking is strong, and offers an explanation for why the ore body is not as large as the areal extent of the parent rock and equivalent climatic conditions. The Weipa bauxite lies directly to the west and down gradient of the highest elevations in the Coen Inlier, which is up to almost 800 m asl (Figure 9.11). The elevated hinterland centred along the Great Divide (Chapter 5) could have been the topographic driver for enhanced fluid flow in this region, hence controlling the location and extent of the Weipa ore resource.

Gas, the 'greener' hydrocarbon

Australia was self sufficient in energy before the arrival of the internal combustion engine, after which the country became an importer of petroleum (Chapter 4). The discovery of oil in Victoria's Bass Strait in the 1960s provided relief from imports for a time. Production of Australian oil peaked in 2000–01, and has since declined by 5% per year. Australia now imports oil, which cost the country $20 B in 2010–11. Globally, around 50% of annual oil production is consumed by the transport industry; however, in Australia, this sector uses 70% of about 1 million barrels per day total. The disproportionate use in Australia probably reflects the nature of commerce (no trade barriers over a large area) and the large distances between population centres (Chapter 1).

In contrast to oil, Australia has significant natural gas endowment provided by the giant gas fields of the North West Shelf (Box 9.3). The US Energy Information Administration rates Australia 29th in global resources with known conventional reserves at 164 trillion cubic feet (tcf), equivalent to 25 billion barrels of oil. The natural gas industry has shown remarkable growth in both the domestic and export sectors over the past few decades, and this is projected to continue. More recently, natural gas is being produced from Australia's coal fields as coal-seam gas. These significant resources are being progressively developed in eastern Australia, with Queensland currently 80% reliant on coal-seam gas for its domestic gas supply. The addition of coal-seam gas's known resources of 42 tcf (Chapter 4) gives a total resource of 206 tcf.

Unconventional resources include tight gas, shale gas and gas hydrates. No definitive gas hydrates have been identified in Australian waters. The reservoirs of tight gas and shale gas resources tend to be deeper (1.5–3.5 km) than the typical coal-seam gas resources, so their development is less contentious with respect to their impact on groundwater. These unconventional resources are not currently being produced in Australia, although there are tight gas projects in the planning stage.

What is natural gas?

Natural gas is composed of simple hydrocarbons, mainly methane (CH_4). Pure methane gas is colourless, odourless and lighter than air. Depending on the source of the gas, it may contain minor quantities of heavier C_2–C_5 hydrocarbons: ethane (C_2H_6), propane (C_3H_8), butane (C_4H_{10}) and pentane (C_5H_{12}). Natural gas with a high concentration of methane is known as a 'dry' gas (e.g. coal-seam gas) and that with a high proportion of C_2–C_5 hydrocarbons is known as a 'wet' gas. A 'lean' gas falls between the two. Other constituents may or may not be present, such as the more complex liquid hydrocarbons (C_{6+}), in addition to nitrogen (N_2), carbon dioxide (CO_2) and hydrogen sulfide (H_2S). A 'sour' gas contains more than four parts per million H_2S and is characterised by an increasingly foul or rotten-egg smell. Australian natural gas is generally 'sweet' due to its low H_2S content.

Inorganic and organic sources have been inferred for non-hydrocarbon gases such as CO_2. A mantle and/or igneous origin for the CO_2 is likely where the natural gas has CO_2 content of 5% or higher. Low levels of helium are generally found (less than 0.5%); however, if the natural gas reserve is large enough, helium can be extracted economically—as happened in 2010 with Australia's first helium extraction plant in Darwin.

Liquid hydrocarbons, if present in high enough concentrations, may separate as 'condensate' from the natural gas when the gas is brought to lower pressures and temperatures at the surface. The remaining natural gas can itself be turned into a liquid, mainly through cooling to lower temperatures. LNG contains mainly methane and occupies about 1/600th the volume of natural gas in the gaseous state. Therefore, LNG is more convenient for transportation over long distances using cryogenic sea vessels or cryogenic road tankers. Liquefied petroleum gas (LPG) comprises mainly propane and butane, and ethane is widely used as a petrochemical feedstock.

Natural gas as an energy source has significant environmental benefits over both coal and oil in terms of its lower greenhouse gas and other emissions. This aspect is of considerable advantage in the further promotion of natural gas use and Australia's energy future. Natural gas remains a cheap energy source in Australia compared with North Asia and Europe. However, wholesale gas prices have generally trended upwards in the past few years, especially in Western Australia as Australia becomes increasingly engaged in the global LNG market.

Petroleum exploration and development requires careful preparation and planning, utilising expensive infrastructure, Bass Strait, Victoria.

Formation and distribution of gas

Australia's conventional gas reserves and resources are distributed unevenly around the continent. More than 90% of the reserves and resources are located offshore from northwest Western Australia (Carnarvon and Browse basins) and in the Timor Sea to the north of Australia (Bonaparte Basin), remote from population centres in the south or east of the country. The largest onshore accumulation of conventional gas reserves and resources occurs in the Cooper and Eromanga basins in northeast South Australia and southwest Queensland. It is this source that currently supplies the bulk of the domestic eastern Australian gas market (SA, ACT, NSW and Qld). Victoria and the emerging Tasmanian market are dominantly supplied from the Gippsland Basin, offshore from southeast Victoria.

Environment and price pressures are increasing the utility and marketability of natural gas, including its use in vehicular transport. Action Bus, Australian Capital Territory.

Image courtesy of Bidgee, Wikimedia Commons

The source rocks for Australia's gas range from the Ordovician in the Amadeus Basin to the late Cretaceous–early Cenozoic in the Bass and Gippsland basins. The major source rocks were derived predominantly from land plants and were deposited in intracratonic or passive margin settings (Chapter 4).

Although source and maturity are the primary control on the composition of the natural gas, secondary alteration processes are also important and can affect the economic value of the resource. If the reservoir depths are shallow (<1500 m), such as in the Gippsland Basin, the Carnarvon and Browse basins of the North West Shelf and the onshore Bowen Basin in Queensland, in-reservoir biodegradation of the gas by microbes can increase the dryness. The microbes can flourish because temperatures are low enough and groundwater flow is sufficient at these reservoir depths. Natural gas can also show differential fractionation along the migration path, or within the reservoir. The bias towards methane leakage relative to longer chain hydrocarbons has resulted in stacked reservoirs, where the shallowest reservoir tends to have the drier gas.

Unlike coal resources, most of Australia's largest conventional gas resources are inconveniently located with respect to the nation's population and therefore demand centres (Figure 9.12). However, the 'stranded' nature of these resources has led to the development of a significant new export industry for Australia, LNG, with Australia rapidly racing towards being one of the world's largest exporters (Box 9.3). Furthermore, the limits of conventional gas in the more populous eastern states are being

AUSTRALIAN LNG—BRINGING ONCE-STRANDED RESOURCES TO THE GLOBAL ENERGY MIX (BOX 9.3)

Image courtesy of Woodside Energy Ltd

Goodwyn A gas platform, North West Shelf Project, Western Australia.

Traditionally, Australian reserves of gas have been far in excess of what has been needed for domestic usage. Much of the gas discovered in Australia's exploration history was discovered during the search for oil, including a string of giant gas fields discovered along the North West Shelf in the 1970s and 1980s. Many of these fields were considered to be 'stranded', being located in areas where the gas could not easily be developed at that time because of remoteness and water depth. The discovery of gas was viewed as an exploration failure or a very poor second prize, at best. The development of the LNG industry has changed that picture, linking the gas resources of northwest Australia with the energy-hungry markets of north Asia.

The industry is now in a period of rapid growth—in addition to the two currently operating projects (North West Shelf project and Darwin), another five are under construction or committed for development (Gorgon, Pluto, Wheatstone, Ichthys and Prelude, the world's first floating LNG project). These projects will help triple Australia's existing LNG export capacity and will mean that by 2016 Australia could rival Qatar for the top spot as the world's largest exporter of LNG.

The innovation of floating LNG technology will further unlock the smaller and more remote gas resources on the North West Shelf. Furthermore, gas is favoured in the global energy mix for its flexibility and its potential, when replacing other fossil fuels, to lower emissions of greenhouse gases and local pollutants.

Image courtesy of Woodside Energy Ltd

North Rankin A platform, North West Shelf Project, Western Australia.

overcome by the identification of large, readily extracted coal-seam gas resources close to major cities (Figure 9.12).

Coal-seam gas: getting a bigger bang for the coal buck

Coal-seam gas is naturally occurring methane (with or without CO_2) that has been absorbed onto the grain faces and micropores of coal or held within fractures or joints (called cleats) with water. The gas is generated by both microbial digestion of the coal and by the thermal maturation process of coalification. Unlike conventional gas, where the source rock generates the gas that then migrates upwards to a reservoir, coal-seam gas remains bound up in the coal seams. Conventional gas trapped in a reservoir generally flows to the surface under pressure. Coal-seam gas requires dewatering of the coal seams to reduce the pressure and allow the gas to flow or be forced to the surface. The permeability of coal seams is very low, and the fracture/joint network controls the fluid flow and recovery of the gas. The flow rate depends on many factors and changes with time, as the ongoing development and recovery of gas (and water) can alter the permeability framework of the coal seams.

Since the start of the 21st century, there has been a rapid commercial rush to develop Australia's unconventional coal-seam gas resources, which are located relatively close to major population centres in eastern Australia and the associated infrastructure (Figure 9.12). The development of these gas sources has alleviated the need for gas to be piped or transported as LNG from western and northern Australia to meet growing domestic demand. Indeed, the coal-seam gas resource is large enough to support a new export industry. Whereas production began in the United States in the 1970s, exploration for coal-seam gas in Australia only started in 1976. The first commercial coal-mine methane operation commenced at Moura in Queensland in 1996, largely for hazard reduction in underground coal-mining operations. There are two LNG processing facilities under construction near Gladstone, based on coal-seam gas resources, which will require a significant infrastructure investment in excess of $30 B.

The economics of gas

Natural gas is converted to LNG for transportation in speciality ships destined for overseas markets—currently Japan, South Korea, China and Taiwan, with other shipments to single spot markets including Spain, Turkey, India and the United States. Although only 14th in the world for resources, Australia is 4th largest exporter of LNG (Table 9.1). Global LNG trade is characterised by two distinct import markets, the Asia–Pacific and the Atlantic. The vast majority of Australia's LNG exports are delivered to economies in the Asia–Pacific, and these countries account for more than half of world LNG trade. As with the other bulk commodities, circumstances in China will play an increasingly important role in the development of Australian markets. China imported around 5.5 Mt of LNG in 2009, which was 64% higher than imports in 2008. The Bureau of Resources and Energy Economics predicts that China will import 13 Mt of LNG in 2012! This continued growth is a direct result of increased capacity for

LNG tanker at Karratha gas plant loading terminal, Western Australia.

Image courtesy of Woodside Energy Ltd

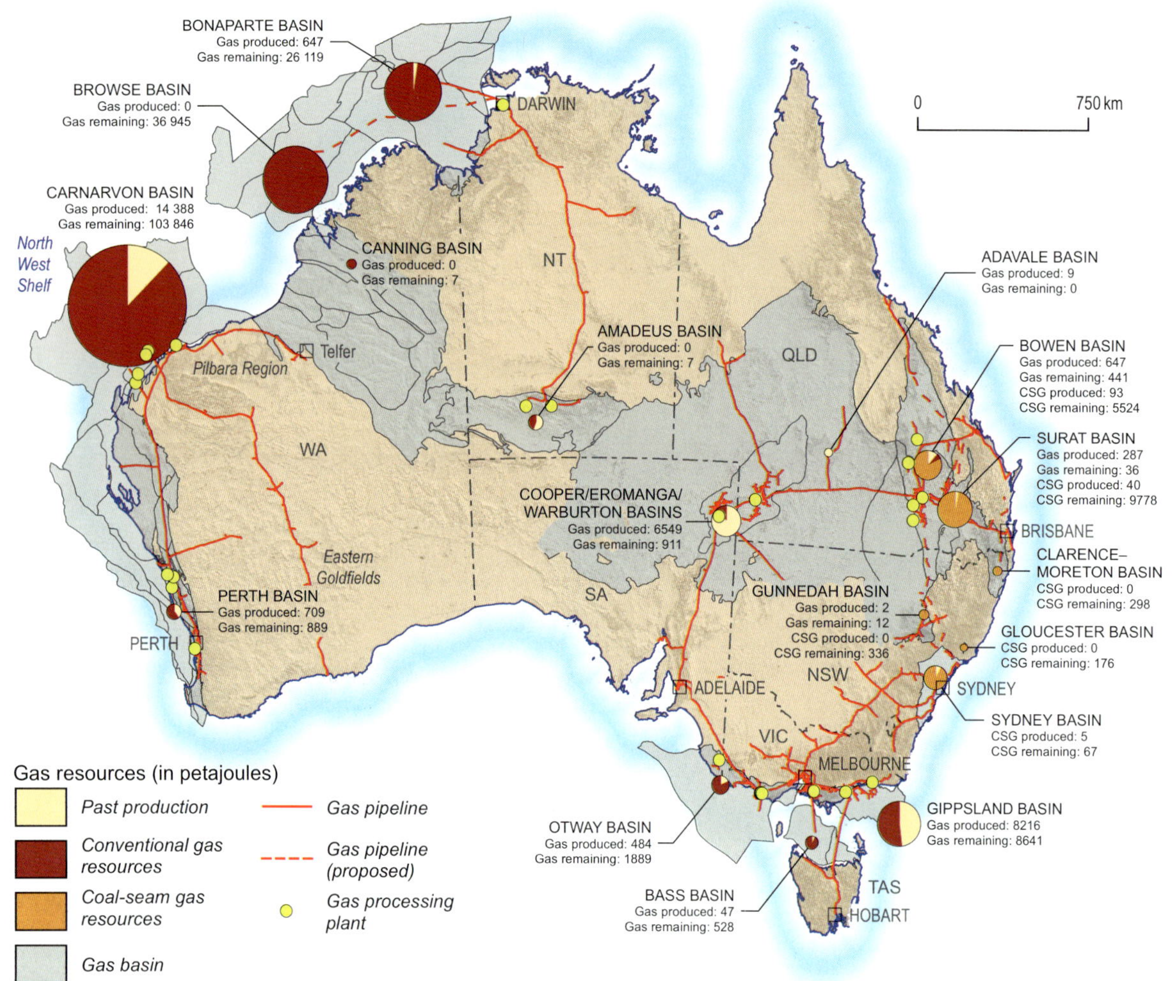

Figure 9.12: Map of Australia's gas resources and associated infrastructure. Note the vast gas resources of the North West Shelf, hosted by the Carnarvon, Browse and Bonaparte basins. The gas pipelines from the Pilbara coast power major mining centres such as Telfer to the east and the gold and nickel fields of the Eastern Goldfields (Chapter 8). The coal-seam gas resources are smaller, but are located closer to the major east-coast population centres, obviating the need for a transcontinental pipeline. (Source: Geoscience Australia and ABARE, Australian Energy Resource Assessment 2010)

gas-fired electricity generation. The fall in LNG prices has also encouraged China to purchase additional spot cargoes, with an immediate impact on prices.

The North Rankin gas field on the North West Shelf is located in moderately deep water some 135 km offshore and 1500 km from the nearest major domestic gas market. Exporting the LNG was considered as the key commercial option for the developers. The North West Shelf project is Australia's largest resources project to date, involving some $25 B of capital expenditure and operated by Woodside Petroleum Ltd and its international consortium partners. Substantial capital investment is required for LNG, which is sometimes beyond the financial resources of local developers.

The first LNG deliveries from the North West Shelf to Japanese buyers were made in 1989. From that time to 2004, production was around 7.5 Mtpa. Export LNG volumes increased to around 11.7 Mtpa following the commissioning of the fourth processing train in 2004. With the completion of a fifth LNG processing train, production capacity increased to 15.9 Mtpa. The $43 B Gorgon project is coming to fruition more than 30 years after its discovery (Box 9.3). The scale of Gorgon means that it will become Australia's biggest resources project, a modern Snowy Mountains Hydroelectric Scheme, which will add an estimated $64 B to Australia's GDP and employ around 10 000 workers.

Additional LNG export volumes are expected in the near future from a number of new ventures, including Greater Gorgon, Pluto, Pilbara and the Browse, all in Western Australia. The Greater Gorgon fields include the expansive Jansz field, and currently represent 25% of Australia's total conventional gas resource. The project partners, Chevron, Shell and ExxonMobil, will construct a 5 Mtpa LNG production plant on Barrow Island, to the east of the resource.

Marriage of the giants: adding value to the bulk commodities

Many of Australia's large bulk commodities are located within a few hundred kilometres of the coast. This relative proximity of resources to ports has driven the development of infrastructure (rail and port) to handle the vast tonnages of these bulk resources. Newcastle is the world's largest export port for coal. The trains running from the Hamersley iron ore province in the southern Pilbara are some of the longest in the world, delivering their wares to the adjacent coast in the north. In many instances, however, the different resources (iron ore, bauxite, coal, gas) are separated from one another by half a continent—such as iron ore in South Australia and coal in New South Wales. In this example, bringing the resources together has relied on shipping, using Australia's coastal waters and ports (Chapter 6).

There has been ongoing political discussion supporting the concept that Australia, as a developed country, should add more value to its raw materials. It was recognised more than 100 years ago that new industries could develop where metalliferous resources could be further refined and transformed by marrying them with energy. Examples include iron ore and coal for steel, iron ore and gas for hot briquettes, and bauxite and coal/gas for aluminium.

Despite its many positive aspects, significant hurdles exist that constrain value adding. Many of the resources companies are not geared up to produce elaborately transformed manufactures and, even if they were, Australia would face high input tariffs in many countries. Implementing newer technologies to upgrade raw materials is not always straightforward, especially on the vast scale required by these industries.

Bauxite + energy (coal or gas) = aluminium

Bauxite is a mixture of hydrated aluminium oxide minerals, including gibbsite, $Al[OH]_3$, boehmite, γ-AlO[OH] and diaspore, α-AlO[OH], together with variable impurities such as silica, iron oxide and titanium-bearing minerals. The Hall–Héroult Process, invented in 1886 to turn bauxite to aluminium, is very energy intensive. It uses around 15.7 kWh of electricity per kilogram of recovered metal. A source of relatively cheap electricity was therefore critical for growing the industry in Australia (*Did you know?* 9.3).

A relative newcomer to the aluminium industry, Australia has secured a prominent position as the world leader in mining and refining of bauxite, alumina and aluminium. From an initial capacity of 20 000 tpa in 1955, at the Bell Bay alumina plant, Australia's production developed quickly. Australia's 6.2 Bt economically demonstrated resource base of bauxite provides a world-class resource for the industry, which consists of 5 bauxite mines, 7 alumina refineries, 6 primary aluminium smelters, 12 extrusion mills and 2 rolled product (sheet, plate and foil) mills. The Australian aluminium industry alone directly employs more than 12 000 people. Bauxite, alumina and aluminium exports were worth nearly $10 B in 2011 (Figure 9.6).

Did you know?

9.3: Energy and aluminium production

Aluminium (Al) is widely used because of the metal's strength, light weight and resistance to corrosion. Earth's crust is rich in aluminium, but due to its reactivity, the metallic state does not exist in nature. The chemical bonds of the metal in its main ore, bauxite, are strong, so extracting Al from its mineral crystals is very energy intensive. Bauxite is first processed using a wet caustic leach to obtain alumina (Al_2O_3), a white granular material. Electrolytic reduction in a molten bath of cryolite ($NaAlF_6$) smelts the alumina to form the metal.

Electricity supply for the smelting process is a major component of production cost but varies depending on the energy source. Smelters are located near sources of plentiful and inexpensive electric power; in mainland Australia, this is generally produced from coal. Al produced from scrap requires one 20th of the energy required for primary production, leading to a high degree of recycling of the metal.

Aluminium ingots ready for export.

Most of the export income from this resource is derived from the refined product—alumina or aluminium metal—rather than the raw bauxite from the mines. This was made possible due to Australia's capacity to supply relatively cheap energy from its enormous coal resources. The Gladstone refinery in central-east Queensland uses the bauxite from Weipa and the coal from the inland Bowen Basin. These factors have led to Australia leading the world in bauxite refining and being fourth largest producer of primary aluminium.

Growth in refining has slowed in recent years, partly due to uncertainty in the future cost of energy. For the future, this presents an irony, as aluminium is such a light metal that the ongoing energy savings from its use in transport infrastructure create a strong demand.

Iron + energy (coal or gas) + other constituents = steel

Steel is an alloy of iron and carbon, with other metals added to meet different functional needs. Steel making is also energy intensive, and Australia has a lot of iron and a lot of energy. Australia's steel making was built on the marriage of the vast coal resources with the iron ore.

In 1915, iron ore from the Middleback Ranges in South Australia became the raw material for the integrated Broken Hill Proprietary Company Limited (BHP) iron and steel industry at Newcastle, New South Wales. The iron ore was shipped to the Newcastle coalfields (Chapter 6). The development of the steel industry was facilitated by BHP metallurgists and chemical analysts who relocated from the Broken Hill lead and zinc operations, with technical assistance from the United States. The transfer of these scientific skills was crucial to the rapid development of sound operating practices for open-hearth steel making under local conditions.

Integrated steelworks were developed at Port Kembla, New South Wales, in 1928, then Whyalla, South Australia, in 1941 and later at Kwinana in Western Australia in 1968. In each case, the steel making was different because of the wide variety of cokes and ferruginous feeds. The result was an industry that adapted its practices to suit the input conditions. Australia also mines metals such as nickel, manganese, chromium and tungsten (Appendix 8.1), which are used in specialist steel manufacture by alloying or galvanising with zinc, or electroplating with tin.

The steel industry is a major employer, and the scale of operations means that steel firms are often a dominant influence in the community in which they are located, such as Port Kembla and formerly Newcastle. This can make communities vulnerable to the vagaries of the market, such as tariffs and exchange rates. There have been major job losses in the industry in Australia, and around the world (except in South Korea and China).

Iron ore + gas = hot-briquette iron

BHP Billiton Ltd attempted to upgrade iron ore to hot-briquette iron by marrying the vast Hamersley Basin iron ore resources with the gas from the adjacent North West Shelf. The Boodarie hot-briquette iron plant located 7.5 km southwest of Port Hedland on the Pilbara coast was completed in 1999 and was planned to produce 2.3 Mt of briquettes per year. The energy-intensive process used hydrogen and carbon monoxide to remove oxygen from the ore, leaving a more than 90% iron briquette for export. A pipeline brought the gas from the west, and the railway brought the iron ore from the south, meeting at Boodarie. The finished briquettes were loaded onto ships at Port Hedland via a vast conveyor-belt network. However, the operation was not financially viable and the plant was closed in 2004, having cost more than $2.5 B.

Unearthing our past and future

The bulk commodities of coal, iron, aluminium and LNG account for more than 40% of Australia's export earnings ($126 B in 2011), sustaining the nation's economic success and the lifestyle of the Australian people. A cornerstone of the Australian economy since the gold rushes, mining was pivotal in shaping mid-19th century Australia. The importance of the resources sector has increased markedly since the mid 20th century, with accelerating export income from the bulk commodities (Figure 9.1). The industrialisation of Asia has provided the demand, driving infrastructure investment in remote regions of Australia. Advances in technology, combined with massive economies of scale and sound policy, have enabled access to the resource and helped to satiate the growing market to which Australia is well located geographically.

Australia's long geological history, fringing passive margins, overall landscape stability and relatively limited amount of deformation in the past 200 Myr have formed and preserved a vast quantity of high-quality resources that are the physical basis for the bulk commodity industry. Competition from other bulk commodity suppliers with a similar Gondwana geological heritage—for example, southern Africa and South America—has not prevented Australian exports from growing rapidly. Australia's educated workforce, system of government and legal framework have provided a sound, stable administrative foundation that allows the geological legacy to be utilised for societal and national benefit. The relatively small domestic population relative to the size of the resource wealth has meant that local demand is readily met, allowing the surplus to be exported.

Commentators talk of Australia's boom period of the early part of the 21st century as being unprecedented in its longevity. What to do with this windfall is a vital question to be answered by the Australian people. In the next chapter, we will look at another of Australia's energy resources, deep heat.

Image by Mining Photo

Aerial close-up of caustic tanks and pipework at bauxite alumina refinery, Gladstone, Queensland.

© Getty Images [G McConnell]

Bauxite reclaimer machine, Weipa, Queensland.

Bibliography and further reading

Resources from a stable base

Australian Government Department of Foreign Affairs and Trade 2009. *Composition of trade Australia*, DFAT, Canberra.

Australian Government Department of Foreign Affairs and Trade 2011. *Trade in services Australia 2010*, DFAT, Canberra.

Australian Government Department of Resources, Energy and Tourism 2011. *Energy in Australia 2011*, DRET, Canberra.

Bureau of Resources and Energy Economics 2012. *Resources and energy statistics, December Quarter 2011*, BREE, Canberra.

Centre for Social Responsibility in Mining 2010. *Indigenous employment in the Australian minerals industry*, Centre for Social Responsibility in Mining, University of Queensland, Brisbane.

Cleary P 2011. *Too much luck: the mining boom and Australia's future*, Black Ink, Australia.

Crabb P 1997. *Murray–Darling Basin resources*, Murray–Darling Basin Commission, Canberra.

The Economist 2011. The next golden state: a 16–page special report on Australia. *The Economist* 28 May – 3 June.

Garnaut R 2010. Growth, cycles, climate and structural change: two hard decades ahead, address to the Academy of Technological Sciences and Engineering—Australia 2030 Conference, Sydney, 11 November 2010.

Grant A, Hawkins J & Shaw L. 2005. Mining and commodities exports. In: *Economic roundup*, spring 2005, Australian Treasury, Canberra, 1–15.

Langton M 2010. The resource curse: still the lucky country? *Griffith Review* 28.

Roarty M 2011. *The Australian resources sector—its contribution to the nation, and a brief review of issues and impacts*, Parliamentary Library background note, Parliament of Australia, Canberra.

Simon J 2003. Three Australian asset-price bubbles. In: *Asset prices and monetary policy*, Richards A & Robinson T (eds), Reserve Bank of Australia, RBA Annual Conference Volume, 8–42.

Sinclair W 1976. *The process of economic development in Australia*, Longman Cheshire Australia (reprinted 1985).

Sykes T 1998. *Two centuries of panic: a history of corporate collapses in Australia*, Allen & Unwin, Crows Nest.

Coal and gas

Bureau of Resources and Energy Economics. *Resources and energy statistics, 2011 September Quarter 2011*, BREE, Canberra.

Australian Government Department of Resources, Energy and Tourism 2011. *Australia's offshore petroleum industry* DRET, Canberra.

Boreham CJ, Hope JM & Hartung-Kagi B 2001. Understanding source, distribution and preservation of Australian natural gas: a geochemical perspective. *The APPEA Journal* 41, 523–547.

BP 2011. *Statistical review of world energy June 2011*. bp.com/statisticalreview

Campbell I 2009. An overview of tight gas resources in Australia. *PESA News* 100, 95–100.

Geoscience Australia & Australian Bureau of Agricultural and Resource Economics 2010. *Australian energy resource assessment*, Geoscience Australia & ABARE, Canberra.

Muir W 2009. Gassing around: operations in the coal-seam gas industry. *PESA News* 103, 79–81.

New South Wales Minerals Council 2010. *Key industry statistics*, New South Wales Minerals Council, Sydney.

Iron ore

Harmsworth RA, Kneeshaw M, Morris RC, Robinson CJ & Shrivastava PK 1990. BIF-derived ores of the Hamersley province. In: *Geology of the mineral deposits of Australia and Papua New Guinea*, Hughes FE (ed.), Australasian Institute of Mining and Metallurgy, Melbourne, 617–642.

Miyano T & Klein C 1983. Conditions of riebeckite formation in the iron-formation of the Dales Gorge Member, Hamersley Group, Western Australia. *American Mineralogist* 68, 517–529.

Morris RC & Kneeshaw M 2011. Genesis modelling for the Hamersley BIF-hosted iron ores of Western Australia: a critical review. *Australian Journal of Earth Sciences* 58, 417–452.

Taylor D, Dalstra HJ, Harding AE, Broadbent GC & Barley ME. 2001. Genesis of high-grade hematite ore bodies of the Hamersley Province, Western Australia. *Economic Geology* 96, 837–873.

Taylor KG & Konhauser KO (eds) 2011. Iron in Earth surface systems. *Elements* 7, 83–120.

US Geological Survey 2010. *Mineral commodity summaries—iron ore 2010*, US Geological Survey.

Bauxite

Eggleton RA, Taylor G, Le Gleuher M, Foster LD, Tilley DB & Morgan CM 2008. Regolith profile, mineralogy and geochemistry of the Weipa Bauxite, northern Australia. *Australian Journal of Earth Sciences* 55, S17–S43.

Geoscience Australia 2012. *Australian mines atlas*, Geoscience Australia, Canberra.

Hao X, Leung K, Wang R, Sun W & Li Y 2010. The geomicrobiology of bauxite deposits. *Geoscience Frontiers* 1, 81–89.

Kew GA, Gilkes RJ & Mathison CI 2008. Nature and origins of granitic regolith in bauxite mine floors in the Darling Range, Western Australia. *Australian Journal of Earth Sciences* 55, 473–492.

Roach IC (ed.) 2004. *Regolith 2004*, proceedings of the Cooperative Research Centre for Landscape Environments and Mineral Exploration Regional Regolith Symposium, November 2004, CRC LEME, Canberra, 350–354.

Taylor G, Eggleton RA, Foster LD & Morgan CM 2008. Landscapes and regolith of Weipa, northern Australia. *Australian Journal of Earth Sciences* 55, S3–S16.

Marriage of the giants

BHP Billiton 2005. Boodarie iron fact sheet. www.bhpbilliton.com/home/investors/news/Documents/HBIFactSheet.pdf

10

Deep heat—Australia's energy future?

Australians love energy. Almost all facets of our modern life depend on it. For much of Australia's European history, our major energy sources have been from hydrocarbons. These, however, are non-renewable and come with increasing environmental and other concerns. In a carbon-constrained future, where will Australia's energy come from? What will power us into the next century and beyond? The answer is literally beneath our feet—our radioactive heritage. Australia is endowed with uranium (U), thorium (Th) and resultant thermal energy. The energy generated by the natural breakdown of radioactive elements is immense and can be captured not only by fission of U and Th in nuclear reactors, but by the use of geothermal energy, using Earth's in-situ heat from this radioactive decay to generate electrical power. Both have potential to supply energy for Australia for thousands of years, particularly geothermal energy—it is renewable and environmentally friendly, and Australia has vast thermal resources, which, if harnessed, could power our future.

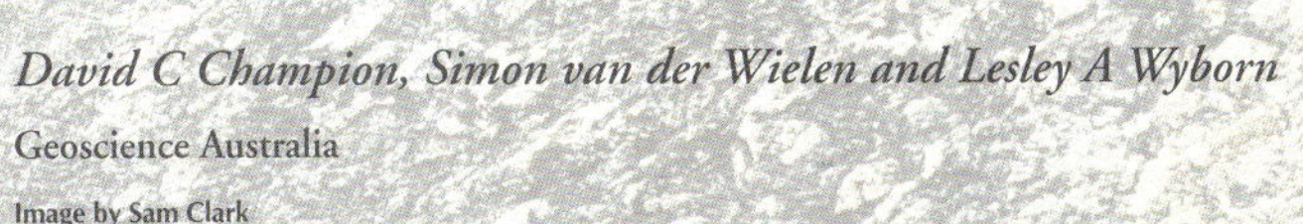

David C Champion, Simon van der Wielen and Lesley A Wyborn

Geoscience Australia

Image by Sam Clark

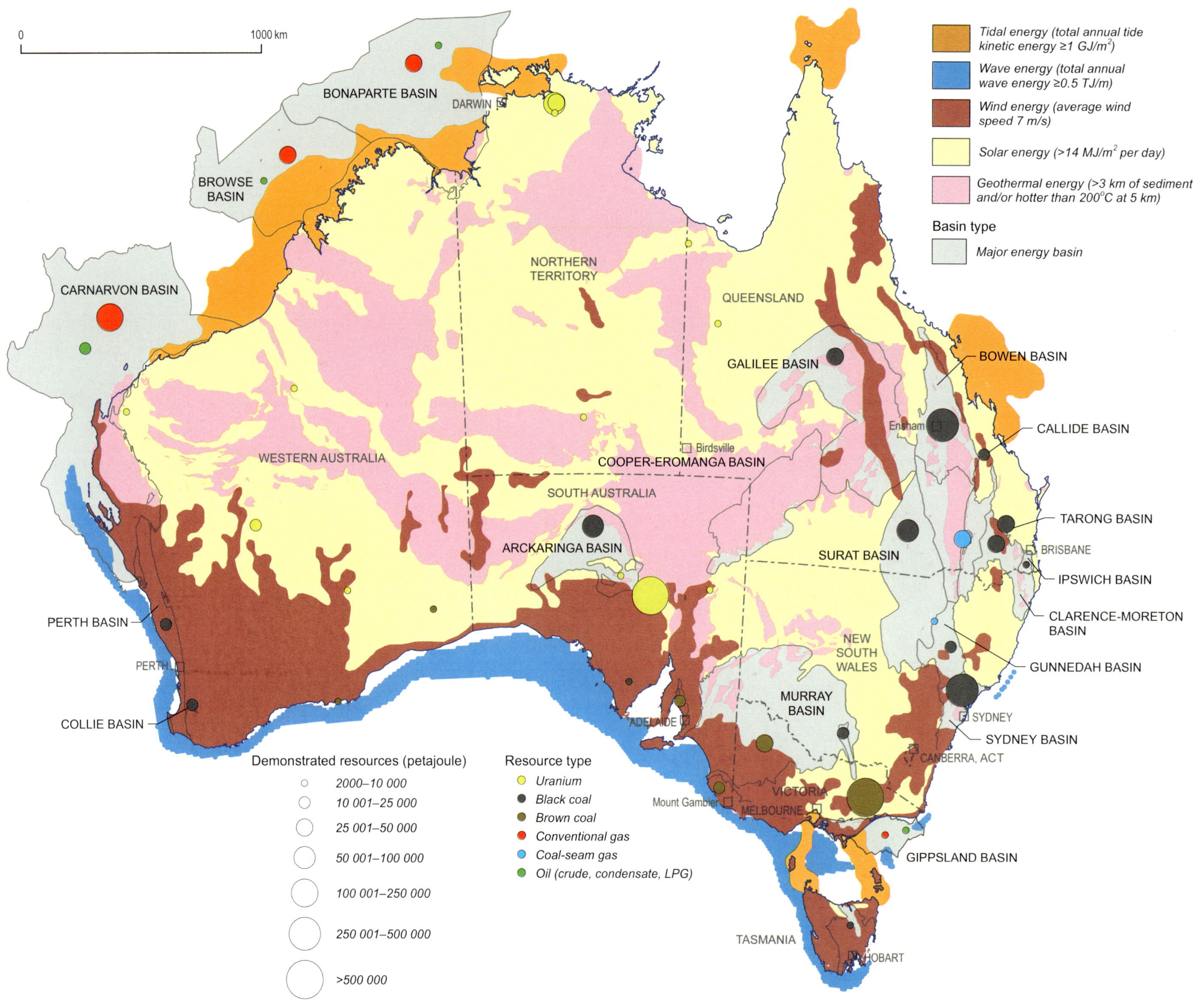
0
1000 km
Tidal energy (total annual tide kinetic energy ≥1 GJ/m²)
Wave energy (total annual wave energy ≥0.5 TJ/m)
Wind energy (average wind speed 7 m/s)
Solar energy (>14 MJ/m² per day)
Geothermal energy (>3 km of sediment and/or hotter than 200°C at 5 km)
Basin type
Major energy basin
BONAPARTE BASIN
DARWIN
BROWSE BASIN
CARNARVON BASIN
NORTHERN TERRITORY
QUEENSLAND
GALILEE BASIN
BOWEN BASIN
Ensham
CALLIDE BASIN
Birdsville
WESTERN AUSTRALIA
COOPER-EROMANGA BASIN
SOUTH AUSTRALIA
TARONG BASIN
ARCKARINGA BASIN
SURAT BASIN
BRISBANE
IPSWICH BASIN
CLARENCE-MORETON BASIN
PERTH BASIN
PERTH
NEW SOUTH WALES
GUNNEDAH BASIN
MURRAY BASIN
COLLIE BASIN
ADELAIDE
SYDNEY
SYDNEY BASIN
CANBERRA, ACT
VICTORIA
Mount Gambier
MELBOURNE
GIPPSLAND BASIN
TASMANIA
HOBART
Demonstrated resources (petajoule)
2000–10 000
10 001–25 000
25 001–50 000
50 001–100 000
100 001–250 000
250 001–500 000
>500 000
Resource type
Uranium
Black coal
Brown coal
Conventional gas
Coal-seam gas
Oil (crude, condensate, LPG)

Future energy

The Australian people, and the Australian way of life, are strongly dependent on energy. It helps us to overcome the very things that define and characterise us—a relatively small, widely dispersed population in a large, mostly hot and dry country. This energy moves us around our cities and the country, cools us over our long, hot summers, warms us in winter, helps feed and entertain us and makes the internet possible, as well as allowing us, and the nation, to make a living. There is almost no facet of our modern life that does not depend, to some extent, on access to energy in its different forms.

This is especially true of high-grade (concentrated) energy, such as fossil fuels. Many of the great steps in the world's modernisation and economic expansion have been driven by this energy source. The switch to coal drove the industrial revolution, while oil and electricity drove the great advances made in the 20th century.

Australians also benefit from the flip side of energy use—energy supply (see *Did you know?* 10.1). Our geological heritage has resulted in us having abundant energy resources—particularly coal (black and brown), oil, gas and uranium (Figure 10.1). Australia is a large producer of energy which is used domestically and exported to other countries (Chapter 9). Australians also rely heavily on these resources to provide us with the lifeblood of modern living—electricity.

Energy is, therefore, both a necessary ingredient and a strong contributor to the nation's wealth and living standards—and there is a strong positive correlation between energy consumption and gross domestic product (GDP). The flip side to these benefits is that our economy and our way of life are vitally dependent on continuing access to reliable and sustainable energy. Fossil fuels—coal, oil and gas—are under threat from a variety of concerns. In a carbon-constrained future, where will Australia's energy come from? What will power us into the next century and beyond? A possible solution is at hand—or more literally 'at our feet'. Australia's geological heritage has left us with a radioactive endowment that can be exploited as nuclear and geothermal energy—an endowment that may power Australians into the future.

Beyond fossil fuels?

Why does Australia need to contemplate alternative future energies? Fossil fuels have served us well and continue to do so. Why won't this continue? The simple answer is that in the short term they will; in the long term, however, they probably will not play such a dominant role. Australia's (and the world's) continued use of coal, oil and gas is threatened by a triumvirate of related concerns:

- *increasing energy consumption* and demand for energy resources, which is placing increasing pressure on supply, affecting availability and affordability
- *depletion of economic resources,* as fossil fuels are non-renewable, their continued long-term usage is not sustainable and the rate of discovery of new resources, especially oil, is generally decreasing

Figure 10.1 (opposite): Major energy resources of Australia (excluding hydro and bioenergy), showing demonstrated current coal, gas, oil and uranium energy resources for the Australian continent. Potentially favourable areas for tidal, wave, wind, solar and geothermal energy are also shown, as are localities mentioned in this chapter. (Source: Geoscience Australia & Australian Bureau of Agricultural and Resouce Economics, 2010)

© Getty Images [D McCardle]

Melbourne skyline at night, Victoria. Most of the electricity powering Melbourne is generated by burning brown coal from the Gippsland Basin.

- *environmental concerns,* as the burning of fossil fuels produces pollutants that have, or may have, short- and long-term environmental implications for the world; greenhouse gases, such as CO_2, and their effects on the world climate, are of chief concern.

Australia is not immune to the long-term implications of depleting hydrocarbon resources. Although Australia has large reserves of coal and gas (Figure 10.1; Chapters 4 and 9), we are increasingly reliant on imported crude oil and refined petroleum products. Environmental concerns are more urgent. Australia is a world leader in greenhouse gas emissions—per capita, because of Australia's small population, one of the highest in the world, and certainly the highest in the Organisation for Economic Cooperation and Development (OECD) countries. Paradoxically, it is Australia's abundant coal resource, and reliance upon it, that is largely responsible for the high carbon footprint (Box 10.1).

Numerous strategies have been proposed to reduce greenhouse gas emissions, including carbon mitigation and switching to cleaner hydrocarbons (Box 10.1). For example, Australia's CO_2 emissions can be reduced significantly by switching to gas-fired electricity generation. The non-renewable nature of fossil fuels and the requirements for long-term sustainability, however, mean that sooner or later Australians will require a switch to cleaner, renewable, or near-renewable, energies. Environmental concerns are only driving this switch more rapidly.

Of course, not just Australia, but all countries, will have to modify their energy mix. This has a number of implications for Australia as an energy supplier, including impacts on the monetary and strategic value of the energy resources (Chapter 9).

ENERGY AND AUSTRALIA'S GREENHOUSE GAS EMISSIONS (BOX 10.1)

Australia is a world-class producer of greenhouse gases, with CO_2 being a major contributor. Studies into why our per-capita emissions are high show a strong relationship to our geological heritage:

- Abundant coal: Electricity generation in Australia is dominated by coal rather than more greenhouse-friendly generation elsewhere (e.g. solar, hydroelectric, nuclear). Electricity generation is responsible for more than half our energy-related emissions.
- Abundant space: Australia, a large country with a very low population (Chapter 1), has high road transport-related emissions. Surprisingly, a significant part of this relates to intra-city travel. The great Australian dream of our own block of land (the ubiquitous 'quarter-acre', about 1000 m^2) has led to sprawling, transport-inefficient cities.

Population densities for Australian cities are low (Figure B10.1), with correspondingly large areal footprints. They follow a trend similar to those for cities in the United States and Canada, with little significant change in density with changes in the population of the city. In contrast, Asian (and other world) cities show a strong positive correlation between population and density, such that more populous cities have much greater densities of people. This tends to result in more efficient transport networks, including public transport, and less transport-related emissions (although additional factors are also important).

Strategies for reducing energy-related greenhouse gas emissions in Australia have been suggested, including:

- Having our (coal) cake and eating it—mitigation processes, such as carbon capture and storage (Chapter 11), which will allow us to continue utilising our extensive fossil fuel resources by disposing of greenhouse gases.
- A gassy cake? Different key ingredient—switching to 'cleaner' fossil fuels. For example, switching from coal to gas for electricity generation will substantially decrease emissions, by 40% or more for CO_2, and significantly more for nitrogen, sulfur and particulate matter. This comes with an added energy-security advantage in that we can use Australia's large (and increasing) gas resources (Chapters 4 and 9).
- A brand new (yellow?) cake—switching to alternative cleaner energies (e.g. wind, tidal, solar, geothermal or nuclear energy).

Aerial view of Melbourne, Victoria, suburbs.

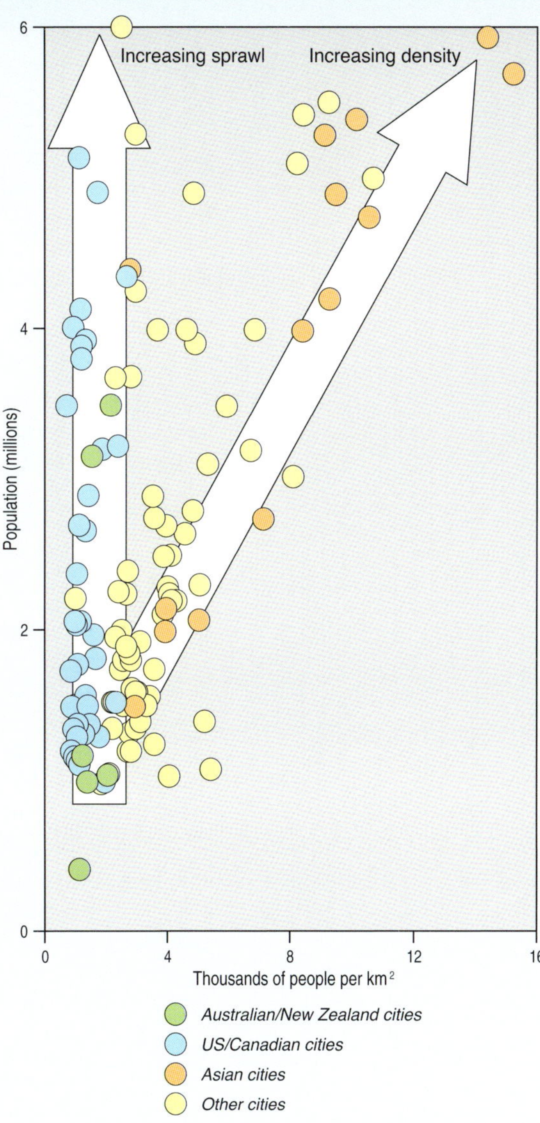

Figure B10.1: City populations versus people density for cities around the world. Population and density data (compiled in 2007) are from the City Mayors Foundation web site (www.citymayors.com/statistics/largest-cities-density-250.html).

© Getty Images [B Wickham]

Bayswater coal-fired power station, Lake Liddell, Hunter Valley, New South Wales.

Future energy resources and Australia

The World Energy Council suggests that future energy supplies will need to meet what it calls the three As:

- accessible (affordable for all)
- available (secure, reliable and sustainable in the long term)
- acceptable (meeting environmental as well as social targets).

Given the future strategic value of indigenous energy sources, another 'A' can be added to this list: Australian (home-grown) energy—the most secure energy accessible.

Australia, by virtue of its geological heritage (Chapters 4, 5 and 6), is potentially well placed to make use of many alternative energies, especially for electricity generation. Australia's hot, arid environment suggests that solar energy should be a significant potential energy source. Australia also has large areas where tidal, wave and wind energy may be viable (Figure 10.1). These alternative energies certainly meet a number of the 'A' requirements. Whether or not they have the potential to fully meet Australia's peak- or base-load energy requirements is less certain.

There are other indigenous energy options for Australia that utilise the immense energy generated by the naturally occurring breakdown of radioactive elements. These are options that overseas experience shows can make significant contributions to a nation's energy demands.

Australia: the hot continent

Red on the inside—a radioactive-powered future?

Radioactive decay—the breakdown of unstable (parent) isotopes of certain elements by transformation of the nuclei and formation of new (daughter) isotopes—releases large amounts of energy, typically in the form of heat. It is a major energy source for planet Earth, provided largely by the decay of just three elements, uranium (U), thorium (Th) and potassium (K)—known collectively as heat-producing elements (HPEs). Radioactive decay is also an energy source that can be exploited for human use. The breakdown of one atom of U or Th produces more than one million times as much energy as that released by the combustion of a molecule of petrol.

Australia is a hot country, but Australia's heat is not restricted to the sun-baked surface alone (Chapter 1). The Australian continent, through its geological heritage, is endowed with well-above-average concentrations of the HPEs, such as U and Th.

The continent has large U and Th resources, as well as 'hot' regions of crust characterised by high heat flow (red on the inside). Together, these have the potential to significantly underpin Australia's energy requirements well into the future, either by fission in nuclear reactors or as geothermal energy (i.e. using Earth's in-situ heat to generate power). The latter, in particular, has the potential to supply large amounts of 'green' energy for Australia's future.

Australia's uranium, thorium and thermal endowment

U and Th resources

Australia has around 33% and 20% of the world's current economic demonstrated resources of U and Th, respectively (Figure 10.2). This high percentage of the world's U is skewed by the supergiant Olympic Dam mine in South Australia, which on its own contains about 30% of the world's known U resource (Chapter 8). Significant other U resources occur elsewhere in South Australia, and also in the Northern Territory and Western Australia (Figure 10.3). Th is more evenly distributed, with significant deposits found in all states (Figure 10.3). The Th data are almost certainly not a true indication of total resources, however, as there is little current world demand for this metal, and little current exploration. Available figures are based on the few deposit types where Th occurs as a by-product.

Australia's thermal endowment

Heat contained within Earth's crust is also a valuable resource of energy and, like U and Th, is unevenly distributed. Surface heat-flow and (5 km) depth-temperature maps both show significant geographical variation in Australia's thermal energy endowment (Figure 10.4). Consideration of world surface heat-flow data shows that, although the measured heat-flow range observed in Australia is not very different from that in other continents, there are a significant number of elevated heat-flow values measurements (>60 mWm^{-2}; Figure 10.5). A notable feature is the concentration of high-heat

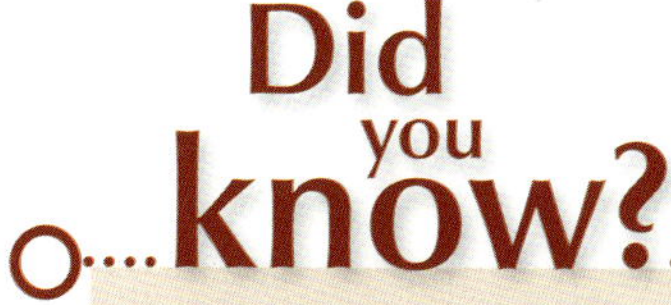

10.1: Energy in Australia

Australia is a large producer and consumer of energy, ranking 20th in total consumption and 15th on a per-capita basis. Australians spend about $50 B on energy every year. Australia is the 9th largest energy producer in the world, with 17 700 PJ produced. More than 60% of this energy is exported, generating significant earnings ($57.5 B in 2009–10).

The energy industry directly employs some 11 000 people and indirectly many others. Our reliable cheap energy also benefits other industries. Mining, agriculture and manufacturing utilise around 31% of our domestic energy consumption, while approximately 25% is used by transport industries (mostly road transport). Around 30% of the total energy budget is used to generate electricity (chiefly by burning coal).

Image by Mining Photo

Excavator bucket loaded with coal, Ensham coal mine, Queensland.

regions within Proterozoic basement terranes, especially in the central part of Australia—the Central Australian Heat Flow Province. This is, in many ways, a surprising result that illustrates the anomalous elevated thermal endowment of some Australian regions. The majority of high heat-flow regions in the world are within active tectonic regimes (e.g. plate margins, characterised by active volcanism and significant advective heat-flow into the crust), or in young crustal regions that have recently experienced active tectonics. Australia is not only dominated by older crust (mostly 3500 Ma to >100 Ma) but occurs within the middle of the Australian Plate (Figure 2.1), well away from active tectonic margins (Chapter 2). For such regions, heat flow is expected to be reduced and certainly not the high heat flows evident within parts of the Australian Proterozoic. Just how anomalous the latter is in Australia is well illustrated by comparisons with Proterozoic-aged rocks elsewhere (Figure 10.5).

The relationship to granites

To explain why Australia has such a radioactive heritage, the distribution of the three elements U, Th and K in the Australian continent needs to be explained—not just where these elements are concentrated in the crust but when they got there. In many respects this is a story about granites—the 'engine room' of the U, Th and K world. Not only do granites make fabulous scenery and great bench-tops, but these same rocks are largely responsible for enriching the upper crust in U, Th and K. Granites are the end product in a chain of processes that have allowed some of the rarest elements in the solar system to become concentrated at crustal depths shallow enough for people to exploit. Importantly, the U, Th and K contents of granites are variable, and some granites (and associated rocks) are much more strongly endowed with U, Th and K than most. As we shall see, these are not just confined to specific regions of Australia but are also associated with specific geological ages. We have suggested that Australia's U, Th and thermal endowment relates ultimately to granites and similar rocks. Is there a spatial correlation between granites and Australia's U, Th and heat resources? The simple answer is 'yes', but with some qualifications. The best correlation is with granites that themselves have above-average levels of U, Th and K, the so-called 'hot' or high heat-producing (HHP) granites. U best illustrates this relationship.

Within Australia, there is a strong spatial association of U mineralisation with U-enriched granites (and volcanic rocks) (Figure 10.6). This is

Remarkable Rocks, 500 Ma granites from Kangaroo Island, South Australia.
Granites, and their extrusive equivalents, are primarily responsible for the enrichment of the upper continental crust in the heat-producing elements potassium, thorium and uranium.
Image by Jim Mason

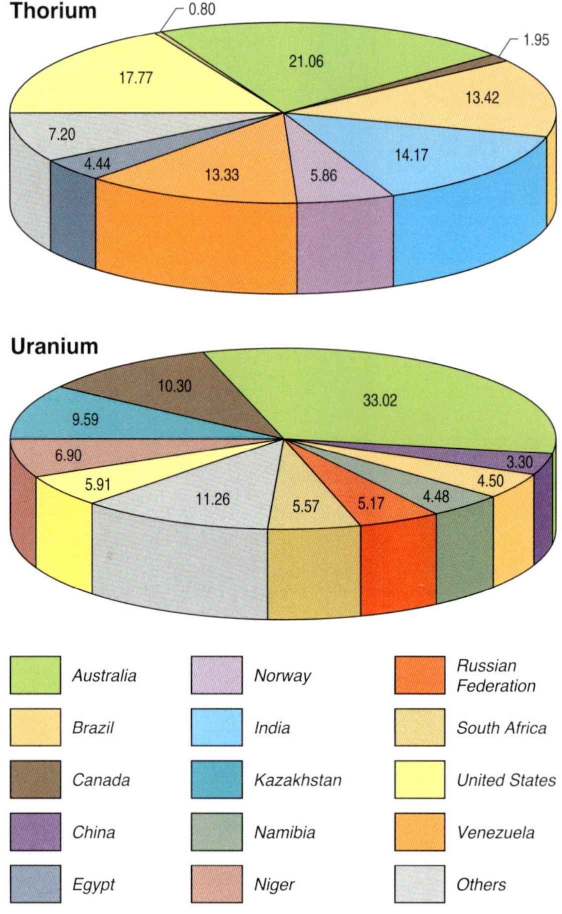

Figure 10.2: Australia's percentage of world economic resources for Th and U.

the case even where the mineralisation may be many millions, even billions, of years younger than the granites (Box 10.2). This is another example of a mineral system (Figure 8.5), where a younger mineralising event has simply tapped an old source rock enriched in U, concentrating the metal to form a new deposit (e.g. Beverley and Four Mile in South Australia). There is a weaker correlation between Th mineralisation and granites. This has more to do with the geochemical behaviour of Th and the nature of Australian Th deposits than the lack of any relationship (Box 10.2). Th deposits in Australia are dominantly placer-style beach-sand deposits (i.e. concentrations of heavy minerals such as the Th-bearing monazite). They are formed by a process not unlike that for alluvial gold. In such deposits, heavy minerals such as monazite, although ultimately derived from granites and related rocks, have been transported often great distances, so that the spatial association with the original Th source is difficult to establish.

Around 80% of Earth's present thermal energy budget is dominated by radiogenic heat from radioactive decay. It is reasonable, therefore, to assume that elevated upper crustal temperatures, in relatively tectonically inactive continents like Australia (Chapter 2), are primarily a response to elevated radiogenic heat production within the crust of those continents. Indeed, such a correlation has, at least locally, been demonstrated. Regions of high heat flow in Australia are either known to be, or are thought to be, associated with the presence of high heat-producing granites. Notably, the converse is not true. Not all high heat-producing granites are associated with elevated surface heat-flow or predicted high temperatures at 5 km or, for that matter, U mineralisation (Figure 10.6). Additional processes are required—in particular, the requirement for overlying insulating rocks, such as coal and shale, in sedimentary basins. These act as thermal blankets, retarding the escape of thermal energy from the buried high heat-producing granites, providing optimum conditions for elevated temperatures in old, tectonically stable continents, such as Australia.

Given Australia's endowment of U, Th and K, it is really not surprising that the continent also contains regions of anomalously hot crust. What in Australia's geological heritage has been responsible for such endowment?

Why Australia?

Crustal zonation

The question of why Australia is enriched in the heat-producing elements is complex and contentious. At the simplest level, the general mechanism for enriching upper continental crust in U, Th and K reflects the magmatic processes that formed the crust itself. Planetary differentiation processes, formation of the core, the mantle and, especially, the crust, have resulted in a strongly zoned Earth with regard to heat-producing elements, with the greatest concentrations in the continental crust. The latter, however, is itself strongly zoned. The majority of the crustal budget of U, Th and K resides in the upper approximately 10 km of crust. In fact, estimates suggest that around 30% of Earth's total budget of Th and U resides in this thin layer of Earth, with Th and

U concentrations more than 100 times those of the primitive mantle composition. This zonation is dictated by the geochemical behaviour of the heat-producing elements (Box 10.2; Figure 10.7).

Not just where, but when?

Although the continental crust is broadly zoned, it should be stressed that the crust is in reality very heterogeneous. In Australia, there are significant geographical variations in U and Th distribution and heat flow—the supergiant Olympic Dam deposit (Figure 10.3), for example, is one area enriched in U. This heterogeneity reflects both a variety of processes and the cumulative and largely non-destructive nature of continental crust. Continental crust is persistent and can be built up over time. In the Georgetown region of north Queensland (Figure 10.4), for example, a geologist can walk a day-traverse from rocks 1600 Ma in age over rocks 430 Ma, 320 Ma, 280 Ma, 130 Ma and 2 Ma in age, stopping to have lunch on river sediments being deposited today. The concentration of heat-producing elements is very uneven in this small area, with the 320 Ma granites being particularly enriched (see later). In broad terms, the Proterozoic rocks in the central third of Australia are characterised by unusually high heat-flows (Figure 10.4). In trying to explain Australia's heat-producing element endowment, we need to focus not just on the processes responsible for crust formation and crustal zonation, but also on the more difficult question of how such processes may have varied through time. Only then can we begin to explain the irregular U, Th and K enrichment within the Australian continent.

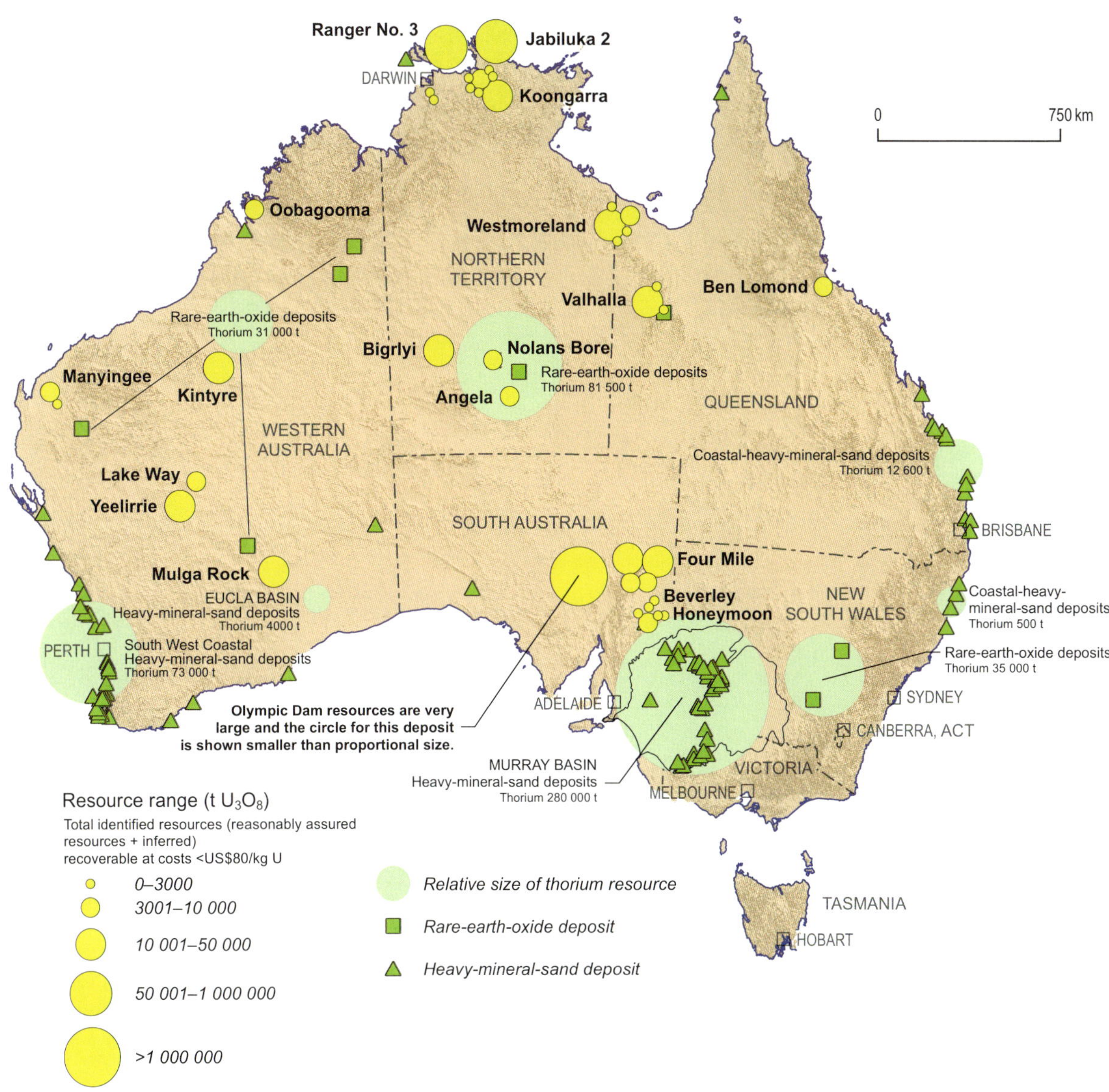

Figure 10.3: Distribution of Australian U and Th deposits and prospects. Note that the U prospects are almost all located in the older western two-thirds of the continent (Chapter 2), while the Th prospects tend to be located in the Murray Basin and other relatively young geological (coastal) settings. (Sources: modified from Department of Resources, Energy and Tourism, 2011; Miezitis et al., 2012)

CONCENTRATION OF HEAT-PRODUCING ELEMENTS IN THE UPPER CRUST (BOX 10.2)

The geochemical behaviour of elements is ultimately controlled by their atomic characteristics, such as atomic radius and electronegativity. Potassium (K), an alkali metal, has a low ionic potential and a large ionic radius. In magmatic systems, it commonly behaves as an incompatible element (i.e. it is excluded from many minerals and preferentially favours the melt). Thorium (Th) and uranium (U) are part of the actinide series (inner transition elements), which have relatively small ionic radii and are highly charged, often with multiple oxidation states—for example, U can vary from U^{3+} to U^{6+}. This combination makes these elements also largely incompatible in most common minerals, such that, like K, melts are generally higher in Th and U (Figure 10.7).

This incompatible behaviour in most magmatic systems is the main reason why Th and U and, to a lesser extent, K, are concentrated in the continental crust (produced largely by partial melting of the mantle), and then in the upper continental crust (by partial melting of lower continental crust) (Figure 10.7). The multivalent nature of U also makes the element more susceptible to changes in oxidation state; U (as U^{6+}) is very mobile under oxidising conditions. This is why U is significantly more susceptible to surficial processes and remobilisation than Th, and why significant amounts of U appear to be 'missing' from the weathered surface of Australia (see *Did you know?* 10.2). In contrast, the more immobile nature of Th (and its incompatibility) is why this element is commonly found in placer deposits. It is geologically easier to concentrate Th-rich minerals, such as monazite, by weathering and subsequent fluvial and marine hydraulic processes than by mobilising the element Th itself.

There is a strong spatial association of U mineralisation with areas of high heat-producing (HHP) granites (and volcanic rocks) in Australia (Figure 10.6). Importantly, this appears to be the case even where the actual age of U mineralisation may be significantly younger, highlighting its often mobile nature (Figure 10.6). A spectacular example of this is in the calcrete-U deposits of Western Australia, such as the soon-to-be-mined Yeelirrie deposit. The U in the calcrete was deposited within a salt-lake environment more than 2.5 Gyr after the first concentration of U in the upper crust by the emplacement of high heat-producing granites. Very young processes of landscape (Chapter 5) and groundwater (Chapter 7) evolution have simply made use of the available U, concentrating it even further to form potentially economic deposits.

The mobile nature of U has also been exploited for in-situ recovery U mining, where the U is leached in-situ, using either an acid or alkaline leaching solution, in conjunction with an oxidising agent, which are pumped through the ore body, obviating the need for conventional extractive mining techniques (open-cut or underground mining). The pregnant fluids are pumped to the surface, where the U is extracted.

In-situ recovery uranium mining at the Beverley mine, South Australia (Figure 10.3), showing well fields with both injection and recovery wells.

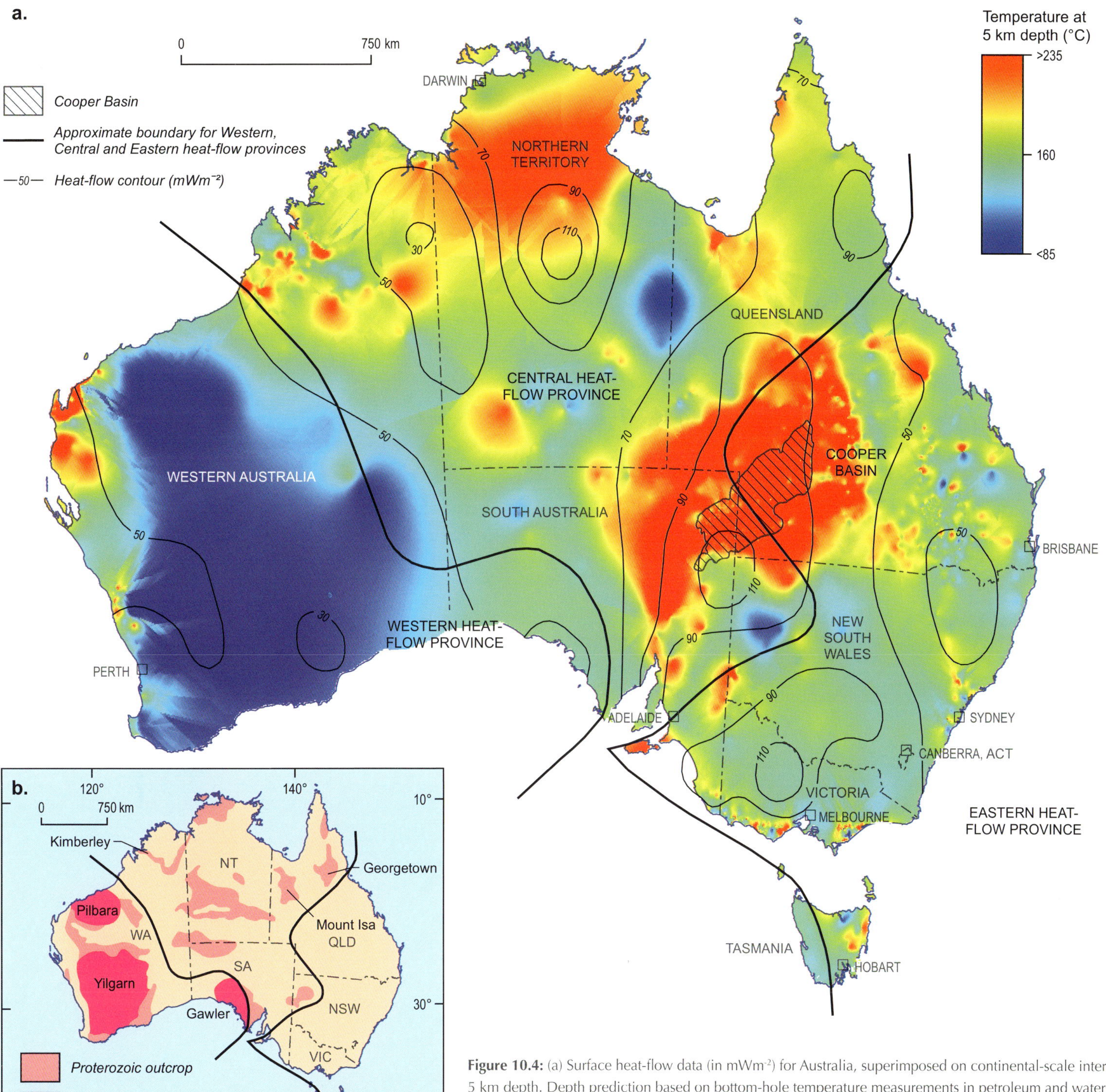

Figure 10.4: (a) Surface heat-flow data (in mWm^{-2}) for Australia, superimposed on continental-scale interpretation of the temperature field at 5 km depth. Depth prediction based on bottom-hole temperature measurements in petroleum and water bore holes. (Source: Gerner & Holgate, 2010). Black lines delineate the Western, Central [Australian] and Eastern heat-flow provinces. (Source: Mike Sandiford (cited in Tyler, 2006); Weber et al., 2011). (b) Western, Central [Australian] and Eastern heat-flow provinces superimposed on exposed Precambrian basement.

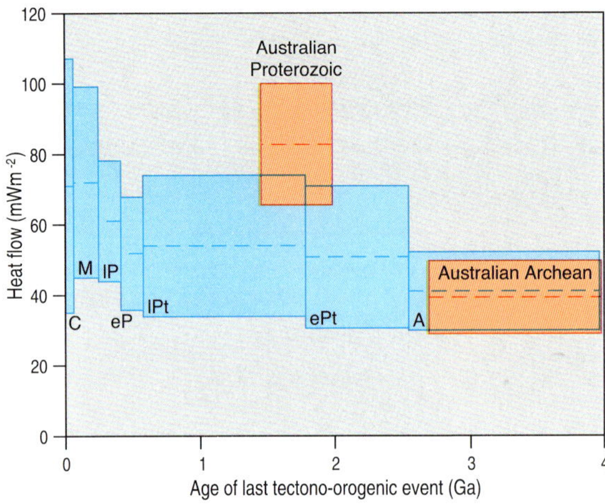

Figure 10.5: Histogram showing global continental heat-flow data, by geological age, compared with heat-flow data from Proterozoic and Archean terranes in Australia. The Australian Proterozoic is a standout in terms of heat flow. Coloured boxes are the age range and mean (±1 s.d.) of heat-flow measurements from provinces around the world (blue) and Proterozoic and Archean provinces (orange) in Australia. (Sources: Morgan, 1984; McLaren et al., 2003) A = Archean; ePt = Early Proterozoic; lPt = Late Proterozoic; eP = Early Paleozoic; lP = Late Paleozoic; C = Cenozoic; M = Mesozoic

The role of granites—vertical and horizontal zonation

The processes that have resulted in an upper continental crust enriched in U, Th and K are primarily magmatic. The behaviour of U and Th in magmatic systems is generally well understood. These elements mostly behave incompatibly, meaning that they prefer to be with the melt, not the solid (Box 10.2). A simple analogy is an ice block. When it starts to melt, Th and U are among the first in the water; when it freezes again, they are among the last out of the water. If liquid water is separated from melting ice (or the melt from the solid, as happens when a magma forms or starts to solidify), those elements that love the melt, such as Th and U (and also K), are concentrated within the melt and depleted in the solid (Figure 10.7; Box 10.2).

Of course, Earth is more complex than an ice cube, but this simple analogy largely holds (Figure 10.7). The continental crust is, in its simplest form, a heterogeneous collection of partial melts of Earth's mantle—melts that have progressively scavenged U and Th, and less so K, from the mantle and concentrated them within the crust. The crust itself has been, and continues to be, modified by a variety of processes—processes that redistribute U, Th and K. These include those readily observable, such as the veneer of weathering and sedimentation that have produced the current Australian landscape (Chapter 5), and those less observable, but nonetheless important, such as magmatism. The latter is chiefly responsible for the strong vertical crustal zonation in U, Th and K. Just as mantle melts scavenge these elements from the mantle and enrich the crust, melting of lower and middle continental crust is the dominant process in mobilising U, Th and K into the upper continental crust, although differentiation processes subsequent to partial melting, such as crystal fractionation, also contribute (Figure 10.7; Box 10.2). The more feldspar- and quartz-rich nature of continental crust dictates that the bulk of this transfer will be by felsic magmatism (i.e. granites and related rocks). Felsic magmatism is also responsible for much of the regional variation observed in U, Th and K, governed by the overall abundance of granites within a region, *and* the actual U, Th and K contents within those granites. Not all granites are equal; it is those high heat-producing granites with elevated U, Th and K that are most closely associated with U mineralisation or high heat flow in Australia (Figure 10.6).

To better understand why, and when, the Australian crust became enriched in heat-producing elements, a better understanding of felsic, especially high heat-producing, magmatism is required. Only when we know the where and when of such magmatism can we begin to isolate the potential mechanisms responsible for the heat-producing element enrichment in Australia.

The heat-producer's guide to Australia

Where and when?

Granites, and other felsic magmatism products, are not only widespread in Australia but span a prolonged age range—more than 3.5 Gyr (Figures 10.8a and 10.8b), from the oldest

Paleoarchean rocks (*ca* 3.48–3.25 Ga) in Western Australia to very young granites in eastern Australia (<30 Ma). Their distribution through time is not uniform. Calculated outcrop surface areas show major peaks in the amount of felsic magmatism during the Neoarchean, Paleoproterozoic and Mesoproterozoic, and during the Middle to Late Paleozoic (Silurian to Permian and Early Triassic; Figure 10.8b). Very similar time peaks are obtained from available U–Pb age data for igneous rocks (Figure 2.20). To explain Australia's radioactive heritage, we need to focus on the high heat-producing granites. Can we identify where these are and when they formed?

It is fact that certain ages of felsic magmatism in Australia, such as the Mesoproterozoic, are characterised by elevated concentrations of U, Th and K. To demonstrate this, we combined Geoscience Australia's extensive whole-rock geochemical database with the remotely sensed gamma-ray data (Figure 2.5). This analysis provides a uniform and unbiased estimate of the concentration of heat-producing elements at the surface across the entire Australian continent (see *Did you know?* 10.2).

Results from geochemical and gamma-ray data (Figures 10.9 and 10.10) confirm that high heat-producing magmatism in Australia is not homogeneously distributed in space or time. There are strong localisations within Western Australia, South Australia, the Northern Territory and northern Queensland, and in Archean, (Paleo- to) Mesoproterozoic, and some Carboniferous–Permian rocks. Just as importantly, there are obvious age periods that appear to have little or no high

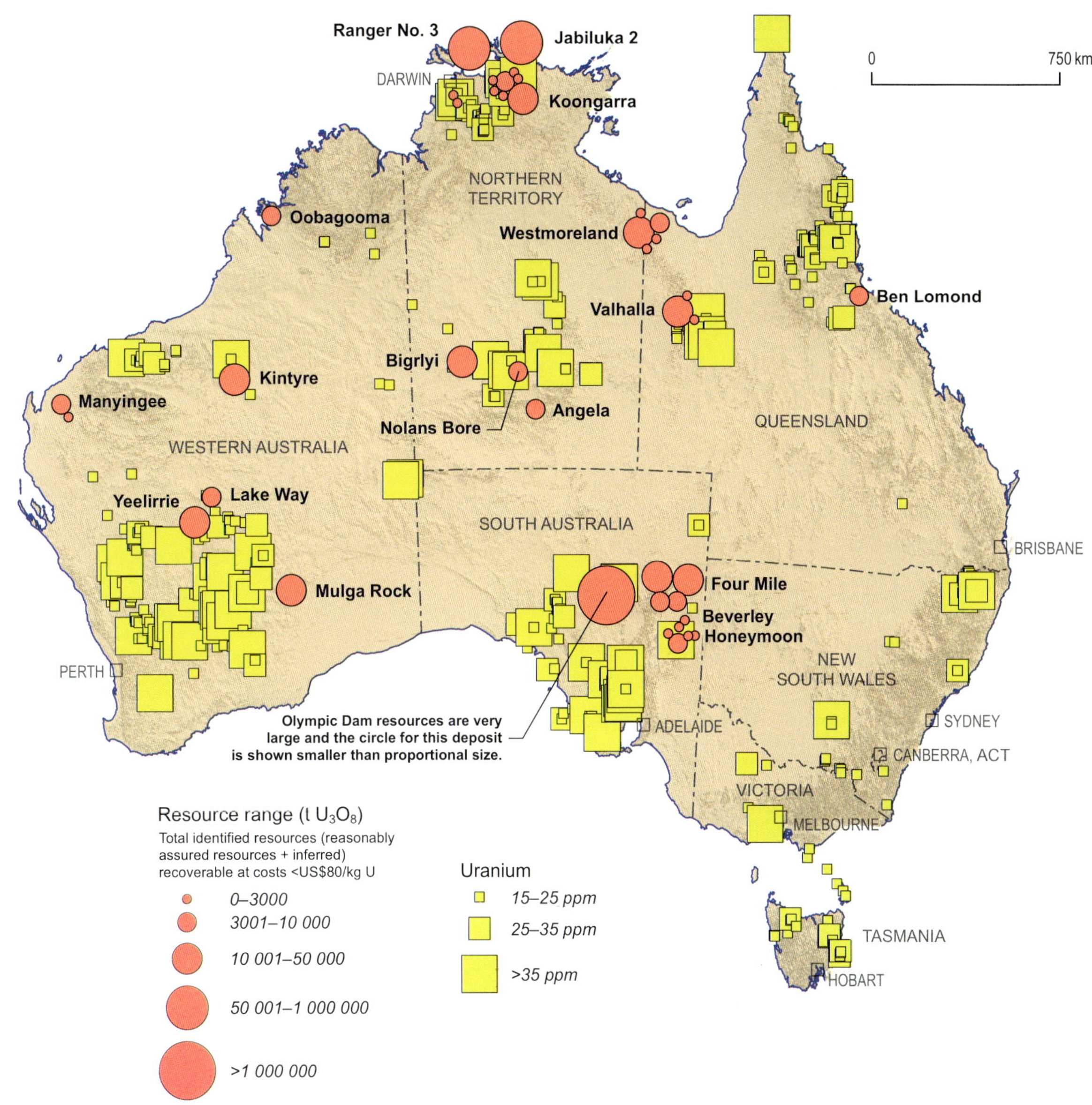

Figure 10.6: Distribution of Australian U deposits in relation to occurrences of granites known to have at least 10 ppm U. Note the close spatial relationship between U-rich granites and U deposits.

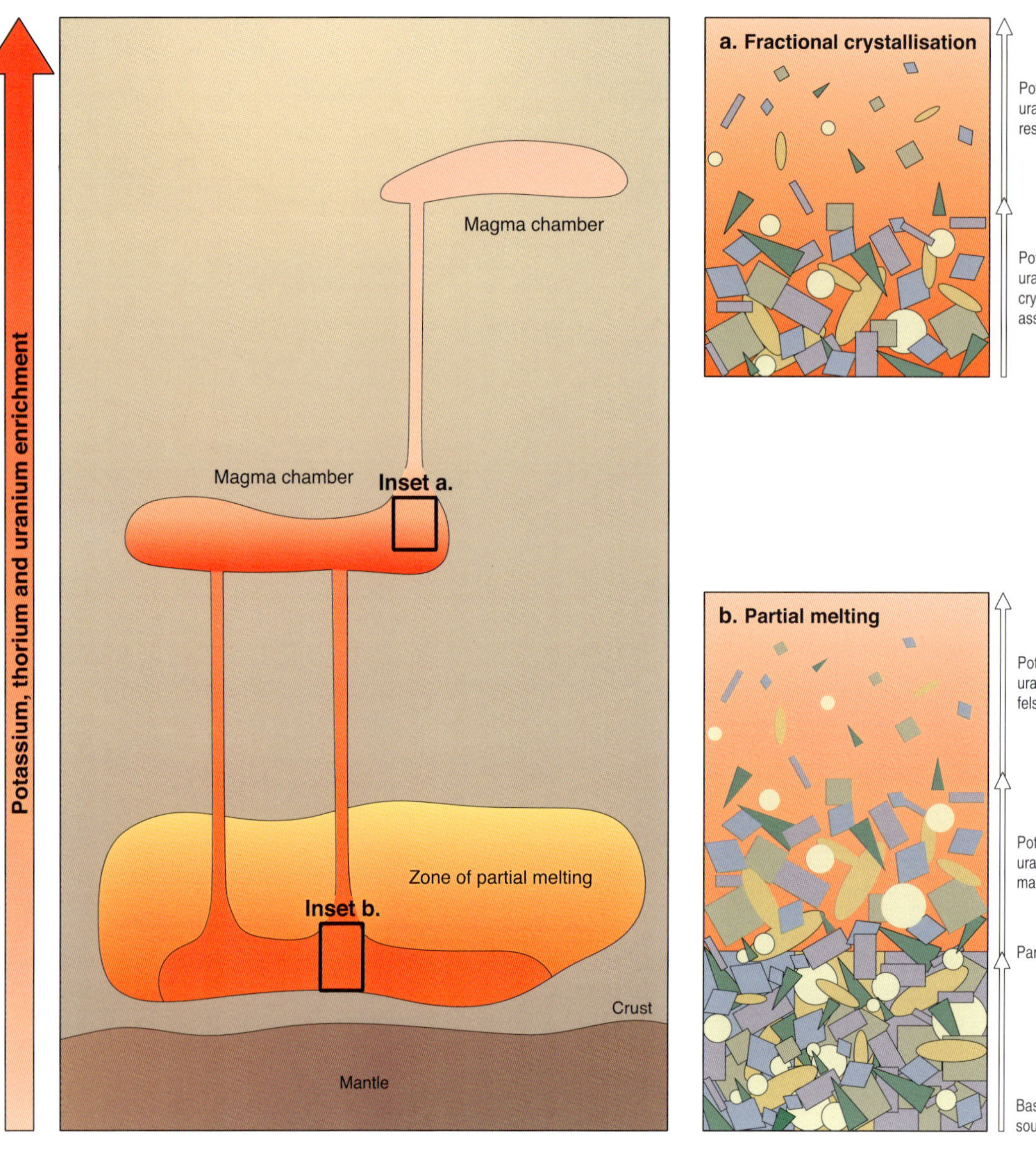

Figure 10.7: Partial melting and fractional crystallisation are two magmatic processes that mobilise the heat-producing elements (HPE) U, Th and K. The cartoon illustrates the effects of both partial melting of the lower crust and fractional crystallisation of a magma. Both processes result in a melt and a solid component, but because U, Th and K commonly behave incompatibly, they prefer (i.e. are concentrated within) the melt component. When rocks undergo partial melting (they never fully melt), they usually produce a U-, Th- and K-enriched melt and a residue (the unmelted bits) depleted in these elements. Similarly, when magma undergoes crystal fractionation, the solid component (the early formed minerals) is depleted in Th, U and K, and the remaining melt is enriched. For either process to work, the melt component must be separated from the solid. This is the norm for partial melting. Partial melts are buoyant and eventually move upwards away from the source, carrying much of the U, Th and K with them, and producing their pronounced zonation in the continental crust. Similar melt–solid separation also occurs with crystal fractionation. Unlike partial melting, crystal fractionation processes can occur at all crustal levels, from near the zone of melting to high in the upper crust.

heat-producing magmatism—for example, the Early to Middle Paleozoic (Cambrian to Devonian)—despite an abundance of felsic rocks of this age (Figure 10.8b).

Figure 10.10 also highlights other points regarding secular changes in Th (and U) in Australia:

- The Proterozoic Eon is the most enriched time period for Th (and U).
- Average Th increases from the early Archean through to the Mesoproterozoic.
- There is a pronounced decrease in Th from the Neoproterozoic onwards (although the dataset for the Neoproterozoic is small).
- A significant portion of felsic magmatism is strongly enriched in Th (and U) relative to modern arc environments.

Figures 10.8, 10.9 and 10.10 show why granites are so important in transporting U, Th and K into the upper crust. This is easily understood by looking at estimates of the actual amounts of U and Th (in tonnes) contained by the granites of each age period. For example, using average U or Th contents (Figure 10.10), combined with outcrop surface areas (Figure 10.8) and making assumptions about rock densities, it can be shown that the Silurian–Devonian granites of Australia contain an estimated 2800 Mt of Th and 600 Mt of U for every kilometre of depth extent. Combining granites of all ages gives an estimate of nearly 15 000 Mt of Th and around 3000 Mt of U; these amounts increase if we assume average granite thicknesses of 2–4 km (and constant concentrations of U and Th). Although these figures are only estimates, they clearly give an indication of the very large amounts of the U, Th and K being transferred to the upper continental crust, and the important role of high heat-producing magmatism.

The Neoarchean to Mesoproterozoic heat-producing element bonanza

It is also worth noting that the peaks of high heat-producing magmatism do not coincide with the peaks of felsic (or even total) magmatism in Australia (Figure 10.8b). This is not an unexpected result—just on simple mass balance considerations, Th and U are insufficient in the continental crust to produce large volumes of high heat-producing granites. These same constraints are what make the enriched periods even more remarkable. Interestingly, it appears that, at the continental scale, the high heat-producing granites tend to closely follow after peaks in granite magmatism (Figure 10.8b). This is evident for the Carboniferous and Mesoproterozoic peaks but also (see below) true for the Neoarchean. It is evident from the data that huge amounts of U, Th and K were mobilised during the Neoarchean, Paleoproterozoic and, to a lesser extent, the Middle and Late Paleozoic. This raises some interesting additional questions that need consideration when trying to explain these enrichments in Australia. In particular, what role, if any, does

Bald Rock, a large Early Triassic granite dome, south of Stanthorpe, Queensland.

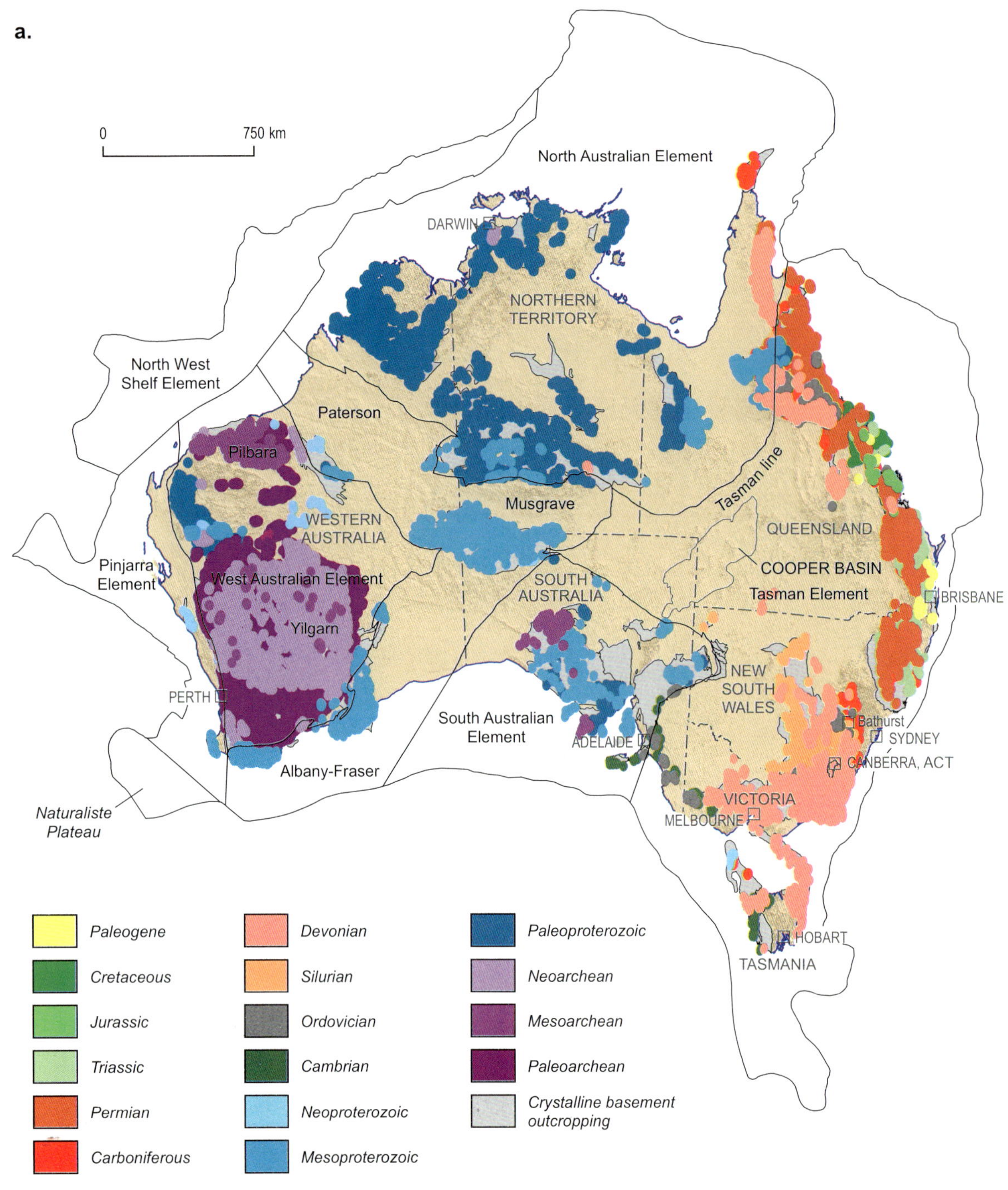

Figure 10.8: (a) Distribution of granites and their ages within Australia, superimposed on crustal elements (Figure 2.10). The granite distribution is derived from the Surface Geology Map of Australia (Figure 2.7), with a 100 m buffer applied. Note the overall trend of magmatism younging to the east, as well as the large areas of cover (in light brown) masking basement rocks (Chapter 5).

this prior voluminous magmatism play? Is it a necessary precursor for the following high heat-producing magmatism, for example, which brings significant U, Th and K into the crust? Perhaps it is not a necessary precursor but simply part of a common evolutionary cycle—for example, high heat-producing granites are often considered (rather nebulously) as post-tectonic or post-collisional. It does suggest, however, that prior history (and not just contemporaneous processes) might have played an important role in producing high heat-producing granites—this is certainly the case for some high heat-producing magmatism.

How and when to get a high heat-producing granite

To better understand the endowment of the heat-producing elements U, Th and K in Australia, we need to explain satisfactorily the enrichments in the felsic magmatism for the Late Archean, Mesoproterozoic and Carboniferous—the three big peaks in Australia's high heat-producing magmatism—and why they occur when they do. This includes taking into consideration prior magmatism and the related geodynamics of such magmatism. There are additional factors to consider. High heat-producing granites, by definition, are more enriched in U, Th and K than average granite. How to explain these enrichments? There are two obvious solutions—either the enrichment process was much more efficient for these granites, or the starting concentrations of U and Th (and K) were already elevated. Both can produce magmatism with elevated heat-producing elements, but could either process produce large areas of such high heat-producing magmatism?

The answer is 'yes'. The high heat-producing magmatism observed in Australia illustrates the efficiency of the two processes—partial melting of rocks already enriched in the heat-producing elements to produce the Archean high heat-producing granites (Figure 10.11); and extreme differentiation, increasing the heat-producing elements by processes within the melt, such as crystal fractionation, to produce the Carboniferous high heat-producing granites (Figure 10.12).

This is not the whole story. These processes do not take into account (geodynamic) drivers, nor do they satisfactorily explain the causes of the prolonged high heat-producing magmatism evident in the Australian Proterozoic. Additional factors must be involved, and these are discussed below (under 'Proterozoic enrichment').

Archean enrichment

Although Archean granites occur in South Australia and the Northern Territory, they are most abundant in the Pilbara and Yilgarn cratons of central and southern Western Australia (Figure 10.8a). Relatively widespread high heat-producing granites of Mesoarchean age (*ca* 2.95–2.85 Ga) occur in the Pilbara Craton, making them the oldest such granites in Australia and, in fact, some of the oldest on Earth. The extremely voluminous high heat-producing granites in the Yilgarn Craton are Neoarchean (*ca* 2.655–2.63 Ga) and form one of the largest exposed occurrences of Archean high heat-producing granites in the world.

The elevated heat-producing element contents in these granites (Figure 10.11c) are largely a function of their being derived by partial melting

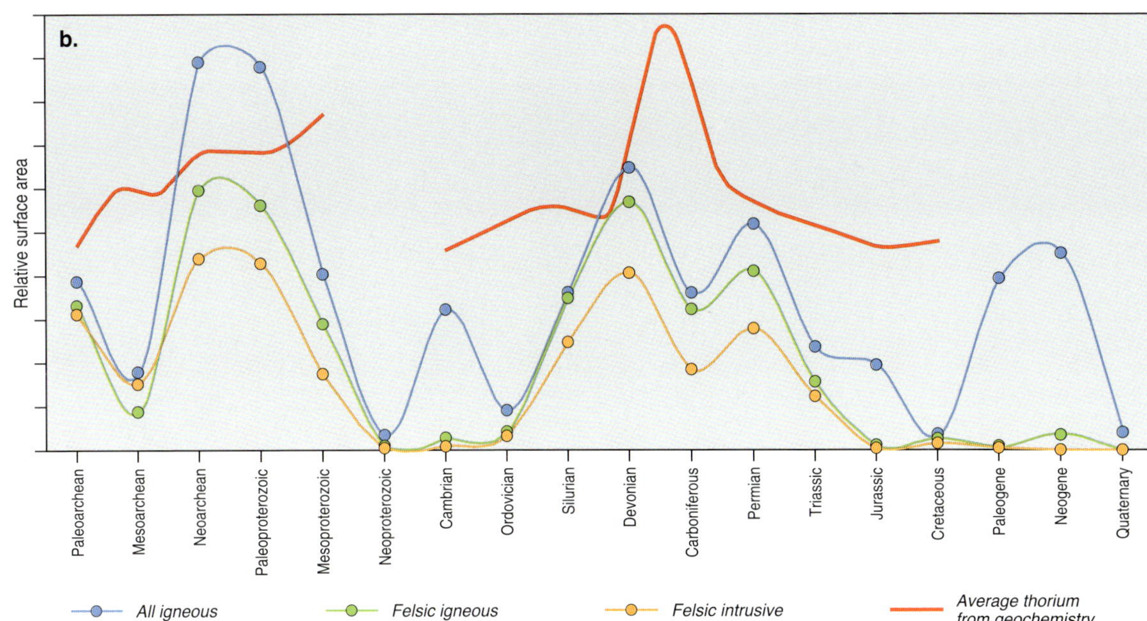

Figure 10.8: (b) Relative surface area of all magmatism, felsic magmatism and felsic intrusive magmatism, by age. Data calculated from surface outcrop based on the Surface Geology Map of Australia (Figure 2.7). Archean eons are readjusted to correct for over-representation of the Paleoarchean in the Surface Geology Map. Peaks in Australian felsic magmatism occur in the Neoarchean to the Mesoproterozoic, and in the Middle to Late Paleozoic (Silurian to Permian and Early Triassic). Compilations of available geochronology for magmatic ages in Australia (Figure 2.20) show very similar, although more precise, curves. Note that the X-axis here is not linear—refer to Figure 2.20 for the linear scale. The average Th content of granites by age (Figure 10.10) is also shown for comparison. The scale for the Th line is not shown.

(Figure 10.7) of granitic rocks that already had elevated (above crustal average) heat-producing element contents (Figure 10.11b). Subsequent further enrichment of the heat-producing elements by crystal fractionation processes played only a minor role. The appearance of K-rich granites is a feature of many Archean terranes around the world, especially in Neoarchean terranes (2.8–2.5 Ga). These granites are generally the last to appear and are interpreted as the first widespread partial melting of a granitic crust.

One of the major requirements for producing large amounts of high heat-producing magmatism in this way is a much larger volume of source rocks (3–5 times larger). This is exactly what is seen in the Archean of Australia and elsewhere. Archean terranes, although complex, largely have a simple bimodal granite–greenstone geology, comprising supracrustal rocks dominated by basalt and other

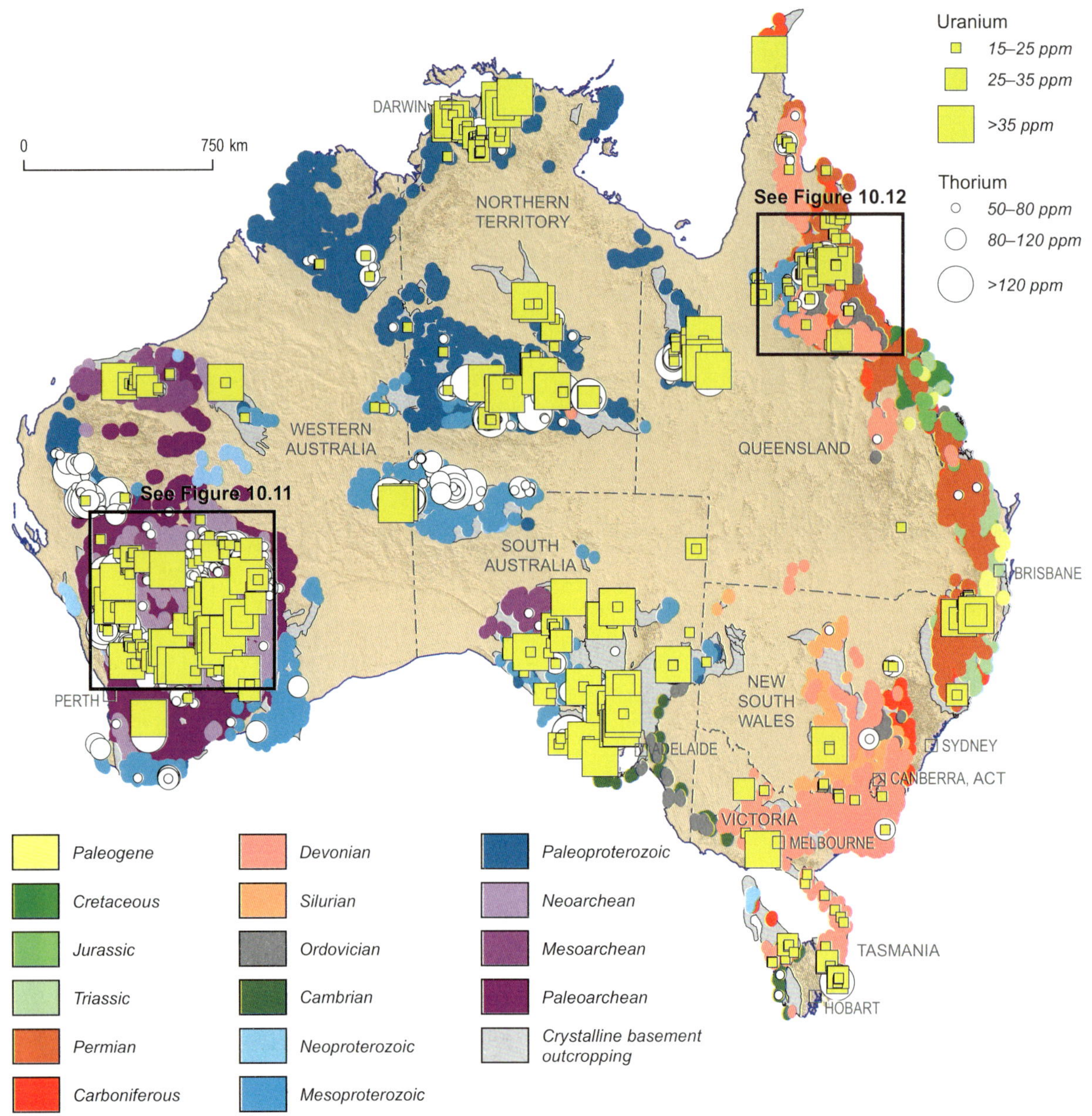

Figure 10.9: Using geochemical data (n = 20 000) to map the distribution of high heat-producing granites in Australia. The figure shows the distribution of granite geochemical data that have either elevated U or Th contents, plotted on top of buffered granite outcrop (coloured by age). Of note is the non-uniform distribution of such elevated heat-producing element samples, with a strong localisation within Archean, (Paleoproterozoic to) Mesoproterozoic, and some Carboniferous–Permian rocks (Figure 10.8a). Granite age colours are as for Figure 10.8a. Location areas for Figures 10.11 and 10.12 are shown by the boxes.

volcanic rocks (greenstones) and voluminous granites (Chapter 2). More importantly, the majority of Archean granites belong to what is known as the tonalite–trondhjemite–granodiorite suite (TTGs), a class of granites distinguished by a characteristic sodic chemistry (high Na and low–moderate K), consistent with derivation by partial melting of basaltic source rocks. These rocks dominate most Archean terranes by volume, including those in Australia, and are ideal source rocks for the Archean high heat-producing granites.

The path to Archean high heat-producing granites, therefore, results from a simple three-step process (Figures 10.11b and 10.11c):

- partial melting of Earth's mantle to produce basaltic rocks
- partial melting of the basaltic rocks to produce tonalite–trondhjemite–granodiorite
- partial melting of tonalite–trondhjemite–granodiorite to produce high heat-producing granites.

This process is well illustrated by the granites from the Yilgarn and Pilbara cratons (Figure 10.8a). The end result is a progressive build-up in U, Th and K at each stage (Figure 10.11). This increase comes with a corresponding decrease in melt volume, culminating in a rock enriched in the heat-producing elements—these are Archean high heat-producing granites. Think of a pyramid, with the high heat-producing granites at the apex. A vast amount of mantle partially melted to produce the basalts, which then partially melted to produce a smaller volume of tonalite–trondhjemite–

granodiorite, which in turn partially melted to produce even less volume of high heat-producing granites. At each stage of volume reduction, there was a corresponding increase in heat-producing element concentration. This is recorded within the Yilgarn Craton, where the high heat-producing granites are one-quarter as abundant as the tonalite–trondhjemite–granodiorite, which is about the expected ratio. To get an idea of the absolute scale of these processes, we can look at surface areas of the high heat-producing granites in the Yilgarn Craton (Figure 10.11a). These are somewhere between the known 30 000 km^2 and the estimated 100 000 km^2. These numbers can probably be tripled when volumes and the third dimension are considered. Interestingly, these high heat-producing granites are thought to be located in granite domes beneath many of the deposits in the Eastern Goldfields of Western Australia (Chapter 8).

In detail, the process is more complex, and additional factors, such as multiple sources, come into play. A feature of Australian tonalite–trondhjemite–granodiorite magmatism is progressive heat-producing element enrichment with decreasing age; that is, younger tonalite–trondhjemite–granodiorite tend to have higher U and Th contents than older ones (a trend that effectively continues into the high heat-producing granites). This probably reflects a component of older tonalite–trondhjemite–granodiorite (i.e. not just basaltic rocks) in the source. Such slightly more enriched tonalite–trondhjemite–granodiorite are common in both the Yilgarn and Pilbara cratons, and they have helped contribute to the high U, Th and K contents within the Archean high heat-producing granites. It is likely that the latter also have a variety of components in their source. Importantly, although the broad process is understood, the geodynamic drivers responsible for these processes are not. There is a general, but not universal, consensus that tonalite–trondhjemite–granodiorite were produced in an arc environment (Chapter 8), not unlike those seen today, although possibly on a hotter Earth (Box 10.3). What environment was responsible for producing the Archean high heat-producing granites, especially the large and widespread amounts in the Yilgarn Craton (Figure 10.11), is much more uncertain.

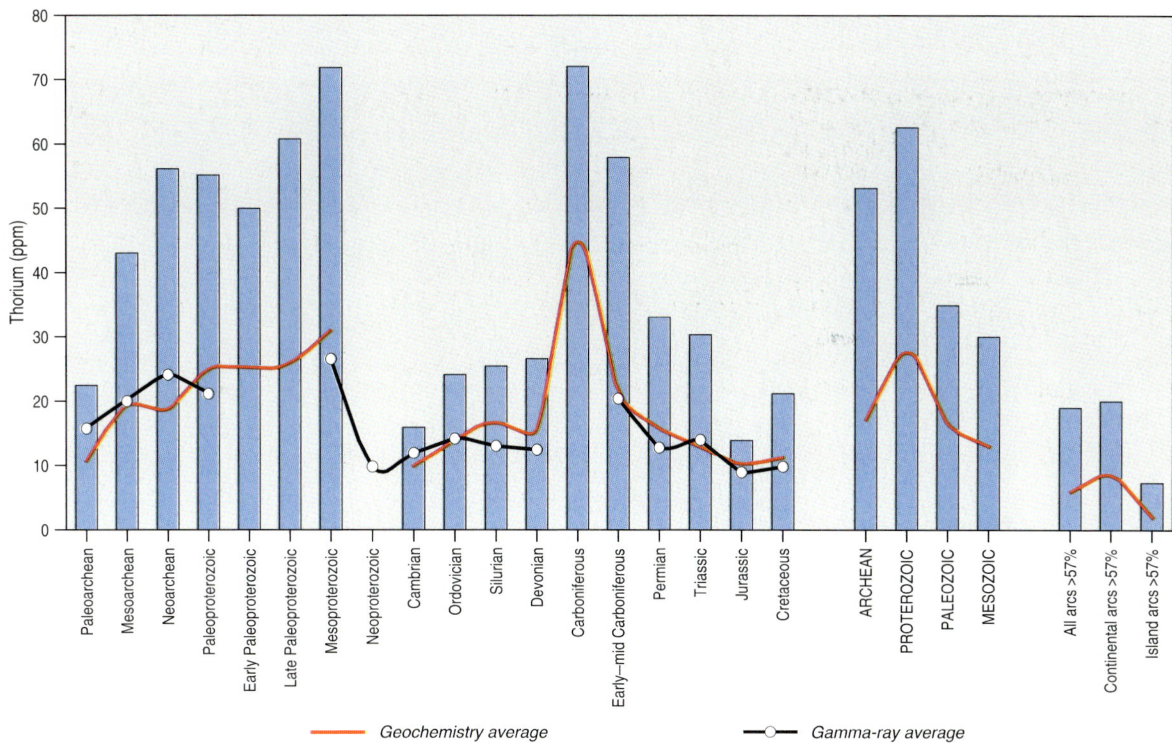

Figure 10.10: Average Th contents of Australian felsic intrusive rocks through time, as calculated from geochemical (red line) and gamma-ray (black line) data. Note the very good correspondence between the average Th calculated from the geochemistry and the average Th calculated from the gamma-ray data. Vertical bars show the 90th percentile values (from geochemical data), which emphasise the elevated values in the Neoarchean and Carboniferous, with the Mesoproterozoic being particularly elevated. Average Th contents for intrusive and extrusive rocks from modern-day island arcs, continental arcs, and island and continental arcs are also shown for comparison; note their much lower Th levels. (Sources: Geoscience Australia; GEOROC database, Mainz University)

HOT EARTH: AUSTRALIA'S HOTTEST REAL ESTATE (BOX 10.3)

Australia was even hotter in the past. Paradoxically, the effects of radioactive decay work in both directions of time. The concentrations of radiogenic isotopes of U, Th and K increase the further back in time we go—for example, the ^{235}U isotope was about twice as abundant at 4.5 Ga when Earth was formed. These changes probably had a number of important effects on Earth and Australia.

Earth's mantle was hotter in the past. This secular change in mantle temperature is commonly used to suggest that Archean geodynamics were different from those operative today, and may be the reason why Archean rocks such as the tonalite–trondhjemite–granodiorite (TTG) suite are largely confined to the early Earth. A further consequence is that heat flows would have been higher (20–30%, assuming similar geology).

Areas of high heat flow today (Figure 10.4) may well have been significantly higher in the Proterozoic, affecting the behaviour and response of the crust (by weakening it), and possibly accounting for the episodic rifting and magmatism we see in Australian Paleoproterozoic and Mesoproterozoic terranes (Chapter 2). These higher geothermal gradients might even have been responsible for generation of some of the Proterozoic granites.

Contrary to expectations, Australia's 'hottest' real estate is not in the cities of Sydney or Melbourne, but in a largely forgotten piece of arid Australia, some 200–300 km northwest of Broken Hill, in the Mount Painter and Mount Babbage inliers of South Australia (Figure B10.3a). Granites (and felsic volcanic rocks) in these areas include both

Thermal springs , Witjira–Dalhousie National Park, South Australia.

Mesoproterozoic and Paleozoic ages, although it is the former, the rocks of the Moolawatana Suite, that are particularly enriched in Th and U. Granites of this suite can be subdivided on the basis of their geochemistry into two subgroups (Figure B10.3b), both of which are examples of high heat-producing granites, but the Mount Neill Subgroup is the standout, with calculated heat production values between 10 μWm^{-3} and 130 μWm^{-3} (average of 25 μWm^{-3}).

We are not entirely sure how the extreme enrichments in the Mount Neill Subgroup were produced. It quite possibly reflects some form of alteration process, although the group of enriched elements (U, Th, rare earth elements and niobium—Nb) are also suggestive of magmatic processes. The high positive correlation between Nb (usually an immobile element in most alteration environments) and Th, for example, indicates that, if it is an alteration feature, it is an unusual one. High heat-producing regions, such as the Mount Painter and Mount Babbage inliers, offer another example of the enigma that is represented by the Paleoproterozoic and Mesoproterozoic in Australia. It is perhaps no surprise that nearby are the Paralana Hot Springs and U deposits, both local historical workings, as well as the operating Beverley Mine and the Four Mile prospect in the adjacent Lake Frome region. Even though the U deposits are significantly younger, it is likely that the U was sourced—directly or indirectly—from the Mount Painter or similar granites.

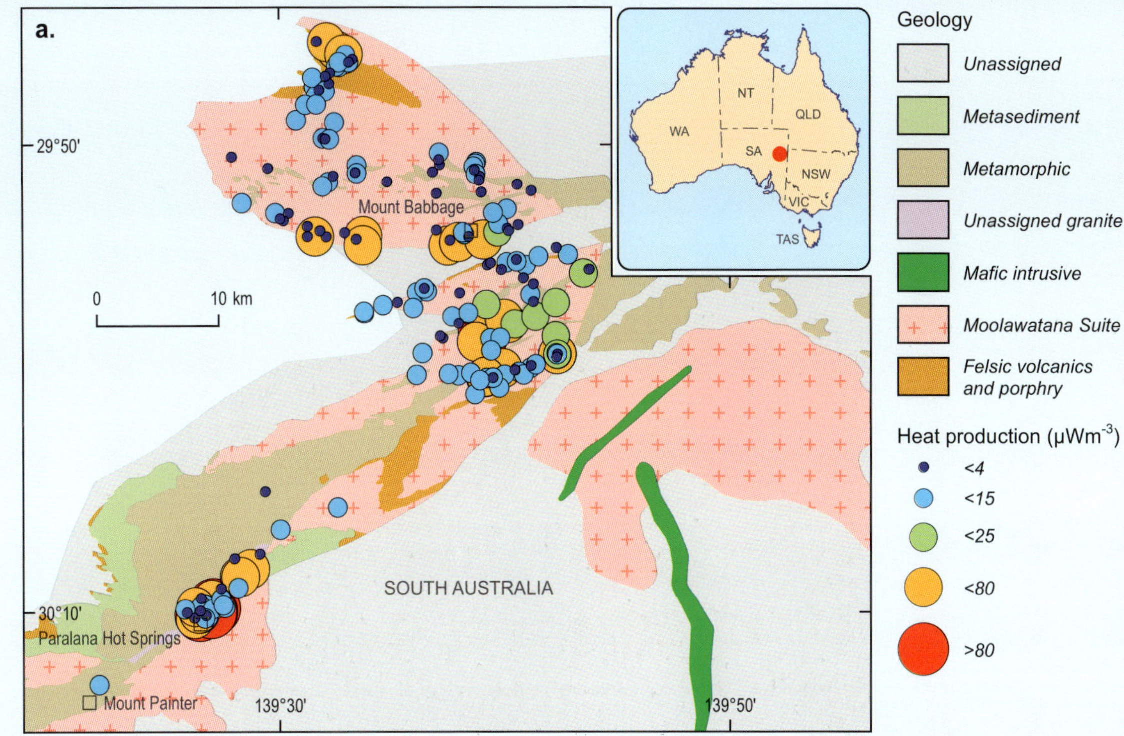

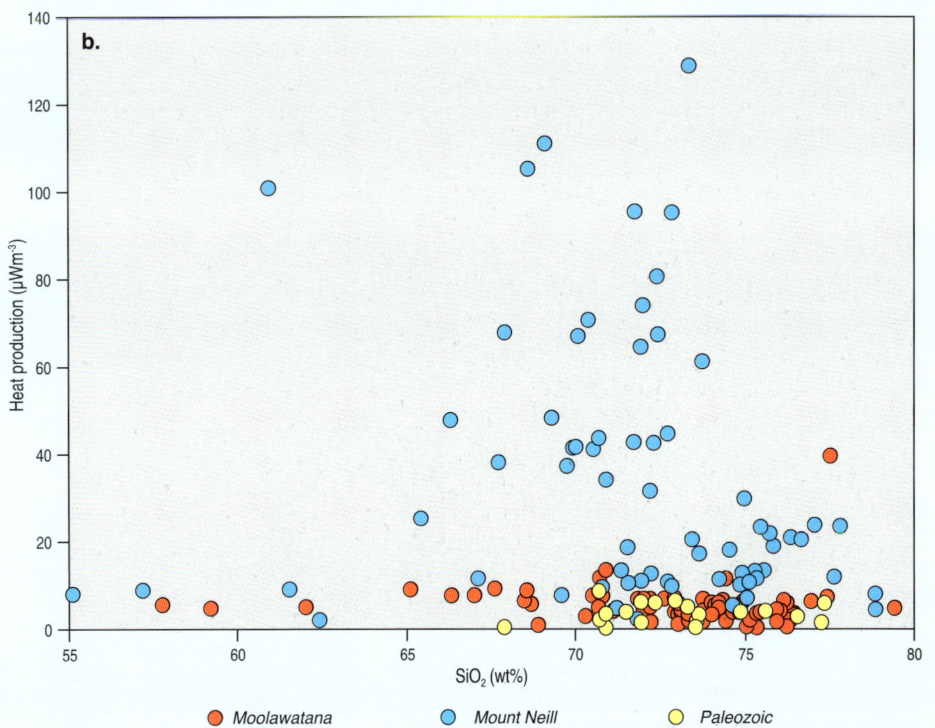

Figure B10.3: Interpreted solid geology of the Mount Painter region of South Australia. (a) Geochemical sample points ranked by heat production values for granites and related rocks of the Mount Painter and Mount Babbage inliers. (b) Heat production values for the Moolawatana Suite. Heat production is a measure of the rate of heat generation in these rocks (in μWm^{-3})—that is, the total heat production from U, Th and K.

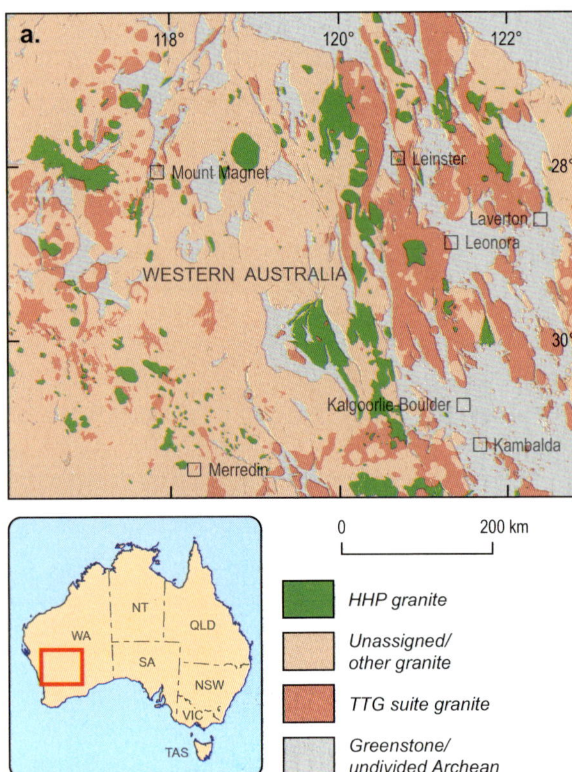

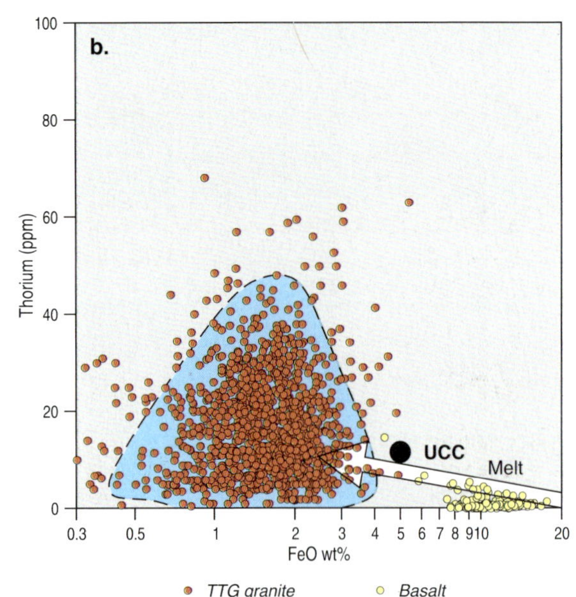

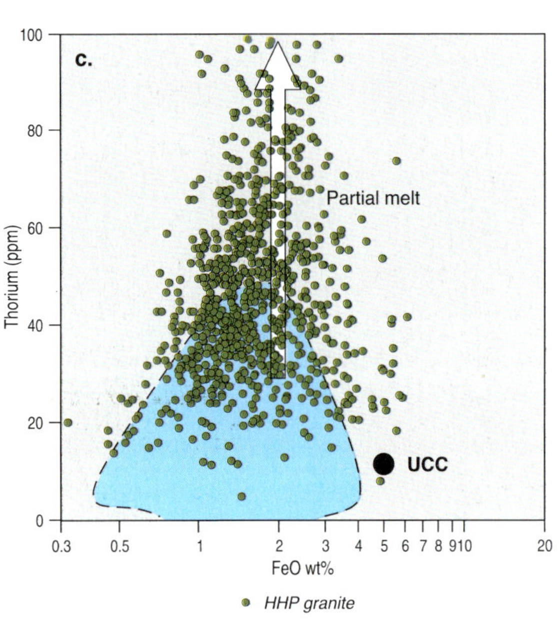

Figure 10.11: Evolution of the high heat-producing (HHP) granites of the Archean Yilgarn Craton, Western Australia. (a) Simplified geological map of part of the Yilgarn Craton, showing the abundant tonalite–trondhjemite–granodiorite (TTG) suite granites (*ca* 3.0–2.655 Ga; in dark red) and the less abundant younger high heat-producing granites (*ca* 2.655–2.63 Ga; in green). Grey regions are the greenstones that are dominated by mafic (basaltic) lithologies. (b) Plots of total Fe (in wt% FeO) vs Th (ppm) for the Archean basalts and tonalite–trondhjemite–granodiorite granites of the Yilgarn Craton. Basaltic source rocks (grey) undergo partial melting to form the ubiquitous and voluminous Archean tonalite–trondhjemite–granodiorite suite (red). (c) The latter then underwent partial melting to produce the younger high heat-producing granites (green). The average Th content of the upper continental crustal (UCC) is also shown, highlighting the strongly elevated Th of the high heat-producing granites in the Yilgarn Craton. Note the lower Th with lower FeO in the high heat-producing granites, indicating that crystal fractionation is not a significant process in these granites (compare with Figure 10.12b).

The Carboniferous period

The only voluminous high heat-producing granites emplaced in Australia in the last 500 Myr are the 345–300 Ma Carboniferous granites (Figures 10.8b and 10.10). These granites occur in a belt that runs inland of the east Australian coast, from around Bathurst (NSW) to far northern Queensland. They also continue sporadically westward and include the granites beneath the Cooper Basin, which form part of some of Australia's most promising current geothermal exploration regions (Figure 10.4). Unlike the Archean high heat-producing granites, the elevated U and Th in the Carboniferous granites are largely a function of extensive crystal fractionation processes. In simplest terms, the observed elevated concentrations have largely occurred after partial melting, driven by separation and removal of U–Th-poor minerals, such as quartz and feldspar, from an increasingly U–Th-enriched magma (Figure 10.7).

These crystal fractionation processes are best illustrated by the widespread Carboniferous granites within the Georgetown–Mt Garnet–Herberton area of north Queensland (Figure 10.12). Here, granites show a range of heat-producing element concentrations, from slightly higher than upper crustal averages to strongly enriched (*ca* 10 times upper continental crust). These large heat-producing element variations are consistent with crystal fractionation, as suggested by positive correlations between Th contents and various indices of differentiation, such as decreasing iron (FeO); compare Figure 10.12 with Figure 10.11. Given the high levels of the heat-producing elements in these granites, it is not surprising that there is a spatial association with both U mineralisation, hot springs and naturally occurring blue topaz—the latter is an excellent indicator of high-percentage crystal fractionation and elevated U and Th, with the colour being due to natural radiation damage.

As for partial melting, the end results of high-percentage crystal fractionation, such as the Carboniferous high heat-producing granites, are volumetrically minor relative to the starting volume. Nevertheless, locally, large areas of such granites can be produced; there is somewhere between around 2500 km^2 and 3500 km^2 for those outcropping in north Queensland (Figure 10.12).

Despite the suggested different processes responsible, Figures 10.11 and 10.12 show similar patterns of enrichment in heat-producing elements. Although this means that discriminating between the two processes can be difficult, this is of only minor concern as both processes—partial melting of enriched source rocks and high-percentage crystal fractionation—rarely operate independently of each other. This is largely because the great majority of high heat-producing granites have a dominant crustal component—that is, their source rocks have had enough prehistory to be already variably enriched in U, Th and K, certainly above average mantle values.

Proterozoic enrichment

The most enriched time period in Australia for the formation of heat-producing elements is the Proterozoic, especially the approximately 400 Ma time period from the Late Paleoproterozoic culminating in the Mesoproterozoic (Figures 10.10 and 10.13). The latter was not only when the most granites enriched in heat-producing elements were formed across the largest area, but also the time when major U mineralisation, such as the giant Olympic Dam deposit, occurred (Chapter 8).

While some of these heat-producing element enrichments in these Proterozoic granites may reflect other processes (Box 10.3), they largely reflect either episodes of magmatic reworking and/or extensive crystal fractionation. This alone does not explain the trend of increasing heat-producing elements from the Archean to a peak in the Mesoproterozoic. Nor does it explain the episodic nature of Proterozoic high heat-producing magmatism, which is in distinct contrast to the Archean and Carboniferous. Additional factors must be involved. This is also apparent from consideration of the general processes discussed for the Archean and the Carboniferous granites. If it is assumed that high heat-producing granites, such as those in north Queensland, simply result from partial melting of a crustal source, coupled with extensive high-level fractionation, then why is there such concentrated high heat-producing magmatism in certain regions? For example, within north Queensland, the high heat-producing granites represent about 30–40% or some 3000 km^2 of magmatism. Why aren't there more of these regions in the Paleozoic of Australia? The Devonian period, for example, includes the major Paleozoic magmatic event in eastern Australia, extending from Tasmania to far north Queensland (Figure 10.8a). Why are there few significant areas of Devonian high heat-producing granites?

Did you know?

10.2: Mapping the movement of the radioactive elements K, Th and U

The recently calibrated airborne gamma-ray spectrometric data, covering >90% of the Australian continent (Figure 2.5) provides unbiased information on the concentration of the radioactive elements (K, Th and U) within the top 50 cm of Australia's surface. When combined with geological information, such as from the Surface Geology Map of Australia (Figure 2.7), these data can be used to quantify the effects of Australia's well-developed regolith on the location of these radioactive elements. This is because K and U are relatively mobile in surficial landscape and regolith processes (Chapter 5), processes reflected in the near-surface materials measured by the gamma-ray data.

Calculations based on this methodology suggest that there may be more than 10 Mt of U missing from the top metre of Australia alone (assuming a Th:U ratio of 4). This is more than five times as much U as in the supergiant Olympic Dam deposit (Chapter 8)! Some of this U ends up within young surficial U deposits such as Yeelirrie in Western Australia.

Image by John Wilford

3D image of the gamma-ray data draped over the digital elevation model. Colours: K (red), Th (green) and U (blue). Note mobile K on hill slopes and U in the valley floor.

Image by Adrian Yee

Naturally radiated blue topaz from the 'Blue Hills' area, Mt Surprise, Queensland.

Similarly, the simple but very efficient basalt–TTG–HHP granite cycle, which dominates the Archean in Australia, appears to cease effectively some time in the 500 Myr period between the last abundant Archean rocks and the first abundant Proterozoic rocks in Australia (Figure 10.13a). Certainly, it is never again the dominant process in Australia—sodic granites, such as tonalite–trondhjemite–granodiorite, are only a very minor component of the (outcropping) Proterozoic magmatism (Figure 10.14a). Why is this? What has changed? The dominant early Proterozoic granites in Australia are in fact not sodic, but overwhelmingly potassic (high K) (Figure 10.14b), with moderate to high levels of the heat-producing elements. This chemical change is reflected in heat-flow and other temperature data, which show that Archean cratons in Australia, and elsewhere (Figures 10.4 and 10.5), are characterised by low heat-flows, in distinct contrast to much of the Australian Proterozoic. This supports the idea that perhaps a fundamental Earth-system process might have changed in the period between the Archean and the Paleoproterozoic.

Additional questions are raised when considering geodynamic controls, which are ultimately responsible for producing high heat-producing magmatism. Most continental crustal growth and (mafic to) felsic magmatism today occurs within subduction zone environments. Geochemical data for both intrusive and extrusive rocks from modern subduction zone environments show that they appear unable to produce significant amounts of high heat-producing magmatism (Figure 10.10). They can and do, however, produce precursors for this process. Does this mean that other geodynamic environments were more dominant at times during the past—perhaps reflecting either secular changes through time, or the role of tectonic cycles such as supercontinent formation and breakup (Chapter 2)? Another possibility is that there have been secular changes in U, Th and K availability, such that geodynamic processes have not changed, but merely their chemical signatures. Although a conclusive answer to these questions may not be known, it is possible to place some constraints on likely geodynamic models by investigating Australia's Proterozoic granites.

Australian Proterozoic granites—geodynamic models

Although there is a wide variety of Proterozoic granite types in Australia (Figure 10.14a), the majority share a number of important features.

They are overwhelmingly felsic and potassic in composition, consistent with being derived from partial melting of predominantly crustal sources. Again, partial melting of source rocks with elevated heat-producing elements, as well as crystal fractionation, probably caused the elevated heat-producing element contents in the Proterozoic granites (Figures 10.13 and 10.14). The chemistry suggests that most Australian Proterozoic granites were not generated within continental or island arcs (Figure 10.10). An arc environment is not ruled out, however, as a backarc could have been responsible for some enriched Proterozoic granites. The situation is complicated by the uncertainty about when modern-style tectonic processes actually commenced. Estimates for the latter range from early Archean through to late Neoproterozoic, although many agree that something similar to modern-style tectonics were operative by the Late Archean. Certainly, there is evidence of features not unlike modern arcs in the Australian Proterozoic, making this a plausible scenario, and most geodynamic models for the Australia Proterozoic invoke arc and backarc processes.

The other often-suggested origin for many Proterozoic granites in Australia, especially those of Mesoproterozoic age, is that they are anorogenic in nature—that is, they were generated within an intraplate environment, well away from any arc influence. Certainly, the heat-producing element–enriched nature of many Mesoproterozoic granites is consistent with such an interpretation—that is, repeated crustal melting driven by crustal extension and high thermal gradients, often with associated mantle-derived magmatism. There is

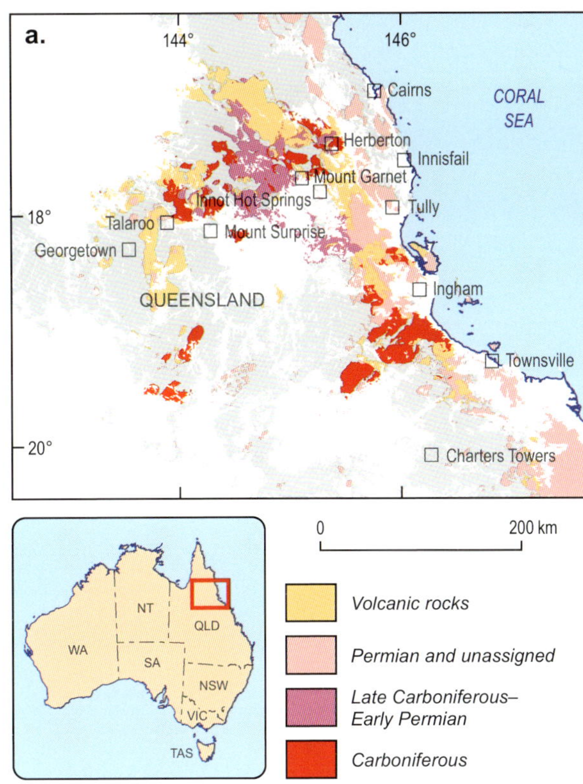

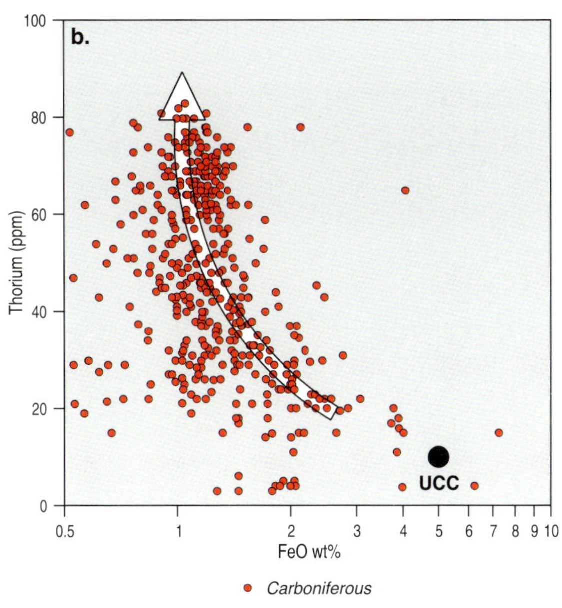

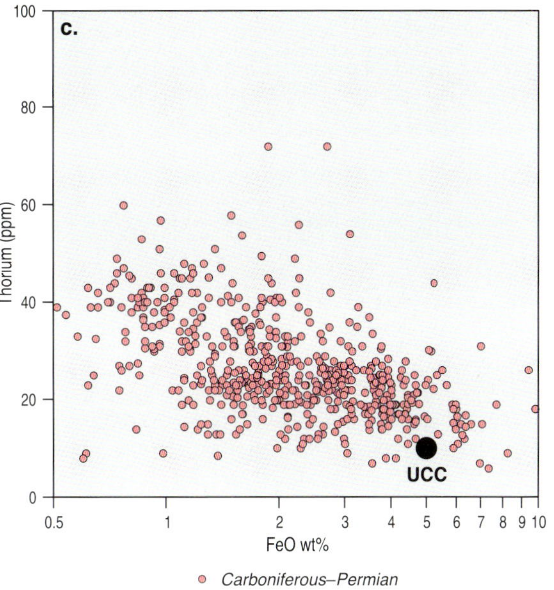

Figure 10.12: Evolution of high heat-producing (HHP) granites in the Carboniferous of north Queensland. (a) Distribution of Carboniferous high heat-producing granites (*ca* 345–310 Ma; in red), and very Late Carboniferous–Early Permian granites (*ca* 305–295 Ma; in pink) in north Queensland. (b) Plots of total Fe (wt% FeO) vs Th (ppm) for Carboniferous high heat-producing granites (red). (c) Very Late Carboniferous–Early Permian granites (pink). Although granites of both ages have undergone crystal fractionation, it occurred to a higher degree in the Carboniferous granites, resulting in their much higher Th contents. Note the strong increase in Th with decreasing FeO—a good indicator of crystal fractionation. Notably, granites of both ages plot above the average Th content of upper continental crustal (UCC), indicating that partial melting of an enriched source also played a role in generating these high heat-producing granites.

Image by Richard McDowell

Elachbutting Rock, a massive Archean (*ca* 2640 Ma) granite outcrop, northeast of Mukinbudin, Western Australia.

also other evidence for this. For example, mafic large igneous provinces—regions of voluminous and widespread mantle-derived magmatism commonly interpreted as intraplate and related to mantle plumes—do occur. An example is the *ca* 1780 Ma Paleoproterozoic Hart Large Igneous Province, in the Kimberley region of Western Australian (Chapter 2), which is contemporaneous with potassic granites. Certainly, multiple plume events could explain the episodic nature of high heat-producing magmatism in the Australian Proterozoic. The presence of large igneous provinces and related rocks, however, does not negate the operation of subduction zones, continental arcs and backarcs. The Hart Large Igneous Province, for example, has been suggested to be contemporaneous with subduction.

There is no reason, of course, why both interpreted geodynamic models are not at least partly correct —that is, both modern-style tectonic processes and large igneous provinces were operative. This is indeed what is proposed by models of the supercontinent cycles, from their formation, driven by subduction and collision, to their fragmentation, driven by repeated rifting and a common association with large igneous provinces and mantle plumes (Chapter 2; Figure 10.13). Interestingly, the greatest period of Proterozoic high heat-producing magmatism in Australia (1.9–1.5 Ga) overlaps with the period of formation of the Nuna Supercontinent, and its subsequent breakup (Figures 2.8 and 10.13). This immediately raises an obvious question: Could the processes that produced Australia's Proterozoic high heat-producing magmatism and much of the country's heat-producing element and thermal energy endowment be directly related to supercontinents?

Supercontinent breakup—the answer?

Several lines of evidence offer support for the supercontinent argument. Firstly, there appears to be a global occurrence of anorogenic (within-plate) magmatism in the Mesoproterozoic, which appears to have peaked around 1400 Ma, although orogenic events were also occurring at this time. Secondly, this global anorogenic magmatism is dominantly confined to regions of Paleoproterozoic crust. This picture has many similarities to what is observed in Australia, where Mesoproterozoic granites are characterised by elevated U, Th and K and appear to be largely 'anorogenic'. They also seem to be strongly localised to predominantly Paleoproterozoic crust, although there is isotopic and geochronological evidence that Archean crust is also involved, at least in some Australian cratons. Studies of Mesoproterozoic magmatism in Australia, and elsewhere, have concluded that the felsic magmatic compositions relate to the composition of their lower crustal protolith.

Supercontinent evolution can explain, at least partly, why the Australian Mesoproterozoic granites are so enriched in the heat-producing elements. The periods of extension and episodic rifting related to supercontinent breakup in the Late Paleoproterozoic and Early Mesoproterozoic (Figure 10.13) would have been ideal for the two processes—crustal reworking and fractionation—that lead to high heat-producing magmatism. There is some evidence that such an environment may also have been conducive to emplacement of enriched mantle melts, also contributing to the heat-producing element enrichment.

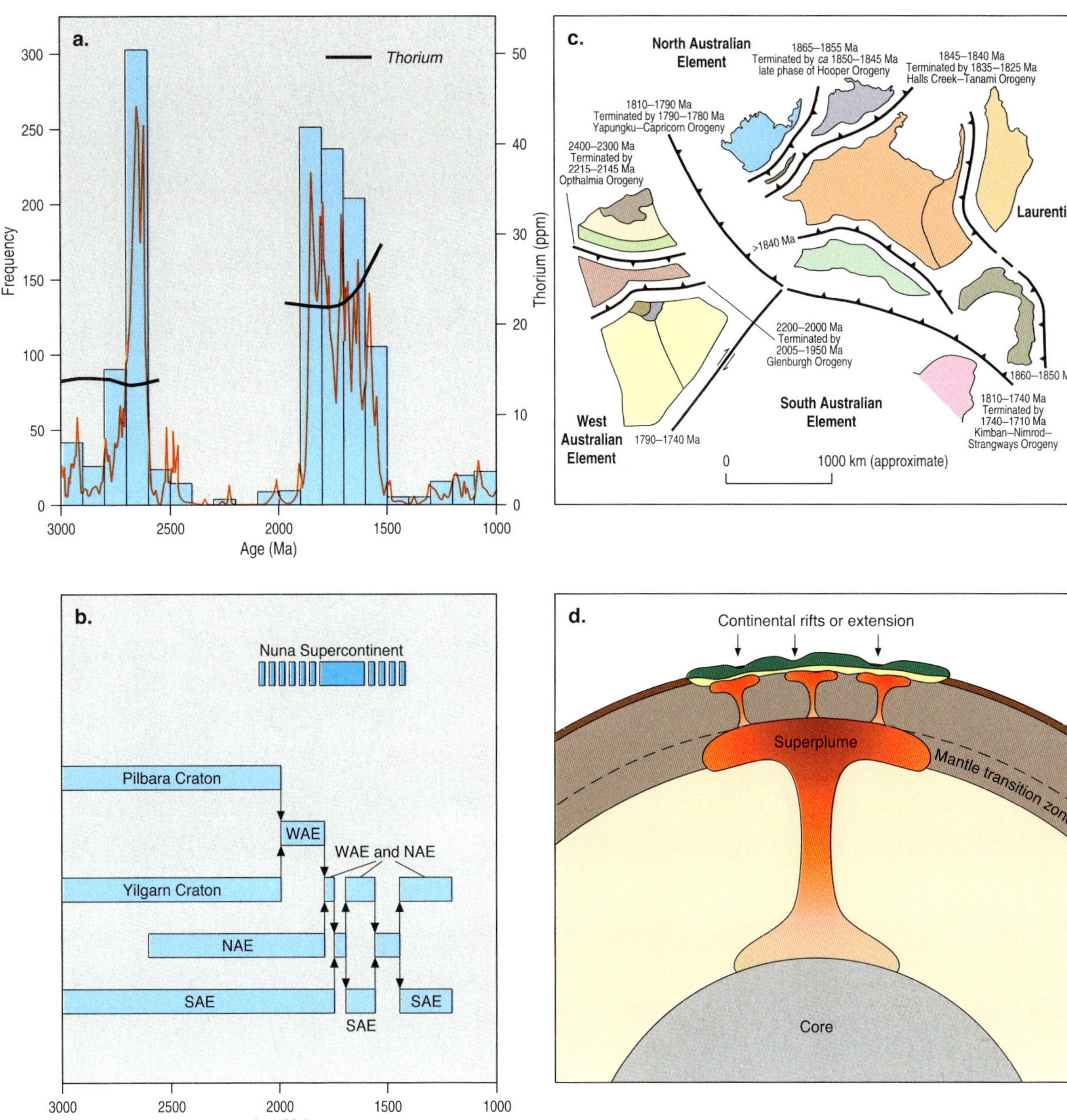

Figure 10.13: Possible supercontinent cycle in the Australian Proterozoic and its relationship to high heat production. (a) Age histogram for Australian magmatism for the period 3000–1000 Ma, as determined by U–Pb SHRIMP zircon ages. (Figure supplied by K Sircombe, Geoscience Australia, pers. comm., 2011). Average Th contents (Figure 10.10) are superimposed in black. (b) and (c) One possible tectonic reconstruction for Australia during this time (see Chapter 2). The main period of Proterozoic magmatism corresponds to the timing of Nuna Supercontinent formation and subsequent breakup. (d) One possible model for the breakup of supercontinents is via formation of superplumes (Santosh et al., 2009), which may be generated by the build-up of heat in the mantle due to thermal insulation beneath the assembling supercontinent. The build-up of heat causes extension and rifting within the supercontinent, and eventual breakup. NAE = North Australian Element; SAE = South Australian Element; WAE = West Australian Element

This is only part of the story. There is an apparent increasing trend in heat-producing element contents from the Neoarchean to the Mesoproterozoic (Figure 10.10), indicating a level of heat-producing element enrichment not seen in modern arc environments and demanding some explanation. One possible solution lies in the Archean geology of Australia and the tonalite–trondhjemite–granodiorite–high heat-producing (TTG–HHP) sequence observable in those regions (Figure 10.11). Perhaps many of the Proterozoic crustal blocks in Australia were themselves originally comprised a significant amount of tonalite–trondhjemite–granodiorite (and high heat-producing granites derived from them), resulting in a variably but relatively enriched crust. This and the subsequent melting of an increasingly hybrid crust in the Proterozoic could readily explain the increasing heat-producing element trend observed from the Archean to the Mesoproterozoic in Australia. Our suggested model (Figure 10.13) to explain why the Australian Mesoproterozoic granites are so enriched in the heat-producing elements (Figure 10.5), and also the apparent increasing trend in heat-producing element contents from the Neoarchean to the Mesoproterozoic is, therefore, via the following two-stage process. The first stage is formation of Australian Paleoproterozoic crustal blocks and subsequent assembly into part of the Nuna Supercontinent (Figure 10.13b). The Paleoproterozoic crust has a significant tonalite–trondhjemite–granodiorite component, representing either modified pre-existing Archean crust and/or new Paleoproterozoic crust formed by tonalite–trondhjemite–granodiorite magmatism. The elevated U and Th contents in the Paleoproterozoic granites are largely a function of reworking (partial melting) of this enriched crust.

The second stage involves periods of extension and episodic rifting related to Nuna Supercontinent breakup in the Late Paleoproterozoic and Early Mesoproterozoic (Figure 10.13b). This tectonic environment would have facilitated crustal reworking and fractionation. This, coupled, with an already enriched crust, resulted in the abundant episodic high heat-producing magmatism in this period.

Is there any evidence for tonalite–trondhjemite–granodiorite heritage in Australia? Although some Proterozoic blocks do contain remnants of

Tasman Island from Cape Pillar in Tasman National Park, Tasmania. The dolerites that make up this region are mafic rocks related to the break-up of the supercontinent Gondwana.

Archean or Paleoproterozoic tonalite–trondhjemite–granodiorite crust, these are at best fragmentary. Direct outcrop evidence is rare. The proposed tonalite–trondhjemite–granodiorite crust would be either of Archean or perhaps an early Proterozoic age, and there is significant indirect evidence for late Archean to very early Proterozoic rocks in many Proterozoic blocks in Australia. Such evidence includes the presence of inherited older zircons, which can be directly dated (Box 2.4), and isotopic signatures of the Proterozoic granites, which both suggest that older crust is much more extensive than indicated by surface geology. Also, from studies of global Paleoproterozoic magmatism, tonalite–trondhjemite–granodiorite commonly occurred up to about 1900 Ma, although there is little evidence to support or refute this in Australia.

The unusual nature of Proterozoic granites in Australia has generated numerous earlier hypotheses regarding their origin, leading to significant controversy over the geodynamic environments they were formed in. Even the question of whether the process of plate tectonics, as it currently operates, was identical in the Paleoproterozoic and the Mesoproterozoic (let alone the Archean) has been discussed. Here, a middle road is proposed, where the original crustal enrichment (in high heat-producing elements)—reflecting the dominance of tonalite–trondhjemite–granodiorite suite granites in the Archean and early Proterozoic—is a relict of processes that might only have operated in these early Earth times. The subsequent supercontinent formation and breakup, with tectonic processes more akin to those operating today, would have provided the ideal environment to exploit any pre-existing crustal endowment and produced the widespread high heat-producing magmatism observed in the rocks today.

Interestingly, this model also potentially answers some of the questions raised earlier. The efficient tonalite–trondhjemite–granodiorite and high heat-producing granite cycle ended (mainly) some time in the latest Archean or early Proterozoic, but not before it had 'conditioned' the Australian crust. The switch from Archean-style geodynamics to something more akin to modern-day tectonics and the Nuna Supercontinent cycle allowed this preconditioned (variably heat-producing element-enriched) crust to be repeatedly exploited—that is, melted to produce high heat-producing granites. The lack of preconditioning is perhaps why many younger regions do not have evidence

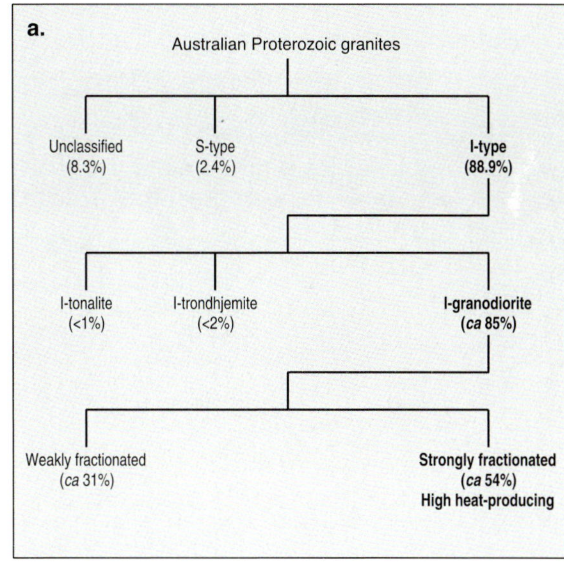

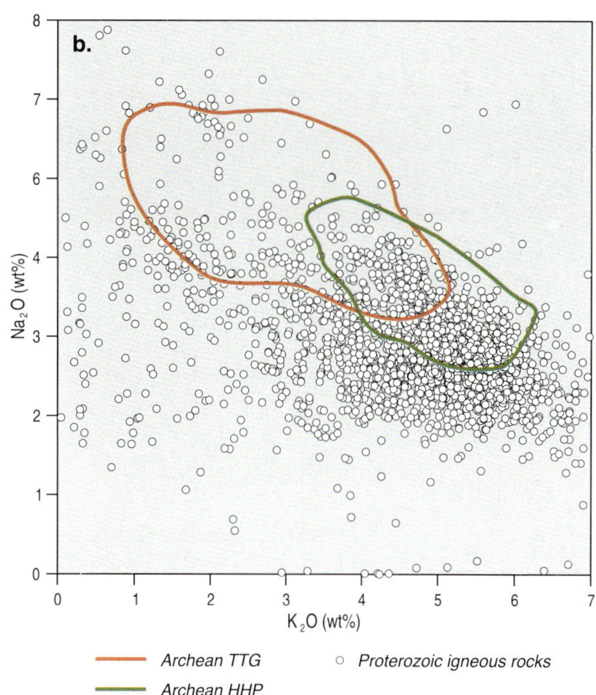

Figure 10.14: (a) Breakdown of Proterozoic granite types in Australia (based on the work of Budd et al., 2001). (b) K_2O versus Na_2O plot for Australian Proterozoic granites. Fields for Australian tonalite-trondhjemite-granodiorite (TTG) (red field) and Archean high heat-producing granites (green field) are superimposed. Both figures show that the majority (>54%) of Australian Proterozoic granites are potassic (high K), often with elevated U and Th. Sodic rocks (high Na), such as tonalites and trondhjemites—similar to the ubiquitous tonalite–trondhjemite–granodiorite found in Archean-aged terranes worldwide—are uncommon in the Australian Proterozoic (<5%).

Image courtesy of ANSTO

The Open Pool Australian Lightwater (OPAL) research reactor at Lucas Heights, Sydney, New South Wales.

for high heat-producing magmatism—for example, within the voluminous Devonian magmatism of eastern Australia (Figure 10.8). It may well be that subsequent magmatic reworking (partial melting) of this now-conditioned Devonian crust will produce high heat-producing magmatism in the future. Certainly, this is what is seen in the Carboniferous of eastern Australia, although the largest occurrence of high heat-producing granites of that age is in north Queensland, within crust of Paleoproterozoic to Mesoproterozoic age. This does tend to confirm the earlier suggestion that both the prior history of continental crust and contemporaneous processes play an important role in generating high heat-producing granites.

Regardless of the process that ultimately produced the voluminous high heat-producing magmatism, clearly Australia's heat-producing element endowment owes a lot to its old geological heritage.

Radioactivity—a potential energy and heat source for Australia

The radioactive decay of elements such as U, Th and K releases large amounts of energy, related to the conversion of very small amounts of mass to energy via Einstein's famous equation ($E = mc^2$). This energy can be harnessed and used in a variety of ways, such as nuclear and geothermal energy. Although both largely utilise the nuclear–thermal energy pathway, the approaches and methodologies are very different. Nuclear energy relies on directly capturing the energy of nuclear fission by the use of fissile radioactive material as fuel in nuclear power stations to generate power. Geothermal energy, on the other hand, uses Earth's natural heat, largely generated by radioactive decay, to produce energy. Australia's antiquity and long geological history mean that the continent is endowed in the heat-producing elements and also has regions with high heat flows. This old geology, therefore, has equipped the nation with alternative energy sources that may have a significant role to play in a non-fossil fuelled future, and not just in Australia.

Nuclear energy—proven energy producer

Nuclear energy has been used as an energy source in some countries for six decades. It is the major source of electricity in some countries, such as France and Japan, and worldwide contributes around 13–16% of total electricity generation. Although there are both upsides and some significant downsides to its use, it is a reliable, relatively cheap, proven, large-scale energy source that can be used for base-load energy requirements (in many ways comparable with fossil fuels). Australia does not currently use nuclear energy, and its current and future value is through exporting U to selected countries that do use it. This not only provides the obvious benefits, such as maintaining Australia's mining industry, developing the country and providing export earnings, but maintains mutual relationships with other countries—all of which greatly benefit from access to reliable energy resources.

Fundamentals

Nuclear power plants rely on the thermal (heat) energy that is generated by nuclear fission (i.e. by splitting atomic nuclei with neutrons). The fission process results in formation of smaller atomic nuclei, various atomic particles and significant amounts of energy. The energy release associated with fission is very large—around a million times more than that released during chemical reactions (Figure 10.15). A kilogram of U, for example, generates as much energy as burning about 10 tonnes of oil. Such energy density is one of the great attractions of nuclear energy. Current commercial thermal nuclear reactors use U as the fissile fuel. Electricity generation in nuclear reactors has many similarities to more conventional electricity generation, in that the generated thermal energy is converted to mechanical energy that is used to drive turbo-generators (mostly by steam or superheated water), either directly or by use of heat exchangers. Although there are a variety of nuclear reactor designs, the majority in use today (called light water reactors) employ ordinary water to drive the turbines, including pressurised and boiling water.

Current usage, resources and potential

Currently, there are more than 400 U-fuelled nuclear reactors in the world. A large number of these are located in countries to which Australian exports U, including France, Germany, Japan, the United Kingdom and the United States. Current reactors require about 68 000 t of U (mine and secondary sources) per year. With the current recoverable U resource (Figure 10.2), this equates to about 80 years of supply—of which Australia contributes about 25–30 years worth. Uranium forms a significant part of Australia's total energy production (*ca* 27% in 2009), with annual export earnings bringing in more than $1 B. Based on 2009 reserve figures (>1223 kt) and prices of about US$40 per pound of U_3O_8, Australia has about US$120 B worth of current in-ground value, which is equivalent to more than 20 years of Australian wheat exports. The current U reserves contain the equivalent of more than 110 years of total energy use in Australia (based on current usage). Of course, Australia does not currently use nuclear energy, but these figures illustrate how Australia contributes to powering other countries.

These figures become more significant when the indicated potential of new-generation reactors—by better use of available U, and use of Th as nuclear fuel—is considered. New reactor designs, such as fast neutron reactors, with their forecast ability to use the much more abundant ^{238}U isotope, will not only extend nuclear fuel supply but will also greatly increase the amount of energy able to be extracted. This will in turn greatly increase the net energy worth of Australia's U resources, such that Australia's current U resource could contain more than 5000 years worth of energy (Table 10.1). Finally, Australia's significant Th resources (Figure 10.2) and their energy potential—some 2000 years worth (Table 10.1)—will also add significantly to Australia's energy inventory if, and when, Th-based fuels become commercially viable. Thorium has a tantalising potential, having a number of natural advantages over U, not the least its being four times more abundant in Earth's crust. It also has a number of disadvantages, however, which have hampered its commercial development thus far (Box 10.4). The tremendous potential of ^{238}U and Th needs to be tempered by the observation that fast neutron and other advanced reactors have yet to be used for commercial electricity generation.

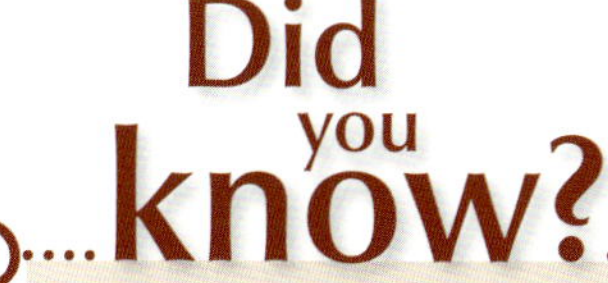

10.3: Radioactive chic

Radioactive elements were all the rage in the early 20th century; they were also very valuable. The *Adelaide Register*'s correspondent Seb. George wrote excitedly on 21 April 1913 about radium and its benefits to South Australia:

> *An ounce of radium would find all the motive power necessary to propel a battle ship until it became obsolete, and could then be utilised to drive one of a newer type, and so on for at least 2000 years. A ton of radium would be sufficient, scientists tell us to drive all the power plants of the world … South Australians should make the most of their opportunities while they have the chance. Europe is now waking to the fact that the largest producing fields of radioactive ores are here, and if we are unable to realise and take advantage of the great gift Nature has bestowed upon us, before long a foreign element will appear and deprive us of our birth right.*

It was a further 62 years before the immense U resource at Olympic Dam was discovered (Chapter 8).

Autunite, calcium uranyl phosphate.

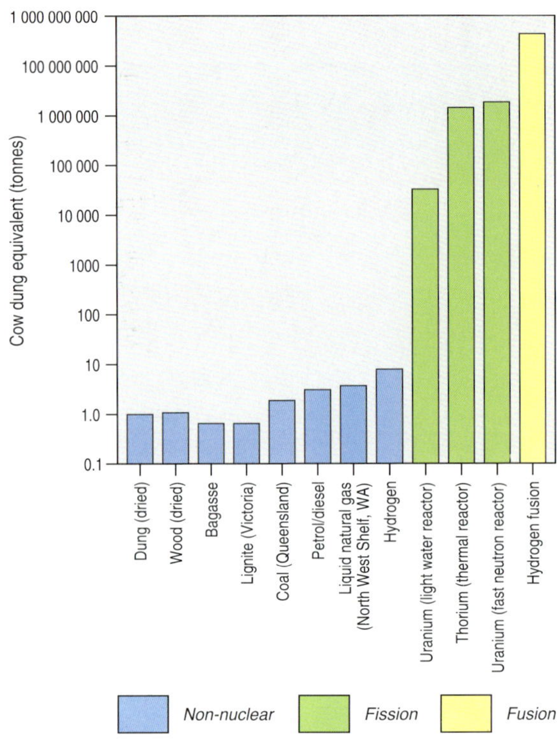

Figure 10.15: Energy densities of selected modern and historic fuels, expressed as the tonnes of cow dung required to be burnt to produce the equivalent energy from one tonne of each fuel type. Note the very large energy densities of nuclear fission and fusion fuels.

Table 10.1: Deep heat and Australia's nuclear energy potential

	Contained energy in petajoules (calculated years of supply)				
	Australian resources (1000 tonnes)	Natural uranium (0.56 PJ/t)	Natural uranium ^{235}U in LWR (0.5 PJ/t)	Natural uranium in FNR (*ca* 28 PJ/t)	Natural thorium in thermal reactor (*ca* 22 PJ/t)
U	1163	651 280 (113)	581 500 (101)	32 564 000 (5642)	
Th	489				10 953 600 (1898)
U top metre	29 900	16 744 000 (2901)	14 950 000 (2590)	837 200 000 (145 045)	
Th top metre	172 000				3 852 800 000 (667 498)

Calculations of the contained and recoverable energy contents, based on current known U and Th resources and various nuclear reactor types (current and in development). Australia's U and Th inventory, in the top 1 m of the crust, calculated from gamma-ray data, and the potential energy content are shown for comparative purposes only—this will, of course, never be mined.
FNR = fast neutron reactor; LWR = light water reactor; PJ/t = petajoules per tonne

Uranium and Th, however, are not renewable energy sources, and, like oil, their discovery and production rates will eventually peak. When this will occur in Australia is not easy to predict. Australia's economic demonstrated resources of U have been strongly increasing over the past 30 years (Figure 10.16)—a trend that will generally continue, given that the current outlook for future discoveries is optimistic, with numerous prospects currently being evaluated, and given the large areas of U-enriched Australian crust.

Geothermal energy—'green' deep heat

Like nuclear energy, geothermal energy has been used as an energy source for decades, both for electricity generation and also by direct use of the heat in a range of low- to medium-temperature domestic, industrial and commercial heating applications (Box 10.5). Current figures show that, although forming a significant percentage of electricity generation in a number of countries (e.g. New Zealand, Iceland, the Philippines), on a worldwide basis it forms a low percentage (<1%) of energy used for electricity generation. This is, in part, a reflection of the nature of modern conventional geothermal power plants, which currently exploit only geographically restricted volcanic geothermal resources. Geothermal energy potential (and exploration) in Australia is based on non-volcanic geothermal resources, and is in its infancy. If, and when, successful, this energy source has an almost unlimited potential that not only could supply Australia for thousands of years, but comes with the added benefits of being both renewable and environmentally friendly, with very low emissions.

Fundamentals

Geothermal energy relies on the natural thermal (heat) energy within Earth. To use this thermal energy for electricity generation, geothermal power plants employ similar processes to more conventional electricity generation—that is, the captured heat energy, in the form of steam or superheated water, is used to drive turbo-generators, either directly or in binary systems by use of heat exchangers. Current technological and economic considerations effectively restrict electricity generation to water temperatures of more than 100°C.

Geothermal energy—the Australian way

There are a number of approaches to harnessing geothermal energy for electricity generation. These approaches depend on the nature of the geothermal resource, as well as its accessibility and other economic factors. In simplest terms, to use this energy economically and efficiently, elevated heat levels at shallow crustal levels (6 km or less) are required. In most geological regimes, purely conductive crustal heat flow (i.e. by conduction only) is not sufficient to generate thermal anomalies conducive to economically recoverable electricity generation. Accordingly, high-grade thermal resources need to be targeted.

Most modern exploration is aimed at either hydrothermal (hot water) or 'hot rock' resources. Hydrothermal resources comprise naturally occurring hot water or steam present in fractured or permeable rock. Such resources are found at a variety of crustal levels, from very shallow (hundreds of metres) to relatively deep (to 4 km or more). These resources can be further subdivided by the nature of the heat source for the water, either 'volcanic' or 'non-volcanic' (Figure 10.17).

In a volcanic hydrothermal resource, the heat is generated by thermal advective–convective processes related to high-level magmatism. They produce elevated near-surface thermal anomalies and moderate- to high-temperature waters by the movement of hot fluids—namely magmas and waters heated by those magmas. Volcanic hydrothermal resources are almost exclusively confined to regions of active tectonics, usually on plate margins, and nearly all modern geothermal plants are in such active tectonic regions (e.g. New Zealand, Japan, the United States and Iceland). Nearly all current economic geothermal energy is derived from volcanic hydrothermal resources.

Australia's current intraplate position means that active tectonic regions are largely absent (Figure 2.1). There are, however, examples of geologically young, but largely inactive, volcanism in eastern Australia, such as Mt Gambier in South Australia and western Victoria (Chapter 2). In Australia, this means that geothermal power plants targeting hydrothermal systems will have to rely on non-volcanic resources—'hot sedimentary aquifer' systems. In such systems, elevated water temperatures are related to elevated geothermal gradients, commonly due to the presence of high heat-producing granites and other rocks with high heat-producing element contents (Figures 10.17 and 10.18). Water temperatures in these non-volcanic hydrothermal systems are typically lower than those in volcanic systems (per given depth) and, therefore, will only be economic (for electricity generation) at significantly greater crustal depths (2–4 km).

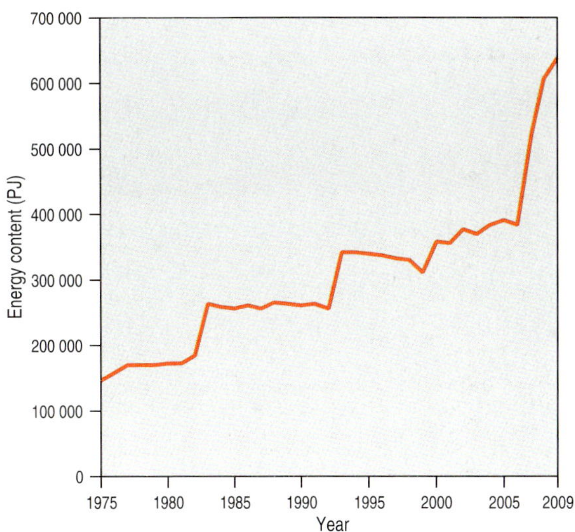

Figure 10.16: Australian U resources expressed as energy content in petajoules (PJ). Note the increasing demonstrated reserves of Australian U. Each of the step-like increases, at 1983, 1993 and 2007, relates to increased exploration at the supergiant Olympic Dam deposit. (Source: Geoscience Australia and Australian Bureau of Agricultural and Resource Economics, 2010)

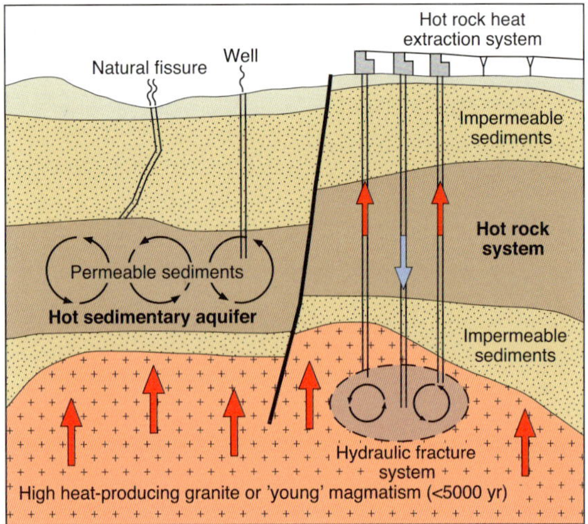

Figure 10.17: Non-volcanic geothermal systems in Australia. The Australian continent is not near a plate margin and does not have active tectonic or volcanic regions. Thus, geothermal power plants in Australia will have to rely on non-volcanic geothermal systems, targeting thermally anomalous regions—typically within or close to high heat-producing granites. Red arrows in granite symbolise high temperatures in the granite under insulating sediments. Red and blue arrows in drill holes show upward extraction of hot water and downwards flow path of cooled water (thermal energy extracted). (Source: Geoscience Australia & Australian Bureau of Agricultural and Resource Economics, 2010)

The other major type of geothermal target is the 'hot rock' resource, which is typically within or close to high heat-producing granites (Figure 10.4). Importantly, although high heat-producing granites are the common heat source for such resources in Australia, it is the combination of such rocks with overlying insulating lithologies, such as sediments, that provides optimum conditions for the elevated temperatures needed for geothermal exploitation. Such insulating rocks, through their low thermal conductivity, slow down the rate of conductive heat loss, effectively trapping the heat below them (Figure 10.17). The thicker or more efficient the insulating layer, the greater the amount of thermal energy trapped (i.e. higher temperatures). For this reason, identified hot rock resources in Australia, and elsewhere in the world, are typically within geological regions characterised by voluminous high heat-producing granites (the best hot rocks) overlain by thick sedimentary basins.

As hot rock resources rely on elevated geothermal gradients, they are typically at much deeper crustal levels (to 4–5 km in Australian geothermal prospects) than hydrothermal resources (especially volcanic types). Exploration and testing of geothermal resources at these depths is not only very expensive but also technically difficult. This is one reason why exploitation of hot rock resources in Australia remains at the proof-of-concept stage. The current economics of drilling deep holes, especially in remote Australia, dictate that only those regions with anomalous high-grade thermal energy are potentially viable.

Current usage, resources and potential

Worldwide production of electricity from geothermal energy between 2004 and 2007 was <1% of total world electricity generation. Even when renewable energies alone are considered, geothermal only contributes around 1.5–2%, which is significantly lower than hydroelectric (around 85–90%), but an order of magnitude higher than wind, solar and tidal combined (<0.2%). Electricity generation from geothermal energy in Australia is currently very low, limited to a small (80 kW net) power plant at Birdsville (Qld).

The Australian and world figures, however, mask the very large potential of geothermal energy, particularly for non-volcanic geothermal resources (assuming that they eventually prove to be economic). The presence of considerable high heat-producing granites in Australia, especially those of Proterozoic age, coupled with the extensive cover of sedimentary basins (Figure 2.9) suggests that Australia is well suited for this type of geothermal energy. In an attempt to semi-quantify these resources, Geoscience Australia undertook a conservative resource analysis of the Australian continent. Using temperature-depth estimates based on drill-hole measurements, the analysis showed that those parts of the upper 5 km of Australia's crust with temperatures of 150°C or more contained the equivalent of 3.3 Myr of Australia's annual energy consumption.

Of course, all this energy will never be extracted, and thus far none has been in Australia, but it does give an idea of the potential. If only 1% could be extracted, there would be more than 30 ka of energy

THORIUM IN AUSTRALIA: A RADIOACTIVE MENACE OR FUTURE ENERGY SOURCE? (BOX 10.4)

Australia has abundant thorium (Th) resources. Although currently valueless, Th, like uranium (U), can be used as a nuclear fuel, and so may be of potential economic importance. Thorium has some natural advantages over U. It is not only about four times more abundant but is also all potentially usable in reactors—unlike U where only about 1% is used (mostly the ^{235}U isotope). This makes Th a tantalising prospect for nuclear energy—one that many countries have investigated.

Unfortunately, Th has a number of features that have hindered its use, not least the presence of strong gamma radiation–emitting daughter products, which make usage difficult and expensive. Current indications suggest that widespread use of Th in nuclear reactors will not occur in the short to medium term.

The combination of lack of world demand, the highly radioactive nature of Th, and its future potential value poses a dilemma for Australia, particularly for our Th deposits. These occur within both heavy mineral sand and rare earth deposits (Figure 10.3). Legislated Australian state mining codes require companies to rehabilitate mines to environmental standards, including acceptable levels of radiation. The Th in such deposits must be diluted to safe, low concentrations before or during dispersal. Although this ensures that environmental standards are met, it comes with the cost that it effectively destroys any future potential for Th within the deposit. In effect, our current safe dispersal of the Th may be burying a future strategic energy resource (Table 10.1) as well as export earnings.

Sand mining, sand dredge, New South Wales.

Did you know?

10.4: Nuclear energy

Australia's abundant energy resources, especially coal and the cheap electricity generated from it (Chapter 9), have meant that there has been little economic or energy security impetus for nuclear energy in this country. Fossil fuels, however, are non-renewable, and Australians will have to seriously consider alternative energy sources—possibly sooner rather than later, given greenhouse gas and climate change concerns.

Nuclear energy is one such alternative energy source, certainly one capable of meeting a nation's electricity requirements and one that is, at least as far as greenhouse gas emissions are concerned, environmentally friendly. Nuclear energy, however, also has some well-documented negatives. Whether we are willing (or not) to venture down this path is a decision that the Australian people may well have to face in the years ahead, but at least the geological legacy of this continent has provided the people with a range of choices.

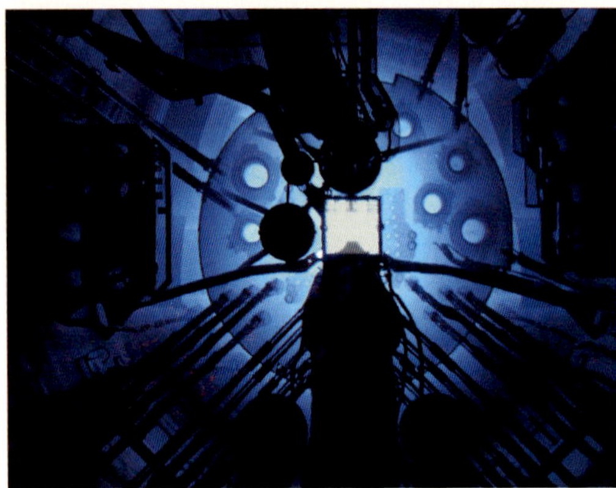

Image courtesy of ANSTO

Looking down into the OPAL reactor, Lucas Heights, New South Wales.

Table 10.2: Geothermal energy: a truly massive 'green' energy source

Geothermal region or project	Contained energy in exajoules[a] (calculated years of supply)	
	Total energy	Usable energy
Geothermal in Australia	19 000 000 (*ca* 3 290 000)	19 000 (3300)
Australian projects (all)	3025 (*ca* 520)	18.2–90.8 (15-3)
Hot rock projects	2404	14.4–72.1
Hot sedimentary aquifer projects	621	3.7–18.6

Estimates of the total and recoverable energy contents of geothermal resources within Australia (Budd et al., 2008) and current geothermal projects (company data). Total years of energy supply for Australia are also shown (in parentheses). Usable energy calculated for geothermal projects assumes 5% or 25% recovery and 12% efficiency. Similar figures are given for the total Australian resource estimate but assuming only 1% extractable. The renewable nature of the energy is not factored into the calculations.
[a] one exajoule (EJ) = 1000 petajoules (PJ)

supply—even a conservative one-thousandth gives about 3000 years (Table 10.2). This does not include the direct-use technologies where temperatures of less than 150°C can be utilised (Box 10.5), nor does it consider the renewable nature of the resource. When this is considered in conjunction with other advantages of geothermal energy, such as zero emissions, 24-hour base-load availability, high reliability, no storage requirement, ability to combine it with other renewable energies (such as solar), and load-following ability, the truly great potential of this energy for powering Australia's future is realised. This is best illustrated when inferred geothermal energy data for current Australian geothermal prospects are taken into consideration (Table 10.2). One of the most advanced hot rock projects in the world is in the Cooper Basin region in South Australia (Figure 10.4) where Geodynamics Ltd has drilled into a Carboniferous high heat-producing granite at depths between 3.8 km and 4.8 km. The company plans to deliver power from a 25 MW plant by late 2013, and much larger production by 2018.

The only working geothermal plant in Australia is in Birdsville, Queensland, where a small low-temperature plant supplies 80 kW of power, or about 25% of that town's electricity needs. The plant utilises 98°C water extracted at around 1.3 km depth from the Great Artesian Basin. Other towns within the basin also utilise this hot water resource. Quilpie has no requirement for hot water systems, extracting its hot water (55°C) directly from the basin. The water is cooled for the cold water supply.

Is the past the key to Australia's energy future?

Energy drives the modern world and underpins our current way of life. The industrial age was fuelled by access to reliable high-grade energy sources, coal and oil, which drove global economic expansion and modernisation. There is a strong correlation between energy consumption levels and GDP. Australia is a large consumer of energy, currently 20th in the world. Australia is well endowed with traditional energy resources (e.g. coal, gas, U) and is now a large energy producer (9th in the world). Australia also benefits from energy exports (more than $70 B in 2011). Energy, therefore, strongly contributes to the nation's wealth and living standards, but increasingly it is recognised that these are dependent on access to cheap energy.

To secure Australia's future prosperity, the nation must have energy security—that is, accessible, secure, affordable and reliable energy into the future. This future will need to be achieved against three competing concerns: increasing annual energy consumption, decreasing resources and increasing environmental concerns from fossil fuel use. The need for energy security and environmental safety will both drive the switch to more sustainable energy types, preferably from indigenous energy sources. Although the hot and arid Australian continent is ideally situated to use alternative energy sources, with solar a potential renewable source, some commentators suggest that such sources may not provide for Australian needs nor meet peak energy loads.

Uranium, Th and K are anomalously concentrated within the continental crust—in particular, in the high heat-producing (HHP) felsic igneous rocks. Fortuitously, Australia is endowed with above-average concentrations of these radioactive elements. There is a strong spatial association of high heat-producing igneous rocks with both U mineralisation and major heat-flow anomalies. The distribution of

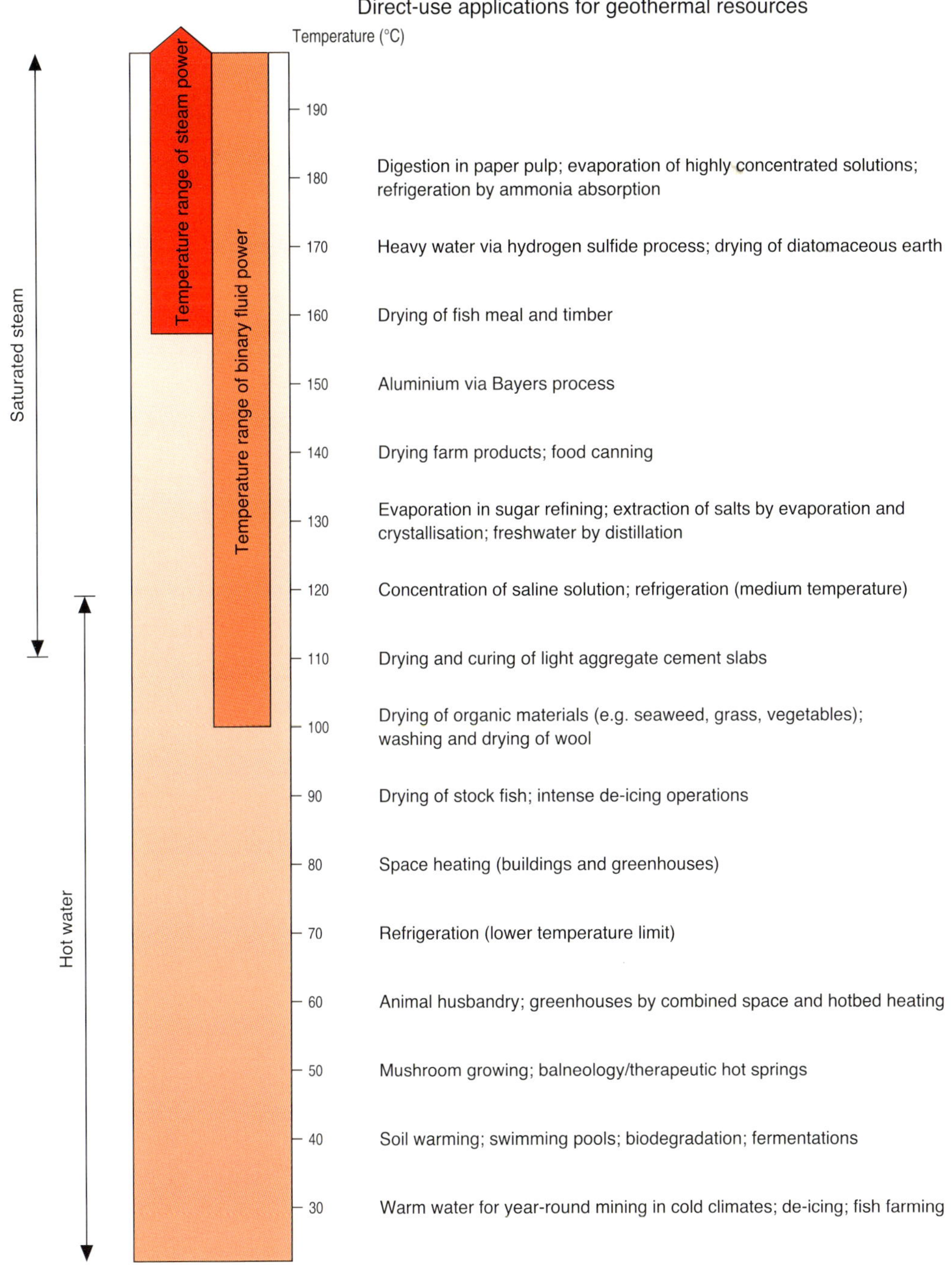

Figure 10.18: Examples of direct-use applications for geothermal energy resources. The higher the temperature of the resource, the greater the number of possible uses. It is possible to structure a series of industries to use thermal resources in a series of cascading applications based on the available temperature following the previous application. (Source: modified from Lindal, 1973)

GEOTHERMAL ENERGY—DIRECT USE IN AUSTRALIA (BOX 10.5)

Although people often think of geothermal energy for generating electricity, it can also be used directly for heating applications. Direct use is typically associated with lower temperature geothermal resources (temperatures <150°C; Figure 10.18). Direct use of geothermal energy normally requires accessing heated groundwater that is naturally circulating through sedimentary strata or fractured rock (a hydrothermal system). This type of geothermal resource is found wherever sufficient quantities of groundwater come into contact with heated rocks, which in Australia means areas with high heat-producing granites, or sediments overlying these rocks. More importantly, the lower temperature requirements for direct use mean such resources are more widespread (in contrast with the higher temperatures required for electricity generation). Australia's Great Artesian Basin, for example, has large, low-temperature hydrothermal systems, with aquifers containing groundwater that comes to the surface as warm as 98°C (Chapter 7). Towns within the basin do make use of this resource; in the western Queensland town of Quilpie, for example, the use of hot bore water (>50°C extracted from *ca* 900 m depth) has meant that there is no need for hot water systems; instead, the townsfolk have cooling tanks to supply cold water.

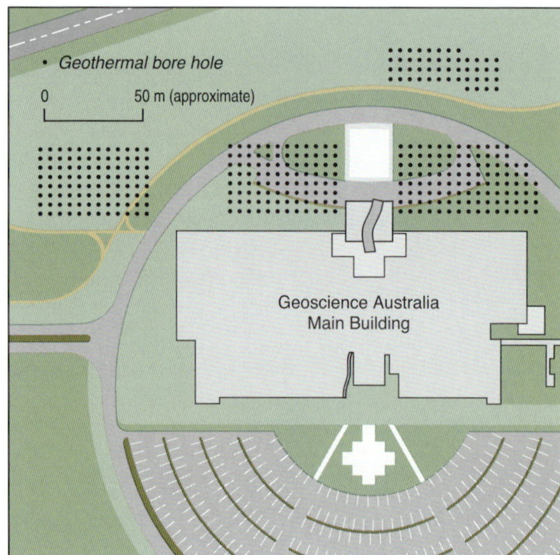

Figure B10.5: The Geoscience Australia building in Canberra, Australian Capital Territory. Air conditioning in the building is undertaken via a ground-source heat pump system, which effectively uses Earth as a heat sink to cool the building in summer or a heat source to warm the building in winter. The heat pump system utilises 352 boreholes, each drilled to 100 m depth—where the ground temperature is a fairly constant 17°C. The heated or cooled water is used by heat exchange units to heat or cool the building before being returned to the bore field.

There are many other likely locations in Australia where warm groundwater is present and could be used for direct-use applications. Direct use of geothermal energy in Australia currently includes heating buildings, spa developments, artesian baths and swimming pools.

There is still significant potential for additional domestic, industrial and commercial applications. These include uses in agriculture for greenhouse heating, crop drying and aquaculture, and in industrial processes such as concrete curing, milk pasteurisation, chemical extraction, refrigeration, drying organic materials, desalination, wool processing and pre-heating of water in coal-fired power stations. Cascading use, whereby the same water is used in successive processes at progressively lower temperatures, is also possible within a single geothermal operation, improving efficiency and economic feasibility.

At the lowest end of the temperature spectrum (Figure 10.18), ground-source heat pumps can be used almost anywhere in the world to provide heating and cooling for buildings. Although used extensively in Europe and the United States, there has been only limited take-up in Australia. The largest ground-source heat pump system in the Southern Hemisphere, however, is in Australia—in Canberra (Figure B10.5).

high heat-producing magmatism in Australia is not uniform, with most in the central third of the continent. In the west and east, where most people live (Chapter 1), they are more localised. Time-wise, the first high heat-producing rocks appear in the mid-Archean and continue to the Mesozoic. Peaks in high heat-producing magmatism occur in the Neoarchean, Mesoproterozoic and Carboniferous, with the greatest concentrations from *ca* 1880 Ma to 1500 Ma. Well over half of Australia's heat-producing element budget was carried into the upper crust by these and earlier (Archean) granites, meaning that the heat-producing element budget is a relic of our continent's distant past.

This anomalous enrichment puts Australia in a unique position for energy security. Not only does the nation have the ability to use Earth's in-situ heat to generate power over wide areas, it also has significant resources of U and Th in conventional deposits. Combined, both resources have the ability to meet base- and peak-load electrical power requirements, and the potential to significantly underpin Australia's energy requirements well into the future. Geothermal energy, in particular, has the potential to provide environmentally friendly energy for hundreds or thousands of years.

It is almost ironic that the key to Australia's energy future is significantly a result of our continent's ancient geological heritage. Only now are we realising this legacy and its full potential.

Image courtesy of Geodynamics Limited

Venus shines on the steam flow at Habanero 3, Cooper Basin, South Australia.

Processing plant, Ranger uranium mine, Northern Territory.

Bibliography and further reading

Energy and energy resources in Australia

Australian Government Department of the Environment, Water, Heritage and the Arts 2008. *Energy use in the Australian residential sector 1986–2020*, DEWHA, Canberra.

Australian Government Department of Resources, Energy and Tourism 2011. *Energy in Australia 2011*, DRET, Canberra.

Budd AR, Holgate FL, Gerner E & Ayling BF 2008. *Pre-competitive geoscience for geothermal exploration and development in Australia: Geoscience Australia's Onshore Energy Security Program and the Geothermal Energy Project*, Geoscience Australia Record 2008/18, Geoscience Australia, Canberra, 1–9.

Dopita M & Williamson R (eds) 2009. *Australia's renewable energy future*, Australian Academy of Science, Canberra.

Geodynamics Limited. Stock market releases and website information. www.geodynamics.com.au/IRM/content/home. html

Geoscience Australia 2012. *Australia's identified mineral resources 2011*, Geoscience Australia, Canberra.

Geoscience Australia 2011. *Energy Security Program achievements—towards future energy discovery*, Geoscience Australia, Canberra.

Geoscience Australia & Australian Bureau of Agricultural and Resource Economics 2010. *Australian energy resource assessment*, Geoscience Australia & ABARE, Canberra.

Gurgenci H & Budd AR (eds) 2008. *Proceedings of the Sir Mark Oliphant International Frontiers of Science and Technology Australian Geothermal Energy Conference*, Geoscience Australia Record 2008/18, Geoscience Australia, Canberra.

Lambert I, Jaireth S, McKay A & Miezitis Y 2005. Why Australia has so much uranium. *AusGeo News* 80, Geoscience Australia, Canberra.

McKay AD, Carson L, Huleatt MB, Miezitis Y, Porritt KR, Sait R, Kay P, Whitaker A, Towner R & Sexton M 2010. *Australia's identified mineral resources 2010*, Geoscience Australia, Canberra.

Mernagh TP & Miezitis Y 2008. *A review of the geochemical processes controlling the distribution of thorium in the Earth's crust and Australia's thorium resources*, Geoscience Australia Record 2008/05, Geoscience Australia, Canberra.

Miezitis Y, Carson L, Lambert IB, Mckay AD & Jaireth S 2012. Classification of Australia's thorium resources and their deposit types. International Atomic Energy Agency (IAEA) Technical Meeting on World Thorium Resources, Thiruvananthapuram, India, 17–21 October 2011.

Heat flow

Gerner EJ & Holgate FL 2010. *OZTemp—interpreted temperature at 5 km depth image*, Geoscience Australia, Canberra.

Jones TD, Kirkby AL, Gerner EJ & Weber RD 2011. *Heat-flow determinations for the Australian continent: release 2*, Geoscience Australia Record 2011/28, Geoscience Australia, Canberra.

Kirkby AL & Gerner EJ 2010. *Heat-flow interpretations for the Australian continent: release 1*, Geoscience Australia Record 2010/41, Geoscience Australia, Canberra.

Morgan P 1984. The thermal structure and thermal evolution of the continental lithosphere. *Physics and Chemistry of the Earth* 15, 107–193.

Pollack HN, Hurter SJ & Johnson JR 1993. Heat-flow from the Earth's interior: analysis of the global data set. *Reviews in Geophysics* 31, 267–280.

Tyler J 2006. Hot-rock energy steaming up. *TAG Newsletter* 140, 22–25.

Weber RD, Kirkby AL & Gerner EJ 2011. *Heat-flow determinations for the Australian continent: release 3*, Geoscience Australia Record 2011/30, Geoscience Australia, Canberra.

Geology

Budd AR, Wyborn LAI & Bastrakova IV 2001. *The metallogenic potential of Australian Proterozoic granites*, Geoscience Australia Record 2001/12, Geoscience Australia, Canberra.

Champion DC & Smithies RH 2007. Three billion years of granite magmatism: Palaeo-Archaean to Permian granites of Australia, presentation given at the Sixth International Hutton Symposium, Stellenbosch, South Africa, July 2007. www.ga.gov.au/image_cache/GA11433.pdf

McLaren S 2008. Hot southern land. *Australasian Science* August 2008, 25–27.

McLaren S, Sandiford M, Hand M, Neumann N, Wyborn L & Bastrakova I 2003. The hot southern continent: heat-flow and heat production in Australian Proterozoic terranes. In: Geological Society of Australia Special Publication 22, 151–161.

McLaren S, Sandiford M & Powell R 2005. Contrasting styles of Proterozoic crustal evolution: a hot-plate tectonic model for Australian terranes. *Geological Society of America* 33, 673–676.

Minty BRS, Franklin R, Milligan PR, Richardson LM & Wilford J 2010. *Radiometric map of Australia*, 2nd edn, scale 1:15 000 000, Geoscience Australia, Canberra.

Neumann N, Sandiford M & Foden J 2000. Regional geochemistry and continental heat-flow: implications for the origin of the South Australian heat-flow anomaly. *Earth and Planetary Science Letters* 183, 107–120.

Palme H & Jones A 2003. Solar system abundances of the elements. *Treatise on Geochemistry* 1, 41–61.

Rudnick RL & Gao S 2003. Composition of the continental crust. *Treatise on Geochemistry* 3, 1–64.

Santosh M, Maruyama S & Yamamoto S 2009. The making and breaking of supercontinents: some speculations based on superplumes, superdownwelling and the role of tectosphere. *Gondwana Research* 15, 324–341.

Windley BF 1993. Proterozoic anorogenic magmatism and its orogenic connections: Fermor Lecture 1991. *Journal of the Geological Society* 150, 39–50.

Other sources of energy

American Nuclear Society 2005. *Fast reactor technology: a path to long-term energy sustainability*, position statement 74, American Nuclear Society, November 2005. www.ans.org/pi/ps/docs/ps74.pdf

Intergovernmental Panel on Climate Change 2007. *Climate change 2007: synthesis report*, IPCC, Geneva.

International Atomic Energy Agency 2005. *Thorium fuel cycle—potential benefits and challenges*, IAEA–TECDOC–1450, IEAE, Vienna.

International Energy Agency 2007. *Nuclear power*, IEA Energy Technology Essentials number 4, IEA, Paris.

International Energy Agency 2010. *Key world energy statistics 2010*, IEA, Paris.

Lindal B 1973. Industrial and other applications of geothermal energy. In: *Geothermal energy: review of research and development*, Armstead HCH (ed.), UNESCO (Earth Sciences) 12, 135–148.

Mackay D 2009. *Sustainable energy—without the hot air*, version 3.5.2, 3 November 2008. www.withouthotair.com

Smil V 1994. *Energy in world history*, Westview Press, Boulder.

United States Department of Energy 2002. *A technology roadmap for generation IV nuclear energy systems*, December 2002, US DOE Nuclear Energy Research Advisory Committee & Generation IV International Forum.

World Energy Council 2007. *Deciding the future: energy policy scenarios to 2050—executive summary*, WEC, London.

World Energy Council 2009. *World energy and climate policy: 2009 assessment*, WEC, London.

World Nuclear Association 2009. *Thorium*. www.world-nuclear.org/info/inf62.html

11 Advance Australia Fair

Does civilisation only exist by geological consent? The previous 10 chapters in this book suggest that this statement might be true, by showing the manifold ways in which the geology of the Australian continent, and its marine extensions, has shaped Australia and continues to shape the people. This concluding chapter looks briefly at the opportunities and challenges that Australian geology creates for the future of this nation.

Chris Pigram,[1] *Richard S Blewett,*[1] *Keith M Scott,*[2,3] *Phil L McFadden,*[4] *David C Champion,*[1] *Phil R Cummins*[1,2]

[1]Geoscience Australia; [2]Research School of Earth Sciences, Australian National University; [3]CSIRO Earth Science and Resource Engineering; [4]consultant

Lucky country?

The social commentator Donald Horne coined the phrase 'The Lucky Country' to describe Australia—a phrase he used ironically, despite its common misquotation today. Nevertheless, in terms of geology, the country has been lucky, as the continent is well endowed with nature's gifts—gifts that have fundamentally shaped the fortunes of the Australian people, as shown in earlier chapters of this book. In navigating the 21st century and beyond, Australians will be influenced by factors such as demographics, politics and economics, but the underlying legacy of Australia's geology will continue to shape the nation.

Current and future societal challenges are and will be varied (Box 11.1), but they are intimately linked with the stewardship of the environment, the sustainable availability of resources (food, fibre, water, energy and minerals) and community safety (Figure 11.1). Successfully meeting these challenges will require an even deeper knowledge and understanding of Earth as a system of complex interrelationships.

The global population is rising rapidly. Standards of living are also rising, making increasing demands on the finite resources of Earth. Australia is not immune to this and, in this developed nation, Australians are high per-capita consumers of resources (Chapter 1). With a 2% population growth rate in recent years, the projected population of Australia is estimated to be more than 35 million by the middle of the 21st century. Such a population increase will have a substantial impact on water, energy and food security, with strong competition between environmental concerns and economic growth. Increasing populations also add to the level of exposure to natural hazards, which may increase in magnitude and frequency with climate change. Many of these challenges have solutions, if somewhat partial, rooted in Australia's geological heritage and in the Earth system (Chapter 2).

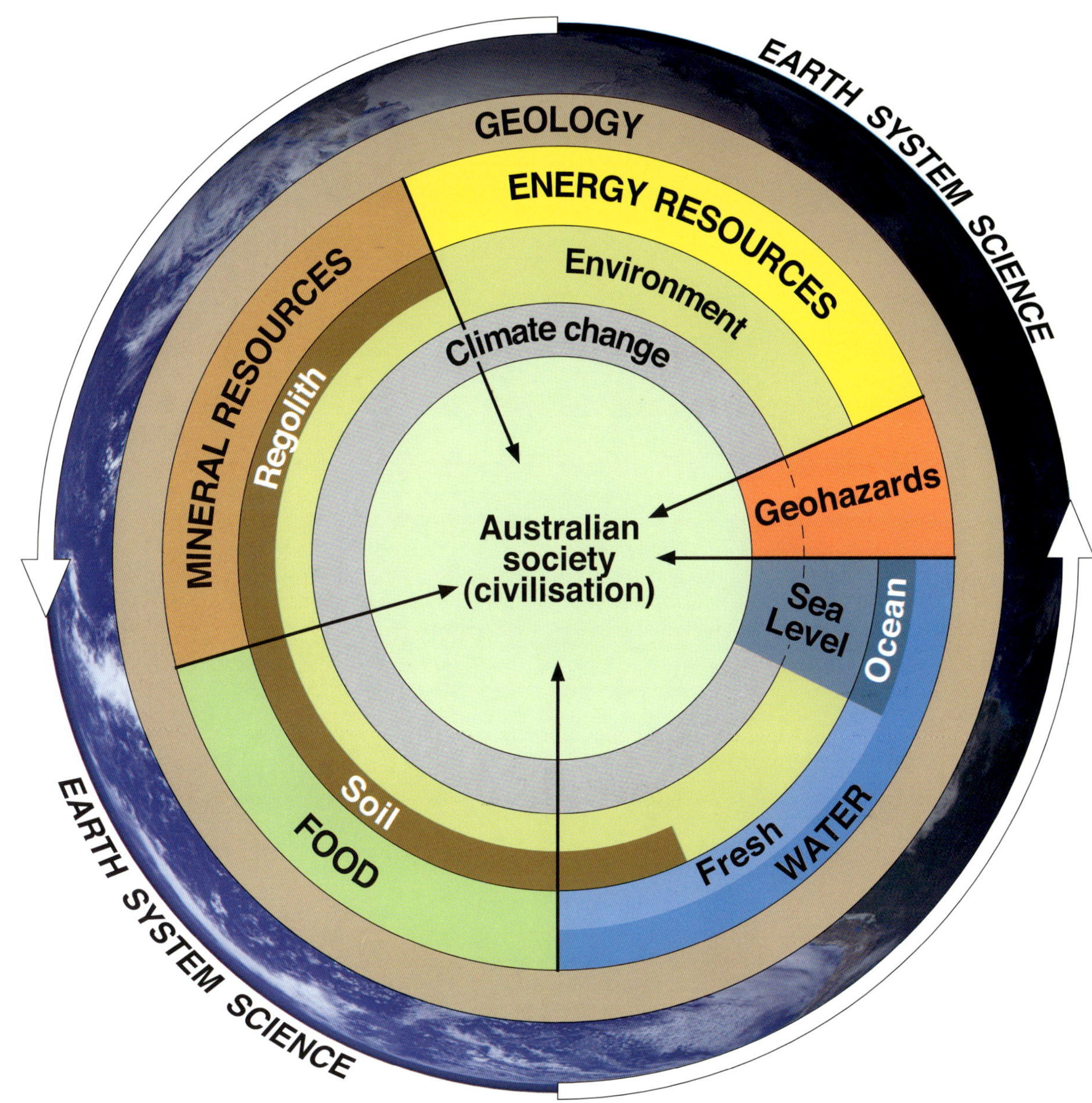

Figure 11.1: Interrelationship between sustainable civilisation and the Earth system.

Opposite: Irrigating bean and sunflower crops.

© Getty Images [Radius Images]

© Getty Images [T Blackwood]

Pejar Dam, one of the dams supplying water to Goulburn, New South Wales. Photograph taken in 2006 at the height of the drought.

Food, fibre and water

As a large country with a diversity of soils and climate zones, Australia supplies food to more than 60 million people. The actual available land for 'agricultural Australia' is not that large; it is similar in size to the area of France, but with much lower productivity. Australia's geological history has left the continent with a deep regolith and a landscape that is old, flat and red (Chapter 5). As a consequence, most Australian soils tend to be fragile, prone to salinity, and depleted in trace elements essential for growing food. The carrying capacity of the land is low and not necessarily amenable to easy 'fixes', such as the addition of fertilisers. Furthermore, past agricultural practices, especially the large-scale clearing of native vegetation, have degraded the environment, leading to elevated water-tables and widespread salinity (Chapter 7). Overgrazing and overclearing, coupled with drought and strong winds, result in dust storms and large losses of valuable topsoil. The growing urban sprawl around Australia's largest cities is adding to the problem as more of the nation's fertile soils close to the consumers become locked away beneath concrete (Chapters 5 and 10).

Water security is vital to civilisation. Human history is littered with once-thriving societies that foundered due to lack of water. Australia is a dry continent with a large arid zone. Our pattern of settlement and development has been, and is, strongly linked to rainfall distribution. Despite this, Australians are large household consumers of water, using more per capita than all other western countries except New Zealand, Canada and the United States. Given Australia's projected future population increases, will the high use of water be sustainable? Although there is a heavy reliance on surface water resources in Australia, the continent has abundant groundwater resources. These are, however, not always located where the demand is, nor are they necessarily of high potable quality, nor are they all renewable. In places, they have been poorly managed, with excessive groundwater use being associated with land subsidence and seawater incursion into freshwater coastal aquifers (Chapter 7).

Competition for water, and how this precious resource is allocated between different interests, is a significant challenge for governments and regulators. One example is balancing the activities of the coal-seam gas industry with the desire

FUTURE CHALLENGES (BOX 11.1)

- Australia is the driest inhabited continent, and the soils are thin and fragile. There is ongoing demographic and economic pressure to produce more food and fibre. The nation's challenge is to effectively and sustainably manage the soils and water (particularly groundwater) of Australia in an environment of growing population and climate change. With the increasing likelihood of groundwater being mined, we must learn from our experience of past resources exploitation to develop, and give real meaning to, the concept of sustainable yield for groundwater use.
- Australia has abundant energy resources. The nation's challenge is to determine the balance of its energy resources in an increasingly carbon- and water-constrained world in ways that best meet the needs of a sparsely populated country in a region that is rapidly growing and industrialising.
- Australia's mineral resources are large, but they are finite. The nation's challenge is to ensure an ongoing pipeline of discovery and development to meet the needs of the domestic economy and the demands from a growing neighbourhood.
- Australia has reduced the risks to people of natural hazard events by nearly a factor of 100 in the past 150 years. The economic costs, however, have been increasing steadily. This increase is driven mainly by increases in exposure (population and value of assets). The prospect of increasing risks from climate change, the interdependency of the world economy and the growing importance of Australia's role in the Australasian region all indicate that natural hazards will continue to be a significant challenge to the nation.
- There is much we do not know about the Australian continent. Many of the nation's challenges involve the environment, sustainable availability of resources and community safety; these are linked to Earth system processes. Meeting the challenges will mean that earth scientists will need to study their field in the context of a larger and more intricate Earth system.

Image courtesy of NASA

Coal-fired power station, Traralgon, Victoria.

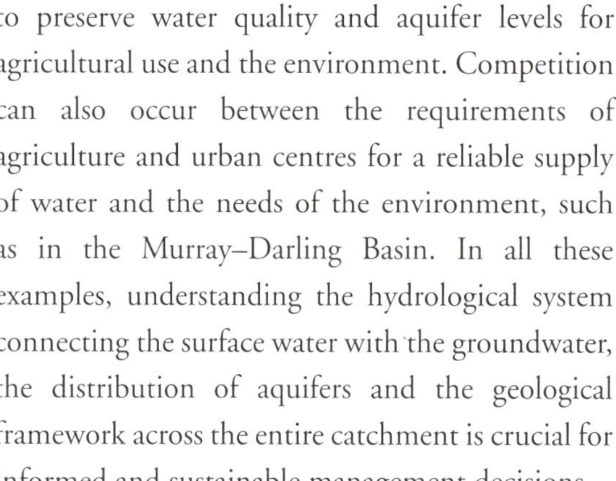

to preserve water quality and aquifer levels for agricultural use and the environment. Competition can also occur between the requirements of agriculture and urban centres for a reliable supply of water and the needs of the environment, such as in the Murray–Darling Basin. In all these examples, understanding the hydrological system connecting the surface water with the groundwater, the distribution of aquifers and the geological framework across the entire catchment is crucial for informed and sustainable management decisions.

Images of town dams reduced to a muddy puddle, and trucks carting water, highlight the vital need for water security. One response by state governments around the country was to build, or put plans in place to build, desalination plants. Technology has driven down the costs and improved the efficiency of desalination. Nevertheless, the process is a large consumer of energy, and the saline by-product needs to be disposed of in ways that do not impact marine ecosystems. Where will this energy for desalination come from in ways that do not pollute? How can the saline by-product be disposed of sustainably? These are questions that go to the heart of the interrelated energy–carbon–water system of Earth.

Energy and power

Economic growth, and the development of all nations, has relied on access to cheap energy. A remarkable outcome of this development has been the transformation of lives of billions of people from poverty to wealth, and long and healthy life spans. Energy is all around us; it is the *power* that can be generated from it and made to do *work* that has transformed humanity. Most of our power is derived from energy-dense fossil fuels: coal, oil and gas (Chapters 4 and 9). Atmospheric pollution (greenhouse gases) from burning fossil fuels is an externality not generally factored into the costs of these energy sources. Nations will increasingly be required to lower their carbon emissions through efficiency measures and alternative sources of energy that produce less CO_2 (Box 11.2, Chapter 10).

Australia is a high per-capita producer of CO_2, with most coming from electricity generation and transport. Australia's geology has ensured that vast amounts of high-quality coal were deposited in Permian basins close to where the nation's main population centres were subsequently established millions of years later. It is no surprise, therefore, that about 70% of Australia's electricity is generated by burning coal (Chapter 9). Australia is also a

CARBON DIOXIDE CAPTURE AND STORAGE (BOX 11.2)

The energy sector produces around 76% of Australia's CO_2 emissions (417 Mt of CO_2-equivalents in 2010), of which electricity generation makes up 36%. A carbon dioxide capture and storage (CCS) strategy involves (1) capture of the CO_2 generated from a stationary source point (e.g. power station); (2) transport of the CO_2 to the site of injection through pipelines or, on a smaller scale, by road, rail or ship; and (3) storage by injecting the CO_2 into a porous, permeable reservoir (e.g. in sandstone) that is at sufficient depth (generally >800 m) to keep the CO_2 in a supercritical, fluid state (Figure B11.1). Potentially, such a strategy could reduce emissions into the atmosphere from CO_2 source locations (fossil fuel power generation and industrial processing) by more than 80%.

These reservoirs generally contain saline water or other fluids, such as oil, natural gas and naturally occurring CO_2. To be suitable for CO_2 storage, the reservoir must be overlain by an impermeable sealing rock such as mudstone, which will prevent the escape of fluids. This is critical because the injected CO_2 is buoyant and will act like any naturally occurring gas, gradually rising from where it is injected until it meets the sealing rock. The CO_2 will be trapped in the reservoir by dissolving into the formation water and, over time, and depending on the geochemistry of the rock, will form carbonate-rich minerals within the reservoir.

The National Carbon Mapping and Infrastructure Plan (2010) confirmed that Australia's sedimentary basins have sufficient storage space for large-scale projects. However, although in some cases there is a good match of emission source with proximal storage sites (e.g. Gippsland and Perth basins), in many places suitable pore space is either not available or is some distance from the emission point (e.g. Sydney and southeast Queensland).

There are currently two CCS projects in Australia, both of which are located in petroleum basins with gas production. In the Otway Basin, western Victoria, CO_2 sourced from a natural CO_2-rich gas field nearby has been injected and stored 2 km underground. The reservoir lies within a Cretaceous rift sequence in the Waarre Formation, which also hosts significant commercial gas deposits. The project has demonstrated the feasibility of injection and storage of CO_2 in Australia, as well as the ability to model CO_2 plume movement, and monitor its chemical and physical changes within the reservoir. Extensive monitoring of the CO_2 throughout and after injection has proven the project to be very successful to date.

The Gorgon CCS project, located in the Carnarvon Basin, Western Australia, involves the removal and geological storage of reservoir CO_2 from the natural gas fields in the offshore Greater Gorgon area. The CO_2 will be injected into the Dupuy Formation, deep below Barrow Island. This formation was deposited in a deep marine slope setting and is sealed by a regional deltaic unit, the Basal Barrow Group Shale. The sedimentological pattern of an excellent reservoir-regional seal combination is repeated in several places across the North West Shelf in Jurassic to Late Cretaceous sequences (Chapter 4). This makes the offshore shelf a highly prospective area for geological storage for future liquefied natural gas projects in the region.

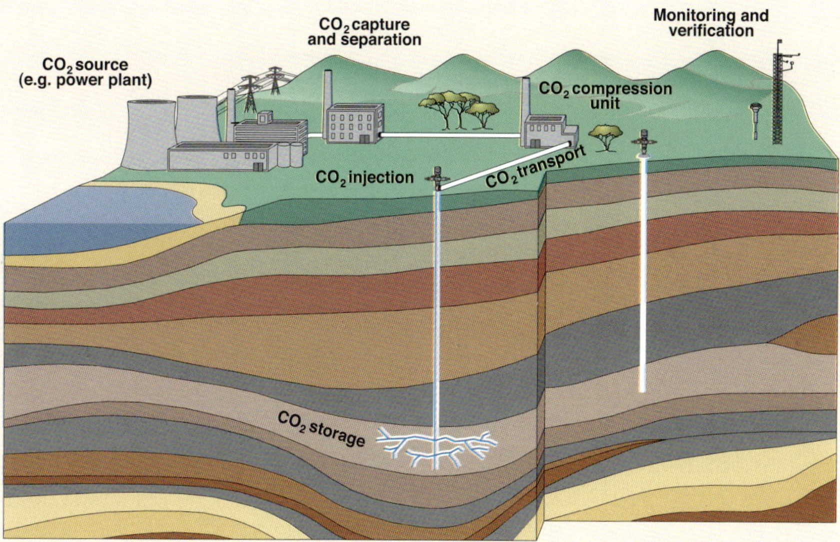

Figure B11.2: The process of capturing carbon dioxide from a power plant, compressing it and injecting it deep underground in rocks. (Source: CO2CRC)

large country with a dispersed population largely living in expansive sprawling cities, making both long- and, particularly, short-distance transport of goods and people a necessity of modern living.

Reducing Australia's dependence on coal will be a challenge economically (coal is readily available and cheap) and a challenge technologically (what to replace it with). This challenge applies not just to the Australian people but to the many export industries that depend on access to cheap power, such as aluminium refineries. One 'geosolution' that might satisfy both challenges is carbon capture and storage, which captures emissions at the source and sequesters them in nearby basins (Box 11.2). Australia would benefit enormously if carbon capture and storage works economically as planned. The economy is highly dependent on coal export earnings, which are forecasted by the Bureau of Resources and Energy Economics at $53 B for 2011–12. With carbon capture and storage, future importing nations would have both a supply of energy and a means of disposing of their CO_2 waste.

The attributes of a good basin for carbon capture and storage are similar to those of oil- and gas-producing basins—that is, a porous and permeable reservoir covered by a seal at a suitable depth to keep CO_2 in supercritical-liquid form. The main coal basins of New South Wales and Queensland do not have these attributes. Most of them are tight and impermeable, making injection and storage of fluid problematic. As a result, most current research on carbon capture and storage in Australia is testing the main oil- and gas-producing basins in Western Australia and Victoria. The search is on, however, for other suitable basins, such as the Surat Basin in Queensland, as well as improvements to the viability of the technology.

Australia's geology also bequeathed the nation another abundant source of energy—both conventional and unconventional gas. Australia is on track to be the second largest exporter of liquefied natural gas (LNG). Burning gas rather than oil or coal has the benefit of lower CO_2 emissions, so that it is commonly viewed as a transitional source of energy before more widespread usage of non–CO_2 producing options in the future.

In the United States, energy security concerns and the development of new technology have seen the emergence of a major new industry in unconventional (shale) gas, which is predicted to make the country a net energy exporter. The development of these gas resources has not been without its critics, because the gas is extracted by a hydraulic fracturing process that increases the permeability of the tight reservoirs. Regulators in the United States are dealing with conflicting interests: those of energy security and new jobs, and protection of the environment and human health.

Australia has been slower than the United States to unlock its unconventional gas resources. The coal basins of eastern Australia also host major coal-seam gas resources (Chapter 9). The increasing demand for domestic energy, jobs and export earnings will likely mean that these unconventional gas resources will be further developed. Unlike offshore gas, some of the best coal-seam gas prospects are located close to communities or beneath valuable farmland. Communities are concerned about loss of or damage to land, pollution and changes to the groundwater aquifers. Meeting these community concerns requires a much greater understanding of the three-dimensional geology of the basins, as well as the interrelationship between the groundwater and the gas resources.

Geology has also bequeathed Australia the world's largest reserves of uranium (U), as well as significant reserves of thorium (Th). Partly because of the easy availability of alternative energy sources, Australia does not currently use its nuclear potential to generate electricity; rather, U is exported to other countries for their domestic power generation (Chapter 10). The merits of nuclear power are as much about politics and community as they are about geology. Australia is also fortunate to have an abundance of Th. Thorium has numerous advantages over U, including the difficulty in producing nuclear weapons using Th. Thorium is three to five times more abundant than U, and can generate more energy per unit mass. Unfortunately, there are no commercial-scale nuclear reactors that can use Th. Although there is no use for it, Th is a by-product of some mineral-sand operations. The Th by-product is diluted with gangue sand and buried, rendering it uneconomic, but safe.

A less contentious source of radioactivity is contained within many of the Proterozoic and some Carboniferous granites of Australia. These are especially well endowed in heat-producing elements and, when buried beneath insulating basins, they are ideal sources of carbon-free geothermal energy, such as in the Cooper Basin (SA). The location of many geothermal plays is now known, but the process of economically extracting the heat and getting it to market poses continuing geological, geochemical and engineering challenges. The potential of geothermal energy in Australia is enormous. If only 1% of the geothermal energy contained in the upper 5 km of Australia's crust could be extracted, there would be more than 30 ka of supply at current usage. Even a more realistic 0.01% gives more than 3 ka of supply. Technological and engineering advances have yet to enable this vast resource to be effectively tapped, but, when they do, Australia might be able to power itself in a low-carbon and sustainable way for thousands of years. Even in the use of truly renewable energy (e.g. hydroelectric, solar, tidal and wind), Australia's position on the globe provides it with a range of climates and latitudinal variation for all of these to potentially make greater contributions to energy requirements in the future.

Although Australia has an abundance of energy, it is short of oil. So is the rest of the world. The world's 'peak oil' may indeed have passed in 2005; Australia's oil production peaked in 2000–01. World oil production appears 'stuck' at around 75 million barrels per day, meaning that high oil prices will remain if demand keeps rising, as it is forecast to do. All is not lost for Australia's oil. The vast basins fringing the southern margin of the continent are largely untested, as are the basins in the remote eastern frontiers, (Chapter 4). Australia also has significant demonstrated resources of unconventional hydrocarbon sources, such as shale oil. These resources remain undeveloped, mostly because of economic and environmental constraints.

The future is difficult to predict, but we can be certain of increasing domestic and overseas demands for energy. Only time will tell what mix Australia chooses. The Australian people are indeed lucky to have such a favourable geological legacy of energy choices.

Mineral resources

The economic growth in China and India, and concerns of Japan and Korea about resource security over the past decade have had a major impact on Australia's economy and trade relationships. Demand for Australian minerals and energy resources has created a boom for the resources sector, and led to closer economic links with Asia, particularly China (Chapter 9). National export earnings from minerals and energy amounted to $190 B in 2011, and much of the manufacturing export income was also resources related (e.g. refining and smelting).

A positive consequence of this demand has been a strong domestic economy, despite the difficulties facing many OECD countries. A negative consequence has been the impact on other sectors of the economy. Demand for Australian resources, along with relatively high interest rates (by international standards), has driven the Australian dollar above parity with the United States dollar (US$1.03 average for 2011). The Australian Treasury predicts that the Australian dollar is likely to remain high for a number of years. As a result, the non-resources sectors of the economy, which have been suffering since the mid 2000s, will probably continue to do so as Australian goods and services remain expensive on world terms.

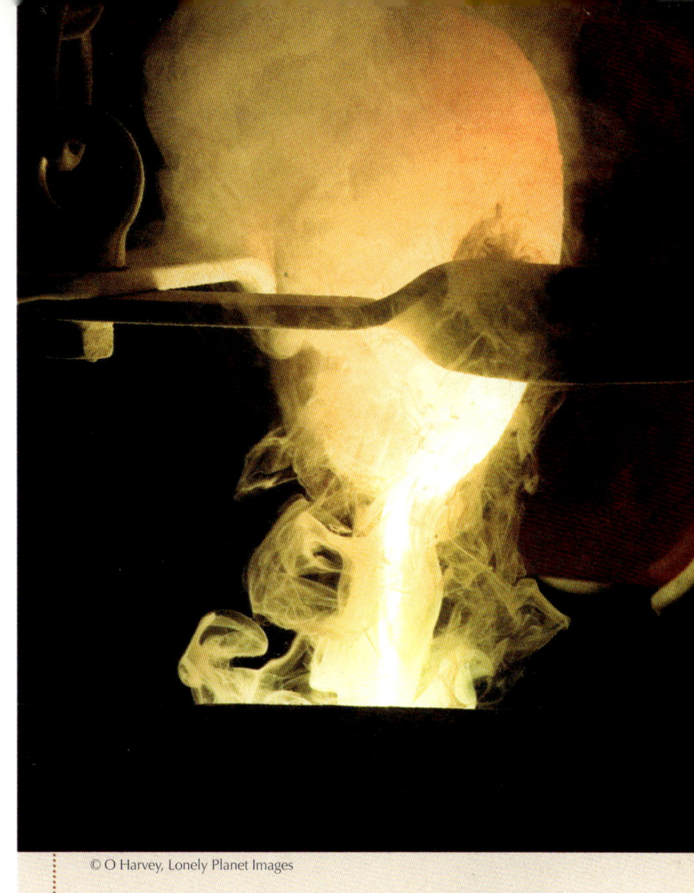

Demonstration of gold pouring at the Australian Prospectors and Miners Hall of Fame, Kalgoorlie, Western Australia.

The minerals industry is a large consumer of energy, so rising energy prices will place additional cost pressures on resource development. Take, for example, the proposed open-cut development at the Olympic Dam mine in South Australia. This is planned to remove an estimated 14 Gt of material over 40 years, which will require an enormous amount of energy to achieve. Rising commodity prices mean that lower grades of ore material (e.g. magnetite iron ore) may become economic. Mining low-grade ore bodies also means that more material must be handled for a given amount of ore. Similarly, mining ore bodies that are deeper in Earth, depending on the methods used, will also require more energy use for a given amount of ore. One of the reasons it is difficult to estimate resource life in the future is that it is highly dependent on the commodity price.

Despite these energy concerns, there is an ongoing need to ensure the steady replacement of the resource inventory. A strong economy provides jobs for citizens and generates taxes that governments use to pay for services. The Australian Treasury estimates that Australia has around 70 years average supply, at current production rates, of the key export resources of iron ore, black coal, oil, gas and gold. With many of these known resources experiencing increasing production rates, their life span is becoming shorter. For many commodities, the rate of extraction now far exceeds the rate of new discovery, despite heavy investment in exploration. Does this mean that peak minerals have been, or soon will be, reached? If so, will this long run in economic growth continue, and how will this affect Australian society?

Where will the new resources come from to replace the exhausted ones? Until quite recently, most mineral deposits discovered in Australia were either exposed at the surface or had some form of surface expression that could be easily detected by geological, geochemical or geophysical exploration methods (Chapter 8). The minerals industry now talks about the *tyranny of depth* because any future large mineral deposits are most likely to be discovered at depth. All of the major mineral provinces in Australia are partially covered by regolith (Chapter 5), and many of these fertile terranes are covered by basins on their margins. There is no reason to suppose that resources will only be found at the current erosion level. With around 80% of the continent covered by regolith, there remain many potential resources. The problem is where to start looking.

Mineral discovery is an expensive 'game of chance', and the odds of success are long (1 in 1000 or less). Successful mineral discovery under cover will need to work on much shorter odds, and require much better prediction before drilling. This can be achieved by changing our thinking on mineral deposits as being symptoms of a system—a mineral system. The mineral system approach asks different questions from those traditionally asked. One question is: why is this mineral deposit there, as opposed to somewhere else in time and space? This approach also benefits from scale—because a

Bauxite/alumina refinery, Gladstone, Queensland.

Image by Mining Photo

Fault scarp generated by the 1968 Ms 6.8 Meckering earthquake, Western Australia. The event produced a 37 km-long surface rupture with a maximum vertical displacement of approximately 2 m (Gordon & Lewis, 1980).

mineral system is many orders of magnitude larger than a mineral deposit, it has a much greater chance of being identified. Thus an explorer can look for the larger footprints of the system (like the spent fluid outflow zones), rather than just the relatively small mineral deposit (Chapter 8).

Exploring beneath regolith cover requires thinking about geology in four dimensions (3D space and time). Successful exploration will depend on a better understanding of the architecture of a given mineral system and how it evolved through time and space, as well as across scales. This is no easy task; it will rely on modelling, testing and numerical simulations. Dealing with uncertainty will also be a challenge. There have been a number of attempts to construct 3D maps of various regions of Australia. In the future, these will become more refined and realistic, and start to incorporate the time dimension with palinspastic reconstructions. Only then can we hope to lower the odds and reduce the risks for successful mineral discovery.

It is also becoming increasingly evident that, just as we need energy security, we also require mineral security. This is especially true for the class of minerals—variably known as strategic or critical minerals—that are vitally important for maintaining and protecting our modern way of life (e.g. electronics, communications, medical, transport industries). It only strictly applies to those elements that are (or will be) vulnerable to availability or supply restrictions, and are not easily substituted. Copper, for example, although extensively used in modern societies, is readily available and, having diverse suppliers, is not vulnerable to supply restrictions; it therefore is currently not a strategic mineral. On the other hand, the rare earth elements—which have a wide range of uses, including long-life batteries, magnets in wind turbines and hybrid cars, and glass polishing for lenses—suffer from numerous potential shortcomings. Although they are relatively common (contrary to what the name suggests), there are not many large economic deposits of rare earth elements. The supply is vulnerable because most of the current production is dominated by one country. This is partly because many rare earth elements reside in minerals that have deleterious associated elements (e.g. Th), making processing economically and environmentally challenging. From a strategic perspective, however, Australia has resources that can provide some measure of rare earth element security.

Hazards at home and abroad

Compared with most of Australia's neighbours, and because of its stable geological position, Australia enjoys a relatively benign natural hazard environment. Extreme events, such as tropical cyclones, floods, severe storms, bushfires, earthquakes and tsunamis, however, can threaten Australia's wellbeing. Indeed, natural hazards cost Australia more than 1% of its gross domestic product annually, with some years experiencing large-scale natural disasters that drive this cost far above the average.

Tropical cyclones are often regarded as the most devastating weather-related hazard, causing severe winds and storm surge that can have major impacts over large areas. Cyclone Tracy destroyed the city of Darwin in 1974, an event still ingrained in the psyche of many Australians, and one that changed the building codes. Although the area is sparsely populated, there are large infrastructure facilities located adjacent to the northwest coast, such as the gas production on the North West Shelf and iron ore export from the Pilbara Craton. These are significant sites in terms of export-earning capacity of the nation, and are vulnerable to the impacts of tropical cyclones, as well as tsunamis.

Floods are, on average, the most costly of Australia's natural hazards, accounting for roughly 30% of the total loss due to all hazards. No wonder Australia was called by Dorothea Mackellar 'the land of droughts and flooding rains' (Chapter 1). Although Australia is the driest inhabited continent, periods of heavy rainfall—typically in La Niña years—can turn vast areas of the arid centre into an inland sea for weeks or months at a time. More short-duration, rapid-onset floods tend to occur in the northern and eastern coastal regions, with several of the main population centres along the eastern seaboard having historical experience of more than one major urban flooding event. A positive effect from these flood events has been the replenishment of parched aquifers and dams, and the flushing of river systems.

Bushfires have been an intrinsic element of the Australian environment since long before humans arrived, and Aboriginal fire-stick agriculture further shaped the environment to one dominated by fire-adapted flora (Chapter 3). Many native plants contain volatile oils and readily burn in the hot, dry conditions that prevail during the Australian summer, particularly in El Niño years. Such fires cause considerable loss of life and property on the urban fringes of cities in Australia's south, where homes are in close proximity to the bush. Occasionally, such fires penetrate farther into suburban areas, where the loss of life and property can be severe.

Victorian bushfire aftermath, 2009.

Geological hazards that pose a threat to Australia's wellbeing include earthquakes and tsunamis. Due to continental Australia's position remote from active tectonic plate margins, earthquake activity is much lower than in neighbouring countries. This low rate of earthquake activity has led to a sense of complacency among Australians. However, that complacency was shattered by a magnitude 5.6 earthquake in Newcastle on 28 December 1989. This resulted in 13 deaths, 160 injuries and an estimated $4 B total economic loss. It changed the building code in Australia to one with an appropriate seismic loading standard. However, Australian cities still contain a substantial number of pre-1989 buildings, particularly brick buildings that are not designed to withstand earthquakes. Thus, the fear remains that, if an earthquake as large as, say, the 1968 Meckering (WA) earthquake (magnitude 6.8) were to occur near one of Australia's major population centres, the impact could be catastrophic. Australians can hope that the chances of such an event occurring are sufficiently low that there will be time to replace most of the older buildings before it occurs. This is a clear example of how humans tend to judge risk from the context of their own life span: 'We have never seen that happen before, so it won't happen here'. Things that seem rare in human experience are often common in the geological time-scale, thereby underlining the ongoing need to significantly improve our understanding of geological processes and their resulting risks to us.

Australia's distance from subduction zones to the north and east have mitigated the impact of large historical tsunamis (Chapter 6). However, the recent occurrence of large earthquakes, such as the 2004 Sumatra–Andaman and 2011 Tohoku earthquakes in subduction zones that were not considered to support such large events, demonstrates the danger of relying on the immediate historical record for assessing hazard. If subduction zones such as those off Java or in the Puysegur Trench south of New Zealand (Figure 2.1) can support earthquakes much larger than historical experience suggests, then the northwestern and southeastern coasts of Australia could face a considerable tsunami threat. Fortunately for Australians, these, and other subduction zones, are at least 2.5 hours tsunami travel time away, which is enough time for the Australian Tsunami Warning System to alert coastal populations.

How might climate change influence the impacts from weather-related hazards in the future? Most climate models predict that southeastern Australia will become hotter and drier, leading to a higher risk of bushfires. The models also suggest that tropical cyclone tracks will move further south (Chapter 6), exposing larger populations that have lower standard building codes to potentially devastating cyclonic events. Rising sea-levels could increase coastal inundation extremes, creating severe impacts on the Australian community (housing and infrastructure), much of which is

Devils Marbles,
Northern Territory.
Image by Jim Mason

Image by Jim Mason

Broome, Western Australia.

located not far above the coastal fringe. Sea-level rise could also impact on coastal groundwater aquifers. Generally, the picture for weather-related hazards may be one of more severe events, although there is too much uncertainty to make precise predictions about just how severe extreme events could be and how often they might occur.

Knowing the unknown Australia

The area of Australia's jurisdiction and stewardship, including the Australian Antarctic Territory and vast marine regions, covers nearly 5% of the planet. This large area comes with a responsibility to describe, understand and manage sustainably. Indeed, each new major marine survey discovers new geology and even new species. On land, new fossils are regularly being identified and described, new stratigraphy is established and new maps are made. The largest scale geological maps for many regions in Australia are at a scale of 1:250 000. We have only just begun to scratch through the surface, with current 3D geological maps being rudimentary. Government agencies around the country have invested heavily in deep-sounding techniques, such as passive seismic, seismic reflection and magnetotellurics, as well as more conventional potential field geophysics (magnetics and gravity). Despite this impressive array of maps, there is much Australian geoscientists do not know. Things can be described and empirical connections made, but do we really understand Earth system processes?

Fragile Earth

What is the greatest threat facing humankind? In the opinion of the celebrated science communicator and geologist Professor Iain Stewart, it is 'our collective inability to appreciate that our cherished human economy relies on a well functioning *earth economy*'. The first Australians see themselves as land custodians, rather than owners. Later Australians appreciate the intense beauty of the landscape, but they do not understand its fragility. As with most of what defines Australia, this beauty and fragility are deeply rooted in the continent's tectonic history, its current tectonic setting and the consequent geology.

Predicting the future is inherently uncertain. Nevertheless, Australians have a responsibility to care for their environment—society depends on this! What is certain about the future is that many things we take for granted will change. The 21st century will be shaped by the finite nature of Earth and its resources, just as industrialisation shaped the 19th and technology the 20th centuries. Although Earth is hospitable at the moment, it is nevertheless indifferent to us. Because the past is the key to the future, we know that eventually it will be lethally indifferent (to its human passengers).

Geoscientists bring a unique perspective to many of society's challenges. These challenges are intimately linked with the stewardship of the environment, the sustainable availability of resources (food, fibre, water, energy and minerals) and community safety (Box 11.1). As a geoscientific community, we can help wider society by explaining how things change, why they change and how to adapt to change. To do this effectively, we need to remember that we must not simply document our science but communicate it, in multiple ways, and show that, as well as the problems, the solutions to the challenges of the future reside within the realm of geoscience.

Bibliography and further reading

AuScope 2011. *The Australian Earth Observatory—infrastructure roadmap*. www.auscope.org.au/userfiles/file/Sidebar%20files/AEO_AuScope_Nov_2011.pdf

Australian Academy of Science 2011. *Searching the deep Earth: the future of Australian resource discovery and utilisation*, proceedings of the Theo Murphy High Flyers Think Tank 2010, Canberra, 19–20 August 2010, Australian Academy of Science, Canberra.

Australian Government Department of Resources, Energy and Tourism 2010. *Energy in Australia 2010*, DRET, Canberra.

Blainey G 2001. *The tyranny of distance: how distance shaped Australia's history*, revised edn, Pan Macmillan Australia Pty Ltd.

Carbon Storage Taskforce 2009. *National carbon mapping and infrastructure plan—Australia: concise report*, Australian Government Department of Resources, Energy and Tourism, Canberra.

Durant W 1926. *The story of philosophy: the lives and opinions of the great philosophers*, Simon & Shorter, New York.

Flannery T 2011. *Here on Earth: a sustainable future*, Text Publishing Company, Melbourne.

Garton P & Gruen D 2012. The role of sovereign wealth funds in managing resource booms: a comparison of Australia and Norway, presentation at the Third Annual Asia Central Bank and Sovereign Wealth Fund Conference, Australian Treasury.

Gordon FR & Lewis JD 1980. The Meckering and Calingiri earthquakes October 1968 and March 1970. Western Australia Geological Survey Bulletin 126.

Gruen D 2011. The macroeconomic and structural implications of a once-in-a-lifetime boom in the terms of trade, address to the Australian Business Economists, 24 November 2011.

Hoekstra AY (ed.) 2003. *Virtual water trade*, proceedings of the International Expert Meeting on Virtual Water Trade, IHE, Delft, the Netherlands, 12–13 December 2002, Value of Water Research Report Series No 12.

Horne D 1964. *The lucky country: Australia in the sixties*, Penguin Books, Melbourne.

Murray J & King D 2012. Oil's tipping point has passed. *Nature* 481, 433–485.

Prime Minister's Science, Engineering and Innovation Council 2010. *Challenges at energy–water–carbon intersections*, PMSEIC, Canberra.

State of the Environment 2011 Committee 2011. *Australia: state of the environment 2011*, Australian Government Department of Sustainability, Environment, Water, Population and Communities, Canberra.

Vogfjord KS & Langston CA 1985. Source parameters of the Ms 6.8 Meckering, Australia earthquake of October 14, 1968: 'a chip off of the old block', *EOS* 66, 963.

Index

An 'f' following a page number indicates a figure; 'i' refers to an image and 'm' to a map.

If a topic is discussed in the text and expressed visually on the same page, only the page is referenced.

A

B

C

D

E

F

G

H

I

J

K

L

M

N

O

P

Q

R

S

T

U

V

W